Introduction to Computational Health Informatics

Chapman & Hall/CRC
Data Mining and Knowledge Series
Series Editor: Vipin Kumar

For more information about this series please visit:
https://www.crcpress.com/Chapman--HallCRC-Data-Mining-and-Knowledge-Discovery-Series/
book-series/CHDAMINODIS

Introduction to Computational Health Informatics

By Arvind Kumar Bansal (principal)

Kent State University

Javed Iqbal Khan

Kent State University

S. Kaisar Alam

President & Chief Engineer at
Imagine Consulting Services

CRC Press
Taylor & Francis Group
Boca Raton London New York

CRC Press is an imprint of the
Taylor & Francis Group, an **informa** business

A CHAPMAN & HALL BOOK

CRC Press
Taylor & Francis Group
6000 Broken Sound Parkway NW, Suite 300
Boca Raton, FL 33487-2742

International Standard Book Number-13: 978-1-4987-5663-1 (Paperback)
International Standard Book Number-13: 978-0-367-43478-6 (Hardback)

Library of Congress Cataloging-in-Publication Data

Names: Bansal, Arvind Kumar, author. I Khan, Javed I. (Professor of
computer science), author. I Alam, S. Kaisar, author.
Title: Introduction to computational health informatics / by Arvind Kumar
Bansal, Javed Iqbal Khan, S. Kaisar Alam.
Other titles: Chapman & Hall/CRC data mining and knowledge discovery
series.
Description: Boca Raton : CRC Press, [2020] I Series: Chapman & Hall/CRC
data mining and knowledge discovery series I Includes bibliographical
references and index.
Identifiers: LCCN 2019042593 I ISBN 9780367434786 (hardback : alk. paper) I
ISBN 9781498756631 (paperback : alk. paper) I ISBN 9781003003564 (ebook)

Subjects: MESH: Medical Informatics
Classification: LCC R855.3 I NLM W 26.5 I DDC 610.285--dc23
LC record available at https://lccn.loc.gov/2019042593

Visit the Taylor & Francis Web site at
http://www.taylorandfrancis.com

and the CRC Press Web site at
http://www.crcpress.com

This book is dedicated to all the visionaries who have worked incessantly to free this world from pain and misery.

Contents

Preface

Since the dawn of civilization, doctors and nurses have strived to relieve people from their pain. Continuous improvement in science and information technology has enhanced the efforts of doctors and nurses by giving them better tools to archive, analyze and transmit clinical data. Information technology promises to provide available medical information seamlessly to providers and caregivers so they can optimize their efforts for the best possible care.

In the last two decades, the increasing presence of computer processing has rendered health information widely available. Combined with computational modeling and the development of distributed databases, clinical data is being archived and analyzed using machine learning techniques and data mining, generating a form of knowledge never seen before. This knowledge is improving life-expectancy by better disease management, development of new vaccines, and drugs with reduced development-cycle time.

It is envisioned that, in the future, the seamless integration of information technology, intelligent analysis techniques and medical science will provide quality care for an affordable price by incorporating better clinical data analysis, providing pervasive care, removing duplicate medical treatment and laboratory data analysis and making data available electronically to collaborating healthcare providers.

The flow of information has raised many issues such as data-format standardizations, adoption of technology and the need for intelligent user-friendly interfaces for the end users such as patients, doctors, hospitals, nurses, pharmacies, insurance providers, policy makers and clinical researchers.

Despite the exponential growth of this multidisciplinary field, there has not been a single textbook that provides the computational aspect of health informatics for both software developers and a new generation of "Health Informatics Scientists"—the books written by clinical scientists present the topic from the perspective of a clinical practitioner. There is a need for a textbook in *Computational Health Informatics* that can prepare computer science or information technology students to understand the computational techniques used in health informatics, along with the related medical concepts.

This book describes various computational techniques, including biostatistics, heterogeneous databases, artificial intelligence, signal analysis, bioinformatics, image analysis, data communication for transmission of clinical data and medical images and their application to clinical data analysis, as well as management of electronic health records and their seamless integration to connect healthcare providers. The book also discusses emerging areas of telemedicine, pervasive care, remote monitoring and bioinformatics for the discovery of drugs, including pharmacokinetics and pharmacodynamics.

This textbook is based upon the Computational Health Informatics course that I have been teaching since 2012, first to graduate students and then to senior-level undergraduate students beginning in 2014. The course content evolved along with my understanding of the lack of knowledge and concepts students need to develop software for health informatics. As I started writing the book, the course material also evolved along with my knowledge. I included new material based upon my research on the ongoing evolution of Computational Health Informatics.

Javed, Kaisar and I committed to writing this textbook in 2015. It has taken four long years to come to fruition, due to our other commitments as well as the need to do extensive research of the scattered material available across multiple disciplines. When Leon Sterling came to know about our efforts, he graciously contributed an important chapter about the need for new technology to meet patients' emotional needs and satisfaction before it can be successfully adopted. The book itself has gone through two revisions.

This textbook will assist (1) computer science students to understand concepts needed to develop techniques and healthcare software; and (2) medical students and practitioners to understand the computational background and concepts for healthcare software and data management. The material is sufficient for one semester at a senior or graduate-freshman level course. The book dwells on concepts and techniques; however, specific in-depth algorithms have been avoided. We believe that the knowledge of the concepts and techniques discussed will prepare the students to follow the necessary algorithms.

In my classes, I could cover Chapters 1 through 7, followed by Chapters 8 and 9. Other instructors may find other combinations, including Chapter 10 on Bioinformatics for Drug Discovery, to be useful. Because of the diversity of topics, I recommend sufficient classroom interactions between the instructors and the students.

We hope that this book will provide a solid foundation to generate a new class of medical technocrats who will understand and apply computational methods to facilitate patient-friendly automation in health-care and improve the interpretation of clinical data.

Arvind Kumar Bansal
Kent State University, Kent, Ohio, USA

Chapter Outlines

We assume that students will have a background of two semesters of programming, introductory knowledge of data structure concepts, and some knowledge of statistics and computer networks. The book assumes that students can write at least 300 lines of code for developing projects. The book is divided into 12 chapters, including a concluding chapter. We have explained concepts in simple intuitive language at an abstract level. We have described examples and case studies as needed.

Chapter 1 introduces informatics and data modeling, and describes modeling of health-related information using the computational techniques to archive, retrieve, transmit, and analyze clinical and patient-centric data. It introduces classifications of medical informatics and defines computational health informatics. It describes the components of electronic health records, including medical images and transparent integration of medical data from heterogeneous sources, including healthcare providers, medical data warehouse, patients, pharmacy, hospitals and insurance agencies. It describes the need for secure transmission of medical data between sources and user-friendly human–computer interfaces.

Chapter 2 describes foundational concepts derived from the needs of archiving, retrieving, transmitting and intelligently analyzing data to extract human comprehensible knowledge from clinical data. It describes data abstractions such as trees, graphs, strings and their matching; image modeling, matching analysis techniques; formats to represent images; image compression techniques; basics of probability and statistics needed for data analytics; curve fitting needed for data analysis; concepts in statistics; different types of databases such as relational databases, object-based databases, multimedia databases, temporal databases; knowledge bases and techniques to keep privacy and security in databases; middleware for data communication; basics of human physiology needed for health informatics; and basics of genomics and proteomics.

Chapter 3 describes various artificial intelligent and machine learning techniques that are used for the automation of human–computer interactions, text extraction and summarization, form filling from doctors' natural language dictation about patients' conditions, monitoring patients' conditions, and clinical data analysis to derive new information and knowledge from a huge amount of data generated. It describes artificial intelligent techniques such as heuristic searches, probabilistic reasoning and modeling, deduction and induction; machine learning techniques such as data clustering, regression analysis, neural networks, support vector machines, Markov processes, Bayesian networks, hidden Markov models and data mining. It also describes analysis and clustering of time-series data. It describes ontology and medical dictionaries to understand and compare an assessment of the patients' condition by various specialist doctors. It also briefly describes techniques for automated information extraction, event analysis and summarization from natural language texts.

Chapter 4 discusses organization of healthcare data that removes the duplication of patient records while preserving the privacy of the patients as protected by HIPAA (Health Insurance Portability and Accountability Act). HIPAA prevents providers and insurers from exposing information to others without necessity and the patient's consent. This chapter also describes automatic acquisition of data from medical sensors, conversion of data for automated archiving and retrieval from the heterogeneous healthcare databases, and many popular standards for the exchange of health information electronically over the Internet. It discusses interoperability and transformation of data to make them compatible with heterogeneous databases containing multimedia objects and temporal objects. It also discusses different views of electronic medical records.

Chapter 5 describes various medical imaging techniques needed to derive and analyze medical images such as X-ray, computer-aided tomography (CAT), magnetic resonance imaging (MRI), ultrasound, positron emission tomography (PET), and other nuclear medicine and optics. The analysis of medical images can offer noninvasively significant insights to clinicians. The chapter also describes various formats and compression techniques for medical image archival, retrieval and transmission. Finally, the chapter describes techniques for the application of medical image analysis with a focus on cancer detection and computer-assisted treatment monitoring.

Chapter 6 describes DICOM, the standard for communicating digital images between medical databases. The chapter describes data structures, modeling a medical process using entity-relationship modeling, transmission protocols and various network levels involved in an image transmission. It also describes briefly the security issues in transmitting medical images.

Chapter 7 describes the various signal analysis techniques to understand ECG (electrocardiograms) for analyzing and monitoring heart-related diseases, electroencephalograms (EEG) to understand brain-related diseases and electromyography (EMG) to analyze muscle-related abnormalities. It also describes how various artificial intelligent techniques, described in Chapter 4, can be applied to extract and analyze ECG and EEG. It discusses various applications of computational analysis to identify different diseases related to the heart, brain and muscles.

Chapter 8 describes the application of various artificial intelligent techniques such as clustering, regression analysis, time-series data analysis, neural networks, clustering and data-mining to perform clinical data analysis derived by clinical trials. It discusses statistical and computational techniques to study drug efficacy, survivability and risk analysis. It describes some applications of clinical decision support systems utilizing knowledge-based systems and artificial neural networks. It discusses the techniques to identify and improve clinical processes and biomarkers for the cost-effectiveness of treatments. Finally, three applications of clinical data analytics have been discussed for cancer detection, detection and management of dynamic organ failure and fatty-liver disease.

Chapter 9 discusses the concepts, techniques and some algorithms for remote care, automated monitoring and transmission of signals, biosignal analysis, archiving the derived data for future analysis and managing information security and patients' privacy during data transmission and archiving. Remote monitoring is becoming important to handle the shortage of medical practitioners, to provide elder-care and to identify refractory conditions important in identifying disease-states of patients.

Chapter 10 describes bioinformatics and its application to drug discovery, efficacy analysis of drugs and derivation of drug dosage and toxicity using pharmacokinetics and pharmacodynamics. This chapter discusses biological concepts necessary for explaining bioinformatics, causes of various diseases, genetic diseases, and pathway aberration-related diseases, as well as vaccine development and improvement in the efficacy of drugs. The analysis techniques describe similarity-based search, genome alignment techniques, dynamic programming techniques, SNP (single-nucleotide polymorphism), GWAS (genome-wide association studies) and microarray analysis to identify signaling pathways. This chapter briefly describes the structure of antibodies and computational techniques to improve the binding-affinity of antibodies to improve drug-effectiveness.

Chapter 11 discusses the lack of understanding about the emotional needs of the healthcare providers, caretakers and patients on the part of software developers. The potential for health informatics software to improve health outcomes for patients is enormous. However, the effective utilization of health informatics software depends upon the adoption and appropriation of the software by a wide range of stakeholders, with a wide range of abilities and motivations. The emotional aspect of this interaction is vital. However, software developers are often unaware of the patients' emotional needs, experiences, and physical and emotional impairments, and thus ignore their needs in the developed software. This chapter also describes four case studies where emotional factors have been taken into consideration during software development for healthcare applications.

Chapter 12 describes the evolution of health informatics and its impact in an aging society and the need of the developing world to provide quality care while maintaining the economy. It also discusses issues in developing standards and adaptability. Finally, this chapter describes some future directions in computational health informatics.

There are five appendices at the end of the book that describe various sources or healthcare-related standards, conferences and journals, organizations, databases and companies related to healthcare. These lists are representative subsets and are not meant to be exhaustive. The purpose of the appendices is to provide the needed data-sources for doing research and project-reports needed for the course. Appendix I describes the websites for major standards and formats described in this book. Appendix II summarizes the list of the conferences and journals that were the source of material for this book. These conferences and journals are rich sources for graduate research and course projects. The list is still not comprehensive, and students should also find other sources for research. Appendix III lists major funding and databank agencies, which are a rich source of data and are also involved in policy decisions regarding healthcare. Appendix IV lists some major national and international databases that will be helpful in graduate and undergraduate students' research and projects. The list is certainly not comprehensive and misses many research databanks from individual research groups and universities, yet is a major source of archived data sufficient for research and projects. Finally, Appendix V lists a small representative subset of companies involved in the healthcare industry. It is divided into different classes such as EHR, medical imaging devices and diagnostics, wearable devices and pervasive care and drug discovery.

Classroom Use of this Textbook

Based upon the experience in the Health Informatics course, this textbook is suitable for a one-semester senior-level undergraduate course or freshman-level graduate course. For the graduate-level offerings, the course needs to be augmented by the research articles given at the back of each chapter and various journals and conferences in the area (see Appendix I). A suggested distribution of the effort is given below:

CHAPTER	TIME IN MINUTES	75-MINUTE LECTURES	45-MINUTE LECTURES	SUGGESTED MINIMUM COVERAGE FOR A SEMESTER-LONG COURSE
Chapter 1	150	2.0	3.0	Full
Chapter 2	300	4.0	6.5	Full
Chapter 3	250	3.5	5.5	Full
Chapter 4	250	3.5	5.5	Full
Chapter 5	180	2.5	4.0	At least Sections 5.8 and 5.9
Chapter 6	180	2.5	4.0	Full
Chapter 7	180	2.5	4.0	At least Sections 7.1–7.5
Chapter 8	150	2.0	3.0	At least Sections 8.1, 8.5, and 8.6
Chapter 9	150	2.0	3.0	At least Sections 9.1–9.4
Chapter 10	180	2.5	4.0	Based on class makeup
Chapter 11	75	1.0	1.0	At least Sections 11.1 and 11.2
Total	**App. 2100**	**28 units**	**43 units**	

Acknowledgments

I thank Kent State University for the "Kent State University-Summa Health System Collaborative Research Grant" that started my collaboration with Dr. Jeffrey Neilson (MD), who steered me to computational health informatics from a medical practitioner's perspective. I also acknowledge Jeff for graciously accepting my request to deliver guest lectures in my first offering of a graduate-level course during Fall 2012.

I acknowledge all the researchers in this fast-growing field for their valuable contributions that became an invaluable source of knowledge and learning. I must acknowledge Javed Iqbal Khan, who nudged me to write this textbook with a promise to contribute the chapter on *DICOM*. I acknowledge Kaisar Alam for contributing the chapter on *medical image informatics*. I also acknowledge the acquisition editor, Randi Cohen, for her constant encouragement and support throughout this long process of writing and improving the text. I acknowledge the reviewers who raised the bar with useful comments. I acknowledge Siemens Healthcare, research groups, researchers, publishers and medical practitioners who permitted their copyrighted images and drawings to be included in this book.

My former PhD student, now Dr. Purva Gawde, contributed to the teaching and provision of feedback of the material that was immensely helpful. She also developed an online version of the course material from an earlier unpublished version of this book that was taught by her and myself. Finally, I acknowledge my PhD advisor and friend, Leon Sterling, who graciously contributed a valuable chapter that raises an important issue that any health technology ultimately must be human-friendly and easy to use for adoption.

Arvind Kumar Bansal

Being a longtime worker in medical informatics – medical image processing, high fidelity and complex video communication, information coding, computation for radiation treatment planning, HIPAA to DICOM, HL7, and medical IoT security – I have felt the need for a common compiled source of knowledge in this multidisciplinary rich and vast area of computational health informatics. Each time I started working on a topic, it involved vast self-learning into a seemingly different wilderness of knowledge. There was never an ideal textbook in this highly important area that could prepare students from a computer science and engineering background for the field.

Over the course of a year, discussions along these lines with Prof. Arvind Kumar Bansal eventually resulted in this project. I am glad to see that finally that dream textbook is here. Given the rapid growth in this area and its highly challenging multidisciplinary topical composition, it inevitably has many deficiencies. However, I am hopeful that, with feedback, this project will become perfect in a few years. More importantly, it will now pave the way to allow students and practitioners to delve deeper into the area of medical informatics with much sharper technical tools than has been possible previously.

I gratefully acknowledge the contribution of my advisor, David Y. Y. Yun, who introduced me to the world of medical computing and affirmed the strength of seeking the bigger picture rising above the individual subareas of computing.

Javed Iqbal Khan

I have been in the area of medical imaging informatics and computer-aided diagnosis since early 1990s and would like to thank Arvind for initiating this much-needed project. I believe that this book will fill a conspicuous void and will be very useful to the practitioners in this area.

I would like to thank six individuals whom I consider both friends and mentors: my PhD advisor, Kevin Parker; my postdoc supervisor, Jonathan Ophir (deceased); my former supervisor, Ernie Feleppa; my former supervisor, Fred Lizzi (deceased); Kazi Khairul Islam; and Brian Garra. Finally, I would like to thank my Creator, my parents (deceased), my wife, our two children, my two siblings and all my friends and family. I really appreciate all your encouragements and support.

S. Kaisar Alam

About the Authors

Arvind Kumar Bansal is a full professor of Computer Science at Kent State University. He received both B. Tech (1979) in Electrical Engineering and M. Tech (1983) in Computer Engineering and Science from the Indian Institute of Technology at Kanpur (IITK), India, and PhD (1988) in Computer Science from Case Western Reserve University (CWRU), Cleveland, Ohio, USA. He has been a faculty member of Computer Science at Kent State University, Kent, Ohio, USA, since 1988 and has taught undergraduate and graduate-level courses in the areas of artificial intelligence, computational health informatics, multimedia languages and systems and programming languages. He also directs the "Artificial Intelligence Laboratory" at Kent State University and has been teaching "Computational Health Informatics" regularly since 2012.

His research contributions are in the areas of artificial intelligence, bioinformatics, proteomics, biological computing models, massive parallel knowledge bases, program analysis, ECG analysis, social robotics and multimedia languages and systems. He has published over 75 refereed articles in journals and international conferences. His research has been funded by NASA and the US Air Force. He has also served in many program committees in the areas of artificial intelligence, bioinformatics, logic programming, multimedia, parallel programming and programming languages. In addition, he has been an area editor in the international journal *Tools with Artificial Intelligence* and is a member of IEEE and ACM.

Javed Iqbal Khan is a full professor of Computer Science at Kent State University. He received his B. Tech (1987) in Electrical Engineering from the Bangladesh University of Engineering and Technology (BUET), Bangladesh, and his MS (1990) and PhD (1995) in Electrical Engineering (Computer Track) from the University of Hawaii at Manoa, Hawaii, USA. He has been a faculty member of Computer Science at Kent State University, Kent, Ohio, USA, since 1997. He has regularly taught undergraduate and graduate courses in the areas of Internet engineering, peer-to-peer systems, artificial intelligence, algorithms and networking.

His research contributions are in Internet Engineering, artificial intelligence, automated knowledge extraction, routing and network decision-making with medical data, perceptual enhancement through eye-tracking, cyber infrastructure for medical-image communication, and networking for education. He has published over 100 articles in refereed international conferences and journals and has been in NSF panels, many program committees and the executive committee of IEEE Internet Engineering. He also led a team that designed and implemented two national educational networks as a part of UN-funded project. His research has been funded by World Bank, NSF, DARPA and NASA. As well, he has been a Fulbright scholar and has served as a senior specialist on high-performance education networking in the Fulbright National Roster of experts. He is an associate editor of *International Journal of Computer Networks and Applications* and is a member of IEEE and ACM.

 **S. Kaisar Alam** received his PhD (1996) in Electrical Engineering from the University of Rochester, New York, USA. His research publications and teaching are in signal/image processing with applications to medical imaging. He was a Principal Investigator at Riverside Research, New York from 1998 to 2013 and the Chief Research Officer at an upcoming tech startup in Singapore from 2013 to 2017. He has been a visiting professor at the Center for Computational Biomedicine Imaging and Modeling (CBIM), Rutgers University, Piscataway, New Jersey (since 2013) and an adjunct faculty at The College of New Jersey (TCNJ), Ewing, New Jersey (since 2017). Currently, he runs his own consulting company specializing in medical image analysis and diagnostic and therapeutic applications of ultrasound. He is a Fellow of the American Institute of Ultrasound in Medicine (AIUM) and a senior member of IEEE and has served in the AIUM Technical Standards Committee and the Ultrasound Coordinating Committee of the RSNA-QIBA. He is an associate editor of *Ultrasonics* (Elsevier) and *Ultrasonic Imaging* (Sage). Dr. Alam has been a recipient of the prestigious Fulbright Scholar award.

Visiting Research Faculty, Center for Computational Biomedicine Imaging and Modeling (CBIM), Rutgers University, Piscataway, NJ, USA

Adjunct Faculty, Electrical & Computer Engineering, The College of New Jersey (TCNJ), Ewing, NJ, USA

President and Chief Engineer Imagine Consulting LLC, Dayton, NJ, USA

Introduction

1

Providing health care is a complex task. Multiple medical practitioners specialize in various aspects of health care and collaborate to treat a patient. Many cooperating hospitals provide surgery and health maintenance. Insurance agencies pay and regulate service bills of patients. Government agencies collect demographics-based medical data, analyze them and formulate healthcare policies. Congress uses the statistical data provided by healthcare agencies to formulate privacy laws, laws to avoid and contain threats of contagious diseases and epidemics, and allocate budgets for a healthier society.

A patient goes through multiple specialists and hospitals for treatment and moves around across diverse geographical locations. In recent years, people have become more mobile due to business necessities and the availability of better transport. Asking a patient to go through the same set of tests will duplicate effort resulting in inconvenience, an increase in medical cost and a burden on resources such as beds and medical staff. Human life expectancy has been steadily growing, and is now around 82 years. Certain diseases become more pronounced in old age, and elderly patients need remote long-term health care in home settings to contain medical cost and resources and reduce patient movement. Besides, elderly patients prefer to remain in the comforting environment of their homes.

In the last three decades, computers and their applications have exploded. Computers are good for: 1) archiving and retrieving of data; 2) intelligent analysis for hypothesis formation; 3) efficient and accurate automation of processes that require a huge amount of human resources and are nearly impossible for humans; 4) providing innovative ways to probe into abnormalities; 5) efficiently communicating information across geographical locations; 6) mimicking speech for human-like interaction; 7) activating an automated process in response to the human voice and 8) analyzing complex signals and images for aberrations and abnormalities. Computers have been employed in the healthcare industry in a variety of ways such as radiology image analysis for noninvasive disease detection; analysis of biosignals (electrocardiogram [ECG], electroencephalogram [EEG] and electromyogram [EMG]) to identify diseases related to vital organs (heart, lung, brain and muscles) and in robotic surgery, etc. Every complex medical device has an embedded computer in it, and this trend will grow.

Providing long-term health care in hospitals is expensive, and is not preferred by patients unless essential. Due to the increased cost and the problem of transportation, elderly people are staying at home and need to be monitored remotely. Providing remote care requires periodic monitoring and automated transmission of certain vital signs and physiological data, such as blood sugar and blood pressure, to a medical care provider. Sending medical images requires high-quality, undistorted lossless transmission for an accurate interpretation. Privacy should be maintained during data archiving, retrieval and transmission over a secure line.

The use of computers and automation leads to: 1) improved data management; 2) reduced loss of information; 3) reduced human errors by overburdened medical staff and in data entry, resulting in fewer accidents; 4) streamlining medical care, resulting in reduced wastage and better utilization of medical resources; 5) efficient search and retrieval capability, allowing doctors to use content-based search to compare previous occurrences of similar diseases and their treatment; 6) better visualization for diagnosis; 7) more efficient and accurate analysis of data for disease diagnosis and hypothesis formation; 8) improved long-term archiving and sharing of medical data among healthcare providers; 9) reduced duplication of procedures and lab tests; 10) portability and transparency of data to improve patients' trust; 11) improved intelligent analysis of data for discovering new knowledge related to disease prediction and identification; 12) enhanced policy formulation based upon statistical evidence; 13) improved mobility of people without sacrificing health care

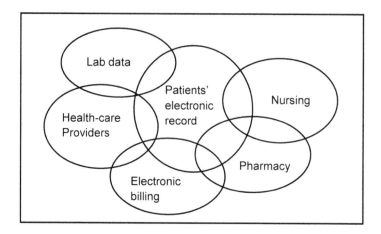

FIGURE 1.1   Types of interacting automation in health care

due to the availability of medical data and 14) reduction of paper consumption. For example, automated image analysis can diagnose malignancies in mammograms and brain MRIs (magnetic resonance imaging). Automated analysis of ECGs can facilitate the work of medical practitioners in the treatment of various heart-related diseases. Automated analysis of EEG can predict an impending epilepsy attack.

There are some disadvantages in automation:

1. Automated interpretation cannot replace the wisdom of human healthcare providers due to inherent limitations in modeling techniques that can affect overall accuracy.
2. The availability of automation encourages excessive and often meaningless data generation. Handling a large amount of data is difficult and error prone.
3. Computer programmers introduce and enforce unnecessary checks and mundane questions in the human–computer interface that take away additional time from healthcare providers, making the automation less attractive. This issue has plagued the adoption of automation tools by medical practitioners.

One problem in the slow adoption of automated healthcare systems is that the doctors and programmers do not understand each other's needs. Healthcare providers find the software overly imposing despite understanding their advantages. Both need to be educated and trained: healthcare providers should understand computational health informatics more, and information scientists should understand the needs of the care providers and patients to make their software user-friendly.

Although there are a few disadvantages, the perceived advantages are significant. Due to automation, healthcare services, pharmacy, medicine dispensing and nursing are getting seamlessly integrated. As shown in Figure 1.1, there are many components of health-automation: patients' electronic records, archiving and analysis of physiological data, improvement in time-management of nursing, pharmacy, medicine dispensing and billing. Computer-based system automation has seamlessly integrated these components in the last decade.

1.1 INFORMATICS

Informatics handles different aspects of information such as modeling a process, digitization of the information, efficient electronic archiving and retrieval of the information, transferring the information, grouping and classifying the information for enhanced data analysis and knowledge extraction, statistical analysis to identify data patterns, analysis of time-series data to identify a trend, learning from the patterns to create simple rules and new medical knowledge, interacting with other information sources to enhance existing knowledge and keeping the information current.

The key factor is the improvement of the overall system efficiency with significant reduction in processing time and required resources. In terms of health care, the resources are: 1) availability of healthcare providers; 2) availability of hospital beds and 3) support personnel hours to handle duplications. The increase in system efficiency will be: 1) improved number of patients treated; 2) faster recovery of the patients due to improved coordination and better diagnosis and 3) increased productivity.

1.2 MODELING HEALTHCARE INFORMATION

Healthcare information from patients is collected using multiple sensor-devices such as an oximeter – a device to measure the saturation level of oxygen in the blood, ECG machines – a device to collect waveforms from heartbeat, EEG machine – a device to collect brain waveforms, spectrometers and other machines. These machines provide raw data and should be interfaced to each other and central computers to exchange information. Different healthcare providers maintain their own data repositories, and the information may be stored using different formats and different coding schemes. For the exchange of information, there must be a common meta-level platform to which other information data formats can be translated.

Raw analog data is first digitized. The digitization of medical data is done using multiple incompatible formats. Software adapters translate data in one format to another. The digitized raw data is denoised, images are enhanced for better resolution and the denoised digital data is structured to gather information. This structured data is stored in an indexed database. Multiple indexed databases are connected using common index keys. This network of information is then analyzed, and data mined to gather knowledge.

One of the major discrepancies between human communication and structured database representation is the lack of natural languages in structured databases. Natural language is comprehensible to humans. However, it suffers from ambiguity and context sensitivity, and its meaning is affected by the background knowledge about a person, trying to interpret the meaning. Two experts may use different set of phrases to communicate similar information. While speaking to a nonprofessional or a beginner, experts simplify the statements to convey similar meaning. There must be means of comparing and finding similarity between two documents having the same meaning. To find similarity between two documents, one must handle *ontology*. *Ontology* uses descriptions of an entity to associate different meanings to the same words or give same meaning to the different words. Different words and phrases having the same meaning can be matched to each other using a dictionary of synonyms and antonyms, or by analyzing the overall meaning.

1.2.1 Data Abstraction

Healthcare domains have multiple entities such as patient, doctors, nurses, hospitals, medications, hospital beds, lab tests, vital organs' signals such as ECG and EEG, administrators, diseases, radiology equipment, physiological test equipment, wearable devices and monitoring devices. Each entity has multiple attributes.

An entity is modeled using only a subset of attribute-value pairs needed to solve the problem. The abstraction uses well-defined data structure to represent the subset of attribute-value pairs. We need data abstraction capabilities in the programming languages to implement the abstract models. It should be easy to implement the required abstractions to develop the corresponding software. A software uses structured data format.

Example 1.1

A patient is modeled abstractly as (*personal information, list of healthcare providers, insurance provider, disease history, list of medicines causing side-effects, symptoms, diagnosis, prognosis*). Each field is further decomposed into many subfields. For example, patients' personal information is modeled as a tuple (*patient's unique identifier such as social security number, name, address, emergency contact*).

Example 1.2

A hospital bed is modeled as a tuple: (*bed-id, location, patient-id, doctor-on-duty, nurse-on-duty, patient's entry-time, patient's condition, patient's lab data, list of medications administered, list of signals monitored*). A hospital department is modeled as an array of beds for inpatient treatment. Each monitoring device is modeled as a tuple (*type of signal, signal output, frequency of signal output, archival format of the signal*).

1.2.2 Raw Data to Information to Knowledge

The raw data coming out of the patient monitoring systems such as ECG machines, EEG machines, echograms and chart showing drug response to a disease is stored in the databases using a structured format. Data is collected using multiple modalities. Each modality is represented using a different data-format. For example, images are stored using "JPEG" format; sketches and figures are stored in "GIF" format; textual information is stored in text formats; video is stored in "MPEG" format; and sound is stored in "WAV" format. This raw data is denoised and enhanced before archiving.

A structured data that conveys some pattern is called *information*. The derived information is further analyzed to: 1) identify associations between different patterns or 2) learn generic rules that connect one or more parameters to the overall pattern. This generic form of rules or association is called *knowledge*. A key goal in the available large amount of clinical data is to extract knowledge that can improve patients' treatments or form public health policies. Generally, the formation of association or rules requires machine learning techniques that employ statistical analysis.

Exchanging information becomes difficult due to heterogeneous structuring formats. To avoid this problem, standardized universal formats are designed by information scientists. *Adapters* and *data-transformers* convert structured data represented in one format to a *universal format* and back to another structured data format.

1.2.3 Inference and Learning

The knowledge can be derived either by: 1) *logical inference* – deriving new derived knowledge rules by combining old knowledge rules or by inductively forming a rule by looking at various examples; 2) *data mining* – identifying common patterns in the data using association rules; 3) *supervised learning* – having prior knowledge of patterns and deriving similarity with the known patterns using intelligent comparison techniques; and 4) *unsupervised learning* – characterizing data elements by some features, and grouping data-elements based upon proximity of feature values using some similarity criteria. Similarity is based on the notion of distance between the feature values. The distance criteria could be the same Boolean value, same fuzzy value and Euclidean distance below some threshold for real and integer values. There are multiple artificial intelligence techniques to derive new knowledge. Collectively, discovery and learning of new knowledge is called *machine learning*.

Derivation of new knowledge has a certain amount of uncertainty due to: 1) the lack of knowledge of all the parameters affecting the outcome; 2) the presence of noise in the measurements; 3) inherent approximation present in artificial intelligence techniques used to derive knowledge and 4) limitations caused by observation, sensor resolution, noise and data processing errors. The level of uncertainty is modeled using probabilistic reasoning with its root in the statistical analysis.

1.3 MEDICAL INFORMATICS

Medical informatics is a broader term. Any informatics related to medical science will qualify as medical informatics. This involves informatics related to patient's health; informatics related to medical discovery such as drug discovery and management; discovery of new surgical techniques, tools and procedures; informatics related to data analytics from the patients' symptoms and lab tests; informatics related to policy decisions based on public health, including gender-based health and policies such as maternity, age-based health and policies such as vaccination, ethnicity-based health and policies. National Library of Medicine (NLM) defines medical informatics as *"providing a scientific and theoretical basis for the application of the computer and automated information systems to biomedicine and health affairs."*

1.3.1 Health Informatics

The United States NLM defines health informatics as "the interdisciplinary study of the design, development, adoption and application of IT-based innovations in healthcare services delivery, management and planning." The overall goal of health informatics is to seamlessly integrate the information related to health services to improve health management in a cost-effective and seamless manner. Health informatics is directly related to efficient patient care and recovery, use of computational and automated devices for disease and patients' recovery management, discovery of new procedures and medications using data analytics to improve health care, to have error-free archiving and transmission of health records to remove duplications and reduce the overall cost of health care.

Health informatics is a multidisciplinary field comprising information science/technology, biomedical technology, communication technology, computer science, social science, behavioral science, management science and statistical analysis for large-scale data analysis and hypothesis verification. The tools include clinical terminologies and guidelines, and information and communication systems, including wireless communication, to improve patient-care delivery by ensuring a high-quality patient-specific data generation, archiving, retrieval and transmission.

Health informatics utilizes: 1) the archiving, retrieval, analysis, transfer and visualization of medical data to improve patient–doctor interaction; 2) machine learning techniques to form a new hypothesis about diseases and drugs, classify data for intelligent analysis and to improve patient monitoring and 3) automation techniques to reduce the time needed to record and analyze the medical data. Automation includes software development and patient-centric human–computer interaction. Since health informatics facilitates the improvement of care providers' effectiveness and improves efficiency of patients' recovery, we should study: 1) the psychology and sociology associated with human acceptance and adoption of new techniques; 2) human perception and comprehension about data collection and analysis; 2) the performance of the system for optimality; 3) maintenance of the privacy of the patient-specific data during information exchange and transmission and 4) emulation of human–human interaction to reduce the unnecessary burden of wasting time of medical practitioners due to enhanced technical complexity.

1.3.2 Clinical Informatics

Clinical informatics is a subfield of health informatics concerned with delivering healthcare service to patients. AMIA (American Medical Informatics Association) defines clinical informatics as "application of informatics and information technology to deliver healthcare services." Clinical informatics involves many subfields such as radiology, ophthalmology, pathology, dermatology and psychology. Clinical informatics involves health signal monitoring, nursing of the inpatients, management of patient–doctor encounters, nursing care, physiological data analysis, radiology image analysis, ECG signal analysis, ophthalmological data analysis, managing the treatment record of the patients, including medications dispensing record and the procedures involved in patients' treatment. There are many further subcategories of clinical informatics such as *dental informatics, pharmaceutical informatics, nursing informatics* and *primary care informatics.* Primary care informatics involves all aspects of family practice, general internal medicine, educating patients, pediatrics, geriatrics and advanced nursing. Dental informatics involves all aspects of dental care, including dental surgery and prosthetics.

1.3.2.1 Nursing informatics

Nursing informatics is a subfield of clinical informatics that integrates nursing science with informatics. The informatics comprises management of records about admitting and discharging patients, patient data collection for archiving, hospital bed management, catheterization, pain management, signal monitoring, medication dispensing, management of therapy charts such as respiratory therapy, patient emergency response, patient recovery analysis, emergency alert system, nurse procedure charting, nursing education, improvement of human technology interfaces and developing models for integrated patient-care management.

1.3.2.2 Pharmacoinformatics

AMIA defines pharmacoinformatics as *all aspects related to using medications by the patients.* Pharmacoinformatics includes all aspects of research, analytics and development of computational techniques, including decision support systems, for prescribing, verifying and dispensing, administering, monitoring and educating the patients and care providers about the medication. Prescription includes the streamlining and automating the process of prescription, administration, verification and billing of the medication such that a medication once prescribed by the physician is automatically checked for side-effects, duplication, and permission by the insurance company before dispensing.

1.3.3 Patients' Privacy and Confidentiality

Each patient is an individual and their state of health is their private information. It cannot be leaked to any other person or organization unless the person needs to know. Even if the person knows the information, the information is for a specific purpose, and it cannot be used for any other purpose.

This regulation is to protect the patients against discrimination by the insurance companies, employers and society. For example, an employer may not employ illegally a patient with cancer; an insurance agency may not insure a person with an existing condition; a community may not allow a person with HIV (Human Immunodeficiency Virus) in the public places.

The regulation that controls this privacy is called *HIPAA* (Health Insurance Portability and Accountability Act) that was passed by US Congress in the year 1996. The HIPAA protects patients by restricting the use of health information held by entities such as doctors, other healthcare providers and their business associates. The privacy rule permits the disclosure of health information needed for patient care, important legal purposes and national security, including epidemics management.

Maintaining patients' privacy and confidentiality has close relationship with how patient-specific data is archived, transmitted, disclosed to third parties not involved in patients' treatment and portability of data. Data must be secured whether it is being archived in computer databases or being transmitted electronically or being shared among a team of medical practitioners associated with the treatment to a patient.

1.4 COMPUTATIONAL HEALTH INFORMATICS

Health informatics has various computational aspects related to individual patient care such as: 1) the collection of data from medical equipment to electronic databases; 2) archiving, retrieval and transmission of patient-related data in an automated computer database so it becomes independent of format of any specific installation; 3) data aggregation and intelligent analysis of data and resource usage for discovering new diseases, medications, treatment's automation and automated patient monitoring; 4) disease diagnosis, disease management and drug distribution; 5) statistical data analysis of patients' recovery data with similar disease; 6) quantitatively studying the effect of medicines before approval for drug administration and 7) statistical data analysis of patient signals such as ECG – heart-related signal, EEG – brain-related signals and EMG – muscle-related signals.

Computational Health Informatics pertains to using computers in health informatics. Various modeling techniques, computational techniques, algorithms and software development to improve the integration of patient-related medical information, improvement of the resource usage for medical care, improvement of medical data handling, improvement of effectiveness of medical practitioners, improvement of efficiency and the automation of health care come within the realm of *Computational Health Informatics*.

Computational Health Informatics is concerned with the development and application of computational techniques for: 1) health-related data collection from the sensors and other sources and their secure archiving in medical databases; 2) retrieval of health-related information from the distributed databases and knowledge bases; 3) secure transmission of health-related information; 4) visualization and intelligent analysis of health-related data for automated diagnosis and discovery of new knowledge; 5) converting the data back and forth from a human comprehensible format to structured data format suitable for archiving in large databases; 6) performing data analysis to improve diagnosis and treatment and 6) remote care of different types of patients and elderly persons. Lately, pervasive health care, bioinformatics and pharmacokinetics have found significant overlap with computational health informatics due to: 1) the extensive use of computational techniques in these fields; 2) discovery of genomic causes of diseases and abnormalities using computational techniques; 3) improved automation in drug and vaccine development and 4) improved remote care using the computational techniques, including intelligent analysis of automatically collected data, automated tracking of patients, and automated secure transmission and archiving of data.

Computational health informatics requires integration of many computer science subfields such as database management, computer algorithms, artificial intelligence and machine learning, signal analysis, image processing, data security and encryption, software engineering and computer networking, including wireless and sensor networks, Internet engineering and embedded computing.

1.4.1 Acceptance and Adoption

There are certain key factors for computational health informatics to succeed: 1) acceptance and adoption by the medical practitioners; 2) compliance with privacy laws; 3) acceptance and adoption by the patients and their care-providers, including close relatives; 4) cost factor to update and upgrade; 5) compatibility with the previous system; 6) support for integration with heterogeneous systems and 7) ease of training.

A key factor in developing health-informatics software is not to make the doctors and medical practitioner slave to the process established by the software designers and programmers for industrial processes. Rather, for better acceptance, adoption and maximization of the time utilization of the medical practitioners, additional effort is needed to develop intelligent software that collects the data from the medical practitioners in a natural form such as voice dictation or handwritten text, automatically convert natural language to structured data format and transform structured data format back to natural language. The intelligent software should also interact with medical practitioners and patients as a human counterpart would. It includes human courtesy and emotional intelligence.

Another key HIPAA is very important because when we transfer the information between the central database to the end user who could be a patient, or a healthcare provider, or a pharmacy, ensure that proper software filters filter out the information not needed by the end user. Due to the privacy constraints, the archived data needs additional security and encryption.

Third important aspect is acceptance by the patients and relatives. Electronic devices are looked with suspicion for many reasons: 1) violation of privacy; 2) presumed lack of response by the human care provider when an alert occurs; 3) fear of failure to operate at a critical time; 4) cumbersome entanglement and interaction with human body in terms of the form factor (weight and size); 5) technical complexity and lack of standardization resulting into a learning curve to operate; 6) lack of human courtesy and human-like interaction and 7) lack of empathy specially for elderly patients. Devices still have a bigger form factor. Recent wireless sensors are better. However, there is no technology to assess the pain and emotion of the patient just by watching the patient.

Technology is expensive to upgrade and integrate with the remaining information system. Because of this limitation, hospitals are slow to upgrade the technology. A new technology also requires training of the staff and patients to use. Unfortunately, due to the lack of standardization and backward compatibility of operations, it is difficult to learn the changes in technology.

1.4.2 Emulating Human–Human Interactions

Emulating human–human interaction is very important because humans are used to interact with emotional intelligence and courtesy, and avoid repeated details and verification. Another important requirement is to maintain the privacy of the patient-specific information in clinical data collection, archiving and transmission. Archiving, processing and transfer of medical data need to maintain patients' privacy. A healthcare provider or an insurance agency should know the patients' condition only on "Need to Know" basis according to the privacy laws of a country.

Traditionally, doctors are used to record dictations on recorders that are converted into a textual form. One will like to employ computers and intelligent techniques to understand doctors' dictations and handwriting, and convert into a structured format that can be easily processed by computers. However, different doctors express the same condition of a patient depending upon their expertise and the intended audience, using different medical phrases and different words. To understand multiple different phrases carrying similar meaning, different types of medical dictionary, thesaurus and cross-language dictionaries are needed along with the natural language understanding, generation and translation software.

1.4.3 Improving Clinical Interfaces

There are multiple ways for a physician to specify the diagnosis. However, textual descriptions do not accurately localize the abnormalities. Information about an organ defect can be conveyed by marking on the appropriate diagram of an organ. There are multiple situations where these visual interfaces are needed such as: 1) heart abnormality; 2) lung abnormality; 3) kidney abnormality; 4) liver abnormality; 5) spine abnormality; 6) neck muscle abnormality; 7) blood flow problems in different parts of the bodies to describe thrombosis (blood clotting); 8) bone fractures and 9) tumors localization in a brain. Such interfaces should automatically be translated to textual description and archived in the multimedia database.

Example 1.3

Figure 1.2 shows a cross-sectional view into a heart showing various chambers and heart valves. A physician when prompted with this visual interface can easily mark multiple heart-related abnormalities on a computer screen.

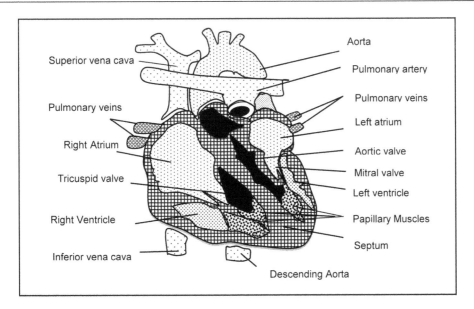

FIGURE 1.2 A cross-sectional view of heart diagram for visual interface (Figure courtesy © Dr. Purva Gawde, part of her PhD dissertation, used with permission).

1.4.4 Privacy and Security

Secure data collection, archiving and transmission facilitates: 1) availability of health care remotely by patients, including elderly care, long-term rehabilitation and chronic disease management; 2) sharing of data between healthcare providers, reducing test-duplications and improving the mobility of people; 3) improved data collection using mobile devices to record transient health conditions that cannot be reproduced; 4) integrated real-time care coordination specially for intensive care unit patients and serious injury patients where paramedics can transmit monitored data to the main hospital in real time and 5) portability of data for a patient.

1.5 MOTIVATION AND LEARNING OUTCOMES

The motivation of this course is to prepare the students who want to develop computational techniques and software in "Computational Health Informatics." Automation of health informatics requires deep understanding and integration of various subfields of computer science along with deeper knowledge of physiological processes and clinical data analysis.

The educators and software developers find a large gap in this fast-evolving automation of health industry where definitions of various subfields are still evolving. The market has created many interdisciplinary jobs that integrate health science, data science, computer science and healthcare management, including the application of computers in hospital management.

This book and the related course in "Computational Health Informatics" will reduce the gap by:

1. providing background knowledge of various clinical concepts that need automation;
2. providing knowledge of various computational concepts and intelligent techniques used for the analysis of clinical data; and
3. providing knowledge of various computational concepts and techniques used in the automation of electronic archiving, retrieval, transfer and analysis of patient-centered data.

This book is for the first course in "Computational Health Informatics" suitable for senior undergraduate students and fresh graduate students. It is also suitable for the researchers in one discipline such as computer scientists to understand the concepts and issues involved in "Computational Health Informatics" or a physician getting educated in understanding computational techniques involved in the automation process of healthcare.

This book prepares the students and the researchers to explore further in this fast-evolving field and does not provide detailed algorithms and software of various approaches described throughout this book. Detailed algorithms can be looked up in the cited research articles, and can be explored further.

1.6 OVERVIEW OF COMPUTATIONAL HEALTH INFORMATICS

This section summarizes various subfields of computational health informatics subsequently described in the following chapters. The scope of computational health informatics keeps evolving. This book describes: 1) automated data collection and medical databases; 2) seamless integration of medical data from heterogeneous sources; 3) various medical knowledge bases related to disease and medical procedures; 4) intelligent and data mining techniques used for data analysis and knowledge discovery; 5) radiology image analysis and transmission; 6) major biosignals such as ECG, EEG and EMG, their role in the related diseases and their computational analysis; 7) analysis of clinical data and the role of decision support systems in helping medical practitioners; 8) techniques and issues used in monitoring patients remotely; 9) computational techniques in analyzing genomic data and 10) pharmacokinetics and related computational techniques for studying the efficacy of administered drugs.

1.6.1 Medical Databases

Medical data can be clinical data such as radiology images of internal organs and their motion, bone fractures, pathology, medication, disease-diagnosis and data related to a large voluntary group of patients being treated for similar abnormalities. Very large databases are used to archive: 1) patient-related information; 2) medication billing and payment data; 3) multiyear EMR (Electronic Medical Record); 4) multiyear medical images such as MRI, computerized tomography (CAT) scans or X-ray images; 5) timed signal and signal analysis data such as ECG, EEG and EMG; 6) videos of surgery or movement of internal organs and 7) actual voice and written notes by the medical practitioners for authentication. These records are used for future reference and diagnosis, to avoid duplication when a patient gets second opinion from another healthcare provider and to save cost. The information is exchanged regularly between the medical organizations and medical practitioners to facilitate the treatment of patients.

Medical databases have huge memory requirement, and the data should be accessed quickly over the Internet. Medical images stored in a medical database should have high resolution for accurate diagnosis.

1.6.1.1 Electronic medical records

Different organizations require a different type of data. For example, a healthcare provider needs patient's history, including diseases, surgeries, medications and their side-effects on the patients, dosages, lab-tests and their results. In contrast, pharmacies only need to know the name of the doctor, medicines and their dosages. The billing department only needs to know the procedure codes that medical practitioner performed, stay-time in the hospital, types of treatment (regular or emergencies; inpatient or outpatient), and insurance companies need to know the procedure codes, name of the patients, any duplication of the procedure codes, and whether the procedure codes are allowed. Medical practitioners should be able to query by date, range of dates and by content-based similarity. This requires innovative information archiving/retrieval techniques.

With the availability of cheap computational power and ever-growing networking of computer, it has become possible to put all the health informatics data into large databases and share the information electronically using a secure computer network either by transmitting or by remotely accessing a central database. This automation of data requires interoperability among multiple types of databases used by health organizations. In addition, database records should be carried over the Internet to the other organization, medical practitioners or the patients.

EMR is a networked database that contains: *personal information of patients, physicians' information, patient's physician-related information, patients' history, information about patient-doctor encounters, including prescription and lab-results, information about patient monitoring, patient-reminders, information about dispensing of the medicine by the pharmacists, list of pharmacies and their information, information about prescriptions sent to pharmacies, information about hospital facilities, information about lab facilities, information about insurance agencies, information about billing, billing-audit related information, Addresses of the entities (patients, physicians, clinics, hospitals etc.), treatment eligibility information, drug-related information.* The information is related to each other using primary keys such as *patient-ids, provider-id, hospital-id, insurance-id, pharmacy-id, encounter-id,* etc. Each of these information contents has multiple fields. For example, patient information includes fields such as *(name, gender, title, occupation, employer, patient-id, social security number, driver license, date of birth, New/repeat, address, phone(s), insurance company, insurance type, emergency contact(s), etc.).* Figure 1.3 shows an interconnection of a subset of these information components related to an electronic health record (EHR) database. The textboxes show different relational tables.

The relational tables are connected through shared-key to other relational tables. For example, the textbox "patient personal information" contains the set of information for each individual patient; the textbox "physician information" contains the set of information for each individual physician; and pharmacy information contains the set of information for each individual pharmacy. "Patient-physician encounter" is a relational table containing all the information about appointment for a patient with a physician. All the edges between the relational tables show the connection between the relational tables using a common field used to identify individual records (tuples) uniquely in at least one of the relational tables.

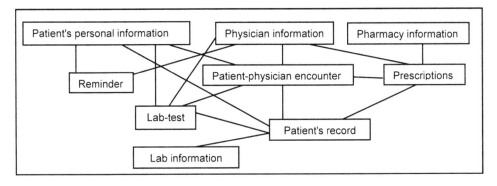

FIGURE 1.3   A small subset of EHR database connectivity

For example, "patient-id" will be a unique id for the relational table "Patient's personal information," and "Physician-id" will be a unique-id for the relational table "physician information." The field "patient-id" connects the relational table "Patient-physician encounter" with the relational table "Patient's personal information" so that related records related to patients can be retrieved. Similarly, the relational table "Patient-physician encounter" is connected to the relational table "physician information" so that physician-related information can be retrieved. Alternately, we can answer a query, what all patients a physician has seen over a period?

Medical databases can be geographically separated as each hospital has its own proprietary information about the patients that cannot be shared with others, other than the patient, without proper authorization as permitted by HIPAA. The cross-references about the same patient-ids are stored in a centralized database so that the records from other hospitals can be retrieved with no duplications as described in Section 3.3.2.

The advantages of EMR are: 1) instant access to integrated data to avoid duplication of the tests; 2) data-archiving for a long period to study the recovery of the patients; 3) analysis of disease specific data to identify patterns of parameters that can cause the diseases; 4) data analytics to study the effectiveness, toxicity (harmful effects) and proper dosage of medications; 5) integration of real-time monitoring of the patients' vital signs, lab results, diagnosis and medicine dissemination and 6) providing remote health care to elderly patients. The overall effect is to reduce the cost of medical care while providing the optimum use of the resources such as patients served per healthcare providers, patients treated per hospital bed and patients served for every nurse.

1.6.1.2 Information retrieval issues

Medical databases contain a huge amount of heterogeneous data. Database records are complex with multiple values, and they are tagged with a unique key (primary-key). Database records are stored in the secondary storage. Images like X-ray have two dimensions; MRI has three dimensions and data characterized by N-features ($N > 1$) have N-dimensions. Computational health informatics deals with images, large-size records and N-dimensional feature space due to the presence of diseases, health conditions and biomarkers that are characterized by N-variables.

Health informatics requires similarity-based comparison to compare to find similar radiology images or similar feature-values in pathology reports to find out similar disease states. Hence, any content-based search mechanism should incorporate similarity-based matching. The search can also be initiated using temporal logic such as "ECG two years before myocardial infarction." Hence, the search should be capable of handling abstract temporal information and match the record time with the query using temporal abstraction.

1.6.1.3 Information de-identification

A patient's treatment is related to patient's background, history and knowledge about the medication and medical procedures. The process of knowledge formation requires statistical analysis. Statistical analysis is not patient specific. Hence, all the individual information that can identify an individual patient needs to be removed from statistical analysis. Such information includes patient's name, patient's address, patient's geographical location, age, gender, ethnicity and patient's medical history. This is important because such information can be unlawfully used against a specific patient by the hospitals, employers, insurance agencies and individual communities. Techniques have been developed to: 1) remove identification of the patient's data before data is statistically analyzed; 2) prohibit unauthorized access and update to patients' data and 3) to ensure the authenticity of data before updating the medical databases. The topic has been discussed in detail in Section 4.10.4.

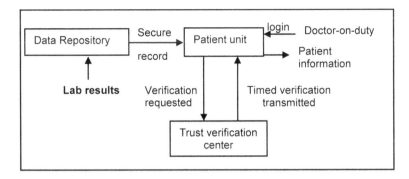

FIGURE 1.4 A schematic of trust-management system for maintaining HIPAA

1.6.1.4 Maintaining patient privacy

HIPAA can be violated by unauthorized access in the database. The database should be protected against 1) illegal accesses; 2) the hackers and 3) unauthorized access within a hospital setting where doctors-on-duty keep changing based upon the shift. If a doctor is not on duty, then HIPAA compliance prohibits him/her to see the patient-specific data.

To comply with HIPAA in a hospital situation where doctor-on-duty keeps changing, a person should be authenticated as the current healthcare provider by a trust-management unit. The doctor-on-duty can open the documents after an authentic key is supplied to him. This privilege is either kept dynamic or it is limited to certain monitors where the doctors can visualize the patient-specific data. The overall schema is shown in Figure 1.4.

1.6.1.5 Standardized medical knowledge bases

Knowledge bases have rules besides basic databases that can derive additional information by combining two or more database relations. Different types of medical knowledge bases interact with each other. Each hospital and provider's office use large standardized knowledge bases of:

1. Procedures for the diagnosis, treatment, surgeries and billing that require efficient compression and transmission over the Internet;
2. Medicines administered to patients in different diseases;
3. Synonyms and antonyms used for automated text analysis so that similarity between two seemingly different texts can be understood; and
4. Images and ECG histories of patients to compare the progression/remission rates of diseases.

1.6.1.6 Automated data collection

There are multiple sensor devices that collect patients' vital sign data, perform automated analysis of lab-data and create radiology images. This data should be automatically archived into a medical database. Each hardware device has a software-based virtual device that controls the hardware operations and collects sensor data from the electronic circuitry. To interface the information between laboratory test results, data collected by virtual devices and hospital information systems are linked using a common clinical interface format. The associated software infrastructure is called *Clinical Enterprise Service Bus* (or *Health Enterprise Service Bus*). It links different virtual devices and medical database

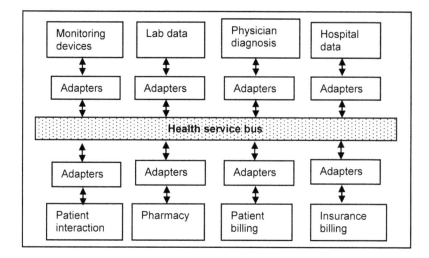

FIGURE 1.5   A schematic for automated data acquisition using health service bus

each having a different standard and format. The adapters transform back and forth the data in common format to the data format of individual virtual device and structured data suitable for archiving in a medical database. There are multiple types of adapters to provide the data transformation as illustrated in Figure 1.5.

1.6.2 Medical Information Exchange

Medical databases are stored in different organizations that use a variety of software developed by various vendors. These software packages are incompatible due to the differences in the data formats. This incompatibility necessitates the use of a common platform and the corresponding software to convert data between various formats using a common platform to support information exchange as follows: *format₁ ↔ common platform ↔ format₂*. To avoid duplication of patients' data, there must be a mechanism to cross-reference the patient records of the same patients archived in different institutions.

Many organizations are involved as end-points of data transfer. It could be a hospital, a clinic, including a remote-clinic, a health-provider's office, a patient's computer or mobile, a mobile computer in the battlefield, a pharmacy, a billing office or an insurance company. All these people are connected to each other over the Internet using a secure communication protocol. This communication pathway is called *Health Information Bus* or *Medical Enterprise Bus*, and is similar to the concept of computer architecture bus inside a computer where different peripheral devices are connected through this common bus.

1.6.2.1 Standards for information exchange

Internet-based transmission is based upon XML (eXtended Markup Language) based tools such as *SOAP* (Simple Object Access Protocol) that embeds the patient-related information to transmit to the destination where the information is again extracted. XML is an intermediate layer Internet-based language for information storage and computation, and is based upon encoding all the nested data into a flat format using user-defined tags. Using XML, complex relational databases – database expressed as a network of two-dimensional relational tables, image databases and complex graph structures can be encoded and transmitted over the Internet with ease, and can be interfaced with other web-based languages such as JavaScript, Java, Python and C#. SOAP is an XML-based communication protocol that acts as an envelope for transmitting the information from a source to the destination.

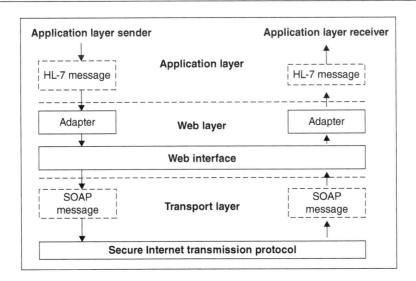

FIGURE 1.6   A schematic of Internet-based medical information exchange using HL7 format

Current technology uses Java-based or XML-based relational/object-based databases. Medical data are transmitted over the Internet using wrapper languages. Many medical XML-based wrappers have been developed for transferring medical data transfer securely between two health organizations or a health organization and a billing organization.

HL7 (Health Level 7) is a clinical markup standard that specifies the structure and semantics of the clinical documents to be transferred over the Internet. Information from a database is embedded in the HL7 format using interface software at the source-end, and is converted back at the destination-end using a software. HL7 is built on top of the SOAP protocol. Providing a common interface and transmission standard provides interoperability between heterogeneous databases/knowledge bases implemented by different vendors across different medical organizations. This interoperability provides ease of information transfer cutting down the cost by avoiding the duplication. The overall transmission layer is shown in Figure 1.6.

1.6.2.2 Types of connectivity

There are different types of communication networks to transmit the data:

1. *Medical Information Bus* (MIB) that is used to send the sensor data to a central monitoring station where it can be analyzed for any emergency condition and visualized by the nurse-on-duty. MIB is used to gather data from the sensors used to monitor the patients post-surgery, in intensive care units and during the surgery. Different sensors are developed by different vendors using their own proprietary data-format. The task of the MIB is to provide a common data-format for information exchange for data coming from different sensors. The advantage of MIB is that devices from different vendors become plug-and-play. The current standard for MIB is IEEE 11073, and it has seven layers: physical layer, data-link layer, network layer, transport layer, session layer, presentation layer and application layer. The communication standard is based upon an intermediate XML message-based language called MDDL – Medical Device Data Language. Each physical device is described as "Virtual Medical Device" (VMD) in MDDL. MDDL codes for: i) medical devices; ii) different types of alerts and iii) units for measurements. More details about clinical interfaces are given in Sections 3.1.3–3.1.5.

2. Communication between different units within the same organization with homogeneous databases is done using HL7 over the Intranet or secure cloud.

3. Communication between different organizations having heterogeneous databases is done using HL7 and various adapters (interfaces). The structured data in a database is converted to HL7 format at the source end, and is retrieved and converted back to structured data at the destination end as illustrated in Figure 1.6.

4. An *enterprise bus* or cloud is used to communicate between different types of organizations such as hospitals, data warehouses, pharmacies, patients, insurance agencies, data analytics centers, academic research centers, healthcare providers outside the hospital network. The advantage of using cloud is that large amount of data, including images can be stored in the cloud, and can be accessed on demand. There are different types of cloud: 1) secure private cloud within the firewall and 2) public clouds. An important issue is to interface the information from secure clouds within the firewall with the public cloud outside the firewall. To protect the privacy and confidentiality of the data going out of secured private cloud, the data must be encrypted sufficiently, and it can only be decrypted at the destination after verifying the encryption key or the password that can be linked to the encryption key. Some of these authentication needs are dynamic. For example, if a healthcare provider has stopped providing care to a patient, then s(he) should not be able to see the patient-related data.

1.6.3 Integration of Electronic Health Records

Multiple hospitals and medical datamarts hold patients' data and have ownership rights on these data. A patient goes to more than one hospital, medical practitioner, or pharmacy and may have more than one insurance company that pays for their bills. All these institutions need to access relevant patient-related information to treat the patient and handle billing information. There are many issues in integrating the records of patients such as: 1) problem of heterogeneity and interoperability; 2) patients being indexed differently by different health organizations; and 3) use of different incompatible software tools from different vendors by the health organizations.

Heterogeneity is caused due to the use of different operating systems, multiple incompatible standards and incompatible data-formats to archive and process data. Due to the implicit incompatibility in the data-format, there is a need for: 1) interface software that converts the data from one format to another to make it understandable across the organizations and 2) the use of middleware like Java (and Java Virtual Machine) or XML-based standard formats that have become the basis of transmitting data from one organization to another organization.

1.6.3.1 Accessing from heterogeneous databases

Another major problem in implementing an integrated record system is that health organizations have different indexing systems to access patients' records in their database. Besides HIPAA privacy restrictions and health organizations' ownership rights, different indexing system makes it impossible to share the record of the same patient across the organizations. There must be one unique index for each patient across multiple organizations to share the patient information. To solve this problem of cross-sharing the patients' information, two types of indexes are created: 1) unique index of a patient and 2) index within the organizations. The unique index is cross-referenced with the index within the organizations.

To access the information from one organization to another, following steps are taken: 1) The index of the requesting organization is converted to the unique index using a cross-reference table stored in a centralized database in a data warehouse; 2) this unique index is converted to the index of the patient in the second health organization supplying the patient information; 3) the information is retrieved; 4) information is converted into a common format and 5) at the requesting organization, information is converted back to the structured format and stored in the local database or viewed. Information can be sent directly between the organizations or through the datamart.

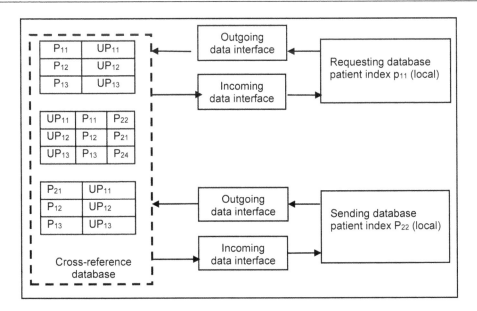

FIGURE 1.7   A scheme of data sharing between two heterogeneous databases

Figure 1.7 shows a schema for information exchange using unique patient index. Two databases correspond to requesting party and sending party. Each database is connected through the incoming and outgoing data interfaces to the cross-reference database that store patients' unique patient-id, while the databases of an individual health-organization stores its local indexing for the patients.

The cross-reference database has two types of tables: 1) a table to look up the unique patient-index given the local-indices and 2) a table to lookup local patient-id given the unique patient-index. Using these tables, the local patient-id from the requesting database is converted to local patient-id of the sending database and vice-versa. The outgoing data interfaces convert the local data format to a common data format such as HL7, and the incoming data interfaces convert the HL7 format back to the local data format.

1.6.3.2 Heterogeneity and interoperability

Institutions use software developed by different vendors. There is no consensus on a standard format between vendors for many reasons such as: 1) competition; 2) evolution of the standards based upon the improvement in the technology; 3) variations in the requirements of different institutions; 4) variations in how different medical practitioners provide the medical information and diagnostics; and 5) privacy laws needed to protect patients' privacies. Attempts are regularly made to develop a consensus format for middle-ware such as *medical procedure code, lab-test code, disease code, medical information exchange format* and *medicine code*. However, these formats also evolve as the systems and procedures continuously change in medical industry, and new information is created.

Another problem is the difference in the perceptions of the software developers and medical practitioners. Current software tools and interfaces are not sufficiently intelligent, and are template-driven. This creates adoptability problems for these tools. These tools are altered based upon the feedback from the medical practitioners. However, this causes multiple incompatible versions of formats and libraries across the medical industry. Incompatibility in documents is also caused by using different words and phrases by doctors to describe the same medical conditions to a different audience (expert vs patient), recommendations and medicines having the same basic chemical/biochemical compound but sold under a different brand name. Wordings also change with a change in the medical domain.

To solve this problem of incompatibility of operating systems, standards, formats, language libraries, the multitude of health vocabulary meaning similar things, intelligent knowledge-based software is needed to interface records and develop dictionaries that provide interoperability, and intelligent software is needed to analyze documents that would identify and match the meaning of sentences and phrases.

1.6.4 Knowledge Bases for Health Vocabulary

We need multiple dictionaries to standardize health vocabulary such as medical procedures, including surgical procedures, medicines, disease states and cross-referencing tables to identify that two names refer to the same entity. There are many popular dictionaries such as LOINC (Logical Observation Identifiers Names and Codes), MedDRA (Medical Dictionary for Regulatory Activities), SNOMED (Systematized Nomenclature of Medicine) and ICD (International Classification of Diseases).

1.6.4.1 LOINC

LOINC is a dictionary of universal code names for medical terminology related to EHRs. The use of universal standardized codes facilitates the electronic exchange of medical information such as medical procedures, including surgical procedures, lab-tests, devices used in lab-tests and clinical observations. Data-exchange standards such as HL7 and IHE (Integrating the Healthcare Enterprise) interface with LOINC codes. More details of LOINC have been described in Section 4.8.3.

1.6.4.2 MedDRA

MedDRA is a tool to encode and communicate the pharmaceutical terminologies related to the clinical tests of drugs, vaccines and drug-delivery devices under development. It covers the tests and outcomes starting from the clinical stage up to the marketing stage. Much of the clinical information such as diseases and disorders, observed signs and symptoms, drug efficacy (effectiveness of the drug), side-effects, adverse events; social relationships and family history are encoded and communicated using MedDRA. These data are pooled, analyzed, compared and verified using standardized data analytics tools. The medical terminology of MedDRA is extensive, and is structured hierarchically.

The advantage of MedDRA is to provide encoding free from language and cultural barriers. The results and outcomes are shared among the clinical researchers. It is a rich, highly specific, hierarchical, medically oriented and rigorously maintained terminology designed to meet the needs of drug regulators and the pharmaceutical industry as a shared international standard.

1.6.4.3 SNOMED

SNOMED CT (Systematized Nomenclature of Medicine – Clinical Terms) describes universal codes for medical terms such as medicine, procedures, microorganisms, diseases, synonyms, anatomy of where a disease occurs, functions and structure of medicines, chemical agents or microorganisms causing a disease, chemical name of the drugs, disease diagnosis, devices and activities used in treating diseases and social relationships associated with disease conditions. SNOMED codes are transmitted over the Internet using exchange standards such as HL7. The coding structure of SNOMED has been discussed in Section 4.8.2.

1.6.4.4 ICD

ICD is a world standard for the classification of diseases, their symptoms, their diagnostics, abnormal findings, origin and spread, complaints and social circumstances. It is supported by WHO (World Health Organization), and is used worldwide to collect statistics of the treatment, symptoms and fatalities caused by various diseases. ICD keeps getting updated as the medicines, treatments and medical procedures get updated. The current version is ICD-10. More details of ICD codes and their structure are discussed in Section 4.8.4.

1.6.5 Concept Similarity and Ontology

Ontology is described using a hierarchical networked structure (directed acyclic graph) involving class, instance, attributes, relations, semantics within a domain, restrictions and events that may affect the entity attributes and relations. Ontology is used to identify equivalency or phrases that convey the similar meaning. However, they may be stated differently depending upon the context.

Given two separate databases from different medical sources, an ontological structure needs to be created to relate the concepts within the databases. Some terms may define a concept that may be subsumed by another term using subclass relationship, while a combination of two terms may define a concept that has not been explicitly stated. This hierarchical (directed acyclic graph) structure is an integral part of *ontology*. Using transitivity between various relationships, two terms are related to each other as illustrated in Example 1.5 and Figure 1.8.

Ontology has been widely used to relate multiple heterogeneous medical databases and knowledge bases to each other by first transforming databases to ontological structure, and then equating two different looking terms using this hierarchical ontological structure.

Example 1.4

The word "growth" in the context of tumor diagnosis will mean malignancy of the tumor, and will be a cause of much anxiety. However, growth within the context of a child will be a healthy and welcome aspect. Multiple medical terminologies have been used within the same and different dictionaries that have similar meaning.

Example 1.5

A cardiologist describes to a physician that a patient has "arrhythmia" (a condition describing an irregular beat-pattern of heart); another cardiologist describes it as "tachycardia"; the third cardiologist describes to the patient as "fast heartbeat;" first cardiologist describes to another cardiologist as "ventricular arrhythmia."

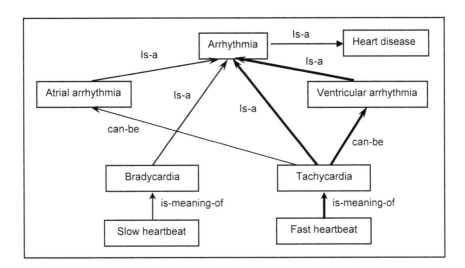

FIGURE 1.8   An illustration of hierarchical structure for the ontology in Example 1.5

A close scrutiny of the phrases shows that all four conditions are related semantically using a hierarchical network. The phrase "fast heartbeat" is related to "tachycardia" using the relation *is-meaning-of*; the entity "tachycardia" is related to the entity "arrhythmia" using the subclass relation *is-a*; the entity "ventricular arrhythmia" is related to the entity "arrhythmia" using the subclass relationship *is-a*; the entity "tachycardia" is related to the entity "ventricular arrhythmia" using a relationship *can-be*. A hierarchical structure with entities as nodes and relations as edges relates all four terms using transitivity of relations as illustrated in Figure 1.8.

In Section 1.6.4, many standards were described. These standards have overlapping domains. One database may use medical codes using one standard such as LOINC while the other may use SNOMED. Ontology will relate the two terms from different databases or match two natural language descriptions. More details about ontology description and their role in integrating information in the heterogeneous databases are described under Section 3.9.

1.6.6 Interfaces

When operating in a heterogeneous environment in a culturally diverse country, the data has to be interfaced to: 1) patient; 2) medical practitioners, including specialists, surgeons, nurses, pharmacists, paramedics and radiologists; 3) billing personnel; 4) appointment and social interaction staff and 5) data analysts and academic researchers.

The data have multiple types such as patients' family history, encounter with the medical practitioners, medication history, diagnosis from the symptoms, history of the symptoms, pre- and post-history radiology images, billing amounts along with codes. The data is exchanged between multiple heterogeneous databases (often placed behind the firewalls) and between medical practitioners using different medical terminology based upon their specialization. This information exchange requires medical interfaces.

Medical interfaces provide: 1) a seamless transition of image and data between heterogeneous databases and 2) interaction between different actors and the data collection system. The second type of interface is a user interface and varies based upon the actor (patient, doctor, nurse), type of disorder (eye disorder, heart disorder, lung disorder, various types of fractures, and brain tumors) and type of intervention (biopsy, surgery, simulation of surgery, etc.).

1.6.6.1 Visual interfaces

The major idea is to provide a natural and effortless interaction with the actor. For example, it would be much easier for a heart specialist to see an image of a heart with arteries and mark the blockage at the exact spot instead of just a textual explanation. Similarly, a patient will like a simple template-based interface to provide the information about him. A surgeon will like to know the exact location of the disorder to plan his surgery and may practice using a simulation with a realistic three-dimensional interface. The image-based interface should have zoom capability and capability to see one or more views so that the region of interest can be accurately located.

1.6.6.2 Natural language interfaces

Physicians and patients will like to understand the diagnosis in natural language, and patients will like to understand the diagnosis in simple terms devoid of complex medical terms. That means there is a need to develop user interfaces that accept and translate clinicians' natural language speech, parse the sentences and translate it to structured data into the database that can be used by computer applications. Similarly, after retrieving the information from a database, the user interfaces need to have the capability to generate equivalent context-aware natural language description.

An important aspect missing in present day user interfaces is the lack of understanding of human–computer interaction and the cognitive aspect of human comprehension in the information presentation and collection. Cognitive science studies human comprehension and performance and will help in developing better user interfaces. The development of interfaces requires middleware tools that integrate well with image-based representation and with the web-based and/or cloud-based databases.

1.6.7 Intelligent Modeling and Data Analysis

Artificial intelligence (or computational intelligence) simulates intelligent reasoning integrating mathematical (such as probabilistic reasoning, heuristic reasoning and statistical reasoning) and computational techniques (search and deductive reasoning techniques and machine learning techniques). The major advantages of computational intelligence are: 1) reduced computation time for hard problems with multiple constraints that cannot be handled easily using traditional algorithmic techniques; 2) pruning the search space to yield solution efficiently; 3) to provide the most plausible solution and explanation by storing the deduction steps; 4) to analyze data and learn from solution steps to improve the performance and 5) to analyze data to discover new knowledge.

Healthcare data analytics means: 1) analysis of lab results for disease diagnosis and recovery; 2) analysis of the lab results for better diagnosis; 3) analysis of the digitized sensor data in the emergency room, during surgery and post-surgery data to identify any alert condition; 4) analysis of medication response patterns to identify the optimum dosage that causes minimal toxicity – harmful effects on the body; 5) analysis of medical images such as MRI, CAT (computer-aided tomography) scan, X-ray, PET (positron emission tomography) scan to identify the abnormalities and diseases in vital human organs with minimum invasiveness; 6) text analysis of doctors' notes to store the relevant medical data in the electronic medical database and 7) development of various user interfaces and adapters that transform data from one format to another format to provide compatibility between distributed databases and knowledge bases.

Artificial intelligence techniques have been applied to predict: 1) the patients' recovery; 2) effective medicine dosage; 3) identify subclasses of heart diseases using ECG analysis; 4) automated mining of the clinical data to identify association patterns; 5) automated understanding of the doctor's speech; 6) automated analysis and matching of textual diagnosis and notes written by the physicians; 7) automated identification of cancer and other malignancies using the analysis of X-rays, CAT scans, MRIs, and ultrasounds, even before the malignancies become painful and harmful; 8) discovery of new biomarkers indicative of complex diseases by data analytic techniques, including statistical analysis; 9) discovery of the genetic causes of disease.

The purpose of using computational intelligence is to: 1) automate the data analytics; 2) provide better accuracy in identifying the causes and outcomes for diseases and recoveries; 3) identify meaningful patterns that will promote health care, including remote care; 4) provide an interface between the end users and the EHR for time-efficient performance; 5) provide compatibility between various data-formats and 6) predict the effect of a treatment on the patient's survivability based on clinical trials.

Intelligent Data Analytics (IDA) utilizes multiple artificial intelligence techniques such as uncertainty-based reasoning, temporal abstractions, probability-based reasoning, Markov models, including Hidden Markov Models (HMM), Bayesian networks, Fast Fourier Transforms (FFT), fuzzy reasoning, clustering, regression analysis, speech recognition, natural language understanding, data mining and visualization tools to automatically (or semi-automatically) identify meaningful patterns embedded in medical data that benefit health management, including diagnosis, patient recovery, prognosis and improving treatment.

1.6.7.1 Hidden Markov model

HMM is a probabilistic abstract transition machine where the state changes probabilistically with time from one state to one of the multiple states. In an HMM, the machine state transitions are not known. Instead, the state transition is probabilistically derived from the measured evidence. The probability of transition is derived by statistical analysis of the large sample set of examples with a known outcome. HMM is described in Section 3.7.3. It has been used to model many phenomena where time-series data involving periodic measurement of values is known. Some applications of computational health informatics are ECG analysis, recovery response to medication, speech recognition, natural language understanding model, gene detection during genome analysis, etc.

1.6.7.2 Uncertainty-based reasoning

Uncertainty-based associates qualitative uncertainty factor with an outcome given an input where the same input may lead to multiple outputs. Uncertainty-based reasoning is important because an observed phenomenon can have multiple outcomes, and all the input parameters or the mapping function between input parameters and the outcome may not be known. Uncertainty-based reasoning is more than rule-based reasoning. It differs somewhat from probability-based reasoning because the uncertainty may not add to 1.0. Probability is based on statistical analysis of large sample size of data.

1.6.7.3 Fuzzy logic

Fuzzy reasoning divides a range of values to a finite and small number of perceptive values. The advantage of fuzzy logic is that it is based upon approximate human perception and cuts down the search space significantly. Fuzzy logic is useful because all the parameters are not known about a phenomenon, and measurements are not always accurate. Fuzzy logic is used during patient–doctor interaction to describe qualitatively medical conditions of a patient. For example, a patient may state that on a scale of 1...10, the pain is eight. It is described in more detail in Section 3.3.3.

1.6.7.4 Bayesian probabilistic network

Bayesian network is used to model a probabilistic phenomenon with multiple input parameters affecting one or more outcomes. A dynamic Bayesian network repeated unfolds the static Bayesian network with time. Dynamic Bayesian network is used to model time-series data involving multiple variables such as the effect of regular periodic medications or treatment of a chronic disease where the medication is given over a long term to contain the progression of the disease.

1.6.7.5 Speech-to-text conversion

Speech-to-text conversion analyzes the speech, separates the speech into individual words separated by embedded silence periods, performs temporal and frequency-domain analysis of sound of individual words and looks up a dictionary to identify the corresponding syllables. These syllables are joined and analyzed further in a context to identify the words. These words are analyzed further to extract the information in the textual form.

Speech to text conversion is a well-developed technology, and has commercially available software. It is used to convert doctor's speech to text form. The text information is further analyzed automatically to extract information related to patient diagnosis and physician's prescription.

1.6.7.6 Text analysis and generation

Traditionally, doctors and nurses write a handwritten note in reporting the case history of a patient. Substituting these notes into computer-generated forms with a limited number of fields will limit physicians' options. Physicians use different health-related vocabulary to explain the same concept based upon their expertise and their assessment of the patients' conditions. Natural language text is also needed to explain the condition to a patient and his relatives in an easy to comprehend manner. This requires that healthcare software should be able to convert natural language summary into structured data into the database, generate natural language from structured data and find equivalence between two textual summaries.

Textual analysis for extracting information requires detection of health-domain specific words and the corresponding values (including fuzzy values). Text generation uses various templates to generate natural language, and equivalence of the two sentences is found using concept similarity and ontology. The technique has been described in Section 3.10.

1.6.7.7 Heuristic reasoning

Most of the resource allocation and scheduling problems are hard to solve algorithmically in a realistic duration. The solution to such problems can be expressed as a state-space search problem where a state is a tuple of variable-value pairs. The solution space is modeled as a huge graph where each state is a node in the graph. Each action that changes the value of one or more variable becomes the edge between the two corresponding states. Because the graph can be huge, any blind search would waste a lot of computational time to find a solution.

An alternative approach is to use a mathematical function to estimate the distance from the current state to the final state and then make sure that next move does not increase the estimated distance. This estimation of the distance using mathematical function is called *heuristics*. The use of heuristics keeps the search very focused and avoids traversal to the nodes that do not lead to final states or increase the distance. This approach reduces the computational time significantly.

In computational health informatics, allocation of any resource such as planning of bed allocation for a maximum number of patients' recoveries, reducing the overall cost of treatment, maximum utilization of physicians, maximum utilizations of equipment are optimization problems and require heuristic reasoning. Heuristics-based intelligent search techniques are described in Section 3.2.

1.6.8 Machine Learning and Knowledge Discovery

Machine learning is a computational intelligence concept used to: 1) identify different classes of the entities such that members in each class behave in a similar manner or have similar attributes; 2) identify a pattern in the data that may predict some known associated property or outcome; 3) learn to improve the solution to a task by shortening the search time in a state-space graph; 4) predict the outcome given the parameters and 5) derive new knowledge by observing and modeling a phenomenon. Machine learning is based upon characterizing an entity by the set of important attributes called *features*, and mapping the feature-values based upon some well-known previous mapping or similarity of feature-values. Similarity analysis is done using a notion of distance between the feature-values.

Many patients may have similar disease symptoms and may go through recovery based upon different set of medications due to the difference in gender, age, ethnic variations or other coexisting disease symptoms. To find the common causes of disease, recovery factors, medication dosages, and a large set of data is analyzed using machine learning techniques.

There are many machine learning techniques that have been employed in computational health informatics. Popular techniques are variants of clustering, regression analysis, decision trees, HMM, probabilistic decision network, neural networks, associative data mining and support vector machines. Various machine learning techniques used in computational health informatics are described in Chapter 3. This section briefly describes few popular techniques to give an intuitive understanding of machine learning.

1.6.8.1 Clustering

Clustering is an automated unsupervised learning technique for the classification of the data elements based upon modeling data in an N-dimensional space where each feature of a data element is a

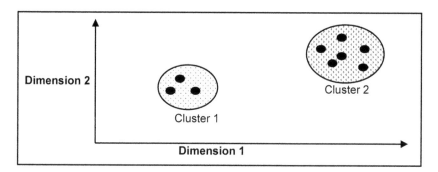

FIGURE 1.9 An example of clusters in two dimensions

dimension. Points are grouped together if the distance between the coordinate-vector for each point in a group is less than a threshold. The underlying assumption is that two entities having similar feature-values have other common attributes and behavior. Clustering has been used to automatically learn the classes of entities exhibiting similar behavior. Many types of clustering techniques have been discussed in Section 3.5.1.

Many notions of distance are used to derive the similarity between two feature-vectors. Popular ones are: *Euclidean distance, Manhattan distance* and *weighted Euclidean distance. Euclidean distance* finds the shortest straight-line path between two points. Given two points in an *N*-dimensional space as $<x_{11}, x_{12}, ..., x_{1N}>$ and $<x_{21}, x_{22}, ..., x_{2N}>$, the Euclidean distance between the points is given by $\sqrt{\Sigma_{i=1}^{i=N}(x_{1i}-x_{2i})^2}$. *Manhattan distance* finds out the sum of the absolute difference of values of the same coordinates between two points. Given coordinate vectors $<x_{11}, x_{12}, ..., x_{1N}>$ and $<x_{21}, x_{22}, ..., x_{2N}>$, the Manhattan distance is given by $\Sigma_{i=1}^{i=N}|x_{1i}-x_{2i}|$. *Weighted Euclidean distance* adds different weights to individual distance components contributed by different dimensions. Given coordinate vectors $<x_{11}, x_{12}, ..., x_{1N}>$ and $<x_{21}, x_{22}, ..., x_{2N}>$, weighted Euclidean distance is given by $\sqrt{\Sigma_{i=1}^{i=N}w_i \times (x_{1i}-x_{2i})^2}$ where w_i is the weight of the *i*th parameter. The rationale is that different parameters have different importance.

Example 1.6

Figure 1.9 illustrates the concept for a popular type of clustering called K-means clustering of nine data elements. Each data element is a vector of two feature-values. Each feature has become a dimension. Thus, we mark the data elements as points in a two-dimensional plane. Two groups of points are close to each other, and the distance between them is below a threshold. These two groups are called clusters. The assumption is that all the points within the same cluster share common properties.

1.6.8.2 Regression analysis

Regression analysis is used to predict the trend based on the curve fitting on a set of data points. Using regression analysis, the dependent variable can be predicted as a function of independent variable in Figure 1.10. The curve is fitted that minimizes the experimental error. The curve can be linear or nonlinear depending upon the problem. Linear curve fitting is called *Linear Regression Analysis*. Linear regression analysis is popular due to simple reasoning involved.

In Figure 1.10, small circles are experimental points. Using these points, a line is drawn such that it has minimum average error distance from all the experimental data. This optimum line is used to predict the value of the dependent variable for a random value of the independent variable within the valid range. Regression analysis has found many applications such as efficacy analysis and toxicity analysis of medication of a treatment regimen; effect of dosage adjustment on the recovery of patients.

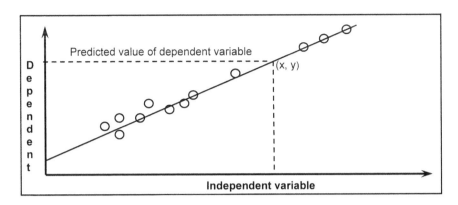

FIGURE 1.10   An example of prediction using regression analysis

1.6.8.3 Decision trees

Decision tree is an automated classification technique in which the decision-making process is based upon a tree traversal using one parameter at a time. The most discriminating parameter is toward the root of the decision tree, and is checked first. Using decision tree, we can classify different disease conditions based upon parameter values as explained in Section 3.5.3.

An entity is modeled as a set of attribute-value pairs. A data element moves down to one of the branches based on the comparison of one of the attribute value (parameter value). The process is repeated until the entity is placed in one of the classes at the end of the tree (leaf-node). This technique classifies a sample set into multiple groups based upon their common properties.

1.6.8.4 Data mining

Data mining is about deriving knowledge, patterns and association between two or more parameters affecting the outcome. It uses statistical analysis techniques to identify the association patterns between different parameters. If the values of one or more independent variables increase, then the values of the dependent variables change in a predictable way.

> **Example 1.7**
>
> A large sample size of patients is given the same medication. Some get side-effects. Suppose that the ethnicity-based analysis establishes that 90% are Afro-Americans, there is an associative pattern between side-effect and Afro-Americans.

In health care, a large amount of data is collected from sensors monitoring the patients, lab-results, medication reports, diagnosis based on the lab results and symptoms. This data is data-mined to: 1) identify a new set of parameter values that cause diseases; 2) derive effective dosage for different class of patients based upon age, gender and ethnicity; 3) identify new disease patterns; and 4) identify biomarkers – biomolecule in internal fluids from the body that indicate the presence of diseases before other detectable symptoms appear. Data Mining is described in Section 3.8.

1.6.9 Medical Image Processing and Transmission

Medical images play a major role in healthcare due to their minimally invasive nature. Without surgery, using various imaging techniques, the images of the internal organs and their functions are captured. These images are used for diagnosis such as bone fracture and deformations; ligament tears;

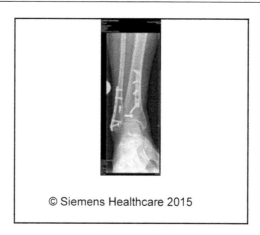

© Siemens Healthcare 2015

FIGURE 1.11   An X-ray of a fractured ankle (Images provided by Siemens Healthcare, used with written permission).

spine injuries; tuberculosis; cancer; hidden cysts and tumors in vital organs such as brain, lungs, heart, liver, pancreas and kidney; malfunctioning internal organs; and state of the fetus during pregnancies. These medical images could be a two-dimensional still image as in X-ray of a bone fracture, a cascaded sequence of images of different slices of an organ used to computationally create a 3D structure as in CAT scan and MRI, or a video of images to record or model the motion such as heart motions to understand the problems in heart wall motion during blood pumping. Before archiving, an image is preprocessed to remove the noise and enhance the image quality using image processing techniques.

Example 1.8

Figure 1.11 shows an X-ray of the fractured ankle that has been operated upon. The X-ray shows if there is any remaining problem in the healing process without performing any surgery. Similarly, Figure 1.12 shows the MRI scan of a brain that can check for any abnormality such as tumor in the brain without performing any invasive surgery. These images can be used to plan future surgery more accurately.

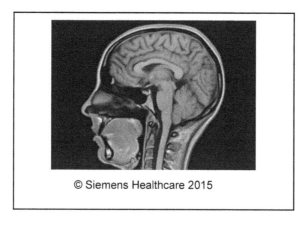

© Siemens Healthcare 2015

FIGURE 1.12   MRI scan of a brain (Images provided by Siemens Healthcare, used with written permission).

1.6.9.1 Image processing techniques

Image processing is used for automated texture analysis and compression of images in X-rays, CAT scan and MRI scan. When an image is transmitted from a hospital to a physician, or from a radiologist to the central database, it needs to be compressed to avoid congesting the transmission line. However, the compression should be such that image is not distorted. Otherwise, disease-related information will be lost. This type of compression is called *lossless compression*. Another important concept in image processing is to compare two images and find out the changes in the textures in the same physical spot: one from the past recording and one from the current recording. A change in texture shows anomalies.

Before performing any image analysis, image quality is enhanced. Improving the quality requires noise removal that involves removal of the spurious pixel intensities in the image. Image analysis can be of many types such as: 1) comparing the image of a patient's organ with an image of a healthy person's organ to identify disease-states; 2) comparing the image of a patient's organ with a past image of the same organ of the same patient to identify the progression of disease or remission of a disease such as tumor.

Before comparing two images, they must be aligned. This process is called *image registration*. After the image registration, intensities and textures of corresponding pixels are compared, and changes are recorded. The extent of enclosed segments of image with similar patterns (texture and/or intensity) is identified. The process to identify homogeneous image-regions with same intensity, color or texture is called *segmentation*. Radiology images are black-and-white images. Hence, segmentation of radiology images requires texture or intensity. A detailed discussion of medical image analysis is given in Chapter 5.

1.6.9.2 Medical image transmission

The images are transmitted using DICOM (Digital Imaging and Communication in Medicine). DICOM is an international standard for the archiving and transmission of clinical images to provide compatibility and interoperability between heterogeneous sources. The standard describes the format to store and exchange medical images and associated information between different units within a hospital and across multiple health-providers.

DICOM interfaces are available for different devices used in radiology such as MRI, CAT scan, X-ray, PET, echogram and photographic films. DICOM stores the personal information of patients in the image-header and supports protecting the patients' personal information using user-specified encryption and authentication. A detailed description of DICOM standard and image-exchange protocols is given in Chapter 6.

1.6.10 Biosignal Processing

Biosignal analysis develops techniques and tools to analyze different types of signals. In health informatics, the signals can be ECG, EEG, MEG and EMG. Signals could also be speech signals coming from the doctor's dictation about patient's condition that needs to be converted into text form to be stored within the database for further matching with past data.

Signal processing and image analysis are important for noninvasive visualization of internal organs, their functioning, their motion and any defect. This information helps in accurate diagnosis, planning the surgery, intervention before the problem becomes acute and monitoring the recovery of the patients.

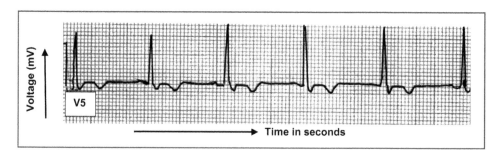

FIGURE 1.13   Repeating P-QRS-T wave patterns of an ECG from a heart patient

1.6.10.1 ECG

ECG is the recording of the electric biosignal responsible for pumping of a heart. Figure 1.13 shows repeated ECG wave patterns. These wave patterns are also called *P-QRS-T* wave pattern. Every time heart compresses and relaxes, it generates one such wave pattern based upon the traversal of electrical signals through the heart.

These electrical signals are collected through electrodes. Based upon the positions of the electrodes, we can infer various problems present in the heart. Heart may pump faster giving more *P-QRS-T* wave patterns or slower giving fewer number of *P-QRS-T* wave patterns. Change in heart-muscle's conditions, blockage in artery or fault in electric signal transmission will give abnormal *P-QRS-T* patterns that are used to predict the corresponding heart-disease.

By analyzing the variations in these wave patterns against the standard ECG of a healthy man, many heart-related diseases are predicted noninvasively without any surgical procedure. ECG is also useful to study the progression of many heart diseases for timely intervention.

In Figure 1.13, the big peak is identified as *R*-wave and occurs when a ventricle (lower chamber of a heart) contracts, and blood is pumped out. The small peak left to the big peak is called *P-wave* and is associated with the contraction of atria – the upper chambers of a heart. There are many issues in analyzing ECG signals such as: 1) removing the initial noise after reading the waveforms from the sensors; 2) automated identification of *P-waves, QRS-waves* and *T-waves*; 3) identifying the variation in the signals' height and width; 4) identifying the number of *P-QRS-T* signals in a second and their frequency of variation with time and 5) identifying the change in slope and spread of waves and the time-dependent pattern change.

Using computational techniques, noise is reduced, and signal-quality is improved. Machine learning and intelligent inferencing techniques such as data clustering, probabilistic reasoning techniques such as HMM, neural networks, Mixed Gaussian models are used to identify different classes of signals and interpret heart diseases associated with an abnormal ECG.

1.6.10.2 EEG

EEG is related to the brain waves. Brain generates different types of waves such as *alpha-waves, beta-waves* and *delta-waves*. These waves have different frequency, and their pattern changes based upon the thought patterns, attention, sleep stages and disease states. Using automated EEG wave patterns, occurrences of diseases such as epilepsy and depression can be predicted. EEG measures the surface electrical signals from the brain.

1.6.10.3 MEG

MEG (Magneto-encephalography) measures the magnetic activity within the brain due to the motion of electric currents. Brain cells interact by generating tiny electrical voltages. The resulting flow of electrical current produces magnetic fields. MEG helmet contains around 300 sensitive magnetometers that measure the generated magnetic fields. MEG is less prone to noise and are better in localizing the effects. MEG is used to discover functional behavior of the brain and cognitive processes in the brain. A combination of EEG and MEG has been applied to locate tumors in the brain to guide neurosurgery.

1.6.11 Clinical Data Analytics

Clinical data analytics is related to computational and statistical analysis of a large amount of data to form a hypothesis about the relationship of various parameters to the outcome such as survivability of a patient in the future before and after the treatment, risk of getting back the disease after a treatment and effectiveness of a treatment or medication in a random clinical trial. It heavily uses statistical analysis, probabilistic analysis, clustering and variations of regression analysis. The statistical analysis depends heavily upon: 1) hypothesis formation; 2) hypothesis verification using various data modeling curves such as Gaussian model and 3) accuracy measurement parameters such as false-positive, false-negative, true-positive and true-negative.

Clinical data analytics has become a keystone for two important tools for modern medical system: 1) evidence-based medicine; and 2) clinical decision support system (CDSS) to assist physicians by using probabilistic rule-based systems where probability is computed using statistical analysis of large sample set.

1.6.11.1 Evidence-based medicine

Patients can be treated using two methods: 1) classical knowledge-based methods used by medical practitioners based upon past practices and symptoms; 2) quantitative evidence-based information. Quantitative evidence-based methods consider quantitative measures of parameters related to the recovery of a patient or a group of patients when a treatment is administered.

Before computational analysis was available, knowledge about medical diagnosis and treatment was passed along by medical practitioners to the next generation. However, every medication (or treatment) has side-effects. Many medicines work on a large group of patients. Yet, they may not always be effective because 1) a specific ethnic group or an individual may have different genetic make-up; 2) an individual may have different reaction to the medicine or 3) an individual may have an excessive toxic response.

Statistical analysis of such information based upon genetic and environmental parameters can group the cases where medicines or treatments work differently. This quantification-based approach is called *evidence-based medicine*. In recent years, evidence-based medicine has become popular due to our capability to computationally analyze the genomic makeup of the individuals for gene-level functionality.

Evidence-based medicine integrates the classical knowledge with the scientific research-based data to provide the best individualized treatment to a patient or a group of patients grouped by ethnic makeup, geographical and environmental variation, age and gender. Computational and statistical analysis is important in evidence-based medicine as the clinical data analysis should be performed on different groups of people based upon age, ethnicity, gender, geographical locations and genomic variations. This analysis requires clustering of observed data to study group-based responses, and trends analysis of patient's recovery and outcome of a treatment for individual patients.

There are both strength and weakness in evidence-based medicine when computational techniques are used. Computational techniques are based upon the premise that all the parameters are well known, are quantifiable, can be accurately measured, prioritized and analyzed. In real world, we do not know all the parameters; we make an incomplete simplified model for the drug development; we make approximations to reduce the overhead of computational costs and we do not know how different parameters are weighted in different situations.

1.6.11.2 Survivability and hazard analysis

Survivability and hazard analysis are about the prediction of the remaining uneventful life of a patient who has gone through an episode of a potentially fatal disease such as cancer, heart attack (myocardial infarction), kidney malfunction and replacement. Survivability analysis is based upon the statistical analysis and regression analysis involving multiple parameters (risk factors and medications) that affect the outcome.

1.6.11.3 Randomized clinical trials

Before a medicine or a medical equipment or a medical procedure is introduced to treat patients, it has to go through a trial on compatible animals followed by a voluntary trial on human volunteers to study the effect of medicine on the specified disease (or abnormality) along with its associated side-effects, adverse reactions and risks on a human body. Effectiveness against the disease and side-effects are measured on a random population to avoid any bias. There are multiple types of bias such as age bias, gender bias, ethnicity bias, climate bias, environmental bias, and inherent biases of the type of patients visiting a clinic. These biases should be minimized. Multiple trials are run concurrently. The results from these trials are combined using a statistical technique called meta-trial. These results are presented to a US agency called FDA (Federal Drug Agency) that is responsible for the approval and regulation of the prescription medications.

1.6.11.4 Clinical decision support system (CDSS)

CDSS' are intelligent rule-based systems that analyze patient-related data to identify and suggest a course of action such as diagnosis, medication, dosage, treatment, including surgery. The system uses content-based information retrieval from EHR, clustering analysis, regression analysis, Bayesian decision network to form probabilistic rules. The actions are suggested based on these rules. It has been observed that healthcare provider's efficiency and accuracy improve while they are using CDSS.

1.6.11.5 Biomarkers discovery

Biomarkers are molecular scale medical signs needed to predict a disease state of a patient that can be observed with minimal invasion to the patient's body. Medical signs are different than medical symptoms. Medical symptoms are easily observed and may not be unique: there is generally many-to-many mapping between disease-states and medical symptoms. However, biomarkers are relatively unique. By looking at the biomarker, a specific disease state, a medical condition or the progression level (or remission level) of a disease can be predicted with high probability. Medical conditions can be a pharmacologic response to a medication or therapeutic intervention such as surgery, blood infusion, blood filtering as in dialysis, etc.

Biomarkers are also used to detect an environmental threat. Biomarkers can also be used to find out the efficacy of an administered drug and other serious side-effects. In a human body, biomarkers can be detected by analyzing body fluids such as blood, urine, sputum, sweat or minimally invasive biopsy.

The computational techniques to identify biomarkers are based upon data mining the proteins or molecules in blood, urine, stool, spinal fluid, sweat or other (mostly fluid) samples that are excreted from the human body and are easily extracted from the body. Microarray analysis and other biochemical and biophysical tests are done on these samples to check if certain protein-content is changing consistently and significantly in the test fluid. Microarray analysis is a bioinformatics technique to find out the variations of genes and proteins associated with different diseases (bacterial or genetic). Two or more microarray results are compared to identify the differences between the quantities of protein expressed in the test sample. This comparison result is then analyzed using computational techniques to identify one or more molecule that can act as biomarkers.

The study is repeated for a larger population of patients to ensure that the biomarker is statistically significant. The computational techniques used in identifying biomarkers involve clustering, regression analysis and other intelligent classification techniques such as support vector machines and the Bayesian network model. The classification techniques are described in Section 3.5; computational aspects of microarray analysis are described in Section 10.2.3.

1.6.12 Pervasive Health Care

Mobile health is related to using mobile devices to access or transmit health-related information remotely. Remote monitoring is also concerned about accessing and transmitting health-related information using the Internet for telehealth and teleconsultation. While mobile health is related to the use of mobile devices, remote monitoring is also used to securely monitor the symptoms of elderly persons and homecare patients, including patients with chronic diseases, over the Internet. The advantage of mobile health is to obtain nursing advice while moving, reduce hospitalization costs, provide patients with comfort in the home environment, and reduce the hazards of traveling, especially for elderly patients.

As Internet connectivity is increasing, smart phones can store and transmit data related to chronic diseases and emergency conditions effectively. Many applications are being developed that can analyze the data locally and send the information about predicted adverse conditions instantly while a patient is moving or residing in his/her home. This facility is also being used to treat the conditions that are potentially fatal and cannot be reproduced easily such as ventricular fibrillation—an irregular heart condition that can be intermittent and can be fatal if untreated.

With the availability of inexpensive sensors to measure ECG, sugar level, and other domestic tests, remote medical centers or elderly patients can send the information to the main hospital for further analysis and for storage in the centralized database. From the computational viewpoint, there is a need for software development for remote data access, data transmission formats for secure medical information delivery in a patient-friendly manner.

1.6.12.1 Patient-care coordination

With so many databases, lab tests and medical practitioners taking care of a patient, the information flow should be timely between the lab-units and medical care providers to ascertain that information is available in a timely way for medical practitioners to provide timely care to a patient. The need for clinical communication between different care providers has become imperative. It is even more important for in-patients and patients in intensive care units where delay in medication or surgical procedures due to the lack of clinical communication can cause fatality or serious side-effects.

The care team of a patient includes nurses, attending physician, surgeons and specialists such as cardiologists, urologists, dieticians, pharmacists and physiotherapists. To compound this problem of coordination, nurses work in shifts, and the incoming nurse should know fully the patients' conditions based upon the recordings of the outgoing nurse.

There are multiple types of information flows in active care coordination of a patient such as 1) nurse-to-nurse information flow; 2) doctor-to-doctor information flow; 3) nurse-to-doctor information flow and vice-versa; 4) doctor-to-lab information flow and vice versa; 5) doctor-to-pharmacist information flow for medication and 6) pharmacist-to-doctor information flow to inform about any side-effect of a drug on the patient or the unavailability of drugs. This requires the development of a different type of EHR system with a high priority of error-free clinical communications. The EHR system should also have associated intelligent software to predict the most probable next state given the current state of the patient to alert a medical care provider in time.

Another important factor in a patient's care coordination is the correct identification of patients, keeping track of her current location within the hospital and the correct identification of the disorder to be treated. The correct identification of a patient requires the use of medical tags for different types of treatments. The tagging can be done using barcode or RFID (Radio-Frequency Identification Device). For example, after getting admitted, a patient's barcode is scanned every time at the different stages of treatment to avoid human errors in communication that may cause incorrect indexing of the generated data or incorrect procedure to be performed on a patient. The barcode is used in lab tests, in-patient treatment, radiology tests such as MRI and CAT scan and surgery.

Patient's care coordination is also important in telehealth for home-based care. One of the major problems in telehealth management is the remoteness of the patient that makes collection of the patient's condition and the dissemination of the instructions challenging. This requires better understanding of the cognition capability and the physical state of the patients and developing cognitively meaningful interfaces that can translate technical instructions to the patients and care providers in an easily comprehensible manner to avoid any mistake.

1.6.13 Bioinformatics for Disease and Drug Discovery

Bioinformatics is the study of biological processes at the gene level and molecular level using computational tools and techniques. Bioinformatics utilizes the computational study of diseases at the genome level to identify the molecular basis of the diseases. Computational studies are complex and involve approximate string matching, dynamic programming, HMM, neural networks, support vector machines, graph-based modeling, graph matching techniques, Bayesian probability, regression analysis along with statistical analysis. As bioinformatics makes an inroad in the computer-assisted development of medicine, health treatment will become more personalized, as evident by *Evidence-based Medicine*.

Bioinformatics has been applied in: 1) the identification of genes and gene-groups whose aberrations cause genetic diseases; 2) the identification of genes whose absence or deletion causes genetic diseases; 3) the study of variations of metabolic pathways or signaling pathways in the presence of specific diseases; 4) the development and study of new class of drugs and their effectiveness and 5) the enhancement of the effectiveness of drug molecules by three-dimensional modeling of drug molecules and analyzing the affinity (binding effectiveness) of the functional part of the molecules.

1.6.13.1 Biochemical reactions pathways

A *biochemical reactions pathway* is the control flow of the molecular transformation (degradation and formation) within our body. There are two types of pathways: *metabolic pathways* and *signaling pathways*. *Metabolic pathways* are chains of biochemical reactions that degrade the molecules received from outside or intermediate molecules into the chemical and biochemical molecules needed by our body. The reactions in metabolic pathways are catalyzed by *enzymes*—a class of proteins used to regulate biochemical reactions.

The cells throughout the body interact with the outside world using a class of molecules called *receptors*. The interactions of these receptors with external entities such as ingested or administered biochemicals, bacteria or a virus start a chain of reactions. These reactions in response to harmful foreign bodies can: 1) identify the external molecule or foreign body; 2) tag and attack the external molecules to protect and 3) enhance the immune system. These chains of reactions are called *signaling pathway* as they are used to transmit information to body's immune mechanism to adopt and prepare the defense mechanism. The signaling pathways utilize molecular interactions to regulate reactions by a dynamic change, shown in a three-dimensional structure of molecules and splicing of the molecules as described in Section 10.1.2. The changes in molecule structures and molecular interactions can be transient.

In bioinformatics, we are interested in: 1) identifying genes and their function; 2) understanding the deviation of the cell function due to the interaction of receptor and external biomolecules and 3) studying various metabolic pathways and signaling pathways and their aberrations. We are also interested in the interactions between human cells and bacterial cells.

1.6.13.2 *Genetic disease discovery*

There are multiple causes of genetic diseases such as: 1) mutations in a nucleotide at a gene level that can result into a different amino-acid changing the three-dimensional configuration and function of the translated protein; 2) insertion and deletion of proteins from the biochemical pathways; 3) mutation in the receptors in the biomolecules affecting the binding with other molecules; 4) lack of sufficient bio-molecules and enzymes affecting the reaction rate of the biochemical pathways and 5) disturbance in the DNA-repair mechanism or protein generation mechanisms. Gene mutations can occur due to changes in environmental conditions such as Ph, salinity, osmotic pressure, temperature, and radiation to name a few.

A disease can be caused either through the pathogenicity that refers to the disease generating cause in the microorganisms, through genetic mutations, or through insertion/deletion of genes through a host pathogen (disease causing agent such as virus or bacteria). Pathogenicity includes production or deletion of chemicals (toxins) that would adversely affect the biochemical environment of the human cells. Many times, there is a delayed link between bacterial infection and diseases. For example, gastrointestinal infection is caused by pathogenic strains of *Escherichia coli*, brain infection can be caused by meningitis strains and HIV is caused by a virus.

The discovery of genetic causes of diseases is based upon: 1) comparing benign and pathogenic (disease causing) bacterial genomes to identify inserted and deleted genes and their functions; 2) comparing the genes of healthy persons and persons suffering genetic diseases; 3) comparison of the pathways and gene-expressions of a human cell in response to an infection causing agent and 4) three-dimensional structure modeling of bio-molecules that can bind with cell receptors to restrict foreign molecules binding to the receptors.

There are many computational techniques that are used in the discovery of the genes responsible for disease. They include: 1) genome comparisons that involve approximate string matching, dynamic programming, graph matching, clustering and use of multiple machine learning techniques such as HMM, probabilistic reasoning and artificial neural networks; 2) modeling three-dimensional structures of bio-molecules using simulation techniques; 3) identifying groups of gene in regulating pathways that requires microarray analysis and clustering and 4) creating multiple databases of biochemical pathways, gene functionality, protein structure and function and protein–protein interactions.

The major problem in the use of bioinformatics to identify the genetic causes of diseases is that the genome data is huge. Computational techniques have to make a lot of approximate models for building genomes from small nucleotide fragments, use statistical techniques to identify gene boundaries, use approximate string matching to compare gene sequences and make approximate three-dimensional models and graph matching techniques. All these techniques use many approximations, heuristics and proba-bilistic techniques leading to inaccuracies in disease discovery. The correct information is also lacking due to the lack of curated data from biochemical laboratories.

1.6.13.3 *Vaccine development*

Vaccine development is based upon two principles: 1) inactivating the harmful bacteria that act as an activator of the immune response 2) activated immune response inactivating bacteria, virus or unhealthy malignant (cancer) cells. The key to identifying a vaccine is to identify the genes that are active part of the harmful bacterial or cancer cells, attenuating their reaction or growth capability significantly or inactivat-ing them. Computational analysis is used to identify the pathogenic islands—the part of the bacteria that causes the harm. The choice of genes to be attenuated has to be carefully designed so the negative effect of the engineered vaccine is minimal while generating the maximum immunogenic response.

In recent times, a variation of the approach is to use recombinant DNA techniques to integrate the epitope part of the pathogen with inactive DNA that can be used as vaccine. An epitope is the part of the pathogen that binds to the human cell in a host–pathogen interaction. It is around 8 amino-acid long. The epitope should come from those proteins that interact with the surface-proteins of the human cells. By deactivating the epitope binding, a pathogen can be made inactive. Recombinant DNA technology splices and integrates genes (or gene-domains) from two species to prepare a protein in a laboratory that does not exist in nature. When injected into the human body, the human immune system generates antibodies against the recombinant protein. These antibodies also can bind to pathogen epitopes making the pathogen inactive. By creating a library of epitopes found in pathogens, generic vaccines can be developed to inactivate multiple pathogens (disease-causing bacteria or virus).

Another approach is to identify those proteins (antigens) that activate some response in human cells. This is called *immunoinformatics* approach. The antigens that can be identified by T-cells (protecting cells against foreign bodies) are the candidates for this approach.

Care is taken so that vaccine does not generate an ill response toward nonpathogenic bacteria or healthy cells and is active in the presence of varying environmental conditions. Some properties for a good protein candidates are: 1) adhesion to the surface protein; 2) lack of similarity with molecules from the host cell and 3) lack of binding with other proteins to avoid cross-protein interaction.

The role of computational tools in bioinformatics is to identify: 1) genes in pathogens causing the diseases; 2) proteins in the pathogens that bind to the cell receptors in host–pathogen interactions and 3) identifying the epitopes in the interacting surface proteins generating human-cell immune response. These tasks require a comparative analysis of the genomes to identify pathogenic genes and to study the three-dimensional configuration of the potential epitopes to estimate their binding affinity for the human cell receptors. Three-dimensional structure modeling should also ascertain that the vaccine retains the three-dimensional configuration and does not alter it configuration when interacting with antigen. Otherwise, its functionality is lost.

Multiple machine learning techniques have been used to classify and separate the epitopes. Some of the popular techniques are HMM, support vector machines, artificial neural networks and dynamic programming for approximate string matching and aligning two sequences to study their similarity.

1.6.13.4 Drug discovery

Drugs can be of many types that include antibiotics and anticancer therapy among others. Antibiotics again could be further classified based upon their action mechanism. Some actions are: 1) attacking the reproduction capability of the pathogens; 2) removing the binding capability of the pathogen; 3) weakening the surface strength of the pathogen, so it can be destroyed by the surrounding environmental factor; 4) reducing their nutrition intake capability. Computational tools used to design new drugs include: 1) chemical structure searching and matching techniques that use graph matching; 2) machine learning techniques for data mining and 3) three-dimensional structure matching techniques for finding the docking potential of the drug surface to the foreign molecule.

Drug discovery also requires the development of multiple databases such as protein–protein interaction databases, drug–protein interaction databases and protein–disease databases. Using three-dimensional simulation studies and pattern-matching techniques to identify the chemical structure of attaching biomolecules (ligand) that will dock to the surface receptors of the foreign body is the key to deliver the chemical needed to activate a foreign body. However, care has to be taken to ensure that the effects are specific to the disease-causing foreign body. Otherwise, human cells can also be destroyed causing serious side-effects.

1.6.14 Pharmacokinetics and Drug Efficacy

Pharmacokinetics is the study of the drug interactions within a human body that includes efficacy, toxicity and side-effects. Bioinformatics is used to predict and design a library of biomolecules that would be effective in treating a disease. Any modeling is an approximation due to: 1) ignoring many known parameters for the ease of modeling; 2) removing many parameters to reduce computational complexity; 3) many unknown parameters; 4) the lack of knowledge of context when a combination of parameters becomes more important; 5) transient parameters whose role changes during a biochemical process at different stages. The approximations used in a computation model cause inaccuracies.

Three-dimensional computation models for molecular structure of the genes, antigens, antibodies and their interactions are approximations. They prune the search space and predict the molecules that can act as drugs. The exact interaction of these molecules needs to be verified by biochemical experiments also called wet-lab experiments. Wet-lab experiments are first tested at the molecular level in test tubes, then modeled in animals such as rats, rabbits and pigs and finally tested on the volunteer patients to estimate their efficacy and side-effects. The study of the effectiveness and correct dosage of a drug is called *pharmacokinetics*.

After a medication is administered, it is absorbed into the cells through multiple mechanisms. The medicine can have side-effects if it remains within the system for a longer period. This phenomenon is called *toxicity*. To reduce toxicity, the ingested biomolecule and derived compounds should be flushed out of the body system shortly after its medicinal effect has taken place. The dosage of a medicine is a function of: 1) how well it reduces the disease symptoms and 2) whether it is below the tolerance level of toxicity. When a drug dosage is incremented, then toxicity also increases. The goal is to increase the efficacy of the medicine, while keeping the toxicity within a tolerance limit.

Drug dosage also differs with age and gender. To study this effect of a drug, clinical data analysis requires computational prediction techniques such as regression analysis. While *pharmacokinetics* reasons about drug availability in the body, *pharmacodynamics* is the study of the effect of drug concentration and the resulting action at a specific site, including its side-effects.

1.6.14.1 Pharmacogenetics

Another interesting emerging field is *Pharmacogenetics*. *Pharmacogenetics* is related to the study of the changes, in effect, of different medicines due to different genomic structure of individuals. The presence or the absence of a gene, variations in the genes causing a different three-dimensional structure of genes and proteins, or the variations in the concentration of certain enzymes and molecules affect related metabolic or signaling pathways. The field will get a boost with the development of more powerful computers and computational techniques. Different medicines are relevant for different persons based upon their genetic makeup. However, the field is still in its infancy.

1.7 SUMMARY

Computational health informatics deals with informatics related to healthcare management of patients. It involves computer science, medical science, health-related information, cognitive science and data analytics. Computer science is used to store, retrieve, transmit, analyze, monitor various signals and provide user interfaces between multiple actors such as patients, doctors, nurses, pharmacists, administrators, hospitals and insurance agencies.

The exchange of information due to electronic databases also provides: 1) longevity of the record for better analysis; 2) avoidance of duplicate lab tests; 3) ease of data access by the healthcare providers and 4) faster availability of data for better patient-care coordination. The major issues in developing EHRs are: 1) interoperability; 2) avoidance of duplication of records between various databases that storage of the information using different formats; 3) secure transmission of data over the web and cloud while keeping patients' confidentiality and privacy and 4) development of user-friendly intelligent interfaces that are easily comprehensible by the end users. Interoperability is needed to exchange the information between different databases. Effective patient-care coordination also needs timely clinical communication between multiple teams involved in the lab tests, diagnosis and service.

Clinical information communication is done at multiple levels: 1) between heterogeneous sensors built by multiple vendors; 2) between homogeneous databases in different units of the same organization; 3) between heterogeneous databases of healthcare organizations; 4) between healthcare organizations and external healthcare providers and 5) between healthcare organizations/healthcare providers to patients and other organizations such as insurance agencies, pharmacies and research organizations. The records are transmitted securely either through clinical messages or by using clouds that can only be accessed after proper authentications.

The archiving of multiple service procedures and transmission of information over the Internet requires standardization of the formats. Many standards have been developed in the last decade for image representation, medical procedure storage and medical information transmission. Some of the popular formats are: 1) DICOM for image storage, retrieval and transmission; 2) LOINC, ICD, SNOMED and MeddRA for medical procedures, drugs encoding and disease encoding; 3) HL7 and IHE for the transmission of information over the Internet.

Artificial intelligence techniques are used to interface patients' and doctors' interactions with the structured data stored in the databases and interfacing the data between heterogeneous databases. Patients and doctors comprehend natural language descriptions and visual interfaces better than template-driven descriptions in the databases. Hence, their interactions, including doctors' notes about patients, are given in natural language and possibly in a verbal dictation to be converted into structured information in the database. In addition, there are different levels of communication to explain the same diagnosis between a doctor and a patient and between two doctors. This requires automated text-generation capability, speech recognition capability, natural language understanding capability and the use of ontology to identify similarity between two texts.

Ontology is used to compare two descriptions of the same entity where in medical terms, entity could be patients' condition, medication to be given, and surgical or clinical procedure to be performed. To compare two terms and phrases, a semantic network along with semantic dictionaries, synonyms and antonyms is used. Semantic network is a directed graph with nodes as entities and edges as the relationship between entities.

An important aspect of medical informatics is the gathering of monitored data of vital organs that provide information about the disease-state or the patients' vitality state with minimal invasion. Radiology images and biosignals provide such information. Major radiology image techniques include X-ray, MRI, CAT scan, PET scan and ultrasound echogram. Analysis of these images facilitates the diagnosis of the disorders in multiple vital organs, including brain, heart, lungs, liver, kidney, spine, blood flow, reproductive system and digestive system. These images have been used not only for diagnosis but also for predicting the growth of the disorder, planning of complex surgeries and prediction of the recovery post-surgery.

Computational analysis of these images involves segmentation of images based upon texture and intensity and separating healthy cell from malignant cell based upon their texture variations. Computational aspect of health informatics involves representing these images in a compressed form with no loss of information, developing indexing schemes for fast retrieval, securely transmitting these images over the Internet, analyzing these images to identify disorders and performing simulation to see how the disease may progress. In addition, techniques are developed to denoise and enhance the images.

Some of the major biosignals are ECG for heart conditions; EEG and MEG for brain diseases and EMG for muscles-related disorders. The extraction and analysis of signals use multiple artificial intelligent techniques to diagnose specific diseases and abnormalities, progression of a disease, prognosis and recovery. Some of the popular computational intelligence techniques are cluster analysis, regression analysis, uncertainty-based reasoning, Bayesian networks, neural networks, fuzzy reasoning, FFT, support vector machines, data mining, rule-based systems and Markov models, including HMM.

The data analytics part requires many artificial intelligence techniques such as clustering, Bayesian network, Markov models, including HMM, probabilistic reasoning and statistical reasoning. There are two related fields, bioinformatics and pharmacokinetics, which interface well. Bioinformatics is about genetics and proteomic causes of disease and the development of biomarkers – the proteins and biochemical molecules whose expression changes as the disease progresses.

1.8 ASSESSMENT

1.8.1 Concepts and Definitions

Acceptance; adapters, adoption; artificial intelligence; artificial neural network; atrium; atrial flutter; atrial fibrillation; barcode; Bayesian network; bioinformatics; biomarker; clinical data analytics, clinical informatics; clustering; CAT; confidentiality; computational health informatics; data mining; decision tree; de-identification; DICOM; drug discovery; drug efficacy; echogram; elderly care; ECG; EEG; EHR; evidence-based medicine; fuzzy reasoning; genomics; health informatics; health information bus; health service bus; heterogeneity; heterogeneous database; HMM; HIPAA; HL7, human–computer interaction; ICD; image compression; image analysis; informatics; information; intelligent interfaces; interoperability; ischemia; K-means clustering; knowledge base; LOINC; lossless compression; lossy compression; magneto-encephalogram (MEG); MRI; Markov model; MedDRA; medical image processing; medical informatics; meta-heuristics; metabolic pathway; microarray analysis; mobile care; monitoring; MRI; natural language interfaces; natural language understanding; neural network; nursing informatics; ontology; patient coordination; patient tracking; pervasive health care; patient-care coordination; pharmacodynamics; pharmacogenetics; pharmacogenomics; pharmacoinformatics; pharmacokinetics; P-QRS-T waveform; privacy; probability-based reasoning, proteomics; regression analysis; remote care; remote monitoring; RFID, risk analysis; secure data; semantic dictionary; semantic net; side-effects; signal processing; signaling pathway; SNOMED; structured data; support vector machine; survivability analysis; text extraction; text generation; toxicity; tracking; transformers, uncertainty; universal format; user interface; vaccine; ventricle; visual interface; XML; virtual device; X-ray.

1.8.2 Problem Solving

1.1 Given a line $Y = 3x + 2$ and points $(1, 6)$, $(0, 1.9)$, $(2, 7.5)$, $(3, 11.3)$, plot the points on a graph, and find out the error in the y axis for each X-value of the point, where the error is defined as the difference of the actual Y-value and the Y-value calculated by substituting in the equation of the line. After calculating the error values, calculate the mean and variance of the error values.

1.2 For these point coordinates, calculate the Euclidean distance, weighted Euclidean distance and Manhattan distance. The weights for the weighted Euclidean distance are 0.3, 0.5 and 0.2 for the first, second and third dimension, respectively. The point coordinates are:

$\{(2, 3, 1), (3, 4, 5)\}$; $\{(1, 5, 7), (5, 6, 2)\}$; $\{(6, 8, 2), (9, 3, 4)\}$; $\{(11, 3, 4), (12, 4, 3)\}$

1.3 Given a set of points that plots the heart rate of the persons against the age with age being the independent variable, plot the points on a graph and intuitively make the clusters such that the maximum difference between any two points in the cluster is less than 8. Then, collect all the points in the clusters and find out the average heart rate for the clusters. If the young people generally have a healthy heart rate, find the outliers for the young people. Outliers are those points that do not fall in a cluster while the value of the independent variables is within the range. A point might lie in two clusters. In such cases, break the deadlock by looking at the value of the independent variables in the cluster. Write the inferences drawn from the cluster in natural language.

$(18, 72)$, $(19, 74)$, $(17, 68)$, $(23, 73)$, $(24, 75)$, $(16, 94)$, $(50, 61)$, $(52, 63)$, $(54, 120)$, $(84, 121)$, $(73, 124)$, $(76, 119)$, $(66, 126)$.

	1	2	3	4
1	0.1	0.3	0.0	0.6
2	0.3	0.1	0.3	0.3
3	0.1	0.2	0.4	0.3
4	0.5	0.0	0.4	0.1

FIGURE 1.14   Probability transition matrix for Problem 1.4

1.4 For the probability transition matrix in Figure 1.14, draw a transition-graph and identify the best transition path from state 1 to state 3. Assume that state 1 is the start state, and state 3 is the final state. A transition graph can be drawn by looking at the cell value, row index and column index. Row index or column index gives the label of a node: row index gives the source node, and column index gives the destination node. A cell stores the weight of an edge with row index as the label of the source index and the column index as the label of the destination. Give all the paths from the initial state to the final state. Path is a sequence of edges from the source node to the destination node. Put loops in parenthesis and star as superscript next to the right parenthesis. For example, (12)* means there is a loop between node 1 and node 2, and 1(21)* 2 means that the path starts from node 1, loops between and node 2 and node 1 and finally comes out of node 2. Here start node is node 1, and the destination node is node 2.

1.5 For the probabilistic transition graph in the Figure 1.15, where an edge weight shows the probability of transition, shows the corresponding matrix representation. A probability transition graph is a directed graph with edges showing the transition. A loop shows a transition to itself in the next time unit. For a missing edge between two nodes, insert "0.0" in the corresponding cell within the matrix. Label the rows and columns using the labels of the vertices in the graph.

1.6 A path is a sequence of edges. The probability of a path is a product (multiplication) of the individual transition probabilities. Derive the transition probability from V1 to V3 using the path V1 V2 V3 and the path V1 V3. By looking at the graph, derive and explain the subset of nodes each node can transition to.

1.7 Given the conditional probability of getting a side-effect of liver damage, given a medicine X is *0.001*, if a medicine is given to *9,240* people in a community, how many people are expected to have liver damage.

1.8 Make an EHR schematics diagram involving *pharmacy, physician's encounter* and *repeat order*. Repeat order is a situation where a patient suffering from a long-term ailment does not need doctor's approval to get a refill for the number of times marked in the prescription. Give all the tables, fields in the tables and the links connecting the tables.

1.9 Extend the EHR connectivity diagram in Figure 1.3 to include the billing to the insurance by a doctor, hospital and the pharmacy. Assume that an insurance company has a name, id, physical address, web address, phone number for contact and a set of policies a person can buy. Assuming that a patient has an insurance policy, update the patient's attribute and its connectivity to the insurance company.

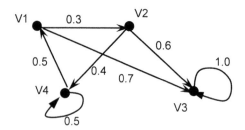

FIGURE 1.15   Modeling transition graph as a transition matrix for Problem 1.5

1.10 Given the following set of 6-tuples that describe (*the name of the patient, age of the patient, gender, disease, medicine, dosage*), give all the repeating patterns and the corresponding association rules, and explain:

(P_1, 10, F, d_1, X, 10 mg); (P_2, 11, M, d_1, X, 10 mg); (P_3, 12, F, d_1, X, 10 mg);
(P_4, 40, F, d_1, X, 30 mg); (P_5, 50, M, d_1, Y, 20 mg); (P_6, 60, F, d_1, X, 30 mg);
(P_7, 55, F, d_1, X, 30 mg); (P_8, 20, F, d_1, X, 30 mg); (P_9, 53, M, d_1, Y, 20 mg);
(P_{10}, 23, M, d_1, Y, 20 mg); (P_{11}, 16, M, d_1, X, 20 mg); (P_{12}, 74, M, d_1, Y, 20 mg).

1.11 The R-waveform of a person's heart at the age of 20 had the duration of 8 ms and a peak of 3.0 mV. Assuming that R-waveform can be modeled as an isosceles triangle, calculate the upward and the downward slopes of the R-waveform.

1.12 An arrhythmic patient's heart is beating 120 beats/min. The waveform being generated is a typical P-QRS-T waveform except that the heart is beating faster. Starting from the SA node, it takes 80 ms to traverse to the AV node from the left-atrium and 82 ms from the right-atrium. This causes 1 mV of P-wave. After a delay of 20 ms, the signal traverses in the ventricle, where first it travels faster in the right-ventricle that causes 2 mV of Q-wave followed by 20 mV of R wave that lasts for 120 ms. Followed by this, there is a small negative S-wave of 2 mV for 5 ms. A T-wave of 5 ms spread of 2 mV occurs after 70 ms of S-wave. The period ends after 25 more milliseconds. Draw two P-QRS-T waveforms and label them.

1.13 Study various diseases of renal (kidney) diseases including renal failure and draw an ontology diagram showing various relationships similar to Figure 1.8.

1.14 Given two hospital systems A and B sharing the information of two patients P1 and P2, who are given a patient-ids I11 and I12 by hospital A and I21 and I22 by hospital B, draw a universal patient indexing scheme to share the information stored by the two hospitals and draw a schematic showing the overall scheme to archive and retrieve information.

1.15 A pharmaceutical company is trying to derive the right dose of a new antibiotic for toddlers. The medicine has to maintain a cumulative concentration of 100 mg in the system to remain effective. However, a concentration of 300 mg will cause serious side-effects and toxicity. If the antibiotics in the system decays at an exponential rate with a half-life of 4 h, give one solution of dosage (amount and equally spaced time duration/day) to maintain the concentration at all time above 100 mg and below 300 mg. State your assumptions.

1.16 In curve fitting during linear regression analysis, the cumulative error is minimized. The individual error ε_i for a single-point P_i is given by *absolute value*($Y_{actual} - Y_{predicted}$). The cumulative error is given as $\Sigma_{i=1}^{i=N} \varepsilon_i$ where N is the number of sampled points. A student claims that he has derived the optimum line that minimizes the curve-fitting error. Give a counter example or use simple calculus to derive the equation of the optimum line to provide a better solution. The equation provided by the student is $Y = 2X + 1$, and the set of points are: (1, 1), (4, 5), (7, 6), (11, 12).

1.17 Many medications are given by body weight. Assume that it requires *five mg* for a *20 pounds* toddler and 10 mg for *80 pounds* boy. How much would it require for *150 pounds* adult. Show the stepwise derivation.

1.18 Write a program that takes the points in (*X-axis value, Y-axis value*) format and the equation of a line that approximately fits these points and performs these operations: 1) identifies the error each estimated point has and 2) calculates the standard deviation of the error values, where the average error is given by $\Sigma_{i=1}^{i=N} |(y_{actual} - y_{estimated})|/n$, and the standard deviation is given by $\Sigma_{i=1}^{i=N} (y_{actual} - y_{estimated})^2/(n-1)$. The number n is the number of data-points, and the vertical bar denotes absolute value.

1.19 Often, distribution plot in a clinical data is modeled using a mixed Gaussian curve. A mixed Gaussian curve combines multiple Gaussian curves with different mean and variance. A general equation for a mixed Gaussian curve is given by the function $f(x) = \Sigma_{i=1}^{i=N} A_i exp^{-\frac{(x_i - \mu_i)^2}{2\sigma_i^2}}$, where A_i is the peak amplitude at the mean μ_i of the ith Gaussian curve, and σ_i is the standard deviation of the ith Gaussian curve. Calculate the value of the function $f(x)$ for various values of x of a mixed Gaussian curve comprising two Gaussian curves having peak amplitude, mean and standard deviation as $(10, 4, 3)$ and $(12, 6, 4)$, respectively. Show a table with 10 points covering both sides of the mixed Gaussian curve.

1.20 An X-ray beam gets absorbed differently in different material: more in bone and less in tissue. If X-ray gets attenuated 20% for every cm of tissue thickness, compute and draw a graph for 0–5 cm thick tissue.

1.21 Write a program that takes three tables joined by index as shown in Figure 1.7, retrieves the information from the second hospital given the patient-name and unique hospital-Id of the first hospital from the first hospital. The data stored could be a number or an image. Images are stored using the URL linked to the file path and the file name.

1.8.3 Extended Response

1.22 Define health informatics and compare it with medical informatics.

1.23 Discuss various components of computational health informatics.

1.24 Discuss the major concern of the physicians and healthcare providers about adopting the automation software being developed by the programmer. Discuss the intelligent technologies to be developed to address this concern.

1.25 Discuss the limitations of intelligent modeling techniques.

1.26 Describe at least five artificial intelligence techniques and their applications in health informatics.

1.27 Explain the functionalities of the interfaces used in health informatics? Why do user interfaces need text generation and natural language understanding capability?

1.28 Discuss various issues involved in retrieving information from medical databases and knowledge bases.

1.29 Discuss the advantages and limitations of *clinical decision support systems.*

1.30 Discuss advantages and disadvantages of evidence-based reasoning? How are they different from traditional clinical practices?

1.31 Describe different types of signals used to study patients' vitality.

1.32 Describe P-QRS-T waveforms in ECG and their relationships to different parts of the heart.

1.33 Explain the differences between EEG and MEG biosignals.

1.34 Define ontology and explain their role in health informatics using an example not described in the book.

1.35 Explain the need of various medical standards used in health informatics? Describe briefly four popular medical standards.

1.36 Explain the need for HIPAA. Explain how HIPAA changes the information-exchanges between patient and the society and medical care provider and the society.

1.37 Explain the importance and approaches taken in data management and information-exchange to maintain privacy and confidentiality in patient-specific information.

1.38 Explain the issues in implementing information exchange in a network of heterogeneous databases. Explain briefly different techniques used to exchange information between heterogeneous databases.

1.39 Explain the issues of remote management of mobile healthcare.

1.40 Explain the role of bioinformatics and pharmacogenetics in healthcare management.

1.41 Read the cited articles on drug–drug interaction and discuss the role of informatics in drug–drug interactions.

FURTHER READING

Modeling Healthcare Information

1.1 *Atlas of Human Physiology*, Taj Book International, Cobham, Surrey, UK, 2009.
1.2 Shortliffe, Edward H., and Cimino, James J., *Biomedical Informatics – Computer Applications in Healthcare and Biomedicine*, 4th edition, Springer Verlag, London, UK, 2014.

Raw Data to Information to Knowledge

1.3 Rowley, Jennifer, "The Wisdom Hierarchy: Representation of DIKW Hierarchy," *Journal of Information Science*, 33(2), 2007, 163–180.

Inference and Learning

1.4 Mckay, David J., *Information Theory, Inference and Learning Algorithms*, Cambridge University Press, Cambridge, UK, 2003.

Medical Informatics

1.5 Bernstam, Elemer V., Smith Jack W., and Johnson, Todd R., "What is Biomedical Informatics," *Journal of Biomedical Informatics*, 43(1), February 2010, 104–110.

Health Informatics

1.6 Coiera, Enrico, *Guide to Health Informatics*, 2nd edition, Chapman and Hall/CRC Press, London, UK, 2003.
1.7 Casey, Ann, "Health Informatics Standards," In *Introduction to Nursing Informatics*, 4th edition, Health Informatics Series, Springer Verlag, New York, NY, USA, 2015, 97–144.
1.8 Hersh, William R., "Medical Informatics – Improving Healthcare through Information," *Journal of American Medical Association (JAMA)*, 288(16), October 2002, 1955–1958.
1.9 Henry, Suzanne B., "Informatics: Essential Infrastructure for Quality Assessment and Improvement in Nursing," *Journal of American Medical Informatics Association (JAMIA)*, 2(3), May/June 1995, 169–182.
1.10 Mantas, John, Ammenwerth, Elske, Demiris, George, Hasman, Arie, Haux, Reinhold, Hersh, William, et al., "Recommendations of the International Medical Informatics Association (IMIA) on Education in Biomedical and Health Informatics – 1st Revision," *Methods of Information in Medicine*, 49(2), January 2010, 105–120.
1.11 Rusu, Mircea, Saplacan, Gavril, Sebestyen, Gheorge, Todor, Nicolae, Krucz, Lonard, and Lelutiu, Cristian, "eHealth: Towards a Healthcare Service-Oriented Boundaryless Infrastructure," *Applied Medical Informatics*, 27(3), September 2010, 1–14.
1.12 Nelson, Ramona, and Staggers, Nancy, *Health Informatics – An Interprofessional Approach*, Elsevier, St. Louis, Missouri, USA, 2014.

Clinical Informatics

1.13 *Informatics Areas: Clinical Informatics*, American Medical Informatics Association, Available at https://www.amia.org/applications-informatics/clinical-informatics, Accessed July 12, 2018.

Nursing Informatics

1.14 Hannah, Kathryn J., Hussey, Pamela, Kennedy, Margaret A., and Ball, Marion J. (editors), *Introduction to Nursing Informatics*, 4th edition, Health Informatics Series, Springer Verlag, New York, NY, USA, 2015.

Patients' Privacy and Confidentiality

1.15 *Health Insurance Portability and Accountability Act of 1996 (HIPAA 96)*, 104th Congress, Public Law, 104–191, Available at https://www.gpo.gov/fdsys/pkg/PLAW-104publ191/pdf/PLAW-104publ191.pdf, Accessed May 31, 2018.
1.16 *Summary of the HIPAA Privacy Rules*, Available at https://www.hhs.gov/hipaa/for-professionals/privacy/laws-regulations/index.html, Accessed May 31, 2018.

Computational Health Informatics

1.17 Buntin, Melinda B., Burke, Matthew F., Hoaglin, Michael C., and Blumenthal, David, "The Benefits of Health Information Technology: A Review of the Recent Literature Shows Predominantly Positive Results," *Health Affairs*, 30(3), March 2011, 464–471.

1.18 Hillestead, Richard, Bigelow, James, Bower, Anthony, Girosi, Federico, Meili, Robin, Scoville, Richard, and Taylor, Roger, "Can Electronic Medical Record Systems Transform Healthcare? Potential Healthcare Benefits, Savings, and Costs," *Health Affairs*, 24(5), September/October 2005, 1103–1117.

1.19 Poissant, Lise, Pereira, Jennifer, Tamblyn, Robyn, and Kawasumi, Yuko, "The Impact of Electronic Health Records on Time Efficiency of Physicians and Nurses: A Systematic Review," *Journal of American Medical Informatics Association*, 12(5), September/October 2005, 505–516.

Overview of Computational Health Infomatics

Acceptance and Adoption

1.20 Jiang, F., Jiang, Y., Zhi, H., et al., "Artificial Intelligence in Healthcare: Past, Present and Future," *Stroke and Vascular Neurology*, 2017, 0: e000101, DOI: 10.1136/svn-2017-000101.

1.21 Avgar, Ariel C., Litwin, A.C., and Pronovost, P.J., "Drivers and Barriers in Health IT Adoptions," *Applied Clinical Informatics*, 3, 2012, 488–500.

1.22 Sassen, E.J, "Love, Hate, or Indifference: How Nurses Really Feel about the Electronic Health Record system," *Computers Informatics Nursing*, 27(5), 2009, 281–287.

1.23 Shortliffe E.H., "Strategic Action in Health Information Technology: Why the Obvious has Taken so Long," *Health Affairs*, 24(5), 2005, 1222–1233.

Medical Databases

1.24 Donaldson, Molla S., and Lohr, Kathleen N. (editors), *Health Data in Information Age*, Committee on Regional Health Data Networks, Division of Healthcare Services, Institute of Medicine, National Academy Press, Washington DC, 1994.

1.25 Lerum, Hallvard, Karlsen, Tom H., and Faxvaag, Arild, "Effects of Scanning and Eliminating Paper-Based Medical Records on Hospital Physicians' Clinical Work Practice," *Journal of American Medical Informatics Association (JAMIA)*, 10(6), 2003, 588–595.

Electronic Medical Records

1.26 DesRoches, Catherine M., Campbell, Eric G., Rao, Sowmya R., Donelan, Karen, Ferris, Timothy G., Jha, Ashish K., Kaushal, Rainu, Levy, Douglas E., Rosenbaum, Sara, Shields, Alexandra E., and Blumenthal, David, "Electronic Health Records in Ambulatory Care – A National Survey of Physicians," *New England Journal of Medicine*, 359(1), July 2008, 50–60.

1.27 Menachemi, Nir, and Collum, Taleah H., "Benefits and Drawbacks of Electronic Health Record Systems," *Risk Management and Health Care Policy*, 4, May 2011, 47–55.

Information De-identification

1.28 Garfinkel, Simson L., *De-Identification of Personal Information*, Internal Report # 8053, Information Access Division, Information Technology Laboratory, National Institute of Technology and Standard, US Department of Commerce, October 2015, Available at http://dx.doi.org/10.6028/NIST.IR.8053, Accessed August 18, 2018.

Maintaining Patient Privacy

1.29 Hu, Jiankun, Chen, Hsiao-Hwa, and Hou, Ting-Wei, "A Hybrid Public Key Infrastructure Solution (HPKI) for HIPAA Privacy/Security Regulations," *Computer Standards and Interfaces*, 32, 2010, 274–280.

1.30 Huang, Hui-Feng, and Lu, Kuo-Ching, "Efficient Key Management for Preserving HIPAA Regulations," *The Journal of Systems and Software*, 84(1), 2011, 113–119.

Medical Information Exchange

1.31 *Introduction to HL-7 Standards*, Available at http://www.hl7.org, Accessed May 26, 2018.

1.32 Hooda, Jagbir S., Dogdu, Erdogan, and Sunderraman, Raj, "Health Level-7 Compliant Clinical Patient Records System," In *Proceedings of the ACM Symposium on Applied Computing*, March 2004, 259–263.

1.33 Shnayder, Victor, Chen, Bor-rong, Lorincz, Konrad, Fulford-Jones, Thaddeus R.F., and Welsh, Matt, *Sensor Networks for Medical Care*, Harvard Computer Science Group Technical Report TR-08-05, 2005, Available at https://dash.harvard.edu/handle/1/24829604, Accessed May 31, 2018.

1.34 Walker, Jan, Pan, Eric, Johnston, Douglas, Ader-Milstein, Julia, Bates, David W., and Middleton, Blackford, "The Value of Health Care Information Exchange and Interoperability," *Health Affairs*, January 2005, W5-10–W5-18.

Integration of Electronic Health Records

1.35 *Integrating the Healthcare Enterprise*, Available at http://www.ihe.net/, Accessed May 31, 2018.

1.36 Sujansky, Walter, "Heterogeneous Database Integration in Biomedicine," *Journal of Biomedical Informatics*, 34(4), August 2001, 285–298.

Heterogeneity and Interoperability

1.37 Martinez-Costa, Catalina, Menárguez-Tartosa, Marcos, Fernández-Breis, Jesuado T., "An Approach for the Semantic Interoperability of ISO EN 13606 and OPENEHR Archetypes," *Journal of Biomedical Informatics*, 43(5), October 2010, 736–746.

1.38 Ryan, Amanda, "Towards Semantic Interoperability in Healthcare: Ontology Mapping from SNOMED-CT to HL7 Version 3," In *Proceedings of the Second Australasian Workshop on Advances in Ontologies*, 72, December 2006, 69–74.

Knowledge Bases for Health Vocabulary

1.39 Aspden, Philip, Corrigan, Janet M., Wolcott, Julie, and Erickson, Shari M., "Health Care Data Standards," Chapter 4, In *Patient Safety: Achieving a New Standard for Care*, National Academy of Press, Available at http://www.nap.edu/download/10863, Accessed August 18, 2018.

Interfaces

Visual Interfaces

1.40 Opoku-Boateng, Gloria A., "User Frustration in HIT Interfaces: Exploring Past HCI Research for a Better Understanding of Clinician's Experiences," In *Proceedings of the Annual Symposium of American Medical Informatics Association (AMIA)*, November 2015, 1008–1017.

1.41 Rind, Alexander, Wang, Taowei D., Aigner, Wolfgang, Miksch, Silvia, Wongsuphasawat, Krist, Plaisant, Catherine, and Schneiderman, Ben, "Interactive Information Visualization to Explore and Query Electronic Health Records," *Foundations and Trends in Human-Computer Interaction*, 5(3), February 2013, 207–298

1.42 Schneiderman, Ben, Plaisant, Catherine, and Hesse, Bradford W., "Improving Healthcare with Interactive Visualization," *Computer*, 46, May 2013, 58–66.

Natural Language Interfaces

1.43 Cawsey, Alison J., Webber, Boney L., and Jones, Ray B., "Natural Language Generation in Health Care," *Journal of American Medical Informatics Association*, 4, 1997, 473–482.

Intelligent Modeling and Data Analysis

1.44 Acampora, Giovanni, Cook, Dianne J., Rashidi, Parisha, and Vasilakos, Athanasios, "A Survey of Ambient Intelligence in Healthcare," *Proceedings of the IEEE Inst. Electrical and Electronics Engineering*, 101(12), December 2013, 2470–2494.

1.45 Lucas, Peter J. F., "Bayesian Networks in Biomedicine and Healthcare," *Artificial Intelligence in Medicine*, 30, 2004, 201–214.

1.46 Owens, Douglas K., and Sox Jr., Harold C., "Medical Decision Making: Probabilistic Medical Reasoning," In *Medical Informatics: Computer Applications in Healthcare*, Addison-wesley, Boston, MA, USA, 1990, 70–116.

Machine Learning and Knowledge Discovery

1.47 Jothi, Neesha, Rashid, Nur A. A., and Husain, Wahidah, "Data Mining in Healthcare – A Review," In *Procedia Computer Science*, 72, 2015, 306–313.

1.48 Tomar, Divya, and Agarwal, Sonali, "A Survey on Data Mining Approaches for Healthcare, *International Journal of Bioscience and Biotechnology*, 5(5), 2013, 241–266.

Medical Image Processing and Transmission

1.49 Deserno, Thomas M., Handels, Heinz, Maier-Hein, Klaus H., Mersmann, Sven, Palm, Christoph, Tolxdorff, Thomas et al., "Viewpoints on Medical Image Processing: From Science to Application," *Current Medical Imaging Reviews*, 9, 2013, 79–88.

Biosignal Processing

1.50 Liang, Hualou, Bronzino, Joseph D., Peterson, Donald R. (Editors), *Biosignal Processing: Principles and Practices*, 1st edition, CRC Press, Boca Raton, FL, USA, October 2012.

1.51 Yao, Shanshan, Swtha, Puchakayala, and Zhu, Yong, "Nanomaterial-Enabled Wearable Sensors for Healthcare," *Advanced Healthcare Material*, 7, 2018, DOI: 10.1002/adhm.201700889.

Clinical Data Analytics

1.52 Byar, David P., "Why Databases Should Not Replace Randomized Clinical Trials," *Biometrics*, 36, June 1980, 337–342.

1.53 Darlenski, Razvigot B., Neykov, Neyko V., Vlahov Vitan D., and Tsankov, Nikolaï K., "Evidence-based Medicine: Facts and Controversies," *Clinics in Dermatology*, 28(5), October 2010, 553–557.

1.54 Every-Palmer, Susana, and Howick, Jeremy, "How Evidence-Based Medicine is Failing Due to Biased Trials and Selective Publication," *Journal of Evaluation and Clinical Practices*, 20, 2014, 908–914.

1.55 Georgiou, Andrew, "Data, Information and Knowledge: The Health Informatics Model and its Role in Evidence-Based Medicine," *Journal of Evolution in Critical Practice*, 8(2), June 2002, 127–130.

1.56 Sheridan, Desmond J., and Julian, Desmond G., "Achievements and Limitations of Evidence-Based Medicine: The Present and Future," *Journal of American College of Cardiology*, 68(2), July 2016, 204–213.

Survivability and Hazard Analysis

1.57 Lee, Elisa T., and Go, Oscar T., "Survival Analysis in Public Health Research," *Annual Reviews Public Health*, 18, 1997, 105–134.

Randomized Clinical Trials

1.58 Kennedy, Harold L., "The Importance of Randomized Clinical Trials and Evidence-Based Medicine: A Clinician's Perspective," *Clinical Cardiology*, 22, 1999, 6–12.

1.59 Spieth, Peter M., Kubasch, Anne S., Penzlin, Ana I., Illigens, Ben M-W., Barlinn, Kristian, and Siepmann, Timo, Randomized Controlled Trials – A Matter of Design," *Neuropsychiatric Disease and Treatment*, June 2016, 12, 1341–1349.

Clinical Decision Support System (CDSS)

1.60 Agency for Healthcare Research and Quality. Health Information Technology. Clinical Decision Support (CDS). https://healthit.ahrq.gov/ahrq-funded-projects/clinical-decision-support-cds. Accessed April 12, 2017.

1.61 Fox J, and Thomson R. Clinical Decision Support Systems: A Discussion of Quality, Safety and Legal Liability issues, In *Proceedings of the American Medical Informatics Association Symposium*, San Antonio, TX, USA, November 2002, 265–269.

Biomarkers Discovery

1.62 Karley, Dugeshwar, Gupta, Deepesh, and Tiwari, Archana, "Biomarkers: The Future of Medical Science to Detect Cancer," *Journal of Molecular Biomarkers and Diagnosis*, 2(5), 2011, DOI: 10.4172/2155-9929.1000118.

Pervasive Healthcare

1.63 Bozadzievski, Trajko, and Gabbay, Robert A., "Patient-Centered Medical Home and Diabetes," *Diabetes Care*, 34, April 2011, 1047–1053.

1.64 Alemdar, Hande, and Ersoy, Cem, "Wireless Sensor Networks for Healthcare: A Survey," *Computer Networks*, 54(15), October 2010, 2688–2710.

1.65 Eren, Halit, and Webster, John G. (editors), *Telehealth and Mobile Health – The E-medicine, E-health, M-health, Telemedicine and Telehealth Handbook*, Chapman and Hall/CRC Press, July 2017.

1.66 Istepanian, Robert S. H., Laxminarayan, Swamy, and Pattichis, Constantinos S. (editors), *M-health – Emerging Mobile Health Systems*, Biomedical Engineering International Book Series (editor: Micheli-Tzanakou, Evangelia), Springer Verlag, 2006.

1.67 *mHealth New Horizons for Health through Mobile Technologies*, Global Observatory for eHealth Series, World Health Organization, 3, 2011, Available from: http://www.who.int/goe/publications/goe_mhealth_web.pdf, Accessed May 31, 2018.

1.68 Pantelopoulos, Alexandros, and Bourbakis, Nikolaos G., "A Survey of Wearable Sensor Based Systems for Health Monitoring and Prognosis," *IEEE Transactions on Systems, Man, and Cybernetics – PART C: Applications and Reviews*, 40(1), January 2010, 1–12.

Patient Care Coordination

1.69 Paré G, Jaana M, and Sicotte C., "Systematic Review of Home Telemonitoring for Chronic Diseases: The Evidence Base," *Journal of American Medical Informatics Association*, 14, 2007, 269–277.

1.70 Singer, Sara J., Burgers, Jako, Friedberg, Mark, Rosenthal, Meredith B., Leape, Lucian, and Schneider, Eric," Defining and Measuring Integrated Patient Care: Promoting the Next Frontier in Health Care Delivery, *Medical care Research and Review*, 68(1), June 2010, 112–127.

Bioinformatics for Disease and Drug Discovery

1.71 Bansal, Arvind K., "Bioinformatics in Microbial Biotechnology – A Mini Review," *Microbial Cell Factories*, 4(19), June 2005, DOI: 10.1186/1475-2859-4-19.

1.72 Bansal, Arvind K., "Role of Bioinformatics in the Development of Anti Infective Therapy," *Expert Review of Anti-Infective Therapy*, 6(1), 2008, pp. 51–63.

1.73 Fuller, Jonathan C., Burgoyne, Nicholas J., and Jackson, Richard M., "Predicting Druggable Binding Sites at the Protein–Protein Interface," *Drug Discovery Today*, 14(3/4), February 2009, 155–161.

1.74 Hirschhorn, Joel N., Lohmueller, Kirk, Byrne, Edward, and Hirschhorn, Kurt, "A Comprehensive Review of Genetic Association Studies," *Genetics in Medicine*, 4(2), March/April 2002, 45–61.

1.75 Martin-Sanchez, F., Iakovisis, I., Nørger, S., Maozo, V., Groen, P. de, Lei J. Van der, et al., "Synergy between Medical Informatics and Bioinformatics: Facilitating Genomic Medicine for Future Health Care," *Journal of Biomedical Informatics*, 37(1), February 2004, 30–42.

1.76 Tenenbaum, Jessica D., "Translational Bioinformatics: Past, Present and Future," *Genomics, Proteomics and Bioinformatics*, 14(1), February 2016, 31–41.

Biochemical Reactions Pathways

1.77 Bild, Andrea H., Yao, Guang, Chang, Jeffrey T., Wang, Quanli, Potti, Anil, Chasse, Dawn, et al., "Oncogenic Pathway Signatures in Human Cancer as a Guide to Targeted Therapies," *Nature*, 439(19), January 2006, 353–357.

1.78 Berg, Jeremy M., Tymockzo, John L., Stryer, Lubert, *Biochemistry*, 5th edition, (editor: Freeman, W. H.), New York, NY, USA, 2002.

1.79 Michal, Gerhard, and Schomburg, Dietmar (Editors), *Biochemical Pathways – An Atlas of Biochemistry and Molecular Biology*, 2nd edition, John Wiley and Sons, Hoboken, NJ, USA, 2012.

1.80 Rosenthal, Mirium D., and Glew, Robert H., *Medical Biochemistry: Human Metabolism in Health and Disease*, 1st edition, John Wiley and Sons, Hoboken, NJ, USA, 2009.

1.81 Silva, Doutor P., *A General Overview of the Major Metabolic Pathways*, Available at http://homepage.ufp.pt/pedros/bq/integration.htm, Accessed August 17, 2018.

Genetic Disease Discovery

1.82 Chautard, Emilie, Thierry-Mieg, Nicolas, Ricard-Blum, Sylvie, "Interaction Networks: From Protein Functions to Drug Discovery. A Review," *Pathologie Biologie*, 57, 2009, 324–333.

1.83 Hirschorn, Joel N., Lohmueller, Kirk, Byrne, Edward, Hirschhorn, Kurt, "A Comprehensive Review of Genetic Association Studies," *Genetics in Medicine*, 4(2), March/April 2002, 45–61.

1.84 Hunter, David J., "Gene-environment Interactions in Human Diseases," *Nature Reviews Genetics*, 6, 2 April 2005, 87–298.

Vaccine Development

1.85 Plotkin, Stanley A., and Plotkin, Susan L., "The Development of Vaccines: How the Past Led to the Future," *Nature Reviews*, 9, December 2011, 889–893.

1.86 María, Ribas-Aparicio R., Artuo, Castelán-Vega J.A., Paulian Monterrubio-López G., Gerardo, Aparicio-Ozores," The Impact of Bioinformatics on Vaccine Design and Development," *INtechopen*, September 2017, DOI: 10.5772/intechopen.69273, Available at https://cdn.intechopen.com/pdfs/55711.pdf, Accessed August 17, 2018.

1.87 Soria-Guerra, Ruth E., Nieto-Gomez, Ricardo, Govea-Alonso, Diana O., Rosales-Mendoza, Sergio, "An Overview of Bioinformatics Tools for Epitope Prediction: Implications on Vaccine Development," *Journal of Biomedical Informatics*, 53, February 2015, 405–414, DOI: 10.1016/j.jbi.2014.11.003.

Drug Discovery

1.88 Bakail, May, and Ochsenbein, Francoise, "Targeting, Protein-Protein Interactions, A Wide Open Field for Drug Design," *Computes Rendus Chimie*, 19(1–2), January/February 2016, 19–27.

1.89 Buchan, Natalie S., Rajpal, Deepak K., Webster, Yue, Alatorre, Carlos, Gudivada, Ranga C., Zheng, Chengyi et al., "The Role of Translational Bioinformatics in Drug Discovery," *Drug Discovery Today*, 16(9/10), May 2011, 426–434.

1.90 Katsila, Theodora, Spyroulias, Georgios A., Patrinos, George P., and Matsoukas, Minos-Timotheos, "Computational Approaches in Target Identification and Drug Discovery," *Computational and Structural Biotechnology Journal*, 14, 2016, 177–184, DOI: 10.1016/j.csbj.2016.04.004.

1.91 Modell, Ashley E., Blosser, Sarah L., and Arora, Paramjit S., "Systematic Targeting of Protein-Protein Interactions," *Trends in Pharmacological Sciences*, 37(8), August 2016, 702–713.

1.92 Percha Bethany L., and Altman, Russ B., "Informatics Confronts Drug-Drug Interactions," *Trends in Pharmacological Sciences*, 34(3), March 2013, 178–184.

Pharmacokinetics and Drug Efficacy

1.93 Pleuvry, Barbara J., "Pharmacodynamics and Pharmacokinetic Drug Interactions," *Anesthesia and Intensive Care Medicine*, 6(4), 2005, 129–133.

Fundamentals

2

Chapter 1 Introduction

Computational health informatics is an interdisciplinary science borrowing information from medical science including physiology and anatomy, statistics and probability and computer science, including image processing, artificial intelligence and data science. This chapter builds up the needed background in the areas of data modeling and multidimensional search, image and multimedia modeling, statistical concepts, probability, databases, biosignals and human physiology.

2.1 DATA MODELING

2.1.1 Basic Data Structures

A *sequence of data-elements*, denoted within two angular brackets "<" and ">," is modeled as a *linked-list*, *an array* of data-elements or a *vector* of data-elements. A *vector* can be extended at runtime. The elements within an array or vector are indexed, providing faster access. The data-elements in a linked-list are accessed sequentially: all the elements before the desired element have to be traversed first. The advantage of a linked-list is in: 1) representing sparse data structure having too many null (or zero) elements and 2) problems that require many insertions and deletions of elements. Insertions and deletion at a specific indexed-location in an array and vector are slow due to shifting of the following data-elements after the operation. A *record* of the form $(x_1, x_2, ..., x_N)$ is modeled as a tuple (or struct) having N ($N > 1$) fields depending upon the programming language being used. A two-dimensional matrix is modeled as a two-dimensional array. An *N-dimensional space* used in clustering and regression analysis is modeled as N-dimensional array. A *sparse-matrix* is modeled as an array of linked-lists where each cell of a linked-list stores a *(column, value)*.

A *straight line* is of the form $Y = aX + b$, where the slope a and the Y-intercept b are real numbers (or integers). In computational informatics, regression analysis data uses a line-segment that is limited in range between $X_{lower-bound}$ and $X_{upper-bound}$. This straight-line equation along with the range information can be modeled as a quadruple of the form *(lower-bound, upper-bound, a, b)*.

Often, medical data are visualized using *pie-charts*. A pie-chart is modeled as a sequence of *circular-segments* quantifying different constituting components. Each circular-segment is modeled as a nested-tuple of the form *(identifier, (start-angle, interval, color))*. Pie-charts are used for population-based classification.

Example 2.1: Pie chart

Information about types of patients in a hospital is modeled as a pie chart: *20% children, 30% youth-females, 30% youth-males and 20% senior citizens.* The corresponding pie chart will be a sequence of four triples: *<(children, (0, 72, color₁)), (youth-females, (72, 180, color₂)), (youth-male, (180, 288, color₃)), (senior-citizen, (288, 360, color₄))>.*

32	17	4	18
210	98	49	90
40	95	47	80
33	35	43	65

a: Macro-block intensity

2	1	1	1
7	4	2	3
2	3	2	3
2	2	2	3

b: Interval maps

FIGURE 2.1   Computing histogram for an image macro-block: (a) Macro-block intensity; (b) Interval maps

2.1.1.1 Histograms

A *histogram* is a statistical summarization that contains frequencies of similar values within a short interval for a range having multiple intervals. Since histogram compacts the information, it is used for quick-retrieval of multimedia entities and multidimensional entities having similar content.

A *histogram* is modeled as a vector of triples of the form (*lower-bound, upper-bound, frequency*), where a *lower-bound* is the smallest value permissible in a bucket (interval), and *upper-bound* is the largest value permissible in the bucket. An interval is a pair of the form (*lower-bound, upper-bound*). If all the buckets have the same size, then the information about lower-bound and upper-bound is dropped from individual tuples. *Equal-interval* histograms are modeled as a *vector of frequencies* starting from the first interval.

Example 2.2: Histogram

Consider an intensity-map of the pixels in a *4 × 4* segment of a radiology image in Figure 2.1. To make a histogram consisting of eight intervals, intensity-values are divided into eight buckets each having *32* values: interval #1 contains intensity values *0–31*; interval #2 contains intensity values *32–63*; interval #3 contains intensity values *64–95* and so on. Figure 2.1b shows the mapping of the intensity values to the corresponding intervals. The frequency of the same interval values in this mapping gives the histogram. The overall histogram is modeled as a vector of eight frequencies: <*3, 7, 4, 1, 0, 0, 1, 0*>.

2.1.1.2 Records in database

Records in databases are modeled as a tuple of the form (*primary-key value, field$_1$ value, field$_2$ value, field$_3$ value, …, field$_N$ value*) ($N \geq 1$). A *primary-key* is a unique field in a relational-table, and is used to access the corresponding tuple using some indexing scheme such as *hash-functions*. A tuple has M fields ($M \geq 2$) possibly of different data types. The major criterion in choosing a proper data structure to represent a relational table is based on the ease of data maintenance, data-access and update. A field could be a simple data item such as an integer, a real number, a string; alternately, it could be a complex information such as an image file, a data file and a video file. Data files have different format representations depending upon the data types. For example, images are represented as .JPEG or .GIF format; video is represented as .MPEG or .AVI format; and binary is represented as .bin format.

Example 2.3: Database record

Let us a consider a patient-record containing his social security as primary-key with additional information: name, age, date of birth, weight and height. It will be modeled as (*Social-security-no, name, age, DOB, weight, height*).

2.1.2 Modeling N-Dimensional Feature-Space

For archiving and accessing points in an N-dimensional feature-space, the simplest technique is to model the space as N-dimensional matrix, where each dimension in the array represents a dimension in the data-space. Figure 2.2 shows modeling of a two-dimensional feature-space.

	#1	#2	#3	#4
#1	0	5	0	8
#2	0	0	7	0
#3	9	0	0	0
#4	0	0	11	0

FIGURE 2.2 An illustration of a two-dimensional matrix

An N-dimensional space is modeled as: 1) an N-dimensional array or 2) a hash-table of linked-lists of tuples indexed by a hash-function as illustrated in Figure 2.3. In the second approach, applying a hash-function on index-value/primary-key is many-to-one mapping resulting into collision of records. To resolve the collisions, the colliding data-tuples are chained using a linked-list. Each cell in the chain contains the information (*index-tuple/primary-key, data-item*). Data-items can be a value, a tuple or a pointer to a complex data structure or file. The major advantages of a hash-table scheme are: 1) index-space is not fixed; 2) it is suitable for mapping sparse matrices and 3) it provides near constant-time data-access.

Example 2.4: Hash-function for data-access

A two-dimensional matrix ($N = 2$), given in Figure 2.2, has been mapped using a hash-table and a simple hash-function $\Sigma_{i=1}^{i=N} index_i / size\,(hashtable)$, where $index_1$ denotes the row number, and $index_2$ denotes the column number. The hash-function adds all the index-values and divides by the hash-table size that is a prime-number for even distribution of tuples in the hash-table. Even distribution reduces the collision. Since there are 5 nonzero data-items in the table, table-size has been chosen as the next higher prime number 7. The corresponding hash-function is (*sum-of-all-indices modulo 7*). The index-pairs are $(1, 2) \longmapsto 3$; $(1, 4) \longmapsto 5$; $(2, 3) \longmapsto 5$; $(3, 1) \longmapsto 4$; $(4, 3) \longmapsto 0$.

The corresponding hash-table and the stored data-items are illustrated in Figure 2.3. Each cell stores a triple: (*row-index, column-index, data value*). The additional overheads in this scheme are: 1) the application of the hash-function on the index-values of the data items and 2) sequential search in the cells. However, a good choice of hash-function will evenly distribute the data-items limiting maximum collision-factor to three minimizing sequential search.

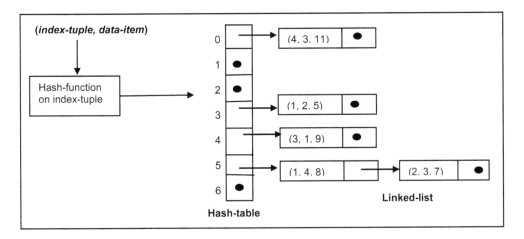

FIGURE 2.3 Hash-table-based storage of N-dimensional sparse data-space

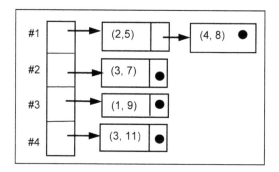

FIGURE 2.4   Representing matrix in Figure 2.2 with only nonzero elements

Two-dimensional space is a special case: images are modeled as two-dimensional bit-map matrices. A two-dimensional matrix can also be represented as an array (or vector) of a linked-list (or vector) of pairs of the form (*column number, data-item*) as illustrated in Figure 2.4. Since only nonzero entries are stored, and indexing is done on row number, and column numbers need to be stored within the cell along with the data-value. This scheme saves memory-space because it stores only nonzero entries. The search time does not increase significantly because the number of elements within each row is sparse.

Example 2.5: Representing a sparse matrix

Consider the matrix in Figure 2.2. The first row has two nonzero entries; rows two, three and four have one nonzero entry each. Each linked-list is a sequence of pairs of the form (*column-index, data-value*). Row #1 has two tuples: (*2,5*) and (*4,8*). Row #2 has one tuple (*3,7*); row #3 has one tuple (*1,9*) and row #4 has one tuple (*3,11*).

2.1.2.1 *Proximity of multidimensional data*

Each N-dimensional data is plotted as a point in an N-dimensional space. There are multiple techniques to estimate distance between these N-dimensional data-points such as *Manhattan distance, Euclidean distance, Minkowski's distance, weighted Manhattan distance* and *weighted Euclidean distance*. Given two N-dimensional tuples $(x_1^1, x_2^1, ..., x_n^1)$ and $(x_1^2, x_2^2, ..., x_n^2)$, *Manhattan distance* is defined as the cumulative absolute-magnitude of value-differences between the corresponding fields in each tuple as described in Equation 2.1. The vertical bars denote absolute-magnitude. *Euclidean distance* is defined as the square-root of the sum of square of the difference between the values of the corresponding fields in both the tuples as shown in Equation 2.2. *Minkowski's distance* is a generalization of *Euclidean distance*, and is defined by Equation 2.3 where vertical bars denote absolute-magnitude, and q is a positive integer.

$$Manhattan\ distance = \sum_{i=1}^{i=n} \left| x_i^1 - x_i^2 \right| \tag{2.1}$$

$$Euclidean\ distance = \sqrt{\sum_{i=1}^{i=n} \left(x_i^1 - x_i^2 \right)^2} \tag{2.2}$$

$$Minkowski's\ distance = \left(\sum_{i=1}^{i=n} \left(\left| x_i^1 - x_i^2 \right| \right)^q \right)^{\frac{1}{q}} \tag{2.3}$$

Weighted distance is based upon giving variable weights to each dimension based upon their importance. In feature-based reasoning, certain features are more important than others. Important features are given more weight. Weights are also given to normalize the variations caused by different units of measurements in different fields of a tuple. A weighted measurement of distance multiplies each of the N factors corresponding to different dimensions by a weight w_i $(1 \leq i \leq N)$ as shown in Equations 2.4 and 2.5. The cumulative sum of all the weights adds up to 1.0.

$$weighted\ Manhattan\ distance = \sum_{i=1}^{i=n} \left(w_i \times \left| x_i^1 - x_i^2 \right| \right) \tag{2.4}$$

$$weighted\ Euclidean\ distance = \sqrt{\sum_{i=1}^{i=n} w_i \times \left(x_i^1 - x_i^2 \right)^2} \tag{2.5}$$

2.1.3 Modeling Graphs

Graph-based reasoning is very important for modeling real-world phenomenon in health informatics. Many data-modeling and inferencing techniques in artificial intelligence model data as graphs. For example, Markov-models, Hidden Markov Models, Bayesian networks, EHR database, knowledge representation, ontology, data mining and Internet-based search are based upon graph-based modeling, and are extensively used in health informatics.

A graph is a pair of the form (set of vertices, set of edges). Each vertex has its own attributes, list of incoming edges and a list of outgoing edges. Each edge has its own attributes including edge-label, edge-weight, source-node and destination-node.

A graph is classified as an *undirected graph*, an *undirected weighted graph*, a *directed graph* or a *directed weighted graph*. A *weighted graph* has a weight associated to each edge in the graph. A *directed graph* has directed edges in one direction: the presence of the directed edge shows connectivity only from the source-node to the destination-node. An *undirected graph* has all the edges traversed in either direction. A graph may have *cycles* in it. A *cycle* is a path starting from a node and reaching the start-node again traversing no edge in the path twice. A *directed acyclic graphs* (DAG) has no cycle.

A graph is modeled using one of the three techniques: 1) a matrix; 2) a set of nodes and connecting edges where nodes represent the entities and the connecting edges represent the relationships between the corresponding entities and 3) a set of edges as 5-tuple of the form (*source-node, destination node, edge-weight, relationship, tuple of other attributes*). In the absence of edge-weight, relationship or the attributes, the corresponding field is dropped.

2.1.3.1 Modeling graphs as matrices

Graphs can be modeled as a two-dimensional matrix such that row indices and column indices are used to model vertices (nodes) in the graph, and the matrix-cells (i, j) correspond to the edge connecting the vertex "i" to the vertex "j." The value inside the cell shows the weight of the corresponding edge. The absence of an edge between two nodes in a graph corresponds to a value of *0.0* in the corresponding cell. The matrix is symmetrical around the diagonal for undirected graphs and asymmetrical for directed graphs.

An *adjacency matrix* is used to model graph connectivity. The values in the cells of the adjacency matrix are 0 or 1. If the two nodes i and j are disconnected, then the corresponding cell (*row i, column j*) has a value 0. Otherwise, the cell value is 1. The value between a node i to itself is *0*. A matrix for a weighted graph can be mapped to an adjacency matrix by mapping (*value* > 0) → 1 in the corresponding cell.

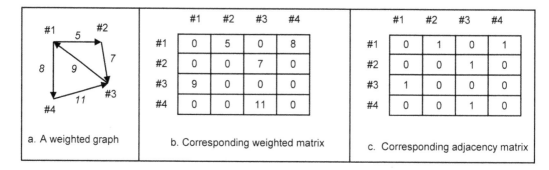

FIGURE 2.5 Graph- and matrix-based representations: (a) A weighted graph; (b) Corresponding weighted matrix; (c) Corresponding adjacency matrix

Matrix-based representation is very useful in solving problems because nodes can be easily accessed using indexing, and connectivity between any two nodes can be modeled using matrix-based searches. The adjacency matrix requires less memory for large-sized matrices.

Example 2.6: Matrix-based representation of a directed graph

Figure 2.5a shows a directed weighted graph. Node #1 has two directed edges to nodes #2 and #4; node #2 has one outgoing edge to node #3; node #3 has one directed edge to node #1 and node #4 has one directed edge to node #3. Figure 2.5b shows the corresponding matrix with the corresponding cells showing edge-weight. A source-node is shows as a row number, and the destination-node is shown as a column number. Figure 2.5c shows the corresponding adjacency matrix.

2.1.3.2 Modeling graphs as a set of vertices

Given a graph with m vertices, the graph can be modeled as a vector (or array) of size m such that each cell of the vector (or array) holds a sequence of tuples containing the destination-nodes, weight and relationship of the outgoing edges from the source-vertex. The embedded sequence is modeled by a vector or a linked-list depending upon the data structures allowed in a specific language. Abstractly, the representation is modeled as a sequence of tuples where each tuple is of the form ($vertex_i$, <($vertex_j$, $weight_{1j}$), … ($vertex_k$, $weight_{ik}$)>) where $vertex_j$ and $vertex_k$ denote the destination-nodes from the source-node $vertex_i$. In the presence of a specific relationship associated with an edge, the relationship also becomes a part of the tuple.

Example 2.7: Modeling graphs as sets of vertices

The graph in Figure 2.5a can be modeled as shown in Figure 2.6. Vertex 1 has two edges: (*1*, *2*) with weight *5* and (*1*, *4*) with weight *8*; Vertex 2 has one edge: (*2*, *3*) with weight *7*; Vertex 3 has one edge: (*3*, *1*) with weight *9* and Vertex 4 has one edge: (*4*, *3*) with weight *11*. Figure 2.6 is the same as Figure 2.3. Not surprisingly, because the edges correspond to nonzero entries in the matrix representation.

2.1.3.3 Modeling graphs as a set of edges

Given a graph with m vertices and n edges, each edge is modeled as a pair (*source-vertex index, destination-vertex index*). A hash function $\mathcal{H}$ is designed so $\mathcal{H}$(*source-vertex index, destination-vertex index*) maps to a *hash-table index*. Since applying the hash-function leads to possible collision, each hash-table cell

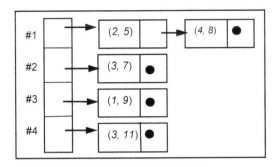

FIGURE 2.6 Modeling graph as a vector of vertices connected to outgoing edges

points to a linked-list. Each cell in the linked-list stores a tuple (*source-vertex index, destination-vertex index, edge-weight, relationship*). Figure 2.7 illustrates the modeling.

Example 2.8: Modeling graphs as a set of edges

Consider the graph in Figure 2.5a. There are five edges: (*1, 2*) with weight *5*; (*1, 4*) with weight *8*; (*2, 3*) with weight *7*; (*3, 1*) with weight *9* and (*4, 3*) with weight *11*. Since there are five edges, we can take a hash-table size of *7*, which is a prime number > 5. The choice of a prime number will facilitate even distribution with less collision. The hash function is $\Sigma_{i=1}^{i=N} index_i / size(hashtable)$. Using this hash function, the edge (*1, 2*) ↦ *3*; edge (*1, 4*) ↦ *5*; edge (*2, 3*) ↦ *5*; edge (*3, 1*) ↦ *4* and edge (*4, 3*) ↦ *0*. Only keys (*1, 4*) and (*2, 3*) collide. Figure 2.7 shows the corresponding data allocation in the hash-table with key. Figure 2.7 is the same as Figure 2.4 because a graph corresponds to the nonzero entries in a matrix.

2.1.4 Trees for Database Search

A *tree* is a *directed acyclic graph* where there is a root node, and the path moves from the root node toward the leaf node. Tree is an important structure in representing data for searching because of the branching based upon value-comparison. A *binary tree* is the simplest tree used in searching a single-dimension sequence. The search time in a binary tree is logarithmic. For eight elements, $\log_2(8) = 3$ comparisons are needed; for *32* elements, five comparisons are needed.

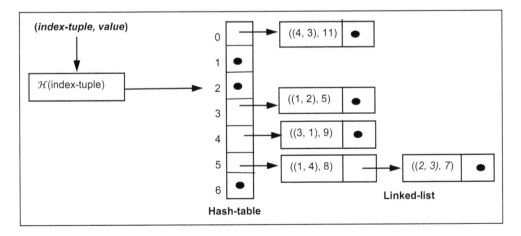

FIGURE 2.7 Modeling graph as a set of edges

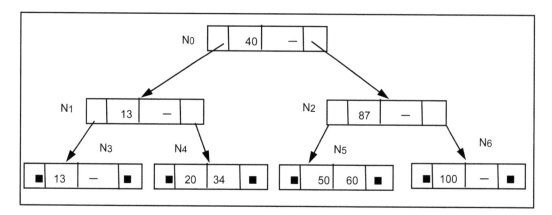

FIGURE 2.8   Interval based search in single dimension

2.1.4.1 Interval-based search

A point-based binary tree can be easily extended to interval-based binary tree by: 1) keeping the points and intervals sorted in a monotonic order; 2) making the binary tree using points and the lower-bound of intervals and 3) keeping the information about the upper-bound of the interval in the node that correspond to an interval. Interval-based trees are very useful in temporal reasoning where a time has to be searched in a time interval in the database, or two-time intervals are matched for a common interval.

> **Example 2.9: Binary tree for interval-based search**
>
> Consider a data set consisting of the points and intervals as {5, 100, [20, 34], 40, 87, 13, [50, 60]}. Sorting all the points and minimum of the intervals in monotonic ascending order organizes the dataset to {5, 13, [20, 34], 40, [50, 60], 87, 100}. The binary tree is organized using the set {5, 3, 20, 40, 50, 87, 100} as shown in Figure 2.8. The nodes N_4, N_5 have the maximum value in the slot; the point nodes leave the maximum value slot empty. There can be many variations to reduce the space wastage if the number of intervals is sparse.

2.1.4.2 Limitations of binary trees

Binary trees have these major limitations: 1) inability to model multidimensional region or N-dimensional space involving multidimensional data-values; 2) branching is based upon single value inequality (greater-than or less-than) comparison in integer, real or string domain; 3) It does not support the multidimensional keys and proximity-based search and 4) the organization of a tree with values in internal nodes does not support page-based organization and retrieval of large-size records from the secondary storage required in the databases.

2.1.4.3 B+ trees for database access

A *B+ tree* is an N-ary tree that has at most N children where $(N > 2)$. It holds the data-elements (or pointer to the data-elements) at the leaf-nodes and uses nonleaf nodes only for the comparison of the key-values. Branching factor b (number of children for an internal node) follows the constraint $\lceil N/2 \rceil \leq b \leq N$ where $\lceil N/2 \rceil$ denotes the ceiling of $N/2$. If the numbers of children become less-than the value $\lceil N/2 \rceil$ after deletion of an element, then neighboring nodes are merged, and tree is restructured by rearranging the ancestor-node. If the number of children exceeds N, then the corresponding node is split, a new ancestor node is added, and the tree is restructured. Restructuring the tree may lead to more

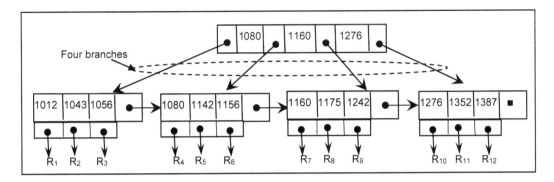

FIGURE 2.9 An illustration of B+ tree for retrieving database records

node-splits (or node-merges). Large value of the branching factor N makes the tree flatter more suitable for page-based storage of tree. Keeping the pointers to the records at the leaf-node makes it suitable for searching large records stored in secondary storage, and loading multiple records at a time using page-based loading from the secondary storage to RAM (Random Access Memory). Figure 2.9 illustrates a quaternary *B+ tree* ($N = 4$).

Example 2.10: B+ tree

Consider a B+ tree that has maximum branching of four. It organizes 12 key-values representing the patient-ids. The key values are *1012, 1043, 1056, 1080, 1142, 1156, 1160, 1175, 1242, 1276, 1352, 1387*. Each key is associated with a big record. The B+ tree is inside the memory (RAM), and the records are stored in secondary storage. The nodes with $N = 2$ will have one key-value; the node with three branches will have two key-values. The *key-values(left-subtree)* are less than the *key-values(right-subtree)*. One such modeling of this information using B+ tree is given in Figure 2.9. The root node has three keys and four branches. To derive the exact branch at the root-level, at most three comparisons are made. At the leaf-node level, at most three comparisons are made. The tree is flatter than a binary tree due to a large branching factor that facilitates its storage in the same page of RAM.

2.1.4.4 PATRICIA tree – fast string-based search

Trie (Retrieval) is a tree used to search the strings. PATRICIA tree is a storage optimized trie. PATRICIA stands for "Practical Algorithm to Retrieve Information Coded in Alphanumeric." PAT-tree is an acronym for *PATRICIA tree*. The tree is arranged so that most frequent mutually exclusive prefixes occur first, and the corresponding branch to the child is labeled by the prefix. The process is repeated until the last character from the string is consumed. Pat-tree is useful to retrieve radiology images or pathology reports based on the caption name that may include patient-name, disease-name and date of image-collection.

Example 2.11: PATRICIA tree

Consider these strings: "ababc, abc, abd, mabc, ababd, bcbcd, bcbbc, bcd, md." The analysis shows that prefix "ab" occurs four times, prefix "bc" occurs three times and prefix "*m*" occurs two times. The three terms also do not share any common prefix. The root node has three branches: prefix "ab" as leftmost branch, prefix "bc" as second branch and prefix "*m*" as the rightmost branch. The remaining suffixes are further split based on the frequency analysis and mutual exclusiveness of the prefix to avoid ambiguity. The resulting tree is shown in Figure 2.10.

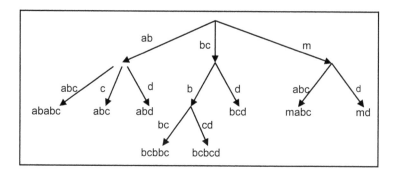

FIGURE 2.10 A PAT-tree for faster string-based search

2.1.5 Spatial Trees for Multidimensional Data

Spatial trees are used to model points in N-dimensional space. Any entity characterized by N-features (*N* > 1) is modeled using *spatial trees*. Each feature (or attribute) becomes a dimension, and the object is mapped to a point in N-dimensional space.

Broadly, the spatial trees are classified as: 1) trees that model an N-dimensional space based upon coordinate-values; 2) trees indexed upon Euclidean distance between the N-dimensional points. In the first category, there are multiple trees such as quad trees; K-D (K dimensional) tree and R-tree (Rectangular tree) and its variants such as R+ tree, SS (Similarity Search) tree, and SR (Spherical Rectangular) tree. Under the second category, there are VP-tree (Vantage Point tree) and its variants such as MVP (Multiple Vantage Point) tree. VP-tree and its variants are indexed based upon distance in feature-space, and are useful for similarity-based searches in queries involving high number of dimensions. In addition, *Hilbert curves* map two-dimensional images as a single dimensional sequence of binary values so the proximity of pixel-distances in the images are nearly preserved. *Hilbert curve* and R-tree have been combined to have *Hilbert R-tree* for modeling the image-regions. The variants of spatial trees have been discussed in the corresponding sections.

Search on spatial trees is used to find multimedia objects similar to an object in N-dimensional space. Spatial trees are very useful in health informatics. Pathology test data, electronic health records and multimedia objects, including radiology images, are modeled using N-features. For example, tumor region(s) in a radiology image is modeled using two-dimensional trees; tumor in MRI is modeled as a three-dimensional tree; pathology report having *K* (*K* > 1) different attributes are modeled as K-dimensional trees. A similarity-based search is used to identify similar radiology images or pathological data in a database and study the corresponding prognosis, treatment and recovery rate.

2.1.5.1 Quad tree

Two-dimensional regions can be represented using quad-trees where ROI (region of interest) or quadrant inside ROI is progressively divided into $2^2 = 4$ quadrants (see Figure 2.11) until the segment is homogeneous and has only one attribute. Further splitting of the homogeneous region stops, and the leaf-node is labeled with the corresponding attribute-value.

Each node has four children except the leaf-node. Each nonleaf node stores the information within the region besides the pointers to the four children. The information related to a region can be a single value or a set of attribute-values.

Example 2.12: Quad tree

Figure 2.11 illustrates a quad tree used to represent the shaded region in a two-dimensional space. There are three types of quadrants: 1) shaded quadrant denoted by S; 2) unshaded quadrant denoted

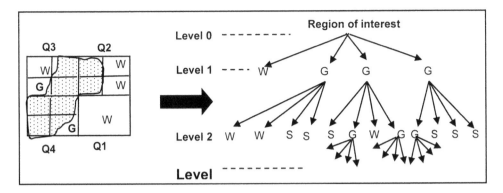

FIGURE 2.11 A quad-tree representation to represent two-dimensional region of interest

by W and 3) partially shaded quadrant containing a set of attribute-values {S, W}, denoted by G. The ROI becomes the root node. In the example, ROI has multiple attribute-values. Hence, it is split into four quadrants.

The first quadrant is unshaded. Hence, it is marked "W," and is treated as a leaf-node. Second, third and fourth quadrants have set of attribute-values {S, W} and are further split. Further splitting of the second quadrant separates the second-level of quadrants to single-attribute values W, W, S, S. These second-level quadrants are treated as leaf-nodes and splitting stops. However, second-level quadrants two and four of the first-level third quadrant have a set of attribute-values {S, W}, and are further split into third-level quadrants. Similarly, the fourth level-1 quadrant has its three quadrants as shaded. Hence, they are treated as leaf-nodes.

It is not always possible to get 100% of a quadrant with just one attribute. Hence, to improve computational efficiency, splitting is stopped when the area of quadrant with the attribute is above a threshold. With medical images, the attribute can be "having the same texture or same intensity value." The corresponding three-dimensional analog of a quad-tree is an *oct-tree* where each three-dimensional block is divided into $2^3 = 8$ *octants*. Each node in the oct-tree is split into eight octants. The splitting continues until an octant has a unique attribute.

2.1.5.2 K-D (K-dimensional) tree

K-D trees represent K-dimensional data as a variant of binary trees. K-D tree divides an N-dimensional space, progressively in two halves at every level. At the ith level ($1 \leq i \leq number\text{-}of\text{-}dimensions$), the ith dimension-value divides the data space. At every level, the median of the ith dimension-value is picked up as the root node to balance the tree. In this scheme, Level #1 belongs to the first dimension, the level i belongs to the ith dimension and the level n belongs to the nth dimension. Dimensions allocation repeats after the nth label: the $(n + 1)$th level is allocated by first dimension-value again.

Example 2.13: K-D tree

Consider the six two-dimensional data-points expressed as triples of the form (*first dimension-value, second dimension-value, data-value*): *(1, 2, 5), (1, 4, 8), (2, 3, 7), (2, 8, 5), (4, 1, 9), (4, 3, 11)*. The median of the first-dimension values is *2*. The triple *(2, 3, 7)* is picked up as the root node. This divides the data space in two parts: {*(1, 2, 5), (1, 4, 8)*} and {*(2, 8, 5) (4, 1, 9), (4, 3, 11)*}. At the next level, second dimension is used. In the left subtree, either of the two data-elements can be chosen as the root. In the right subtree, the median for the second-dimension coordinate-values is *3*. The triple *(4, 3, 11)* is chosen as the root node of the subtree. The final K-D tree is shown in Figure 2.12.

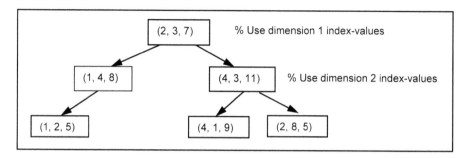

FIGURE 2.12   An example of K-D tree representation of two-dimensional data-elements

Searching a K-D tree is based upon selecting the next value from the feature-vector and using for comparison with the value of the children-node at the next level. This equals changing the dimension. After the last value in the tuple is used, the comparison starts again with the first value of the tuple. That means after $(K - 1)$ levels, the comparison with the same dimension repeats.

The condition for periodic search of every dimension can be relaxed by storing the index of the dimension a node is splitting the data sequence. The information about keeping the equality on left sub-tree or right subtree can also be relaxed by keeping the information about this choice in the node.

2.1.5.3 R (rectangular) tree

An *R-tree* uses *minimum bounding rectangle (MBR)* to envelope data-points or pixels having common attributes such as similar intensity values and stores the coordinates of the rectangle in the information node of a tree. The rectangles can be of varying size as shown in Figure 2.13a. MBRs and smaller rect-angles are included in bigger rectangles in a nested tree as shown in Figure 2.13a and 2.13b. An R-tree shows progressive splitting of a complex region into smaller subregions. Each internal node of a subtree contains more than one subregions. Leaf-node contains the MBRs.

R-tree is formed by first enclosing the clusters using an MBR. The MBR is characterized by {(*minimum x-coordinate, minimum y-coordinate*), (*maximum x-coordinate, maximum y-coordinate*)} of the MBR. Two or more MBRs are joined to form a bigger rectangle. A fill-factor of 30%–40% is a good threshold to identify a rectangular space. After identifying the rectangles, the corresponding coordinate-points are identified. The coordinate of the bigger rectangle is derived by the minimum and maximum of the coordinates of all the enclosed MBRs.

The process is continued until only one encompassing rectangle remains. The embedded sibling rectangles may be overlapping. However, tree is organized in such a way to minimize the overlaps. Insertion in an R-tree is guided by two factors: 1) least enlargement of the MBRs or the enveloping rectangles 2) least increase in

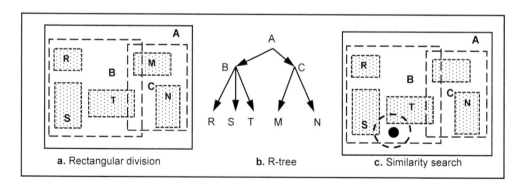

FIGURE 2.13   R-tree and similarity search: (a) Rectangular division; (b) R-tree; (c) Similarity search

the overlap between the rectangles. Each region is labeled for better inference and indexing. The siblings of a node show the nearest regions.

Example 2.14: R-tree

Let us take the example in Figure 2.13. Assume that the coordinates for the MBRs are: S – ((*10, 20*), (*30, 60*)), T – ((*40, 20*), (*60, 40*)), R – ((*10, 80*), (*30, 100*)), M – ((*50, 70*), (*80, 100*)), and N – ((*90, 20*), (*120, 65*)). The MBRs *R, S* and *T* are combined to form the rectangular region *B*, and the MBRs *M* and *N* are combined to form the rectangular region *C*. The coordinates for the rectangle *B* are ((*10, 20*), (*60, 100*)). The coordinates for the rectangle *C* are ((*50, 20*), (*120, 100*)). The top-level rectangle *A* is combined by joining *B* and *C*. The coordinates for the rectangle A are ((*10, 20*), (*120, 100*)).

The advantages of R-tree are: 1) it models a region having multiple clusters of data-points without storing the empty spaces between the clusters; 2) it supports querying the inclusion of a data-point in a specific cluster; 3) it supports query about finding similar clusters; 4) given a data-point, it supports identifying all other points with similar feature-values within a distance as illustrated by the circular region in Figure 2.13c. The advantages of using MBRs are: 1) easy to check the inclusion of a point in a rectangular region by comparing minimum and maximum coordinate-values and 2) quick localization of the cluster of points using interval-based inclusion. R-tree and its variations are used extensively in similarity-based search of multidimensional keys to find out the closest match.

The variants of R-trees used in image representation are R+ trees, SS-tree (Similarity Search tree), SR (Sphere/Rectangle) tree and *TV Trees*. *R+ tree* is like R-tree except the overlap between the regions corresponding to the siblings in R-trees is minimized. The rectangles that show overlap are counted in more than one subtree. SS-tree uses spheres instead of rectangles, and is described in Section 2.1.5.4. SR-Tree is another variation of SS-tree and R-tree that uses intersection of spheres and rectangles and benefits from the advantages of both volume saving property of rectangles and faster similarity-based search property of spheres.

TV-tree is a height-balanced R-tree with two additional properties: 1) only few most significant dimensions are used for indexing and 2) substituting an active dimension by less significant dimension for indexing if the more significant dimension has same value in many subtrees. The dimensions in TV-tree are prioritized based upon their significance in feature-space. This property of dimension reduction of TV-tree has also been used in developing variants of K-D trees.

2.1.5.4 SS (similarity search) tree

SS-tree uses bounding spheres instead of MBRs and use embedded spheres as regions instead of embedded rectangular regions as shown in Figure 2.14. The center of the sphere is the centroid of the points in the sphere. The spherical regions are characterized by (*centroid-coordinates, radius*). The centroid of the circles is calculated as the average of the coordinate-values of the individual points in the sphere as illustrated in Figure 2.14.

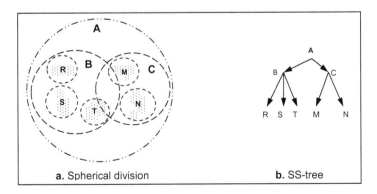

FIGURE 2.14 An illustration of SS-tree: (a) Spherical division; (b) SS-tree

Example 2.15: SS-tree

Figure 2.14 illustrates the corresponding SS-tree for the Example 2.14. The centroid of a bounding sphere is derived using the mean of the coordinates, and the radius for a sphere is derived as the maximum of all the distances from the center to the points in the bounding sphere. Here, the sphere-radius is calculated by $\sqrt{(x_{max} - x_{min})^2 + (y_{max} - y_{min})^2}$, and the centroid is computed by the mean value $\left(\frac{(x_{min} + x_{max})}{2}, \frac{(y_{min} + y_{max})}{2}\right)$. If the density of the cluster is uniform, the centroid and radii are: R – ((20, 90), 28.3), S – ((20, 40), 44.7), T – ((50, 30), 28.3), M – ((65, 85), 42.4) and N – ((105, 42.5), 54.1).

If the number of data-points is equal in each cluster, the centroid the sphere B can be calculated using the mean of the centroids of the spheres R, S and T. The centroid of the rectangle B is calculated as (30, 53.3). The radius is given by the maximum of the (*distance between the centroid of the enclosing sphere and individual sphere + radius of the individual sphere*). The radius of the sphere B is calculated as max(<63.3, 61.3, 59.0>) = 63.3. Similarly, the coordinates and the radius of the sphere C are calculated as (85, 63.8) and 29.2. The coordinates of the root node and the corresponding radius are given by (57.5, 58.6) and 91.8, respectively.

SS-tree is suitable for: 1) grouping data space having multiple K-means clusters and 2) efficient similarity-based reasoning because points in proximity are included in the bounding sphere at the leaf-level. SS-tree also reduces the memory requirement at each node by storing two values: (*centroid coordinates and radius*) compared to four values in R-tree: minimum coordinates and maximum coordinates. One criticism of an SS tree is that the bounding volume requires more space reducing the search efficiency for higher-dimensional data.

2.1.5.5 *VP (vantage-point) tree*

VP tree uses indexing based upon Euclidean distance. There are many variations of indexing based upon Euclidean distance such as GH-tree (Generalized Hyperplane Tree), FQ (Fixed Queries) tree, GNAT (Geometric Near-neighbor Access Tree) and MVP (Multiple Vantage Point) tree.

VP-trees are used for identifying and retrieving images and other high-dimensional that are similar to a query. Given a set of data-points, the data-points having distance below a threshold-value from the vantage-point (reference-point) are placed in left subtree, and the remaining data-points are placed in the right subtree. The simplest form of VP-tree has a branching factor of two. The node in a binary VP-tree is (*Node, Median distance, left-subtree-pointer, Right-subtree pointer*). VP-trees can be generalized to be M-ary VP-tree with branch factor > 2; each branch represents a different distance in sorted order. A VP-tree is illustrated in Figure 2.15 and explained in Example 2.16.

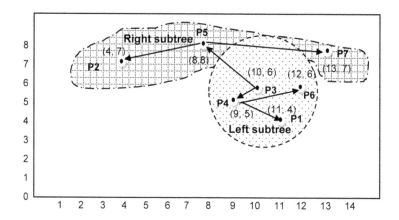

FIGURE 2.15   An illustration of a VP-tree

Example 2.16: VP tree

Consider a cluster of two-dimensional points P_1 *(11,4)*, P_2 *(4,7)*, P_3 *(10,6)*, P_4 *(9,5)*, P_5 *(8,8)*, P_6 *(12,6)* and P_7 *(13,7)*. Let the initial distance-threshold be *2.5*. The point P_3 is compared with all the other points to compute the distance. The distances are: P_3P_1 – *2.2*; P_3P_2 – *6.1*; P_3P_4 – *1.4*; P_3P_5 – *2.8*; P_3P_6 – *2.0*; P_3P_7 – *3.1*. Based upon the comparison with the threshold, left subtree will consist of point-set {P_1, P_4, P_6}, and the right subtree will consist of the point set {P_2, P_5, P_7}.

Let us assume that the next mid-point is P_4. The process is repeated again for the set P_1, P_4 and P_6 with P_4 as the root of the left subtree with a threshold of *2.3*. This will put the point P_1 as the left leaf and the point P_6 as the right leaf. Assume that the point P_5 is the mid-point in the right subtree, and the threshold value is *4.5* at P_5. The distance P_5P_2 is *4.1*, and the distance P_5P_7 is *5.1*. Hence, the point P_2 is the left child, and the point P_7 is the right child. The overall VP-tree with a branching factor of two is shown in Figure 2.15. However, a VP-tree can be generalized to have branching factor > 2. Here, every subtree of the vantage point will have a range of distance from the vantage-point.

Given M-points, the distances between all the points are calculated, and the median-point that almost equally divides the population is chosen as the root node. The process is repeated recursively, until the set of points in a subtree are consumed. Given a query Q, a VP-tree is searched, starting from the root node, using a user-defined similarity threshold δ following these rules: 1) return the current-node having *distance(Q, current-node)* $\leq \delta$; 2) go to the left subtree if *distance(Q, current-node)* $\leq$ threshold median-distance stored in the current node; 3) otherwise, go to the right subtree. The process is repeated until the leaf-node is compared. The direct distance between root and left-child(left-child(root)) on left subtree may be shorter than the cumulative sum of distance between the left-child(root) and left-child(left-child(root)) due to multidimensional nature of the space and triangular inequality which states that sum of the two sides of a triangle is greater than the length of the third side.

MVP (Multi-Vantage Point) tree is a variant of VP tree with multiple vantage points instead of a single vantage point. It has two advantages over VP-tree: 1) keeps the pre-computed pair-wise distance between the vantage-point and the data-points; 2) uses more than one vantage-point to partition the space. MVP-tree reduces the number of distance computations.

2.1.6 Trees for Multidimensional Database Search

Database search requires higher branching-factor and keys to be stored in the leaf for pointing to large records in the secondary storage in page-based organization. A database search is supported by B+ trees. Multidimensional search requires multidimensional trees as explained in Section 2.1.5. A multiattribute indexing method should satisfy four major properties: 1) good storage utilization; 2) small number of disk accesses to answer a query; 3) incremental reorganization with the growth of data and 4) ability to handle range searches, including interval matching. K-D tree is not good for partial queries based upon a smaller subset of features. An example of such a partial query would be to search for patients getting radiation-therapy on a specific date.

To support multidimensional queries involving large data-records, many hybrids of B+ trees and multidimensional trees have been developed. Some of the popular trees are K-D-B tree (K-dimensional B+ tree) and its variants.

2.1.6.1 K-D-B tree and variants

K-D-B tree merges the multi-attribute properties of K-D tree and efficient page-based access of B+ tree. K-D-B tree inherits: 1) the property of higher branching factor from B+ tree; 2) placement of keys at the leaf-node along with the pointers and periodic check on different dimensions from K-D tree. Each node is in a box-shaped region. Each region is divided into multiple mutually exclusive box-shaped regions traversed using K-D tree-based traversal. Branching factor of the node is guided

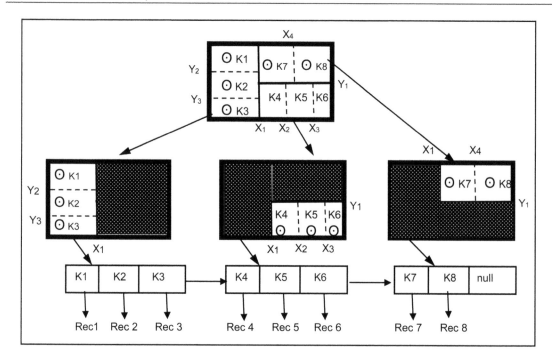

FIGURE 2.16   A schematic of a KDB+ tree

by the property of the B+ tree. At the leaf-node, the actual multi-attribute key-values and the corresponding pointer to the data-record (stored in the secondary storage) are saved. A K-D-B tree is illustrated in Figure 2.16.

K-D-B tree has these major problems: 1) continuous uploading and modification of data-set can make the K-D-B tree lopsided in certain dimensions adversely affecting the balancing properties and space utilization and 3) searching for similar data is not straightforward and may require traversing multiple branches because K-D tree is not explicitly indexed on distance; 3) insertion of a data is complex and requires node-splitting.

Interval in K-D-B tree is implemented similar to intervals discussed in a binary tree in Section 2.1.4.1. The nodes associated with interval-based attributes carry the maximum value of an interval. Interval-based comparison is used to: 1) check the inclusion of a point-value within an interval and 2) discover the common interval between the query interval and a recorded interval for temporal inference as explained in Section 2.7.2.

hB (holey Bricks) tree and BKD tree are two popular variants of K-D-B tree to improve its performance. An hB tree is actually a dynamic acyclic graph that groups multiple leaf-nodes of the K-D tree into one node to access the pages on the secondary storage for better space utilization. It has two properties: 1) index-nodes are organized as K-D trees and 2) node splitting due to insertion of a record may require more than one attributes. An hB tree solves the problem of space utilization in a static tree. However, if the tree is continuously updated dynamically, then its performance also deteriorates. However, it has better performance than K-D-B tree.

BKD tree improves the performance and the space utilization over hB trees. A BKD tree is a set of $\log_2(N/M)$ trees where N is the number of data-elements, and M is the number of data-elements in a secondary storage block. Each tree is laid out on the disk like K-D-B tree. All the $\log_2(N/M)$ trees are probed simultaneously to answer a query. The internal nodes are stored in blocks, and traversal is done breadth-wise in the block to build the K-D tree. The scheme has two advantages: 1) update and data-access can be done concurrently because data-access uses a different tree than data-update; 2) it has near 100% space utilization.

2.1.7 Time-Series Data

Time-series data is a periodic discrete sampling of continuous data. There are multiple applications of time-series data such as: 1) studying disease progression, 2) long-term treatment of chronic diseases, 3) recovery after a surgery, 4) managing age-specific diseases and 5) efficacy analysis of new drugs. The nature of time-series data includes big data, multidimensionality of data and frequent updates.

2.1.7.1 Representing time-series data

There are multiple ways to represent time-series data such as: 1) representing data in a time-domain; 2) representing data using coefficients of Fourier transform (frequency-domain); 3) representing data using coefficients of wavelet-transform (both time- and frequency-domain); 4) hidden Markov models summarizing the transition patterns in the data.

The data also needs to be compressed in time-domain for the storage and the computational efficiency. Any compression technique somewhat distorts the data due to inherent loss of information. A popular approach is to segment the sampled data either in periodic fixed-size intervals or adaptive meaningful intervals of varying size. The values are stored either as absolute-value or as the difference from the previous value. Under the assumption that time-series data are continuous, and do not change abruptly, a difference-based data representation saves memory space. However, it is less robust against data corruption. Actual values are unaffected by the corruption of a data-element in the sequence.

A popular technique to compress time-series data is *adaptive piecewise aggregate approximation*. In this technique, complete period is segmented into chunks of time that depends upon the shape of the time-series data. Sampled data in each segment are analyzed statistically. The statistical aggregation can be mean, median, variance of the sampled values or mean of the slope of the curves within the segment. After computing the aggregate-value, the time-series data within a segment is represented by this value. The approach is illustrated in Figure 2.17. Each segment is characterized by the 4-tuple (*segment-start, segment-end, mean, standard-deviation*).

Another approach to compress the time-series data is to express sampled data-points with a large segment using a straight line. This technique is called *piecemeal linear representation* (*PLR*). Time-series data becomes a sequence of straight-line equations. Actual values are approximated by first going to the

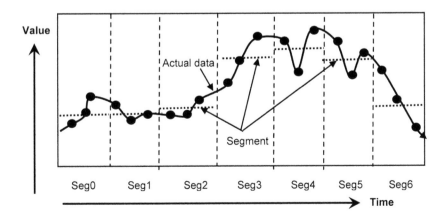

FIGURE 2.17 Compressing time-series data using segmentation and aggregation

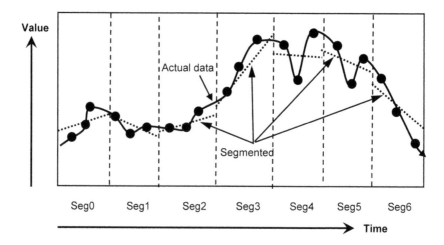

FIGURE 2.18   Compressing time-series data using segmentation and linear curve fitting

relevant segment, and then use the line-equation to retrieve the value of the attribute for specific-time. A line-equation is characterized by 4-tuple of coefficients (*segment-start, segment-end, slope, y-axis-intercept*). The approach is illustrated in Figure 2.18.

Another approach is to identify the salient-points in the time-series data, retain the salient points in the compressed data and perform segmentation between the salient points. This reduces the effect of data distortion due to segmentation-based compression.

2.1.7.2 Indexing structure

For content-based retrieval in spatial-domain, data has to be modeled as $(n + 1)$ dimensions with time being one dimension. For just time-domain search, data is modeled as single dimensional tree. For multidimensional query, a popular data representation for indexing the time-series data is a variation of R-tree (see Section 2.1.5), R+ tree based upon MBR (minimum bounding rectangle) or variations of R+ tree. Inside this structure, the actual data-points, aggregate-values such as mean and standard deviation, or coefficient of the linear segments is stored. For spatial-domain retrieval, data-points other data structures such as KDB+ tree or combinations of other multidimensional trees (see Section 2.1.5) are also used.

To reduce the number of pointers in large number of time-series data, additional variations of trees have been utilized such as: 1) ISAX (Indexable Symbolic Aggregate ApproXimation) tree; 2) TPI (Time-series Pattern Index) tree; and 3) TSR (Time-Series R) tree.

2.1.7.3 ISAX-based indexing

ISAX file takes the interval compressed representation of time-series data as explained in Figure 2.17 and further compresses it by mapping the values into a finite number of symbolic values varying from $0...(2^N - 1)$ using N-bits. The real-valued data is represented by one of the values between $0...(2^N - 1)$ with a cardinality of 2^N. The data can be further compressed by ignoring the lower-order bits at the cost of irreversible information loss. The distances between symbolic values need not be the same. Rather, it is decided by the density of the values. More dense segments have shorter distance for better resolution. These segment boundaries are coded into a lookup table for future retrieval of information. This sequence of symbolic value becomes the signature of the time-series data, and is used to store, retrieve and match two time-series data. The signature can also be mapped to a unique file-name for the ease of retrieving similar data. Two time-series data with different cardinality can be approximately matched by matching the most significant bits and ignoring the least significant bits.

Example 2.17: ISAX indexing

Consider a sequence of interval means for the time-series data as <*2.3, 5.0, 11.2, 6.7, 9.8, 13.2, 12.4, 6.9*>. The range varies from *2.3* to *13.2*. These values are mapped into eight symbolic values *0…(2³ − 1)* using three bits. Assume that the distance between the symbolic intervals is equal. Thus, there are eight value-segments of size *(13.2 − 2.3)/8 = 1.36*. The value-intervals mappings are: *2.3…3.6 → 0; 3.7…5.0 → 1; 5.1…6.4 → 2; 6.5…7.7 → 3; 7.8…9.1 → 4; 9.2…10.5 → 5; 10.6…11.9 → 6; 12.0…13.2 → 7*. Based upon this mapping, the data-series is represented symbolically as <0_8, 1_8, 6_8, 3_8, 5_8, 7_8, 7_8, 3_8> where the subscript shows the maximum number of possible intervals. In a binary form, the information can be represented as <*000, 001, 110, 011, 101, 111, 111, 011*>. The information can be further compressed by ignoring the least significant bits. Thus, the information can also be coded as <*00, 00, 11, 01, 10, 11, 11, 01*> or <*0, 0, 1, 0, 1, 1, 1, 0*>. The signature can be hashed to generate a file-name for the quick retrieval of the data-series.

TPI-index

TPI tree is an extended grid-based tree that reduces the number of pointers to the block containing multidimensional data. The index contains three types of information: 1) frame identifier block (FID Block); 2) dynamic range table (DRT) and 3) hierarchical directory block (HD blocks). A *frame-identifier block* is of the form (*pattern-number, frame-number*). *DRT* is a two-dimensional table that carries the min-max range of the values for every dimension for every region-vector. *Hierarchical directory* is a two-dimensional table that maps a region-vector to the corresponding subdirectory. At the leaf level, a region-vector maps to the corresponding FID block-pointer.

TSR-tree

TSR tree is an enhanced R-tree that also stores minimum bounding time-series by selecting maximum and minimum of the all the time-series data.

2.1.8 Trees for Spatiotemporal Access

Another important aspect of modern healthcare is capturing and reasoning about the data that involves remote care and mobile care. In remote care, the geographical information is also stored in the tree. K-D-B trees can be used for fixed location-based queries by storing the GPS information as one attribute. However, for mobile care, the problem becomes complex as the location of the patient keeps changing with time. Such a condition occurs in many situations such as: 1) a patient is moved from one location to another location within a hospital; 2) an elderly patient is walking or driving and 3) paramedics are carrying a patient in an ambulance. The mobile reasoning about patients' location requires time as a parameter.

2.1.8.1 Time-parameterized R-trees

A *time-parameterized R-tree (TPR tree)* is a dynamic R-tree that changes with time based upon: 1) the insertion of an object in the R-tree; 2) movement of an object in the R-tree that updates the R-tree based upon the estimated velocity and time-horizon needed for the query and 3) deletion of an object. Given, the velocity of an object, a velocity-rectangle is drawn around the object, having a size (velocity × time-horizon) and taking its *x*-axis projection and *y*-axis projection. The rectangle enclosing the object is adjusted to include the velocity-rectangle. An object is inserted so overlap between the rectangles is minimized, and rectangle gets minimal enlargement. After an object is deleted, the enveloping rectangle is readjusted so its size is minimal. If the MBRs become full, then objects are reinserted. An illustration of TPR tree is shown in Figure 2.19. The time-horizon is assumed to be one second.

Example 2.18: Time parameterized tree

Figure 2.17 shows two MBRs R_2 and R_3 and the enclosing root node R_1. The MBR R_2 contains three objects: moving objects MO_1 and MO_2 and the stationary object SO_1. The MBR R_3 contains one moving object MO_3 and one stationary object SO_2. The moving objects MO_1, MO_2 and MO_3 have the corresponding velocity-box VB_1, VB_2 and VB_3, respectively. The moving objects are periodically sampled, and the MBRs are altered as shown in Figure 2.19b.

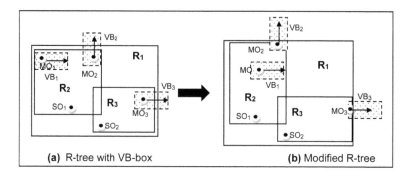

(a) R-tree with VB-box   **(b)** Modified R-tree

FIGURE 2.19   A schematic of temporal parameterized R-tree: (a) R-tree with VB-box; (b) Modified R-tree

2.2 DIGITIZATION OF SENSOR DATA

Computational health informatics is concerned about storage, retrieval, analysis, transmission and security of a large amount of data collected from sensors. Sensors collect analog data to be converted into a digital format for processing in computer memory, archiving in the secondary storage, and transmission to other computers. Digital data is transformed back to analog data for rendering using digital-to-analog conversion. Analog data used in health informatics are: speech, ECG, EEG, EMG, ultrasound echogram, X-ray images, MRI and CAT-scan images, biopsy images, photo images of digestive systems taken from the capsule cameras, teleconsultation images, remote-care images, surgical images and video of pumping heart, etc. Different standardized digital formats are needed to store different media such as biosignal, sound waveform and light waveform. The archived digital representation needs data-compression to reduce memory and bandwidth requirement.

The advantages of digital data are: 1) it can be archived, processed and transmitted electronically and 2) digital error can be detected and corrected using parity-bits during archiving and transmission. However, digital representation suffers from inaccuracies due to quantization error because it has limited resolution based upon the number of bits used in modeling the analog signal. Hence, analog $\rightarrow$ digital $\rightarrow$ analog conversion does not recover the original waveform. The quantization error decreases with an increased number of bits used to model analog signal. However, higher number of bits also increases the memory space requirement and consumes higher bandwidth during data-transmission.

2.2.1 Analog to Digital Conversion

Analog form changes continuously in real-time, while the digitized data is sampled discrete values represented in groups of quantized binary digits (*0* or *1*). Analog-to-digital conversion of a waveform is based upon sampling the amplitude of the waveform at a high frequency with a regular period and converting the sampled value into binary-form. The arrangement of sequence of binary values along with the meta-information such as sampling rate, number of channels and information structure becomes the basis of standard digital format. Figure 2.20 shows a schematic of sampling an analog signal for digitizing.

The quality of reproducibility of the original analog waveform from the converted digital waveform depends on two factors: 1) number of bits used to represent an amplitude; 2) sampling rate. Faster sampling results into better reproduction of the analog waveform during digital-to-analog conversion for better human perception. However, faster sampling results in more sampled data. Higher the number of samples means a higher number of stored bytes requiring a higher amount of memory and higher processing time.

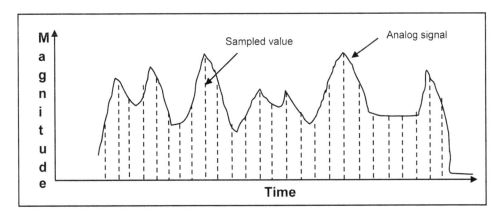

FIGURE 2.20   Sampling analog signals for conversion to digital data

2.2.1.1 Standardized sound format

A popular format for sound waveform is .WAV format. A schematics of .WAV format is given in Figure 2.21. The format stores the information about sampled digitized amplitude of the sound of multiple channels since there may be more than one channel of sound for stereoscopic rendering of audio.

The sampling rate is decided based upon whether the sound contains human speech or high-frequency high-fidelity music. Male speech hovers around 1 kHz, and female frequency hovers around 2 kHz. According to famous *Nyquist's criterion*, the sampling rate should be at least twice as much to avoid losing any information. Normally for capturing human speech, the sound is sampled over 4000 times per second.

The information in the popular sound format such as "WAV" format includes three types of blocks: 1) descriptor block; 2) format block and 3) data block. The *descriptor block* contains the information about the format-tag and the block-size. The format block contains the information about: 1) number of channels; 2) sampling rate; 3) audio format; 4) bits per sample to store digitized amplitude; 5) byte rate per second and 6) bytes per sample for alignment. The data block contains the digitized sampled data with multiple interleaved channels.

2.2.1.2 Error correction and preprocessing

In digital representations, bit-errors are introduced during A/D conversion, archival over a long period in electronic memory and data-transfer inside the computer or over the data-link due to degradation of storage media or the presence of spurious noise. Data errors are detected and corrected using *parity-bits*. Parity-bits are additional information-bits that are computed as a function of information being protected. In the case of an error, information-bit patterns change, and the parity function does not yield the correct parity-bits. Comparison of the original parity-bits and the derived parity-bits detects and corrects the error. There are two parity-bit schemes: *error detection scheme* and *error-correction schemes*. *Error correction schemes* require more parity-bits and cause additional transmission overheads.

Chunk Descriptor	fmt Subchunk						Data	
RIFF <Chunk size> WAVE	fmt	subchunk-size	Code-format	Channels	sampling-rate	Blockalign		

Channel 1   Channel 2

FIGURE 2.21   .WAV format organization

The digital data is also preprocessed to reduce noise and enhance the quality of data before storing into digital archives. The digitized data is stored in standardized media-formats to facilitate software development and data-transmission.

2.2.2 Digital Representation of Images

Images are stored in a digital format. Radiology images are black-and-white images as more resolutions can be packed in the same memory space, and eyes have higher density of black-and-white detecting rods at the back of a retina than cones that detect colored pixels.

Each digitized black-and-white image is a collection of pixels (*Pic*ture *ele*ment). Each pixel is modeled as an 8-bit byte. Each byte can capture *256* levels of intensity. A colored image uses *3* bytes, one byte for each color, for one pixel for levels of transparency. The number of colors represented by three bytes is $2^{24} = 16.58$ million colors that are good enough to capture most of the real-world colors. A high-resolution image would require additional memory and transmission overhead.

In the real world, a phenomenon keeps occurring in real-time. Both eyes and cameras (including digital camera) take snapshots of a real-time phenomenon. A video is modeled as a sequence of fast snapshots where the brain cannot separate during rendering giving the perception of motion.

Enormous amounts of bytes are required in medical videos such as pumping heart behavior, high-resolution MRI that is a sequence of two-dimensional image slices or transmission over the Internet for real-time diagnosis and multiexpert sharing and collaboration.

2.2.2.1 *Proximity preserving image representation*

Images have two dimensions. Images have to be mapped from two-dimensional space to single-dimensional space to search in one-dimensional space. However, simply splitting images row-wise or column-wise will be inaccurate due to the distortion of distances between the neighboring pixels. Row-wise splitting moves away two pixels in adjacent rows; column-wise splitting moves away two pixels in adjacent columns. To maintain the proximity of the pixels in two-dimensional plane, two types of mappings are used: 1) Z-curve and 2) Hilbert-curve as shown in Figure 2.22.

In a Z-curve, the pixels are traversed in the *first-order Z-pattern* to reduce the distance distortion; in a Hilbert-curve, the pixels are traversed in the *first-order square pattern* reducing the distance distortion. Both curves use the nested patterns for coding images: first-order patterns are grouped to form second order-patterns that are grouped to form third-order patterns and so on. The empirical study shows that the images stored using Hilbert-curve perform better in deriving image-similarity.

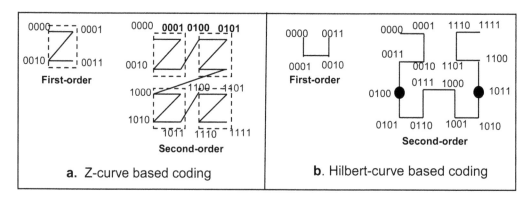

FIGURE 2.22   Mapping two-dimensional images to proximity preserving sequences: (a) Z-curve–based coding; (b) Hilbert-curve–based coding

2.2.2.2 Standardized image formats

The images are stored in RGB colors and transparency for every pixel along with the transparency factor. For colored images, RGB information could be *16* bits each or at most *32* bits each for true color. For black and white pictures such as X-rays images or MRI images, only intensity-values are stored. The three popular formats for images are: JPEG (Joint Photo Expert Group) for photographs, GIF (Graphical Information Format) for drawings, and PNG (Portable Network Graphics) for better portability. JPEG has many variations. Variants of JPEG2000 are used for medical images.

Digital formats for colored images use four bytes of information for each pixel: one for each constituent RGB-color and the last one for transparency. Some global information is also stored such as the *name of the format, number of pixels, palette of colors* used in the image and *mapping of the color-indices in the palette to actual color values.*

Palettes are constructed by: 1) grouping the similar color-values (color-values separated at most by some threshold value δ) together, 2) storing the mapped colors into a table, and 3) associating a color-index with each triple of mapped color. Palettes of colors are used because any specific image uses only a small region of color space, and our eyes cannot separate small variations in the color-space. The use of palettes compresses the pixel coding to just one byte improving the space utilization.

There are many ways to compress the images: 1) the use of palettes, 2) the use of a *16 × 16* (or *4 × 4*) macro-block and averaging out the intensity and color values in the macro-block and 3) the use of transformation functions on bytes to reduce the number of bits on more frequently occurring colors. The palette-based coding substitutes three-bytes of RGB color of a pixel by one byte of color-index in the palette resulting in significant memory savings. Significant compression is achieved by averaging the individual pixel information at the macro-block level. However, the original color information is lost.

Example 2.19: Palette-based coding

Consider a *3 × 3* block of an image in Figure 2.23a. Assume that the value of δ is ±*16*. A color will assume the previously assigned color if its value is with the δ-range. Each cell in the matrix shows four bytes of information. Figure 2.23b shows the color-bucket mapping. After color mapping, only four combinations are left. A palette table is shown in Figure 2.23b, and corresponding color-code is shown in Figure 2.23c.

2.2.2.3 Standardized video formats

Video information contains both the image data and audio data. A video is a sequence of still frames with associated audio. Our brain perceives the objects in the frames as moving provided there is a spatiotemporal continuity, and the frames are rendered at *30* frames/second or more. Video format stores the information in two ways: 1) storing video and audio data separately using a time-stamp for synchronization and 2) storing video and audio data in an interleaved manner one frame at a time. The first

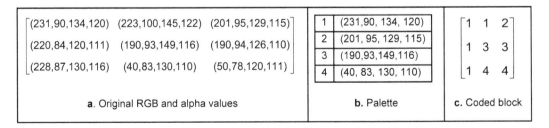

FIGURE 2.23 An illustration of palette-based coding: (a) Original RGB and alpha values; (b) Palette; (c) Coded block

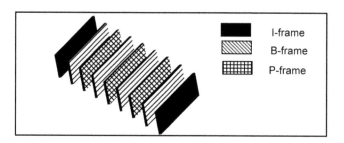

FIGURE 2.24   An illustration of sequence of different types of frames in MPEG

scheme is used by *MPEG format*, and the second scheme is used by *AVI format*. *MPEG and AVI* are two popular formats to archive the video. *MPEG format* is the popular format for transmitting videos over the Internet due to its extensive synchronization and compression capabilities.

Time-stamping is a scheme that associates the same sequence number of a frame with the corresponding frame of image and the sound in a video. After associating the sequence-number, image-frames are buffered and transmitted separately using multiple channels. At the receiver-end, these packets are disassembled, arranged and sorted using the time-stamps. The sorted frames are rendered in the same order as the original display.

A video format holds additional information such as: 1) frame-rate, 2) time-stamps, 3) buffer-size, 4) number of macro-blocks, 5) frame-type and 6) motion-vector. The displacement of a macro-block in following frame captures motion of objects, and is called *motion-vector*. *Frame*-type and *motion-vector* are specific to *MPEG* format.

MPEG (Moving Pictures Expert Group) format has three types of frames: *I-frame, P-frame* and *B-frame* as illustrated in Figure 2.24. An *I-frame* is an image frame that contains the color information of individual macro-blocks in an image; A *P-frame* stores the displacement of macro-blocks from the previous frame (either I-frame or P-frame); A *B-frame* stores the bidirectional displacement of the block with respect to the preceding frame and the following frame. The advantages of motion-vectors are: 1) it takes one byte of information for only those macro-blocks that have been displaced compressing the image with no loss of information and 2) the frames lost during transmission are rebuilt by looking at the previous frames or following frames.

2.2.3 Image Compression

There are two types of compression techniques: 1) lossless compression and 2) lossy compression. Information is not lost during transformation in lossless compression, and images are nearly recovered to the state before the transformation except the quantization error. However, in lossy compression, information is lost during image transformation, and the original image cannot be restored after the compression. For visual perception, lossy compression is sufficient. However, it is inaccurate for the automated detection of abnormalities. Lossy compression saves more memory space and is preferred over the lossless compression where disease encodings and detection are not involved.

There are many techniques to perform lossless compressions. Some of the popular techniques are: 1) Huffman coding; 2) predictive coding; 3) palette-based color indexing and 4) removing interframe redundancies by using motion-vectors. First three techniques are used for single images, and the last technique is used in video compression. Huffman coding is described in Section 2.2.3.1. Palette-based color indexing is described in Section 2.2.2.2.

Predictive coding is based upon predicting the value of a pixel based on neighboring pixels and past changes. It is a two-step process: 1) predicting the value using neighboring pixels and 2) taking the difference between the actual value and predicted value to calculate the prediction-error. This prediction-error value is small and is encoded. The image is reconstructed using the prediction-error and the predicted value.

2.2.3.1 Huffman coding

Huffman coding is a statistical coding technique that classifies the bit-patterns in a file by the frequency of occurrences. After the classification, the file is transformed so frequently occurring patterns are transformed to smaller variable-sized and mutually exclusive binary sequences based upon their frequency of occurrences. The size of the transformed bit-sequence is smaller for higher frequency of occurrence. The intensity-bytes are replaced by these mapped bit-sequences. A variable-length unambiguous code is generated for each pattern.

The coding has to be unambiguous. Otherwise, a bit sequence would be interpreted ambiguously as two or more different patterns during decompression. For example, the coding-elements in the set {*111, 110, 11, 01*} are proper candidates for mapping as no combination of them can be interpreted as the other. However, {*0, 11, 011*} are not mutually exclusive.

2.2.3.2 Segmentation and image compression

The image analysis for compression requires identification of a *region of interest* (ROI). An ROI is identified by deriving homogeneous segments with the required texture property. Segments are derived by choosing few seed-points and growing the region until the neighboring pixels have different texture. Segment analysis is described in Section 5.8.1 under medical image analysis.

2.2.3.3 Compression in digital image formats

GIF format uses a lossless compression algorithm to compress the information, and uses a palette of colors to store the colors. JPEG format uses a lossy compression algorithm. Image coding format contains: 1) tag to mark the start of the information, 2) number of bytes needed for the image, 3) global palette telling the number of colors used in the image in the sorted order of color importance to reduce the number of bits used for storing the information, 4) number of pixels used to define a macro-block, 5) length and width of the image, 6) number of blocks in the images needed to store the color, 7) tag to mark the start of each macro-block and 8) list of the color information for each macro-block. The list of color information for each block includes: 1) local color palette and 2) size of the sub-macro-blocks and list of colors used in sub-macro-blocks.

2.3 APPROXIMATE STRING MATCHING

String matching reasons about: 1) whether a string is substring of another string and 2) the alignment of two similar strings to expose the maximum amount of latent similarity hidden due to misalignment. For example, identifying that a string "ttc" is a substring of the string "lltcytcyycyyttcwcxyz" in linear time is an important problem. It can be used in content-based matching to match a sound pattern, or to match a shape pattern or a macro-block pattern. Similarly, matching two patient-names "Arvind" and "Arawind" for similarity requires string alignment.

Approximate string matching has many applications in health informatics because different words with the same meaning are written and pronounced differently even in the same language by different cultures. For example, "color" in American English is the same as "colour" in British English. Often, in a note by doctors, the words are misspelled, and they need to be interpreted correctly. For example, the incomplete word "ischmic" written by a doctor on a note can be read using an *online character recognition system* (OCR system) correctly. Yet, it cannot be matched with any word in the medical database without correcting the word using a spell-checker. A spell-checker uses approximate string matching and discovers the nearest meaningful word in a medical-terms' dictionary using insertion, deletion, substitution or a combination of the three operations. Approximate string matching requires five operations: *matching, insertion, deletion, substitution* and *transposition* of characters.

2.3.1 Hamming Distance

Formal metric is needed to find the similarity of two strings. For the strings of equal sizes, *Hamming distance* is used to measure the similarity. *Hamming distance* finds out the number of matching positions in two strings without realigning the strings. For binary strings, bit-wise exclusive-OR gives those positions where the positions do not match. Counting these positions gives the *Hamming distance*. The number representation of regular binary representation is not compatible with *Hamming distance*. For example, in binary representation, the numbers *10* and *01* have a difference of *1*. However, the Hamming distance is *2*. Hamming distance is used widely in parity-error detection and error-correction.

Example 2.20: Hamming distance of a string

Consider the binary codes *01110011* and *10111010*. The Hamming distance between them is *4* because the positions *1, 2, 5* and *8* do not match.

2.3.2 Edit-Distance

For unequal size strings, *edit-distance* is used to measure dissimilarity. *Edit-distance* is the minimum number of steps taken to transform one string into another using *insertion, deletion, substitution* of a character by another character and *transposition* of two characters. The operation *insertion* places a character at a specific position; the operation *deletion* removes a character from the substring; the operation *substitution* is a combination of deletion followed by an insertion of the desired character or vice-versa. Two types of metrics have been used popularly to measure the *edit-distance* between two strings: 1) *Jaro–Winkler distance* and 2) *Levenshtein edit-distance*.

Example 2.21: Edit-distance

An OCR system reads a doctor's hand-writing about the diagnosis of "ventricular arrhythmia" as "vetrcullas arhyhma." The approximate string-matching inserts characters "*n*" and "*i*" delete the additional character "*l*" and substitutes the last character "*s*" by "*r*" to transform the word "vetrcullas" to "ventricular." Similarly, characters "*r*," "*t*," and "*i*" are inserted in the misspelled word "arhyhma" to make it "arrhythmia." The operation *transpose* exchanges two characters. The operations *substitution* and *transposition* are derived operations; the operations *insertion* and *deletion* are kernel. *Substitution* is modeled as *a deletion* followed by an *insertion*. *Transposition* is modeled as two substitutions.

2.3.2.1 Jaro–Winkler distance

Jaro–Winkler distance is a modification of Jaro's distance that considers common prefix in two strings being compared. Jaro's distance, denoted by d_j, is described by Equation 2.6, and Jaro–Winkler distance, denoted by d_w, is described by Equation 2.7. Jaro's distance is based upon matching character. It also takes into account the distance between the positions of the corresponding matching characters. Two characters are matching if they are the same and the position-distance between them is less than or equal to $floor\left(\frac{maximum(length(s_1),\ length(s_2))}{2}\right) - 1$.

In Equation 2.6, the variable m is the number of matches, and the number t is half the number of transpositions. *Transposition* is defined as the shuffle of the misaligned characters to align them. The distance d_j is zero in the absence of matching characters. The term t is zero for the matching characters in the same order. Otherwise, d_j is equal to the average of the three components $\left(\frac{m}{|s_1|}, \frac{m}{|s_2|}, \text{ and } \frac{m-t}{m}\right)$. Each of three components is less than or equal to *1.0*. The value of the distance d_j varies between *0.0* and *1.0*. The value is *1.0* when the strings are identical and there is no transposition.

$$d_j = \begin{cases} 0 & \text{if } (m = 0) \\ \dfrac{1}{3}\left(\dfrac{m}{|s_1|} + \dfrac{m}{|s_2|} + \dfrac{m-t}{m}\right) & \text{otherwise} \end{cases} \tag{2.6}$$

Jaro–Winkler distance uses two additional variables: 1) length l of the common prefix at the start of the strings and 2) a scaling factor f to adjust the score. Empirically, the value of the scaling factor f is *0.1*. The Jaro–Winkler distance is equal to *1.0* for identical strings.

$$d_w = d_j + \left(l \times f(1 - d_j)\right) \tag{2.7}$$

Example 2.22: Jaro–Winkler distance

Consider two words "Arvind" and "Arawind." The length is *6* for the first string and *7* for the second string. Prefix "Ar" matches in the words. The distance-threshold for shuffling the characters is *floor(max(6, 7)/2)) – 1 = 2*. The characters "A," "R," "I," "N," and "D" match in both the words. Since the matched characters are in the same order, the value of the variable t is 0. The *Jaro distance* is $\frac{1}{3}\left(\frac{5}{6} + \frac{5}{7} + \frac{5-10}{5}\right) = 0.85$. The Jaro–Winkler distance is *0.85 + 2 × 0.1 × (1 – 0.85) = 0.88*.

2.3.2.2 Levenshtein edit-distance

Levenshtein edit-distance is the minimum number of edits needed on one string to transform into the second string. The edit operations are: 1) insertion; 2) deletion and 3) substitution. There are two variations to handle the cost of substitution. The first variant defines *substitution* as a kernel operation and assigns it the cost 1. The second variant defines *substitution* as a *deletion* followed by an *insertion*. Hence, the second variant assigns a cost of two for each substitution.

The minimum edit-distance can be computationally calculated using dynamic programming technique where each character of the first string S_1 is represented as a column of a two-dimensional matrix, and each characters of the second string S_2 is represented as a row of the matrix. Given two strings S_1 and S_2 of the size m and n, respectively, a $(m + 1) \times (n + 1)$ matrix is drawn with column-values varying from $0...m$ and the row-values varying from $0...n$. The matching starts from the top left-hand side corner, and *Levisdist(i, j)* is calculated using Equation 2.8. The cost is *0* if $S_1(i)$ and $S_2(j)$ match; otherwise, the cost is *1* for insertion, deletion and substitution using the first variant. The cost for substitution is *2* for the second variant. After filling in the values, the distance is found in the cell(m, n).

$$\begin{cases} Levisdist(i, 1) = i; \\ Levisdist(1, j) = j; \\ \\ Levisdist(i, j) = \min \begin{cases} Levisdist(i-1, j)+1 \\ Levisdist(i, j-1)+1 \\ Levisdist(i-1, j-1)+\text{cost} \end{cases} \end{cases} \tag{2.8}$$

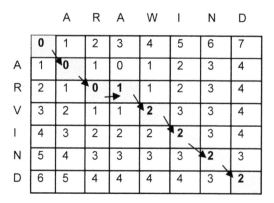

FIGURE 2.25   Deriving *Levenshtein edit-distance* using dynamic programming

Example 2.23: Levenshtein edit-distance

Consider the strings "Arvind" and "Arawind." The edit operations to transform "Arvind" to "Arawind" require: 1) insertion of the character "a" after the character "r" and substitution of the character "v" by the character "w." Hence, the edit-distance is 2. For the second variant, the edit-distance would be 3: one for the insertion of the character "a," and the two for the substitution of the character "v" by "w."

Figure 2.25 describes the calculation of the edit-distance using a two-dimensional matrix. If the insertion/deletion and substitution cost is *1*, the overall distance is *2*.

2.3.3 Applications of Approximate String Matching

Approximate string matching has been extended for: 1) similarity analysis of sequences of nucleotide and amino-acid in bioinformatics 2) speech recognition to match differently spoken sequences of phonemes. A sequence of nucleotide is a linear unfolded structure of a gene, and a sequence of amino-acids is a linear unfolded structure of a three-dimensional protein.

   Approximate string matching has also been used to: 1) glue together the fractions of subsequences of nucleotides derived from wet labs to identify the complete genome, 2) to estimate the functionality of genes in a newly sequenced genome and 3) to identify the conserved subdomain of genes across species under the assumption that conserved domains map to an important function of the gene. A similar operation is performed in matching and understanding the spoken words because different cultures and regions speak the same words differently. These words and phrases are represented as a sequence of phonemes and are aligned and compared to reduce the error in the identification of the words and phrases. A variation of approximate string matching is *dynamic time-warping* used to match two time-series data and has been described in Section 2.7.2.1.

2.3.3.1 Dynamic programming

Dynamic programming is a matrix-based variant of approximate string matching technique for aligning the characters in two similar nucleotide or amino-acid sequences using three operations: insertion, deletion or substitution. The rows represent one sequence, and the columns represent another sequence. The overall matching function is defined as given in Equation 2.9. A mismatch is penalized, and a match is rewarded by adding a positive similarity-score that is looked-up from a standardized biochemical matrix. For matching amino-acid sequences, BLOSUM (BLOcks SUbstitution Matrix) or PAM (Point Accepted Mutation) matrices are used that show similarity-score for different pairs of amino-acids.

$$match\_score(i, 0) = 0 \ for \ 0 \leq i \ \leq m$$

$$match\_score(0, j) = 0 \ for \ 0 \leq j \leq n$$

$$match\_score(i, j) = max \begin{pmatrix} match\_score(i - 1, j - 1) + similarity\_score(char(i), char(j)) \\ match\_score(i - 1, j) + gap\_penalty(char(j)) \\ match\_score(i, j - 1) + gap\_penalty(char(i)) \end{pmatrix} \quad (2.9)$$

The alignment pattern is generated by identifying the bottom rightmost column and moving one step in three directions: up, left, and diagonal-up-left, along the diagonal, tracking the maximum value until the cell (0, 0) is reached. The gap-penalty is user-defined based upon statistical analysis.

2.4 STATISTICS AND PROBABILITY

Data analytics in health informatics uses machine learning techniques to discover new patterns from a large sample-size of data. Machine learning techniques exploit statistical analysis and probability to study the trend, correlation between various features, prediction of future occurrence and learning the behavior.

Statistical analysis and probabilistic reasoning are two interrelated key-stones essential for computational intelligence and machine learning based analysis. The training of artificial neural networks, Markov models, SVM (Support Vector Machine) is based on statistical analysis. Markov models, HMM (Hidden Markov Models), probabilistic networks and quantitative handling of uncertainty are based upon probabilistic reasoning. Conditional probability and Bayes' theorem are used in probabilistic reasoning. Probability calculation is based upon statistical analysis.

A simple example of the probabilistic reasoning is: what is the probability that a patient having a specific biomarker value for cancer actually has cancer? This is an important decision because if the probability is high, then the patient may have to go through an invasive biopsy. Otherwise, other noninvasive lab-test or radiology test can be done. Another example is to derive the probability of recovery for a treatment regimen. Another example is to identify the probability of a disease occurring in an ethnic group at a particular age.

2.4.1 Statistics

Statistics is a population-based abstraction and analysis technique to understand the overall patterns within the population. Statistics can derive the patterns, understand variations in a pattern, derive deviation from the expected patterns and derive the probability of an event. Statistical analysis involves metrics such as mean, median, mode, variance, standard deviation and correlation. Statistical data are modeled using different types of probabilistic curves such as Gaussian distribution, Poisson distribution, binomial distribution and multinomial distribution. These distributions are well-studied and are used to classify the outcome.

2.4.1.1 Basic metrics

The simplest statistical metrics are *mean, median, mode, range* and *variance. Mean* is the average of all the values in a population. It is derived by adding all the values in the population divided by the population-size. *Median* is the middle-value in a population. For a population having $2m + 1$ elements ($m > 0$), m values are less than or equal to the median value, and m values are larger than or equal to the median value; and $(m + 1)$th element is the median value. For a population having $2m + 2$ element ($m > 0$), first m elements are less than or equal to the median value and the last m elements are greater than or equal to median value, and median value is derived as an average of the $(m + 1)$th element and $(m + 2)$th element in the sorted sequence. *Mode* is decided by the maximum frequency of a value in the population space. *Range* is decided by the difference of the largest value and the smallest value. *Variance* is used to calculate the extent of dispersion of values in the range. Larger variance means larger dispersion. The variance is defined as the average of the sum of the square of the difference of data-values and the mean. For very large population, variance, denoted by σ^2, is given by Equation 2.10, where y_i is the value of the ith element, μ is the mean, N is the number of data-elements and σ is the standard-deviation. For small sample-space, Equation 2.11 gives better results for the variance σ^2.

$$\text{variance for large population } \sigma^2 = \frac{\sum_{i=1}^{i=N}(y_i - \mu)^2}{N} \tag{2.10}$$

$$\text{variance for small population } \sigma^2 = \frac{\sum_{i=1}^{i=N}(y_i - \mu)^2}{N-1} \tag{2.11}$$

Example 2.24: Mean and standard deviation

Blood pressure varies due to ingested food, physical activities and mental state. A blood-pressure average for two weeks is a better indicator of a patient's hypotension condition. For these blood-pressure samples, compute the average and variance of the (*systolic pressure, diastolic pressure*): (*90, 70*), (*100, 65*), (*80, 60*), (*102, 72*), (*75, 55*), (*90, 65*), (*95, 65*), (*90, 60*), (*100, 80*), (*110, 75*), (*100, 75*), (*95, 65*), (*90, 65*), (*88, 62*).

Each of the two dimensions is computed separately under the assumption they are independent, and there is no correlation between them. The average of systolic pressure is *93* (rounded to the nearest integer), and the average of the diastolic pressure is *67* (rounded to the nearest integer). Since the sample size is smaller, the denominator *13* is used for calculating the variance. The variance for the systolic pressure is *1,079/13 = 83.0*, and the variance for the diastolic pressure is *578/13 = 44.5*. The corresponding standard-deviations are: $\sqrt{83} = 9.1$ and $\sqrt{44.5} = 6.7$, respectively.

2.4.1.2 Correlation

Another important factor in a statistical analysis is the *correlation* of two variables x and y. If a variable y varies in the same direction as an independent variable x, then the variable y is *positively correlated* with variable x. It is used to study the relationship between independent variables and dependent variables in regression analysis. *Independent variables* are controlled parameters in an experiment. *Dependent variables* are the measured outcomes. The metrics to study the *correlation* of two variables x and y is called *covariance*. Covariance measures the spread of the points in a two-dimensional plane and is defined as the average deviation of all the points from the centroid given by (μ_x, μ_y) as shown in Equation 2.12.

$$covariance = \frac{\sum_{i=1}^{i=N}(x_i - \mu_x)(y_i - \mu_y)}{N-1} \tag{2.12}$$

Covariance is used in computing the most optimum curve that minimizes the overall error that fits the experimental data as described in Section 2.4.5.1.

Another related definition is the *correlation-factor*. It is defined as the ratio of the covariance and the product of the standard deviations of the corresponding two variables as shown in Equation 2.13. A *correlation-factor*, denoted by the Greek letter ρ, varies from *−1.0* to *+1.0*. A value between *0.0* and *+1.0* means *positive correlation*; y changes in the same direction as x. Correlation-factor value = *0* means that the two variables x and y are not correlated; change in the value of the independent variable x does not affect the value of y. A value between *−1.0* and *0* means independent variable is negatively correlated with the dependent variable; dependent variables changes in a direction opposite to the changes in the independent variable(s). A value near *+1.0* indicates strong positive correlation, and a value near *−1.0* indicates strong negative correlation.

$$correlation\ factor\ \rho = \frac{covariance(x, y)}{\sigma_x \sigma_y} \tag{2.13}$$

Example 2.25: Covariance and correlation factor

A scientist postulates there is a strong correlation between sodium intake and hypertension. A person's blood pressure was measured against the sodium intake. The sodium intake was measured in mg/day. The result is summarized as (*sodium-intake, systolic-pressure*): (*1000, 110*), (*1200, 120*), (*1500, 125*), (*1800, 130*), (*2000, 140*), (*2500, 150*). Calculate the covariance and the correlation-factor, and infer.

Sodium intake is denoted by the independent variable x, and the systolic pressure is denoted by the dependent variable y. The mean μ_x is *1666.7* mg/day. The mean μ_y is *129.2*. The standard deviation σ_x is *550.2* mg/day. The standard deviation σ_y is *14.4*. The covariance is *7765.5*. The correlation factor is 7765.5/(550.2 × 14.4) = 0.98 showing strong positive correlation between excessive sodium and blood pressure.

2.4.2 Probability

Probability is the ratio of the occurrence of an outcome divided by the number of times an experiment is repeated. Since, the number of outcomes can never exceed the number of experiments, probability varies between *0.0* and *1.0*. Probability is deeply related to statistical analysis and the sample-size. The computation of accuracy in probability depends upon a large sample-size. Given the probability of an event A occurring is $p(A)$, the probability of the event not occurring is $1.0 − p(A)$. Given a possible set of mutually-exclusive events, the sum of their probability is equal to *1.0*. The probability of an impossible event is *0.0*.

An event may be a composite such as: 1) two events occurring together; 2) sequence of events; 3) two or more mutually-exclusive events. Given two events A and B that do not change the total number of possibilities, the probability of occurring together is $p(A) \times p(B)$, where $p(A)$ is the probability of event A, and $p(B)$ is the probability of event B. If the events A and B are disjoint, then the total probability is given by $p(A) + p(B)$. If the events A and B overlap, then the probability is $p(A) + p(B) − p(A \cap B)$, where $A \cap B$ is the overlap region. If the probability of an event is $p(A)$, then the probability of it occurring M times in a sequence of N is given by $^N C_M\, p(A)^M\, (1 − p(A))^{(N−M)}$.

The conditional probability of an event B to occur after an event A has occurred is denoted as $p(A|B)$. The probability of two events A and B occurring one after another can be represented as $p(A, B) = p(B) \times p(A|B)$. For two independent events A and B, the probability $p(A|B)$ reduces to probability $p(A)$, and the probability $p(A, B)$ reduces to $p(A) \times p(B)$.

Example 2.26: Conditional probability

The probability of a person going for swimming in glades is *5%*. The probability of getting flesh-eating bacterial infection from swimming in glades is *1%*. What is the overall probability of a person going for swimming and contracting the bacteria.

There are two events: 1) swimming denoted by *S* and 2) contracting flesh eating bacteria denoted by *B*. Probability of all the actions is denoted by *p(S, B)*. Using the chain rule of conditional probability, *p(S, B) = p(S) × p(B|S) = 0.05 × 0.01 = 0.005*.

2.4.2.1 Bayes' theorem

Bayes' theorem derives conditional probability *p(A|B)* given the absolute probability of each event is *p(A)* and *p(B)*, and the conditional probability *p(B|A)*. Bayes' theorem can be stated by Equation 2.14. The theorem states that the probability *p(A|B)* can be computed by knowing *p*(A), *p*(B), and *p*(B|A). The equation is derived from the equality *p(A, B) = p(A) × p(B|A) = p(B) × p(A|B)*. Bayes' theorem has been used extensively in probabilistic reasoning and deriving the probability of a rare event given the pattern.

$$p(A|B) = \frac{p(A) \times p(B|A)}{p(B)}$$

(2.14)

Baeyer's equation is a powerful tool to derive the probability of a cause given the probability of an effect. The generalized formula for the cause and effect reasoning using Bayes' formula is given by Equation 2.15.

$$P(cause|effect) = \frac{p(effect|cause) \times p(cause)}{p(effect)}$$

(2.15)

Example 2.27: Bayes' theorem

The probability of the occurrence of biomarker given a cancer is *20%*. The probability of cancer in a community is *3%*. The probability of raised biomarker level in the community is *10%*. Give the probability that cancer is there if the biomarker is raised above a threshold.

Let the occurrence of cancer be denoted as the *cause*, and the raising of the biomarker be the *effect*. Then *p(biomarker)* is *0.10*, *p(cancer)* is *0.03*, and *p(biomarker|cancer)* is *0.2*. Thus, p(*cancer|biomarker*) = *0.03 × 0.2/0.1 = 0.06 (6%)*.

2.4.3 Probability Distribution Functions

Population samples are modeled using multiple probability distribution functions that have been studied well. Popular distribution-functions used in medical informatics are: 1) Gaussian distribution, 2) Poisson distribution, 3) binomial distribution, 4) multinomial distribution and 5) exponential distribution.

2.4.3.1 Gaussian distribution

Gaussian distribution is a symmetric bell-shaped continuous distribution around the mean μ. A distribution may also be a half-Gaussian distribution shown by the right-hand side of the bell-shaped curve or the left-hand side of the bell-shaped curve. The probability density of a Gaussian distribution is given by Equation 2.16, where μ is the mean, and σ is the standard deviation. Most values in the normal distributions are centered at the mean μ, and the probability decreases as the data move away from the mean.

$$Gaussian\ distribution\ function\ f(x|\mu, \sigma) = \frac{e^{\frac{-(x-\mu)^2}{2\sigma^2}}}{\sigma\sqrt{2\pi}} \tag{2.16}$$

If the standard deviation is large, then the bell curve is more spread. Conversely, a smaller standard deviation makes the bell curve more focused around the mean.

The probability that a value belongs to the population is tested by checking the distance from the mean. The Z-score indicates how many standard deviations away a sample point is from the population mean. Z-score is defined as $(x - \mu)/\sigma$ where x is the value of the data-point, μ is the mean and σ is the standard deviation. The Z-score for a specific data-sample having multiple data-points is given by Equation 2.17, where $\bar{x}$ is the mean of the sample, n is the sample size and μ is the population mean.

$$Z\text{-}score = \frac{\bar{x} - \mu}{\sigma/\sqrt{n}} \tag{2.17}$$

2.4.3.2 Bivariate Gaussian distribution

A phenomenon depending upon two or more variables is modeled using multivariate Gaussian distribution. An example of multivariate Gaussian distribution is the study of cancer caused by two or more environmental factors and genetic makeup. Another example of multivariate analysis is the ECG analysis to detect cardiac diseases based on frequency patterns (temporal property) and the morphological properties. Variables may be correlated. Hence, the distribution function involves both variance and covariance. A bivariate Gaussian distribution is given by Equation 2.18 where X and Y are the variables, μ_X and μ_Y are the corresponding mean values, σ_X and σ_Y are corresponding standard deviations and ρ is the covariance.

$$f(X, Y, \mu_X, \mu_Y, \sigma_X, \sigma_Y, \rho) =$$

$$\frac{1}{2\pi\sigma_X\sigma_Y\sqrt{1-\rho^2}}\exp\left(-\frac{1}{2(1-\rho^2)}\left[\frac{(x-\mu_X)^2}{\sigma_X^2} + \frac{(Y-\mu_Y)^2}{\sigma_Y^2} - \frac{2\rho(x-\mu_X)(y-\mu_Y)}{\sigma_X\sigma_Y}\right]\right) \tag{2.18}$$

The exponential part and denominator both contain the covariance ρ.

2.4.3.3 Other distributions

Poisson distribution is a discrete distribution for $k = 0, 1, 2, \ldots$ and uses a parameter λ as shown by Equation 2.19. In medical informatics, it is applied to study: 1) the probability of occurrence of a rare event such as genetic mutation in a population and 2) an event occurring during a period of time such as a number of childbirths in a maternity word. The mean μ of a Poisson function is equal to λ, and the variance σ^2 is also equal to λ.

$$Poisson\ distribution\ function\ p(k) = \frac{\lambda^k}{k!}e^{-\lambda} \tag{2.19}$$

Binomial distribution is a discrete distribution for $k = 0, 1, 2, \ldots$ and uses two parameters n and k based upon combinatorics of selecting k elements out of n elements. It is given by Equation 2.20 where ^nC_k

is the number of ways to pick k ($k > 0$) entities out of n entities. Binomial distribution function is useful in understanding written text to model the distribution of words and topics inside a document.

$$binomial\ distribution\ function\ f(n, k) = {}^{n}C_{k}\, p^{k}\left(1 - p\right)^{n-k} \tag{2.20}$$

$$combination\ {}^{n}C_{k} = \frac{n!}{(n - k)! \times k!} \tag{2.21}$$

Binomial distribution is easily extended to multinomial distribution having k parameters each having a probability $p_i (1 \leq i \leq k)$. The multinomial distribution function f is given by Equation 2.22.

$$f\left(n, k, x_1, x_2, \dots x_k\right) = \frac{n!}{x_1! x_2! \dots x_k!}\, p_1^{x_1}\, p_2^{x_2} \dots p_k^{x_k} \tag{2.22}$$

The exponential distribution function is defined by Equation 2.23. The probability distribution function is 0 for the negative value of the parameter λ.

$$f(x; \lambda) = \begin{cases} \lambda e^{-\lambda x} & if\ x > 0 \\ 0 & if\ x = 0 \end{cases} \tag{2.23}$$

2.4.4 Hypothesis and Verification

A *hypothesis* is a testable statement about a relationship between independent and dependent variables. A *null hypothesis* denoted by $\mathcal{H}_0$, is a statement that can be validated or invalidated. An *alternate hypothesis,* denoted by $\mathcal{H}_1$, is another statement that holds if the *null hypothesis* is invalidated. An example of a *null hypothesis* is that a new drug developed is more effective than the existing drug. A corresponding alternate hypothesis would be "newly developed drug is not more effective than the existing drug."

A *hypothesis* is expressed as a function of some random variables used in a distribution curve. For example, if the distribution curve is Gaussian, then the random variables are the mean μ and the standard deviation σ. Based upon the number of parameters in a hypothesis, it is divided into two categories: 1) *nonparameterized hypothesis* and 2) *parameterized hypothesis*. A *nonparameterized hypothesis* has no parameters from the distribution in the statement of the hypothesis. A *parameterized hypothesis* contains at least one parameter from the distribution. A *simple parameterized hypothesis* specifies all the parameters from the distribution. A *composite parameterized hypothesis* does not specify all the parameters from the distribution.

A *null hypothesis* is tested using various statistical distributions and techniques. There are two types of error in testing a null hypothesis: 1) Type I error and 2) Type II error. Type I error occurs when a null hypothesis ($\mathcal{H}_0$) is wrongly rejected. Type II error occurs when a null hypothesis $\mathcal{H}_0$ is considered true, whereas, in fact, it is invalid.

2.4.4.1 Confidence intervals and margin-of-errors

A *confidence interval* is the region of a distribution curve that contains the true value of the observed data with a user-asserted *confidence-level*. A *confidence-level* is measured as the certainty that the sampled point is within the distribution curve. A *confidence-level* decides the probability that a sample point is within the *margin-of-error* that is decided by the tolerance between the population mean and the sample mean. Higher confidence-level means smaller *margin-of-error*, and vice-versa. Standard confidence levels are 95%, 98% and 99%. Z-scores for the standard confidence-levels are stored in a lookup-table. For a symmetric two-tailed bell-shaped Gaussian curve, the *margin-of-error* is given by Equation 2.24.

$$margin\ of\ error = \frac{Z_{\mu/2} \times \sigma}{\sqrt{sample\ size}} \qquad (2.24)$$

By rearranging the equation, sample-size is given by Equation 2.25.

$$sample\ size = \frac{Z_{\alpha/2}^2 \times \sigma^2}{(margin\ of\ error)^2} \qquad (2.25)$$

Here, $Z_{\alpha/2}$ is the Z-score for two-tailed symmetrical Gaussian distribution and is looked up from a standard Z-score table. $Z_{\alpha/2}$ is *1.96* for the *95%* confidence-level, *2.33* for the *98%* confidence-level and *2.57* for the *99%* confidence-level. Equation 2.23 states that the *margin-of-error* increases with larger variance or smaller sample-size.

2.4.4.2 Hypothesis testing

To test a hypothesis, limited size samples are picked, and the results are inferred based upon the sample-size. The sample-size is chosen carefully based upon the population-size, confidence interval, margin-of-error and standard deviation. Hypothesis testing is based upon knowing the confidence-interval of the distribution. Under the assumption that the normalized area under the distribution is *1.0* and the condition that the hypothesis is true, the probability of an element x belonging outside the confidence interval is α where α is the area lying outside the confidence interval. This probability value is called *p-value*. The probability $p(x \in X \mid \mathcal{H}_0) = \alpha$, and the probability of type I error is α. If the *p-value* $> \alpha$, the null hypothesis $\mathcal{H}_0$ is not rejected because some of the area belongs within the confidence-interval. The areas of the distribution curve outside the various confidence-intervals have been calculated and are used to test a hypothesis.

2.4.5 Curve Fitting

An important aspect of health informatics is the analysis of data to derive the correlation between independent and dependent variables. An experiment gives a large number of data-points that measure the values of the dependent variables for a value of the independent variable. The relationship between the independent variable and dependent variable is important for trend analysis and to predict the value of the dependent variable for a new value of the independent variable or vice-versa. One way to understand the correlation is by plotting a curve with a known equation of the form $y = f(x)$ where y is a dependent variable, x is an independent variable and "f" is the function. The function could be a linear-curve in the form of a straight line, a quadratic function, or a higher-order function.

Example 2.28: Curve fitting

A drug company derives the dosage necessary for patients. However, patients' age and weight vary. The dosage also varies with age and weight. In an experiment, a drug company varies the amount of a drug to check patients' positive response to a medicine and records their age and weight. The plot gives the response of a limited set of patients. To predict the dosage for a random patient, a linear plot is drawn connecting these points.

The curve fitting is based upon error-minimization of the predicted value derived from the curve and the actual observed value. However, the number of sampled points is limited. The best curve based upon the limited number of points minimizes the errors if an additional number of data-points are added.

Curve fitting is done using limited sample-size, and is only an estimate of the actual phenomenon. Higher-order functions may fit the curve in a limited sample-space, only to realize later, they are not actual fits when more data arrives. Linear curves (straight lines), exponential curves and quadratic curves have been used extensively in medical informatics for different purposes.

Linear curve fitting is common in regression analysis to see the correlation of an independent variable on a dependent variable. Exponential curve fitting has been utilized in studying the population-based phenomenon such as cell decay, drug efficacy and effect of a treatment. Quadratic equations have been used to model the functioning of the organs under a treatment. For example, Mayo Clinic has used quadratic curves to estimate the functioning of the kidney (*Glomerular filtration rate*).

2.4.5.1 Fitting a straight line

The simplest form of curve fitting is done using the straight-line equation $y' = \beta x + \infty$, where y' is the estimated value of the variable y based upon the curve fitting; β is the slope of the fitted line and ∞ is the intercept of the line on the y-axis. The error ε_i for each individual point is given by $(y_i - y'_i) = (y_i - (\beta x_i + \infty))$. Cumulative error $E = \sqrt{\sum_{i=1}^{i=N} \varepsilon_i^2}$ and is described by Equation 2.25. By equating the partial differentiation $\frac{\partial E}{\partial \beta} = 0$, it can be shown that given n-points $x_1, ..., x_N$ ($N \gg 1$) for a large sample size, the slope β is given by Equation 2.26, and minimizes the overall error E. The lines pass through the mean value $\bar{y} = \Sigma y_i/N$ and mean value $\bar{x} = \Sigma x_i/N$. Using the slope β and that optimum line passes through $(\bar{x}, \bar{y})$, the intercept-value α is calculated. This information of linear curve fitting is used extensively in linear regression analysis as explained in Section 4.1.1.

$$cumulative\ error\ E = \sum_{i=1}^{i=n}(y_i - (\beta x_i + \alpha))^2 \tag{2.26}$$

$$fitted\ line\ slope\ \beta = \frac{cov(x,\ y)}{var(x)} \tag{2.27}$$

Example 2.29: Linear curve fitting

Consider that a drug company is trying to derive the concentration of the medication after eight hours of administering the medication in adult males. The medicine is to be given to patients three times in *24* hours. The researchers have determined that medication concentration has to be *4.0* µg/ml (microgram/milliliter) of blood for the drug to be effective. The medicine distributes evenly in the blood system within an hour of oral-administration. An adult male has around *4.5* liters (*4500* ml) of blood in the body. The readings for different administered dosages are given as a pair (*administered dosage, concentration in blood after eight hours*). Equate the line and determine the medication that should be given to an adult male such that after eight hours, the concentration is *4.5 microgram/ml*. The company conducted a randomized clinical trial and collected the average drug concentration remaining within the body eight hours after the oral administration. The data is: (*10 mg, 1 µg/ml*), (*15 mg, 1.3 µg/ml*), (*20 mg, 1.5 µg/ml*), (*25 mg, 1.85 µg/ml*), (*30 mg, 2.14 µg/ml*).

Based upon the data, a linear plot is made with administered medication on x-axis and remaining concentration on y-axis. The mean of the administered medication is $\bar{x} = \Sigma x_i/5 = 20.0\ mg$. The mean of the remaining concentration is $\bar{y} = \Sigma y_i/5 = 1.56\ \mu g/ml$. The corresponding variance and covariance are *250.0* and *14.2*. The slope β is *14.2/250.0 = 0.06*. A value of ∞ is calculated using the equation $\bar{y} = 0.06$ $\bar{x} + \alpha$. Knowing the value of the mean $\bar{x} = 20$ and $\bar{y} = 1.56$, the value of the intercept α is calculated as *0.36*. Hence, the equation of the line is $y = 0.06\ x + 0.36$. Note that the line segment is meaning less for $x < 0$ because the *dosage* ≥ 0 as shown in Figure 2.26.

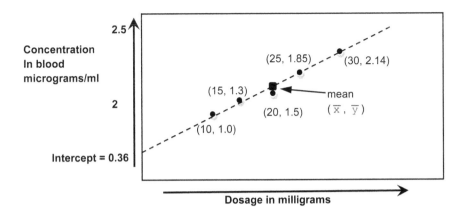

FIGURE 2.26   An illustration of curve fitting using mean-square error minimization

2.5 MODELING MULTIMEDIA FEATURE SPACE

Images are characterized by features such as color, texture, shape and intensity. Sounds are characterized by attributes such as amplitude, frequency, patterns of frequencies and energy level. Speech has many attributes such as frequency-patterns and frequency-range used in different phonemes (basic unit of utterances that make up a spoken word) and envelope of the phonemes. Videos are modeled as sequences of image-frames. Medical images are an integral part of noninvasive patient monitoring for disease detection. This section describes various features of images.

Regular black-and-white images mostly use eight-bit representation giving 256 levels of intensity. Medical MR (magnetic resonance) images use 12-bit representation having 4096 intensity levels for better resolution. Lower values show darker colors.

Images are modeled by: 1) region-based attributes such as intensity, color and texture and 2) periphery-based attributes such as contour. The regions are divided into smaller parts called *macro-blocks*. Each macro-block represents a block of pixels. A typical size of macro-block is *16 × 16* pixels (or *4 × 4* for high-resolution images). To compress the images and to reduce the computational overhead, statistical functions such as mean, median and maximum are used to aggregate the attributes within the macro-blocks. After associating the aggregate attributes within the macro-blocks, individual attributes of pixels are lost, and a macro-block behaves like a unit data-element. The homogenized attributes of the macro-blocks are arranged in form of a binary matrix. Transformations are applied to the matrix for further compression.

2.5.1 Texture Modeling

Image texture is defined as a repetitive pattern of intensity variation among the neighboring pixels in a region. The emphasis is on the similarity of the patterns, variations in intensity-levels and the repetition of the patterns.

Image texture plays a major role in the identification of the abnormal cells such as cancerous cells in different organs and cirrhosis of the liver. Healthy cells are spherical or oval shaped and have uniform well-defined texture. Diseased cells grow irregularly and have a heterogeneous structure. Texture analysis of images plays a major role in identifying the unhealthy cells in radiology images and biopsy images.

Texture analysis is classified into these major categories: 1) histogram, 2) gradient-based analysis, 3) run-matrices, 4) two-dimensional Hurst operator, 5) cooccurrence matrix also known as a SGLD (Spatial

Grey Level Dependence) matrix, 6) LBP (Local Binary Pattern), 7) autoregressive models, 8) Gabor filters and 9) wavelets. First six techniques use statistical analysis. The seventh technique is model-based, and the last two techniques use transformation functions.

2.5.1.1 Histogram as texture

As described in Section 2.1.1.1, a histogram is a vector of tuples of the form $<(interval_1, frequency_1),$ ..., $(interval_N, frequency_N)>$. An image has two or three dimensions, and MRI images have 4096 intensity-levels. The interval-buckets are distributed based upon the intensity-intervals. By knowing the histogram and the interval-size, percentile-based distribution of an image-matrix is computed. Information about the highest intensity available in specific percentile is kept. This statistical summarization characterizes images for similarity-based analysis and image separation.

> **Example 2.30: Histogram-based texture modeling**
>
> Consider the *4 × 4* gray-color image-matrix given in Figure 2.27. The corresponding histogram using interval-size 100 showing only nonzero frequencies is given as a sequence of triples of the form (*lower-bound, upper-bound, frequency*). The histogram is <(*2000, 2100, 4*), (*2100, 2200, 3*), (*2200, 2300, 2*), (*2300, 2400, 2*), (*2400, 2500, 1*) (*3100, 3200, 1*), (*3200, 3300, 3*)>.
>
> The corresponding mean of the intensity-values, derived using the histogram, has been calculated as $\frac{1}{N}\sum \frac{(interval_i^{min} + interval_i^{max})}{2} \times frequency_i$, where $interval_i^{min}$ denotes the lower-bound of the intensity in the *i*th interval, $interval_i^{max}$ denotes the upper-bound of the intensity in the *i*th interval and $\frac{Interval_i^{min} + Interval_i^{max}}{2}$ denotes the average intensity in the *i*th interval. The mean value is *2450*. The standard deviation is computed as $\sqrt{\frac{1}{N-1}\sum \left(\frac{(interval_i^{min} + interval_i^{max})}{2} - \mu\right) \times frequency_i}$. The standard-deviation is *430.5*. The top *10%* (two pixels) have approximate highest intensity value as *3250* – the mid-point of their interval. The calculation of the mean, variance and percentile based upon histogram introduces some quantization-error due to the larger size of the intervals.

2.5.1.2 Gradients as texture

Gradient is the maximum positive difference in the intensity-value in a specific direction when compared to its adjacent neighbors. Assume that the directions are coded as: *0* for vertical-down shown by the arrow "↓"; *1* for diagonal-down-left shown by the arrow "↙"; *2* for horizontal-left shown by the arrow "←"; *3* for diagonal-up-left shown by the arrow ↖; *4* for vertical-up shown by the arrow ↑ ; *5* for diagonal-up-right shown by the arrow "↗"; *6* for horizontal-right shown by the arrow "→" and *7* for diagonal-down-right shown by the arrow "↘." The information is further summarized by mapping the

2054	2083	2076	2124
2162	3205	3207	2210
2063	3256	3143	2176
2311	2422	2312	2296

FIGURE 2.27   A 4 × 4 image-intensity matrix to illustrate histogram-based texture

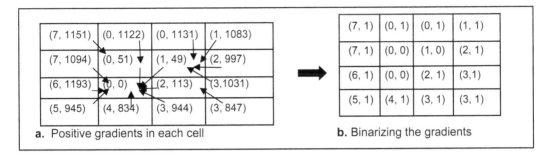

FIGURE 2.28 Gradient-based texture analysis: (a) Positive gradients in each cell; (b) Binarizing the gradients

actual gradient-magnitude to a binary *0* or *1*. A gradient-magnitude above a threshold is mapped to *1*. Otherwise, it is mapped to *0*. The histogram of this summary acts as a texture. The scheme is illustrated in Figure 2.28.

Example 2.31: Gradient-based texture analysis

Let the threshold value be 100. Using binary mapping, the image in Figure 2.28a is translated to the matrix in Figure 2.28b. The histogram of positive gradients in Figure 2.25 is < (0, 1), (1, 1), (2, 2), (3, 3), (4, 1), (5, 1), (6, 1), (7, 2)>.

2.5.1.3 Run-length matrices

Run-length matrix stores in each cell a run of the cells having the same intensity-index in one of the eight directions: *horizontal-right, horizontal-left, vertical-up, vertical-down, diagonal-right-down, diagonal-left-down, diagonal-right-up, diagonal-left-up*. There are eight run-matrices, one for each direction. Each matrix is of the size: *number of gray-color levels within a gray-color range × size-of-the maximum possible run*. The gray-color values are put into buckets by ignoring *m* lowest order bits. The advantages of dividing intensity-range into buckets are: 1) reduction of noise-effect and 2) improvement in the space utilization by using just one byte. The space utilization is further improved by removing those rows from the matrix that do not belong to the gray-color range in the image. Two run-matrices are illustrated in Figure 2.29.

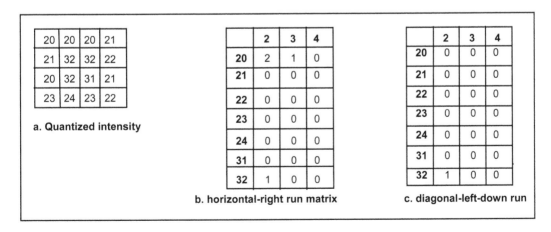

FIGURE 2.29 An illustration of run-length matrices: (a) Quantized intensity; (b) Horizontal-right run matrix; (c) Diagonal-left-down run

Example 2.32: Run-length matrices

The matrix in Figure 2.29a shows the bucket-index of the gray-color intensity for a 4 × 4 macro-block of MRI by removing four lowest order bits from *12*-bit gray-color intensity-values. The range of the bucket-index varies from *20* to *32*. The run-matrices in Figure 2.29b and 2.29c add an additional column to display the exact bucket-indices present in the range. The *horizontal-right run* is given in Figure 2.29b, and the *diagonal-left-down run* is given in Figure 2.29c. In Figure 2.29b, there are two horizontal-right runs of the form <*20, 20*>, one horizontal-run of the form <*32, 32*> and one horizontal-right run of the form <*20, 20, 20*>. In Figure 2.29c, there is only one diagonal-left-down run <*32, 32*>. Other six run-matrices can also be created by analyzing the image-matrix in Figure 2.29a.

2.5.1.4 Hurst operator

Hurst operator takes a circular area of a radius ρ around a pixel and calculates a histogram of number of pixels having the same intensity, their average distance from the central pixel and their range (*maximum value – minimum value*) for each distance. The pixels with similar intensity in the histogram bucket are treated as a single-data-points and given the same label. The slope and intercept of a Hurst plot between the *log*(*range*) vs. *log*(*distance*) are good parameters to match the textures. Other useful statistical parameters are the mean value $\mu_X(\rho)$ and standard deviation $\sigma_X(\rho)$ of the intensity values of the neighboring pixel as described in Equations 2.28 and 2.29, respectively. The variable X is the central point, and ρ is the specified radius. The set of neighboring points is denoted by the symbol S, the intensity of a neighboring point $q \in S$ is denoted by *intensity*(q) and N is the number of neighboring points within the radius ρ. Double summation is due to the two dimensions in the image.

$$\mu_X(\rho) = \frac{1}{N} \sum_{q \in S} \sum intensity(q) \tag{2.28}$$

$$\sigma_X(\rho) = \sqrt{\left(\frac{1}{N} \sum_{q \in S} \sum \left(intensity(q) - \mu_X(\rho) \right)^2 \right)} \tag{2.29}$$

Example 2.33: Hurst operator

Let us consider a radius of two in Figure 2.30. Each cell the in the figure gives two values: label number/intensity value. The points having the same intensity have the same label. Table 2.1 shows the label number of the points having the same intensity, number of pixels having the same intensity and the average distance of the points (from the center-point) having the same intensity. The calculated mean and the standard deviations are *114* and *1.54*, respectively.

		3/110		
	2/112	1/118	2/112	
3/110	1/118	0/ 128	1/118	3/110
	2/112	1/118	2/112	
		3/110		

FIGURE 2.30 Neighboring pixels in a radius of 2 around center pixel

TABLE 2.1 Pixel quantities and average distance

PIXEL LABEL	NUMBER OF PIXELS	EUCLIDEAN DISTANCE
0	1	0.0
1	4	1.0
2	4	1.41
3	4	2.0

2.5.1.5 Cooccurrence (SGLD) matrix

Cooccurrence matrix captures the texture by using a position operator of a pair of cells involving the pixel-cell and one of the adjacent neighbors around. For example, the pair of image-cells can be chosen as (x, y) and $(x + 1, y)$. If the intensity of the image cell (x, y) is I, and the intensity of the image cell $(x + 1, y)$ is J, then the value of cell (I, J) in the cooccurrence matrix is incremented by 1. Hence, the cell (I, J) of the cooccurrence matrix contains the cumulative sum of all possible combinations of the intensity-pairs (I, J) in the image-matrix involving a cell and its immediate right neighbor. The cooccurrence matrix is normalized by dividing every entity by the number of pairs to give cooccurrence probabilities.

Example 2.34: Cooccurrence matrix

Consider the matrix in Figure 2.31a having four levels of gray: *0, 1, 2, 3*. Hence, occurrence matrix will be a *4 × 4* matrix. There are only two adjacent value-pairs *(1, 1)*. Hence, the cell *(1, 1)* in the cooccurrence matrix contains the value *2*. Other entries are filled similarly. The sum of the numbers in the cooccurrence matrix is *12*. Hence, division by *12* gives the probabilities of cooccurrence in Figure 2.31c.

Cooccurrence matrix has been used to measure many texture-related parameters such as *correlation, energy, entropy, and contrast. Correlation* measures gray-level dependency in the image. *Entropy* and *energy* are measures of homogeneity of the image. Uniform cooccurrence matrix has lowest energy. *Contrast* measures the intensity variation in the image. The equations for the four parameters are given in Equations 2.30–2.33.

$$energy \ (uniformity) = \sum_{i=0}^{i=n-1} \sum_{j=0}^{j=n-1} p^2(i, j) \tag{2.30}$$

$$entropy = \sum_{i=0}^{i=n-1} \sum_{j=0}^{j=n-1} p(i, j) log_2 p(i, j) \tag{2.31}$$

(a) Gray image matrix	(b) Co-occurrence matrix	(c) Probability matrix		

$$\begin{bmatrix} 1 & 1 & 2 & 3 \\ 2 & 1 & 3 & 1 \\ 0 & 2 & 1 & 0 \\ 2 & 3 & 1 & 1 \end{bmatrix} \qquad \frac{1}{12}\begin{bmatrix} 0 & 0 & 1 & 0 \\ 1 & 2 & 1 & 1 \\ 0 & 2 & 0 & 2 \\ 0 & 2 & 0 & 0 \end{bmatrix} \qquad \begin{bmatrix} 0 & 0 & 0.083 & 0 \\ 0.083 & 0.167 & 0.083 & 0.083 \\ 0 & 0.167 & 0 & 0.167 \\ 0 & 0.167 & 0 & 0 \end{bmatrix}$$

FIGURE 2.31 An illustration of cooccurrence matrix: (a) Gray image matrix; (b) Cooccurrence matrix; (c) Probability matrix

$$contrast = \sum_{i=0}^{i=n-1} \sum_{j=0}^{j=n-1} (i-j)^2 p(i,j) \tag{2.32}$$

$$correlation = \sum_{i=0}^{i=n-1} \sum_{j=0}^{j=n-1} \frac{(i-\mu_x)(j-\mu_y) \times p(i,j)}{\sigma_x \sigma_y} \tag{2.33}$$

where the *mean* μ_x, *mean* μ_y, *standard deviation* σ_x, and *standard deviation* σ_y are given by Equations 2.34–2.37, respectively. The term $p(i,j)$ is the value in the cell (i,j) of the probability matrix.

$$mean\ \mu_x = \sum_{i=0}^{i=n-1} i \sum_{j=0}^{j=n-1} p(i,j) \tag{2.34}$$

$$mean\ \mu_y = \sum_{j=0}^{j=n-1} j \sum_{i=0}^{i=n-1} p(i,j) \tag{2.35}$$

$$standard\ deviation\ \sigma_x = \sqrt{\sum_{i=0}^{i=n-1} (i-\mu_x)^2 \sum_{j=0}^{j=n-1} p(i,j)} \tag{2.36}$$

$$standard\ deviation\ \sigma_y = \sqrt{\sum_{j=0}^{j=n-1} (j-\mu_y)^2 \sum_{i=0}^{i=n-1} p(i,j)} \tag{2.37}$$

Example 2.35: Computing texture related parameters

Given the probability matrix in Figure 2.31, the energy is calculated as *0.14;* the entropy is calculated as *−2.92* and the contrast is *1.83.* The mean μ_x and μ_y are *1.58* and *1.33,* respectively. The values of σ_x and σ_y are *0.80* and *0.99,* respectively. The calculated value of the correlation is *−0.14.*

2.5.1.6 Local binary pattern

In *local binary pattern* (LBP), the intensity of a pixel is compared against the predetermined number of neighboring pixels to derive the texture pattern. The simplest form of local binary pattern uses *3 × 3* block of pixels around the center pixel to derive information. In a *3 × 3* block, there are eight adjacent neighbors of the center pixel with the radius $\rho = 1$. The comparison with the intensity of a neighboring pixel is mapped to "1" if the intensity of the neighboring pixel is greater than or equal to center pixel, otherwise "0" is assigned. This process is done for every pixel. This comparison creates a byte associated with the center pixel. The exact position of each bit location derived from comparison operation is based upon some convention such as the counterclockwise traversal of neighbors. After assigning this pattern byte with every pixel, a histogram having 256 bins is created that retains the texture information of the image.

The simplest form of the LBP computation is when the radius = *1*, and the immediate adjacent pixel-values are compared as shown in Figure 2.32. Increasing the radius involves pixels beyond adjacent pixels and increases the computational complexity.

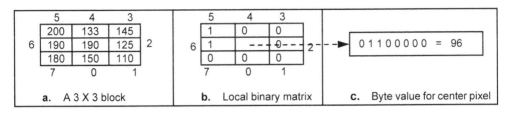

FIGURE 2.32 Assigning texture-pattern value to a pixel using LBP: (a) A 3×3 block; (b) Local binary matrix; (c) Byte value for center pixel

Example 2.36: Local binary pattern

Figure 2.32a shows a 3×3 block showing the intensity values of individual pixels. The comparison of the center pixel-intensity with other pixel-intensities derives the corresponding bits. Assuming that leftmost bottom pixel is mapped to the most significant bit of the byte, and assignment is done in a clockwise manner, the corresponding byte value is *01100000*. This value is mapped to central pixel.

2.5.1.7 Gabor filters

Gabor filter is a transformation function that retains the local spatial and frequency-domain information about the objects. The output of a Gabor filter retains the orientation and scale characteristics of an image, and is unaffected by illumination changes. Gabor filter enhances the low-level features of images such as edges, peaks and contours, and is good for analyzing the rough contours present in malignant cells. Gabor filter is obtained by modulating the amplitude of a sinusoidal wave with a wavelength λ and a bivariate Gaussian distribution function with standard deviations σ_x and σ_y as described in Equation 2.38.

$$\Phi(x,\ y,\ \lambda,\ \alpha) = \frac{1}{2\pi\sigma_x\sigma_y} exp^{-\frac{1}{2}\left(\frac{x'^2}{\sigma_x^2} + \frac{y'^2}{\sigma_y^2}\right)} exp^{j\frac{2\pi x'}{\lambda}} \tag{2.38}$$

where the pair $(x,\ y)$ denotes the coordinates of a pixel; α is the orientation of a Gabor filter and σ_x and σ_y are the standard deviations of the x-coordinates and y-coordinates, respectively. The values of transformed coordinates x' and y' are given by $x' = x\ cos(\alpha) + y\ sin(\alpha)$ and $y' = -\ x\ cos(\alpha) + y\ sin(\alpha)$. Gabor filter-based texture analysis has been applied to detect cancer cells in biopsies.

2.5.1.8 Wavelets

A *wavelet-transform*, given by Equation 2.37, decomposes a signal into a combination of basis-functions, using a time-dilation factor a (also called *scaling*) and a translation factor b, to move the window of raw waveforms. Equation 2.39 shows the application of a wavelet function $\psi(t)$ on a signal $x(t)$ in time-domain. Wavelet-transform is discretized by modeling time-dilation factor as 2^m and translation factor b as $2^m \times l$, where m and l are positive integers.

$$W_{a,\ b}\left[x(t)\right] = \frac{1}{\sqrt{a}} \int_{-\infty}^{+\infty} x(t)\psi * \left(\frac{t-b}{a}\right) dt \tag{2.39}$$

where $\psi*(t)$ is the complex conjugate of $\psi(t)$, a is the dilation factor, and b is the translation factor.

After applying a wavelet-transform, a wave is decomposed into smaller wavelets. Wavelet-transforms use the scaling (dilation) to pick up waveforms with different frequencies. Larger scales are used to match lower frequencies. Smaller scales are used to match signals with higher frequencies. Wavelet-transform has been used in the ECG waveform analysis to identify R-waveform by matching the shape of the transformed signal to the shape of the wavelets as described in Section 7.2.2.1.

2.5.2 Shape Modeling

Images of tumors are detected using texture analysis of radiology images as described in Section 5.9. The shapes of the radiology images are compared with the shape of the tumor-image from the past of the same patient using shape-based analysis to derive the growth (or remission) of the tumor. Shape-based analysis is also used to identify different organs in radiology images.

Two-dimensional shapes of tumors in a radiology-image are modeled as irregular polygons. Three-dimensional shapes such as human brain, heart, lung and liver are modeled as sequences of slices in CAT scan and MRI where each slice is a two-dimensional polygon. There are multiple techniques to model irregular polygons. All are based upon representing polygons as a sequence of tuples. Modeling a bitmap image as a polygon significantly reduces the memory requirement.

2.5.2.1 Contour-based techniques

There are four popular techniques to model shapes: 1) using a centroid and distance from centroid to the peripheral point at a regular angle-interval; 2) drawing concentric circles with regularly spaced radii and identifying the intersection point of the periphery with the concentric circles; 3) drawing tangent from a peripheral point to the next nearby peripheral point, and measuring the angle of the tangent with respect to previous the previous tangent in the polygon and 4) chain-coding the macro-blocks on the periphery using a set of eight directions.

The first technique is based upon 1) identifying the evenly distributed points on the periphery; 2) deriving the centroid $(\bar{x}, \bar{y})$ of the peripheral points; 3) identifying the peripheral points at regular angular difference; 4) computing the radial distance between the centroid and these peripheral points and 5) representing the object as a vector of these radial distances as described in Example 2.37 and Figure 2.33.

> **Example 2.37: Contour modeling using centroid**
>
> Let us take an irregular shape in Figure 2.33. The centroid of the image is shown as a big dot. The object is modeled as $((\bar{x}, \bar{y}), <\delta_1, \delta_2, \delta_3, \delta_4, \delta_5, \delta_6, \delta_7, \delta_8, \delta_9, \delta_{10}, \delta_{11}, \delta_{12}>)$, where $(\bar{x}, \bar{y})$ is the centroid, and δi is the distance of the ith intersection-point from the centroid.

The second technique is based upon: 1) drawing equidistant concentric circles from the centroid and identifying the intersection point on the periphery and 2) arranging the polar-coordinates as described in Example 2.38 and Figure 2.34.

> **Example 2.38: Contour modeling using concentric circles and centroid**
>
> In the example in Figure 2.34, the irregular-shaped contour is mapped using six concentric circles. Starting from the innermost concentric circle, their radii are denoted as R_1, R_2, R_3, R_4, R_5 and R_6. The

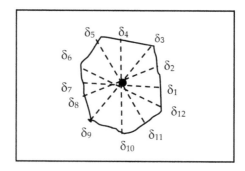

FIGURE 2.33   Modeling an object by vector of distances from the centroid

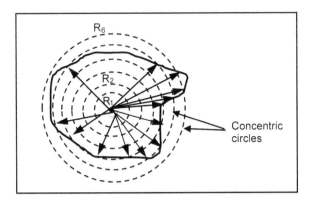

FIGURE 2.34   Modeling an object as a polygon using a vector of intersection points from the centroid

concentric circles intersect the contour at different places. The vector V_i joining the centroid to the intersecting point is modeled as (M_i, θ_i) where $M_i \in \{R_1, R_2, R_3, R_4, R_5, R_6\}$, and θ_i is the angle between the vector V_i and the last vector $V_{(i-1)}$. The angle θ_1 is the angle between the last vector V_N and V_1. There are 13 intersecting points. Hence, the contour is modeled as a vector of 13 tuples: $< (R_4, \theta_1), (R_5, \theta_2), (R_6, \theta_3), (R_6, \theta_4), (R_5, \theta_5), (R_5, \theta_6), (R_5, \theta_7), (R_4, \theta_8), (R_4, \theta_9), (R_4, \theta_{10}), (R_5, \theta_{11}), (R_5, \theta_{12}), (R_4, \theta_{13})>$. In addition, the representation contains the information about the centroid $(\overline{x}, \overline{y})$. The use of radial-coordinates makes the representation rotation-invariant.

The third technique uses a tangent-based modeling when the curve suddenly changes beyond a threshold value as shown in Figure 2.35. The polygon is modeled as a sequence of pairs of the form (*point-coordinates, angle between the previous and the current tangent*). The angle is considered positive in the counterclockwise direction and negative in the clockwise direction.

Example 2.39: Contour modeling using tangents

Let us consider the image in Figure 2.35. There are ten points as marked where the curvature changes significantly. Let the starting point be (x_1, y_1). The polygon corresponding to the irregular shape is modeled as a sequence $<((x_1, y_1), \theta_1), ((x_2, y_2), \theta_2), ((x_3, y_3), \theta_3), ((x_4, y_4), \theta_4), ((x_5, y_5), \theta_5), ((x_6, y_6), \theta_6), ((x_7, y_7), \theta_7), ((x_8, y_8), \theta_8), ((x_9, y_9), \theta_9), ((x_{10}, y_{10}), \theta_{10})>$.

The chain-coding technique divides $360°$ into eight equal directions: 1) vertical-down; 2) diagonal-down-left; 3) horizontal-left; 4) diagonal-up-left; 5) vertical-up; 6) diagonal-up-right; 7) horizontal-right and 8) diagonal-down-right. These directions are coded from 0 to 7. The peripheral macro-blocks are identified.

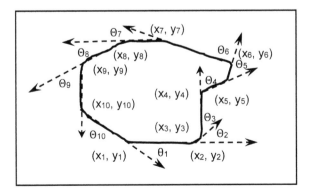

FIGURE 2.35   Polygon-based modeling of irregular shapes using tangent angles and coordinates

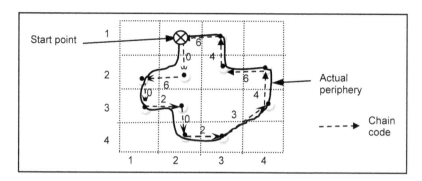

FIGURE 2.36 A chain-code–based modeling of an image region

Starting from one of the peripheral macro-blocks, the remaining peripheral macro-blocks are traversed one adjacent macro-block at a time. The chain code is represented as a pair of the form (*coordinate of the start-point, sequence of directions for one unit cell traversal*).

Example 2.40: Contour modeling using chain-ode

Figure 2.36 shows an irregular shape consisting of *16 4 × 4* macro-blocks. Assume that the left most top corner *(2, 1)* is the starting point. The traversal is done in the counter clockwise direction. The directions are codes as: 1) *vertical-down* → *0*; 2) *diagonal-down-right* → *1*; 3) *horizontal-right* → *2*; 4) *diagonal-up-right* → *3*; 5) *vertical-up* → *4*; 6) *diagonal-up-left* → *5*; 7) *horizontal-left* → *6* and 8) *diagonal-down-left* → *7*. A chain-code based upon counterclockwise traversal is ((2, 1), <0, 6, 0, 2, 0, 2, 3, 4, 6, 4, 6>) where the first field (2, 1) is the start-point, and the sequence <0, 6, 0, 2, 0, 2, 3, 4, 6, 4, 6> is the contour-map. The cell (4, 4) has been ignored because the image covers a very small part of the cell.

2.6 SIMILARITY-BASED SEARCH TECHNIQUES

The queries involve exact-match queries, partial-match queries, multiattribute queries involving inequalities, Boolean queries, range-based queries, temporal queries, including combinations of time-intervals, time snapshots, and complex queries, formed by combining queries using logical operators. The queries are based on textual search, temporal search, multimedia search including medical images, and search involving K-dimensional feature-vectors. In multimedia databases, similarity queries play a major role as the providers want to know: 1) prior similar events that match the current conditions and 2) similar images of abnormal conditions, prognosis and therapy.

Multimedia queries also need similarity-search because each time, the feature-values associated with multimedia objects vary either due to change in the ambient condition such as equipment, lighting condition, exposure time in the case of medical images, sensor noise that modifies the pixel intensities; patients' movement during image capture changing the image; the deterioration or healing of the internal organs changing the overall image.

Multimedia objects are also accessed using captions. However, a query may contain only part of caption or misspelled caption requiring similarity-based matching on captions. The search returns a set of near matches that are sorted on similarity scores to identify best-match(es).

Complex queries have N-attributes ($N \geq 1$). Each attribute represents one dimension in K-dimensional space. An image has two-dimensions or three-dimensions. Time is a separate dimension. To retrieve data-entities based upon similar (or partial) conditions, attributes or images, similarity search in K-dimensional space is needed.

The database should support archival that ensures similarity-based search on multidimensional attributes. For similarity-based search on images, the pixels in images are arranged so the intensities of neighboring pixels in two-(or three) dimensional plane can be compared. Similarity search requires that values around the query-values be identified and ranked in the descending order of matching. Similarity search has many issues: 1) arrangement of the features (multiple dimension) data so the distance between the given value and the result is within a threshold; 2) assignment of the weights to features with varying significance; 3) dimension-reduction from the index-structure to make the search more manageable.

Examples of similarity-based searches are: 1) similar medication and dosage administered to patients in a disease-condition; 2) similar image patterns in the medical repository to identify the previous diagnosis; 3) interval-based matching to identify the patterns with similar diagnosis for laboratory results; 4) matching the current condition with some time-interval in the past when a patient went through similar trauma.

2.6.1 Matching Query and Database Entity

In a similarity-based search, a feature-vector has multiple dimensions, and an image is indexed by feature-vectors or histograms. Based upon the similarity score, few best matches are sought. To perform similarity search in K-dimensional space, four major classes of tree structures and their variations are used: 1) K-D tree and its variations; 2) R+-tree and its variations such as SS-tree; 3) VP-tree and its variations and 4) trees integrating spatial trees and B+ trees such as K-D-B tree and its variation BKD tree as discussed in Section 2.2.

Each query is mapped to a K-tuple feature-vector. These feature-vectors contain K-values: one for each dimension. To perform similarity search, a threshold value τ is needed so the Euclidean distance between the feature-vector corresponding to the query feature-vector and the feature-vector of the database entry is less than τ. Mathematically speaking $\sqrt{\sum_{i=1}^{i=k} (x_i^q - x_i^d)^2} \leq \tau$ where the query feature-vector is $(x_1^q, ..., x_k^q)$, the database feature-vector is $(x_1^d, ..., x_k^d)$ and τ is the distance threshold. Another variation of the similarity-search is weighted similarity where different dimensions in the feature vector are given different weight: more significant dimensions are given higher weights. Mathematically speaking, the equation is written as $\sqrt{\sum_{i=1}^{i=k} w_i(x_i^q - x_i^d)^2} \leq \tau$, where w_i is a positive weight associated with the ith dimension.

2.6.2 Tree Traversal Techniques

2.6.2.1 Traversing R+ tree

Given a point (x, y) as query and the similarity-search distance δ, the tree traversal in R+ tree is based upon comparing (x, y) with the minimum and maximum coordinates of the current rectangle. If the query-coordinates are within the region that is (($minimum$ x-$coordinate$ $(rectangle) < (x - \delta)) \bigwedge ((x + \delta)$ $< maximum$ x-$coordinate(rectangle))) \bigwedge (minimum$ y-$coordinate(rectangle) < (y - \delta) \bigwedge ((y + \delta) < maximum$ y-$coordinate(rectangle)))$, then the smaller embedded rectangles under the current nodes are explored in an iterative manner. The traversal continues until the smallest region containing the query point is identified.

In the case, the condition is not completely satisfied, the rectangle containing the coordinate (x, y) may not have all the nearest-points, and the neighboring rectangles are also explored as shown in Figure 2.13c. The neighboring rectangles can be traversed: 1) by chaining the neighboring rectangles or 2) using backtracking to reach the neighboring nodes that contains any part of the interval $((x - \delta) < x < (x + \delta); (y - \delta) < y < (y + \delta))$. This can be done by backing up to the ancestor-node and checking other sibling nodes that contain at least one of the four corners of the rectangle within them. The complete MBR also might be embedded in the rectangular region given by the coordinates $(x - \delta, y - \delta)$ and $(x + \delta, y + \delta)$. In that situation, whole MBR has to be returned.

2.6.2.2 Traversing SS-tree

Traversing an SS-tree is similar to traversing R+ trees with an exception that the rectangles are substituted by the spheres. The comparison is given by $((((x\text{-}coordinate(sphere\text{-}centroid) - radius(sphere)) < (x - \delta)) \land ((x + \delta) < x\text{-}coordinate(sphere\text{-}centroid) + radius(sphere)) \land (y\text{-}coordinate(sphere\text{-}centroid) - radius(sphere)) < (y - \delta)) \land ((y + \delta) < y\text{-}coordinate(sphere\text{-}centroid) + radius(sphere))$. In the situation, there is partial inclusion, the neighboring spheres are also investigated.

2.6.2.3 Traversing K-D and K-D-B trees

K-D trees and K-D-B trees have no direct way to map coordinate-based traversal to the proximity and have to resort to using stack and backtracking to search the nodes that satisfy the proximity-condition $distance(query\text{-}tuple, node\text{-}tuple) \leq \delta$ where δ denotes the similarity-threshold. If $dimension_i(query\text{-}tuple) - dimension_i(node\text{-}tuple) > \delta$, then the traversal to the left-subtree is pruned since all the nodes in the left-subtree have smaller values for the ith dimension. Similarly, $dimension_i(query\text{-}tuple) - dimension_i(node\text{-}tuple) < \delta$ then the traversal to the right subtrees is pruned because all the nodes in the right subtree will have larger values for the ith dimension. Otherwise, both the trees are traversed, and pruning is possible only at the node level because checking the values of the ith dimension is not directly related to the overall distance. All those nodes that satisfy the proximity-condition $distance(query\text{-}tuple) - distance(node\text{-}tuple) \leq \delta$ are returned as possible matches.

2.6.2.4 Traversing VP trees

A VP-tree is traversed using a user-defined similarity-threshold δ and the proximity-radius $\rho^{current}$ stored at the current-node. First, the $distance(query\text{-}tuple, current\text{-}node\text{-}tuple)$ is computed using one of the distance metrics. If the computed distance is less than δ, the node is included in the set of similar nodes. After testing the current-node, left- and right-subtrees are traversed as follows. If the $((distance(query\text{-}tuple, current\text{-}node\text{-}tuple) < \delta + \rho^{current})$, then only the left subtree is traversed. If the $(distance(query\text{-}tuple, current\text{-}node\text{-}tuple) \geq \rho^{current} - \delta)$, then only the right subtree is traversed. If neither of the conditions are satisfied, then both the subtrees are traversed. The tree-traversals to left and right subtrees are not mutually exclusive. Both left and right subtrees can be traversed depending upon satisfaction of the conditions. Hence, the search is done recursively.

Example 2.41: Traversing VP tree

Consider Figure 2.15. The distance metrics being used is Euclidean distance. A user defines similarity-threshold δ as *1.2*, and searches for data similar to the tuple *(9, 4)*. The search starts at the root node P_3 *(10, 6)*. The proximity-radius ρ^{curr} at P_3 is *2.5*. The value of $distance((9,4), P_3)$ is *2.2* (> *1.2*). The query *(9, 4)* is not within the similarity-range of P_3. Hence, P_3 is not included in the similarity-set S. Yet, $distance(current\text{-}node, query) < 1.2 + 2.5$. Hence, left subtree is chosen for traversal. Similarly, $distance(current\text{-}node, query) \geq 2.5 - 1.2$. Hence, right subtree is also traversed.

The query probes the left-child P_4 and computes $distance((9,4), P_4)$. The distance between P_4 and the query is *1.0* < similarity-distance threshold of *1.2*. The node P_4 is included in the set S. The value of new ρ^{curr} becomes *2.3*. The $distance(P_4, query) < 1.2 + 2.3$. Hence, the left subtree of P_4 is also traversed. Since the $distance(P_4, query) \geq 2.3 - 1.2$, the right-subtree of P_4 is not traversed. Next, the point P_1 in the left subtree of P_4 is probed. The $distance(P_1, query)$ is *2.0* > similarity-distance threshold *1.2*. Hence, the point P_1 is rejected.

The query probes the P_5 in the right subtree of P_3. The distance between the point P_5 and the query is $\sqrt{(9-8)^2 + (4-8)^2} = 4.1 > 1.2$. Hence, P_5 is not included in the set S. The proximity-radius ρ^{curr} at P_5 is

4.5. The *distance(P_5, query)* $< 1.2 + 4.5$. Hence, the *left-subtree(P_5)* is traversed. The *distance(P_5, query)* $\geq 4.5 - 1.2$. Hence, the *right-subtree(P_5)* is traversed. The traversal to the node P_2 derives the proximity distance as $\sqrt{(9-4)^2 + (4-7)^2} = 5.8 > 1.2$. Hence, P_2 is rejected. The traversal to the node P_7 derives the proximity distance as $\sqrt{(9-13)^2 + (4-7)^2} = 5.0 > 1.2$. Hence, the point P_7 is rejected. Finally, the point P_4 is returned.

2.7 TEMPORAL ABSTRACTION AND INFERENCE

Time-based data collection and analysis are very important in health informatics to compare the current conditions with past data. Patients' clinical data are periodically taken, recorded and analyzed. Time is modeled as a sequence of points (snapshots) or time-intervals. Time is abstracted as a discrete measurable attribute where the distance between two adjacent instances is expressed as a time-interval. This interval can be dilated for finer analysis, or compressed for better space utilization. Temporal information is recorded in database using two techniques: 1) object-versioning and 2) attribute-versioning. *Object-versioning* keeps the objects occurring at the same instance linked-together using pointers or a common temporal key. *Attribute-versioning* is used to study the history of a single attribute over a period. It contains the *time-stamp*, *time-unit* and the *data-value*.

2.7.1 Modeling Time

The diagnosis and actions are based upon *linear time*. However, the activities can be concurrent, and the events can be *partially ordered*. *Total order* means that transitivity relation holds between the events: $event_2 \succ event_1 \bigwedge event_3 \succ event_2$ implies $event_3 \succ event_1$, where the symbol "$\succ$" denotes "comes later." However, in *partial-ordering,* the transitivity of events does not hold: if $event_2 \succ event_1 \bigwedge event_3 \succ event_2$ does not imply $event_3 \succ event_1$. Partial-ordering – based modeling is called *branching time.* For example, a patient is admitted in the ICU (Intensive Care Unit) of a hospital, and is being continuously monitored. At the same time, many lab tests are being done on him. Here, monitoring and lab-tests are partially ordered.

The time is linear and is measured in discrete quantum steps. However, within an interval, the activities may repeat with some periodicity. This kind of periodic time modeling is called *circular time* modeling. For example, a patient taking medication for maintaining diabetes can be modeled using circular time. He takes insulin in the morning, lunch time and evening every day.

2.7.2 Time Interval-Based Matching

Time-unit can be enlarged to summarize the time-series data. However, at a later time, queries may be posed with different time-units that involve the summarized data. A time-based query can be: 1) an instance-based query; 2) a time-interval-based query and 3) a content-based query asking when an event occurred. The time itself can be stated as: 1) an absolute value such as a specific date and 2) a relative time (or time-interval) with respect to a reference time using both "before" and "after" the reference time.

Time-stamp can be recorded in a patient-record in two ways: 1) transaction time and 2) time of event occurrence. Event-time is not monotonous because of simultaneous occurrence of multiple events and partial ordering among the events as explained in Section 2.7.1. Another issue is the loss of information in the summarized data because summarization loses the resolution and pattern of change in the archived feature-values.

TABLE 2.2　Interval-based matching for information retrieval

REQUESTED TIME	ARCHIVED TIME	MATCHING CONDITION → ACTION
τ	T	$((\tau > T - \delta)) \bigwedge (\tau < T + \delta))) \rightarrow$ Retrieve
Before τ	T	$(\tau < (T + \delta)) \rightarrow$ Retrieve
After τ	T	$(\tau > (T - \delta)) \rightarrow$ Retrieve
τ	$T_1: T_2$	$((\tau > T_1 - \delta)) \bigwedge (\tau < T_2 + \delta))) \rightarrow$ Retrieve
Before τ	$T_1: T_2$	$(\tau < (T_2 + \delta)) \rightarrow$ Retrieve
After τ	$T_1: T_2$	$(\tau > (T_1 - \delta)) \rightarrow$ Retrieve

In retrieving values, the requested time and the archived time-interval need to be matched. Requested time can be a time-instance or time-interval that can be expressed absolutely or relatively with respect to a reference-time. A relative-time is first converted into an absolute time before comparing. The unit of the time should also be the same, and the notion of proximity should also be defined for similarity-based matching. For example, if a radiology report about spine injury is being searched that was taken in the last ten years. However, a radiology report is available from 12 years back. The information should be retrieved despite it being over ten years old. It gives more information than not having any information.

A simplified version of a time-interval based matching process has been summarized in Table 2.2.

Assume that the symbol δ denotes the proximity threshold. Often, proximity threshold is explicitly stated by the care-provider. In Table 2.2, a time-interval is represented as *start-time: end-time*, the symbol "$\bigwedge$" denotes logical-AND and the symbol "$\bigvee$" denotes logical-OR. It is assumed that the requested time precedes the current-time.

Often, two time-series data have to be matched for similarity. For example, two ECG recordings of the same patients have to be compared to analyze similarity in patterns of irregular beat-patterns. Another example would be matching patient's recovery in response to a combination therapy.

Given two time-series data, they may have similar patterns. Yet, they may look different due to: 1) quantization error caused by time compression; 2) scaling caused by the use of different units; 3) presence of other data interspersed between the similar patterns; 4) difference in sampling pattern of discrete-time and 5) compression techniques of two time-series data. Figure 2.37 illustrates the differences in two time-series data sharing the same pattern P_1 and P_2 that may not give the best match if matched sequentially.

Example 2.42: Matching time-series data

Figure 2.37 illustrates two time-series data that have two matching patterns P_1 and P_2. For the convenience of representation, the time-stamps for series #1 are denoted by T_{1i}, and time-stamps for series #2 are denoted by T_{2j}. The values of i and j vary from $0...9$, A, B, C to avoid confusion. The scaling factor in both the time-series is the same. However, the event-time of P_2 is somewhat shifted due to the insertion of some additional data in P_1 during the time-interval T_{15}–T_{16}.

In addition, the values in the patterns are different due to the difference in sampling-time despite having the same analog signal. There is also some difference in signal-values between T_{1A}–T_{1C} due to the quantization-error. The time-series data #1 is represented as the sequence $<(T_{11}, 8), (T_{12}, 8), (T_{13}, 13), (T_{14}, 13), (T_{15}, 11), (T_{16}, 11), (T_{17}, 15), (T_{18}, 17), (T_{19}, 12), (T_{1A}, 6), (T_{1B}, 6), (T_{1C}, 6)>$; the time-series data #2 is represented as the sequence $<(T_{21}, 8), (T_{22}, 8), (T_{23}, 16), (T_{24}, 12), (T_{25}, 12), (T_{26}, 14), (T_{27}, 17), (T_{28}, 13), (T_{29}, 9), (T_{2A}, 5), (T_{2B}, 5), (T_{2C}, 5)>$.

Simple sequential matching of values will not identify the similarity in Figure 2.37. This is quite a complex task, and multiple techniques have been used to match time-series data that include clustering, Markov model, sequence alignment algorithms such as DTW (<u>D</u>ynamic <u>T</u>ime <u>W</u>arping) algorithms. Clustering and Markov model have been discussed in Chapter 4. Dynamic programming technique for the

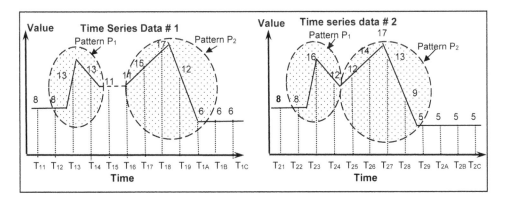

FIGURE 2.37   An illustration of issues in matching time-series data

sequence alignment has been discussed in Section 2.3.3. Dynamic time warping is a variation of sequence alignment using dynamic programming technique described in Section 2.3.3.1.

2.7.2.1 Dynamic time warping

Dynamic time warping takes two continuous and monotonic time-series data and matches them using a variation of dynamic programming technique. The matrix for dynamic programming is arranged so left-hand bottom corner corresponds to the first data-element in both time-series data, and the right top-corner corresponds to the last data-element. Equation 2.8 is modified to incorporate difference in magnitude as cost, and tracing the DTW path from $(1, 1)$ to (m, n) along the diagonal and picking the cell with minimum value as described in the equation. The symbols v_i^1 $(1 \le i \le m)$ and v_j^2 $(1 \le j \le n)$ denote the ith value in the time-series #1 and jth value in the time series #2, respectively. Equation 2.40 is used to match time-series data. Dynamic time warping is used in matching speech phrases and matching disease recovery trend by analyzing the patterns of lab-tests.

Example 2.43: Dynamic time-warping

Consider the time-series data in Figure 2.37. The corresponding matrix for DTW is shown in Figure 2.38. The DTW path is derived by joining the cells with minimum scores along the diagonal. The DTW path is shown by the shaded cells. The value inside the cells ideally should be zero for perfectly matching patterns. However, sampling-time variation, time-shift due to embedded values and quantization-error add to the matching score. Following the DTW, path maps the discrete time-sequence and gives the optimum pattern match.

$$match\_score(0, j) = \infty \; where \; 0 < j < n$$

$$match\_score(i, 0) = \infty \; where \; 0 < i < m$$

$$match\_score(i, j) = min \begin{pmatrix} matching\_score(i - 1, \; j - 1) + distance(v_i^1, \; v_j^2) \\ matching\_score(i - 1, \; j) \; + \; distance(v_i^1, \; v_j^2) \\ matching\_score(i, \; j - 1) \; + \; distance(v_i^1, \; v_j^2) \end{pmatrix} \qquad (2.40)$$

In this example, the time-interval $T_{11}–T_{15}$ maps to the time interval $T_{21}–T_{25}$. After the first matching pattern, there is a break between the time-interval $T_{15}–T_{16}$ in the time series #1 as shown by horizontal DTW path. Similarly, the time-interval $T_{16}–T_{1C}$ maps to the time-interval $T_{25}–T_{2B}$. After matching the second pattern, there is another nonmatching time-interval in the time-series #2. The matching pattern

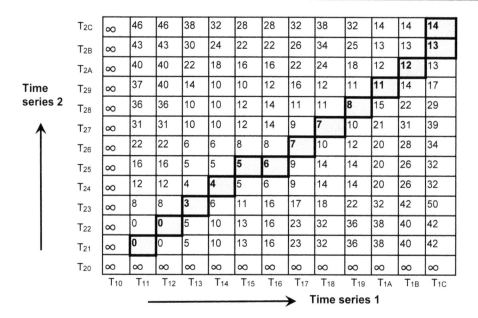

FIGURE 2.38   Dynamic matrix to derive DTW path for time-series data matching

P_1 corresponds to the time-interval T_{11}–T_{15} in the time-series #1 and the time-interval T_{21}–T_{25} in the time-series #2. The matching pattern P_2 corresponds to the time-interval T_{16}–T_{1C} in the time-series #1 and the time-interval T_{25}–T_{2B} in the time series #2.

2.7.3 Temporal Analysis

The analysis includes the prediction of progress in the recovery of a patient's condition, discovering how the disease affects patients, deriving the efficacy of a medicinal dosage, archiving patients' condition for future retrieval and analysis. To summarize the values, multiple statistical techniques are used such as: 1) average of the values within the larger interval, 2) range of values in the form (*maximum, minimum*); 3) median within the interval and 4) Gaussian distribution expressed as a pair (*mean, variance*).

Example 2.44: Temporal analysis

A patient's blood pressure is being measured every ten minutes while he is recovering after an outpatient surgery. The recovery period is two hours. During the recovery, patient's temperature shot up two times and went down 3 times. His blood pressure is given as follows. The first element is time of measurement. The second term is systolic pressure, and the last term is diastolic pressure. His blood-pressure variation has to be recorded while summarizing the data.

<(2:00 PM, 90, 62), (2:10 PM, 92, 62), (2:20 PM, 70, 55), (2:30 PM, 90, 60), (2:40 PM, 92, 65), (2:50 PM, 100, 66), (3:00 PM, 140, 95), (3:10 PM, 95, 70), (3:20 PM, 100, 72), (3:30 PM, 104, 74), (3:40 PM, 70, 50), (3:50 PM, 140, 96), (4:00 PM, 120, 80)>.

The data can be summarized using hour interval at *2:30 PM* and *3:30 PM* using a combination of (*duration*, (*maximum systolic, minimum systolic*), (*maximum diastolic, minimum diastolic*), (*average systolic, average diastolic*)). The summarized data would be <(2:00 PM – 3:00 PM, (100, 70), (66, 55), (89, 62)), (3:00 PM – 4:00 PM, (140, 70), (96, 50), (128, 90))>.

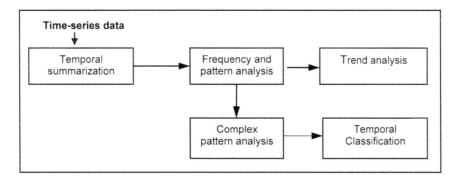

FIGURE 2.39   A schematic of temporal abstraction and classification

The process of temporal abstraction facilitates: 1) efficient storage of patients' conditions for future comparison; 2) discovery of temporal-pattern changes in patients' conditions such as recovery rate, deterioration rate, emergency alert when the monitored signal violates the threshold; 3) discovery of knowledge in terms of event-based causality.

Temporal-pattern analysis is integrated with knowledge-based reasoning as illustrated in Figure 2.39. First, a time-series data is compressed using proper time-interval and unit. Next, frequency and pattern-analysis are performed on the time-series data.

After identifying the patterns, complex pattern analysis is done to identify repeats of pattern sequences or trend in pattern changes. Complex pattern analysis and trend analysis themselves require previous knowledge stored in knowledge bases or machine learning techniques. This derived information is archived in the knowledge base.

2.7.4 Knowledge-Directed Temporal Analysis

Medical practitioners are interested in finding an embedded pattern in time-series data, trend analysis, variations after an event such as medication-administration and the effect of medications. Such investigation pertains to chronic disease management where multiple time-stamped data show some pattern that can be correlated to medication or environmental conditions or dietary inputs. They are also interested in knowing the time-interval of an event. This form of abstraction requires the integration of aggregating multiple time-stamped data samples to a time-interval and intelligent inferencing techniques described in Chapter 3.

Example 2.45: Integrating temporal abstraction and intelligent inferencing

A stroke patient is in ICU. His ECG and blood pressure are continuously being monitored. After a medication to dissolve blood clots is given to him, his heart rate suddenly drops, and his blood pressure drops. The physicians will like to know the duration and extent of heart rate drop, duration of blood pressure level drop, and the delay in time after the medication is administered when the event of blood pressure drop occurs. Such an analysis will require temporal abstraction and intelligent inferencing technique.

Example 2.46: Temporal abstraction

A patient is being prepared for surgery by administering general anesthesia. The administration of anesthesia causes his blood pressure to drop. Fast administration of anesthesia will cause patient to go into the comma. The anesthesiologist will like to monitor the heart rate and the blood pressure and measure the rate of blood pressure drop and change in the heart rate. The measurement of rate of change will require temporal abstraction.

2.8 TYPES OF DATABASES

This section describes the basics of databases to benefit those students (or practitioners) who lack the background in different types of databases. The knowledge gained in this section will be useful in Chapter 4 that discusses the organization of healthcare data. There are many types of databases in use in health industry: *relational databases, object-based databases, multimedia databases, temporal databases, distributed databases* and *knowledge bases*. These databases and knowledge bases are connected in an intricate way that is still evolving.

2.8.1 Relational Database

Two-dimensional table is the simplest form to organize data: entities are rows and attributes are columns. For example, rows could be *patient-id,* and columns could be *patient-names, date of birth, address, primary-care physician,* etc. This form of table is called a *relational table,* and rows are called *records.* A row is modeled as a tuple of the form (*primary-key, attribute₁-value, ..., attributeₙ-value*). The number of entities is equal to the number of rows, and the number of attributes is equal to the number of columns.

Primary-key is used to index an individual record using either a tree-based search scheme or a hash-table. Each primary key is uniquely associated with the index of the corresponding record it represents. By knowing the index or the point of the record, the corresponding record is retrieved. Figure 2.40 illustrates a hash-function-based information retrieval.

A *relational database* is a network of *relational tables* joined by common attribute(s). The splitting of a single large relational table in multiple connected relational tables, removes duplication in attribute values, helps in the storage utilization, and updates to the database because different relational tables are stored separately. Besides, a smaller part of the database is updated making the retrieval and update faster.

Example 2.47: Relational database

Consider a patient–doctor encounter database. *Patient-id* is the primary-key to retrieve the information about the patients, and *doctor-id* is the key to retrieve information about the doctors. However, a patient has many doctors for different ailments and for second opinions besides a family practitioner. Similarly, a doctor sees many patients with different ailments. There is a need for a relational table where both the keys, patient-id and doctor-id, can be joined to list information about the patient–doctor encounter. Figure 2.41 gives a simple schematic of relational database for the patient–doctor encounters.

There are three tables: *patient-information table, doctor-information table* and *patient-doctor encounter table.* Patient-information table contains patient-specific information such as *patient-id, patient-name,*

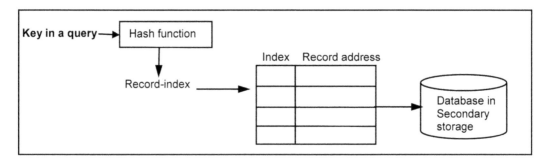

FIGURE 2.40   Retrieving information from a simple relational table

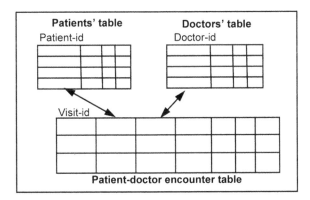

FIGURE 2.41   A simple example of relational database

patient's address, patient's diseases, patient's primary-care physician, patient's allergies, patient's pharmacy, patient's medication, and so on with the primary key as *patient-id*.

The doctor's relational table contains the information such as *doctors-id, doctor's name, doctor's hospital association, doctor's address, doctor's expertise, doctor's patients, doctor's availability,* and so on with the primary key as *doctor-id*.

The third relational table will have a *visit-id* as the primary key and will contain the attributes such as *patient-id, doctor's id, appointment time, type of encounter, diagnosis, suggested lab-tests, referrals to other doctors, drugs prescribed, drugs ordered to pharmacy, billing code, etc.* The field's *patient-id* and *doctor-id* are used to retrieve the information from patients-table and doctors-table. This table is also used to answer statistical queries related to doctors and patients.

2.8.1.1 Limitations of relational databases

Relational databases cannot capture: 1) complex relationship between objects or 2) nested relationship between objects. Many types of entities such as human organs comprise different parts that may have different diseases, and one disease may affect many other parts from the body. Object-based modeling of human organs and other nested entities is natural in object-oriented databases.

2.8.2 Object-Based Databases

Object based programming supports: 1) notion of nested class, 2) inheritance, 2) generic template of a class that can generate multiple active objects at runtime and 3) information hiding. The visibility of data and code can be hidden to promote modularity of a large software. Object-oriented programming supports complex nested relations, multimedia objects that involve multiple components and interoperability in text analysis involving the tree-structure of the ICD (International Classification of Diseases) codes. For example, eye is made of iris, retina, cornea and vitreous chamber. Each of them can have different diseases. All these diseases are eye-related diseases. Each component has subcomponents, and each subcomponent has multiple related diseases. There is an implicit tree-structure of subclasses: a disease in a component of an eye is also a disease in the eye.

Object-oriented programming has been extended to handle object-oriented databases. Methods are described to model a class of an organ such as eye, heart, lung, brain, etc. The same organs of each patients use the same methods described in the class-library of the organ. Similarly, multimedia objects are easily described as an object-instance of a multimedia class.

2.8.2.1 Types of object-based databases

There are two types of object-based databases: 1) OODBMS (Object Oriented Database Management System) supported by full object-oriented programming and 2) ORDBMS (Object-based Relational Database Management System). ORDBMS extends *relational database management systems* to include: 1) the definitions of complex structured types and user-defined abstract data types; 2) inheritance of methods and subtypes for nested relations; 3) typing to include reference to a file storing a bit-level information for multimedia objects; 4) user-defined methods for user-defined types and 5) operators for structured types that can override built-in operators. Each object is given a unique *object-id*. An object could be a tuple, a table, or URL of multimedia objects. The advantage of *object-id* is that an object can be stored anywhere without changing the program to access the data. Using inheritance and subtype allows the same method for subtypes of an object-class.

2.8.3 Multimedia Databases

Multimedia databases store multimedia objects in the databases. A multimedia object has multiple components such as image, sound, video, text and their combinations. Multimedia objects are complex because:

1. An object can have many media contents to be stored separately in different formats. However, contents are rendered simultaneously. This requirement necessitates synchronization and time-stamping the frames and data at regular periodic intervals for accurate rendering;
2. For medical diagnosis, the archived objects related to disease and diagnosis need to be restored with minimum loss of information to reduce false-positive or false-negative diagnosis;
3. Minimum noise should be introduced in the media during analog-to-digital conversion, archiving, retrieval, rendering and transmission; and
4. Transmitted media should be encrypted and secured to maintain the confidentiality of the content and the patient identification.

The major issues in these multimedia databases are: 1) indexing for quick retrieval of large size of files; 2) content-based similarity searches to retrieve similar multimedia objects for the diagnosis; 3) mechanisms for transmission of huge media files over the Internet in near real-time; 4) creating summary files for large media albums and clips for the ease of human interaction and comprehension and 5) automated comparison of sequence of images to derive progression (or remission) of diseases such as cancer and tumor growth.

2.8.4 Temporal Databases

A temporal database contains time-stamped data. Real-time databases have timing-constraints and have time-sensitive information. Databases for monitoring critical patients and ICU patients are temporal real-time databases: the signals such as ECG and oxygen level in the blood are continuously monitored, and an automated alarm is raised instantaneously after life-threatening conditions are observed or predicted. In this section, we study the characteristics of temporal and real-time databases.

In all temporal databases, a time-stamp is associated with every event or recording to: 1) establish a chronology of events, 2) perform temporal reasoning about disease progression (or regression) in the future, 3) project the rate of change of patient's condition, 4) study the effect of drug-dosage and interaction of drugs with other drugs when taken together or within short duration and 5) manage the ambulatory patients, patients in emergency room, critical inpatients and ICU patients. Two types of time are recorded in a temporal database: 1) actual time of an event and 2) time of recording in the database (also called *transaction time*). Actual-time of an event is not monotone because many monitoring activities and physiological tests can be done concurrently. However, for the database maintenance and recovery, we need a monotonic order provided by the *transaction-time*.

Time-stamping could be absolute based on a clock, or relative to an event such as "taking medicine before meals," and "taking medicine after meals." Multiple activities are grouped together as one atomic activity with a deadline for the event-start and the deadline for the event-finish. Periodic activities are time-stamped using a global time-interval and a pair (*time-quantum, relative offset*) as illustrated in Example 2.48.

Example 2.48: Examples of time-quantum

Consider a case of antibiotic administration for 10 days starting January 16, 2016. The time-interval is *01/16/2016* to *01/25/2016*. The time-quantum is an element within the set {*1, 2, ..., 10*}, and the relative offsets are members of the fuzzy set {*morning, afternoon, night*}.

Using time-stamped information in database, various artificial intelligence techniques are used to reason for the outcome. Some of the AI techniques are *dynamic belief network, Markov process, and HMM* (*Hidden Markov Model*). These techniques are described in Chapter 3.

2.8.4.1 Queries in temporal databases

Time is included as a column in the medical record to facilitate temporal access. Temporal databases have two types of queries: 1) time-based query that includes a specific time or a time-interval and 2) attribute-based queries. *Interval based queries* (*time-slice*) also use temporal logic such as "before a time-point," and "after a time-point," retrieving all the history and performing statistical operations on it. Matching a specified time-interval in the query and a recorded time-interval requires using interval-based temporal logic as described in Table 2.2.

2.8.4.2 Issues in temporal databases

Since the temporal databases have two types of queries, *attribute based and temporal,* preferring one type of query over the other effects the retrieval efficiency for other types of queries. Another problem is that event-time need not be unique due to multiple concurrent events occurring together. Hence, transaction-time is used for the temporal keys. The use of transaction-time sacrifices accuracy.

2.8.5 Knowledge Bases

Knowledge bases are a collection of logical rules and an embedded database. Queries are answered by applying logical rules on the database. *If-then-else rules* and *uncertainty-based reasoning* also provide limited intelligent reasoning capability. Associated *certainty-factors* use probabilistic analysis (or knowledge extracted from an expert) to apply a rule probabilistically when there are more than one outcomes. Logical rules can also be updated like data when a new pattern is identified. Logical rules are interpreted using *forward or backward chaining* described in Sections 3.3.1 and 3.3.2. Knowledge bases incorporate multiple types of constraints including *semantic constraints* on the database. Examples of the constraints are: 1) a child has to be above the age of 12 to take a specific medicine and 2) people in a specific community have to go through a medical diagnostic more frequently for a specific disease due to the environmental conditions.

2.8.6 Distributed Databases and Knowledge Bases

Medical databases are highly distributed. Each database has its own format, their coding that span across multiple library-versions. These heterogeneous databases bridge the incompatibility gaps using translators. In addition, many issues arise from data-distribution such as: 1) temporal compatibility between locally processed and remote data, 2) timely availability of remote data, 3) reliability of remote data against corruption, 4) even distribution of workload, 5) freshness of the available data and 6) security and reliability of a communication channel. Time-stamping is one way to manage the distributed multimedia databases involving transmission of multiple streams such as sound-stream, image-stream and their combinations in a video-stream.

The use of cloud makes the data availability at distributed sites. However, cloud-based storage and retrieval of data have many security concerns. Data cloud providers are also considered semi-trustworthy. While protecting the data against outsider hackers, they themselves are curious about data.

2.9 MIDDLEWARE FOR INFORMATION EXCHANGE

2.9.1 eXtended Markup Language (XML)

Extended Markup Language extends HTML (Hyper Text Markup Language). XML supports user-defined tags besides built-in tags. The advantages of user-defined tags are: 1) support of complex structure, including nested relations in a database; 2) modeling of graphs for visualizing complex biomolecules, metabolic pathways and signaling pathways; 3) building and retrieving complex multimedia objects with its components interspersed across different URL sites; 4) modeling of computations and 5) modeling of complex drawings. Together, these models are sufficient for transmitting health-related information over the Internet. There are three types of health-related information: 1) information that needs not be stored in a specific medical process, 2) background information that does not change or changes very slowly and 3) information that keeps changing continuously. The third type of information needs to be transmitted frequently.

Example 2.49: XML-based coding of medical information

The example in Figure 2.42 describes the information about the heart condition of a patient "John Doe" using a *patient-id*. It describes heart as a nested multimedia object with many attributes. Attribute-values are constants or a URL (Universal Resource Locator) that stores an image or a video.

```
<patient>
            <patient_id> 123 </patient_id>
            `<patient_name> John Doe </patient_name>
            <Heart>
                <SA_node> <attributes of SA node> </SA_node>
                <Left_atrium> <attributes of Left_atrium> </Left_atrium>
                <right_atrium> <attributes of right_atrium> </right_atrium>
                <left_ventricle> <attributes of left_ventricle> </left_ventricle>
                ...
                <abnormality> (disease ischemia) (duration_years 2) (severity high) </abnormality>
                <heart_video>  https://www.CCF.org/john_doe/heart_video.mpg </heart_video>
            </Heart>
            ...
</patient>
```

FIGURE 2.42   An illustration of XML-based coding of heart-related information

```
<soap:Envelope
        xmlns:soap = "http://www.w3.org/2001/12/soap-envelope"
        soap:encodingStyle = "http://www.w3.org/2001/12/soap-encoding">
        <soap:Header>
        ...
        </soap:Header>
        <soap:Body>
                ...
                <soap:Fault>
                        ...

                </soap:Fault>

        </soap:Body>

</soap:Envelope>
```

FIGURE 2.43   An illustration of the template for SOAP-based message transmission

2.9.2 SOAP and Message Envelope

SOAP (Simple Object Access Protocol) is an XML-based high-level protocol for transmitting complex information between two information resources over the Internet. It puts encrypted information in a message envelope that contains the information about *origin of the message, destination of the message* and *public key to restore the message at the destination.* The overall template is illustrated in Figure 2.43.

    <?xml version = "1.0"?>.

    The header tag is called "<soap:Header>," and the body tag is called "<soap:Body>". The message is embedded between the tags "<soap:Envelope>" and "</soap:Envelope>." The "<soap:Body>" block's information includes: 1) the URL of the destination and 2) an optional embedded "<soap:Fault>" block to handle the faults. There are four elements in the "<soap:Fault>": 1) "<faultcode>" – code to detect the faults; (2) "<faultstring>" – an explanation of the fault; (3) "<faultfactor>" – what caused the fault and (4) "<detail>" – application-specific description.

2.10 HUMAN PHYSIOLOGY

Our body is a complex integration of multiple organs. Various hormones control biochemical activities. Enzymes control reaction-rates of metabolic activities. Immune system continuously protects cells from foreign bodies. New cells are constantly produced; dead cells are regularly cleared. Each cell reacts to environmental changes using molecular sensors. Lungs absorb and oxygenate blood, and heart pumps out oxygenated blood to the cells. Each cell generates energy for various functions, including reproduction using oxygen. Any malfunction in this integrated body may manifest as a disease that can be identified by analyzing: 1) visible symbol; 2) genetic changes; 3) changes in the biochemical concentration, changes in the protein-levels in the body-fluids, and changes in functioning human organs monitored using biosignals or biopsies.

    Medical treatments use monitoring vital signs for the diagnosis of various diseases. Noninvasive techniques have minimal trauma and the associated cost. The mechanisms include radiology images, ultrasound images, biosignals such as ECG, EEG, EMG, biomarkers, levels of dissolved biochemical in internal fluids such as blood, urine, sweat and spinal fluids. Cell samples are also collected using a biopsy. The collected cells are cultured to magnify the number of cells and study abnormalities.

2.10.1 Modeling Human Body

There are many organs in our body such as brain, heart, lung, liver, kidney, stomach, pancreas and pituitary gland. Each organ has one or more body-functions. Major body-functions are: 1) energy generation, 2) reproduction, 3) transforming ingested molecule to proteins and other needed biomolecules and chemicals, 4) growth of the body, 5) production and regulation of hormones for controlling physiological functions, 6) production of RBC (Red Blood Cells) to carry oxygen, 7) production of T-cells, antibodies and WBC (White Blood Cells) to protect against foreign bodies, 8) protecting against injury, 9) fighting external infections, 10) identifying foreign bodies and 11) cleaning up the dead cells from the body.

2.10.1.1 Cardiopulmonary system

Heart and lung system that oxygenates blood is called *cardiopulmonary system*. It is very important for the cells throughout the body to receive nutrients, oxygen and other biomaterial necessary to fight the foreign body, and repair body.

Lung comprises small air-sacks called *alveoli t*hat expose the blood to oxygenated air and facilitates the exchange of oxygen and carbon di-oxide (CO_2) from the blood. Lung absorbs inhaled oxygen, oxygenates oxygen-depleted blood and removes carbon dioxide from the blood. During exhalation, carbon dioxide retrieved from the oxygen-depleted blood is released. oxygen-depleted blood enters from the right chamber of the heart to both the lungs, gets exposed to oxygenated air stored in lung-sacs and oxygen-rich blood enters the left chamber of the heart again. Heart pumps blood to different parts within a body and collects blood from different parts of the body. Concurrently, oxygen-depleted blood is carried to lungs, and oxygen-rich blood is carried from lungs to heart.

On an average, 3.44 liters of blood passes through human-lungs every minute that increases by approximately 7% in the presence of an atmosphere with 5% elevated CO_2 level than the normal level. Increase in the CO_2 level decreases the level of oxygen in blood. Lack of oxygen enhances the heart pumping rate to get more oxygen from lungs and to provide the required oxygen to the body. An increased heart-beat and increased blood flow causes stress on the heart muscles and blood vessels. Lack of oxygen in the blood also depletes the level of oxygen in tissue cells and results in a condition called *hypoxia* that may cause unconsciousness and death.

The amount of CO_2 in the atmosphere, the rate of breathing, time of holding the air, the capacity and flexibility of *alveoli* control the level of gas exchange in the lung. Increase in the CO_2 level in the atmosphere causes a condition called *hypercapnia* where the concentration of CO_2 in blood increases, and the concentration of oxygen in blood decreases.

Hemoglobin in the pumped blood carries oxygen to various cells. Oxygen reacts with sugar and carbohydrates inside the cells to generate energy. Blood also contains nutrients and circulating cells to protect against injury and protection against invasion by external microorganisms. Clogging of blood vessels due to cholesterol, a coagulating biomolecule in the blood, increases the blood pressure.

2.10.1.2 Liver

Liver is a triangular organ below the diaphragm on the right-hand side. It has two lobes: left lobe and right lobe. It works with pancreas, gallbladder and intestine to digest, absorb and process food. It is responsible for multiple functions such as: 1) filtering blood coming from the digestive tract; 2) generating bile juice for digestion of ingested carbohydrate, proteins and fat; 3) production of blood-related

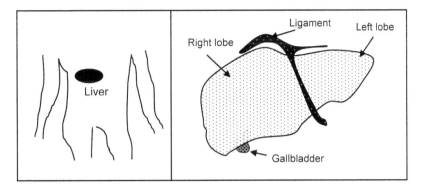

FIGURE 2.44 Location and structure of the liver

protein-components, proteins responsible for clotting the blood at the site of injury, and albumins related to water regulation in the body; 4) removal of dead RBC (Red Blood Cells); 5) metabolism of nitrogenous and carbohydrate materials absorbed through the intestine; 6) detoxifying chemicals and metabolizing drugs and 6) conversion of carbohydrate and fat to glucose for glucose regulation in the body. Figure 2.44 shows the location and structure of a liver.

Major diseases of the liver are: 1) liver cancer; 2) hepatitis – inflammation of the liver; 3) liver failure due to infection, genetic disorder or excessive alcohol; 4) infection caused by gallstones; 5) cirrhosis – long-term damage to the liver causing it to malfunction and 6) hemochromatosis – excessive iron deposit in the liver damaging it.

2.10.1.3 Kidney

Kidneys are bean-shaped organs on either side of the body. The kidney is made of millions of tiny filters called *nephrons*. Kidneys are located below the rib cage behind the belly and are connected to the blood-vessels. Kidneys filter out excess undesired fluid and water-soluble chemicals into the blood. The filtered blood goes to heart for oxygenation, while the unsought fluid is excreted as urine through a bladder – a funnel-shaped sack used to keep urine before drainage. Urine contains many chemicals not required by the body. Two major components of urine are *urea* and *ammonium*. Kidneys are essential for electrolyte balance, regulation of ion concentration, pH level balance, blood pressure management, water level balancing and production of many hormones. They also regulate extracellular fluid volume and ensure enough plasma to keep blood flowing to vital organs. Two important biomarkers excreted by a kidney are *creatinine* and *blood urea nitrogen* (BUN). Kidneys are adversely affected by diabetes, high blood pressure and heart disease.

2.10.2 Heart

Heart is a complex pumping muscle with four chambers and four valves as shown in Figure 2.45. Right-side of a heart is separated from the left-side of the heart using a thick muscle called *a septum*. Heart chambers are covered by thick muscles. The upper chambers are called *atria*, and the lower chambers are called *ventricles*. *Right-ventricle* is connected to lungs via *a pulmonary artery* to carry oxygen-depleted blood to lungs. *Right-atrium* is connected to the rest to the body via two major veins that carry oxygen-depleted blood to heart. *Left-atrium* is connected with the lungs via *pulmonary veins* that carry oxygen-rich blood from the lungs to the heart. *Left-ventricle* is connected with the body through the *aorta* – the major artery connecting a heart to the rest to the body.

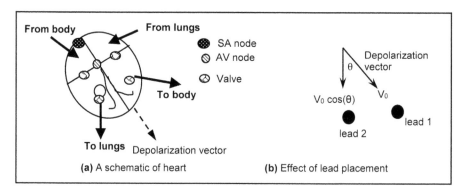

FIGURE 2.45   Basics of heart and electrical signal recording: (a) A schematic of heart; (b) Effect of lead placement

Four valves are: 1) *mitral valve* between the *left-atrium* and the *left-ventricle* that allows oxygen-rich blood to flow from the left-atrium to the left-ventricle; 2) *tricuspid valve* between the *right-atrium* and the *right-ventricle* that allows oxygen-depleted blood to flow from the right-atrium to the right-ventricle; 3) *aortic valve* between the *left-ventricle* and the body that allows oxygen-rich blood to flow from a heart to the body and 4) *pulmonary valve* between the *right-ventricle* and the lung that sends oxygen-depleted blood from the right-ventricle to the lungs.

Oxygen-rich blood comes from lungs to the *left-atrium* flows through the *mitral valve* to the left-ventricle, and from *left-ventricle* to the rest to the body using *aortic valve*. The oxygen-depleted blood comes from rest of the body to the *right-atrium*, flows through the *tricuspid valve* to the right-ventricle and flows to the lung through the *pulmonary valve*. The process repeats almost every second using a cycle of: 1) atrium compression that sends blood from atria to the corresponding ventricles; 2) ventricle compression that sends oxygen-rich blood to body and oxygen-depleted blood to lung concurrently and 3) relaxation of the heart muscle that concurrently fills right-atrium by oxygen-depleted blood coming from the body, and left-atrium by oxygen-rich blood coming from the lungs.

2.10.2.1 Depolarization of the heart-cells

The compression of heart-chambers is achieved using a cycle of electrical *depolarization* and *repolarization* that involves Ca^{++} ion (calcium ion), Na^{+} ion (sodium ion) and K^{+} ion (potassium ion) exchanges within the heart cells. *Depolarization* starts from the top right-hand side of the right-atrium called *SA node* and gradually progresses through atria and then to ventricles. Inside the ventricle, after a little delay, the electric pulse spreads through a three-dimensional network of electric-charge carrying fibers. During the depolarization of atria, the upper chamber compresses. During the depolarization of ventricles, lower chamber compresses. Depolarization is followed by repolarization when all the cells get back into the original relaxed state. As the atria relaxes, the oxygenated blood from lung enters the left atria, and the deoxygenated blood from the body enters the right atria. The cycle repeats almost every second.

2.10.2.2 ECG

The recording of the electrical signal during the depolarization and repolarization cycle from the chest-surface near a heart using one or more leads is called ECG (Electrocardiogram). Each cell in the heart is involved in the electrical signal. The overall vector sum of the electrical signals for each cell decides the vector-direction of the electrical signal. The angle this vector makes with the leads' positions changes the amplitude of the signal in different leads using the cosine law $V = V_0 \, cos(\theta)$ as shown in Figure 2.45b.

The first lead *Lead₁* is aligned with the depolarization-vector, and reads the electrical voltage V_0 that is measured in millivolts. The second lead *Lead₂* is at an angle of 30° from the depolarization-vector, and measures the value $V_0 \cos(30)$. Thus, the same ECG signal measures differently in multiple leads based upon their orientation with respect to the depolarization-vector. A 12-lead reading for ECG signals is common. Many times, in emergency rooms, 3-lead ECG reading is used. The interpretations of ECG readings are described in Section 7.1. Deviations of the ECG readings from the ECG of a healthy heart show abnormalities, and have been correlated with various cardio-pulmonary diseases.

2.10.3 Brain and Electrical Activity

The brain is made of around hundred billion *neural cells* connected to each other through *dendrites*. Different parts of a brain are responsible for different functions such as vision, hearing, balancing, emotion, language, speech, muscle-action, memory, cognition, perception, etc. These parts are connected through complex information pathways. MRI (Magnetic Resonance Imaging) of a brain has shown that different parts in a brain light up during human activities.

The brain generates different types of waves that vary based upon the mental and physical activities such as sleeping state, resting state, active state, anxious state, epilepsy, change in emotional state, eye blinking and muscle motions. EEG (Electroencephalography) is a recording of brain waves using multiple electrodes placed on the scalp in an organized pattern as explained in Section 7.5.1.

2.10.3.1 Brain waveforms

Based upon current understanding, there are seven types of brain waves: *alpha, beta, gamma, delta, theta, mu* and *SMR*. *Alpha waves* are neural oscillations in the range of 7.5–15 Hz and occur during closed-eyes and relaxed state, and are also associated with REM state of dreaming and lapse of attention. *Beta waves* are between 12.5 and 30 Hz, and are further subcategorized as: 1) *low beta waves* in the frequency range 12.5–16 Hz; 2), *mid beta waves* within the frequency range 16–20 Hz and 3) *high beta waves* between the frequency range 20–30 Hz. *Beta waves* are associated with walking, active state and anxiety level. *Delta waves* have frequency between 0 and 4 Hz, have highest amplitude, and are related to deep sleep. *Delta waves* are associated with the release of many hormones, including growth hormone. *Delta wave* disruption indicates metabolic disorder, sleep disorder, dementia and other mental disorders. *Theta waves* occur within the frequency range of 4–10 Hz and are associated with the movement activities such as running, jumping, walking. They change in response to various administered drugs. *Gamma waves* are in the frequency range of 25–100 Hz, and are associated with cognition, consciousness and elevation of mood. *Mu waves* are a synchronized pattern of electrical activity within the range of 7.5–12 Hz, and are most prominent when the body is resting. *SMR waves* (Sensory Motor Rhythm) are in the frequency range of 13–15 Hz, and occur during the period of inactivity.

2.11 GENOMICS AND PROTEOMICS

Genomes comprise *chromosomes*. Chromosomes comprise *genes*. Genes are the basic coding units of cells responsible for high-level functions. A gene may have multiple functions. Different types of genes are responsible for different functions. Gene's differentiation and specialization occur during the period of fetus development.

2.11.1 Genome Structure

Genes are a double-stranded sequence of four types of nucleotides: *A* (*adenine*), *T* (*thiamine*), *G* (*Guanine*) and *C* (*Cytosine*). Genes have a three-dimensional dynamic structure. Two strands complement each other: *A* is a complement of *T*, and *G* is a complement of *C*. Complements are held together by *hydrogen-bonds* between a hydrogen atom and an electronegative atom. *G-C bases* have three hydrogen bonds between them, and *A-T bases* have two hydrogen bonds between them. Genes are modeled as a sequence of the four nucleotides for computational studies.

Genes are divided into two parts: *regulating region* and *coding region*. *Regulating region* decides how fast and how much genes will be copied and translated into protein, and *coding region* is responsible for *protein creation*. In bacterial genes, *regulating region* immediately precedes the coding region. However, in Eukaryotes such as yeast, fungus, mammals, including humans, plants, *regulating* region can be far away from the coding region. RNAs (Ribosomal Nucleic Acid) are involved in multiple activities such as: 1) carrying a template of gene for protein creation; 2) carrying amino-acids needed for protein creation and 3) building proteins from the RNA-based templates of a gene. In addition, microRNA, a special class of RNA, are responsible for regulating the protein production. Defects in protein creation machinery cause many diseases.

2.11.1.1 Chromosomes

A *chromosome* is a packing unit in a genome. A chromosome consists of multiple genes. Generally, bacterial genomes have only one chromosome with some exceptions. Higher-level organisms have multiple chromosomes. Humans have 46 chromosomes. These chromosomes occur in pairs except *sex-chromosomes* that are different in males and females. The combination pattern of parental sex-chromosomes decides the gender of a child.

2.11.1.2 Proteins

Some genes are transformed into a sequence of amino-acids, and a complex three-dimensional dynamic structure of amino-acids matures to become *proteins*. Transcription genes are responsible for creating the mRNA (messenger RNA) based template for a gene that is converted into protein by rRNA (ribosomal RNA) – a small protein making factory. Three nucleotides of a gene form a *codon*. These *codons* map to a unique amino-acid – a basic building block of a protein. Since there are *4* nucleotides, $4^3 = 64$ possible codons occur that map on to 20 naturally occurring amino-acids in a human body. Many codons map on to the same amino-acid because the third position in many codons don't care.

*M*essenger RNA (m-RNA) carries a complemented single-strand imprint of the corresponding DNA encoding of a gene. These templates are created using transcription genes that slide over single gene to generate the corresponding m-RNA version of the gene. Another type of RNA molecules called t-RNA carry amino-acids needed for protein creation. The t-RNA molecules deposit the associated amino-acid in a peptide building unit called ribosomal RNA in a pattern guided by the m-RNA. The amino-acids are added in a chain-form, the sequence of amino-acid matures over a period and assumes a 3D folded-structure called *protein*.

Proteins are made of multiple domains. A domain is a single-function unit within protein. Proteins also bind to other proteins or genes or biomolecules through exposed patterns of amino-acids called *receptors*. The binding of a protein to another molecule makes a joint molecule that changes the 3D configuration of one or both molecules and alters their functionality.

2.11.1.3 Protein functions

A protein may have multiple functions, and its function changes based upon the change in 3-dimensional configuration. A protein consists of various segments of two types of regions: *hydrophilic* and *hydrophobic*. As the name suggests, hydrophilic regions have the affinity for water and strongly interact with water;

TABLE 2.3 Functions of amino-acids

NAME	CODE	CLASSIFICATION	FUNCTION
Alanine	Ala/A	Non-polar	Sugar metabolism, produces antibodies
Arginine	ARG/R	+ve charged	Neurotransmitter; increases endurance and reduces fatigue; detoxifier; involved in DNA synthesis
Asparagine	Asn/N	Polar	Excitory neurotransmitter
Aspartic acid	Asp/D	−ve charged	Increases stamina; protects live; DNA and RNA metabolism; immune function
Cysteine	Cys/C	Polar	Protection against pollution, radiation and ultraviolet; skin-repair
Glutamate	Glu/E	−ve charged	Neurotransmitter; involved in DNA synthesis
Glutamine	Gln/Q	Polar	Maintains normal and steady blood sugar level; muscle strength; provides energy to small intestine
Glycine	Gly/G	Non-polar	Slices DNA to produce different amino-acids
Histidine	His/H	+ve charged	Treats anemia. Found in high concentration in hemoglobin
Isoleucine	Ile/I	Non-polar	Essential for the synthesis of hemoglobin
Leucine	Leu/L	Non-polar	Skin, bone and tissue healing
Lysine	Lys/K	+ve charged	Component of muscle protein; needed in the synthesis of enzymes and hormones
Methoinine	Met/M	Polar	Breaks down fat; reduces muscle degeneration
Phenylalanine	Phe/F	Aromatic	Healthy nervous system. Boosts memory and learning
Proline	Pro/P	Non-polar	Critical for cartilage, tendon and ligaments and strong heart muscle
Pyroglutamic	Glp/U		
Serine	Ser/S	Polar	Maintains blood sugar level, boosts immune system
Threonine	Thr/T	Polar	Formation of collagen; prevents fat deposit in liver; helps in antibodies production
Tryptophan	Trp/W	Aromatic	Neurotransmitter serotonin synthesis; Helps sleep-aid and antidepression
Tyrosine	Tyr/Y	Aromatic	Increases energy, mental clarity and concentration antidepression
Valine	Val/V	Non-polar	Essential for muscle development

hydrophobic regions do not have an affinity for water and avoid water interaction. Out of 20 amino-acids, some are hydrophilic, some are hydrophobic and some are polar.

Table 2.3 shows the various functions of amino-acids within a human-body. Proteins are also used in a body for: 1) regulating biochemical reactions within the body in the metabolic pathways; 2) protecting against foreign invading biomolecule by blocking the interaction of foreign biomolecule; or 3) making foreign biomolecules inactive by binding to their receptors.

2.11.2 Biochemical Reactions Pathways

There are multiple biochemical reactions in a human body. The chain of these reactions is called a *pathway*. Two types of pathways are very important for the cell-function: 1) pathway involved in transforming a biomolecule to another biomolecule. These pathways are called *metabolic pathway* and 2) pathways involved in cell reproduction, DNA repair, cell response to environment change, sending messages to other cells, protecting a body by destroying foreign attacking microorganisms, improving the body's immune system and cleaning up the dead cells. These pathways are called *signaling pathways*. Understanding both these pathway mechanism is very important for disease management and treatment.

2.11.3 Gene Mutations and Abnormalities

Genes continuously go through mutations due to interactions with environmental factors such as chemicals, pH variations, pressure, radiation, heat-stress, interaction with foreign micro-organisms and other biochemical reactions. The body's gene-repair mechanism protects genes from mutation. However, repair is not foolproof, and probabilistically certain mutations go unrepaired causing their defective translation to proteins, and changes in gene-interaction with other genes and proteins. These alterations result in the lack of gene-expression (transcription and translation), and are responsible for genetic variations that show up often as genetic defects or genetic diseases. A small mutation in specific can cause serious defects. Sickle-cell anemia and many cancers are caused by such genetic defects.

Genetic transference from parents also plays a major role in genetic aberrations. Genes are also altered due to interactions with other foreign bodies such as viruses and may cause permanent aberrations manifested as diseases. Computational analysis of genes studying these aberrations and their effect on higher level genetic makeup (phenotype) is helping to develop better vaccines and medicines.

2.11.4 Antigens and Antibodies

Antigens are foreign biomolecule generally proteins or polysaccharides that are part of a harmful bacterium or virus that invade and stress a human body. A foreign biomolecule is recognized by a special cell called a T-cell in the body that identifies and tags the foreign biomolecule. *Antibody* is a special Y-shaped biomolecule. The two tips of the Y-shaped antibody bind to antigens to make them ineffective as shown in Figure 2.46.

The end of the Y-shaped antibody acts as the defense mechanism against foreign molecules by: 1) binding to the receptors of the molecules inside the proteins not permitting foreign molecules to bind with the receptors effectively stopping the alteration in the 3D confirmation of the proteins that will disrupt the healthy pathway of biochemical reactions; alternatively, 2) binding to the receptor of the foreign molecules making them ineffective against the receptors of the body protein involved in health reactions.

The important thing is the how tight and effective this binding is. This requires computational model to discover the best configuration of the antibody that would have strong binding. This is done using two techniques: 1) 3D computational model and 2) affinity analysis. Three-dimensional modeling approximates the three-dimensional folding structure of proteins using simulation. Affinity analysis derives the strength of binding of amino-acids under a specific protein configuration.

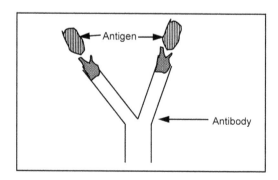

FIGURE 2.46   Antigen and antibody interaction through protein binding

2.12 SUMMARY

This chapter has discussed many background concepts needed to understand the interdisciplinary field of "Computational Health Informatics." The major fields discussed are: 1) fundamental data structures to access and search multidimensional and multimedia objects used in healthcare, including mobile objects; 2) digitization, recording format, compression sensor data; 3) approximate string matching; 4) statistics and probability needed for inferencing; 5) modeling of multimedia attributes using textures and shapes; 6) techniques for similarity-based search; 7) techniques for temporal abstraction and inferencing; 8) basic properties of different types of databases; 9) XML language and SOAP protocol for secure transmission of medical data over the Internet; 10) basics of human physiology and 11) basics of genome structure and function needed for computational analysis of genome for disease and biomarker identification.

Records in a database are modeled as tuples with one field unique to the record. That field is called *primary-key*. Biosignals are modeled as two-dimensional data with one dimension as time. Histograms are the frequency patterns in different intervals. Histogram is used to identify similar multimedia objects or entities based upon content-based matching.

An *N-dimensional feature-space* is modeled using an N-dimensional matrix that is implemented as a multidimensional array. *Sparse-matrices* have a very small percentage of occupancies of nonzero data items. A two-dimensional *sparse-matrix* is modeled as an array of linked-lists. The use of linked-lists is cost-effective due to the storage of a small number of nonzero data-items. It improves space utilization and retrieval time. Sparse N-dimensional spaces are efficiently modeled using a hashed-array, where each element of the hashed-array is a pointer to a linked-list of the cells of form (*N-dimensional-index, value*). A *hash-function* is used to derive the index. The mapping from the N-dimensional matrix cell to hashed-array index is not unique; collision occurs. This collision is resolved using sequential traversal of linked-lists of colliding data-elements. With a careful design of the hash-function and the size of the hashed-array, collision is reduced to a very small number, generally limited to two or three collisions.

For comparing the distance between the points in N-dimensional space, Manhattan distance, Euclidean distance, Minkowski's distance, and their weighted versions are used. The computation of weighted distance is based upon assigning different weights to the individual absolute difference for each dimension component in the overall distance.

Graphs are used to model many processes and probabilistic finite states machines used in medical processes and signal analysis. Graphs are modeled as matrices: rows and columns represent the nodes in a graph, and the cell value (*i, j*) holds the weight of an edge connecting node *i* to node *j*. A graph can be *undirected* or *directed*. In an *undirected graph*, the movement is bidirectional. In a directed graph, the direction is one way only from source-node to destination-node. A graph has a cycle if starting from one node and not repeating any intermediate node, the same node can be reached again. Tree is a special kind of directed acyclic graph.

Efficient retrieval of exact data, given a key-value, can be done using trees capable of handling multidimensional attributes. Multiple types of trees are used for retrieving information based upon multidimensional keys. They can be grouped as multidimension-based, region-based, prefix-based, and distance-based trees. Multidimension-based trees include K-D trees and its variants. Region-based trees include R-tree and its variants such as quad trees, R+ trees, SS-trees, SR-trees, TV-trees, TPR trees, etc. Prefix-based trees include PAT tree and its variants. Distance-based trees include VP-trees and its variant such as MVP trees. In addition, for the fast access to a record, hashing technique uses a hash-function to map primary-key to a near-unique index used to retrieve the record.

B+ trees and its integration with multidimensional-trees are used to retrieve data from the large databases. B+ trees store the key-values at the leaf nodes for efficient space utilization in the form of a page in the secondary storage. Since one page at a time is retrieved from the secondary storage to the RAM, this scheme also reduces undesired latency-time during data-retrieval.

For mobile care, the GPS locations of the objects keep changing. Hence, R-trees keep changing dynamically. Dynamic R-trees are constructed by combining an additional rectangle derived by combining velocity-vector and time before R-tree is recomputed, and the previous R-tree. The resulting R-trees are also adjusted when a node leaves or enters an MBR (Minimum Bounding Rectangle).

Digital data include a combination of image, sound, text, and video to archive, transmit and reproduce real-world information at a different time and space. Analog data captured by the sensors are first quantized by sampling them at discrete periodic intervals to convert into a digital data. High-resolution data have low quantization error. However, they require more memory and are slow to transmit due to excessive data. Z-curve and Hilbert's curve are two techniques that are used to map two-dimensional image to a sequence of binary data with minimal distortion. Instead of representing image data as a sequence of rows, these techniques use Z-pattern or square-pattern to traverse neighboring data before traversing data that are further apart.

For medical images, the information should be accurately recorded, transmitted in real-time and reproduced to give the same perception as the actual event. All multimedia contents are stored differently: image is stored in "MPEG," "JPEG", or "GIF" formats and variants; sound is stored in "WAV" format; video is stored in "AVI" or "MPEG" format; and text is stored in "TXT" format. Image files and sound files are large, and references to their file locators are embedded inside the XML document during transmission. Parity-bits are used for error-detection and error-correction.

Image formats are compressed using *lossless* or *lossy* compressions. Lossless compression provides actual recovery at the destination with no loss of information. Lossy compression provides a lot more compression, but loses information. Medical images require actual images for the diseased part at the destination and prefer lossless compression for the region of interest that may have abnormality. Other parts are compressed using lossy compression. There are multiple ways to compress image data such as the use of macro-blocks and motion-vectors, palettes of used colors, the use of transformation based upon color-frequency and predictive coding. Images are divided into multiple segments of homogenized segments that are labeled for analysis of their properties. These segments are associated with healthy and abnormal cells in the radiology images.

String matching takes two strings that vary somewhat in content and/or length, aligns them and calculates a similarity-score. If the similarity-score is below a threshold, then the strings are considered similar. String matching uses two types of techniques: Huffman coding and edit-distance. Huffman coding works on equal-length strings and uses number of nonmatching positions. Edit-distance also can handle unequal lengths using insertion/deletion of characters. There are two popular approaches to compute edit-distance: Jaro–Winkler distance and Levenshtein distance.

Dynamic programming is used to match two unequal approximately similar strings. It is based upon computing the matching-score of a cell (i, j) using the maximum of three cells $(i - 1, j)$, $(i - 1, j - 1)$ and $(i, j - 1)$ and progressively moving along the diagonals. A variant of dynamic programming is called dynamic time-warping that is used to match time-series data to identify time-intervals having similar patterns.

Statistical reasoning is used in many AI techniques such as neural networks, Markov model, hidden Markov model, probabilistic reasoning and verification of a hypothesis. Statistics is based upon studying aggregation of various attributes in a large sample set. Some of the aggregation functions are mean, median, mode, standard deviation and covariance.

Probability is strongly related to statistics because it is based upon computation of the percentage of an event occurring in a large sample set. Variable dependence (or independence) plays a major role in computing joint-probability of a multivariable event to occur. Conditional probability of an event to occur changes when the variables are dependent. Bayes' theorem equates absolute probabilities to conditional probabilities of two or more variable events, and has been used in many AI techniques including cause-and-effect based reasoning.

There are many well-explored probability distribution functions that are used to model the behavior of large sample sets and hypothesis verification. The popular distribution functions used in medical informatics are Gaussian, Poisson, binomial, multinomial and exponential. A null hypothesis is a verifiable statement that uses a probability distribution function and a confidence-interval.

Curve fitting is used to predict the values and study the trend analysis. The analysis using curve fitting to study the change in value of the dependent variable as independent variable changes. Regression analysis could be modeled using a linear curve or a nonlinear curve. A linear curve is popular due to its simplicity. In linear regression analysis, a straight line is fitted that passes through the average mean value of x-coordinates and y-coordinates of the sample points with a slope given by the ratio cov(x, y)/var(x) and minimizes the error between the actual value of y-coordinates and the estimated y-coordinates.

Multimedia objects such as image and sound are modeled using multiple characterizing features. Two types of characterizations are popular for modeling medical images: *texture* and *shape*. Texture modeling is based upon finding out the local patterns that vary. They are derived by a combination of comparing the intensity values of the neighboring pixels, their aggregations and their directional variations. There are also popular transformation-based techniques such as Gabor filter and wavelet-transform that extract different image characteristics in frequency and time-domain.

Shape-based modeling uses boundary-based modeling or Fourier transforms. Medical images use boundary-based modeling to study the shape-variations of a region of interest such as tumor. Boundary-based shape modeling use: 1) distance from the centroid of the region; 2) change in the direction of a curve and 3) approximations based upon polygons to model an irregular shape.

Most of the similarity-searches involving media objects require more than one feature-vector. A *similarity-search* uses variations of K-D tree, VP-tree and R-tree (R+ tree). K-D tree is a binary tree capable of searching K-dimensional space by rotating the dimensions periodically. R-tree divides a region into two regions (or more in N-dimensional space) progressively until only one significant cluster is left in the region.

Similarity-search is done by finding out the node with the smallest distance to the neighboring node during the traversal of the tree using the key-value. Besides KD-tree, there are PAT-trees that are used for fast string search and can be used for caption-based mapping of multimedia objects. Two-dimensional images have to be mapped to single dimensional images while preserving the proximity of pixels in a two-dimensional plane. Z-curve or Hilbert's curve based pixel-traversal are used to reduce the proximity distortion while transforming two-dimensional pixel-data to a sequence of binary values.

Temporal data recording and temporal query are quite important in health informatics to identify similar episodes in the past, or to compare the same abnormality from the past to find out the growth or remission. *Time-series data* analysis is also needed to predict alert conditions in case of emergency-care patients, patient recovering post-surgery, and patients in intensive care units. Time can be modeled as linear time, circular time or time-intervals.

Time-series data is first compressed using larger time-intervals before archiving to reduce the amount of data to be archived. Compression uses aggregation functions like mean of the values within an interval, fitting a linear curve between the points within an interval. However, compression leads to loss of information. Matching time-series data requires temporal logic and dynamic time-warping. *Temporal logic* is used to match two time-intervals or to check inclusion of a point-time in a time-interval. *Dynamic time warping* is used to compare two time-series data to identify common patterns. It is a variant of dynamic programming for matching two sequences of data.

Medical databases include distributed, temporal and multimedia databases and use both indexing based upon patient-id or similarity-based search on various contents to find out similar entries for data analytics. The databases are distributed because different medical organizations are owners of data, and they collaborate in data sharing. The databases are temporal because they hold time-stamped data such as patient monitoring and patient-recovery data where patient related signals and patient-states change. The databases have multimedia objects such as MRI, CAT-scan, PET scan, X-rays of different body parts, echograms, ECG, and EEG. Most images are black-and-white and are intensity based.

The databases are implemented as relational, object-relational and object-based databases. Relational databases are limited as they cannot model complex objects and medical datatypes needed in medical databases. The advantage of object-oriented databases is that it is supported by class inheritance and information hiding. Class inheritance allows the method reuse in subclasses. Medical databases have to be very secure to protect against unauthorized and fraudulent access or corruption. The access to medical databases should be guided by HIPAA policy and should be role-based at the record level to support only authorized access. To keep the database secure, medical databases are encrypted, fragmented and have role-based keys to give a role-based access to medical practitioners.

A major issue in multimedia databases is quick retrieval of multimedia using content-based similarity-match. Multimedia systems also need a synchronization mechanism for appropriate rendering at the destination. Multimedia, like any other patients' data, need to be encrypted and deidentified to provide confidentiality and privacy to a patient.

Temporal databases can have queries based upon time-stamps and temporal intervals or other attributes. Queries based upon time-stamps require indexing based upon time and temporal logic to match the time. However, time-based indexing causes limitations in efficient access based upon other attributes and vice-versa. Another major problem is that the actual times-stamps for events need not be unique because multiple events can occur concurrently. For the database recovery, time should be monotonic. Hence, transaction-time (time when an event is recorded into the database) is used for indexing that may not be accurate.

Medical databases are distributed, and the retrieval from distributed databases causes disparity of time-stamps and delays due to information transmission. There are many issues that need to be addressed in distributed databases such as information loss, corruption, time-synchronization, and freshness of data. Knowledge bases use a layer of if-then-else rules on top of databases to combine data in a meaningful way to answer complex queries.

XML is an Internet-based middleware language that can be used for transmitting medical data including multimedia images and database structures over the Internet. Since the tags can be user-defined, they can be extended to model: 1) nested databases 2) complex human organs; 3) other complex objects; 4) embedded computations; 5) interoperability and 6) languages for domain-specific object representation and manipulation. SOAP is an XML-based protocol to transmit XML information by providing an envelope to embed the information. It has a header that contains the information about origin, title, and type of information and a body that contains the information.

The human body has many vital organs and emanating signals. This book discusses signals associated with the heart (ECG), signals associated with the brain (EEG and MEG), and briefly signals associated with muscles (EMG). Other techniques in data analytics to identify diseases of livers, kidneys and other organs are related to identifying biomarkers in blood, spinal fluid and urine.

The heart is a four-chamber pulsating muscle-complex that takes oxygen-depleted blood from a body, transmits it to the lungs to absorb oxygen, receives oxygenated blood from the lungs and pumps blood into the body again at regular intervals approximately every second. The pumping of blood is controlled by a cycle of electrical activity called *depolarization* and *repolarization* of upper and lower chambers of the heart. During *depolarization* of upper chambers (atria), blood moves through valves to the lower chambers (ventricles). During the *depolarization* of ventricles, oxygenated blood is pumped from the left-ventricles into the body, and oxygen-depleted blood is pumped from the right-ventricles to the lungs. During the *relaxation* of atria, oxygenated blood moves from the lungs to the left-atrium, and the deoxygenated blood moves from the body to the right-atrium. The process repeats almost every second in a healthy heart.

ECG is the recording of the electrical signal of the heart cycle of depolarization and repolarization at the skin surface. The voltage is picked up by one or more leads. Different leads pick up different voltage based upon their orientation to the cumulative depolarization-vector. The amplitude is given by the cosine law $V = V_0 \, cos(\theta)$, where θ is the orientation of the lead with respect to the depolarization-vector V_0.

The brain has seven types of waves: alpha, beta, gamma, theta, mu, SMR and delta. These different brain waves are in the range of 0–100 Hz and are related to sleep patterns, resting, activity, motion, cognition and many disease states such as epilepsy and Parkinson's disease. Analysis of EEG can describe sleep disorders, epilepsy, cognitive disorders and motor control issues.

Human genome is made of multiple chromosomes. Each chromosome is a group of genes. Genes are the basic units of functions in a human body. Genes are double-stranded structures of four nucleotides (A, T, G, C), which are tightly bound in pairs (A $\Leftrightarrow$ T; G $\Lleftarrow\Rrightarrow$ C) through hydrogen bonds. Genes are translated to proteins using three different types of RNA molecules: mRNA, t-RNA, and rRNA.

Proteins have multiple functions in the body and are one of the three basic building blocks (proteins, fat and carbohydrates) of a human body. Proteins in a human body are made of 20 naturally occurring amino-acids that have multiple functions. Proteins are also involved in biochemical reactions, defense of the cells, and interactions with other molecules. Proteins have multiple domains and may have multiple functions based upon their dynamic three-dimensional configurations. Their configuration is very important for the protein-function and is altered by interactions with other molecules.

A biochemical reaction pathway is a chain of biochemical reactions involved in some common functionality. There are two types of pathways: *metabolic pathways* and *signaling pathways. Metabolic pathways* are involved in decomposition of ingested molecules and transformation into chemicals and biochemical molecules needed by the body. *Signaling pathways* are used to protect and repair the body against foreign microorganisms or adverse environmental conditions.

Antibodies are Y-shaped proteins whose tips dock to the foreign bodies or the human protein receptor breaking the interaction between foreign molecules that disrupt the regular biochemical reaction pathways by binding to proteins involved in catalyzing the biochemical reactions.

2.13 ASSESSMENT

2.13.1 Concepts and Definitions

Acyclic graph; adjacency matrix; alpha wave; alternate hypothesis; amplitude; analog-to-digital conversion; amino-acid; antibody; antigens; approximate string matching; AV-node; AVI format; B-frame; B+ tree; Bayes' theorem; beta wave; binomial distribution function; biochemical pathways; bioinformatics; biomarker; biosignals; BKD tree; brain; brain waveforms; cardiopulmonary system; chromosome; circular time-model; conditional probability; confidence interval; confidence level; contrast; correlation matrix; co-occurrence matrix; correlation; covariance; curve fitting; delta wave; depolarization; depolarization vector; digitization; directed acyclic graph; directed graph; distributed database; dynamic programming; dynamic time warping; ECG; edit-distance; EEG; EMG; energy; entropy; error correction; error detection; Euclidean distance; evidence matrix; exponential distribution function; feature-vector; Gabor filter; gamma wave; Gaussian curve; Gaussian distribution function; gene; gene mutation; genome; GIF format; gradient; graph; Hamming distance; hash function; hashing; hB tree; heart; Hilbert curve; Hilbert R-tree; HIPAA; histogram; HMM; Huffman coding; human physiology; Hurst operator; hypothesis testing; I-frame; image compression; image formats; image texture; indexing; interval based search; initialization matrix; Jaro–Winkler distance; JPEG format; knowledge base; K-D tree; K-D-B tree, KDB+ tree traversal; kidney; Levenshtein distance; liver; local binary pattern; local search; lossless compression; lossy compression; lung; Manhattan distance; margin-of-error; Markov model; matrix; mean; median; metabolic pathway; middleware; minimum bounding region (MBR); Minkowski's distance; mode; MPEG format; multimedia database; multimedia query; multinomial distribution; MVP tree; nucleotide; null hypothesis; Nyquist criterion; object-oriented database; object-relational database; object versioning; P-frame; p-value; partial order; PAT tree; pathway; Poisson distribution; probabilistic distribution function; probabilistic reasoning; protein; Quad tree; regression analysis; relational database; repolarization; R-tree; R+-tree; R+ tree traversal; run-length matrix; SA-node; sampling rate; SGLD tree; shape modeling; signaling pathway; similarity search; simulated annealing; SOAP; spatiotemporal data structure; sound formats; sparse matrix; spatial tree; SS-tree; SS-tree traversal; standard deviation; statistical reasoning; string matching; summarization; synchronization; temporal abstraction, temporal analysis; temporal database; text-summarization; texture analysis; theta wave; time modeling; time stamping; time versioning; time-series data; total order; TPR tree; tree; Trie, TV tree; variance; video format; VP tree; VP-tree traversal; WAV format; wavelet; wavelet transform; weighted Euclidean distance; weighted Manhattan distance; XML; Z-curve; Z-score

2.13.2 Problem Solving

2.1 Given the following directed graph in Figure 2.47, give the corresponding matrix representation and the corresponding adjacency matrix. Represent the matrix as an array of linked-lists.

2.2 Given a list of three-dimensional tuples: (23, 12, 4), (10, 24, 6), (45, 6, 14), (90, 83, 12), (65, 34, 2), (10, 76, 45), draw a balanced K-D tree. Then show the mechanism to search for the value (90, 83, 12).

2.3 Given these triples showing the X–Y coordinates of the cluster-centroids and the radius of the clusters, give an R-tree so only one cluster is covered minimally by the leaf-nodes of the R-tree. The triples are: (1, 1, 2), (6, 8, 3), (20, 13, 2), (15, 6, 2), (16, 16, 2).

2.4 Make an almost balanced K-D tree for the following three-dimensional dataset: (12, 34, 45), (34, 20, 17), (24, 40, 2), (37, 11, 19), (45, 20, 63), (35, 18, 23), (24, 76, 11), (33, 55, 12). Explain the rationale for the tree and the logic for balancing.

2.5 Search K-D tree in Problem #1 for the value (33, 15, 34). Show the steps.

2.6 Given a small subset of histograms containing 8 elements each, devise a similarity-based matching scheme assuming equal weight given to each field, and try it out on the following database of histograms for the query <3, 5, 1, 8, 6, 9, 4, 2>. Use Euclidean distance for matching, and index the database using K-D tree. Use traversal algorithm you developed earlier in Problem 2.2, and give top two values. Then, compare the histogram sequentially with each element in the database, and find out the top two solutions. Improve the algorithm you developed in Problem 2.2, so that it can traverse both left and right subtree optimally improving the best match but not traversing all the nodes. Give the top two matches. Compare your algorithm with optimal K-D tree traversal algorithms in research literature.
Database:

<3, 2, 1, 6, 6, 7, 4, 2>, <4, 5, 1, 8, 6, 9, 3, 2>, <3, 5, 7, 8, 6, 9, 4, 1>
<3, 5, 1, 8, 6, 9, 4, 20>, <3, 2, 1, 8, 6, 10, 4, 2>, <3, 5, 1, 8, 3, 4, 4, 2>
<7, 7, 1, 8, 6, 9, 6, 2>, <3, 6, 2, 8, 6, 9, 4, 3>, <3, 5, 1, 8, 6, 9, 6, 2>
<3, 9, 1, 8, 6, 12, 4, 2>, <1, 5, 1, 8, 6, 9, 4, 2>, <7, 5, 1, 8, 6, 9, 4, 7>

2.7 Given a set of clusters ($45 \leq x \leq 55, 23 \leq y \leq 30$), ($55 \leq x \leq 58, 40 \leq y \leq 45$), ($45 \leq x \leq 55, 32 \leq y \leq 38$), ($10 \leq x \leq 25, 13 \leq y \leq 20$), ($30 \leq x \leq 35, 40 \leq y \leq 45$), ($5 \leq x \leq 15, 10 \leq y \leq 20$), give an R+ tree and an SS-tree assuming that the centroids for the regions are (48, 26), (57, 42), (50, 35), (15, 15), (32, 42), and (10, 15), respectively. Make sure that the circles do not overlap and cover the cluster region completely.

2.8 Give a VP-tree for the following set of points: (12, 24), (11, 23), (8, 18), (4, 14), (7, 10), (15, 12). Explain your reasoning.

2.9 Give a PAT-tree for the searching the database with following names: "Manning," "Mark," "John," "Johnny," "James," and "Margret." Assume that each name is associated with a .jpg file to be retrieved from the database.

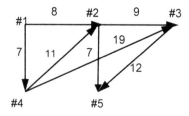

FIGURE 2.47   Graph for Problem 2.1

2.10 Write a generic program to search and insert an element in N-dimensional K-D tree. Check out your program on the following three-dimensional data: (*3, 10, 5*), (*8, 18, 6*), (*5, 15, 11*), (*12, 23, 7*), (*19, 14, 24*), (*34, 5, 17*), (*19, 25, 4*). Use Euclidean distance to measure the nearness of a data element to existing data in the tree. Store the location of all the data that are less than absolute distance threshold $\tau = 3$ in an array and sort the array in the end to give the best matching similar value. In the case, no elements are found within the distance threshold, insert the element in a location based upon search.

2.11 Read at least two articles on similarity-search techniques on R-trees and search for the nearest clusters in the radius of three units from the position (*14, 5*). Show the method.

2.12 Given a table of ten patients' records with the primary-key as patient-ids: "P123," "P456," "P017," "P789," "P127," "P230," "P761," "P089," "P193," "P549," give a hashing function using the size of table as 13 that would make the index mapping nearly unique with minimal collision, and show the hash-table entries.

2.13 For the pair of frames in Figure 2.48, assuming that the first frame is a reference frame, give the motion-vectors that would create the second frame using the first frame and the motion-vectors. The transmitted data is a quadruple: (*Frame Row, Frame Column, Motion-vector*). Only nonzero motion-vectors are transmitted. Give the motion-vector to be transmitted in form of quadruples: (*frame-row, frame-column, frame-shift in X-direction, frame-shift in Y-direction*). Explain your answer.

2.14 Given the 5×5 block in Figure 2.49 for intensity pattern of pixels, for each pixel, except first row, first column, last row and last column, find out the local binary pattern value and show the matrix. For, the first row, imagine a row above it by extrapolating the values of the row by subtracting the difference between the second and first row elements. For the last row, add an extra imaginary row below the last row by adding the difference between the last row and the last row elements. Repeat the process for the first and last columns. Give the overall LBP (local binary pattern).

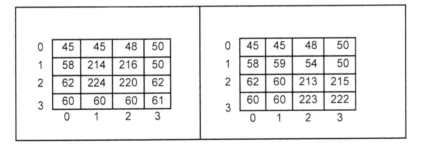

FIGURE 2.48 Intensity patterns of pixels for Problem 2.13

120	156	178	185	190
126	177	180	200	222
132	166	175	191	200
145	184	194	210	240
160	190	210	220	255

FIGURE 2.49 Intensity pattern for Problem 2.14

2.15 Given a single strand of a gene fragment *ATTGCAGCTACTGCTTCG*, give the complimentary strand, break it up into a group of three nucleotides called *codons* starting from the first nucleotide, and give the corresponding amino-acid sequence.

2.16 Do the histogram analysis of the two tables in problem #6, show the histogram, and match the histograms. Find out the percent similarity as total matching count/total count.

2.17 Given a sequence of sampled values <*35, 45, 57, 63, 157*>, and convert them into binary representation.

2.18 Given a polygon made of the following sequence of coordinate pairs, compute its centroid. Then find out the vector of distance from the centroid to the coordinate pairs and the slope each point makes with respect to the centroid. The points are: *(4, 0)*, *(6, 3)*, *(8, 7)*, *(4, 12)*, *(2, 5)*.

2.19 Compute the mean and standard deviation for the following sample of values: *3, 3, 4, 7, 1, 9, 8, 6, 5*.

2.20 Given these pairs of independent and dependent variables, find out the covariance. The value pairs are: *(2, 4)*, *(3, 5)*, *(4, 7)*, *(6, 9)*, *(8, 11)*, *(11, 13)*, *(12, 16)*.

2.21 For the following points, use the equations of the slope for optimum curve fitting in Section 2.1.6, find the covariance, variance, and equation of the optimum line for linear regression analysis. After computing the equation of the optimum line derive the individual error for each actual point by taking the absolute value $|Y_{actual} - Y_{estimated}|$. The points are: *(2, 3)*, *(3, 4)*, *(6, 8)*, *(10, 13)*, *(14, 18)*, *(20, 24)*, *(21, 25)*.

2.22 For the following string pairs of the same size, compute the Huffman distance and mark the matching positions. The string pairs are: "Arvind" and "Arwind"; "Pail" and "Pale"; "Right" and "Write"; "X-ray" and "Xrays".

2.23 For the following string pairs, calculate the Levenshtein edit-distance if the cost of insertion or deletion is "1," and the cost of substitution is "2" (deletion followed by insertion). Align the two strings in the pairs so the edit-distance is minimum. Assume that lowercase characters are same as uppercase characters. The string pairs are: ("X Ray" and "X-ray"), (Colorblind and Colourblind), (Electrocardiogram and ECG), (ECG and EEG), (Typhoid and Tiefoid). Show the optimum alignment and the edit-distance.

2.24 Repeat Problem 2.23 using Jaro–Winkler criteria for edit-distance.

2.25 Write a general-purpose program using dynamic programming technique and Equation 2.4 to align two strings. Try this program on these two sequences of amino-acids: "AGGCTAAG" and "AGCGCATAAG."

2.26 For the following sequence of three hourly reading of body temperature of a child suffering from infection for four days, summarize the data to explain to a physician what his temperature has been in the last four days. Summarize the data first and show the summarization (hint: read Section 2.4)

1: <6AM: 100, 9 AM: 102, 12 PM: 104, 3 PM: 104, 6 PM: 102, 9 PM: 103, 12 AM: 102, 3 AM: 100>

2: <6AM: 102, 9 AM: 102, 12 PM: 104, 3 PM: 104, 6 PM: 104, 9 PM: 104, 12 AM: 102, 3 AM: 102>

3: <6AM: 102, 9 AM: 103, 12 PM: 105, 3 PM: 104, 6 PM: 103, 9 PM: 102, 12 AM: 102, 3 AM: 102>

4: <6AM: 100, 9 AM: 101, 12 PM: 103, 3 PM: 104, 6 PM: 102, 9 PM: 101, 12 AM: 102, 3 AM: 100>

2.27 Write a program in C#, Java, C++ or Python to create a hash-table that reads several triple of the information of the form (*patient-id, social security number, name, date of birth, Insurance company, pharmacy*). A hash-table is an array of pointers to linked-lists. The information about the patient is accessed using patient-id. Name of the person is used to provide robustness if the data-entry clerk makes an error with the patient-id. Name should be accepted as first name followed by last name. Use simplified version of Jaro–Winkler or Levenshtein edit-distance to match the name and accept the name if edit-distance is below a threshold. Support following queries: 1) What is the social security number of the patient? 2) What is the insurance company of the patient?, and 3) what is the date of birth of the patient?

2.28 Now a day, antibiotic-resistant bacteria are becoming widespread. If there is a probability that a bacterium gets resistant to an antibiotic "A" is 0.02, and the probability that the same bacteria got resistant to another antibiotic that kills bacteria using different independent mechanism is 0.05. What is the probability that bacteria would get resistant to combination therapy? Suppose that the treatment is given ten times, what is the probability a resistant strain will develop anytime between first and tenth treatment? Explain your reasoning.

2.29 A person's heart pumps at 70 beats/minute in a clear unpolluted field, while he is walking. He needs 3.5 liters of blood being pumped into the body every minute to maintain the level of oxygen. Assume that a person travels to a metro city where pollution causes CO_2 level to rise by 15%. What will be the new heart beat and the blood being pumped/minute in the body if for very 5% increase in CO_2, blood requirement increases by 7%?

2.30 A 2520×1960 image is being compressed using 4×4 macro-blocks. Each macro-block will have a single color that is the median of the colors in the macro-block. Four bytes can present each macro-block: RGB value and transparency. Each pixel requires four bytes. Estimate the percentage of compression and the size in kb (kilo bytes) of data a compressed image will have.

2.31 Given a nested representation of a tumor as a quad-tree of the form ((1, ((1, tumor), (2, tumor), (3, healthy), (4, healthy))), (2, tumor), (3, ((1, tumor), (2, healthy), (3, healthy), (4, tumor))), (4, healthy)), draw and color the tumor.

2.32 A person suffering from lung infection has many white patches in his left lung in Figure 2.50. Represent these patches using an R+ tree and explain your rationale. Assume that the lung size is 8×4 inches.

2.13.3 Extended Response

2.33 Explain linear regression analysis and the equation for line that minimizes the overall error.

2.34 Describe various techniques used to compress multimedia images.

2.35 Explain the role of macro-blocks and palette in image-compression.

2.36 Read review and tutorial articles on standard JPEG, JPEG 2000, and MPEG 4 and compare their compression capabilities.

2.37 Describe ways to retrieve multimedia images from a multimedia database.

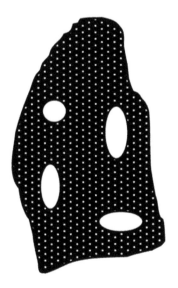

FIGURE 2.50 Modeling infected lung image using R+-tree for Problem 2.32

2.38 Describe the tree-search mechanisms to handle multidimensional keys used in multimedia databases.

2.39 Read various similarity-based search schemes in published articles and discuss.

2.40 Describe different models of time and give two examples for each type of modeling from the healthcare domain.

2.41 Explain the limitations of relational databases and how object-oriented databases remove the limitations of relational databases.

2.42 Describe the issues in secure medical databases? Explain various techniques to ensure the security of medical databases.

2.43 Explain data mining and its role in medical data analytics. Read two well cited survey articles about the role of data mining in medical data analytics, and discuss.

2.44 Explain Markov process and Hidden Markov model, including their matrix-based representation using clear examples.

2.45 Explain the role of heuristic in intelligent search.

2.46 Explain K-means clustering using a simple example.

2.47 Explain the mechanism of cardiopulmonary mechanism for oxygenating the blood.

2.48 Explain the different types of brain-waves and their roles.

2.49 Explain the advantages of object-oriented databases over relational databases.

2.50 Explain the importance of temporal abstraction in temporal databases.

2.51 Explain the information loss during the summarization of time-series data.

2.52 Explain the structure of heart and role of electrical activity for pumping of a heart.

2.53 Read tutorial articles on cardiopulmonary diseases and explain using schematic diagrams of the effects on lung and heart under these conditions: 1) asthma; 2) hypertension; 3) COPD (Cardiac Obstructive Pulmonary Disease); 4) pulmonary embolism; 5) excessive smoke inhalation containing CO (carbon monoxide), CO_2 (carbon dioxide), and NO (nitric oxide).

2.54 Describe different types of pathways and the mechanism through which antibodies block the invasion by foreign biomolecules.

2.55 What are the different means of compressing images and videos in different formats? Explain the answer abstractly instead of a specific format.

2.56 Explain various data structures used for multimedia image representation. How are these data structures compared for similarity-based search?

FURTHER READING

Data Modeling

2.1 Ballard, Chuck, Herreman, Dirk, Schau, Don, Bell, Rhonda, Kim, Eunsaeng, and Valencic, Ann, *Data Modeling Techniques for Data Warehousing*, Technical Report # SG24-2238-00, International Technical Support Organization, International Business Machine (IBM), Available at http://eddyswork.synthasite.com/resources/Data%20Modeling%20Tech%20For%20Data%20Warehouseing.pdf, Accessed June 2, 2018.

2.2 Johnson, Stephen B., "Generic Data Modeling for Clinical Repositories," *Journal of American Informatics Association (JAMIA)*, 3(5), September/October 1996, 328–339.

2.3 Kowalski, Gerald J., and Maybury, Mark T., *Information Storage and Retrieval Systems – Theory and Implementation*, Kluwer Academic Publishers, Boston, USA, 2002.

Basic Data Structures and Modeling Graphs

2.4 Brass, Peter, *Advanced Data Structures*, Cambridge University Press, New York, NY, USA, 2008.

2.5 Goodrich, Michael T., Tamassia, Roberto, and Goldwasher, Michael H., *Data Structures and Algorithms in Python*, John Wiley and Sons, 2013.

2.6 Samet, Hanan, *Foundations of Multidimensional and Metric Data Structures, Morgan Kaufman Series in Data Management*, Morgan Kaufman Publishers, San Francisco, CA, USA, 2006.

2.7 Schaffer, Clifford A., *Data Structures and Algorithms (C++ Version)*, edition 3.2, Department of Computer Science, Virginia Tech, Blacksburg, VA USA, Available at https://people.cs.vt.edu/shaffer/Book/C++3e20120102.pdf.

Modeling N-dimensional Feature Space

2.8 Chang, Edward Y., *Foundations of Large-Scale Multimedia Information Management and Retrieval: Mathematics of Perception*, Springer, New York, NY, USA, 2011.

2.9 Gaede, Volker, and Günther, Oliver, "Multidimensional Access Methods," *ACM Computing Surveys*, 30(2), June 1998, 170–231.

Trees for Database Search

2.10 Brass, Peter, *Advanced Data Structures*, Cambridge Press, New York, NY, USA, 2008.

2.11 Dao, S., Yang, Q., and Vellaikal, A., "MB+ Tree: An Index Structure for Content Based Retrieval," Chapter 11 of *Multimedia Database Systems: Design and Implementation Strategies* (editor: Nwosu, K. C., Thuraisingham, B., and Berra, P. B.), Kluwer Academic Publishers, Boston, MA, USA, 1996.

Spatial Trees for Multidimensional Data

2.12 Askitis, Nikolas, and Zobel, Justin, "B-Tries and Disk-based String Management," *The VLDB Journal*, 18(1), 2009, 157–179, DOI: 10.1007/s00778-008-0094-1.

2.13 Beckman, Norbert, Kriegel, Hans-Peter, Schneider, Ralf, and Seeger, Bernhard, "The R-tree: An Efficient and Robust Access Method for Points and Rectangles," in *Proceedings of the ACM SIGMOD Int. Conf. Management of Data*, Atlantic City, NJ, USA, 1990, 322–331.

2.14 Benteley, J. L., "Multidimensional Binary Search Trees in Database Applications," *IEEE Transactions on Software Engineering*, SE-5(4), July 1979, 330–340.

2.15 Kamel, Ibrahim, and Faloutsos, Christos, "Hilbert R-tree: An Improved R-tree Using Fractals," In *Proceedings of the 20th International Conference on Very Large Data Bases*, September 1994, 500–509.

2.16 Kurniawati, R., Jin, J. S., and Shephard, J. A., "SS Tree - An Improved Index Structure for Similarity Searches in High Dimensional Feature Space," In *Proceedings of the Conference on Retrieval for Image and Video Databases V*, San Hose, CA, Feb 1997, SPIE Proceedings Series, 3022, 110–120.

2.17 Samet, Hanan, *Foundations of Multidimensional and Metric Data Structures, Morgan Kaufman Series in Data Management*, Morgan Kaufman Publishers, San Francisco, CA, USA, 2006.

Trees for Multidimensional Database Search

2.18 Lomet, David B., and Salzberg, Betty, "The hB-tree: A Multi-attribute Indexing Method for Good Guaranteed Performance," *ACM Transactions on Database Systems*, 15(4), December 1990, 635–658.

2.19 Nievergelt, Jürg, Hinterberger, Hans, and Sevcik, Kenneth C., "The Grid File: An Adaptable, Symmetric Multikey File Structure," *ACM Transactions on Database Systems*, 9(1), 1984, 38–71.

2.20 Procopiuc, Octavian, Agarwal, Pankaj K., Arge, Lars, and Vitter, Jeffrey S., "Bkd-Tree: A Dynamic Scalable kd-Tree," In Eighth International Symposium *Advances in Spatial and Temporal Databases*. Lecture Notes in Computer Science. 2750: 46–65. DOI: 10.1007/978-3-540-45072-6_4.

2.21 Robinson, John T., "The K-D-B Tree: A Search Structure for Large Multidimensional Dynamic Indexes," In *Proceedings of the ACM SIGMOD International Conference on Management of Data*, Ann Arbor, MI, USA, April/May 1981, 10–18.

Time-Series Data

2.22 Camerra, A., Palpanas, T., Shieh, J., and Keogh, E., "iSAX 2.0: Indexing and Mining One Billion Time Series," *2010 IEEE International Conference on Data Mining*, Sydney, NSW, 2010, pp. 58–67. DOI: 10.1109/ICDM.2010.124.

2.23 Chatzigeorgakidis, Georgios, Skoutas, Dimitrios, Patroumpas, Kostas, Athanasiou, Spiros, and Skiadopoulos, Spiros, "Indexing Geolocated Time Series Data," In *Proceedings of the 25th ACM SIGSPATIAL International Conference in Advances in Geographic Information System*, Redondo Beach, CA, USA, November 2019, DOI: 10.1145/3139958.3140003.

2.24 Kim, Young I., Park, Youngbae, and Chun, Jonghoon, "A Dynamic Indexing Structure for Searching Time-Series Patterns," In *Proceedings of 20th International Computer Software and Applications Conference (COMPSAC '96)*, Seoul, South Korea, August 1996, 270–275, DOI: 10.1109/CMPSAC.1996.544176.

Trees for Spatiotemporal Access

2.25 Saltenis, S., Jensen, C. S., Leutnegger, S. T., and Lopez, M. A., "Indexing the Positions of Continuously Moving Objects" In *Proceedings of the ACM SIGMOD International Conference on Management of Data*, New York, NY, USA, 2000, 331–342.
2.26 Tao, Yufei, Papadias, Dimitris, and Sun, Jimeng, "The TPR*-Tree: An Optimized Spatio-Temporal Access Method for Predictive Queries," In *Proceedings of the 29th VLDB Conference*, Berlin, Germany, 2003.

Digitization of Sensor Data

2.27 Sonka, Milan, Hlavac, Vaclav, and Boyle, Roger *Image Processing, Analysis and Machine Vision*, 4th edition, Cengage Learning, Stamford, CT, USA, 2015.

Analog to Digital Conversion

2.28 Walden, Robert H., "Analog-to-Digital Converter Survey and Analysis," *IEEE Journal on Selected Areas in Communications*, 17(4), April 1999, 539–550.

Digital Representation of Images

2.29 Hudson, Graham, Léger, Alain, Niss, Birger, and Sebestyén, István, "JPEG at 25: Still Going Strong," *IEEE Multimedia*, 24(2), Apr.-June 2017, 96–103.
2.30 *Joint Photographic Experts Group*, Available at http://www.jpeg.org/index.html, Accessed June 6, 2018.
2.31 Taubman, David S., and Marcellin, Michael W., "JPEG2000: Standard for Interactive Imaging," *Proceedings of the IEEE*, 90(8), August 2002, 1336–1357.

Image Compression

2.32 Bethencourt, J., Sahai, A., and Waters, B., "Ciphertext-Policy Attribute-Based Encryption," In *Proceedings of IEEE Symposium on Security and Privacy*, Oakland, CA, USA, May 2007, 321–334.
2.33 Langdon, G., Gulati, A., and Seiler, E., "On the JPEG model for Lossless image compression," In *Proceedings of the 2nd Data Compression Conference*, 1992, 172–180.
2.34 Oh, Han, Bilgin, Ali, and Marcellin, Michael W., "Visually Lossless Encoding for JPEG2000," *IEEE Transactions on Imaging Processing*, 22(1), January 2013, 189–201.
2.35 Wallace, Gregory K., "The JPEG Still Picture Compression Standard," *IEEE Transactions on Consumer Electronics*, 38(1), February 1992, 18–34.

Approximate String Matching

2.36 Cohen, William W., Ravikumar, Pradeep, and Feinberg, Stephen E., "A Comparison of String Metrics for Matching Names and Records," *KDD Workshop on Data Cleaning and Object Consolidation*, 3, Washington DC, USA, August 2003, 73–78.
2.37 Winkler, William E., "Overview of Record Linkage and Current Research Direction," *Technical Report Statistics #2006-2*, Research Report Series, Statistical Research Division, US Census Bureau, Washington DC, USA, 2006, Available at http://www.census.gov/srd/papers/pdf/rrs2006-02.pdf, Accessed on June 1, 2018.
2.38 Zaro, M. A., "Advances in Record Linkage Methodology as Applied to the 1985 Census of Tampa Florida," *Journal of the American Statistical Association*, 84, 1989, 414–20.

Statistics and Probability

Statistics

2.39 Bluman, Alan G., *Elementary Statistics: A Step by Step Approach*, 9th edition, McGraw Hill Education, New York, NY, USA, 2014.

Probability

2.40 Bertsekas, Dimitri, *Introduction to Probability*, 2nd Edition, Aethna Scientific, Nashua, NH, USA, 2008.

Curve Fitting

2.41 Rigalleau, Vincent, Lasseur, Catherine, Raffaitin, Christelle, Perlemoine, Caroline, Barthe, Nicole, Chauveau, Phillippe, Combe, Christian, and Gin, Henri, "The Mayo Clinic Quadratic Equation Improves the Prediction of Glomerular Filtration Rate in Diabetic Subjects," *Nephrology Dialysis Transplantation*, 22(3), March 2007, 813–818.

Modeling Multimedia Feature-Space

Texture Modeling in Medical Image Analysis

2.42 Barry, Brian, Buch Karen, Soto, Jorge A., Jara, Hernan, Nakhmani, Arie, and Anderson, Stephan W., "Qualifying Liver Fibrosis through the Application of Texture Analysis of Diffusion Weighted Imaging," *Magnetic Resonance Imaging*, 32, 2014, 84–90.
2.43 Casttellano, G., Bonhilla L., Li, L. M., and Cendes F., "Texture Analysis of Medical Images," *Clinical Radiology*, 59, 2004, 1061–1069.
2.44 Depeursinge, Adrien, Foncubierta-Rodriguez, Antonio, Ville, Dimitri V. D., and Müller, Henning, "Three-Dimensional Solid Texture Analysis in Biomedical Imaging: Review and Opportunities," *Medical Image Analysis*, 18, 2014, 176–196.
2.45 Chandrasekhar, Ramachandran, Attikouzel, Yianni, and DeSilva, Christopher J. S., "Texture Analysis of Mammograms Using Two-Dimensional Hurst Operator," In Proceedings of the 13th International Conference on Digital Signal Processing, Santorini, Greece, July 2002, 97–100.
2.46 Fogel, I. and Sagi, D., "Gabor Filters as Texture Discriminators," *Biological Cybernatics*, 61(2), June 1989, 103–113.
2.47 Kachouie, Nezamoddin N., and Fieguth, Paul, "A Medical Texture Local Binary Pattern for TRUS Prostate Segmentation," In *Proceedings of the 29th Annual IEEE International Conference of the EMBS*, Lyon, France, August 2007, 5605–5608.
2.48 Wei, Datong, Chan, Heang-Ping, Helvie, Mark A., Sahiner, Berkman, Petrick, Nicholas, Adler, Dorit D., and Goodsit, Mitchell M., "Classification of Mass and Normal Breast Tissue on Digital Mammograms: Multiresolution Texture Analysis," *American Association of Medical Physics*, 22(9), September 1995, 1501–1513.

Shape Modeling

2.49 Loncaric, Sven, "A Survey of Shape Analysis Technique," *Pattern Recognition*, 31(8), 1998, 983–1001.
2.50 Shu, Xin, and Wu, Xiao-Jun, "A Nodel Contour Descriptor for 2D Shape Matching and Its Application to Image Retrieval," *Image and Vision Computing*, 29, 2011, 286–294.
2.51 Yuan, Zhanwei, Li, Fuguo, Zhang, Peng, Chen, Bo, "Description of Shape Characteristics through Fourier and Wavelet Analysis," *Chinese Journal of Aeronautics*, 27(1), 160–168.

Similarity-Based Search Techniques

Nearest Neighbor Search Methods

2.52 Abbasifard, Mohammed R., Ghahremani, Bijan, and Naderi, Hassan, "A Survey on Nearest Neighbor Search Methods," *International Journal of Computer Applications*, 95(25), June 2014, 39–52.
2.53 Bozkaya, Tolga, and Ozsoyoglu, Meral, "Indexing Large Metric Spaces for Similarity Search Queries," *ACM Transactions on Database Systems*, 24(3), September 1999, 361–404.
2.54 Friedman, Jerome H., Baskett, Forest, and Shustek, Leonard J., "An Algorithm for Finding Nearest Neighbors," *IEEE Transactions on Computers*, 24(10), October 1975, 1000–1006.
2.55 Fu, Ada W., Chan, Polly M., Cheung, Yin-Ling, Moon, You S., "Dynamic VP-tree Indexing for N-nearest Neighbor Search Given Pair-wise Distance," *The VLDB Journal*, 9(2), July 2000, 154–173.
2.56 Petrakis, Euripides G.M., and Faloutsos, Christos, "Similarity Searching in Medical Image Databases," *IEEE Transactions on Knowledge and Data Engineering*, 9(3), May/June 1997, 435–447.

2.57 Yianilos, Peter N., "Data Structures and Algorithms for Nearest Neighbor Search in Generic Metric Spaces," In Proceedings of the fourth annual ACM-SIAM Symposium on Discrete algorithms (SODA '93), Austin, TX, USA, January 1993, 311–321.

Tree Traversal Techniques

2.58 Drozdek, Adam, *Data Structures and Algorithms in C++*, 4th Edition, Cengage Learning, Boston, MA, USA, 2013.
2.59 Manolopoulos, Yannis, Nanopoulos, Alexandros, Papadopoulos, Apostolos N., and Theodoridis, Yannis, *R-Trees: Theory and Applications*, Spring-Verlag, London, UK, 2010.

Temporal Abstractions and Inference

Modeling Time

2.60 Allen, J. F., "Towards a General Theory of Action and Time," *Artificial Intelligence*, 23, 1984, 123–154.

Temporal Analysis

2.61 Batal, Iyad, Sacchi, Lucia, Bellazzi, Riccardo, and Hauskrecht, Milos, "A Temporal Abstraction Framework for Classifying Clinical Temporal Data," In *Proceedings of the Annual Symposium of American Medical Informatics Association*, San Francisco, CA, USA, November 2009, 29–33.
2.62 Combi, Carlo, and Sahar, Yuval, "Temporal Reasoning and Temporal Data Maintenance in Medicine: Issues and Challenges," *Computers in Biology and Medicine*, 27(5), 1997, 353–368.

Time Interval Based Matching

2.63 Müller, Meinard, "Dynamic Time Warping," Chapter 4, In *Information Retrieval for Music and Motion*, Springer, Heidelberg, Germany, 2007.

Knowledge Based Temporal Analysis

2.64 Sahar, Yuval, "A Framework of Knowledge-based Temporal Abstraction," *Artificial Intelligence*, 90, 1997, 79–133.
2.65 Stacey, Michael, and McGregor, Carolyn, "Temporal Abstraction in Intelligent Clinical Data Analysis: A Survey," *Artificial Intelligence in Medicine*, 39(1), January 2007, 1–24.

Types of Databases

Relational Database Systems

2.66 Ramakrisnan, Raghu, and Gehrke, Johannes, *Database Management Systems*, 3rd edition, Publisher: McGraw-Hill Higher Education, New York, NY, USA, 2003.

Object-Based Databases

2.67 Dietritch, Suzanne W, and Urban, Susan D., *Fundamentals of Object Databases: Object-Oriented and Object Relational Design*, Morgan & Claypool Publishers, San Rafael, CA, USA, 2011.

Multimedia Databases

2.68 Lu, Guojun, *Multimedia Database Management Systems*, Artech House, Boston, MA, USA, 1999.

Temporal Databases

2.69 Keogh, Eamonn, Chu, Selina, Hart, David, and Pazzani, Michael, "An Online Algorithm for Segmenting Time Series," In *Proceedings of IEEE International Conference on Data Mining (ICDM)*, November/December 2001, San Jose, CA, USA, 289–296.
2.70 Orphanou, Kalia, Stassopoulou, Athena, Keravnou, Elpida, "Temporal Abstraction and Temporal Bayesian Networks in Clinical Domain," *Artificial Intelligence in Medicine*, 60(3), March 2014, 133–149.
2.71 Özsoyoglu, Gultekin, and Snodgrass, Richard T., "Temporal and Realtime Databases – A Survey," *IEEE Transactions of Knowledge and Data Engineering*, 7(4), August 1995, 513–532.

Distributed Databases and Knowledge Bases

2.72 Haug, Frank S., and Rahimi, Saeed K., *Distributed Database Management Systems: A Practical Approach*, Wiley, Hoboken, NJ, USA, 2010.

Middleware for Information Exchange

2.73 Chinnici, Roberto, Moreau, Jean-Jacques, Ryman, Arthur, and Weerawarana, Sanjeeva (editors), Web Services Description Language (WSDL) Version 2.0 Part I: Core Language, Available at https://www.w3.org/TR/wsdl/, Accessed May 27, 2018.

2.74 *Efficient XML Interchange (EXI) Format 1.0*, 2nd edition, February 2014, Available at https://www.w3.org/TR/2014/REC-exi-20140211/, Accessed May 27, 2018.

2.75 Fabians, Benjamin, Ermacova, Tatiana, and Junghanns, Philipp, "Collaborative and Secure Sharing of Healthcare Data in Multi-clouds," *Information Systems*, Elsevier, 48, 2015, 132–150.

Human Physiology

Modeling Human Body

2.76 Betts, J. Godon, Desaix, Peter, Johnson, Eddie, Johnson, Jody E., Korol, Oksana, Kruse, Dean, et al., *Anatomy and Physiology*, OpenStax, Rice University, Houson, TX, USA, Available at https://d3bxy9euw4e147.cloudfront.net/oscmsprodcms/media/documents/AnatomyandPhysiology-OP.pdf, Accessed June 1, 2018.

2.77 Roberts, Alice, *The Complete Human Body – The Definitive Visual Guide*, 2nd edition, Dorling Kindersley Limited, New York, NY, USA, 2016

Heart and ECG

2.78 Garcia, Tomas B., *Introduction to 12-Lead ECG: The Art of Interpretation*, 2nd edition, Jones and Bartlett Learning, Burlington, MA, USA, 2013.

Brain and Electrical Activity

2.79 Britton, Jeffrey W., Freay, Lauren C., Hopp, Jennifer L., Korb, Pearce, Koubeissi, Mohamad, Livens, William E., Pestana-Knight, Elia M., St. Louis, Erk K, *Electroencephalography (EEG): An Introductory Text and Atlas of Normal and Abnormal Findings in Adults, Children and Infants*, American Epilepsy Society, 2016, Available at https://www.ncbi.nlm.nih.gov/books/NBK390354/pdf/Bookshelf_NBK 390354.pdf, Accessed June 1, 2018.

2.80 Carter, Rita, Aldridge, Susan, Page, Martyn, and Parker, Steve, *The Human Brain Book: An Illustrated Guide to its Structure, Function, and Disorders*, DK Limited, New York, NY, USA, 2009.

Genomics and Proteomics

Genomic Structure and Pathways

2.81 Campbell, A. M., and Heyer, Laurie J., *Genomics, Proteomics and Bioinformatics*, 2nd Edition, Benjamin Cummings, San Francisco, CA, USA, 2006.

2.82 Thangadurai, Devrajaan, and Sangheetha, Jeyabalan, *Genomics and Proteomics: Principles, Technologies and Applications*, CRC Press, Boca Raton, FL, USA.

Antigens and Antibodies

2.83 Abbas, Abdul K., Litchtman, Andrew H., and Pillai, Shiv, *Cellular and Molecular Immunology*, 8th edition, Elsevier Saunders, Philadelphia, PA, USA, 2014.

Intelligent Data Analysis Techniques

3

Artificial Intelligence (computational intelligence) emulates human intelligence using computational techniques. The major advantages are in reducing the computational overhead of solving hard problems using mathematically well-defined guesses (heuristics), optimizing the solution under complex constraints, learning to solve a problem using multiple mathematically sound techniques such as decision trees for classification, prediction using regression analysis, classification using clustering and statistical analysis and pattern recognition.

Healthcare databases have a huge amount of disease-related data. The data includes pathology reports, radiology reports, vital signs' data, ECG signals, patient monitoring data, data related to physician–patient encounter, drug dosage data and patient-survey data. The data needs to be automatically analyzed to identify various conditions associated with these diseases. Different administered medications and regimens help patient recover. The dosage of these medications needs to be controlled based upon age-group, gender, patient's history of drug-reactions and drug-toxicities. During recovery, patients have multiple side-effects, and various drugs have different levels of toxicities to be analyzed and minimized. In addition, administered drugs interact with each other causing additional side-effects.

Patients' biosignals such as ECG, EEG and EMG indicate many disease conditions. These biosignals need to be analyzed automatically using intelligent techniques. X-rays, MRI, CAT-scan and PET-scan need to be analyzed automatically to identify abnormalities such as tumor growth, malignant lumps in breast, brain tumors, kidney stones, fractures in bones, herniated discs in spines and much more. Different types of cells are counted automatically from blood samples in pathology tests. Automated detection and analysis are necessary because the pattern is not so obvious to human experts, and the sample-size is too large for experts to handle. Besides, the summarization of health records facilitates data mining new information about abnormal conditions or disease conditions. With growing resolution in data, intelligent data-analytics is making the drug administration more accurate.

Intelligent techniques are also needed to provide human-friendly interfaces. Humans use natural languages to describe patients' conditions to be translated into structured information to be archived in databases. Healthcare practitioners use different clinical terms to describe the same disease conditions. An ontological dictionary is needed to identify the equivalent terms in medical domains. Doctors and nurses use handwritten notes and voice recordings to store patient conditions. These notes are translated to structured information in databases. Spoken language translation to text often has word-sense ambiguity that needs to be resolved. The information in an unstructured text needs multilingual translation when transmitted across the countries.

There are many subareas of artificial intelligence used in healthcare informatics. The major ones are: 1) heuristic search technique; 2) inferencing techniques based upon mathematical logic, uncertainty-based reasoning, probabilistic reasoning and fuzzy reasoning and 3) machine learning techniques.

3.1 DIMENSION REDUCTION

Data is mapped in N-dimensions based upon feature-vectors. Each dimension corresponds to a feature from the feature-vector. Features characterizing an entity are presumed to be orthogonal and do not affect each other. It is assumed that the noise caused by dimensions on each other is minimal. Perfect independence of variables implies zero covariance. However, the coordinate systems may not always be orthogonal. Certain features may contribute significantly to the resolution and identification of an object-class. Ignoring the features (dimensions) that contribute little to the identification of object-classes will reduce the computational overhead and facilitate visualization of the clusters. The purpose of dimension-reduction is to transform the coordinate systems that: 1) maximizes the variance of the individual classes; 2) minimize the covariance between the individual classes (or dimensions) to remove interference; 3) retain significant features.

To derive proximity, the values in different dimensions are normalized to take care of: 1) incompatibility in units of values associated with the feature-set in two feature-vectors and 2) uneven effect of the change in feature-vectors by different parameters on the outcome. Those feature vectors that affect outcomes marginally are assigned smaller weights. A feature is dropped from the consideration to improve computational efficiency if its weight is less than a threshold-value.

Retaining significant features and dropping less-significant features in data analysis is called *dimension reduction*. The popular techniques are *principal component analysis, linear discriminant analysis* and *independent component analysis*.

3.1.1 Principal Component Analysis

Principal component analysis is based upon transforming the current coordinate-system $X_1, ..., X_N$ ($N > 1$) to a system of new orthogonal coordinate-system $X'_1, ..., X'_M$ ($M \geq 1$ and $M < N$) such that the points have largest variance along the axis X'_1 then along X'_2 and so on as shown in Figure 3.1.

Each new dimension X'_i ($1 \leq i \leq M$) is expressed as a linear combination of original dimensions $X_1, ... X_N$. After few major dimensions, the variance becomes small along the remaining dimensions, and the remaining dimensions are ignored. Principal component analysis is suitable for data with normal Gaussian distribution.

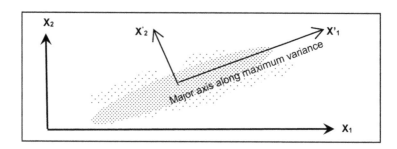

FIGURE 3.1   Transformation of coordinates-system in PCA to get axis of maximum variance

An N-dimensional data element X is modeled as $(x_1, ..., x_N)^T$. The coordinate transformation is performed using Eigenvector-based analysis. Given a vector of values in the old-coordinate system and a $N \times N$ matrix of transformation coefficients, denoted by P, the corresponding vector of values in the new coordinate system is written as $Y = PX$, where $Y = (y_1, ... y_M)^T$ is the transformed coordinate-values for the data-point. The problem is reduced to deriving the transformation matrix P.

3.1.2 Linear Discriminant Analysis

Linear discriminant analysis is based upon separating two or more classes of objects using distance, mean, variance and covariance. LDA aims at maximizing the separation between the classes and minimization of covariance of the samples within the same classes. The classes of objects are separated by transforming the coordinate system such that the sum of distance between the points within the same clusters is minimized, and the sum of the distance between the points of the different clusters is maximized. It is suitable for Gaussian distribution.

Scatter-matrix is an estimate of the covariance matrix for multivariate normal distribution. LDA uses two types of scatter-matrices: (1) within-class scatter matrix denoted by Σ^w and (2) between-class scatter-matrices denoted by Σ^b. LDA searches for a group of basis-vectors, which make different class samples, and have the smallest within-class scatter and the largest between-class scatter. The main objective of LDA is to find a projection matrix that maximizes the ratio of the determinant of Σ^b to the determinant of Σ^w. Two scatter-matrices are given by Equations 3.1 and 3.2.

$$\Sigma^b = \sum_{i=1}^{i=k} N_i(\mu_i - \bar{\mu})(\mu_i - \bar{\mu})^T \tag{3.1}$$

$$\Sigma^w = \sum_{i=1}^{i=k} \sum_{j=1}^{j=N_i} (x_{ij} - \mu_i)(x_{ij} - \mu_i)^T \tag{3.2}$$

where x_{ij} is a data-point in ith class, μ_i is the mean of the ith class, $\bar{\mu}$ is the overall mean of all the classes, N_i is the number of data-points in the ith class and k is the number of classes.

3.1.3 Independent Component Analysis

Independent component analysis uses a higher order of moments, and is suitable for non-Gaussian distributions. Like PCA (principal component analysis), the goal is to derive a transformation matrix to identify a new set of the coordinate system that has fewer dimensions, and has minimum dependence on each other. ICA has been applied to separate different speech signals, analysis of EEG, analysis of MRI, and biological image processing.

3.2 HEURISTIC SEARCH TECHNIQUES

Most problems can be modeled as a *state-space* problem, which is a graph of multiple states. A *state* is a tuple of common feature-values modeled as a node in the graph. A *move* takes the solution-process to next state from the current state, and is modeled as edge of the state-space graph. AI-based techniques, move toward the goal in a focused way using *heuristics* and ignore other states that regular breadth-first search or depth-first search and their variants would normally traverse. Using these techniques, a solution (*goal-state*) is reached faster than algorithmic techniques.

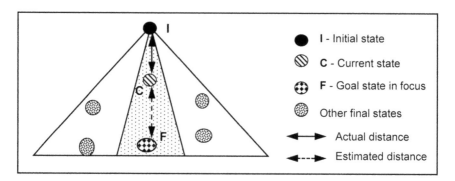

FIGURE 3.2   An illustration of a heuristics-based and goal-state search

Exploring and focusing on the goal-state so that unnecessary parts of state-space are not explored is called *heuristics*. *Heuristics* means intelligent guessing the distance between the current-state and the goal-state and to move in a focused manner. The use of an appropriate heuristics-based search reduces the computation overhead significantly. However, unlike breadth-first search, it does not always guarantee reaching the goal-state. The process of a heuristics-based state-space search is illustrated in Figure 3.2.

The bigger triangle shows the overall search space that is modeled as an acyclic directed graph. The shaded circles denote different states. The search starts at an initial state I and moves toward a known goal-state F. There may be other goal-states of which the search-process may be unaware. The distance between the initial state I and the current-state C is the actual traversed distance; the distance between the current-state C and the goal-state F is estimated based upon a mathematical function. The shaded triangle is the actual focused search-space.

3.2.1 Goal-Directed Global Search

There are many techniques that use heuristics such as: 1) best-first search and 2) A* algorithm. In a best-first search, the next state is picked based upon the path with the shortest distance to the goal-state F from the current-state C. In A* algorithm, global minimum distance $I \rightarrow^* F$ is estimated based upon the sum of the traversed-distance $I \rightarrow^* C$ and the *estimated-distance*(C, F); there is no commitment to the current best estimated path. The notation $\rightarrow^*$ denotes zero or more edges making a path from the source node to the destination node.

3.2.1.1 Best-first search

Best-first search estimates the distance from the children-states to the goal-state and adds the *distance*(*current-state, child-state*) to the corresponding estimated distances. If the current-state has m ($m > 1$) children, a sequence of m distances is formed. This sequence is sorted in the ascending order, and the child-state that gives the smallest cumulative distance (*distance*(*current-state, child-state*) + *estimated distance*(*child-state, goal-state*)) is selected as the next current-state. The process repeats. Since the best state is based upon estimated-distances and does not take into account the actual traversed distance, it is not optimal. Another problem is the size of the stack to keep the *fringe* (the sequence of alternate possibilities) keeps growing.

One approach to solve the storage overhead is to limit the alternate states in the fringe to K ($K > 1$) best alternative candidates. The value K is a user-defined parameter. Each time the search moves to a new state, additional distances of the form d_i = *traversed-distance*(*current-state, child-state*) + *estimated-distance*(*child-state, goal-state*) are included in the priority-queue modeling the fringe in an ascending order. The process repeats until the goal-state is reached.

There is no overhead of storing the alternatives in the *greedy best-first search algorithm*. Every time, the best child candidate with minimum distance d_i is picked; other candidates are discarded. The major problem in this scheme is the early commitment. Dropping other paths in favor of the committed paths rules out the possibilities of searching a better solution even if it exists.

3.2.1.2 A* search

A* search considers three types of distances: 1) *actual-distance(start-state, current-state)*, 2) *actual-distance(current-state, child-state)* and 3) *estimated-distance(child-state, goal-state)*. The sum of all these three distances is sorted after every move. The state with the shorted overall distance is picked. A* search is optimal: a solution is found if it exists provided the heuristic is optimistic. The problem with A* is that it needs extra memory space to store the previous travels to unexplored alternative states in the fringe.

3.2.2 Local Search – Hill Climbing

In many search problems, goal-state(s) cannot be located in advance during the traversal, making distance estimation very difficult. A goal-state may be missed depending upon the search-terrain. To solve the first problem, local search schemes are used that do not estimate the distance, but calculate the goodness of the current state after every move using an evaluation function under the assumption that it is moving toward the goal-state by improving the goodness of the current state. This type of search is called a *local search* or *hill-climbing search*. In a local search, the best move is picked as the maximum of the goodness-values of the next reachable states. However, as shown in Figure 3.3, the hill-climbing search suffers from many problems: 1) local maxima, 2) plateau and 3) ridges.

Local maximum is a state where all the next probable states are worse than the current state. However, local maximum is not the goal-state and is not globally the best state. *Plateau* is a state where all the next probable states are similar to the current state, and the search gets stuck on the plateau. Ridges are those points in the search-space that look close to the goal-state. However, goal-state is not reachable from them as illustrated in Figure 3.3.

3.2.3 Combining a Local and Stochastic Search

To get out of local maxima, there has to be a mechanism to: 1) jump out to another part of the search-terrain; 2) keep a view of multiple parts of the search-terrain and jump to the part of the terrain with high probability of reaching the goal-state; 3) start with multiple threads and later focus on the part of the terrain where the evaluation function improves consistently faster; 4) probabilistically keep some alternate moves with low probability while doing a local search, and in the case, the solution gets stuck in local maxima, use the alternate moves and 5) progressively keep evolving the search-space using genetic algorithm.

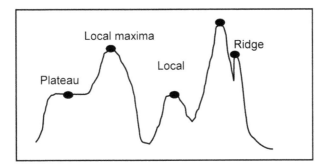

FIGURE 3.3   Hill climbing local search issues

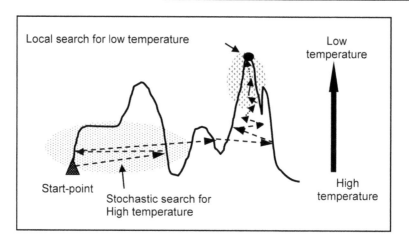

FIGURE 3.4 Probabilistic search strategy in simulated annealing

There are four very popular techniques and their combinations that use these approaches: 1) stochastic search, 2) genetic algorithm, 3) simulated annealing and 4) nature-inspired metaheuristics. Stochastic search uses random jumps when a local maximum is identified; genetic algorithm uses an evolution-based approach; simulated annealing keeps some bad moves with low probability to get out of local maxima to another unrelated part of the terrain; and nature-inspired metaheuristics uses an interleaved combination of repeated randomization and focusing to reach the goal-state.

3.2.3.1 Simulated annealing – stochastic + local search

Simulated annealing uses a notion of temperature that is slowly lowered until the temperature becomes 0. Each state is evaluated using an energy-function. The evaluation score is called *energy-difference*. After making a move, the next state is selected with probability 1 if $\Delta E = E_{next-state} - E_{cur-state} > 0$. Otherwise, the next-state is picked with a probability $e^{\Delta E/KT}$, where K is a empirically derived user-defined constant, and T is the current temperature. The initial temperatute T_0 is very large, and the probability of accepting the next randomly selected move is $e^0 = 1$. Hence, the next-state is picked randomly. However, as the temperature gets lower, the probability of picking a bad solution moves toward 0, and the search becomes focused. The overall scheme is illustrated in Figure 3.4.

> **Example 3.1: Simulated annealing**
>
> Consider an elderly patient suffering from multiple chronic diseases. Each disease requires one or more medications. These medications may interfere with each other. A doctor will like to know what is the optimum dose that will improve and/or maintain the condition of the patient by keeping the side-effect level to the minimum. Assume that the multiple drug–drug interactions are known, and a heuristic function can estimate the cumulative goodness based upon predicted side-effects and drug-interference. A simulated annealing search can identify the right dosage with minimal predicted side-effect for the administered drugs.

3.2.4 Genetic Algorithms

Genetic algorithm simulates human genetics. There is a pool of partial solutions that can solve sub problems of a task. Evolutionary algorithm progressively goes through a sequence of operations: *selection, cross-over* and *mutation*. The solution is modeled as a sequence of digits or binary numbers. The

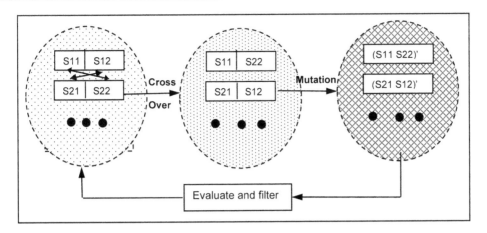

FIGURE 3.5 A schematic of various operations in a genetic algorithm

different solutions are picked probabilistically: good solutions are picked with higher probabilities, and bad solutions are picked with lower probabilities. After picking the solutions, they are paired according to a strategy, and cross-over produces new solutions. Cross-over takes between two or more parents. The process of cross-over takes partial solution from each parent. The new solutions are randomly mutated. The new solutions after mutations are tested using a *fitness-function*. Solutions having a value less than a threshold value are rejected. The process is repeated improving the pool of solutions. The whole process is illustrated in Figure 3.5 using two parents. However, many variants employ three of more parents from the pool.

In Figure 3.5, cross-over takes place between two samples across well-defined segments. The edges show the probability of selection. The probability of selection is based upon the quality of the parent in terms of nearness to the goal-state or the goodness value after applying the fitness function.

Example 3.2: Genetic algorithms

An elderly person is suffering from three diseases. He takes seven medications. The medications, if taken together, go through drug–drug reaction and cause side-effects. If the medicines are taken separately, two hours apart, there is no side-effect. However, it is very difficult for an elderly patient to remember multiple medications. Hence, the medication timings should be minimized while keeping the side-effects low. This problem is a typical candidate for genetic programming where different combinations are tried, and each time a new combination of time-delay is tried based upon previous combinations. Better solutions are kept, and others are discarded.

3.2.5 Nature-Inspired Metaheuristics

There are two important concepts in any heuristic search process: 1) randomization to move around in the search space so that a search does not get stuck in a local maxima and 2) focusing to prune the search-space and reduce computational overhead. Excessive randomization slows down the search process, and too much hill-climbing can lead to a local-maxima.

A new class of algorithms has been developed to handle these issues. These algorithms are nature inspired and are based upon repeated interleaving of diversification (randomization of search) and focusing (hill-climbing). Some of the popular ones are *ant-colony search, bee-search, firefly-search, swarm-based optimization*. Metaheuristic searches are good for optimization.

3.2.5.1 Ant colony optimization

Given a scenario of having multiple paths to a goal-state, an ant-colony search is a population-based path minimization technique that automatically identifies the path with the minimum cumulative sum of the edge-weights to reach from the start-node to the goal-state. The whole state-space is modeled as a graph where each move is an edge, and each intermediate state is a vertex. The scheme is inspired by ants' movement strategy in nature.

The scheme spawns multiple threads (ants) to explore the state-space. All paths are initially explored randomly with equal probability. Each thread (ant) leaves a decayable token (pheromone concentration) at the traversed edge. The pheromone level decays over time with an exponential decay-rate γ. Over time, more threads probabilistically pass through the same set of shorter-length edges connecting the start-state and the goal-state because the overall decay of the token (pheromone) is less in shorter paths. The use of pheromone decay at the edges has two advantages: 1) it makes the search population-based and 2) the approach allows relearning in the event the goal-state is altered.

The probability of picking an edge depends on four parameters: 1) desirability d where d is inversely proportional to the edge-weight w, 2) decay factor γ, 3) cumulative pheromone concentration $C_{ij}(t)$ at an edge and 4) pheromone deposition rate δ as shown in Equations 3.3 and 3.4. The factors α and β are user-defined *influence-factors* derived by experimentation. The decay of concentration at an edge (i, j) is given by the equation $C_{ij}(t) = C_{ij}(0)e^{-\gamma t}$ that reduces to the linear Equation 3.4 for small values of γ.

$$P_{ij} = \frac{C_{ij}^{\alpha} \times d_{ij}^{\beta}}{\sum\limits_{i,j=1}^{i,j=N} C_{ij}^{\alpha} \times d_{ij}^{\beta}} \tag{3.3}$$

$$C_{ij}(t+1) = (1 - \gamma)C_{ij}(t) + \delta\, C_{ij}(t) \tag{3.4}$$

3.2.5.2 Bee search and firefly search

Bee search and *firefly search* have two differences from *ant colony optimization*: 1) Bee search and firefly search are not based upon the decay of the pheromone concentration and 2) *bee search* and *firefly search* are based upon communicating the strength of the goal-states to other threads.

Bee search and *firefly search* also differ from each other. Information communication in *bee search* is limited to a distance and is good for finding multiple local maxima such as multiple clusters of malignant cells in an organ's MRI based upon texture analysis. Firefly search is based upon broadcasting the information that weakens as the distance from the goal-state increases.

Firefly search can be made global-search by strengthening the signal-propagation and reducing the decay factor, or local-search by weakening the signal strength and increasing the decay factor. Firefly search can be utilized for intelligently identifying specific organ-parts during image analysis by knowing the desirable landmarks of an organ and multiple threads working on the different part of the organ-contour.

3.2.5.3 Particle swarm-based optimization

Swarm-based optimization is inspired by the movement of a flock of birds. The flock of birds moves together in a specific direction without colliding. Each bird is equivalent to a separate local thread performing a stochastic search locally on a different part of the terrain. However, each thread is influenced by the mean of the best-found values by all the threads and its own best and keeps moving toward the centroid of the best positions of all the threads. The overall motion combines global communication and mass scale trajectory adjustment and has both a stochastic and deterministic components in the search: the stochastic component is present in a local search and the deterministic component is present in following the current best by the flock.

3.3 INFERENCING TECHNIQUES

The knowledge is represented in the form of a set of rules. Each rule is of the form: $X_1 \ op_1 \ X_2 op_2 \dots op_N X_n \rightarrow y$, where $X_1, \dots, X_n$ are relations (or negation of a relation), $op_1, \dots, op_N$ are logical operators, and y is the consequence. The rules are grouped together based upon the same relational name and the same number of arguments. The argument is a constant, or a variable. The use of variables makes the rule generic. More than one example can match the same generic rule.

There are two inferencing processes: 1) induction: given multiple examples, how to express them as a rule with variables and 2) deduction: to find out that an information can be derived using a set of rules. A new fact may be inserted at runtime. The deduction process is further divided into: 1) backward chaining and 2) forward chaining.

3.3.1 Forward Chaining

In forward chaining, after a new fact is added, it is combined with various rules to derive new applied facts. The new facts are repeatedly combined with the set of existing rules until new applied-facts cannot be derived. These applied-facts become part of the database. When a query is asked, the query is tested against the database. Indexing and matching are used to retrieve the facts and match. A failed retrieval or matching results in failure, and a "false" is returned.

The problem with forward chaining is that a new fact is combined with all the facts in the database using all the possible rules in the knowledge-base. That makes it computationally expensive. The advantage of forward chaining is that it derives all possible facts, and is comprehensive; there is no possibility of missing a solution. Forward chaining is good for monitoring because new information is immediately assimilated, and any abnormality can be derived. However, due to the processing of all the rules multiple times, it is inherently slower than backward chaining when the rule-base is large.

Example 3.3: Forward reasoning

Consider the following three forward chaining rules as follows.

1. *has_glucose_level_above(A1C, 200) → has_person(severe, diabetes).*
2. *has_person(severe, diabetes) → reduce_calorie_intake_below(1400).*
3. *has_person(severe, diabetes) → give(insulin).*

Suppose that a person is tested for her glucose level every six months with an average glucose level (A1C) value 220. The fact will match the left-hand side of the rule #1, and trigger rule #1 resulting into a new fact to be added in the database "has_person(severe, diabetes)." This added fact will trigger rules #2 and #3, suggesting that the person should eat only *1400 calories/day* and start taking insulin.

3.3.2 Backward Chaining

In backward reasoning, a query is divided into a conjunction or disjunction of subqueries that are individually solved and combined with the result of other subqueries based upon the logical operator connecting them as illustrated in Figure 3.6a. The deduction process is based on building an AND-OR tree where the sibling nodes are connected using a logical operator (logical OR, logical AND or negation). The leaf-nodes of an AND-OR tree are the lookup results from the database facts. The splitting process is based upon matching a query to the left-hand side of a rule and passing the binding-values to the right-hand side of the rule. A value of *true* at the root node means that the query has succeeded. Otherwise, the query fails.

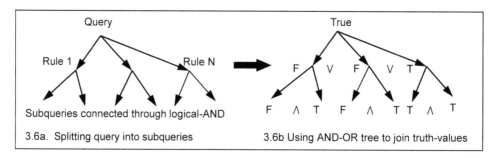

FIGURE 3.6   An illustration of a backward chaining system: (a) Splitting query into subqueries; (b) Using AND-OR tree to join truth-values

In Figure 3.6b, the first rule fails because one of the subqueries returns *false*. The second rule fails because one of the subqueries returns *false*. However, the third rule succeeds returning *true*. The advantage of backward-chaining is that the search is focused; only part of the search space is traversed. This approach is good for asking specific queries and developing explanation systems.

Example 3.4: Backward chaining

Consider these rules:

1. *has_diabetes (X, severe) ← has_glucode_level(X, W) ∧ W > 200.*
2. *has_diabetes(X, moderate) ← has_glucose_level(X, W) ∧ W > 140 ∧ W = < 200.*
3. *has_diabetes(X, Y) ← has_glucose_level(X, Z) ∧ Z > 140.*
4. *has_glucose_level("John", 180).*

For the above rules, a query *has_diabetes("John", Y)?* matches with the left-hand side of the rules *1* or *2* and get reduced to the right-hand sides of the rules. The subquery *has_glucose_level(john, W)* will match with the fact *has_glucose_level("John", 180)* to get the value of *W = 180*. This fails the rule *1*. However, rule *2* succeeds returning the value of *Y* as *moderate*. The backward reasoning keeps splitting the query into smaller subqueries until the corresponding facts are looked-up.

3.3.3 Fuzzy Reasoning

Sensor-based measurement gives real values. However, real values have large possibilities, can have noise and may be imprecise. In addition, when we parameterize the features of an entity, we may not pick all the features. Humans take care of these problems using approximate mapping. Even in the absence of external sensors, fuzzy values can be used to describe a patient's condition. For example, a doctor may ask a patient to describe her pain on a scale of one to ten. Without identifying the features, all the feature-vectors have been mapped to approximate perceptive value between one and ten. These values are called *fuzzy values*, and reasoning is called *fuzzy reasoning*.

Quantitative values are transformed to fuzzy values during fuzzy reasoning. The fuzzy values may be quantified again after the reasoning. The advantage of fuzzy values is that the use of a smaller number of classes prunes the search space improving computational efficiency. In addition, patient-specific information collected by medical practitioners is also fuzzy. For example, a nurse may write her bed-side note that: 1) "patient was somewhat drowsy"; 2) "patient was having excessive pain"; 3) "Blood pressure was somewhat high."

Mapping of quantitative values to fuzzy values uses a linear or a Gaussian model. The minimum and maximum intervals of neighboring fuzzy classes overlap. The knowledge of the range and the type

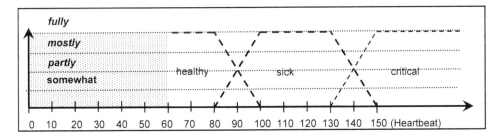

FIGURE 3.7   An illustration of mapping quantitative value to fuzzy value

of modeling is used for the quantification of fuzzy-values. Figure 3.7 shows a linear model of fuzzy classification of a patient's condition. A heart patient can be *critical, sick* or *healthy*. The sickness is being judged by the fast heart-beat rate.

Example 3.5: Fuzzy reasoning

If the heart-beat is above *150,* then the health-state is *critical.* The criticalness diminishes linearly between *150* and *130* and becomes zero at *130.* The sickness decreases from *150* slowly and becomes *fully sick* at *130.* It remains *fully sick* until *100* and then diminishes linearly between *100* and *80.* The healthy heart starts from heart-beat *100* and becomes "fully healthy" at *80* and remains in that state until *60.* There are five subclasses given by the fuzzy set {*not, somewhat, partly, mostly, fully*}. A conversion table between fuzzy values and quantification is given in Table 3.1.

The whole range from *60–150* is reduced to eight intervals using fuzzy modeling. These seven intervals are assigned a truth value with *fully healthy* being *1.0* and *fully critical* being *0.0* for fuzzy reasoning. To reason about truthness, these eight cases can be assigned a truth-value as illustrated in Table 3.1.

3.3.3.1 Fuzzy logic

A fuzzy truth-value varies from *0.0* to *1.0*: *false* being *0.0* and *truth* being *1.0*. The fuzzy logic implements *fuzzy-AND* and *fuzzy-OR,* which differs from Boolean-AND and Boolean-OR. Unlike, Boolean logic that has only two values: *true* or *false,* fuzzy logic can also handle a *partial truth* and a *partial false. Fuzzy-OR* is defined as the maximum of two truth-values. *fuzzy-AND* is defined as minimum of two truth-values. *Fuzzy negation* is defined as $(1 - truth\_value(x))$. An alternate way of defining *fuzzy-AND* of variables x and y is using the multiplication $truth\_value(x) \times truth\_value(y)$. Similarly, fuzzy-OR can be also be defined as $(1 - (1 - truth\_value(x)) \times (1 - truth\_value(y)) = truth\_value(x) + truth\_value(y) - (truth\_value(x) \times truth\_value(y))$.

TABLE 3.1   A lookup table for fuzzy values and quantifiable heart rate

	CONDITION	HEART RATE	TRUTH-VALUE
1	Fully healthy	60–80	1.0
2	Mostly healthy	80–85	0.9
2	Partly healthy and partly sick	85–95	0.7
3	Somewhat healthy mostly sick	95–100	0.6
4	Fully sick	100–130	0.5
5	Mostly sick somewhat critical	130–135	0.4
6	Partly sick partly critical	140–145	0.2
7	Mostly critical	145–150	0.1
8	Fully critical	>150	0.0

3.3.4 Uncertainty-Based Reasoning

Logical reasoning assumes that all the parameters are known, and they can be tested for Boolean true or false. In reality, all the parameters are not known in the real-world reasoning. Besides, the number of evidence variables is limited. Hence, with the known parameters, one cannot ascertain for certainty truthness. This is the basis of uncertainty. When the conditions match, the outcome can be estimated to occur with some certainty. This quantification is called a *certainty-factor*. In uncertainty-based reasoning, each rule also contains a certainty-factor of truthness.

Example 3.6: Uncertainty-based reasoning

1. *is_coughing(X)* → *has_disease(X, 'common-cold', 0.3).*
2. *is_coughing(X)* → *has_disease(X, 'strep-throat', 0.7).*

Consider the above two forward-chaining rules with an uncertainty factor. It says that a "if a person is coughing, then there is 30% certainty that he has common-cold and 70% certainty that he has strep-throat."

3.3.5 Inductive Reasoning

Inductive reasoning is based upon generalization of multiple examples to form a common logical rule based upon known samples to label a set of points that exhibit a common trait among the attributes. Unlike clustering where a group label is unknown beforehand, the group-label is known in inductive reasoning. After the logical rule is formed, it's applied to classify the new data-points and predict the behavior of entities based upon their attribute values. Inductive reasoning is probabilistic because all the parameters affecting the classification may not be known.

Example 3.7: Inductive reasoning

One hundred samples of a specific type of breast cancer are characterized by the presence of lumps in mammograms having specific texture-pattern P and size-range R in *90%* of the cases. The attributes are statistically consistent. A logical rule is derived that "a lump within the breast with texture-pattern P and size-range R is cancerous." A physician can deduce that a lump in a breast with texture pattern P and size within the range R is cancerous with high confidence and recommend a biopsy for invasive investigation.

3.4 MACHINE LEARNING

Learning is assimilation of new knowledge in a way that reduces the number of steps and improves the accuracy and efficiency to perform a task. Learning is further classified as: 1) memorization of the information such as facts, traveled search path to lookup in the future that reduces the processing time; 2) generalization from examples to form generic rules; 3) classification of the data items to different groups based on major common attributes; 4) summarization of data and text; 5) association between two sets of attributes to find some common pattern; 6) prediction by discovering a function that maps a set of input parameters to the outcomes and 7) analyzing the images and identifying recognizable patterns. Memorization is the simplest form of learning. After solving a problem multiple times, humans memorize the moves to make and prune out the remaining state-space. The memorization of the landmark states and the corresponding moves reduces the execution time of the problem solving significantly.

In health informatics, learning has many applications such as: 1) finer classification of diseases; 2) identification of abnormal growths in the organs; 3) analysis of the progression of disease; 4) understanding the speech patterns and phrase patterns of different expert doctors; 5) data mining the large databases for finding the association between various disease symptoms and biomarkers; 6) automated correlation of biosignals in ECG, EEG and EMG with various diseases; 7) predicting the location of a patient being tracked in pervasive care; 8) analyzing the activities of the remote care patients, including fall detection; 9) identification of the genetic diseases and 10) predicting the efficacy and the dosage of the medicine.

There are three mechanisms for automated learning: 1) *supervised*; 2) *reinforced*; and 3) *unsupervised* learning. Following sections describe each technique.

3.4.1 Supervised Learning

In supervised learning, individual examples are associated with class-labels using preexisting knowledge. Each example is modeled as a tuple of attributes. An expert may train apprentices, or automated software to diagnose a specific disease using known changes in vital signs, biomarkers and lab-test data. The important aspect of supervised learning is the presence of training samples with known correct labels. After some examples, the learner or the automated software starts associating new data with specific labels. The class-label is derived by performing the similarity-search on feature-vectors to identify the best match. The association process is continuously refined based upon the training samples until the class-label generated by the classifier is the same as the actual label with a high degree of confidence.

Some of the popular techniques for training are: 1) support vector machines (SVM); 3) regression analysis and 4) inductive reasoning. These techniques have a common property: the presence of the learning samples with known labels for the output. All these techniques start with some initial value and are further refined based upon the feedback until they output the correct label on the sample test within an acceptable tolerance limit.

3.4.2 Reinforced Learning

Reinforced learning is an automated pattern analysis technique where an automated intelligent software derives the same search path to find the same goal-state. Based upon the frequency of occurrence of the intermediate-states in multiple common search paths, a subset of intermediate-states and the path between the intermediate-states and the path from intermediate-states to goal-state are memorized.

The difference between supervised learning and reinforced learning is that there is no training session in reinforced learning. Rather, it is based upon statistical analysis of the patterns in the sequence of actions taken to reach the goal-state. It is more like on the job training where an apprentice observes the same sequence of operations by an experienced surgeon and learns the technique.

3.4.3 Unsupervised Learning

Unsupervised learning is used when no known labels are associated with data being analyzed. Data-items are classified using some common combination of attribute-values. For example, given a set of different dosages of medications, different patients from different ethnic groups, age and gender may respond differently in terms of recovery periods and side-effects. Hence, dosage, age, gender and ethnicity become the attributes.

The basic property of unsupervised learning is that the number of classes and their labels are unknown. These labels need to be derived based upon some Boolean classifier, multivalued classification or interval-based classification. An important aspect of unsupervised learning is to identify the set of parameters used to characterize the data-elements. Incomplete or incorrect characterization of data-elements may lead to erroneous grouping and learning.

Examples of unsupervised learning are clustering and decision trees. Clustering is based upon the notion of proximity (see Section 2.1.2.1) of data-points in N-dimensional space. There is no priority between various parameters under proximity based upon Euclidean distance or weighted Euclidean distance. Decision-trees use the notion of entropy and information gain to prioritize the parameters and derive a small set of logical rules from a large sample-set.

3.5 CLASSIFICATION TECHNIQUES

3.5.1 Clustering Techniques

Clustering techniques map data-elements in an N-dimensional space and discover the subregions in N-dimensional space that are in proximity based upon some notion of distance between the data-elements or the density of the data-elements within a region. The data-points in a cluster are modeled in multiple ways. Some of the popular techniques are: 1) grouping in a spherical shape in an N-dimensional feature-space so the Euclidean distance between the data-points within a cluster is below a threshold; 2) data-points within a cluster have a density of occurrence higher than a threshold; 3) tree-shaped cluster where each node in the tree is a subcluster.

> **Example 3.8: Application of clustering**
>
> Heart disease is a large group. Under this group, there are many smaller groups such as: 1) valve-based disease, 2) heart-muscle–based disease, 3) artery-blockage–based disease, 4) atrial disease and 5) ventricular disease. *Atrial diseases* are related to atrium enlargement, atrial fibrillation and atrial flutter. *Ventricular diseases* are ventricular fibrillation, ventricular flutter, ventricular tachycardia and ventricular hypertrophy. These diseases are identified using feature-vectors derived from ECG analysis. Some of the features are: presence/absence of P-waves, height of P-wave, width of P-wave, width of QRS wave, height of QRS wave; pulse rate, blood pressure, blood-flow, and electrolyte levels. Each feature becomes a dimension in an N-dimensional feature-space. The set of feature-values is plotted as a data-point in the N-dimensional space.

3.5.1.1 Hard-clustering vs soft clustering

There are two types of cluster modeling: *hard-clustering* and *soft-clustering*. A data-point can belong to only one cluster in a *hard-clustering model*. A data-point can belong to multiple clusters with different probabilities in a *soft-clustering model*. The soft-clustering model is useful when data-points exhibit properties of more than one group. A data-point is called an *outlier* if it does not belong to any cluster. A spherical cluster is characterized by two attributes: the *centroid* and the *radius* of the cluster.

Figure 3.8a describes an example of a hard cluster, and Figure 3.8b gives an example of a fuzzy cluster (soft cluster). The probability of inclusion in a soft-cluster is decided by the nearness to the centroid with the centroids of the individual clusters. The probability of inclusion is given by $\left(1 - \frac{distance(P,\ cenrtroid(A))}{distance(P,\ centroid(A)) + distance(P,\ centroid(B))} \right)$. As the distance between a point and the centroid increases, the probability of belonging to that cluster goes down. The sum of the probabilities of a point belonging to multiple clusters adds to *1*.

In soft-clustering, two clusters may overlap. Soft-clustering can be derived using *Gaussian mixture models,* where each Gaussian distribution corresponds to a single cluster, and two Gaussian distributions may overlap. A point in the overlapped regions belongs to both the clusters.

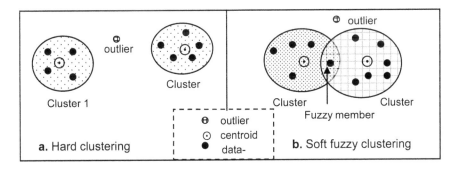

FIGURE 3.8 (a) Hard-clustering vs. (b) soft-clustering

Example 3.9: Soft-clustering

Consider Figure 3.8b. The clusters A and B are overlapping. The point P is in the overlapped area and belongs to both the clusters. Assume that the distances of the point P from the *centroid*(A) and *centroid*(B) are *4* units and *6* units, respectively. The probability of point P being in cluster A is *(10 − 4/10) = 0.6* and in cluster B is *(10 − 6/10) = 0.4*.

There are many techniques to derive clusters. Some of the popular ones are: K-means clustering, hierarchical clustering, fuzzy C-means clustering, density-based clustering, spanning-tree–based clustering, and their variants. The following sections will discuss the basics of these popular techniques.

3.5.1.2 K-means clustering

K-means clustering is used to identify regularly shaped hard-clusters. Cluster-identification is based upon establishing K seed-points that act initially as the centroids of the spherical clusters. Initial radii of the clusters and K seeds ($K > 1$) are chosen randomly, manually based upon inspection of the distribution or based on density analysis of the macro-blocks within the region to be analyzed.

For performing initial density analysis, the whole data space is divided into multiple blocks much greater than K by dividing each dimension in N-dimensional space such that $\prod_{i=1}^{i=N} D_i > K$ where D_i is the number of divisions of the value-intervals in ith dimension. This division divides the space in $\prod_{i=1}^{i=N} D_i$ blocks. The number of data-points in each block is counted, and the blocks with highest frequency above a threshold not in proximity are picked for seed-points. The centroid of these selected blocks becomes the seed-points. To avoid overlap between the clusters, the radius is chosen as half of the smallest distance between the chosen seed-points. Once the seed-points and radius r are chosen, the process is of identifying the points that are r distance away. This requires approximately $O(P \times K)$ comparisons where P is the number points, and K is the number of centroids. After identifying the number of points in each cluster, the outliers – points outside the clusters – are counted. Too many outliers show a bad choice of seed-points or radius. In case of a large number of outliers, new sets of k seed-points are chosen until the number of outliers is below a threshold. For smaller dimensions such as two-dimensional data-spaces, seed-points can also be identified by visually inspecting the data-points on a computer screen.

After identifying the seed-points, the clusters are derived using an iterative process. The data-elements are placed in a cluster with a centroid within the radius r after every iteration. After placing the data-points in each cluster, the mean of x-coordinates and y-coordinates of points within a cluster derives a centroid. The old centroid is substituted by the new value (*x-mean, y-mean*). The membership of each point within the cluster is computed again. During this process, some points in other clusters or outliers may become part of the new cluster, while some points in the old version of the cluster become outliers. This change in the set of points alters the *x-mean* and *y-mean* again. The iteration is repeated until the distance between the old centroid and new centroid is below a threshold. This process is repeated for every seed-point.

Example 3.10: K-means clustering

Consider a set of 17 data-points in two dimensions for analysis with three seed-points. The seed-points and radius are derived using some initial analysis. The data-points are *(2, 3)*, *(2, 4)*, *(1, 2)*, *(1, 1)*, *(6, 5)*, *(6, 6)*, *(7, 6)*, *(7, 7)*, *(5, 7)*, *(12, 8)*, *(9, 9)*, *(11, 10)*, *(1, 9)*, *(10, 11)*, *(8, 2)*, *(10, 10)*, *(11, 12)*.

X-coordinate interval varies from *1* to *12*, and Y-coordinate interval varies from *1* to *12*. The intervals are divided into 3×3 blocks to provide enough possibilities for random variation of coordinates of seed-points. There are three X-intervals: X_1, X_2 and X_3, where $0 < X_1 < 4$; $4 < X_2 < 8$ and $8 < X_3 \leq 12$. Similarly, there are three Y-intervals: Y_1, Y_2 and Y_3, where $0 < Y_1 < 4$; $4 < Y_2 < 8$ and $8 < Y_3 \leq 12$. There are nine blocks X_1Y_1, X_1Y_2, ..., X_3Y_2 and X_3Y_3. The number of data-points in these blocks are: X_1Y_1: 4, X_1Y_2: 0, X_1Y_3: 1, X_2Y_1: 1, X_2Y_2: 5, X_2Y_3: 0, X_3Y_1: 1, X_3Y_2: 0, X_3Y_3: 6.

Based upon this analysis, there are three seed-points: centers of X_1Y_1, X_2Y_2 and X_3Y_3. The coordinates of the seed-points are *(2, 2)*, *(6, 6)* and *(10, 10)*. The smallest distance between the centroids is *5.6*. The radius is less than or equal to half the distance. Let us take *2.7* to avoid any overlap. Now the distances from each data-point to the three centroids are computed. Those data-points with a distance less than *2.7* to a centroid are put in the corresponding cluster. The centroid *(2, 2)* corresponds to *Cluster₁*; the centroid *(6, 6)* corresponds to *Cluster₂*; and the centroid *(10, 10)* corresponds to *Cluster₃*. In the first iteration, *Cluster₁* = {*(1, 1)*, *(1, 2)*, *(2, 3)*, *(2, 4)*}, *Cluster₂* = {*(6, 5)*, *(6, 6)*, *(7, 6)*, *(7, 7)*} and *Cluster₃* = {*(9, 9)*, *(10, 11)*, *(10, 10)*, *(11, 12)*}. The outliers after the first iteration are: {*(1, 9)*, *(8, 2)*, *(12, 8)*}. After taking the means of x-coordinates and y-coordinates of data-points in the clusters, the coordinate-pairs for the new centroids are: *(1.5, 2.5)*, *(6.5, 6)* and *(10, 10.5)*. In the next iteration, all the clusters remain the same, and the computation terminates. Figure 3.9 shows three K-means clusters. A centroid need not be one of the data-points as shown in Figure 3.9.

After the new points are inserted in the old data-set, then there can be two options: 1) recompute the clusters; 2) incrementally enhance the clusters by inserting a new point in one of the existing clusters. For large data sets, the second option is chosen. The choice of seed-points and radius is critical in automated clustering. Otherwise, many important clusters may be missed.

For very large data sets, the number of clusters in the dataset is large. Each cluster is characterized by the pair (*centroid, radius*). Clusters of clusters are formed to accommodate all the clusters. This further compresses the overall data and helps in finding additional interesting cluster-patterns in the dataset.

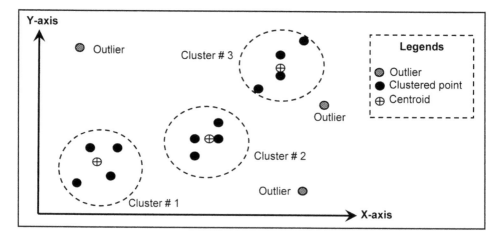

FIGURE 3.9   An example of K-means clusters and outliers

3.5.1.3 *Hierarchical clustering*

Clusters can be grouped using a hierarchical tree. The leaf-nodes are the smallest clusters and uses the distance between clusters to form a tree of clusters. Smaller clusters are treated as the children nodes of the resulting bigger cluster in the hierarchy. Grouping is continued until one big cluster is achieved. This becomes the root-node of the tree.

Using a *greedy algorithm,* the nearest clusters are joined incrementally. There are many variations to derive the distance between two clusters: 1) minimum distance between points across the clusters; 2) maximum distance between two points across clusters; 3) distance between the centroids. In the following illustration, the distance between the centroids has been used to measure the distance between two clusters.

The centroid of two joined clusters is calculated by knowing the number of data-points in each joined cluster and using the equation $\bar{x}^{joint} = \frac{\bar{x}_1 \times n_1 + \bar{x}_2 \times n_2}{(n_1 + n_2)}$ and $\bar{y}^{joint} = \frac{\bar{y}_1 \times n_1 + \bar{y}_2 \times n_2}{(n_1 + n_2)}$, where $(\bar{x}_1, \bar{y}_1)$ is the centroid for the first cluster; $(\bar{x}_2, \bar{y}_2)$ is the centroid of the second cluster; n_1 is the number of points in the first cluster; and n_2 is the number of points in the second cluster. The distance between the centroids of joint-cluster and individual clusters is given by $\sqrt{(\bar{x}^{joint} - \bar{x}_i)^2 + (\bar{y}^{joint} - \bar{y}_i)^2}$. The process is repeated with the new set of clusters until there is only one element in the set. The hierarchical clustering becomes a tree with edges showing the distances between the centroids of the clusters and joint-clusters. The tree-structure is known as *dendrogram* and is illustrated in Figure 3.10.

Example 3.11: Hierarchical clustering

Consider the clusters in Figure 3.10. All computed data are rounded to first decimal place. There are six clusters with centroids {(8, 3), (10, 14), (1, 1), (3, 7), (13, 17) and (6, 8)}. The clusters have 54, 23, 65, 72, 44, and 30 points, respectively, in the clusters.

The distance between the centroids is calculated using $O(N^2)$ operations. The distance is sorted in ascending order. The smallest distance is between the centroids (3, 7) and (6, 8). Using the equation $\bar{x}^{joint} = \frac{\bar{x}_1 \times n_1 + \bar{x}_2 \times n_2}{(n_1 + n_2)}$ and $\bar{y}^{joint} = \frac{\bar{y}_1 \times n_1 + \bar{y}_2 \times n_2}{(n_1 + n_2)}$, the centroid of the joint cluster is $((72 \times 3 + 30 \times 6)/102$, $(72 \times 7 + 30 \times 8)/102) = (3.9, 7.3)$ with a distance of 1.0 from (3, 7) and 2.2 from (6, 8).

The new set becomes {(8, 3), (10, 14), (1, 1), (3.9, 7.3), (13, 17)} with 54, 23, 65, 102, and 44 points in the clusters, respectively. The next nearest centroids are (10, 14) and (13, 17).

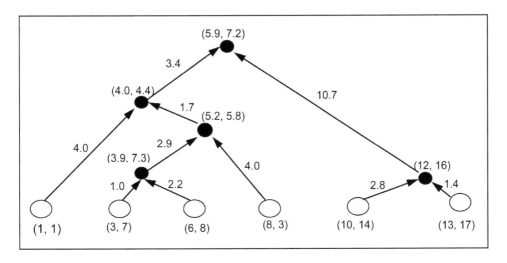

FIGURE 3.10 An example of hierarchical clustering

The joint-centroid for them is $((23 \times 10 + 44 \times 13)/69, 23 \times 14 + 44 \times 17)/69 = (12.0, 16.0)$. The distances of this joint centroid from $(10, 14)$ and $(13, 17)$ are 2.8 and 1.4, respectively. The new set becomes $\{(8, 3), (12, 16), (1, 1), (3.9, 7.3)\}$. The next shortest distance is between $(8, 3)$ and $(3.7, 7.3)$. The new joint centroid is $((54 \times 8 + 102 \times 3.7)/156, (54 \times 3 + 102 \times 7.3)/156) = (5.2, 5.8)$.

The distances of the centroid of this joint-cluster from $(8, 3)$ and $(3.7, 7.3)$ are 4.0 and 2.9, respectively. The new cluster set is $\{(5.2, 5.8), (12, 16), (1, 1)\}$ with 156, 69, and 65 data-elements, respectively. The next smallest distance is between centroids $(5.2, 5.8)$ and $(1, 1)$. The new joint centroid is $((156 \times 5.2 + 65 \times 1)/221, (156 \times 5.8 + 65 \times 1)/221) = (4.0, 4.4)$. The distances of the centroid of this joint-cluster from $(5.2, 5.8)$ and $(1, 1)$ are 1.7 and 4.0, respectively. The new cluster set is $\{(4.0, 4.4), (12.0, 16.0)\}$ with 221 and 69 data-elements, respectively. The root of the dendrogram is $((221 \times 4.0 + 69 \times 12.0)/290, (221 \times 4.4 + 69 \times 16)/290) = (5.9, 7.2)$. The root of the dendrogram has distances from $(4.0, 4.4)$ and $(12.0, 16.0)$ as 3.4 and 10.7, respectively.

3.5.1.4 Incremental K-means clustering

For very large datasets, after the clusters are formed, additional new points are inserted incrementally to the nearest clusters based upon their distances from the cluster-centroids. After adding the points in the corresponding clusters, the centroids of the corresponding clusters are recomputed by using the equation $\mu_i^{new} = \frac{m \times \mu_i^{old} + x_i}{n + 1}$, where m is the number of data-elements in the old cluster, μ_i^{old} is the previous mean in the ith dimension, μ_i^{new} is the new mean in the ith dimension and x_i is the value of the ith coordinate of the inserted data-point. Incremental K-clustering is illustrated in Figure 3.11.

Three very large clusters are shown in the shaded circles. The symbol "$\oplus$" shows the centroid of these clusters. The new points are absorbed in the clusters nearest to them, and the centroids and radii of the updated clusters are recomputed.

Example 3.12: Incremental K-means clustering

Assume that the *cluster #1*, the *cluster #2* and the *cluster #3* are characterized by $(3.0, 3.0), 2.0, 100)$, $((6.0, 5.0), 1.9, 80)$ and $((9.0, 8.0), 1.8, 110)$, where the first field in the tuple shows the coordinates of the centroid, second field shows the radius of the cluster and the third field shows the population of the data-points in the cluster. The coordinates for the three new points P_1, P_2 and P_3 are $(2.0, 4.8), (5.5, 7.0)$ and $(10.7, 7.2)$, respectively. The point P_1 is closest to the *cluster #1*, and the distance from the centroid of the *cluster #1* is 2.05. Point P_2 is nearest to the *cluster #2*, and the distance from the centroid of

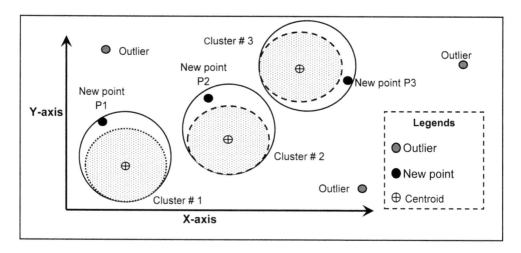

FIGURE 3.11   An illustration of incremental K-means clustering

the *cluster #2* is *2.06*. Point P_3 is closest to the *cluster #3*, and the distance from the centroid of the *cluster #3* is *1.87*. The new centroid for *cluster #1* is calculated as $((3.0 \times 100 + 2.0)/101, (3.0 \times 100 + 4.8)/101) = (2.99, 3.01)$ with a radius $\sqrt{(2.99 - 2.0)^2 + (3.01 - 4.88)^2} = 2.05$. Similarly, the new coordinates and radii of the centroids of *cluster #2* and *cluster #3* are calculated.

3.5.1.5 *Density-based clustering*

Density-based clustering derives irregular-shaped clusters having data-points density above a threshold in any unit area within a cluster. It can derive objects and entities having irregular shapes such as characters in text recognition, arteries, fractures in the bones, trachea, folding in the brain, heart, intestines and other internal organs.

Density-based clustering is based upon identifying the blocks having a common attribute such as uniform texture or color and iteratively joining the adjacent blocks having a very similar density pattern. Figure 3.12 illustrates density-based clustering to derive irregular shapes. Figure 3.12a shows the two irregular clusters, and Figure 3.12b shows the use of a density-based cluster to identify the character "A."

3.5.1.6 *Spanning-tree–based clustering*

Another approach to form clusters is based upon exploiting a *minimum spanning-tree*. This scheme is suitable for smaller data set. In this technique, a minimum spanning-tree is derived that connects the data-elements; data-elements are treated as vertices in the tree. Any two vertices in a *minimum spanning-tree* are connected using a single path of the smallest possible weight. Discovering a minimum spanning tree for N data-elements uses an algorithm having time-complexity of $O(N^2)$. First, the distance of each point against every other point is calculated using $O(N^2)$ algorithm, then the set of weighted-edges with distance-as-weight is sorted. The iterative process starts with the set S that containing all the edges between every pair of vertices. The set S is sorted in ascending order of edge-weight starting with the smallest edge-weight.

The output T is a set of subsets modeling the connected subtrees. Initially, the set T is empty. Each time a new edge is picked, such that at least one of the two nodes in the edge is not in the same subset of the set T. An edge that has both its vertices in the same subset is deleted from the consideration. Two separate subsets in the set T are joined into one larger subset if the two end-vertices of the next edge are included in two subsets of T. After picking an edge from the set S, the edge is deleted from S. The process is repeated until all the edges in the set S are consumed, or the set of vertices in the set T covers all the data-elements.

After finding the minimum spanning-tree, all the edges above a threshold value are removed from the tree. This breaks the tree into a forest (*a set of smaller well-connected trees*), each having edges with weights less than the threshold value. The data-elements in each smaller tree form a cluster.

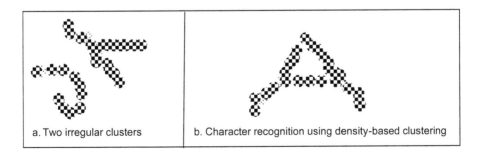

| a. Two irregular clusters | b. Character recognition using density-based clustering |

FIGURE 3.12   Density-based clustering

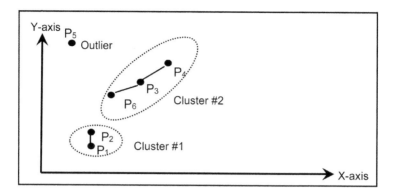

FIGURE 3.13 An illustration of minimum spanning-tree–based clustering

Example 3.13: Spanning-tree–based clustering

Consider a data set $P_1 = (2, 3)$, $P_2 = (2, 4)$, $P_3 = (4, 7)$, $P_4 = (5, 8)$, $P_5 = (1, 9)$, $P_6 = (3, 6)$. The threshold value is 2.0 to prune the edges. The corresponding spanning-tree–based clustering is illustrated in Figure 3.13.

The sorted set S of the weighted edges is $\{(P_1P_2, 1.0), (P_3P_4, 1.4), (P_3P_6, 1.4), (P_2P_6, 2.2), (P_1P_6, 3.2), (P_2P_3, 3.6), (P_3P_5, 3.6), (P_5P_6, 3.6), (P_1P_3, 3.6), (P_3P_5, 3.6), (P_4P_5, 4.1), (P_2P_4, 5.0), (P_2P_5, 5.1), (P_1P_4, 5.4), (P_1P_5, 6.1)\}$.

The iterative process starts by picking the smallest edge $(P_1P_2, 1.0)$. The set S becomes: $\{(P_3P_4, 1.4), (P_3P_6, 1.4), (P_2P_6, 2.2), (P_1P_6, 3.2), (P_2P_3, 3.6), (P_3P_5, 3.6), (P_5P_6, 3.6), (P_1P_3, 3.6), (P_3P_5, 3.6), (P_4P_5, 4.1), (P_2P_4, 5.0), (P_2P_5, 5.1), (P_1P_4, 5.4), (P_1P_5, 6.1)\}$, and the set of subtrees T becomes $\{\{(P_1P_2, 1.0)\}\}$. Next, the edge $(P_3P_4, 1.4)$ is picked. The set S becomes $\{(P_3P_6, 1.4), (P_2P_6, 2.2), (P_1P_6, 3.2), (P_2P_3, 3.6), (P_3P_5, 3.6), (P_5P_6, 3.6), (P_1P_3, 3.6), (P_3P_5, 3.6), (P_4P_5, 4.1), (P_2P_4, 5.0), (P_2P_5, 5.1), (P_1P_4, 5.4), (P_1P_5, 6.1)\}$, and the set T becomes $\{\{(P_1P_2, 1.0)\}, \{(P_3P_4, 1.4)\}\}$. Next, the edge $(P_3P_6, 1.4)$ is picked. The set S becomes $\{(P_2P_6, 2.2), (P_1P_6, 3.2), (P_2P_3, 3.6), (P_3P_5, 3.6), (P_5P_6, 3.6), (P_1P_3, 3.6), (P_3P_5, 3.6), (P_4P_5, 4.1), (P_2P_4, 5.0), (P_2P_5, 5.1), (P_1P_4, 5.4), (P_1P_5, 6.1)\}$. The vertex P_3 is already in the subset $\{(P_3P_4, 1.4)\}$. Hence, the edge $(P_3P_6, 1.4)$ is inserted in this subset, and the set T becomes $\{\{(P_1P_2, 1.0)\}, \{(P_3P_4, 1.4), (P_3P_6, 1.4)\}\}$. Again, the next smallest edge $(P_2P_6, 2.2)$ is picked from the set S. The set S becomes $\{(P_1P_6, 3.2), (P_2P_3, 3.6), (P_3P_5, 3.6), (P_5P_6, 3.6), (P_1P_3, 3.6), (P_3P_5, 3.6), (P_4P_5, 4.1), (P_2P_4, 5.0), (P_2P_5, 5.1), (P_1P_4, 5.4), (P_1P_5, 6.1)\}$. The vertex P_2 is in the subset $\{(P_1P_2, 1.0)\}$ in T, and the vertex P_6 is in the subset $\{(P_3P_4, 1.4), (P_3P_6, 1.4)\}$ in T. Hence, both these subsets are joined, and the weighted-edge $(P_2P_6, 2.2)$ is inserted in this larger subset.

The set T becomes $\{\{(P_1P_2, 1.0), (P_3P_4, 1.4), (P_3P_6, 1.4), (P_2P_6, 2.2)\}\}$. Next, the weighted $(P_1P_6, 3.2)$ is picked from the set S. The set S becomes $\{(P_2P_3, 3.6), (P_3P_5, 3.6), (P_5P_6, 3.6), (P_1P_3, 3.6), (P_3P_5, 3.6), (P_4P_5, 4.1), (P_2P_4, 5.0), (P_2P_5, 5.1), (P_1P_4, 5.4), (P_1P_5, 6.1)\}$. The weighted edge $(P_1P_6, 3.2)$ is discarded because both the vertices are in the same subset of T. Similarly, the weighted edge $(P_2P_3, 3.6)$ is discarded. The process is repeated. After the insertion of the weighted edge $(P_3P_5, 3.6)$, all the data-elements are covered, and the algorithm to derive the minimum spanning-tree terminates. The set T is $\{\{(P_1P_2, 1.0), (P_3P_4, 1.4), (P_3P_6, 1.4), (P_2P_6, 2.2), (P_3P_5, 3.6)\}\}$. The edges P_3P_5 and P_2P_6 have weights >2.0 and are removed from the set T. The resulting set is $\{\{(P_1P_2, 1.0)\}, \{(P_3P_4, 1.4), (P_3P_6, 1.4)\}\}$. It consists of two clusters $\{P_1, P_2\}$ and $\{P_3, P_4, P_6\}$. The data-element P_5 is an outlier.

3.5.1.7 Clustering time-series data

Clustering time-series data is a generalization of static clustering because feature-values are sampled over a period instead of a data-point. To compare two time-series data, time is discretized. Distance metric is different for short-time series and long-time-series data. For short time-series data having N-intervals, a popular distance-metric is the square-root of the sum of the squares of difference between slopes of two time-series for each unit time-interval as given by Equation 3.5. Clustering is done using K-means clustering as described in Section 3.5.1.2 or incremental K-clustering as described in Section 3.5.1.4.

$$distance = \sqrt{\sum_{i=1}^{i=N}\left(\frac{x_{i+1} - x_i}{t_{i+1} - t_i} - \frac{y_{i+1} - y_i}{t_{i+1} - t_i}\right)^2} \qquad (3.5)$$

where x_i is the value of the ith sample in first time-series, and y_i is the corresponding value for the second time-series data. For the comparison to be meaningful, two time-series data should be aligned so the distance between them is minimized. The alignment is done using dynamic time warping as described in Section 2.7.2.1.

3.5.2 Support Vector Machine

Support vector machines are based upon limiting the regions of known clustered data-points with labels using a combination of multiple $(N - 1)$ lower-dimensional hyperplanes for N-dimensional data-points. A single hyperplane divides the data into two regions on either side of the hyperplane. In a two-dimensional data, hyperplane reduces to a straight line as shown in Figure 3.14.

An optimum hyperplane separates the groups by discovering the nearest points to the hyperplanes on either side such that the sum of the distance between these points, and the separating line is maximum. This sum of the distance is called *margin*. The separating hyperplane is adjusted so that nearest points on either group are equidistant from the hyperplane.

For a two-dimensional plane, separating hyperplane is a straight line modeled as $y = mx + b$. Assume that the points are in first group are $(x_i^1, y_i^1)(1 \le i \le m)$, and the points in the second group are $(x_j^2, y_j^2)(1 \le j \le n)$. Thus, in order to separate the two groups, all the points in the first group should satisfy the condition $y_i^1 - mx_i^1 \le B$, and all the points in the second group should satisfy the condition $y_j^2 - mx_j^2 \ge B$.

The distance between the separating hyperplanes is derived by identifying the perpendicular lines passing through (x_i^1, y_i^1) and (x_j^2, y_j^2), identifying the intersection points with $y = mx + B$, deriving the Euclidean distances between the nearest-points and the corresponding intersection points, and summing up the distance. Multiple clusters of points are separated using SVM technique using a set of such linear equations.

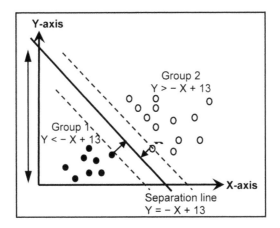

FIGURE 3.14   Separating two regions of data-points using SVM technique

Example 3.14: Support vector machine

In Figure 3.14, there are two groups of points. The groups are separated by the thick black line $Y = -X + 13$. The first group of points, located in the lower left region, is shaded in dark black color, and satisfies the inequality $Y < -X + 13$. The second group of points are shown as unshaded circles, and satisfies the inequality $Y > -X + 13$. The point P_1 is the nearest point from the group 1, and the point P_2 is the nearest point from the group 1. Both P_1 and P_2 are equidistant from the separation line, and the sum of their distances gives the margin of the line. The slope of the perpendicular is $+1$, and it passes through the point P_1 and intersects the separation line. Similarly, the other perpendicular passes through P_2, and intersects the separation line.

3.5.3 Decision Trees

Decision tree is a decision-making approach based upon classification using attribute-values. Each branch of a decision tree corresponds to a predicate such as arithmetic comparison that returns a Boolean value *true* or *false*. A predicate may involve Boolean attributes, a multivalued set, or an interval-based comparison. Combining the predicates along a path from the root node to the leaf node gives a conjunctive (logical-AND) logical rule for the classification. This conjunctive rule puts the data-element into a class at the leaf-node based upon the attribute-values of an entity.

Features are prioritized in the decision tree. Features that clearly separate distinct groups are preferred. Feature selection is done using an information theoretic metrics called *entropy*. *Information gain* is calculated as the magnitude of the difference of entropy after executing predicate using an *attribute-test(predicate)* involving a selected feature and entropy before executing the predicate. Mathematically speaking, information gain is defined as ($entropy^{after} - entropy^{before}$). The information gain is more for the significant features that contribute more to the classification. Higher information gain means that the outcomes are well separated. The attribute-test (predicate) with bigger information gain move toward the root node, and the features giving the smaller-gain move toward the leaf node.

Entropy E is defined as: $-\sum_{i=1}^{i=N} prob(x_i) \times log_2(prob(x_i))$ where x is the variable, x_i is one of the N possible values of the variable x, and $prob(x_i)$ is the probability of x_i. The entropy of a random Boolean variable v with probability p is given as $E(v) = -(p \times log_2 p + (1 - p) \times log_2(1 - p))$. The first term corresponds to successful outcome ($v = true$), and the second term corresponds to unsuccessful outcome ($v = false$). An attribute with k distinct values divides the sample space into k subsets $S_1, ..., S_k$, each with the probability $p(S_i)$ such that $\sum_{i=1}^{i=k} p(S_i) = 1$. Each subset has probability p_i of a successful possible outcome and $(1 - p_i)$ probability of unsuccessful possible outcome. The overall entropy after the test of an attribute-value is given by Equation 3.6.

$$E = -\sum_{i=1}^{i=k} p(S_i)(p_i log_2 p_i + (1 - p_i)log_2(1 - p_i)) \tag{3.6}$$

Example 3.15: Decision trees

Let us consider a condition where a cancer-type is detected using different biomarkers present in blood. Assume there are three biomarkers *A*, *B* and *C*. The presence of high values of these biomarkers above a threshold shows probabilistic presence of the cancer-type. Let us assume that biomarker *A* has *1.0%* probability of being high, biomarker *B* has *0.5%* probability of being high and biomarker *C* has *3.0%* probability of being high. Let us also assume that the probability of biomarker *A* showing cancer is 40%, biomarker *B* showing cancer is 20% and biomarker *C* showing cancer is 30%. Discover the order in which a patient should be tested for cancer. We assume that when the biomarkers are not high, then there is a default 2% probability of having cancer in the remaining population. We also assume that after the biomarker is found high, then a biopsy-test is done to further ascertain the cancer.

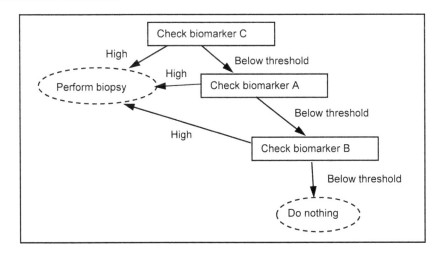

FIGURE 3.15   A decision tree based upon information gain analysis

Let us assume that the general probability of having cancer patients is *2%* in the community. The initial entropy is $-(0.02\ log_2 0.02 + 0.98\ log_2 0.98) = -(-0.11 - 0.03) = +0.14$. The entropy after testing for biomarker *A* is $-(0.01 \times (0.4 log_2 0.4 + 0.6 log_2 0.6) + 0.99 \times (0.02\ log_2 0.02 + 0.98\ log_2 0.98)) = +0.149$. The information gain is *0.009*. For the biomarker *B* the entropy after testing is $-(0.005 \times (0.2 log_2 0.2 + 0.8 log_2 0.8) + 0.995 \times (0.02\ log_2 0.02 + 0.98\ log_2 0.98)) = +0.143$. The information gain is *0.003*. For the biomarker *C* the entropy after testing is $-(0.03 \times (0.3 log_2 0.3 + 0.7 log_2 0.7) + 0.97 \times (0.02\ log_2 0.02 + 0.98\ log_2 0.98)) = 0.162$. The information gain is *0.022* for biomarker *C*. Thus, biomarker *C* should be tried first followed by biomarker *A*. Biomarker *B* is tried at the end. The corresponding decision tree is given in Figure 3.15.

3.5.4 Artificial Neural Network (ANN)

Neural network is a supervised learning model to simulate the functioning of a network of simulated neurons called *perceptrons*. The network has multiple layers. The power of neural network increases with additional layers. A perceptron fires "1" when the input signal is above a certain threshold and emits a signal "0" when the input signal is below the threshold value. The input signal is a cumulative weighted sum of the signals coming from the previous layer. A negative bias is applied to each perceptron to eliminate unregulated firing of neurons due to noise. This negative bias is added to the cumulative input signal before comparing the strength against the threshold value.

There are three layers: 1) input layer that accepts the input signals; 2) one or more hidden layers that process the intermediate signals emitted by the input layer or the previous hidden layer; 3) output layer that outputs the final values. Each perceptron in the current layer is connected to a different set of perceptrons in the next layer. Each connection is associated with a weight between *–1.0* and *+1.0*. Negative weights act as *inhibitors*, and positive weights act as *enhancers*. An inhibitor-connection inhibits the firing of a neuron while enhancer-connection enhances the firing of a neuron.

ANN is modeled as a weighted directed graph with perceptrons as nodes and connections as edges between the nodes. The weights vary between *1.0* and *+1.0*. A node in a neural network is denoted as N_{ij} for the *i*th network layer and *j*th node within the layer. An edge in the neural network is denoted as $(N_{ij} \rightarrow N_{(i+1)k})$. The weight of the edge $(N_{ij} \rightarrow N_{(i+1)k})$ is modeled as $W_{ij \rightarrow (i+1)k}$. The negative bias at a node N_{ij} is denoted as $b_{ij,}$ and the emission is denoted as $E_{ij.}$ Given these notations, the weighted sum of input at a node $N_{(i+1)k}$ including the negative bias $b_{(i+1)k}$ is given by $\sum_{j \in [1...m]} E_{ij} \times W_{ij \rightarrow (i+1)k} + b_{(i+1)k}$.

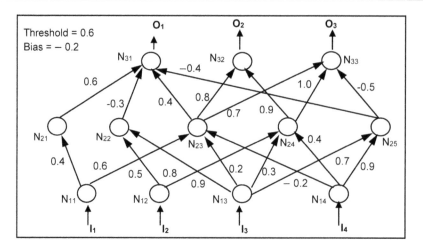

FIGURE 3.16   A simple feed-forward neural network

Perceptron is modeled using a sigmoid function that emits *0* or *1*. Perceptron output can also be modeled using an exponential function $\frac{1}{1+e^{-input}}$, where *input* is the weighted sum $\sum_{j \in [1...m]} E_{ij} \times W_{ij \to (i+1)k} + b_{(i+1)k}$. In the exponential function, output value approaches 0 for negative input and close to *1* for positive input.

Example 3.16: Artificial neural network

Figure 3.16 shows a feed-forward neural network with three layers: 1) input layer; 2) hidden layer and 3) the output layer. All the nodes in the network have a standard bias of −0.2 to take care of noise. There are four inputs I_1, I_2, I_3 and I_4. There are three outputs: O_1, O_2 and O_3. Each output is a function of the four inputs. The functions are adjusted by varying the weights dynamically until the derived output $O_{i(1 \leq i \leq 3)}^{d}$ is very close to the actual output O_i^{actual}.

Assume that the weights have stabilized, and the inputs are $I_1 = 1$, $I_2 = 0$, $I_3 = 1$ and $I_4 = 1$. The inputs at N_{11}, N_{12}, $N_{13,}$ and N_{14} are *1, 0, 1, 1*, respectively. After applying the negative bias of −0.2, the resulting input signals become *0.8, −0.2, 0.8, 0.8*. The nodes N_{11}, N_{13} and N_{14} fire with an emission *1* because their input values are greater than the threshold-value 0.6, and N_{12} emits *0*. The weighted inputs at N_{21}, N_{22}, N_{23}, $N_{24,}$ and N_{25} including the negative bias are *0.2, 0.7, 0.4, 1.3* and *1.4*, respectively. The nodes N_{21} and N_{23} do not fire because their input values are less than the threshold value of *0.6*, while other remaining nodes N_{22}, N_{24} and N_{25} fire and emit the value *1*. The input values including the bias value at the nodes N_{31}, N_{32} and N_{33} in the output layer are −0.5, 0.65 and 0.3, respectively. The node N_{32} fires and emits the value *1*; other two nodes N_{31} and N_{33} emit *0*.

3.5.4.1 Training an ANN

The neural network is a complex regression function that is dynamically adjusted until it outputs the desired values. A neural network first goes through a training process before it becomes active for classification. ANN is trained using a multitude of samples with known outcomes. Training process starts with some ad hoc set of connections based upon the input parameters. The training phase uses: 1) comparison of the actual output value with the derived output value to identify the error content and 2) use this error value to go-back to the previous layer and adjust the weights to reduce the error. If the actual output is *1* and computed output is *0*, then the weights of the previous layer connections are increased, or negative biases are decreased so the overall error for all the output is decreased. Similarly, if the actual output is *0* and the computed output is *1*, then the weights of the previous layer connections are decreased, or negative biases are increased to reduce the overall error.

ANN has been used extensively in health informatics for the identification of diseases based upon the value of the biomarkers, automated classification of diseases using texture patterns, image analysis, clustering of data, ECG waveform analysis, and derivations of genes from nucleotide data.

3.6 REGRESSION ANALYSIS

Regression analysis is based upon finding a function $Y = f(X_1, X_2, ..., X_N)$, where $X_{I (1 \leq I \leq N)}$ is an independent variable, and Y is a dependent variable. Regression analysis is used to predict the values of dependent variables given the value(s) of independent variables. The choice of function is guided by two factors: *simplicity* and *accuracy* of prediction in the future for new values of the independent variables. The simplest form of regression analysis is when the number of independent variable is one, and the regression function is a linear function. In that case, the function becomes a straight line given by an equation $y = mx + B$, which is fitted on the training data-points as described in Section 1.6.6.

Based upon the number of independent variables and the curve used to model the regression function, regression analysis can be modeled as: 1) *linear regression analysis*; 2) *nonlinear regression analysis*; 3) *multilinear regression analysis* and 4) *multi-nonlinear regression analysis*. *Linear regression analysis* uses a straight line for the curve-fitting. *Multilinear regression analysis* combines the effects of more than one independent variable. *Nonlinear regression analysis* uses a nonlinear curve such as quadratic curve, parabolic curve, cubic curve or other higher ordered curve-fitting on the available data-points. Multi-nonlinear regression analysis combines the effect of the nonlinear regression analysis.

Linear regression analysis is much easier to model, and empirical results have shown that the overall error for future test-data is much less for linear function approximation compared to nonlinear function approximations. A variation of regression analysis is called *logistic regression analysis* where the value of the dependent variable is mapped to a set of discrete values. The simplest form of logistic regression analysis is when the value of dependent variable is mapped to two values in a domain *{0, 1}* or *{true, false}*. This form of logistic regression analysis is used in a medical analysis to analyze and predict when a person will have an abnormal condition or healthy condition.

3.6.1 Linear Regression Analysis

Given N data-points $(x_1, y_1), ..., (x_N, y_N)$, and the equation of the fitting-line $y^P = mx + B$, where y^P is the computed value of the dependent variable for a value of the independent variable x. The fitting-line always passes through the mean of the data-points $\left(\frac{\sum_{i=1}^{i=N} x_i}{N}, \frac{\sum_{i=1}^{i=N} y_i}{N} \right)$. Actual experimental value y^{actual} and the predicted value y^P for the same value of the x-coordinate differ by a small random error denoted by the Greek symbol ϵ. Thus, y^{actual} is equal to $y^P + \epsilon$. For each experimental data-point, the magnitude-of-error is denoted by ϵ_i, and is equal to the absolute value $| y^{actual} - y^P |$. To find out the best fitting, the cumulative error from all the points is minimized. The cumulative error is computed as $\sqrt{\sum_{i=1}^{i=n} (y_i^{actual} - y_i^P)^2}$. After the substitution, the sum of the square of the errors is given by $\sum_{i=1}^{i=N} (y_i^{actual} - (mx_i + B))^2$. If we plot the error $\epsilon = | y^{actual} - y^P |$, it will be a Gaussian distribution with some variance.

To derive the value of the slope of the optimum line, partial differential equation is used. Using partial differentiation with respect to the slope m and equating the partial differentiation equal to zero for minima, the slope $m = \frac{covariance(x, y)}{variance(x)}$ minimizes the overall absolute value of errors. The lines pass through the mean value $(\bar{x}, \bar{y})$ where $\bar{x} = \frac{\sum_{i=1}^{i=N} x_i}{N}$ and $\bar{y} = \frac{\sum_{i=1}^{i=N} y_i}{N}$. Substituting the value of m, $\bar{x}$ and $\bar{y}$, the corresponding value of B is derived as $\left(\bar{y} - \frac{covariance(x, y)}{variance(x)} \times \bar{x} \right)$. The equation is further simplified as both variance and covariance have the term "$N – 1$" as the denominator that cancels out. The overall equation after the cancellation is given as:

$$\hat{r} = \frac{\sum_{i=1}^{i=N}(x_i - \bar{x})(y_i - \bar{y})}{\sqrt{\sum_{i=1}^{i=N}(x_i - \bar{x})^2} \times \sqrt{\sum_{i=1}^{i=N}(y_i - \bar{y})^2}}$$

(3.7)

The value of the Pearson's coefficient $\hat{r}$ varies between $+1.0$ and -1.0. The value $\hat{r} = +1.0$ means very strong positive correlation between the independent variable x and dependent variable y; $\hat{r} = -1.0$ means very strong negative correlation. The condition $\hat{r} = 0.0$ means that the dependent variable y is not correlated with the independent variable x, and the independent variable x should be dropped from the set of parameters affecting the dependent variable y.

An important issue in the regression analysis is that how strongly a dependent variable y is correlated with an independent variable x. There are three possibilities: 1) strongly positively correlated; 2) not correlated and 3) strongly negatively correlated. Positive correlation means the value of a dependent variable y increases with the increase in the value of an independent variable x; negative correlation means that the value of a dependent variable y decreases when the value of an independent variable increases. Correlation is derived using Pearson's correlation-coefficient $\hat{r} = \frac{covariance(x, y)}{\sigma_x \sigma_y}$, where $\sigma_x = \sqrt{variance(x)}$ and $\sigma_y = \sqrt{variance(y)}$ are the standard deviations of x-coordinates and y-coordinates of the data-points.

Data-points are more scattered for smaller Pearson coefficient $\hat{r}$ as shown in Figure 3.17 and less scattered for large value of Pearson coefficient $\hat{r}$. The slope of the fitted line in linear regression and the Pearson's coefficient $\hat{r}$ are related by the relationship $\hat{r} = m \times (\sigma_y / \sigma_x)$, where m is the slope of the fitted line.

Example 3.17: Linear regression analysis

Consider a data-set of 12 points $(1, 1)$, $(2, 3)$, $(3, 6)$, $(4, 7)$, $(6, 12)$, $(9, 16)$, $(11, 18)$, $(14, 30)$, $(15, 30)$, $(16, 36)$, $(19, 40)$, $(20, 45)$. The mean of x-coordinates $\bar{x} = \frac{(1+2+3+4+6+9+11+14+15+16+19+20)}{12} = 10.2$, and the mean of y-coordinates $\bar{y} = \frac{(1+3+6+7+12+16+18+30+30+36+40+45)}{12} = 20.3$. The variance of x-coordinates σ_x^2 is $\frac{\sum_{i=1}^{i=12}(x - 10.2)^2}{11} = 46.0$. The variance of y-coordinates σ_y^2 is $\frac{\sum_{i=1}^{i=12}(y - 20.3)^2}{11} = 234.4$. The standard deviation x-coordinates $\sigma_x = \sqrt{46.0} = 6.8$. The standard deviation of y-coordinates $\sigma_y = \sigma_y = \sqrt{234.4} = 15.3$. The covariance is $\frac{\sum_{i=1}^{i=12}(x-10.2)(y-20.3)}{11} = 102.9$. The slope of the fitting line is $covariance(x, y)/variance(x) = 102.9/46.0 = 2.24$. The line passes through the point $(10.2, 20.3)$. Substituting this value in the

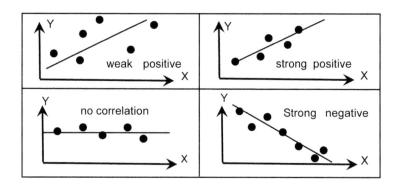

FIGURE 3.17   Correlation type and scattering of points

equation of the straight-line $y = mx + B$, the intercept $B = 20.3 - 2.24 \times 10.2 = -2.55$. The straight-line becomes $y = 2.24x - 2.55$. The Pearson's coefficient $\hat{r} = \frac{covariance(x, y)}{\sigma_x \sigma_y} = 0.99$. That means that the dependent variable y is strongly and positively correlated with the independent variable x.

One of the problems in curve fitting is the randomness of the error. Error cannot be correlated with the values of the independent variable. For example, it cannot increase or decrease with the increase of the independent variable x, show any periodic behavior, or be a function of the independent variable x. A statistical test called t-test is performed on error distribution to ascertain that the residual error ε is random.

3.6.2 Multilinear Regression Analysis

Multilinear regression function with two independent variables is modeled as $Y = m_1 X_1 + m_2 X_2 + B$ representing a combination of two straight lines: one for each independent variable with slope m_1 corresponding to the first straight line and m_2 corresponding to the second straight line. Multilinear regression function is generalized for N ($N > 1$) independent variables as $Y = m_1 X_1 + m_2 X_2 + \ldots + m_N X_N + B$. The assumption in multilinear regression function is that all the variables $X_1, \ldots X_N$ are orthogonal that means that independent variables do not interfere with each other in deciding the value of the dependent variable. Multinonlinear regression analysis involves at least one nonlinear function in modeling the regression function.

3.6.3 Logistic Regression Analysis

The value of the dependent variables in logistic regression analysis is mapped to a discrete set of values. The simplest form of the category contains two values such as $\{0, 1\}$ or $\{true, false\}$. The curve that fits a logistic regression analysis has a transition range. Below the transition range, the value of the dependent variable maps to 0; beyond the transition range, the value of the dependent variable maps to 1. The logistic regression is modeled using Equation 3.8.

$$y = \begin{cases} 1 \text{ if } \beta_0 + \beta_1 x + \varepsilon > 0 \\ 0 \text{ otherwise} \end{cases} \qquad (3.8)$$

where ε is an error given by the logistic distribution that is similar to Gaussian distribution with heavier tails. The parameters β_0 and β_1 are derived using an iterative process. The parameters β_0 and β_1 are related to a logistic function $f(x)$ given by Equation 3.9. The logistic function $f(x)$ is the probability of the dependent variable y being equal to 1 or $true$.

$$f(x) = \frac{1}{1 + e^{-(b_0 + b_1 x)}} \qquad (3.9)$$

3.6.4 Bayesian Statistics and Regression Analysis

In the regular statistical analysis, the parameters are fixed based upon some prior knowledge and hypothesis, and the curve is fitted to minimize the error between the estimated values and actual values. The assumption is that parameters are known and fixed, and there is noise in measuring the actual data that causes error. The noise could be due to many reasons such as sensor error, experiment setup error, quantization error from the analog signal to digital signal, change in the ambient conditions while doing the experiment, variation of hidden variables that have been ignored in the model or human error.

There could be another approach of looking at the data and the parameters that are decided by some hypothesis rooted in prior knowledge. In this approach, the hypotheses (and parameters) are further refined based upon the observed data. The first set of parameters is estimated based upon: 1) a generative model such as probability distribution to simulate data distribution and 2) a prior knowledge, belief or some hypothesis as a starting point. Prior knowledge or belief biases the probability distribution. The hypothesis (or the subset of parameters) is further modified and refined based upon the observed data. If a sufficiently large amount of data is available, the bias introduced by the prior knowledge is reduced. This approach is called *Bayesian statistics*. It is a two-layered approach: 1) estimating the prior hypothesis and the corresponding parameters based upon some prior knowledge and 2) modifying the hypothesis and the corresponding set of parameters based upon the observed data. For example, least square method to minimize error is a prior belief based upon reasoning rooted in Euclidean distance with equal weight given to each dimension. However, it's modified to have a weighted Euclidean distance with different weight given to various dimensions based upon the observed data.

3.6.4.1 Bayesian regression analysis

Bayesian regression analysis predicts the probability distribution of the parameters given a model, the set of data, and a probability distribution of the error terms. If the dataset is denoted by X, and the parameter set in the model is denoted by Θ, the Bayesian formula is written as $p(X = x) \times p(\Theta = \theta \mid X = x) = p(\Theta = \theta) \times p(X = x \mid \Theta = \theta)$. The Bayesian regression analysis derives the *posterior distribution $p(\Theta = \theta \mid X = x)$*. The values of $p(X = x)$, $p(\Theta = \theta)$, and $p(X = x \mid \Theta = \theta)$ are needed to derive the probability $p(\Theta = \theta \mid X = x)$. The probability $p(X = x)$ is a *constant of integration* used to normalize the probability because probabilities should add up to *1.0*. The probability $p(\Theta = \theta)$ is called *prior distribution,* and is based upon the *prior estimates* of the parameters using prior knowledge, and $p(X = x \mid \Theta = \theta)$ is the *likelihood function*. The *posterior distribution* is summarized using multiple criteria such as mean, median, mode and mid-point of the range.

In a single variate linear regression analysis, modeled as $y = mx + b$, there are two parameters in the fitted line: the intercept b and the slope m. Each parameter is represented by a probability distribution function with a *prior estimate*. The probability distribution function is characterized by the mean, variance and the covariance matrix. For a multivariate function of the form $y = a_1 x_1 + a_2 x_2 + \dots + a_m x_m + \beta$, each independent variable and the intercept β will be characterized by a probability distribution function with a mean, variance and covariance.

3.6.5 Issues in Regression Analysis

There are issues with multilinear regression analysis and the data-points. The experimental data-points show the cumulative effect of multiple variables. Often, effects of multiple independent variables cancel each other showing poor correlation while individually they may have strong correlations. The experimental data should be plotted for one independent variable at a time. However, in real-world medical data, it may not be possible to study the effect of just one independent variable at a time. For example, many factors such as loss of vitamins, less blood-flow to the brain and the formation of plaque in the brain-cells may affect cognition-loss during old age. However, there is no way to mask off the effect of other independent variables, and just study the effect of one independent variable.

Another issue in multilinear regression analysis is *overfitting* that is defined as a situation when there is no known correlation between an independent variable and the dependent variable based upon common-sense reasoning. Yet, a curve-fitting shows correlation. For example, an independent variable "staying up late" and the dependent variable "cognition loss" may exhibit a strong correlation without justifiable scientific basis. Despite a strong correlation, the rationale for such outcome is weak.

Another problem with multilinear regression is *co-linearity*. Co-linearity is defined as the case when two independent variables are correlated. They contribute twice in the regression model and make the model faulty. For example, a study to discover the causes of diabetes may use both *BMI* (Body Mass Index) and *weights* as independent variables. Since BMI and weights are correlated, the outcome will be incorrect.

Another problem in modeling multilinear regression analysis is the loss of the independent variable where a major parameter of the outcome is missing. In such situations, the error will be inconsistent, and the model will not show correct correlation.

Another problem is when two or more independent variables affect the dependent variable similarly. However, one of the independent variable effects more than the other. For example, if we are looking at lung cancer of a person who has been smoking for long, had been living in an asbestos house and has been working in a nuclear power plant where he gets exposed to above-normal radiation every day for a prolonged period, then all three independent variables are cancer causing. It is possible that one independent variable may overshadow the effect of other variables. This phenomenon is called *residue confounding.*

3.7 PROBABILISTIC REASONING OVER TIME

It is common and useful to predict the future outcome in clinical science. Based upon the lab test, a doctor can predict the recovery of a patient with certain confidence, predict the progression of disease, predict the onset of a disease based upon the change in biomarkers and can provide intervention even before a predicted disease can become fatal. However, in all these predictions, a probability-factor is involved. Prediction about recovery and progression of a disease requires *temporal probabilistic reasoning.* One problem in the prediction with limited sample of data in time-based reasoning is that it may not be very accurate initially. However, with the increase in the amount of data, the picture become clearer and the prediction can be done with more confidence.

Probabilistic reasoning over time also includes a new dimension of time in inference mechanism. Since the effect on an outcome is seen by the world using measured evidence using sensors, it answers about the probable cause if the sensor values change. For example, the probability that an administered drug is causing the recovery is based upon the change of values of lab-tests and biomarkers. Similarly, we can ask a question: what is the probability of a disease if the values of a biomarker change over a period? Biomarkers get altered because of other reasons too. Disease and biomarkers are probabilistically related.

A general query is the chance of X_t – an event at time T, if the evidence for the events $X_0, \ldots X_{t-1}$ up to time $(t - 1)$ are known. Using a transition matrix an and evidence matrix in first-order Markov processes, the value of $p(X_t)$ can be derived by knowing $p(X_{t-1})$ and the values of the evidence variables $E_0, \ldots, E_t$. The transition matrix gives the conditional probability $p(X_t | X_{t-1})$ of transition from a state X_{t-1} to a state X_t, and evidence matrix gives the probability of evidence E_t in a particular state X_t. The formula for deriving the probability of an event X_t is given by Equation 3.10.

$$p(X_t) = p(X_0) \times \sum_{i=0}^{i=t-1} p(X_i \rightarrow X_{i+1}) \times p(E_{i+1}|X_{i+1}) \qquad (3.10)$$

The term on the left-hand side is the probability of the occurrence of a dependent variable at time t. The first term on the right-hand side is the initial probability at time $t = 0$. The second term is the summation of the products of probability of transition from a state $X_i \rightarrow X_{(i+1)}$ and probability of the state emitting evidence signals $E_{(i+1)}$ at time $t = i + 1$ when it is in the state $X_{(i+1)}$. Sensors measure the evidence signals.

This reasoning is extended further to solve the problems of *predicting* the future, or *smoothening* the results from the past. By using the probabilistic reasoning over time, the effect of noise or sensor errors is reduced from the past readings. Prediction is modeled as finding the probability $p(X_{N+M})$ at time $t = (N+M)$ given we have only sensor values up to the time $t = N$, and only $p(X_0), \ldots, p(X_N)$ and the evidence $E_0, \ldots, E_N$ are known. As the value of M increases, the distance from the current evidence increases, and the prediction accuracy decreases.

There are three major classes of probabilistic reasoning over time: 1) Bayesian decision network; 2) first-order Markov model and 3) hidden Markov model. Hidden Markov model is a special subclass of first-order Markov models where the states and transition between states are only estimated by observing the emitted signals.

3.7.1 Bayesian Decision Network

Bayesian decision network is a probabilistic reasoning technique that is based upon Bayes' formula described in Section 2.3.3. It is used for probabilistic relationship between the input and the outcomes. Bayesian network is represented as a directed acyclic graph where input to a node is an incoming edge. The outgoing edge of a node may connect to another node. The input signal to a node affects the outcome probabilistically, and is represented as *conditional probabilistic table (CPT)* as shown in Figure 3.18.

Each CPT gives the conditional probability based on the event in the corresponding node. The conditional probability table includes the probability for input being *true* or *false*. The conditional probability is of the form $P(X_i|parent(X_i))$, where X_i is a node, and there is an incoming edge from a $parent(X_i)$ to X_i. Given the CPTs for each node, the probability of outcome is given by the product of the chain of conditional probabilities in the path from the initial condition to the outcome.

Example 3.18: Bayesian decision network

Consider a hypothesis that lung cancer is caused by excessive smoking or genetic abnormality or nuclear radiation or asbestos. People suffering from cancer have higher probability of cancer-related death, breathing difficulty and fatigue. There are different probabilities of getting cancer from these causes.

The CPT for cancer combines four possibilities. There are three edges going out of the node cancer. Since there are four inputs for cancer, and each input can have two values: *true* or *false*, there are 16 (2^4) input combinations. There is a conditional probability for each input combination. For example, a query can be: what is the probability of person having fatigue due to nuclear radiation only? It would be given by the product $P(fatigue|cancer) \times P(cancer|nuclear\ radiation) \times P(nuclear\ radiation)$ under the assumption that all four inputs are independent of each other. There may be errors due to sensor readings that contribute to overall uncertainty.

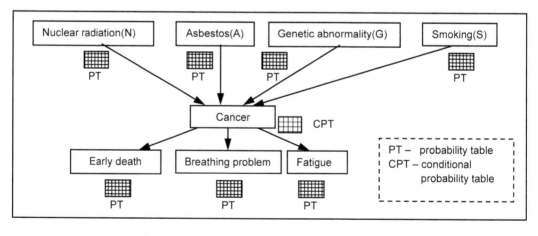

FIGURE 3.18   Bayesian decision network for modeling causes of cancer

3.7.1.1 Temporal Bayesian network

The concept of Bayesian network is extended to temporal Bayesian network (also called dynamic Bayesian network) by reading sensor values periodically. Dynamic Bayesian network is a repeated occurrence of Bayesian network, where each network is based upon a different discrete time. Two adjacent networks are interconnected by a time-step. Dynamic Bayesian networks are used in treatment selection, diagnosis, monitoring and long-term disease management such as diabetes, heart-related diseases, cancer management and kidney-related diseases.

If the current state depends only on the previous state, and evidence depends on the current state, the probability of next occurrence of the phenomenon can be predicted using the probability tables between the current-state → next-state and evidence-probability-table in HMM. We also assume that the transition follows first-order Markov process that is the next state at time $t + 1$ only depends upon the current state at time t. We assume that the state change is only Boolean for simplicity. Multiple values in the state-transition can also be handled. However, it would require all possible mappings from M previous states to M next states. If continuous input in addition to the previous state occurs, the next state also depends upon the inputs and previous state as described in Example 3.19. The probability of the next Boolean state is given by the recursive equation $P(X_{t+1}) = P(X_t) \times P(X_{t+1}|X_t) \times P(E_{t+1}|X_{t+1})$. By unfolding the equation recursively, it can be expressed by Equation 3.11.

$$P(X_{t+1}) = P(X_0) \times \prod_{i=1}^{i=t} (P(X_{i+1} \mid X_i) \times P(E_{i+1} \mid X_i)) \tag{3.11}$$

The equation can be extended to predict the recovery after K units of time. However, the transition-matrix and evidence-matrix will not be accurate because there will be no information available about the time $t + 2$ onwards.

Example 3.19: Application of dynamic Bayesian network

For example, if a person is being given medicine to improve his cancer, and the recovery is being modeled using: 1) image processing of cancerous cells; 2) patient survey of his feelings of pain; 3) outward symptoms of fatigue and 4) biomarkers using pathological tests. This recovery can be modeled as dynamic Bayesian network with drug and drug dosage as input to the state of the cancer with emissions as size of the cancerous lump, fuzzy scale of patient's feeling better and fuzzy scale of outward symptoms.

The drug dosage changes with time, and the recovery state and drug administration are connected through a CPT. Similarly, the recovery state and emissions: lump-size, patient feeling better, outward signs and biomarkers. Each emission and recovery-state are connected through CPT derived using statistical analysis from the past cases. An overall model is given in Figure 3.19. The model can be generalized for any disease by changing the inputs, CPTs and the evidence.

The recovery state at time $t + 1$ is predicted knowing the recovery state at time t and evidence at time $t + 1$ using the Bayesian equation. Assuming finite limited possibilities of drug-dosage and Boolean possibility for drug, there are CPTs for (drug-dosage, drug, previous-recovery state) → next-recovery-state. The number of entries in the CPT is given by the product of the number of possible input values and recovery states. The recovery is modeled using a finite number of states. There is a probability table for each emission for each recovery state. For simplicity, the recovery state and evidence of improvement are modeled using Boolean values.

3.7.2 Markov Model

Many problems can be modeled as a probabilistic finite state nondeterministic machine that changes state with time. A probabilistic finite state machine can transition nondeterministically to one of the probable states in the next time-unit. However, the transition is probabilistic. The machine has a set of initial states

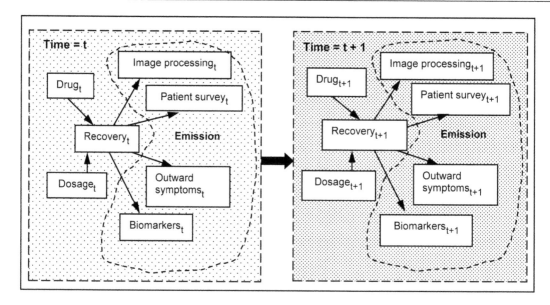

FIGURE 3.19   Modeling recovery as a dynamic Bayesian network

that is a subset of machine-states. The machine can probabilistically be in any initial state at the start. The sum of probabilities of being in all the initial states is *1.0*. Similarly, sum of the probabilities of transition from one state to the next probable states is *1.0*.

3.7.2.1 First-order Markov model

In a generalized Markov process, the transition to the next state may depend upon a sequence of previous states. However, one particular class of Markov process is used more often in the modeling that assumes that the transition to the next state depends only upon the current state, and no other past information is needed. This class of Markov process is called first-order Markov process and is computationally tractable. The probability of transition can be modeled as an $N \times N$ matrix, so the row shows the current state, and the column shows the next state. The value N is number of states in the finite machine. The cell (*row, column*) shows the probability of transition from the current state to the next state. The number of nonzero values in a row shows the number of possible transitions from a state to other states. The initial state probabilities are modeled as a vector where each element corresponds to the probability of being in a probable initial state. There can be more than one start-state.

A first-order Markov model is modeled either as a four-tuple: (*set of states, set of transitions, transition matrix, initialization-vector*). It can also be visualized using a directed weighted graph where the states are the vertices, and the edges are the transitions. The edge-weights in the graph-based representation of the machine represent the probability of transition as illustrated in Figure 3.20.

Example 3.20: Markov model

Figure 3.20 shows a probabilistic transition graph for first-order Markov process and the corresponding transition-matrix and initial state matrix. Note that the graph is an asymmetric directed graph. There are two initial states, state #1 and state #3, each with the initial probability of *0.5*.

A transition sequence in the Markov model is of the form $(S_i \rightarrow S_j)$, $(S_j \rightarrow S_k)$, $(S_k \rightarrow S_m)$ such that the destination from the previous transition becomes the source in the current transition. Similarly, the

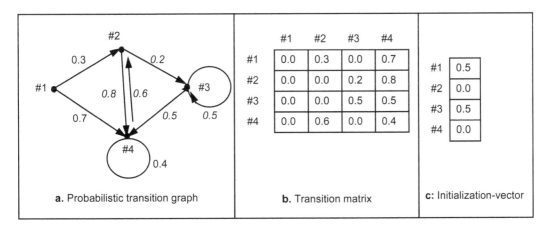

FIGURE 3.20 Modeling a Markov process: (a) Probabilistic transition graph; (b) Transition matrix; (c) Initialization-vector

destination in the current transition becomes the source in the next transition. The overall probability of the transition sequence is given by the multiplication of the individual transition-probabilities: $probability(S_1 \rightarrow S_2 \rightarrow S_3 \rightarrow S_4 \rightarrow S_5) = probability(S_1 \rightarrow S_2) \times probability(S_2 \rightarrow S_3) \times probability(S_3 \rightarrow S_4) \times probability(S_4 \rightarrow S_5)$.

Both Markov model and its variation *hidden Markov model* (HMM described in Section 3.7.3) have been used extensively to model and analyze many complex phenomena such as ECG analysis for heart disease analysis, speech recognition, gene recognition from genome sequences, patient monitoring and recovery.

3.7.2.2 Training Markov models

Markov models are trained by observing large number of state transitions. Probability of transition between two states is defined as the ratio of the number of transitions between two states and the total number of transitions. The assumption is that the transition between the states can be directly observed.

3.7.3 Hidden Markov Model

In the Markov model, there is a clear assumption that the state of the machine is visible to the world, and there is clear one-to-one mapping between the state and the corresponding emission. By looking at the emissions, we can identify the current state uniquely. However, in the real life, it is not the case. Many states give the same evidence, and a state emits more signals measured by different sensors. The state and transition are guessed by the evidence collected by the sensors. The evidence may have measurement inaccuracies due to the surrounding conditions, error in the sensor equipment or limitations in the resolution of the sensors.

There are many-to-many mappings between the set of states and the set of emissions. One state may probabilistically emit multiple emissions. Similarly, the same emission may be emitted by more than one state. Under the assumption that the transitions between states and the mapping between being in a state to the sensor-evidence are not correlated, we have another parameter to consider while modeling: the probability of many-to-many mapping between the set of states and the set of emissions. Because of this probabilistic mapping between the states and emissions, it is difficult to derive the state of the machine given the emission sequences. This interesting subclass of Markov model is called Hidden Markov Model (HMM) because the states and the sequence of transitions of states in the machine cannot be derived with certainty rather only probabilistically.

HMM has been applied in multiple areas related to health informatics such as speech recognition, natural language processing, gene identification, text analysis, ECG analysis for abnormalities and disease diagnosis from various measured parameters such as ECG, lab-tests and visible symptoms.

3.7.3.1 Modeling HMM

Like first-order Markov models, HMM is modeled as a probabilistic transition graph such that the transition to the next state is only decided by the current state. HMM is described as a five-tuple $(\Sigma, \Psi, I, \Delta, E)$, where Σ is the set of states, Ψ is the set of discrete sensor signals, I is the initialization vector, Δ is the transition-probabilities matrix between the states and E is the evidence-probabilities matrix.

The transition-matrix Δ is an $N \times N$ matrix, where N is the number of states in the HMM. Each cell (i, j) $(1 \leq i, j \leq N)$ of the transition-matrix stores the transition-probability from state i to state j. This is similar to Markov model described in Section 3.7.2. The evidence-matrix E is an $N \times M$ matrix, where N is the number of states, and M is the number of emissions. The rows denote the states in an HMM, and columns denote the emissions. Each cell (i, j) $(1 \leq i \leq N, 1 \leq j \leq M)$ stores the probability of a signal-emission.

The graphical representation of the HMM is similar to that of Markov model with an addition: each state also shows the signals it emits. The sum of probabilities of all the emissions for each state is *1.0*.

Example 3.21: Hidden Markov model

The Example 3.20 is modified to include the emission probabilities. The emission-matrix for the corresponding HMM is given in Figure 3.21a. The evidence-matrix is shown in Figure 3.21b. Transition-matrix and initialization-vector are the same as in Example 3.20. The emission matrix shows that there are three emissions: *x*, *y* and *z* with different probabilities. Probability-value *0.0* means that the corresponding state does not emit any signal.

3.7.3.2 Probability of transition sequence

The calculation for the probability of transition sequence is extended to include the emission probabilities also in addition to transition probabilities between the states. The only way to guess the state is based upon

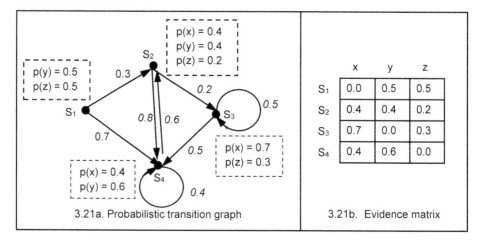

3.21a. Probabilistic transition graph   3.21b. Evidence matrix

FIGURE 3.21   Probabilistic graph for an HMM: (a) Probabilistic transition graph; (b) Evidence matrix

the emissions. Hence, the identification of the states is probabilistic. To take care of this probability, transition probability between states has to be multiplied with the evidence probability to derive the overall probability.

Example 3.22: Probability of transition sequence in an HMM

Starting from one of the initial states #1, the probability of a state transition sequence $(S_1 \rightarrow S_2 \rightarrow S_4 \rightarrow S_4)$ emitting signal "zxyx" would be given as *initial-probability(S_1) × evidence-probability$(z|S_1)$ × transition-probability$(S_1 \rightarrow S_2)$ × evidence-probability$(x|S_2)$ × transition-probability$(S_2 \rightarrow S_4)$ × evidence-probability$(y|S_4)$ × trans-probability$(S_4 \rightarrow S_4)$ × evidence-probability$(x|S_4) = 0.5 \times 0.5 \times 0.3 \times 0.4 \times 0.8 \times 0.6 \times 0.4 \times 0.4 = 0.0023$.*

Similarly, we can reason it in the reverse direction. Given a sequence of emissions $E_1, ..., E_M$, first, we identify how many probable transition sequences are possible, and then compute the probability for each one of them. The transition sequence with highest probability is the *most probable transition sequence*. To derive the transition sequences, we start from the first emission, and identify the states that emit the signal. Transitions are considered only from the valid states derived from the previous emission. The process is repeated after collecting the set of states for the next emission until all the emissions are consumed. The set of transition sequences is given by the Cartesian product of set of states emitting measured signals. From this set, those transition sequences with a zero-transition probability are removed; remaining transition-sequences are probable. The most probable sequence of state-transitions has the highest probability.

Example 3.23: Most probable sequence of state-transitions

Let us consider an emission sequence "zyyx" that was measured using the sensors. The signal z is emitted by the states $\{S_1, S_2, S_3\}$; the signal y is emitted by the set of states $\{S_1, S_2, S_4\}$; the signal x is emitted by the set of states $\{S_2, S_3, S_4\}$. The overall set of possible transitions is given by the Cartesian product $\{S_1, S_2, S_3\} \times \{S_1, S_2, S_4\} \times \{S_1, S_2, S_4\} \times \{S_2, S_3, S_4\}$. There are 81 possible state-transition sequences including sequences with zero probability. Based upon the transition-matrix and evidence-matrix in Figure 3.21, state-transitions $S_1 \rightarrow S_1, S_1 \rightarrow S_3, S_2 \rightarrow S_1, S_2 \rightarrow S_2, S_3 \rightarrow S_1, S_3 \rightarrow S_2, S_4 \rightarrow S_1$ and $S_4 \rightarrow S_3$ have zero probabilities. Hence, transition sequences containing these transitions are deleted.

Applying these facts on the Cartesian product leaves only 14 possible transition sequences: $\{S_1 \rightarrow S_2 \rightarrow S_4 \rightarrow S_2; S_1 \rightarrow S_2 \rightarrow S_4 \rightarrow S_4; S_1 \rightarrow S_4 \rightarrow S_2 \rightarrow S_3; S_1 \rightarrow S_4 \rightarrow S_2 \rightarrow S_4; S_1 \rightarrow S_4 \rightarrow S_4 \rightarrow S_2; S_1 \rightarrow S_4 \rightarrow S_4 \rightarrow S_4; S_2 \rightarrow S_4 \rightarrow S_2 \rightarrow S_3; S_2 \rightarrow S_4 \rightarrow S_2 \rightarrow S_4; S_2 \rightarrow S_4 \rightarrow S_4 \rightarrow S_2; S_2 \rightarrow S_4 \rightarrow S_4 \rightarrow S_4; S_3 \rightarrow S_4 \rightarrow S_2 \rightarrow S_3; S_3 \rightarrow S_4 \rightarrow S_2 \rightarrow S_4; S_3 \rightarrow S_4 \rightarrow S_4 \rightarrow S_2; S_3 \rightarrow S_4 \rightarrow S_4 \rightarrow S_4\}$. If imposing additional restriction that the first state must be a starting state, the transition sequences starting from the states s_2 and s_4 are removed leaving only 10 probable transition sequences: $\{S_1 \rightarrow S_2 \rightarrow S_4 \rightarrow S_2; S_1 \rightarrow S_2 \rightarrow S_4 \rightarrow S_4; S_1 \rightarrow S_4 \rightarrow S_2 \rightarrow S_3; S_1 \rightarrow S_4 \rightarrow S_2 \rightarrow S_4; S_1 \rightarrow S_4 \rightarrow S_4 \rightarrow S_2; S_1 \rightarrow S_4 \rightarrow S_4 \rightarrow S_4; S_3 \rightarrow S_4 \rightarrow S_2 \rightarrow S_3; S_3 \rightarrow S_4 \rightarrow S_2 \rightarrow S_4; S_3 \rightarrow S_4 \rightarrow S_4 \rightarrow S_2; S_3 \rightarrow S_4 \rightarrow S_4 \rightarrow S_4\}$. The maximum probability would give the most probable sequence.

Finding out the most probable transition sequence does not mean that in actual life that transition would occur. However, that is the best modeling for probabilistic transitions that computer science can provide using probabilistic transition. Similarly, the assumption that the next transition depends always upon the last state, and the past states do not affect the next transition is also a simplification that may not always occur in the real-life events.

3.7.3.3 Queries in HMM

Many types of queries can be asked using HMM: 1) what is the highest probability for a specific measured sequence of emissions and the starting state? 2) given the current state of the machine and the transition-matrix, what state the machine would be after n emissions? 3) what is the probability of going from one state to another state with n emissions? 4) given the current state and a sequence of emissions, what is the best probable state from where the transition started?

The first class of query has been discussed in Section 3.7.3.2. The second class of query is also answered by the discussion in Section 3.7.3.2 with a further modification: the transition sequence should start from the current state. All the transition-paths and their probabilities are computed. The path with the highest probability is identified, and the last node in the transition sequence is the probabilistically the optimum final state after the given emissions. The third class of queries is related to first and second class. Only those path sequences are retained that have the last node as the desired node. The probabilities of all the path-sequences ending in the desired end-node are computed, and their average (or median) is taken. If the best probability is required, then the path with the maximum probability is chosen. The fourth class of queries is answered by reverse reasoning starting from the end-state. In each iterative step, the set of previous reachable states are computed, until all the emissions are consumed. The set of the transition-paths are collected, and their probabilities are computed using the equation: $emission\text{-}probability(emission_N | state_N) \times transition\text{-}probability(state_{N-1} \rightarrow state_N) \times emission\text{-}probability(emission_{(N-1)} | state_{(N-1)}) \times \ldots \times transition\text{-}probability(state_0 \rightarrow state_1) \times emission\text{-}probability(emission_0 | state_0)$.

3.8 DATA MINING

Data mining is a computational process to find association or correlation between two independent variables or attributes in a huge amount of data. The rules are represented as logical rules using conjunction (logical AND) and disjunction (logical OR) of variable-values (or interval of values or generalization of concrete values) or their negation. The rules are of the form $X_1\ op_1\ X_2\ op_2 \ldots op_N\ Xn \rightarrow y$, where Xi are the values, $op_i (0 \le i \le n)$ are the arithmetic-comparison or logical operators and y is the outcome.

For example, an association rule may say $(Biomarker_1\text{-}value > Threshold_1)$ or $(Biomarker_2\text{-}value > Threshold_2) \rightarrow renal\text{-}failure$. Another example of data-mining is that $(ethnic\text{-}group = Indian)$ and $(age > 50) \rightarrow higher\text{-}coronary\text{-}disease$. In this example, a statistical analysis of Indians shows that the mean number of people in the age group above *50* years have a higher probability of coronary disease.

Example 3.24: Example of interestingness in data mining

Consider an example of blood-test from a large population. Assume that the average level of biomolecule b be μ mg/ml. The biomolecule level for a large population will assume a Gaussian distribution with a mean μ and a standard deviation σ. The 95% cutoff would occur at $\mu \mp 2\sigma$. Let the level of b for patients having kidney disease be ρ, and $\rho > \mu + 2\sigma$. The value of biomolecule significantly deviates from the mean and is interesting and is relevant as a biomarker for kidney disease.

During data-mining, multiple interesting patterns related to a specific goal may be identified. These patterns are ranked based upon their significance, and most interesting few patterns are selected. There are many aspects to knowledge discovery using data mining. It can be done using unsupervised learning such as clustering for classification, regression analysis to predict the trend and correlation, pattern matching techniques, forming association rules by deriving correlation and text mining.

Data mining has been used in medical informatics in identifying biomarkers, effects (including side-effects) of medications, relationships between various variables in lab-tests and disease conditions and the identification of combinations of biosignals that specify an abnormal medical condition.

3.8.1 Interestingness

An important aspect of identifying patterns for data-mining is interestingness based upon the patterns that deviate from the expected probability. Nine factors have been identified for "interestingness" of a pattern during data mining. The factors are: 1) *conciseness*; 2) *generality*; 3) *reliability*; 4) *peculiarity*;

5) *diversity*; 6) *novelty*; 7) *surprising*; 8) *utility* and 9) *actionable*. *Conciseness* is defined as attribute-value pairs that are much less than the expected values. *Generality* is defined when a pattern contains a relatively larger number of attribute-value pairs in a sample set. A pattern is *reliable* if it occurs consistently across different datasets. A pattern is *diverse* if the elements in the dataset differ significantly from each other, and there is no continuity in attribute-value pairs. A pattern is *novel,* if based upon the prior knowledge, then it could not be derived. A pattern is *surprising* if it is an exception to existing knowledge or rule-set of patterns. Data-mining goals are different, and patterns derived using data-mining are goal-specific; a pattern may have high *utility value* for one goal, and no value for another goal. A pattern is *actionable*, if the identification of pattern allows us to take an action in some domain. Based upon the domain of data-mining, researchers have devised other properties of interestingness.

3.8.2 Associative Learning

Consider a database D that contains N-tuple each containing M ($M \geq 2$) Boolean variables. Associative learning is based upon deriving statistically high significant correlation between a set of Boolean variables as *antecedents* of the form $\{V_i, ..., V_J\}$ and a set of Boolean variables $\{V_{K, ...,} V_M\}$ as *consequent* to form an *if-then rule* of the form $\{V_i, ..., V_J\} \rightarrow \{V_{K, ...,} V_M\}$ in a relational database where many records are stored. The consequents-set shares no variable with the antecedents-set. Boolean variables can be extended to testing the value of a general variable above a threshold using an equality or an inequality test.

For example, *high-sodium* is equivalent to (*sodium-level* $\geq$ *high-threshold*) that returns *true* or *false*. Hence, *high-sodium* is Boolean. Using Boolean variables, we can analyze patient databases in a specific domain to identify useful association rules, for example, $\{high\text{-}sodium, high\text{-}cholesterol\} \rightarrow high\text{-}blood\text{-}pressure$; $\{cancer\text{-}biomarker, big\text{-}lump\text{-}in\text{-}breast\} \rightarrow breast\text{-}cancer$ and so on. There is a difference between association-rule and logical-rule though. Logical-rules implicitly support causality. However, association-rules only support strong correlation.

An association rule is of the form *antecedent* $\rightarrow$ *consequent*. The union (*antecedent* $\bigcup$ *consequent*) forms an association-set. The identification of association rules requires: 1) identification of the optimal number of association sets, 2) pruning of the redundant subsets, 3) splitting of the association-sets using conditional probability of occurrence of the consequent when an antecedent occurs and 4) merging the rules to form minimum number of meaningful rules.

Association rules are used extensively in data-mining. Often, the probability of occurrence of a phenomenon is low, and it is difficult to determine whether two attributes or phenomena are associated, or it is a mere coincidence due to the lack of statistical significance. For example, certain types of cancer are very rare. They are associated uniquely with certain biomarkers. However, higher frequency is needed by traditional algorithms to derive the association rules.

3.8.2.1 Associative rules formation

In the formation of association rules, three properties are very important: *support, confidence-factor,* and *lift-ratio*. *Support* is the fraction of the database tuples that contain the required set of variables. For example, *support*($\{V_i, ..., V_J\}$) is the fraction of the database-tuples that contain the set $\{V_i, ..., V_J\}$. *Confidence-factor* is the ratio of the fraction of tuples containing both antecedent-set $\{V_i, ..., V_J\}$ and consequent-set $\{V_K, ..., V_M\}$ divided by the fraction of tuples containing the antecedents-set $\{V_i, ..., V_J\}$.

Confidence is the conditional probability of consequent occurring when an antecedent occurs. Higher confidence shows higher association between the antecedent and consequent. There needs to be a minimum support threshold denoted as *minsup* for every variable in the antecedent and consequent. Similarly, antecedent-set and consequent-set together should satisfy the minimum *confidence-factor threshold.* If the *minsup* and/or *confidence-threshold* are high, then many useful association rules are pruned out. Conversely, if the threshold values are low, then many weak associations will qualify as association rules. Lower value of *minsup* indicates lower frequency of occurrence of variables, and the association calculated for less frequently occurring variable can be mere coincidence. However, in disease diagnosis, many diseases are rare and the corresponding biomarkers are less frequent.

One problem with the high occurrence of a variable in the database is that it may show a false correlation even when the antecedent and consequent are independent. To verify the independence of antecedent and consequent, a metric called *lift-ratio* is used. *Lift-ratio* is defined as the ratio of *support(antecedent ∪ consequent)* and the product *support(antecedent) × support(consequent).* If the *lift-ratio* is around *1.0*, then the antecedents and consequents are independent. *Lift-ratio* greater than *1.0* shows that the antecedent has increased the probability of the consequent to occur in the same record. A good association rule has *high support, high confidence*, and *lift-ratio > 1.0*. The formal definitions of the three terms, *support, confidence* and *lift ratio* are given in Equations 3.9, 3.10 and 3.11, respectively.

$$support = \frac{frequency(variableSet)}{Total records} \tag{3.12}$$

$$confidence\ factor = \frac{frequency(antecedent \cup consequent)}{frequency(antecedent)} \tag{3.13}$$

$$lift\ ratio = \frac{support(antecedent \cup consequent)}{support(antcedent) \times support(consequent)} \tag{3.14}$$

3.8.2.2 Apriori algorithm

Association rules are derived using many techniques such as *apriori algorithm, Elcat algorithm* and *FP growth algorithm. Apriori algorithm* is a popular algorithm based upon a theorem that says that subsets of a frequent set are also frequent. The algorithm is based upon discovering all the associated set of variables that satisfy the *minsup* cutoff. After identifying the associated set of variables, various subsets of combinations are tried as an *(antecedent → consequent)* rule such that the *confidence-factor* is higher than the minimum threshold, and the *lift-ratio* is > *1.0*.

Apriori algorithm uses three step: 1) progressively identify bigger sets of variables that satisfy *minsup* cutoff, and all its subsets satisfy *minsup* cutoff; 2) removal of all the bigger sets with one or more subsets having support less than the *minsup*; 3) enumeration of various combination of *antecedent → consequent* such that *(antecedent ∪ consequent)* is equal to an association set, and it satisfies the minimum confidence threshold. The bigger sets are derived using a lattice starting from the single variables sets that are progressively joined to form bigger sets as shown in Figure 3.22.

Starting from singleton sets, two variable sets are formed followed by three variable sets. A set is not considered if any of its subset has support less than the *minsup*. The apriori algorithm stops when it cannot form new bigger sets. Multiple rules are joined using two strategies: 1) merging the rules that share the same prefix in the rule consequent and 2) prune the rules if any of its subset does not have high confidence.

Example 3.25: Deriving associative rules using apriori algorithm

Consider Figure 3.22. If the support of subset {*A*} is less than *minsup*, then these bigger subsets {*AB*}, {*AC*}, {*AD*}, {*ABC*}, {*ACD*}, {*ABD*} and {*ABCD*} will not be considered in the formation of the associative rules. For example, given two rules *AB → CD* and *AC → BD*, the merged rule will become *A → BCD*. The newly formed rule also includes *AC → BD*. If the rule *AC → BD* has confidence lower than minimum threshold, then rule *A → BCD* is not formed.

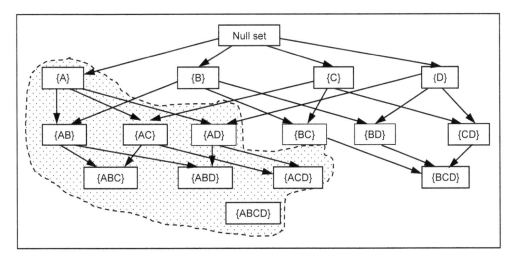

FIGURE 3.22   Use of lattice structure to form bigger subsets

Example 3.26: Data mining a database

Consider the relational Table 3.2. The table shows the data about sodium content and cholesterol content for eight patients. There may be other electrolytes. However, for the sake simplicity, discussion is limited to four variables: *sodium concentration, potassium concentration, cholesterol ratio* and *systolic blood pressure*. Assume that the threshold for high systolic blood pressure is *150*; the threshold for high sodium is *160* nmol/liter; the threshold for high potassium is *5.5* nmol/liter and the threshold for high cholesterol (LDL/HDL ratio) is *6.5*.

Systolic blood pressure occurs during ventricular depolarization at the time of heart contraction to pump the blood in the arteries. The numbers in the table are not the real numbers and have been generated to illustrate the concept. The field *Sod* denotes the sodium-concentration. The field *Pot* denotes the potassium-concentration. The field *Chol* denotes the cholesterol-level. The field *BP* denotes the systolic blood-pressure. The fields *A* to *D* on the right-hand side are the corresponding Boolean values after comparing with the threshold value: the field *A* stores the output of (*value(Sod) > sodium-threshold*); the field *B* stores the output of (*value(Pot) > potassium-threshold*); the field *C* stores the output of "(*value(Chol) > Cholesterol-threshold*); the field *D* stores the output (*value(BP) > BP-threshold*). Bit-value *1* denotes *true*, and the bit-value *0* denotes *false*.

Assume that the value of *minsup* is 70%, and the minimum *confidence-factor* is 85%. The *support* values are: for {*A*}: *71%*, {*B*}: *0%*, C: *86%*, D: *86%*. Only {*A*}, {*C*} and {*D*} meet the threshold criteria. There are three combinations of these subsets: *AC, AD* and *CD*. The *support* for AC is *71% > minsup*. The support for *AD* is *71% > minsup*. The support for *CD* is *86% > minsup*. These can be combined

TABLE 3.2   An example for associative rule mining from relational database

SOD.	POT.	CHOL.	BP	A	B	C	D
140	4.5	6.5	157	0	0	1	1
170	5.0	5.8	160	1	0	1	1
160	4.5	7.2	164	1	0	1	1
200	4.7	6.2	200	1	0	1	1
130	4.9	4.3	120	0	0	0	0
197	5.2	5.5	156	1	0	1	1
220	4.8	6.0	189	1	0	1	1

TABLE 3.3   Associative rule formation

	RULES	CONF	ABOVE THRESHOLD	LIFT	COMMENT
1	AC → D	1.00	Yes	1.16	Good rule. Merged with rule #2 to give A→CD
2	AD → C	1.00	Yes	1.16	Good rule. Merged with rule #1 to give A → CD
3	CD → A	0.83	No	1.16	Not considered
4	A → C	1.00	Yes	1.16	Covered by rule #1 and rule #2
5	A → D	1.00	Yes	1.16	Covered by rule #1 and subsequent merged rule
6	C → D	1.00	Yes	1.16	Good rule
7	C → A	0.83	No	1.16	Not considered
8	D → A	0.83	No	1.16	Not considered
9	D → C	1.00	Yes	1.16	Good rule

to form the set *ACD*. The support for *ACD* is *71% > minsup*. Some possible *antecedent → consequent* combinations, their *confidence-factor*, and *lift-ratio* are given in Table 3.3 for the illustration. The list is not exhaustive. The support for {*B*} is *0*. Hence, all the subsets derivable from the singleton subset {*B*} are left out.

Six rules satisfy the *confidence* and *lift-rule* criteria. The rules are *AC→ D, AD → C, A → C, A→ D, C → D* and *D → C*. The rules *AC → D* and *AD → C* share the same prefix in the antecedent and are merged to give the rule *A → CD*. Rules *A → C* and *A → D* have the same *confidence* as the rule *A → CD* and are already included. Other two rules are *C → D* and *D → C*. Finally, after merging the rules, there are three associative rules: *A → CD; C → D; D → C*.

Translating into the actual pathological terms, the associative rules say: 1) if a person has high sodium concentration, then s(he) has high cholesterol and high blood pressure; 2) if a person has high cholesterol, then s(he) has high blood pressure; 3) if a person has high-blood pressure, then s(he) has high cholesterol.

3.8.3 Temporal Data Mining

Temporal data mining is concerned with deriving interesting patterns and association-rules between time (or time-intervals) and various attributes. Temporal data mining is used to: 1) predict the future values, 2) remove errors or missing values from the past, 3) classify the data into different clusters; 4) correlate temporal behaviors occurring together, 4) data summarization and visualization, 5) patterns of sudden changes and 6) Markov model based upon the transition-patterns of segment values.

3.8.3.1 Identifying meaningful patterns

An *episode* is a larger sequential pattern that has some interesting property and needs to be explored. By arranging the temporal pattern as a subsequence, and progressively increasing the length of the subsequence, episodes can be identified. These patterns are identified using apriori algorithm as explained in Section 3.8.2.2. There is partial ordering among episodes. The episodes that are contained in bigger episodes are ignored.

An episode is relevant within a context. A context may persist for a longer period and may encompass more than one episode. For example, a person with hypothyroid starts losing weight, suffers from arrhythmia and may have liver problems. The analysis of the time-series data has to be done in the context of hypothyroid. Without context, there are many reasons for weight loss, and a meaningful analysis of time-series data is difficult.

3.9 INTEROPERABILITY AND ONTOLOGY

Ontology is defined as the study of beings that means a study of entities, their relationships, and equivalence of the groups of entities and their relationships to other groups of entities. This involves understanding the meaning of the words, discovering relationships between the words and phrases using synonyms, antonyms and class-hierarchies. The equivalences of words and phrases are expressed as a knowledge-base that uses rules and algorithms to traverse up and down the class-hierarchy relationships.

Ontology has major applications in health science in identifying the terms and phrases with equivalent or similar meaning in different contexts. The ontology information has been used in SNOMED CT. The structure-based information expressed in LOINC code is conducive to ontology equivalence. One of the popular models of ontology GALEN (General Architecture of Languages, Encyclopedias and Nomenclatures) connects nursing terminologies, pain terminologies, decision-support knowledge, surgical procedures and anatomy.

Example 3.27: Ontology

Figure 3.23 illustrates ontology using a simple example of a subset of heart diseases. Heart disease is subclassified as 1) *arrhythmia* – irregular heart-beats, 2) *ischemia* and 3) myocardial infarction. Arrhythmia is subclassified as *supraventricular arrhythmia* or *ventricular arrhythmia*. *Supraventricular arrhythmia* is subclassified as *atrial fibrillation* and *atrial flutter*. Myocardial infarction, the death of the muscle cells, is caused by ischemia blockage of blood flow due to deposits in the arteries.

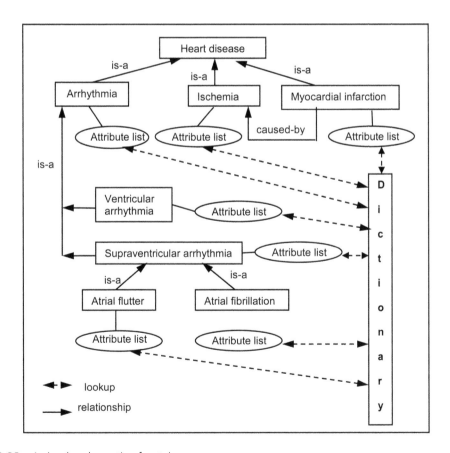

FIGURE 3.23   A simple schematic of ontology

Many medical terms have been used here such as *arrhythmia, ischemia, myocardial, infarction, hypermetropia, supraventricular, ventricular, atrial, fibrillation* and *flutter*. Each word has a meaning and attributes associated with it. Some of these words are synonymous such as *supraventricular* and *atrial*. Some words may be used differently in different contexts. We need a semantic dictionary. Some words have similar meaning needing a synonyms-dictionary.

In Figure 3.23, there are two relationships: *is-a* and *caused-by*. This helps in identifying similar diseases and cause of diseases. For example, a doctor may tell a patient's authorized kin "the patient has a heart-disease," while he may tell an attending nurse "the patient has arrhythmia." Since "arrhythmia" is-a subclass of "heart-disease," the sentence "the patient has a heart-disease" subsumes "the patient has arrhythmia" and the second sentence is medically more accurate.

Similarly, if a doctor determines that a patient has myocardial infarction, then it is understood that patient went through ischemia. The words "supraventricular" and "atrial" are synonymous and can be looked up the synonyms dictionary. The meaning of the medical words "arrhythmia," "atrial," "ventricular," "ischemia," "myocardial" and "infarction" can be looked up in the semantic dictionary.

3.9.1 Entities and Relationships

The basic unit of ontology is the representation of an *entity*. An *entity* could be a meaningful word, a meaningful string or a phrase. Multiple equivalent words and groups are put in an equivalence group. These groups of identical or very similar words or phrases map to a *concept*. A medical-dictionary of concepts exists in a knowledge-base. There are three types of ontological information:

1. *Semantic networks* consisting of multiple semantic types connected through relationships in a hierarchy to form a hierarchical graph. A hierarchical graph is a layered graph where different layers are graphs and are connected through different properties such as inclusion (subclass relationship) or some form of summarization of lower layers into upper layers. The edges represent type of relations between the semantic types. There are multiple types of relationships such as *has-location, is-part-of, is-a* and *includes*.
2. Assignment of semantic types to *concepts*. Each *concept* is an instance of at least one semantic type. However, a concept may be associated with more than one semantic type.
3. *Instances of relations* between actual concepts.

Example 3.28: Semantic-type

For example, we can describe *a semantic-type* as *heart disease*. Multiple concepts are associated with *heart-disease* such as *arrhythmia and ischemia* that are related to other concepts through multiple relations. Another example of a semantic type is *inflammation*. *Inflammation* could occur at multiple locations making it more specific.

3.9.1.1 Unified medical language system

UMLS (Unified Medical Language System) is a data warehouse of medical terms and relationships that contains three types of information: 1) *meta-thesaurus*; 2) *semantic network* and 3) *specialist lexicon*. The meta-thesaurus contains two million biomedical and clinical concepts from 100 different terminologies from various sources such as SNOMED CT. The concepts are connected through relations to form a semantic network. The relationships connect two concepts and may be hierarchical. There are two types of hierarchical relationships: *parent–child relationship* and *broader/narrower relationship*. The concepts are grouped into *semantic-types*. The semantic-types are grouped in *semantic-groups*. Some examples of semantic groups are *anatomy, phenomenon, disorders, chemicals* and *drugs*. Currently, there are 15 such semantic-groups. The matching process uses *subsumption* and *similarity-based approximate matching* to match the concepts. Subsumption means that a minor category is included into a more general category.

3.9.1.2 MedDRA and ontology

Modeling of ontology in MedDRA uses four types of concepts: 1) type of disorders, 2) physiological functions, 3) pathogenic agents that cause diseases and 4) physical structure levels in the body. The concepts could be primitive that cannot be split further, or complex that include other concepts. Complex concepts are built using the generalization of simple concepts or by combining concepts using logical operators. The grouping of individual relationships is called *roles*. Two types of roles are used: 1) relations connecting primitive concepts and 2) relations defining properties of the concepts.

Semantic relatedness is subsumed by association of two concepts or relations to each other. For example, *inflammation* is associated with *ibuprofen*. Semantic relatedness includes *is-a* relationship and similarity based upon some notion of distance.

3.9.2 Similarity Measures

The semantic similarity is classified into two categories: 1) path-based and 2) content-based. Path-based similarity depends upon the weighted distance of the two entities. The links are weighted because the links between the entities do not always represent the same distance. There are multiple approaches to compute path-based similarity. Content-based similarity is based upon the quantitative measurement of the properties of the entities.

3.9.2.1 Path-based similarity

The first approach is given in Equation 3.15. In this approach, similarity is defined as the reciprocal of the *shortest_weighted_distance* between the two nodes N_1 and N_2 in the semantic network:

$$similarity(N_1, N_2) = \frac{1}{shortest\_weighted\_distance(N_1, N_2)} \tag{3.15}$$

The second definition uses the common ancestor of the two nodes and the depth of the nodes as illustrated in Equation 3.16. For a hierarchical tree with a single root, similarity is defined as the ratio of the depth of the first common ancestor and the average depth of the two nodes N_1 and N_2. This means that the nodes N_1 and N_2 are less similar if the first_common_ancestor is further away from the nodes N_1 and N_2.

$$similarity(N_1, N_2) = \frac{depth(first\_common\_ancestor(N_1, N_2))}{(depth(N_1) + depth(N_2) / 2)} \tag{3.16}$$

The third definition uses the shortest path between the two nodes and the depth information. The similarity is defined as the negative log of the ratio of the shortest-path between the two nodes and twice the total depth as illustrated in Equation 3.17.

$$similarity(N_1, N_2) = -log\left(\frac{shortest\_weighted\_distance(N_1, N_2)}{2 \times total\_depth} \right) \tag{3.17}$$

Example 3.29: Path-based similarity

Let us consider the semantic network in Figure 3.23 to heart-related diseases. The shortest weighted distance between *atrial flutter* and *atrial fibrillation* is *2* based upon the shortest path: *atrial flutter* → *supraventricular arrhythmia* → *atrial fibrillation*. The distance from each disease to the first common ancestor *supraventricular arrhythmia* is *1*. The depth of the node *supraventricular arrhythmia* from the root node *heart disease* is *2*. The maximum depth is *3*. Using Equation 3.15, the similarity between *atrial-flutter* and *atrial fibrillation* is *0.5*. Using Equation 3.16, the similarity between *atrial-flutter* and *atrial fibrillation is* $\frac{2}{(3+3)/2}$ = *0.67*. Using Equation 3.17, the similarity between *atrial-flutter* and *atrial fibrillation* is $\left(-log\left(\frac{2}{2 \times 3}\right)\right)$ = $-log\ 0.33$ = *0.39*.

3.9.2.2 Content-based similarity

Given a term T in a corpus of terms making a concept, the information content is described as: $-log(p(T))$ where $p(T)$ is the probability of the term occurring in a text corpus. For the occurrence in text corpus, all the synonyms, all its taxonomic descendants and their synonyms are included in the probability $p(T)$. Given the information content of two nodes T_1 and T_2, the difference of the information content describes the similarity (or dissimilarity) between them. The difference between the information content is given by the factor Another definition of content-based similarity is based upon the information content of the first common ancestor because the information content of the first common ancestor is present in both the terms T_1 and T_2. The information content $c(T_1, T_2)$ is defined by Equation 3.18.

$$c(T_1, T_2) = -log\Big(p\big(first\_common\_ancestor(T_1, T_2)\big)\Big)$$ (3.18)

3.9.3 Ontology Formation

The building of the ontology is based upon: 1) identifying the primitive concepts based upon the similarity of the meaning of the terms in the group, 2) hierarchically connecting the primitive concepts to complex concepts and 3) connecting the concepts using relational networks to form a semantic-network. The connectivity may use "is-a" relationship or logical operators. The first step requires matching of the meaning of the terms based upon a medical dictionary. There are automated tools to facilitate building of such dictionaries. The meanings of many terms are contextually defined: the same term may have different meanings in two different concepts.

Ontology is built using knowledge-based semantic primitives. There are three types of knowledge-based primitives: 1) linguistic meaning, 2) formal mathematical description of the linguistic meaning and 3) computational action to be taken to map the mathematical description to linguistic meaning. The methodology of building uses four steps: 1) building the corpus of medical terms, 2) semantic normalization of the terms in the corpus, 3) formalization of the meaning of the knowledge primitives and 4) the use of knowledge representation languages to make the domain operational.

3.9.3.1 Corpus building

Corpus building requires the use of NLP (natural language processing) tools that automatically analyze the meaning of the medical terms by analyzing the medical texts and building the context using statistical analysis and language dictionary. While analyzing the medical text, the same medical domain is maintained. Some examples of medical domains are: 1) intensive care unit; 2) trauma center; 3) cardiovascular surgery; 4) neurosurgery; 5) obstetrics; 6) diabetic management, etc.

3.9.3.2 Semantic normalization

Semantic normalization of the terms ascertains the consistency of the use of the terms every time, they are used in the medical text. This requires statistical analysis for validation, formalization of the meaning of the terms using some consistent semantic theory that associates distance-based similarity and difference between the meanings of various terms in the corpus. There are four principles to ascertain the formal meaning: 1) similarity with the parent, 2) similarity with the siblings, 3) difference from the parents and 4) differences from the siblings.

3.9.3.3 Concept formation

Each concept refers to a set of terms with well-defined meaning. New concepts are derived using: 1) set theoretic operations: intersection, union and complement; 2) use of logical operators to make complex concepts and 3) the use of inclusion to set up the hierarchy.

3.10 INTELLIGENT INTERFACING TECHNIQUES

3.10.1 Natural Language Understanding

Natural language analysis and understanding play a major role in health informatics both for archiving disease-related information and communicating the information in a human-comprehensible form. In traditional systems that use human interpreters, natural language is used as a means of daily bedside reports of patients, doctor's dictation describing patients' condition and diseases, interpretation of radiology reports, reporting and interpretation of lab-tests. With the availability of integrated databases used for fast information archival, information retrieval and data analysis from the computer, there is a need to interface natural language to structured-data used in a health informatics database.

The use of natural language is more conducive for human comprehension. In contrast, structured data for computer representation keeps changing with different versions of databases and knowledge-bases and is too detailed for human memorization. However, structured data representation is suitable for computer-based information processing and fast data retrieval. This requires natural language to structured-data conversion and vice versa.

The speech or text to structured data conversion requires: 1) speech-to-text conversion; 2) written text to computer understandable format conversion; 3) interactive text recognition; 4) lexical analyzer and natural language parser to verify grammatical correctness of the sentences and phrases; 5) medical phrase and word extractor with the help of context and medical domain-specific dictionary. *Word-sense disambiguation* is done to remove any discrepancy in the meanings of the words and phrases. Medical phrases are converted to structured-data using ontology information to match the medical-data to-be-archived and archived medical data to avoid any duplication. Once the best match is identified, and the similarity is used within the specified threshold, and the archived data is modified to reflect minor adjustments. If the similarity is not within the specified threshold, then the medical phrase contains new information, and it needs to be processed to represent into structured format and archived.

The generation of natural language text requires standardized code dictionaries such as ICD-10, SNOMED CT and LOINC that are looked up to derive human comprehensible names of the archived codes. After looking up the codes, a natural language sentence generator is used to generate human comprehensible text.

3.10.1.1 Characteristics of speech waveforms

Speech waveform is a superimposition of many fundamental frequencies and overtones mixed in different proportions to form a *phoneme*. Multiple phonemes make a word. Words are separated by a small silence. Two sentences are separated by silence. Using the silence as a delimiter, and having a knowledge base of phonemes' features with the features extracted from the speech between two silence period extracts sequence of phonemes that can be looked up in a phoneme-to-word dictionary to derive the words. The features used in the speech recognition are: 1) set of base frequencies; 2) overtones; 3) amount of energy and 4) sequence of slopes of the voice envelopes in a phoneme. Phonemes are divided into different classes that have different time-segments and frequency variations. Six factors affect the spoken speech: 1) Phonetic identity, 2) pitch, 3) speaker style, 4) microphone, 5) environmental noise and room acoustics and 6) emotion-level.

3.10.1.2 Speech recognition techniques

A speech signal is divided into frames of 20–30 ms to analyze the speech. The speech is preprocessed by reconstructing it to the natural form from the transmitted form and removing the noise. The speech is then converted to frequency domain using Fast Fourier Transform (FFT). After the frequency domain is identified, then MFCC (Mel Frequency Cepstral Coefficient) is derived for each frame. MFCC is calculated using Equation 3.19.

$$MFC(frequency) = 2595 log_{10}\left(1 + \frac{frequency}{700}\right) \tag{3.19}$$

The sequence of MFCC for a phoneme is collected. This MFCC can be analyzed either using dynamic programming (see Section 2.1.7) to match the MFCC of the archived word to identify the most similar word. Another approach is to use HMM to derive a possible set of matching words. Each phoneme is statistically analyzed by studying a large domain of related medical text to derive the HMM for different relevant medical words. A word becomes a cascade of HMM. The sequence of MFCC is matched with the archived HMM to derive the most probable phonemes joined to derive the word.

Handling ambiguities in a natural language sentence is a major problem. There are multiple sources of ambiguities: 1) domain; 2) context; 3) use of improper sentence structure; 4) multiple meanings of the words; 5) incorrect spellings. The domain and the context should be established to understand the meaning of a sentence. The sentences are parsed statistically. Different rules used for parsing are applied probabilistically to derive parse trees. The advantage of statistical parsing is that as word-sense is disambiguated, the most probable parse-tree is easily identified.

3.10.1.3 Word-sense disambiguation

One of the major problems in natural language understanding is to derive the meaning of the words exactly as the same word may have different meanings based upon the domain and the context. There are many ambiguities present in using a word such as: 1) different part-of-speech; 2) different meaning; 3) *metonymy* – the use of a word for a concept. An example of metonymy is "The patient was treated by the Hospital H." Here the word "hospital" does not mean the building structure but doctors in the hospital H.

Word-sense can be disambiguated by: 1) logical association of one of the meanings with the meanings of the other words in a specific context; 2) connecting the implied knowledge from previous discourse to the next sentence using some logical or temporal relationship; 3) knowing the domain and the context. Word-sense disambiguation can also be derived by associating concepts with the clusters of words, and identifying the clusters in a sentence or a sequence of statements. For example, a cluster of words {*lung, coughing, blood, chest-pain*} in a text segment refers to the association with the concept "tuberculosis."

The set of associations is derived by analyzing multiple texts in the same context and domain to identify common meaningful words and their frequencies occurring together within vicinity and occurring separately. *Association-factor af* for any two words is defined as the ratio of the frequency of the words occurring in the same context and product of the frequency of the words in the related texts (see Equation 3.20). The maximum value of an *association-factor* is *1.0* when every time the two words occur together in the vicinity within the same sentence or same paragraph of the related text.

$$af(\{word_1, word_2\}) = \frac{frequency(\{word_1, word_2\})}{frequency(word_1) \times frequency(word_2)} \tag{3.20}$$

The concept of the *association-factor* is extended further to *n-grams* – *n* words occurring together in the vicinity. The *n-grams* are associated with a concept, disease or an abnormality with certain probability derived using statistical analysis of medical texts within the same domain and context. The association-factor for n-gram is given by Equation 3.21, where "Π" denotes the product of the terms.

$$af(\{word_1, \ldots, word_N\}) = \frac{frequency(\{word_1, \ldots, word_N\})}{\prod_{i=1}^{i=N} frequency(word_i)} \tag{3.21}$$

Example 3.30: Association-factor of n-grams

Consider the following definition of arrhythmia: "Arrhythmia is a disease of *unregulated refractory beats* that is caused by the presence of *ectopic-nodes* and the lack of periodic *sinus beats*. Arrhythmia is further divided into *supraventricular arrhythmia* and *ventricular arrhythmia*. *Supraventricular arrhythmia* is caused by the presence of *ectopic-nodes* in *atria*, and *ventricular arrhythmia* is caused by the presence of *ectopic-nodes* in the *ventricles*."

This text contains repeated occurrences of set of nontrivial words {*beats, ectopic-nodes, supraventricular, ventricular*}. The disease arrhythmia can be associated with this set. Under the assumption, that all the words in the text are within proximity threshold, the *association-factor* of word "ectopic-nodes" with word "supraventricular" is $(3 + 2)/(3 \times 2) = 0.83$. The association-factor of the word "supraventricular" with "ventricular" is $(2 + 2)/(2 \times 2) = 1.0$. The *association-factor* of all four words in the set is given as $(2 + 3 + 2 + 2)(2 * 3 * 2 * 2) = 9/24 = 0.38$.

3.10.1.4 Multilingual translation

Movement of people is beyond the lingual borders. Although, most of the medical information in the medical databases is stored in English, the information, medical notes and dictations stored in medical databases in other languages are common. A query may be asked in one language to retrieve information from a database containing reports and notes in another language. To retrieve complete medical information, search medical literature in other languages and process them, there is a need of machine translation of the information written in one language into another language.

A popular scheme of machine translation is SMT (<u>S</u>tatistical <u>M</u>achine <u>T</u>ranslation). SMT is a phrase-based translation scheme based upon splitting a sentence into multiple phrases. A phrase is defined as a collection of words occurring together. The words in these phrases are translated into another language having same domain and context, and then shuffled to fit the language grammar of the target language. There are multiple ways to split a sentence into phrases. All possible ways of splitting a sentence into phrases can be modeled as a graph. The overall approach is to prune the search space in such a way that the translated text has maximum probability of preserving the meaning of the sentence. This requires maximization of the overall probability of translating the source sentence to target sentence given by Equation 3.22.

$$probability = max\left(\sum_{i=1}^{i=N} weight_i \times log(feature\_function(target, source))\right) \tag{3.22}$$

The feature-functions include probabilistic relation of source phrases to corresponding target phrases, the reordering model to capture different phrase order in two languages and *word penalty* to penalize the translations that have a length mismatch. The model is trained using a large lingual model, and the weight is derived by minimizing the error that gives the corresponding target phrase. The words may be *simple* or *compound*. A *compound word* is made of many *simple words*. However, different languages have different approaches to combine the words.

Besides the adaptation of specific domain, the analysis of text should also consider matching the *genre* of the text. Some examples of a genre are radiology report, bedside chart, doctors' notes, postsurgical report, lab report, text messages between the health providers and patients or between two health providers. Each genre has a specific writing style, level of short cuts and abbreviations used to express the same information and spelling mistakes. For example, personal text messages are not edited for spelling mistakes.

3.10.2 Intelligent Text-Extraction

Natural language understanding systems are required in: 1) CDS (Clinical Decision System) – software-based semi-automated decision-making system); 2) medical literature analysis for drug side-effects and dosages and developing domain-specific vocabulary; 3) medication administration errors; 4) clinical note analysis to extract disease, dosage period and duration of various medical conditions and 5) radiology report analysis to derive information from experts' notes on radiology images.

The CDS software is triggered by recognizing the key-words and phrases used in a specific procedure. For example, a CDS may analyze various test reports of a patient stored in the database and find out whether a person is suspected of have a chronic disease such as tuberculosis, cancer and diabetes. Another example is the extraction of the temporal information such as the duration of a treatment, administration of the medication, encounter time between the patient and the healthcare providers. Word-sense disambiguation is challenging in physicians' notes due to ill-formed sentences, errors in a character-recognition system and ambiguous or context-sensitive meaning of a large number of medical terms and English words.

Ontology information is needed in natural language understanding system and clinical data extraction for sharing domain-specific information, cross-disciplinary exchange of information, clinical diagnostic modeling system and derivation of relationship between disease-related terms. Ontology can be developed manually. However, manual development of ontology is labor-intensive and cannot be easily updated in the future. Another approach of the ontology development is based upon domain-specific text analysis.

The automated text analysis systems include many components: 1) lexical analysis and identification of sentence boundaries; 2) probabilistic parsing; 3) probabilistic part-of-speech tagger; 4) negation detection and 5) entity and relationship detection using dictionary-lookup. Probabilistic parsing is used to generate many probable parse-trees using probabilistic rules. The words are looked up in the domain-specific entity-class dictionary, generate all possible synonyms and generate best combination of words that matches a concept in the SNOMED CT database. Negation identification is an important factor and can be identified using lexical analysis often. Since, the number of concepts is very large, an indexing scheme is used to match the concepts.

3.10.2.1 Latent semantic analysis

Latent Semantic Analysis is a probabilistic technique to associate a meaning to a document by deriving the probabilities of occurrences of medical-terms within the document. Given a knowledge-base like ICD-10 and a corpus, the probabilities of occurrences of medical-terms are derived using following steps:

1. Common words and punctuations are removed from the documents; only meaningful words are retained in the documents.
2. A two dimensional-table TDM (Term Document Matrix) of *terms* and *documents* is created where rows represent the medical-terms present in the knowledge-base and the document, and columns represent the document-label. Each cell of the table is indexed by (*term, document-label*), and gives the frequency of a term in the document. In identifying the terms, approximate string matching and a synonyms' dictionary are used to take care of typographical errors

and synonyms. Each document is an M-dimensional vector where M is the number of identified medical-terms in the corpus.

3. TDM is transformed to a weighted matrix based upon two concepts: 1) more weight is allocated to the terms occurring locally in the same document and 2) less weight is allocated to the terms that occur globally in multiple documents using a technique called *TF-IDF* (<u>T</u>erm <u>F</u>requency <u>I</u>nverse <u>D</u>ocument <u>F</u>requency) as explained in Example 3.31. TF-IDF weight is calculated by Equation 3.23.

$$term\_weight =$$

$$\frac{term\_frequency\ in\ the\ document}{word\_count\ in\ the\ document} \times log\left(\frac{documents\ in\ the\ corpus}{documents\ containing\ the\ term}\right) \qquad (3.23)$$

4. The resulting weight-matrix is then transformed using SVD (<u>S</u>ingular <u>V</u>alue <u>D</u>ecomposition). SVD is a technique to transform a matrix into a form $X = U\Sigma V^T$, where U and V^T are orthogonal matrix, and Σ is a nonnegative diagonal matrix of singular values. The property of an orthogonal matrix is that all the columns are orthonormal, and $A \times A^T$ is an identity matrix. Two vectors are orthonormal if they are orthogonal and unit-vectors. The singular values in the diagonal are arranged in the descending order, and the top K rows ($K \ll$ *number of terms*) are selected for reducing the total dimension, because the lower singular values are noisy. This K-dimensional vector space is called *Latent Semantic Space*. It is represented as $\hat{X} = \hat{U}\hat{\Sigma}\hat{V}^T$, where $\hat{U}$ is of the form $[U_1, ..., U_k]$, $\hat{\Sigma}$ is the $K \times K$ singular matrix, and $\hat{V}$ is of the form $[V_1, ..., V_k]$.

5. A document D is projected in this K-dimensional *latent semantic space* as $\hat{D} = \hat{\Sigma}^{-1}\hat{U}^T D$, where D is the original document-vector. Similarity between two documents D_1 and D_2 is measured using cosine similarity that is computed as $inner\text{-}product(\hat{D}_1, \hat{D}_2)$.

Example 3.31: Latent semantic analysis

Consider a corpus of 10,000 medical reports related to heart diseases. The domain is heart diseases. The knowledge-base used is ICD-10. All the medical terms related to heart diseases are candidates for the corresponding *TDM*. Consider one document that has the word *ischemia* 15 times. The word count after removing punctuations and common words in the document is 1564. The term *ischemia* occurs in 600 documents. The weight for the term *ischemia* in the weighted-matrix is $\frac{15}{1564} \times log\left(\frac{10,000}{600}\right) = 0.01 \times 1.22 = 0.012$.

3.10.2.2 Topic modeling

A *topic* contains a set of key-words having a specific frequency pattern. However, a document contains only words, phrases and sentences visually. *Topics* are latent (hidden) variables and have to be identified. Identification of distribution pattern of various topics is important for understanding the text and identifying the domain of the textual report.

A *document* can be understood as a fuzzy cluster of *topics* where each *topic* is a soft-cluster of key-words with certain probability. A document can also be modeled as a graph such that topics are the vertices, and relationships between them are edges. Using the subgraphs of topics, a document can be analyzed for the presence of complex themes. Two documents can be matched to each other for similarity by matching the resulting graph.

The documents can also be analyzed to derive the level of co-occurrence of various topics in the documents. After identifying the topic distribution and co-occurrence, it can be applied for information retrieval, document similarity and document exploration.

Topic modeling is a probabilistic unsupervised machine learning technique for discovering the topics in the document and annotates the document-parts with different topics. Topic model analysis: 1) generates the dictionary of key-words associated with a topic in various documents; 2) identifies

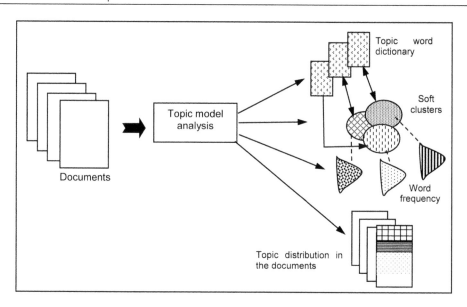

FIGURE 3.24　A schematic of topic model analysis

the topics discussed in a document; 3) identifies the frequency distribution of the key-words in a topic and 4) identifies a cohesive topic distribution within the documents. Mathematically speaking, topic model analysis computes the posterior distribution $p(z, \theta, \beta \mid w)$, where z is the key-word association with the topic, θ is the topic proportion per document and β is the topic distribution in a corpus. There are many techniques to calculate the posterior distribution. The overall scheme is illustrated in Figure 3.24.

Derivation of the clusters is based upon: 1) counting the frequency of interesting words such as domain-specific medical-terms and 2) grouping the words using a machine learning technique. One popular technique for grouping the words is *Latent Dirichlet Allocation* (LDA). *LDA algorithm* has two goals in mind: 1) optimizing the allocation of words into topics based upon maximum likelihood and 2) allocation of topics to the documents. It works iteratively by initially assigning the words randomly to K clusters, and then progressively resampling the words and putting them into another cluster such that the overall probability is maximized. The process is iterative.

3.10.3　Automated Text Summarization

As the automation is increasing, more data is being generated. Patient-related information is embedded in databases and clinical notes. Clinical notes include bedside monitored physiological data such as ECG and oximetry data, pathology reports, radiology reports, surgery reports and daily bedside nursing reports. These reports need to be summarized for the ease of archival and quick review of the patients' disease conditions, diagnostics, treatment and recovery. Clinical reports and records include text, numeric values, narratives, diagrams, images, patient history, temporal information and relational information. The information to be extracted includes: 1) administrative information such as name, insurance and account; 2) background, including social background, substance abuse, genetics background, illness record and drug side-effects; 3) timeline involving patient–doctor encounter, hospitalization, surgery and other procedures; 4) illness record, including symptoms, diagnosis and medication record, including dosage.

The summarization requires: 1) visualization such as pie charts for the ease of comprehension in limited space; 2) time-series data to textual description and 3) concise representation of textual data. A schematic of the overall architecture for summarization is given in Figure 3.25.

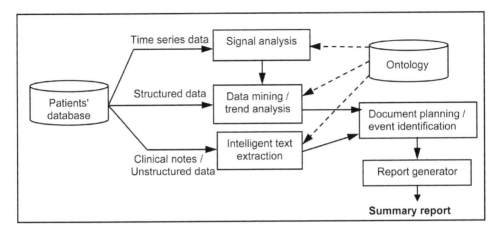

FIGURE 3.25 An overall scheme for summarization

This scheme has five major components: 1) analysis of time-series data such as ECG, oximetry data, 2) intelligent text-extraction system from clinical notes, 3) data-to-text conversion from a database, 4) document planning event identification system and 5) report generation system. In these stages, there is a close association with ontology to derive various concepts and relationship between concepts. Time-series data analysis system finds various features, identifies the patterns and derives the trends.

Text-extraction system uses a natural language understanding system and ontology information to derive various concepts and relationships, and temporal information from the clinical notes and the output of the signal analysis system. Data-to-text conversion system finds the associative rules and patterns in the database and converts them into the textual form. Event analysis system uses temporal information from previous stages and arranges the document into a tree of linked events. The report generation system takes the tree input and generates a user readable report.

3.10.4 Event Extraction from Clinical Text

Clinical notes from the healthcare providers carry important information about the occurrence of the events. These events are very important to build patients' history and perform comparative analysis. There are four types of clinically relevant events: 1) clinical concepts, 2) clinical departments, 3) evidential and 4) occurrences. *Clinical concepts* are defined as the problems, tests performed and treatments. *Clinical departments* are defined as surgery and outpatient. *Evidential* is defined as the source of information that indicates the onset of a problem such as patient's-complaint and appearance of visual symptoms. *Occurrences* are defined as the events that happen to a patient such as "admission" and "follow-up", etc. Each event is characterized by three attributes: 1) type; 2) polarity (positive or negative event) and 3) modality. *Modality* can be "actual occurrence," "conditional occurrence," or "possibility."

Multiple events are related using temporal relationships as described in Sections 2.7.1 and 2.7.2. In the textual form, these temporal relationships can be expressed as "before," "after," "during," "at the same time," "simultaneously," "overlapped," "concurrently," "prior to" *<event 2>*, "post"*<event 1>*, etc.

Events are defined by time-expressions that include a specific time, duration, age, frequency and relevant verbs and nouns. Specific time can be identified by day, date, month and year. Relative time is identified by the words such as "before," "after," "prior," "later," "earlier," "post," "ago," "following," "next," and "beyond." Duration is identified by a number followed by time-unit such as "minutes," "days," and"months," "years." It could also be preceded by words such as "for," "over," "lasting" and "within."

A text is analyzed to identify such temporal words to identify the events. After identifying the anchor points of events, text is analyzed for any additional relationships between the two events in textual proximity.

3.10.5 Case Studies

There are many popular text analysis and extraction systems such as: 1) cTakes (Clinical Text Analysis and Knowledge Extraction System) developed by Mayo Clinic to analyze clinical notes; 2) MedLEE (Medical Language Extraction and Encoding System) developed at Columbia University to process radiology reports; 3) HITex (Health Information Text Extraction System) – a natural language understanding system developed at Brigham Women's Hospital and Harvard Medical School for integrating biology and bedside reports; 4) BioTeks and MedKAT (Medical Knowledge Analysis Tool) developed by IBM to analyze pathology and medical notes; and 5) SymText and Mplus to analyze lung reports to detect pneumonia. In all these systems, there is one common factor. They are domain-specific to derive the meaningful clinical words. In this section, we discuss cTakes and MedKAT as case studies.

3.10.5.1 cTakes system

cTakes system analyzes clinical notes to derive diseases, disorders, signs of diseases and abnormalities, symptoms, procedures, anatomical sites, medications and other semantically viable information to support the clinical research domain. It uses dictionary mapping to update the information and to retrieve and match the information from dictionaries. It is a modular system of pipelined components: 1) sentence boundary detector, 2) tokenizer, 3) normalizer, 4) part-of-speech tagger, 5) shallow parser, 6) negation annotator and 7) entity recognition.

Sentence boundary detector looks for sentence terminators such as periods, question mark or exclamation mark. *Tokenizer* first splits the sentence by using various punctuation marks such as space, colon, semicolon, and then creates a token stream where each token is an internal representation. *Normalizer* annotates each word with their lexical properties such as case, inflection, spelling variants and punctuations. *Part-of-speech tagger* identifies the part of speech using NLP grammar and data dictionaries. The *entity recognition module* uses a dictionary lookup to derive the associated entity and the corresponding concepts. The dictionary also uses the synonyms for similar words and antonyms for words with negation near them. Each discovered entity has the corresponding semantic type and the corresponding attributes that are domain specific.

3.10.5.2 MedKAT system

MedKAT is an NLP understanding system that extracts tumor-related information from semi-structured and unstructured data embedded in pathology reports. The retrieved information includes lymph node status, anatomical site, type of the cancer, and histology – microscopic tissue study. MetKAT has four major components: 1) document ingestion, 2) natural language processing system, 3) concept finding and 4) relation finding.

Document ingestion system identifies sections, subsections, headers and correlations between disjoint texts describing the same tissue. *Natural language processing system* determines the sentence boundary, annotates the medical sentences, tags words with the corresponding parts-of-speech and identifies the contexts. *Concept finder* maps textual sentences to the corresponding concepts. Concept finding is a two-step process: 1) concept-mapper maps a textual phrase to all probable concepts in the ontology and 2) filters. The derived concepts are progressively filtered by many filters such as anatomical site and histology. *Relation finder* discovers the relationships between two classes after identifying the leaf-classes in concept-finder stage. Leaf-classes are the basic classes in the concept-tree derived during concept-finder stage.

3.11 SUMMARY

Many fundamental artificial intelligent techniques are used in the analysis of medical data and data mining such as heuristic search, meta heuristics, machine learning, probabilistic reasoning, including Markov process and HMM (Hidden Markov Model), text analysis, deduction and induction and fuzzy modeling. These techniques are regularly used in analyzing pathological data, patient monitoring data, radiology images and clinical text to diagnose diseases, and drug-related information from clinical notes.

A knowledge-base is a set of cause-and-effect rules that governs a very large set of samples. Two operations are involved with knowledge-bases: 1) induction – forming the rules by generalizing and identifying common patterns from a large sample space and 2) deduction – given a knowledge-base and information (or query) to derive new derived facts that would answer a query.

There are two ways to apply the rules: *forward reasoning* and *backward reasoning.* In forward reasoning, given the preconditions, the corresponding postcondition is derived, and then it is applied iteratively to all the rules until new facts cannot be derived any more. In backward reasoning, a complex query is progressively broken into a group of logically connected simpler subqueries until a subquery can be looked up in the database or can be executed as simple kernel function. The forward reasoning system is computationally expensive and exhaustive and is good for monitoring. Backward reasoning prunes the search space and gives the first solution until prompted to give more solutions.

Fuzzy reasoning is based upon dividing the value range in limited number of classes and performs approximate reasoning on these classes of values. The advantage of fuzzy reasoning is that it is close to human perception of real-world values and improves the computational efficiency.

Uncertainty-based reasoning associates a certainty-factor with every rule instead of Boolean values to model what happens in the real world. Uncertainty-based reasoning is more realistic because all the parameters of an event are not known; some may be hidden. In addition, due to the limited amount of evidence and many-to-many mappings between the internal states and evidence, the exact cause of the evidence is unknown.

The major class of artificial intelligence techniques used in computational health informatics are different machine learning techniques such as cluster analysis, regression analysis, decision trees for classification, artificial neural networks, support vector machines, Bayesian decision network and associative rule learning. Machine learning also uses statistical techniques to learn the patterns. We have also discussed ontology to handle equivalence between multiple ways of representing the same concept based upon the type of communication and the clinical domain. Probabilistic reasoning over time is important to model the progression of disease over a period, effects of medications, ECG analysis, diagnosis of various diseases, text analysis and speech recognition.

Machine learning can be of many types: supervised, reinforced and unsupervised. Learning helps in identification and classification of abnormalities; derive the patterns in the change of vital signs and biomarkers; discovery of correlation and association between multiple disease and related markers. Learning could be: 1) memorizing and indexing a partial path from the initial state to the goal-state; 2) fast mapping of the feature-vectors of the given objects to a class label; 3) associating different actions or attributes in a highly correlated database using pattern analysis. The goal of learning is to improve the time-efficiency and accuracy of doing a task. For the classification problems, it is important to identify the features that can discriminate between different classes of objects.

Metaheuristics or nature-inspired heuristics is based upon iterative applications of two processes: 1) random exploration also called *divergence* and 2) focused search for faster exploration providing good possibilities of solution also called *convergence.*

Cluster analysis is an unsupervised learning technique. It identifies the class of objects that are close together based upon some notion of distance. The distance could be Manhattan distance, Euclidean distance or Minkowski's distance. The distance could be weighted in the sense that certain dimensions are given more weight than others based upon their importance in a classification. There are many techniques for clustering. Some popular ones are K-means, spanning-tree–based clustering, density-based clustering, hierarchical clustering and fuzzy clustering. The points not lying in any cluster are called *outliers*. After the clusters are labeled based upon relevant features, new data-items are mapped to a class based upon the values of the features. Time-series data are clustered using many techniques. One popular technique is to sample the time-series data at equal intervals find the difference between the cumulative slopes in the corresponding intervals and use it as the distance.

The major problem in the identifying clusters is the computational-time overhead and memory overhead. Larger number of features would increase the dimensions during cluster analysis and increase the computational overhead. Less important dimensions are removed to reduce the computational-time and memory overhead. The major techniques are principal component analysis, linear discriminant analysis and independent component analysis. Principal component analysis is based upon transforming the axis such that the data-point distribution has maximum variance along the transformed axis. Linear discriminant analysis is based upon separating the cluster of points such that the distance between the points within the cluster is minimized, and the cumulative distance between the centroids of the clusters is maximized.

Decision trees are used to classify the data set into multiple classes using comparison of the values of different features. Those features that separate the sample space into different mutually exclusive subclasses are preferred. To derive the most important feature, entropy change is used when a new feature is used. The information change is derived as the difference of entropy after a feature-based comparison and the entropy before the use of a feature. The feature that gives the maximum information change is picked up first in the decision tree.

Artificial neural network is a multilayered connection of perceptrons to solve a complex problem. A perceptron is a signal-processing center that accepts multiple weighted inputs from other perceptrons, and fires "1" to the outgoing connections when the sum of the weighted inputs is greater than the threshold value. A negative bias is also added to a perceptron to cancel any change in the output value due to random noise. The perceptrons in a layer are connected to the perceptrons in the next consecutive layer using weighted edges, and whole ANN is modeled as a directed acyclic graph. ANNs are used to predict the outcome for a set of inputs. In the first phase, a set of samples with known outcomes are selected, and the weights are adjusted continuously using a back-propagation of error between the actual value and the generated output value. The process is repeated until the error is minimal below a threshold.

Given a set of data-points, a curve is fitted to connect the points to model a function $y = f(X_1, ..., X_N)$ such that the cumulative error between the actual values and the estimated values is minimum. This fitted curve is used to predict the value of the dependent variable y for a new tuple of the values $(x_1 \in X_1, ..., x_N \in X_N)$ of the independent variables $X_1, ..., X_N$. If the fitted curve is linear, then the regression analysis is called *linear regression analysis* and is modeled as $y = C_1X_1 + C_2X_2 + ... + C_NX_N + D$. For multiple independent variables, the regression analysis is called *multilinear regression analysis.*

Regression analysis has been used to model recovery, disease progression and medication dosage analysis. A regression is *strongly correlated* if the value of the dependent variable changes significantly when the independent variable changes. The correlation is measured using Pearson correlation-coefficient that varies between [− *1.0* ... *+1.0*]. Positive coefficient shows *positive correlation*, and negative coefficient shows *negative correlation*. The choice of an independent variable causally connected to dependent variable is very important. In the presence of multilinear regression analysis, the correlation may not show up correctly due to the cumulative effect of the independent variables.

Support vector machine (SVM) is based upon separating the data-points using a set of hyperplanes. Each hyperplane divides the data-points into two partitions. The slope of the separating line is decided by maximizing the distance between the two points nearest to the separating hyperplane one from each partition. The separating hyperplane is centered between the nearest-points, and the distance between the separating hyperplane and either nearest point is called *margin*.

Inductive reasoning generalizes the common attributes in multiple samples to find probabilistic logical rules. After a logical rule of the if-then form is derived, it is used to derive the outcome, if the conditions are satisfied.

A Bayesian network is a probabilistic reasoning system where multiple inputs are probabilistically combined to give an output, and the input values keep changing. This probabilistic reasoning is used for prediction. Bayesian network is represented as a sequence of directed acyclic graphs. Each internal node in the graph is associated with a conditional probability table between the input values and the output values.

Markov process is a temporal probabilistic finite state nondeterministic automata where the state in the next time-instance depends upon the current state. First-order Markov process is a subclass of Markov process that approximates the next state based upon the current state under the assumption that the immediate previous state contributes maximum to the next state. First-order Markov process reduces the computational overhead significantly. Markov process is modeled using two probability matrices: *transition-matrix* and *initialization-vector*.

HMM (Hidden Markov Model) is a variation of first-order Markov process where the state of the machine is predicted probabilistically by the evidence. This is due to many-to-many mappings between the internal states and the evidence signals. HMM has many applications in medical diagnosis, text analysis, speech understanding, natural language understanding, ECG analysis to detect different waves, gene-detection, recovery analysis and disease detection.

Associative learning derives associative-rules of the form *antecedent* $\rightarrow$ *consequent*. It is useful in data analytics for the discovery of new and sometimes unusual patterns. Apriori algorithm is a popular algorithm to derive association rules. The formation of a rule is based upon forming the associative set of variables and splitting the associative set into two subsets: *antecedent* and *consequent*. Only those variables are selected that have *support* above a minimum threshold. Similarly, only those associated rules are selected that have *confidence* above a threshold. Multiple rules are merged if they share the same prefix. Associated rules may exhibit false correlation, if the support is high. To handle false correlations, *lift-ratio* is used that is a ratio of support of the association set and the product of the support of *antecedent* and *consequent*. A *lift-ratio > 1.0* shows false correlation. Association rules are used in identifying the association of disease, the lab-tests and biomarkers.

Temporal data mining identifies pattern classes that have some interesting property in a temporal sequence of multidimensional data. Medical data is full of temporal time-series data. Temporal data-mining first segments and compresses the time-series data, and then uses the apriori algorithm to identify subsequences having interesting properties such as significant deviation of frequency of occurrence from the mean or some values satisfying user-defined thresholds. Temporal data mining is also affected by the context because certain data-patterns become more significant within a context. Context may persist for a longer period.

Ontology is concerned about the study of entities, their relationships, and equivalence of the groups of entities and relationships. In health science, it has tremendous application as the same concept may be expressed using multiple phrases by different healthcare providers. Ontology requires word-sense disambiguation in a domain and concepts, and traversing up and down the class-hierarchical relationship to derive if two words or phrases are related. UMLS (Unified Medical Language System) is a dataware house of medical terms and relations. It contains a meta-thesaurus, semantic network of entities, their relationships and specialist lexicons. To derive similarity between two concepts, various similarity-measures are used. Similarity-measures are based upon the distance of the two concepts in the semantic network or the information content. Ontology formation is based upon building the semantic network using the relationships and developing the hierarchical tree using "is-a" relationship.

Speech recognition is important for automated human–computer interaction and understanding dictations by physicians and medical-care providers. Speech-to-text conversion requires recognition of phonemes – the basic unit of utterance. Different phonemes have different features such as duration, pitch, energy and slopes of word-envelopes. The speech is divided into multiple 30 millisecond frames which are analyzed. Word boundaries are derived by marking the silence periods in speech because two words are separated by a silence period.

Natural language understanding systems and automated text-extraction play a major role in health informatics because healthcare providers store the reports in natural languages, and the lab reports and diagnosis are given to the patients in natural languages. Natural language understanding is also needed because transfers of medical notes across cultural and linguistic boundaries require automated machine translation from one language to another language.

Automated text-extraction is required to extract relevant clinical terms from the notes for automated database update and form filling. To understand natural languages, the domains of the text and the contexts must be specified because the meaning of the word can be easily disambiguated using domain and context. Document analysis is based upon the identification of the topics and the soft clusters of medical-terms within a domain and a context.

Latent semantic analysis is based upon the creation of a two-dimensional matrix (TDM) of medical-terms and documents using a medical-term knowledge-base and SVD (single value decomposition). Topics in a document are identified using *topic modeling. Topic modeling* is a probabilistic technique to cluster the medical-terms in proximity. A document is modeled as a graph of related topics. Two documents can be compared for similarity either using graph-matching or using inner-product of documents modeled as a vector of weights calculated using TF-IDF technique on TDM.

Event extraction systems search for the temporal anchor points such as dates and event marking annotations such as problem, occurrence and a patient's complaint. Two events are related using the words denoting temporal relationships such as "before," "after" and "during."

3.12 ASSESSMENT

3.12.1 Concepts and Definitions

Antecedent; apriori algorithm; artificial neural network; association rule; backward chaining; Bayesian association; Bayesian decision network; Bayesian statistics; Bayesian regression analysis; centroid; class-hierarchy; classification, clustering; clustering time-series data; collinearity; concept; concept finding; conditional probability table; confidence; consequent; correlation; covariance matrix; cTakes; data dictionary; data mining; decision-tree; deduction; dendrogram; density-based clustering; dimension reduction; disambiguation; dynamic probabilistic network; emission, emission matrix; entity; entropy; Euclidean distance; feed-forward neural network; first-order Markov process; forward reasoning; fuzzy cluster, fuzzy reasoning; greedy algorithm; hard-cluster; hidden Markov model; hidden layer; hierarchical clustering; incremental K-means clustering; independent component analysis; induction; inductive reasoning; information content; information gain; intelligent text-extraction; interoperability; initialization matrix; input layer; K-means clustering; latent semantic analysis; learning; lift-ratio; linear-discriminant analysis; linear regression analysis; logic reasoning; Logistic regression analysis; Manhattan distance; Markov process; MedKAT; Meta-thesaurus; MFCC; Minkowski's distance; minsup; multilinear regression analysis; multilingual translation; natural language translation; natural language understanding; neural network; nonlinear regression analysis; ontology; outlier; output layer; overfitting; path-based similarity; Pearson coefficient, principal component analysis; probabilistic network; probabilistic reasoning; reinforced learning; relation finding; relationship; residue confounding; scatter matrix; semantic group; semantic network; semantic type; similarity; similarity measures; soft-cluster; spanning tree; spanning-tree–based clustering; speech analysis; speech-to-text conversion; speech recognition; speech waveform; supervised learning; support; support vector machine; synonyms; TDM; TF-IDF; temporal data mining; text extraction; text summarization; time-series data; time-series data mining; time-series data segmentation; topic modeling; training hidden Markov model; training Markov model; training neural network; transition-matrix; transition probability; uncertainty-based reasoning; unsupervised learning; word-sense disambiguation.

3.12.2 Problem Solving

3.1 Given the following data-points, derive clusters using K-means clustering, and show your computation. Devise a reasonable technique to identify initial seed points, its radius and number of clusters. The initial radius should not be greater than maximum-interval/N where N is the number of clusters. Find the clusters, centroids, final radii and outliers. The data-points are: {(1, 2), (1, 3), (2, 1), (2, 3), (1, 4), (4, 8), (5, 7), (6, 7), (5, 8), (6, 8), (12, 13), (13, 12), (12, 15), (13, 14), (1, 12), (14, 2)}.

3.2 For the above data-points, use spanning-tree–based method to derive the clusters.

3.3 Figure 3.26 shows the recovery graph of a patient suffering from diabetes II. In diabetes, sugar becomes unregulated. A patient can be cured of diabetes II by weight reduction or taking different medicines such as insulin, Glipizide and Metformin. There are three moves: 1) reduce weight (denoted by RW); 2) change medication (denoted by I for insulin, G for Glipizide or M for Metformin) and 3) change dosage (denoted by DI or DG or DM). A state is defined when one of the three parameters is altered.

The measured sugar level defines the distance from the goal node. If the current sugar level of a patient is *330* mg/dl. The ideal sugar level is *90* mg/dl. Sugar level below *50* causes coma and death. The edge between states contains two types of information: treatment regimen and days to recover. Explain the route, a person should take to reach the goal state in the minimum number of days without getting an excessive low level of sugar. Use the knowledge of heuristics and goal search to answer the question.

3.4 Given these clusters characterized by their centroids and number of points in the cluster, give a hierarchical tree joining the cluster, and show the centroid for each new cluster. The data has been given in the format (*x-coordinate, y-coordinate*): *number of data-point in the cluster*. The centroids are (2, 4): 20; (5, 7): 40, (10, 12): 70 and (14, 16): 100. Show all the calculations.

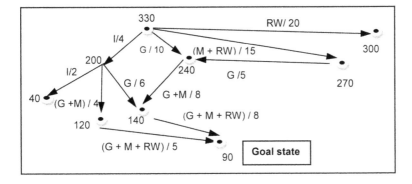

FIGURE 3.26   A state-space graph for the recovery of a diabetic patient

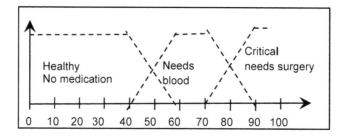

FIGURE 3.27   Defuzzification

3.5 Chest-pain is related to an abnormal heart condition such as ischemia that causes cell injury/ death. Ischemia is caused by the lack of oxygen. A lack of oxygen is caused by blockage in the blood-vessels or the lack of oxygen absorption by the lungs. Often, doctors estimate the level ischemia and heart blockage by the level of chest-pain. A doctor converts this fuzzy value to a quantifiable value using Figure 3.27. If a patient says that on a scale of *0* to *10*, his pain is *6*. Calculate the percentage blockage in heart and the medication suggested by the doctor. Explain and show your computation.

3.6 For the neural network in Figure 3.28, derive the output values. Assume that the negative bias of each perceptron is −0. 2. Also assume that the threshold value for firing of a perceptron is 0.1.

3.7 In a disease *X*, a patient can be in four states: *dead-state, sick-state, recovering-state* and *healthy-state*. The patient is being administered drug every single logical unit of time. From the sick-state, he can go to dead-state with a probability *0.1* and to the recovering-state with a probability *0.9*. From the recovering-state, he can go to the dead-state with a probability *0.02*, to the sick-state with a probability *0.18* and the healthy-state with a probability *0.8*. Assume that the initial state is a sick-state with probability *0.1* or healthy-state with probability *0.9*. Draw a graph showing the Markov process and give the corresponding initial matrix and the transition-matrix. Represent the transition-matrix as an array of linked-lists.

3.8 For the HMM in Figure 3.29, derive the transition and evidence matrices and give the most probable path for the signal emissions *XYZ* and *XXY*. Use the reasoning that a state has zero possibility if the corresponding state does not emit the required signal. Do not compute prob-abilities for such cases.

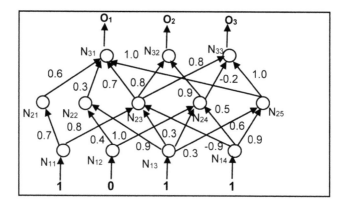

FIGURE 3.28   Neural network for Problem 3.6

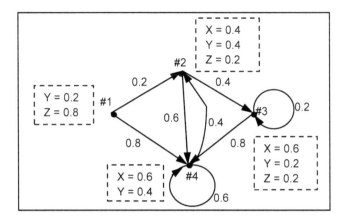

FIGURE 3.29   HMM for Problem 3.5

3.9 Fit a linear curve for these data-points that minimizes the residual error. Calculate the equation of the line and the Pearson coefficient. Using Pearson's coefficient, infer the correlation between independent variable X and dependent variable Y for the dataset {$(0, 4)$, $(2, 8)$, $(3, 10)$, $(4, 9)$, $(7, 20)$, $(9, 26)$, $(12, 30)$}.

3.10 Give a probabilistic network where high blood-pressure is caused due to excessive sodium concentration, cholesterol level and high blood pressure causes heart-failure and brain aneurism.

3.11 Given the Boolean table (Table 3.4) consisting of three variables X, Y and Z, give the corresponding *support, confidence, lift-ratio, association sets* and the minimal set of valid associations rules derived after removing all the redundant association rules. Show the complete process. Assume that minsup is 65%, and the confidence threshold is 70%.

3.12 Draw a semantic network for lung cancer caused by smoking tobacco. Use *is-a* relationship for the hierarchy and other appropriate relationships. Use meta-thesaurus from SNOMED CT for creating the ontology. For the following paragraph, find out the *association-factor* between the words "tobacco," "lung" and "cancer." Show the process.

   "Lung cancer is a deadly disease that is caused by smoking tobacco for a longer period. Tobacco smoking chars the lung sacs reducing the capability to breathe. Tobacco inhalation from the smoke of tobacco being smoked by other persons can also cause cancer."

3.13 Given the following time-series data between two readings of oxygen carrying capability of a patient on two different days measured for 60 minutes at the five-minute intervals, find out the overall distance between 12 samples. Use the distance-measure described in Section 3.2.5 to derive the distance. Assume that the data are aligned and have an equal number of samples.
   Time-series data #1: <85, 90, 92, 89, 88, 84, 87, 93, 83, 91, 92, 94>
   Time-series data #2: < 94, 96, 97, 92, 99, 95, 96, 98, 95, 92, 97, 98>

TABLE 3.4   Association table for Problem 3.8

	VARIABLE X	VARIABLE Y	VARIABLE Z
1	1	0	1
2	1	1	1
3	0	1	1
4	0	0	1
5	1	1	0
6	1	1	1
7	1	1	1

3.14 A drug company tests a new medicine and claims to improve significantly the oxygen carry-ing capability of blood. The drug was tried on six patients having COPD (Chronic Obstructive Pulmonary Disease). The drug was tested on a group of ten patients, each having mean oxim-eter reading of *78*. If the threshold level for hypoxemia is 85, show the effectiveness of the medicine with change in age (hint: first make K-means clusters of the given new reading based upon time-series data and derive the centroids. Finally infer the data).
Time-series data (Age *8*): < *94, 96, 97, 92, 99, 95, 96, 98, 95, 92, 97, 98*>
Time-series data (Age *10*): < *92, 96, 99, 93, 96, 95, 97, 98, 96, 92, 99, 98*>
Time-series data (Age *12*): < *94, 96, 99, 96, 96, 98, 97, 98, 96, 92, 99, 99*>
Time-series data (Age *62*): < *90, 92, 88, 92, 91, 90, 97, 92, 90, 92, 89, 94*>
Time-series data (Age *72*): < *91, 88, 88, 92, 91, 90, 87, 92, 90, 92, 89, 91*>
Time-series data (Age *78*): < *88, 88, 88, 89, 91, 90, 87, 91, 90, 92, 89, 91*>

3.15 Data plotted in a three-dimensional space show following mean and variance. Assuming that there is no need for unit scaling, which two dimensions should be retained during dimension reductions? Explain your answer.
X-axis: mean: 50, variance: 100
Y-axis: mean: 30; variance 10
Z-axis: mean 60, variance 40

3.16 Write a program that uses regression equation to derive the optimum line that passes through a set of given points and predicts the estimated value of the dependent variable for a value of independent variable interactively. Test your program using following initial data for curve fit-ting: (2, 2), (3, 8), (4, 12), (5, 11), (6, 14), (7, 18), (8, 17).

3.17 Write a program that takes the equation of a line and the coordinates of N (N is parameterized) two-dimensional points, finds out the perpendicular distance from the line, and displays the coordinates of the points, distance from the line and intersection point of the perpendicular from the point to the line.

3.18 Given two biomarkers to test liver cancer in blood, an insurance company is prioritizing which two biomarkers to pay for. The probability of the biomarkers A and B to express themselves is 10% and 8% in a population. The default population without elevated markers has *4%* prob-ability of getting cancer. The elevation of biomarker A exhibits the probability of cancer-type X as *50%*, cancer-type Y as *30%* and no cancer as *20%*. The elevation of the biomarker B exhibits cancer-type X as *10%*, cancer-type B as *40%* and no cancer as *50%*. Which biomarker should be tested first for testing for cancer, and why? How is your answer affected if the diagnosis is made more specific to cancer-type X or cancer-type Y. Explain your answer and show your computations.

3.19 Given the value of the coefficients $\beta_0 = -8$, $\beta_1 = 0.2$, and the simplified error value $\epsilon = 0.01$, give the range of values of independent variable x, when the dependent variable y would be *1* (hint: use logistic regression in Section 3.6.3). Use these values to plot the logistic regression function $F(x)$.

3.20 Assume that a person's cognitive function reduces with age after the age of 60 using the equa-tion *cognitive-function = −2 × (age − 60) + 100* on a scale of *0...100*. Assume that the person needs at least 90% of the cognitive capability to be socially active. At what age, he should take medication for cognitive enhancement. Show your computation.

3.21 In Figure 3.23, compute the depth-based similarity between the disease-pairs (*atrial flutter, Myocardial infarction*), (*myocardial infarction, ischemia*), (*ventricular arrhythmia, ischemia*) using different equations. Compare the similarity results and show your computation.

3.22 For the following definition of cancer: "Cancer is an unregulated tissue growth when neighbor-ing cells stop communicating with each other. After the cells stop communicating, the cells go through unprecedented growth. Cancer is characterized by the unusual growth of tumor, which has a different texture than the healthy cells," derive the characterizing set based upon statisti-cal analysis, and then calculate their *association-factor* with each other.

3.23 Using Figure 3.23, show how following sentence-pair can be matched. You can make an assumption about by the set of key words and associations with diseases. You can also assume that a descendant is included in the ancestor's definition using is-a relationship and is very similar.

"John is suffering from heart-disease."

"John has ectopic nodes in atria that is causing a supraventricular arrhythmia."

"John is suffering from supraventricular arrhythmia."

3.24 For the following two short documents, identify the medical terms associated with heart diseases and ECG analysis and show the TDM for medical-terms and phrases vs. paragraphs. Then, compute the weighted matrix using TF-IDF analysis.

Paragraph 1: "Heart disease can be of multiple types: 1) ischemia; 2) disease related to electrical-irregularity; 3) valve-irregularity and 4) heart-muscle problems. Many times, these diseases are interconnected. For example, arrhythmia is a disease caused by electrical irregularity, and Myocardial infarction is a disease caused by the lack of oxygen."

Paragraph 2: "Electrical irregularity problem is started either by: 1) electrical conductivity problem in HIS fibers or 2) the presence of ectopic-nodes in the atria or ventricles. The heart diseases having ectopic-nodes in atria are called supraventricular arrhythmia, and the heart diseases having ectopic-nodes in ventricles are called ventricular arrhythmia."

3.25 For the following text, perform temporal extractions to identify the events and temporal relationships. Assume that the temporal keywords are available to you in a lookup dictionary. Explain the computational process. Make reasonable assumptions about the event-markers, temporal keywords and temporal relationships.

"John has been suffering from prostate-cancer since he was 62 years old. Three years before getting cancer, he had a heart-surgery where two stents were placed in the aorta. He had a surgery to correct prostate-cancer at the age of 64 – two years after he was diagnosed for cancer."

3.12.3 Extended Response

3.26 Explain the process to derive K-means cluster using a set of data-points. Create the two-dimensional data-points yourself using a random number generator.

3.27 Explain the difference between hard clustering and fuzzy clustering using a simple example.

3.28 Explain the process and various embedded components of intelligent text-extraction from clinical notes.

3.29 Explain Bayesian decision network and the role of various probability tables in Bayesian network.

3.30 Explain first-order Markov process and hidden Markov model using a simple example. Clearly show the graph, transition-matrix, and evidence-matrix for the HMM example.

3.31 Explain linear curve fitting and Pearson's coefficient in regression analysis. Also explain the issues involved in multilinear regression analysis.

3.32 Explain the various stages of association rule-mining using a simple example different from the book.

3.33 Explain how spanning-tree is used to cluster the data-items. What are the drawbacks of a spanning-tree–based clustering.

3.34 Explain hierarchical clustering using a simple example not in the book.

3.35 Explain the role of ontology in clinical text analysis to derive entity-relationship network.

3.36 Explain the role of dimension reduction in clustering.

3.37 Explain principle component analysis and linear discriminant analysis and compare the two techniques.

3.38 Explain the role of semantic network in ontology formation.

3.39 Explain the role of word disambiguation in natural language understanding and language translators.

3.40 Read the research literature for similarity measures in "Further Reading" and write a one-page report on various techniques for content-based similarity matching.

3.41 Explain the role of statistical analysis in various machine learning techniques.

3.42 Explain the advantages of incremental K-means clustering over other clustering techniques for a large amount of data.

3.43 For tumor detection in a brain, what type of clustering would be used and why? Explain your answer.

3.44 Study two research articles on HMM in disease detection and the analysis of time-series data and summarize the articles in two pages.

3.45 Use research literature search using *Google scholar* and explain how a biomarker for genetic diseases such as cancer is derived using machine learning techniques.

3.46 Explain the method of latent semantic analysis using a simple two-paragraph example that generates a five by five matrix.

3.47 Read the papers on the back of this chapter and explain topic modeling.

FURTHER READING

Dimension Reduction

3.1 Balakrishnama, Suresh, and Ganapathiraju, Aravind, "Linear Discriminant Analysis – A Brief Tutorial," Technical report, *Institute of Signal and Information Processing*, Department of Electrical and Computer Engineering, Mississippi State University, Available at http://www.music.mcgill.ca/~ich/classes/mumt611_07/classifiers/lda_theory.pdf, Accessed on May 30, 2018.

3.2 Fodor, Imola K., "A Survey of Dimension Reduction Techniques," Technical Report # UCRL-ID-148494, *US Department of Energy – Lawrence Livermore National Laboratory*, May 2002, Available at https://e-reports-ext.llnl.gov/pdf/240921.pdf, Accessed May 30, 2018.

Heuristic Search/Inferencing Techniques

3.3 Russell, Stuart, and Norvig, Peter, *Artificial Intelligence – A Modern Approach*, 3rd edition, Pearson Education, Harlow, Essex, UK, 2014.

Fuzzy Reasoning

3.4 Chen, Guanrong, and Pham, Trung T., *Introduction to Fuzzy Systems*, Chapman and Hall/CRC Press, Boca Raton, FL, USA, 2005.

Machine Learning

3.5 Bishop, Christopher M., *Pattern Recognition and Machine Learning*, 8th printing, Springer Verlag, New York, NY, USA, 2006.

3.6 Vapnik, Vladimir N., "An Overview of Statistical Learning Theory," *IEEE Transactions on Neural Networks*, 10(5), September 1999, 988–999.

Classification Techniques

Clustering Techniques

3.7 Jain, Anil K., Murty, Narsimha M., and Flynn, Patrick J., "Data Clustering: A Review," *ACM Computing Surveys*, 31(3), September 1999, 264–323.

3.8 Hanwell, David, and Mirmehdi, Majid, "QUAC: Quick Unsupervised Anisotropic Clustering," *Pattern Recognition*, 47(1), January 2014, 427–440.

3.9 Knorr-Held, Leonhard, and Raßer, Günter, "Bayesian Detection of Clusters and Discontinuities in Disease Maps," *Biometrics*, 56(1), March 2000, 13–21.

3.10 Saitou, Naruya, and Nei, Masatoshi, "The Neighbor-Joining Method: A New Method for Reconstructing Phylogenetic Trees," *Molecular Biology Evolution*, 4(4), July 1987, 406–425.

3.11 Xu, Rui, and Wunsch II, Donald, "Survey of Clustering Algorithms," *IEEE Transactions of Neural Networks*, 16(3), May 2005, 645–678.

Clustering Time-Series Data

3.12 Liao, Warren T., "Clustering of Time Series Data - A Survey," *Pattern Recognition*, 38(11), November 2005, 1857–1874.

Regression Analysis

3.13 Draper, Norman R., and Smith, Harry, *Applied Regression Analysis*, 3rd edition, John Wiley and Sons, Danvers, Massachusetts, USA, 1998.

Probabilistic Reasoning over Time

Bayesian Decision Network

3.14 Andreassen, Steen, Riekehr, Christian, Kristensen, Brian, Schønheyder, Henrik C., and Leibovici, Leonard, "Using Probabilistic and Decision-Theoretic Methods in Treatment and Prognosis Modeling," *Artificial Intelligence in Medicine*, 15(2), February 1999, 121–134.

3.15 Bueno, Marcos L.P., Hommersom, Arjen, Lucas, Peter J.F., Lappenschaar, Martijn, Janzing, Joost G.E., "Understanding Disease Processes by Partitioned Dynamic Bayesian Networks," *Journal of Biomedical Informatics*, 61, June 2016, 283–297.

3.16 Carlin, Bradley P., and Louis, Thomas A., *Bayes and Empirical Bayes Methods for Data Analysis*, 2nd edition, Chapman and Hall/CRC Press, Boca Raton, FL, USA, 2009.

Markov and Hidden Markov Models

3.17 Russell, Stuart, and Norvig, Peter, *Artificial Intelligence – A Modern Approach*, 3rd edition, Pearson Education, Harlow, Essex, UK, 2014.

Data Mining

3.18 Aggarwal, Charu C., *Data Mining*, International edition, Springer Verlag, Cham, Switzerland, 2015.

3.19 Brown, Donald E., "Introduction to Data Mining for Medical Informatics," *Clinics in Laboratory Medicine*, 28(1), March 2008, 9–35.

3.20 Harrison, James H., "Introduction to the Mining of Clinical Data," *Clinics in Laboratory Medicine*, 28(1), March 2008, 1–7.

3.21 Tan, Pang-Nin, Steinbach, Michael, and Kumar, Vipin, *Introduction to Data Mining*, 2nd edition, Pearson Publishers, Boston, MA, USA, 2018.

3.22 Wu, XinDong, Zhu, ZingQuan, Wu, GongQuin, and Ding, Wi, "Data Mining with Big Data," *IEEE Transactions on Knowledge and Data Engineering*, 26(1), January 2014, 97–107.

3.23 Zaki, Mohammed J., and Meira Jr., Wagner, *Data Mining and Analysis*, Cambridge University Press, New York, NY, USA, 2014.

Interestingness

3.24 Geng, Liqiang, and Hamilton, Howard J., "Interestingness Measures for Data Mining: A Survey," *ACM Computing Surveys*, 38(3), September 2006, Article No. 9, DOI: 10.1145/1132960.1132963.

Associative Learning

3.25 Agrawal, Rakesh, Imielinski, Tomasz, and Swami, Arun, "Mining Association Rules between Sets of Items in Large Databases," In *Proceedings of the ACM SIGMOD International Conference on Management of Data*, Washington, DC, USA, May 1993, 207–216.

3.26 Gonzalez, Graciela H., Tahsin, Tasnia, Goodale, Britton C., Greene, Anna C., and Greene, Casey S., "Recent Advances and Emerging Applications in Text and Data Mining for Biomedical Discovery," *Briefings in Bioinformatics*, 17(1), 2016, 33–42.

3.27 Kumhare, Trupti A., and Chobe, Santosh V., "An Overview of Association Rule Mining Algorithms," *International Journal of Computer Science and Information Technologies*, 5(1), February 2014, 927–930.

3.28 Ordonez, Carlos, "Association Rule Discovery with the Train and Test Approach for Heart Disease Prediction," *IEEE Transactions on Information Technology in Biomedicine*, 10(2), April 2006, 334–343.

3.29 Vaani, Kannika N., Ramaraj, Eswara T., "An Integrated Approach to Derive Effective Rules from Association Rule Mining using Genetic Algorithms," In *Proceedings of the 2013 IEEE International Conference on Pattern Recognition, Informatics and Mobile Engineering*, Salem, India, February 2013, 90–95.

Temporal Data Mining

3.30 Fu, Tak-chung, "A Review of Time-Series Data Mining," *Engineering Applications of Artificial Intelligence*, 24(1), 2011, 164–181, DOI: 10.1016/j.engappai.2010.09.007.

3.31 Keogh, E., Lin, J., Fu, A., and Herle, H.V., "Finding the Most Unusual Time-Series Subsequences: Algorithms and Applications," *IEEE Transactions on Information Technology in Biomedicine*, 10(3), 2006, 429–439.

3.32 Mascovich, Robert, and Sahar, Yuval, "Medical Temporal-Knowledge Discovery via Temporal Abstraction," In *Proceedings of the 2009 Annual Symposium of American Medical Informatics Association*, 2009, 452–456.

3.33 Orphanou, Kalia, Stassopolou, Athena, and Keravnou, Elpida, "Temporal Abstraction and Temporal Bayesian Networks in Clinical Domain: A Survey," *Artificial Intelligence in Medicine*, 60(3), March 2014, 133–149, DOI: 10.1016/j.artmed.2013.12.007.

3.34 Post, Andrew R., and Harrison, James H., "Temporal Data Mining," *Clinics in Laboratory Medicine*, 28(1), March 2008, 83–100, DOI: 10.1016/j.cll.2007.10.005.

Interoperability and Ontology

3.35 Bowman, Sue, "Coordinating SNOMED-CT and ICD 10," *Journal of the American Health Information Management Association*, 76(7), July/August 2005, 60–61.

3.36 Cimino, James J., and Zhu, Xinxin, "The Practical Impact of Ontologies on Biomedical Informatics," *IMIA: Yearbook of Medical Informatics*, 15(1), 2006, 124–135.

3.37 Smith, Barry, Arabandi, Sivaram, Brochhausen, Mathias, Callhoun, Michael, Ciccarese, Paolo, Doyle, Scott, Giband, Bernard, et al., "Biomedical Imaging Ontologies: A Survey and Proposal for Future Work," *Journal of Pathology Informatics*, 6(37), June 2015, Available from http://www.jpathinformatics.org/text.asp?2015/6/1/37/159214, Accessed June 6, 2018.

Entities and Relationships

Unified Medical Language

3.38 Fung, Kin W., and Bodenreider, Olivier, "Utilizing the UMLS for Semantic Mapping Between Terminologies," In *Proceedings of the American Medical Informatics Association (AMIA) Annual Symposium*, Washington DC, 2005, 266–70.

3.39 Humphreys, Betsy L., Lindberg, Donald A.B., Schoolman, Harold M., and Barnett, Octo G., "The Unified Medical Language System: An Informatics Research Collaboration," *Journal of American Medical Informatics Association (JAMIA)*, 5(1), January–February 1998, 1–11.

MedDRA and Ontology

3.40 Gruber Tom, "Ontology," in *Encyclopedia of Database Systems* (editors: Ling, Liu and Tamer, Özsu M.), Springer, Boston, MA, USA, 2009, DOI: 10.1007/978-0-387-39940-9_1318.

3.41 Mozzicato, Patricia, "MedDRA – An Overview of the Medical Dictionary for Regulatory Activities," *Pharmaceutical Medicine*, 23(2), April 2009, 65–75, DOI: 10.1007/BF03256752.

3.42 Bousquet, Cédric, Sadou, Éric, Souvignet, Julien, Jaulent, Marie-Christine, and Declerck, Gunnar, "Formalizing MedDRA to Support Semantic Reasoning on Adverse Drug Reaction Terms," *Journal of Biomedical Informatics*, 49, June 2014, 282–291, DOI: 10.1016/j.jbi.2014.03.012.

Similarity Measures

3.43 Haripse, Sébastein, Sánchez, David, Ranwez, Sylvie, Janaqi, Stefan, and Montmain, Jacky, "A Framework for Unifying Ontology-based Semantic Similarity Measures: A Study in Biomedical Domain," *Journal of Biomedical Informatics*, 48, April 2014, 38–53.

3.44 McInnes, Bridget T., and Pedersen, Ted, "Evaluating Semantic Similarity and Relatedness over the Semantic Grouping of Clinical Term Pairs," *Journal of Biomedical Informatics*, 54, April 2015, 329–336.

3.45 Pivavarov, Rimma, and Elhadad, Noémie, "A Hybrid Knowledge Based and Data Driven Approach to Identifying Semantically Similar Concepts," *Journal of Biomedical Informatics*, 45, 2012, 471–481.

Ontology Formation

3.46 Henegar, Corneliu, Bousquet, Cedric, Louet, Agnes L., Degoulet, Patrice, and Jaulent, Marie C., "Building an Ontology of Adverse Drug Reactions for Automated Signal Generation in Pharmacovigilance," *Computers in Biology and Medicine*, 36 (7–8), July–August 2006, 748–767.

3.47 Leroux, Hugo, Metke-Jimene, Alejandro, and Lawley, Michael J., "Towards Achieving Semantic Interoperability of Clinical Study Data with FHIR," *Journal of Biomedical Semantics*, 8(41), 2017, 1–14.

3.48 Liu, Hongfang, Wu, Stephen, Tao, Cui, Chute, Christopher, "Modeling UIMA Type System Using Web Ontology Language: Towards Interoperability among UIMA-Based NLP Tools," In *Proceedings of the 2nd International Workshop on Managing Interoperability and Complexity in Health Systems*, Maui, Hawaii, USA, 2012, 31–36.

3.49 Marenco, Luis, Wang, Rixin, and Nadkarni, Prakash, "Automated Database Mediation Using Ontological Metadata Mappings," *Journal of American Medical Informatics Association*, 16(5), September/October 2009, 723–737.

Intelligent Interfacing Techniques

Natural Language Understanding

Speech and Language Processing

3.50 Juang, Biing H., and Chen, Tsuhan, "The Past, Present and Future of Speech Processing," *IEEE Signal Processing Magazine*, 15(3), May 1998, 24–48.

3.51 Jurafsky, David, and Martin, James H., *Speech and Language Processing: An Introduction to Natural Language Processing, Computational Linguistics and Speech Recognition*, Prentice Hall, Upper Saddle River, NJ, USA, 2000.

3.52 Leu, Fang-Yie, and Lin, Guan-Liang, "An MFCC-Based Speaker Identification System," In *Proceedings of the IEEE 31st International Conference on Advanced Information Networking and Applications (AINA)*, Taipei, Taiwan, March 2017, 1055–1062.

3.53 Manning, Christopher D., and Schütze, Hinrich, *Foundations of Statistical Natural Language Processing*, MIT Press, Cambridge, MA, USA, 1999.

Word-Sense Disambiguation

3.54 Savova, Guergana K., Coden, Anni R., Sominsky, Igor L., Johnson, Rie, Ogren, Philip V., Groen, Piet C.D., and Chute, Christopher G., "Word Sense Disambiguation Across Two Domains: Biomedical Literature and Clinical Notes," *Journal of Biomedical Information*, 41(6), December 2008, 1088–1100.

Multilingual Translation

3.55 Pecinaa, Pavel, Dušek, Ondřej, Goeuriot, Lorraine, Hajič, Jan, Hlaváčová, Jaroslava et. al., "Adaptation of Machine Translation for Multilingual Information Retrieval in the Medical Domain," *Artificial Intelligence in Medicine*, 61(3), July 2014, 165–185.

Intelligent Text-Extraction

3.56 Manning, Christopher D., Raghavan, Prabhakar, and Schütze, Hinrich, *Introduction to Information Retrieval*, Cambridge University Press, New York, NY, USA, 2008.

Latent Semantic Analysis

3.57 Crain, Steven P., Zhiu, Ke, Yang, S-H, Zha, Hongyuan, "Dimensionality Reduction and Topic Modeling: From Latent Semantic Indexing to Latent Dirichlet Allocation and Beyond," In *Mining Text Data* (editors: Aggarwal, Charu C., and Zhai, Cheng X.), Springer, Boston, MA, USA, 2012, 129–161.

3.58 Gefen, David, Miller, Jake, Armstrong, Johnathan K., Cornelius, Frances H., Robertson, Noreen, Smith-McLallen, Aaaron, and Taylor, Jennifer A., "Identifying Patterns in Medical Records Through Latent Semantic Analysis," *Communications of the ACM*, 61(6), June 2018, 72–77, DOI: 10.1145/3209086.

Topic Modeling

3.59 Blie, David M., Ng, Andrew Y., and Jordan, Michael I., "Latent Dirichlet Allocation," *Journal of Machine Learning Research*, 3, January 2003, 993–1022.

3.60 Torii, Manabu, Wagholikar, Kavishwar, and Liu, Hongfang "Using Machine Learning for Concept Extraction on Clinical Documents from Multiple Sources," *Journal of American Medical Informatics Association*, 18(5), September 2011, 580–587, DOI: 10.1136/amiajnl-2011-000155.

Automated Text Summarization

3.61 Goldstein, Ayelet, and Shahar, Yuval, "An Automated Knowledge-Based Textual Summarization System for Longitudinal, Multivariate Clinical Data," *Journal of Biomedical Informatics*, 61, June 2016, 159–175, DOI: 10.1016/j.jbi.2016.03.022.

3.62 Nenkova, Ani, and McKeown, Kathleen, "A Survey of Text Summarization Techniques," In *Mining Text Data* (editors: Aggarwal, Charu C., and Zhai, ChengXiang), Springer, Boston, MA, USA, 2012, 43–76.

3.63 Portet, François, Reiter, Ehud, Gatt, Albert, Hunter, Jim, Sripada, Somayajuli, Freer, Yvonne, and Skykes, Cindy, "Automatic Generation of Textual Summaries of Neonatal Intensive Care Data," *Artificial Intelligence*, 173(7–8), May 2009, 789–816.

3.64 Scott, Donia, Hallett, Catalina, and Fettiplace, Rachel, "Data-to-text Summarization of Patient Records: Using Computer-generated Summaries to Access Patient Histories," *Patient Education and Counseling*, 92(2), August 2013, 153–159, DOI: 10.1016/j.pec.2013.04.019.

Event Extraction from Clinical Text

3.65 Sahar, Yuval, "Dynamic Temporal Interpretation Contexts for Temporal Abstractions," *Annals of Mathematics and Artificial Intelligence*, 22(1 – 2), February 1998, 159–192.

3.66 Sohn, Sunghwan, Wagholikar, Kawishwar B., Li, Dingcheng, Jonnalagadda, Siddartha R., Tao, Cui, Komandur, Ravikumar E., and Liu Hongfang, "Comprehensive Temporal Information Detection from Clinical Texts: Medical Events, Time and TLINK Identification," *Journal of American Medical Informatics Association*, 20(5), April 2013, 836–842, DOI: 10.1136/amiajnl-2013-001622.

3.67 Wang, Wei, Kreimeyer, Kory, Woo, Emily J., Ball, Robert, Foster, Matthew, Pandey, Abhishek, et al., "A New Algorithmic Approach for the Extraction of Temporal Associations from Clinical Narratives with an Application to Medical Product Safety Surveillance Reports," *Journal of Biomedical Informatics*, 62(C), June 2016, 78–89, DOI: 10.1016/j.jbi.2016.06.006.

Case Studies

3.68 Coden, Anni, Savova, Guergana, Sominsky, Igor, Tanenblatt, Michael, Masanz, James, Schuler, Karin, et al., "Automatically Extracting Cancer Disease Characteristics from Pathology Reports into a Disease Knowledge Representation Model," *Journal of Biomedical Informatics*, 42(5), October 2009, 937–949, DOI: 10.1016/j.jbi.2008.12.005.

3.69 Dugas, Martin, "ODM2CDA and CDA2ODM: Tools to Convert Documentation Forms between EDC and EHR Systems," *BioMed Central Medical Informatics and Decision Making*, 15(40), May 2015, DOI: 10.1186/s12911-015-0163-5, Available at https://bmcmedinformdecismak.biomedcentral.com/track/pdf/10.1186/s12911-015-0163-5, Accessed September 23, 2018.

3.70 Patrick, Jon, Wang, Yefeng, and Budd, Peter, "An Automated System for Conversion of Clinical Notes into SNOMED Clinical Terminology," In *Proceedings of the Fifth Australian Symposium on the ACSW Frontiers*, Darlinghurst, Australia, January 2007, 219–226.

3.71 Xu, Hua, Stenner, Shane P., Doan, Son, Johnson, Kevin B., Waitman, Lamuel R., and Denny, Joshua C., "MedEx: A Medical Information Extraction System for Clinical Narratives," *Journal of Medical Information Association (JAMIA)*, 17(1), January/February 2010, 19–24, DOI: 10.1197/jamia.M3378.

cTakes System

3.72 Savova, Guergana K., Masanz, James J., Ogren, Philip V., Zheng, Jiaping, Sohn, Sunghwan, Kipper-Schuler, Karin C., and Chute, Christopher J., "Mayo Clinical Text Analysis and Knowledge Extraction System (cTakes): Architecture, Component, Evaluation and Application," *Journal of American Medical Information Association (JAMIA)*, 17(5), September/October 2010, 507–513, DOI: 10.1136/jamia.2009.001560.

MedKAT System

3.73 *The MedKAT Pipeline*, Available at http://ohnlp.sourceforge.net/MedKATp/, Accessed on May 30, 2018.

Healthcare Data Organization

4

BACKGROUND CONCEPTS

Section 1.2 Modeling Healthcare Information; Section 1.6.1 Medical Databases; Section 1.6.2 Medical Information Exchange; Section 1.6.3 Integration of Electronic Health Records; Section 1.6.4 Knowledge Bases for Health Vocabulary; Section 1.6.5 Concept Similarity and Ontology; Section 1.6.8 Machine Learning and Knowledge Discovery; Section 1.6.9 Medical Image Processing and Transmission; Section 1.6.10 Biosignal Processing; Chapter 2 Fundamentals

Huge amount of data requires proper organization for efficient archiving, retrieval and updates in real time with no duplication. The problem becomes more complicated in the health industry because there are many sources of heterogeneous data to be reconciled and integrated in real time; an error in data reconciliation may cause an error in medical decisions and judgments by medical practitioners. As the reliance of the medical industry on electronic data increases, this need will become very critical. However, the purpose of creating electronic databases is important because the information can be stored for a long time and transmitted globally to other medical institutions, physicians and patients in real-time and on-demand.

Computational health informatics needs to model a large amount of data used in: 1) Electronic Health Record (EHR); 2) data analytics such as cluster analysis, regression analysis, etc.; 3) storage and analysis of time stamped biosignals such as EEG and ECG signals; 4) medical image storage and analysis for abnormalities such as cancer and 5) video analysis such as in the movement of heart muscles. Healthcare data is classified into three major categories: 1) text such as physicians' notes and human-comprehensible description of a patient's condition and test reports; 2) structured database values in the form of integers, real numbers, and strings and 3) multimedia objects such as video during surgery, physicians' dictations, patient–doctor conversations in telemedicine and high-resolution medical images. In addition, data need to be structured for efficient access and processing. Different sources of data have to be interfaced at multiple levels: 1) device-database interface for automated data acquisition, 2) database-database interface for automated information exchange and 3) human-database interface for automated archival of survey data, diagnosis, physician's notes, instructions and training.

This chapter describes the acquisition, organization and exchange of healthcare–related information. It discusses various issues and techniques needed to reconcile heterogeneous databases from different organizations, issues and techniques related to retrieving and transmitting information from one database to another database. It also describes various standards for medical terms used to provide interoperability between heterogeneous databases and describes the overall architecture to integrate heterogeneous databases and information flow across them. The chapter also describes techniques to incorporate HIPAA into existing medical information archiving, retrieval and transmission to protect patients' privacy and confidentiality. As a case study of EMR (Electronic Medical Record), it discusses a publicly accessible EMR database *OPENEMR*.

4.1 MODELING MEDICAL DATA

The storage of data for long periods of time helps in discovering the history of the patient immediately and cuts down the information retrieval time. The capability of immediate data-communication allows multiple health organizations to collaborate without increasing the cost and with no duplication of the diagnostic tests.

Healthcare information is heterogeneous such as data related to analysis of lab tests, data related to prescription, data related to patient-management inside a hospital, data related to patients' real-time monitoring in an ICU (Intensive Care Units) and data related to real-time collaborative surgery. Some of the data may not require real-time updates such as a database of prescriptions, while others require real-time updates of the corresponding databases such as real-time monitoring in ICU and collaborative-surgery.

4.1.1 Modeling Electronic Health Record

EHR is organized as a complex heterogeneous network of medical databases with automated or semi-automated data-acquisition from the patients' monitoring, physiology labs, radiology labs and mobile healthcare units; and transparent information exchange between databases. These databases should answer temporal and multimedia queries involving different types of high-resolution images, videos and audios across different spoken languages. Health data is organized as graphs at various levels: 1) device-level during data-acquisition; 2) database-level and 3) Internet-level.

The databases are a network of relational databases and object-based databases capable of automated content-based multimedia retrieval and temporal queries involving ontologies. The network can be modeled as a weighted graph if we can use bandwidths (data-transfer rate) as the weights of the edges. To access a node in a graph, a search starts from one node, traverses along the edges toward the desired node to access the information stored in the desired node. The information has to be protected against unauthorized access and should be verifiably authenticated.

4.1.1.1 Modeling multimedia objects

A large part of medical data contains multimedia objects. Some examples are: 1) MRIs; 2) CAT scans; 3) X-ray images; 4) videos of heart-movements; 5) doctors' dictations about the patients' diagnosis; 6) recordings of the medical panel meetings; 7) sequences of images showing disease-progression such as tumors, tuberculosis, etc. The data are accessed to: 1) retrieve patients' history while diagnosing diseases; 2) treat patients based upon similar symptoms; 3) understand the effect of a medication for data analytics; 4) educate medical professionals; 5) assist in guided-surgery; 6) enable automated speech-to-text conversion and 7) transmit the patients' information to authorized medical facilities.

4.1.2 Modeling Biosignals

Biosignals such as ECG and EEG recordings are modeled as two-dimensional data with one dimension being *time* and the other dimension as *amplitude*. An additional field is needed to label different types of waveforms. Thus, biosignals are abstracted as a sequence of triples (*label, time-interval, amplitude*). For the higher sampling rate, the time-interval between two amplitude values is shorter, and the waveform recovered from the sequence of amplitudes at the destination will be more accurate. However, the number of amplitude-values will be much larger requiring additional memory storage, slow data-processing, slow transmission over the communication-link and slow rendering at the other end. In medical monitoring, high sample-rate without losing accuracy is a necessity to accurately reproduce the signal at the destination for proper diagnosis. The standard sampling rate for ECG varies from one to three ms.

4.1.3 Medical Image Compression

Multimedia data contains a huge amount of: 1) text and binary data files, 2) audio files, 3) image files and 4) video files. Disease-related data such as MRIs are high-resolution images. The storage of multimedia data requires huge amount of secondary memory space. The transmission of raw data needs a large bandwidth. Besides, it slows down the data-transmission rate by clogging up the network. This slow-down can create life-threatening situations in time critical operations such as remote instructions to paramedics and collaborative surgery. Hence, data needs to be transmitted in a compressed form. The transmitted multimedia should also be quickly decoded at the destination for appropriate medical actions.

Healthcare multimedia is of multiple types: 1) videos used for training or instructions; 2) videos of multiexpert collaborations; 3) remote encounters of patient and the physician; 4) videos of patients' regions of interest during robotic surgeries; 5) microscopic images of infected regions after cell cultures; 6) black-and-white radiology images such as X-ray, CAT scan, MRI, PET scan, etc.; 7) colored image-reconstructions from black-and-white radiology images for better human perception and 8) video snapshots of patients for posture analysis in remote care. Disease-related images require high-resolution reproducibility of the original image for the disease-area at the physicians' end for better diagnosis.

4.1.3.1 Lossy vs lossless compression

Lossless image-compression must reproduce diagnostically relevant parts of images accurately at the destination. Hence, lossless compression of medical images is required. Each image runs into multiple megabytes depending upon the technique. On an average, an image occupies *5–50* MB of memory space. Medical images need sufficient compression because high-resolution images need a large amount of memory and take excessive time during transmission over the Internet. However, after reconstruction, the information-loss should be negligible.

One approach to solve this seemingly contradictory problem is to divide an image in two parts: 1) region of interest (ROI) containing medically relevant information such as tumors and 2) the remaining region. ROI is compressed using *lossless compression* for the accurate reproduction at the destination. The remaining region is compressed using *lossy compression* to save memory storage and reduce image-transmission time while maintaining the quality of medically relevant part of an image. For multiple slices of images, a motion-vector is computed by performing similarity analysis between macroblocks.

The needs for compression are divided into three cases: 1) local prediagnosis images to be diagnosed locally; 2) images transmitted over the Internet for an expert-opinion and 3) images archived in the multimedia database after diagnosis for future use. The first class of images is not compressed or compressed using lossless compression for future reference. It is important for an expert to see the complete image to identify the regions of interest where possible abnormalities are present. In the second case, the requirements are contradictory: 1) the faster transmission requires lossy compression to reduce the transmission time, while the preservation of information requires lossless compression. Here progressive reconstruction of an image is used. Image quality is slowly improved incrementally by sending the refined image later, while compressed images are sent earlier for faster diagnosis. Eventually, the image quality will be as good as lossless compression. In the third case, after the image has been diagnosed, the regions of interest are marked. Here, the regions of interest are compressed using lossless compression algorithms, and the remaining regions are compressed using lossy compression to reduce the storage requirement.

4.2 AUTOMATED DATA ACQUISITION AND INTERFACES

Hospitals have equipment from different vendors. An equipment communicates automatically with other connected instruments and the required databases. For example, ECG machines monitoring a patient in an ICU is connected to a central computer that displays the ECG signals, stores the signals in a temporal database for the ease of future retrieval of data and analyzes the signals to derive emergency alerts.

To communicate the information between instruments and computers, there should be: 1) a virtual description of the devices based on a software model of the machine, 2) a standardized format of information exchange, 3) a built-in software to translate software commands to hardware signals and 4) adapters to convert the messages sent by the software model of the instrument to the common message-format and vice versa for information exchange.

4.2.1 Virtual Medical Devices

Virtual Medical Device (VMD) is an abstract model of the hard devices that interfaces with other VMDs and the database through information exchange and command transfer. A VMD is modeled by an object-oriented model called Domain Information Model (DIM) that abstractly models the real-world entities. A VMD has two major components: *static model* and *dynamic model*.

4.2.1.1 Static model of VMD

The static model is divided into multiple small definition packages: 1) medical package, 2) alert system package, 3) system package, 4) control package, 5) extended service package, 6) communication package, 7) archival package and 8) patient package.

The *medical package* defines the abstraction of a VMD that collects biomedical signals. The definitions include modes, versions and status of the device. The *alert package* defines various alert codes and descriptions, including their priority levels. The *system package* describes the definitions of the system devices that process the collected biosignal and vital signals. The *control package* provides the commands for remote control of the devices and the measurements. It also provides operating system locks for avoiding racing conditions while collecting the data from the device. *Extended device-package* includes the virtual description of the devices other than the medical devices such as scanners. *Communication package* describes the definitions of device communication controllers. *Archival package* stores the commands to archive biosignal data, vital-signs data and time-framed data into a file. *Patient package* stores the information for monitoring patient and transmitting collected data.

4.2.1.2 Dynamic model of VMD

The *dynamic model* is a client-server model for checking connection of the newly connected devices and data transmission from the DCC (Device Communication Controller) of the monitoring device to the BCC (Bedside Communication Controller) of the connected server. The whole protocol has four operations: 1) detection of the connection event, 2) verification of the compatibility of the VMD definition and the corresponding data-formats, 3) exchange of MDIB (Medical Device Information Base) and 4) data transmission in a compatible data format. The base information exchanged between a device and the server consists of: 1) basic encoding of the message, 2) abstract syntax for the message to be exchanged between devices and 3) services that a host can use to request data from devices. In addition, there are details of transport layer, network layer, data link layer and physical layer for message communication. Various network layers are explained in Section 6.1.

The major advantages of automated data acquisition are: 1) sharing of data by the nurses, 2) freeing the nurses to do other tasks, 3) timely and consistent data recording and 4) accurate and representative results. To reduce the overhead of data storage, multiple readings are taken at different periods, and data is summarized episodically.

4.2.2 Data Interface Model and Technology

The structure that makes the medical devices plug-and-play and facilitates a data-exchange without any interoperability issue are part of the IEEE standard commonly known as *HIB* (*Health Information Bus*) or *MIB* (*Medical Information Bus*). *Health-information-bus* is analogous to the *memory bus* in a computer where data can travel in a standardized format to different destinations starting from the hardware sensors. Some devices that are covered through Health Information Bus (HIB) are: 1) vital signs monitor, 2) ventilator, 3) pulse oximeter – measures the oxygen content, 4) defibrillator – used for heart stimulation during cardiac-arrest, 5) ECG, 6) insulin pump – used for administering insulin to diabetic patients, 7) body composition analyzer – used to estimate the fat content in a body, 8) sleep-apnea breathing equipment and 9) glucose meter – used to measure the glucose in the blood.

MDDL (M̲edical D̲evice D̲ata L̲anguage) is used to send messages from a medical device to personal computers. The message includes the information such as: 1) data value, 2) units of measurements, 3) device definitions and descriptions and 4) alert codes. MDDL is covered by IEEE ISO 11073-1xxxx. Some examples of MDDL codes are: oxygen level, diastolic and systolic blood pressures, mean blood pressure, heart rate, glucose concentration in arterial blood and glucose concentration in general. A schematic is illustrated in Figure 4.1.

4.2.2.1 Adapters

This requires the development of software interfaces. Each destination will have its own format of data representation. Data transformation between multiple data-formats is provided by *adapters. Adapters*

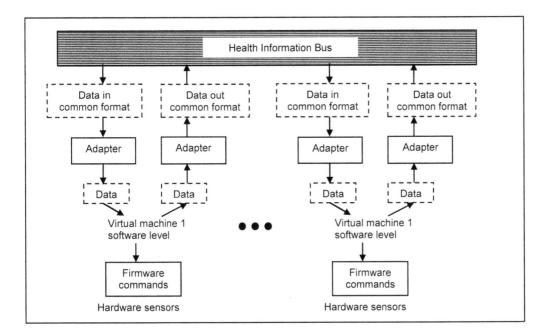

FIGURE 4.1   HIB and medical equipment connectivity

convert the information from one format used in an organization or sensor to the standardized format used in health-information-bus and vice versa. This conversion is necessary for the interoperability of the databases and interoperability between sensors and databases. *Adapters* are also needed to exchange information between two institutions or two heterogeneous units (within the same institutions) using incompatible formats due to hardware provided by different vendors.

Adapters have four components: 1) a source connector, 2) a destination connector, 3) filters and 4) the transformer. A *source connector* listens to the external sources for the messages. It uses polling on ports to find out if a message is available on that port. A *destination connector* connects to an external system to transmit data. This data must have been transformed from the internal representation format of the unit to a common-bus data format. A *filter* determines which incoming message should be accepted based upon a set of rules that filter out certain inquiry types, sources that originate a message, or the extent of the request. A transformer converts the source-format to the destination-format.

4.2.2.2 Transformers

A *transformer* converts a message from the input format to: 1) the format of the HIB (Health Information Bus) when transmitting information from the source-connector to HIB; alternately 2) from the HIB-format to the destination-connector format. A transformer selects the needed data from the message based on the labels in international standards such as ICD, SNOMED, LOINC, DICOM. The separation of these four components provides extra modularity as the change in the internal formats of the units will alter the corresponding transformer.

4.2.2.3 Message flow

There are two types of messages: *synchronous* and *asynchronous*. *Synchronous messages* facilitate immediate automatic response from the server where the source connector is located. Asynchronous messages are needed when there is no immediate need for the response or where the response needs human input such as a response from a patient or a healthcare provider.

Another problem with the information exchange is the semantic aspect of data. The same information is stored in multiple units using varied textual phrases. For example, if the source is a doctor or a patient, the incoming-message contains English language to be parsed, checked for equivalence using synonym's dictionary or using ontology. Often, only reference to a data is sent along with the message to maintain security, and the data is retrieved using a secure channel using proper authentication. The transformers also have access to tools such as natural language processors, dictionary lookup and transformation based upon ontology information.

4.3 HEALTHCARE DATA ARCHIVAL AND INTERCHANGE

Medical databases are patient-centric and contain different types of patient-related information. The databases contain lab tests, radiology reports, images of various organs and bones, videos of moving organs such as heart, database of biosignals and their monitoring during intensive care, patient encounters, patient histories, medication-related information and billing information. Different parts of databases are used by different group of healthcare providers such as: 1) doctor; 2) bedside nurse; 3) radiologist; 4) surgeon; 5) data analyst and clinical researcher; 6) data scientist involved in data mining de-identified data to identify useful rules, new disease conditions and side-effects of medications; 7) pharmacist to dispense the medications; 8) billing department to code the procedures for the billing purpose and 9) insurance companies to verify the procedures and codes and pay the relevant bills. The summarized databases are used by public health institutes to derive demographic-related statistics, suspected spread of epidemics and the corresponding treatments.

4.3.1 Datamarts

Healthcare databases are stored in secured sites called *datamarts* for providing privacy, and retrieved using a *Master Patient Index* (MPI) that is unique to a patient despite the location of data-storage. Different medical providers such as hospitals, physicians, laboratories have different databases about the same patients and are located at geographically diverse places connected through the Internet. Each organization has its own data-storage format provided by different vendors and different local patient-ids for the same patient. These patient-ids for the same patient need to be reconciled to avoid duplications and the mix-up of records. Databases need to be automatically updated based upon the sensor readings.

4.3.2 Data Warehouse

Data warehouses are the centralized database-archives that contain the information about: 1) patient health in standardized formats; 2) patient demographic data such as age sex, race, marital status, address, hospital charges and length of stay; 3) admission and discharge data; 4) discharge disposition such as home, death or nursing facility; 5) payment sources such as "Medicare," "Medicaid," personal or insurance company code; 6) healthcare provider data such as a hospital, physician(s), specialty doctor(s), surgeon and nursing units; 7) billing details such as charges in each invoice, service dates, previous invoices and amount recovered; 8) readmission data; 9) insurance denial data; 10) pharmacy medication details; 11) pharmacy billing data and 12) patient survey information showing satisfaction with the service.

There are multiple approaches to building dataware houses. They can be built using datamarts, and then merged to make a bigger dataware house. Datamarts are smaller functionality-based data-storage units such as financial-datamart, survey-datamart, health-insurance-claims datamarts, pharmacy datamart, patient-care datamart, intensive care-unit datamart, genomic datamart, radiology datamart, pathology datamart, psychological datamart, cardio-vascular datamart and operating-room datamarts. Each of these datamarts has their own architecture connecting different set of database tables. These datamarts are connected using bus architecture to form a data warehouse. The advantages of using datamarts and data warehouse are: 1) sharing of data across health organizations while maintaining required security; 2) de-identification and aggregation of data for data analytics; 3) verifying the procedure and medication errors in patient's treatment and 4) setting up a common policy of data retrieval that maintains data security and integrity.

4.3.2.1 Design and operations

Running of data warehouses requires many operations such as: 1) insertion of new data; 2) secure retrieval of data; 3) maintaining interoperability of data; 4) abstracting low-level data for high level data analytics; 5) analysis of the disease-spread and drug-use trend among population; 6) data mining of radiology and pathology data to derive knowledge and 7) statistical reporting of data to study broader health issues in population and to develop an anticipative healthcare system.

Example 4.1

Health-claim datamart consists of many relational database tables such as *claims, hospital, insurance, patient, disputes and staff, etc. Claims-table* contains claim-related information. *Hospital-table* contains hospital-related information. *Insurance-table* contains insurance-companies-related information. *Dispute-table* contains dispute-resolution information. *Patient-table* contains personal profile of patients. Staff-table contains the information related to staff processing the claims.

Claims-table contains *claim-id, claim-status-code, hospital-id, insurance-company-id, insurance-type, patient-id, last-updates, claim-filing date, processing-stage, claimed amount, claim-recovery date, recovered claim* and the *processing-staff. Hospital-table* contains *hospital-id, hospital-name, hospital*

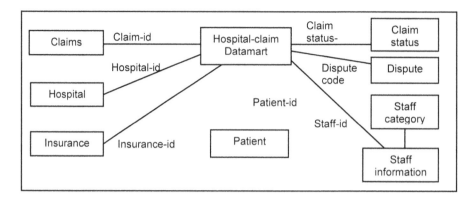

FIGURE 4.2   Hospital-claim datamart

address, contact person and *other details*. *Insurance-table* contains *insurance-company-id, company-name, company-address, contact-name, insurance-type* and *other details*. *Patient-information table* contains *patient-id, patient-name, address, emergency-contact, nearest relative, insurance-related information*. *Disputes-table* contains *dispute-id, dispute-description, dispute-resolution code*. *Staff-information table* contains *employee-id, employee-name, unit-id* and other staff-related details. The overall scheme is shown in Figure 4.2.

These tables are selectively connected to each other using a field that act as primary-index in the supporting table. For example, the primary-key in hospital information is *hospital-id*. The *claim-table* is connected to *hospital-table* through *hospital-id*, and to the table *patient-table* through the key *patient-id*.

4.3.3 Information Clearing House

Information clearing house is a natural resource center that collects, analyzes and disseminates large amount of medical information related to public health. A clearing house acts as the verifier of the claims, authenticates customer-requests, checks the validity of the procedure-codes, sends verification results to appropriate authorities, imposes HIPAA regulations on archived data to produce information that is private and secure to disseminate to the end-party. It has two major abstract functions: 1) convert standard format data in the database into a nonstandard readable or transmittable format after the de-identification (to maintain confidentiality) and disambiguation of data; and 2) take the data in a nonstandard format and convert the information in a standard format to be archived in a database.

4.4 HEALTH INFORMATION FLOW

There are three types of information flow: 1) sensor to a database for the automated archiving of the monitored data; 2) information exchange between heterogeneous databases within the same organization such as pathology labs, radiology labs, ICU, recovery units, outpatient units, etc. and 3) information exchange between geographically separated service-units such as hospitals, pharmacies, healthcare providers, pathology laboratories, datamarts, information clearing houses that provide de-identification and authorization of patient-data, patients through firewalled portals, billing services, insurance providers, data-analytics centers, public health analysis centers and research centers.

4.4.1 Health Service Bus

EMRs are integrated into clinical practices to exchange information between interoperable heterogeneous systems while maintaining the privacy and security. Multiple types of databases reside into a heterogeneous network connected through a *local area network* within an organization. Abstract model of information flow between heterogeneous databases is like the bus architecture used in a computer network. The requested data is converted into a standardized transportable format, transmitted over the Internet and reconstructed at the destination for the visualization to the user according to a permissible subschema.

This information flow through heterogeneous databases is achieved using *Medical Service Bus* (*MSB*) or *Health Service Bus* (*HSB*). *HSB* is built upon *Enterprise Service Bus* (*ESB*) technology. ESB is a message-based middleware to integrate highly distributed and heterogeneous data sources. A generic end-point in ESB consists of: 1) service container; 2) invocation and management framework and 3) an interface. HSB integrates many clinical services, pharmacies, billing services into an integrated system as illustrated in Figure 4.3.

HSB analyzes the contents of messages and routes them. The different end-points connected to HSB are: 1) EMRs belonging to various hospitals, 2) point-of-care accesses like physicians' offices, 3) data warehouses, 4) billing centers, 5) health-data analytics centers, 6) pharmacies, 7) pathology labs and 8) radiology labs. HSB has a common standardized format called HL7. Before submitting an information to the HSB, the data must be converted to HL7 format, transmitted over the Internet using SOAP protocol, and is converted back to the local format of the destination.

4.4.2 Clinical Information Flow

Patients go through a complex set of tasks during a treatment. These tasks are connected to each other through a dependency-graph known as *clinical information flow*. As illustrated in Figure 4.4, a patient going through the cancer treatment has to go through a complex set of tasks.

The information flow involves: 1) initial patient survey; 2) lab tests; 3) patient–doctor encounter; 4) initial diagnosis; 5) radiological image analysis; 6) biopsy and additional lab tests; 7) analysis of the severity of the disease; 8) appointments for the further treatment such as surgical procedures, radiotherapy and chemotherapy if cancer occurs; 9) surgical and other postsurgical procedures; 10) monitoring procedures in a recovery room; 11) patients' discharge procedures and 12) billing. At every step, the corresponding preceding subtasks must be completed before moving to the next subtask. The flow of the tasks also decides the database connectivity and database accesses.

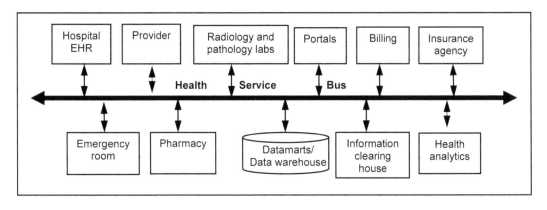

FIGURE 4.3 Health Service Bus

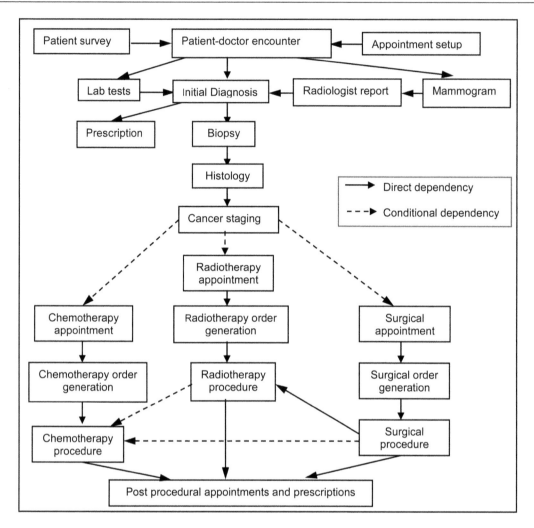

FIGURE 4.4   A schematic of clinical workflow for breast cancer treatment

Example 4.2

Figure 4.4 shows a schematic of a workflow for breast cancer detection and treatment. Initial patient survey and appointment scheduling is followed by a patient–doctor encounter. Based upon the examination, the doctor orders more lab tests and radiology for a better diagnosis. Based upon the lab tests and mammogram (X-ray of a breast), a radiologist or an automated image processing system gives an approximate location of the lump to be investigated further.

A doctor performs the biopsy. The sample is sliced using histological techniques and analyzed in pathological labs for cancerous cells and cancer staging. Cancer staging decides the extent of spread of cancer in the body resulting into the treatment.

Malignant cancers need chemotherapy, the use of chemical compounds to kill the left-over cells after the surgical procedures and radiotherapy, and radiotherapy – the use of radiation from radioactive material to kill the malignant cells. Both these procedures are nonspecific and kill healthy cells as well as cancerous cells. Special orders are generated for surgical procedures, radiology procedures, chemotherapy procedures and radiotherapy procedures. After these procedures, postsurgical disease-management is started that includes regular checkup and prescription of the drugs.

4.4.3 Clinical Interface Technology

Clinical interfaces are used to: 1) display and accept the diagnosis information from physicians, specialists and nurses in a user-friendly and intuitive manner using graphical user interfaces; 2) transform natural language responses from patients to the structured data-format supported by EHR; 3) convert a structured database information in EHR to a patient-friendly response, including human-readable form and 4) transform natural language description of patient's diagnosis, prognosis and monitoring notes by the medical practitioners to an integrated information in EHR. The medical practitioners have different expertise. The output generated must follow their expertise domain. The medical terminology used by the medical practitioners must be converted into the codes used by the international standards for medical terms. Many such standards exist such as ICD-10, SNOMED and LOINC for universal access and information transfer.

After a domain-expert diagnoses the disease or an abnormality, the location, extent and classification of a disease are specified by: 1) filling in of the medical database manually; 2) natural language speech; 3) textual description; 4) specifying the exact location and the extent on a sketch of the organ and 5) natural language voice-dictation. Approach #4 is accurate for localization and the spread of diseases, and other modes are good for narrating the symptoms of the disease.

The framework of translating natural language texts to structured-data consists of: 1) a *parser*; 2) *domain-specific grammar rules*; 3) *XML-based text generator, dictionary of domain-specific reserved words* and 4) *mapping table for the standards* such as *ICD, LOINC and SNOMED*. The scheme has six steps: 1) conversion of natural language texts into lexicons, 2) parsing the stream of lexicons using domain-specific grammar rules, 3) extracting the reserved words from the output using the dictionary of domain-specific words, 4) converting the domain-specific reserved words to the codes in the international standards using the mapping tables and 5) and converting the statements to XML-codes using an ML generator. A schematic is given in Figure 4.5.

The input can be dictations, handwritten notes or a typed natural language text. Dictation is converted to natural language text using an interactive speech-to-text converter. Similarly, handwritten-text is converted to natural language text using an interactive online character recognition system. The interactive converter has

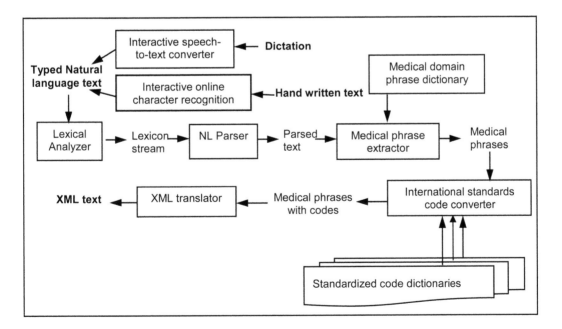

FIGURE 4.5 Automated conversion of natural language text/speech to XML-text

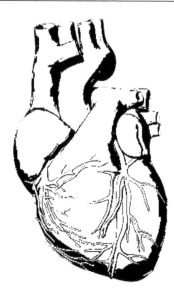

FIGURE 4.6   Visual interface of heart to facilitate marking of the disease area for diagnosis

two additional functions: 1) displaying the converted text to the user and 2) editing the converted text using a speech command. These two functions are essential to maintain 100% accuracy in the medical records.

Natural language text is analyzed using a *lexical analyzer* that converts a text into a stream of lexicons based upon the knowledge of the reserved words and delimiters. The lexicon-stream is passed through a *natural language parser* (NLP) that generates a parsed-text. The generated text is passed through a medical-phrase extractor that mines the text using a dictionary of domain-specific medical phrase-dictionary. These phrases are identified by matching the phrases and keywords used in a medical domain. The extracted medical-conditions, lab-tests, and medical procedures are converted to standardized codes by a software that employs medical knowledge-bases such as LOINC, SNOMED and ICD-10. After the standardized codes are inserted, an XML converter translates the text into an XML-document.

Besides speech-based interfaces, the expert caregivers prefer a detailed visual interface of an organ, as illustrated in Figure 4.6, to mark the locations of abnormalities instead of textual articulations. For example, a neurosurgeon will be greatly assisted by an image of the spinal cord to mark which vertebra is herniated. Similarly, a nephrologist will like a detailed diagram of kidney to mark the site of abnormality in a patient's kidney. The interface should be able to read the markings of the expert physician, and translate it to the textual description of the location along with the abnormality.

4.4.4 XML-Based Representation

XML-based representations of databases store the templates in DTD (Document Type Definition). DTD is a declaration of document structures and attributes of XML entities. The field-names or their synonyms are stored in the DTD. The element values are stored in the remaining XML document. The major advantages of storing database in XML format are: 1) conversion to HL7 format for communication of the patient-record becomes easier and 2) database in XML format can be updated dynamically.

Example 4.3

Consider an in-patient being treated for high sugar-level. His vital signs are monitored. Necessary lab-tests have been ordered to measure blood-sugar and mineral-loss through urine every 6 hours. The patient has entries in the database for making chart for blood-sugar level and minerals such as

```
<?xml version = "1.0">
<!DOCTYPE EMR [
<!ELEMENT EMR ( Patient_Info, Location, Vital_Signs, Observation, Sugar_Level, Lab_Tests, Treatment)>
<!ELEMENT Patient_Info (Patient_Id, Name, DOB, Gender)>
<!ELEMENT Patient_Id (#PCDATA)>
<!ELEMENT Name (#PCDATA)>
<!ELEMENT DOB (#PCDATA)>
<!ELEMENT Gender (#PCDATA)>
<!ELEMENT Location (#PCDATA)>
<!ELEMENT Vital_Signs (Temperature, Heart_Rate, Blood_Pressure, Pulse)>
<!ELEMENT Temperature (#PCDATA)>
<!ELEMENT Heart_Rate (#PCDATA)>
<!ELEMENT Blood_Pressure (#PCDATA)>
<!ELEMENT Pulse (#PCDATA)>
<!ELEMENT Oservation (Observation_Id, Observation_Type, Notes)>
<!ELEMENT Observation_Id (#PCDATA)>
<!ELEMENT Observation_Type (#PCDATA)>
<!ELEMENT Notes (#PCDATA)>
<!ELEMENT Lab_Test (#PCDATA)>
<!ELEMENT TREATMENT (Drug, Therapy, Dosage)>
<!ELEMENT Drug (#PCDATA)>
<!ELEMENT Therapy (#PCDATA)>
<!ELEMENT Dosage (#PCDATA)>
]>
</EMR>
```

FIGURE 4.7   An illustration of DTD for XML description

magnesium, calcium and potassium. The XML version of the database contains the patient profile along with the lab-tests, 6 hourly blood-sugar level tests, the prescribed medication and the dosage. A DTD is given in Figure 4.7. The annotation "!DOCTYPE" defines that the root element is "EMR". The tag "!ELEMENT" defines the nested structure of a record in the "EMR". The tag "#PCDATA" asserts that data is entered.

4.5 ELECTRONIC MEDICAL RECORDS

EMR keeps many types of multimedia and temporal data related to a clinical information such as the lab results, radiology reports, radiology images, prescriptions and medications, history of patients, surgery-related notes, emergency room notes, patient–doctor encounter reports and diagnosis, appointment schedules, progress notes, physical therapy notes, logs of monitoring patients' vital signs, including ECGs, patient history, patient allergies, demographics, physicians' directives for further actions. An overall framework of EMR system is illustrated in Figure 4.8.

The overall clinical information system consists of: 1) Data-acquisition system, including device and clinical interfaces; 2) local EMR databases; 3) knowledge-bases of medical-terms and software, including adapters and transformers and 4) information-exchange system to handle heterogeneity and to exchange information with other datamarts. The patient-specific information, including laboratory results, radiology reports, radiology images, operative notes, clinical information flow, prescriptions, medication administration information, patient history, biosignal analysis and patients' progress notes forms the *medical context system,* which facilitates the diagnosis of the patients' disease-states.

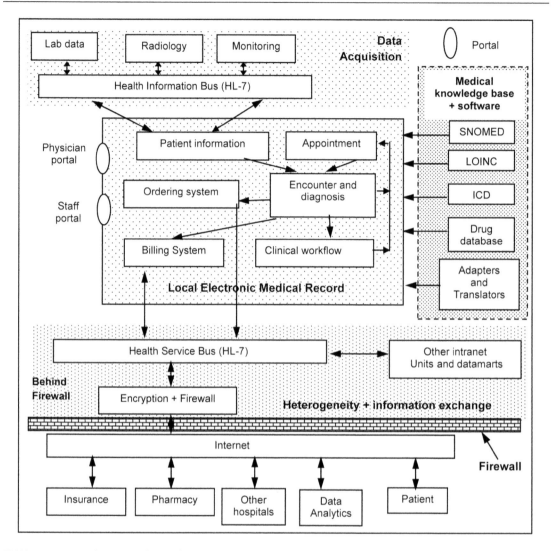

FIGURE 4.8   A schematic of EMR framework

Major technology enablers in a clinical information system are: 1) speech recognition system; 2) text extraction and generation systems; 3) mobile data-access systems; 4) clinical workflow engine; 5) wearable devices, monitoring units and automated lab-data analysis equipment; 6) virtual devices, adapters and transformers; 7) software, including human-machine interfaces, clinical-interfaces, image-analysis software, data-analysis software, data-exchange software and security enhancement software; 8) bus-architecture and Internet-based data communication and 9) query answering system for patient-specific queries (individual and statistical), including temporal and multimedia queries.

Given a patient-related information and her appointment schedule, a patient meets a medical provider. After the encounter, a diagnosis is made. Based upon the diagnosis, a clinical workflow is developed for further investigation and treatment. Based upon this workflow, a new set of appointments is established as shown by the feedback loop between the clinical workflow box and the appointment box.

After the diagnosis, the physician also orders pathological tests, radiology tests, biosignal analysis tests, orders other surgical procedures or orders further investigation by another expert. The order goes over the Internet to other clinics or hospital units using *HIB* in HL7 format. HL7 format is automatically translated to SOAP, as explained in Section 2.9.2.

After the patient–doctor encounter, the physician also orders a list of updated medications to the pharmacy. The order is encrypted and sent through the Internet to the pharmacy where a software decrypts it, and is automatically stored in the database. The database may be in a cloud that can be retrieved by the pharmacist.

A patient–doctor encounter, other lab tests and procedures generate bills handled by a billing system. The billing system uses ICD code-conversion acceptable to the insurance agencies. The electronic invoices are encrypted and sent through the Internet to the insurance companies for the payment. EMR uses multiple technologies such as: 1) database technology described in Section 2.8; 2) structured and unstructured data processing techniques described in Sections 2.1 and 2.2; 3) HIPAA privacy and protection at the database record level described in Section 4.10; 4) semi or automated data entry such as automated data acquisition, voice dictations, handwriting recognition that uses density-based clustering and statistical analysis to recognize characters, direct entry from tablets, and various templates and clinical interfaces; 5) data communication technologies, including clouds and Internet-based communication, adapters and transformers and DICOM; 6) Security technologies such as encrypted storage in cloud, encrypted transmission, strong passwords and using digital signatures for authentication and verification; 7) XML-based wrappers for web services and interoperability and 8) intelligent techniques for content-based information retrieval described in Section 4.7.2, data analysis for disease-discovery, and *clinical decision support systems* (CDSS) to improve operation flow and to help disease-diagnosis by the experts as described in Section 8.5.

Besides structured data, EMR also stores unstructured textual data for: 1) human comprehension and 2) for the verification and legal purposes. The structured data is translated into different languages for international collaboration or treatment across cultural and geographical borders. Automated translation and automated analysis of dictations and text require multiple technologies such as: 1) speech recognition and speech-to-text conversion, 2) natural language understanding described in Section 3.10.1, 3) intelligent text-extraction described in Section 3.10.2, 4) text-summarization described in Section 3.10.3 and 5) image analysis and character recognition based upon density-based clustering. Answering database queries and content-based retrieval require the use of string matching, multidimensional trees, texture and shape modeling, similarity-based search and temporal abstraction, as described in Sections 2.3, 2.5, 2.6 and 2.7.

The major advantages of the EMR systems are: 1) seamless integration; 2) automated report generation; 3) automated data acquisition; 4) reduction in time used in documenting clinical encounters; 5) improvement in appointment and patient–doctor encounters; 6) eliminating the cost of paper charts and trails resulting into cost saving and environmental protection; 7) significantly reducing the time and effort to locate medical records; 8) promoting portability of patient data resulting into inter-organizational and physicians' collaborations, including real-time collaborative surgery; 9) automated charge capture; 10) lower risk of embezzlement; 11) optimization of medical resources; 12) eliminating the duplications of medical tests; 13) better archival and analysis of patient data by keeping a longer history; 14) improvement in treatment using data analytics on big data of patient treatment and recovery; 15) improvement in the economy by using specialists to analyze the lab results in realistic time and 16) increased patients' satisfaction through improved healthcare.

The disadvantages of the EMR system are: 1) increased initial cost of setting up the system and the related infrastructure; 2) additional costs of updating the software systems based upon evolving standards; 3) increased cost of continuous training of existing staff; 4) ongoing costs of vendor support and maintenance; 5) interoperability among heterogeneous software systems developed by various vendors and interoperability between different versions of HL7; 6) existing chasm between the available software developers and medical needs that results in wastage of the time of physicians and medical staff by forcing them to answer nonintuitive mundane questions for verification that results in extra typing time and 7) concerns about data-ownership and control in the data-sharing scenarios. All the disadvantages can be grouped as enhanced cost, additional learning curve, slow adoption due to less than desirable EMR system and human interfaces and issues with data ownerships.

The advantages outweigh the disadvantages in the EMR systems. Besides, as the software developers get trained in developing medical systems and use intelligent techniques in the EMR-human interfaces, the quality of interfaces would improve significantly.

4.5.1 Components of Clinical Information System

Based upon their usage, clinical information systems have been categorized as: 1) ordering system; 2) result reporting systems; 3) clinical documentation systems; 4) data dictionaries for medical-terms; 5) drug databases; 6) clinical pathways consisting of multiple guidelines, protocols and work lists; 7) customer-relation management system; 8) clinical decision support system; 9) data and interaction interfaces, including clinical interfaces; 10) patient survey systems; 11) standardized data transmission formats: DICOM for images and HL7 for patient-specific data and 12) human-EMR interfaces for automated data capture and human-comprehensible data generation.

4.5.1.1 Ordering system

An ordering system consists of three subsystems: 1) order entry system, 2) order management system and 3) order communication system.

An *order entry system* is designed for the use by physicians and clinicians and includes: 1) order set, 2) checking for duplicate orders, 3) drug interactions, 4) allergies, 5) dose-range, specifically for chronic disease management, 6) online medication administration record and 7) work-list for task generation from orders.

An *order management system* uses intelligent systems and knowledge-bases for advanced ordering and decision support systems. It includes; 1) alerting a doctor or pharmacy or patient for reordering the prescription and 2) automated ordering (or prompting) of the procedures and suggested medicines based upon diagnosis and automatically generated clinical pathways using the knowledge bases of clinical pathway templates.

An *order communication system* is related to capturing and transmission of the orders from the physician/hospital to the pharmacy and from the patient to the pharmacy. It consists of: 1) communicating single orders to other hospital units or pharmacy; 2) communicating multidisciplinary orders according to a clinical dataflow to other hospital units; 3) charging for the order; 4) automated refilling the prescription and 5) sending the requisitions.

4.5.1.2 Portals for data-access

There are three major classes of portals to access patient-specific data: *patient portals*, *physician portals* and *employee portals*. *Patients' portals* are used for personal health information, appointment information, calendar and automated time-triggered reminders, online questionnaires and surveys, mail-items, lab-test reports, radiology reports and electronic bills. *Physicians' portals* are used for reference library, secure messaging, results (pathology reports, surgery reports, radiology reports, diagnosis, etc.), prescription management, referral management and hand-held wireless input. *Employee portals* are used for: 1) content management and publishing; 2) workflow tools; 3) billing information management; 4) searching calendar and directories related to a patient; 5) sending medication request to pharmacists; 6) consulting a patient and 7) hand held wireless solution, especially for remote care.

Clinical documentation accessible to authorized physicians and medical practitioners include: 1) clinical flow sheet; 2) patients' charts and history of access to treatment-related information, including people who accessed, time-of-access and location-of-access. The reported results are: laboratory reports, radiology reports, pharmacy reports, diagnosis, ordered prescriptions and clinical workflows.

4.5.1.3 Interfaces

There are multiple interfaces such as: 1) registration process; 2) admission process into in an emergency room, hospital, radiology unit and surgical theater; 3) discharge from an emergency room, hospital, radiology unit and surgical theater; 4) device-interface to the EMR for automated data-acquisition from physiological lab-equipment, vital-signs monitoring equipment, radiology equipment; 5) interactive clinical-interfaces showing various organs for accurate marking of diseased areas; 6) interface for automated information collection of semi-structured information such as dictation, hand-written notes in natural languages; 7) patient-survey interfaces; 8) patient-wellness questionnaires; 9) patient accounting and billing interfaces; 10) appointment-related interfaces and 11) ordering-system interfaces.

The interfaces automatically collect input information and use intelligent translators to convert into a combination of structured and semi-structured data in the HL7 format for the transmission that is again extracted and analyzed using intelligent techniques to structured data for insertion in a database. It also uses a natural language generator to transfer structured data in the database to human comprehensible form.

4.5.2 EMR Views

Each group of users has a different view of a database. This mapping of a database for different group of users is called "views". The database administrators organize data to answer queries posed by different groups without duplicating entries in the database. Duplicated entries will be a serious problem as all the copies of the duplicated entry (possible distributed) will have to be updated simultaneously. These sections describe four major views: 1) patients' view; 2) healthcare providers' view; 3) pharmacists' view and 4) accounting view for billing and claim. However, a multitude of views depend upon the actor's role.

4.5.2.1 Patient's view

The view is specific to every hospital or a healthcare provider. This schematic describes a patient's view of an EMR. Patients can see: 1) lab tests, including images such as X-rays and MRIs; 2) personal profile; 3) doctors' expertise and professional information; 4) their diseases, deficiencies and prescriptions; 5) their allergies and side-effects; 6) surgical procedures performed; 7) appointments and past-encounters; 8) reminders; 9) billing statements; 10) payment information and 11) disputed statements.

The patient-profile table has multiple records each having the information about a patient's account-number, patient-id, date of birth, employer-id, insurance-company, pharmacy-used, current and permanent address, phone numbers, emergency contacts and citizenship. Billing statement's table will include multiple fields such as invoice number, hospital-id, billing date, the amount billed and payment-ids for payment information. Each table has a *primary-key*. Medications will use SNOMED codes. Deficiencies, diseases, allergies and side-effects will use ICD-10 code. Lab-tests will use LOINC code. A simplified schematic of the view is shown in Figure 4.9.

4.5.2.2 Healthcare providers' view

Medical providers have limited access to the patients' disease history and lab-tests restricted by HIPAA policy. Many past records, lab tests, patient–doctor encounters regarding other doctors and unrelated radiology tests are not accessible. In addition, they do not have access to personal data about the patients such as immediate relatives and billing-related information.

Medical providers provide the procedure code, and the remaining billing information is handled by the billing department. The billing department has limited access to the patient's profile, appointments records and messages. This view is regulated using role-based privileges given to the different role-players based on HIPAA requirements. The roles are dynamic and timed.

The nurses in primary physicians' office will have information to vital signs, medications, allergies and side-effects, pharmacy where prescription would be sent, lab tests ordered during a patient–doctor

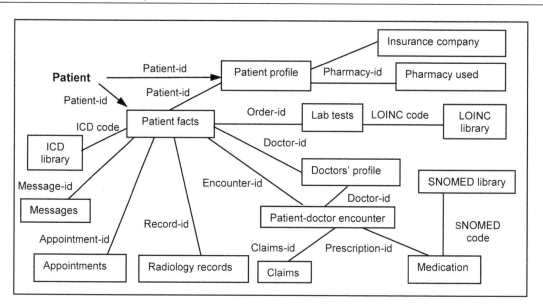

FIGURE 4.9   A schematic of patient's view of the database

encounter. A radiology nurse or a nurse in emergency room or a nurse for an admitted patient may also have access to lab-results or access to radiology-images from the current encounter. The nurse will have restricted access to the patients' diseases, history, including past encounters with the doctor based on HIPAA requirements, and billing information.

A nurse in the hospital for an emergency visit will have more access to personal information as HIPAA requirements are relaxed for emergency conditions. For example, she may get access to radiology reports and other past records based upon the type of emergency. In addition, the nurses for in-patients will have the timed information for monitoring the dosage, vital signs, side-effects to medications, other abnormalities, doctors' notes and recommendations and access to inventory of medications in the hospital. Figure 4.10 gives a schematic of a medical provider's view.

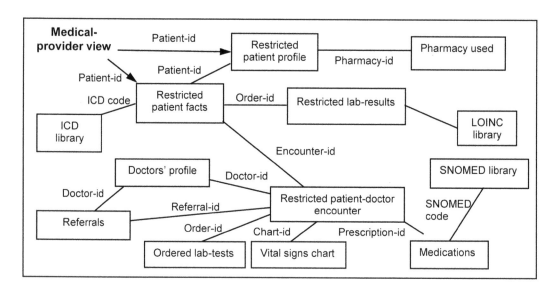

FIGURE 4.10   A medical provider's view of the database with patient's information

4.5.2.3 Pharmacist's view

Pharmacists' view is limited to the dispensing of medications, duration of a prescription, doctors' approval for medications, insurance-plans, pricing of various medications under a plan, generics available for a prescription, prescription date, dosage of medicine, dispensing date, billing for the insurance company, messages from the doctors approving or disapproving a medication, inventory of the medicine, vendor of the medicine, manufacturers, supply schedule, etc.

Each employer (or self-employed) has a *group number* based upon negotiation with the corresponding insurance company. A combination of *group number* and the corresponding *insurance-plan type* sets up the *allowed medication and payments* to be paid by the insurance company. An *insurance-plan* may substitute a brand-name drug by a generic drug, and limit the subsidy paid on a drug. All that information is stored in a table called *Allowed medication + generics + cost*.

A pharmacists' view of the patients' personal information is limited to his/her identification such as the date of birth, social security number, insurance-id, insurance company and the primary insurer on a scheme. Pharmacists do not need to know the emergency contacts, past history, radiology reports of the patients, doctor–patient encounters or claims of the medical providers. A simplified scheme for a pharmacist's view is given in Figure 4.11.

4.5.2.4 Billing and claims view

The claims and billing office is concerned about claiming the cost of the treatment from the insurance companies or the guarantor(s) of the treatment. A treatment is mapped to multiple allowed procedure-sequences performed by medical practitioners. A procedure-sequence is made of multiple procedure codes. Each procedure has the corresponding *ICD-code* allowed by an insurance company. Each procedure-code's price is set up by the insurance company based upon the patient's (or the guarantor's) negotiation with the insurance company and the insurance plan. Medical providers provide the procedures for the disease-code given in ICD-10.

The billing department checks the cost of procedure-codes from the table of allowed procedures-codes corresponding to a disease-code given in ICD-10 and bill for the procedure-codes to the insurance company. Since the billing department is concerned about the claims, it does not have access to patients' medical history and personal profiles according to HIPAA rules.

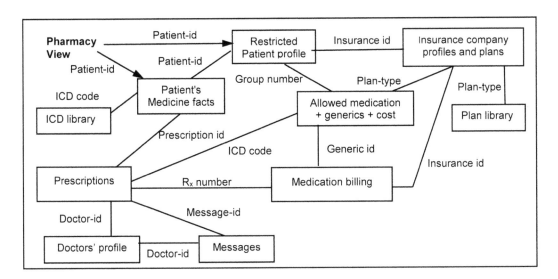

FIGURE 4.11   A simplified view of the pharmacy view of the database about the patient and medication

4.5.2.5 *Case study of OPENEMR*

There are many open source databases such as *Care2X, EHMIS, GNUMed, Medical, MedScribbler, OpenClinic, OpenEHR, OpenEMR, OpenMRS, PatientOS, Ultimate EMR* and *Vista*. These database tools are patient-centric. Available commercial software packages are *AdvancedMD, AetnaHealth, Allscripts, BENCHMARK SYSTEMS, Cerner, chirotouch, CoreCloud, CureMD, eClinicalWorks, ECLIPSE, GE Healthcare (Centricity Practice Solutions), Greenway, HealthFusion, Kareo, MediTouch, MODERNIZING MEDICINE, NextGen, NextTech NueMD, OP, Practice Fusion, PRAXIS, prongoCIS* and *WRS Health*.

OPENEMR is an open-source medical database, used by many small-scale hospitals and clinics in the developing countries, and can become a tool for practicing EMR database in a class setting. The database-schema has tables for storing patients' data, medical-providers' information, patients' medical history, immunizations, automatic notifications, patient-doctor encounters, insurance companies, insurance data, vital sign charts, medical facilities, procedure types, procedure orders, procedure reports, procedure results, prescriptions, drugs, drug inventory, billing, claims, payments, chart tracker, documents, addresses, employer data and list of forms.

Forms are of many types such as such as a patient-survey record, a vital-signs form, a medical-checkup review form, a dictation form, forms for miscellaneous billing options, etc. A patient-survey form contains the information about patient history such as height, weight, fatigue level, insomnia, fever, chills, pain and location, etc. The medical checkup review form contains the information about primary care physicians' findings during a medical checkup such as past surgery, hearing problems, vision problems, nerve conduction problem; etc. The miscellaneous billing form describes various sources of payments depending upon the conditions such as road-accidents, employment-related hazards, etc. Most of the names in the table are self-explanatory.

4.6 HETEROGENEITY IN MEDICAL DATABASES

EHR avoids duplication by allowing access of the patients' records across various medical facilities within a hospital and across hospitals. However, each hospital or medical-provider's clinic has their own local patient-id for the same patient. In addition, each hospital has its own commercial database source and security system. These databases represent data using incompatible data-formats; there is no standardization in data-formats and data structures between different vendors. The only standardizations are in disease codes (ICD-10), lab-procedure codes (LOINC), clinical terms' codes (SNOMED) and transmission of information using HL7. This creates a problem in sharing data across different hospitals and clinics.

To solve the incompatibility problem between the heterogeneous databases, two techniques are used: 1) mapping patient-ids of various hospitals and clinics for the same patient to a common patient-index called *MPI* (Master Patient Index) and 2) providing a VMR (Virtual Medical Record) that allows access to heterogeneous distributed databases located at other facilities using virtualization and MPI. A bigger database schema is created remotely using VMR.

4.6.1 Master Patient Index

MPI is used for heterogeneous patient management systems by sharing information between multiple institutions. Every hospital before transmitting a patient-specific data converts the local patient-id to MPI that is converted back to the local patient-id of the destination hospital. Although, the individual hospitals do not share their local patient-ids, the datamart stores the MPI → local patient-id tables for every hospital to facilitate the conversions. A simplified schematic of MPI is shown in Figure 4.12.

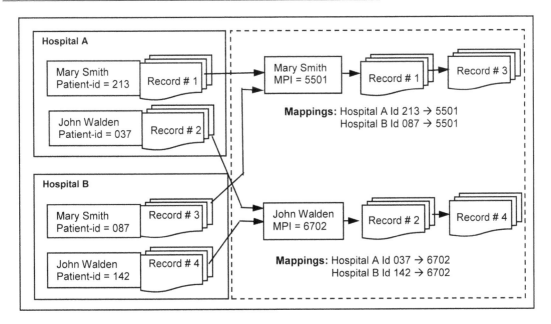

FIGURE 4.12 Mapping different patient-ids from different hospitals to a common MPI

Example 4.4

In Figure 4.12, there are two hospitals A and B. Both have stored the information about two patients "Mary Smith" and "John Walden" in their local databases. The patient-ids of Mary in hospital A and hospital B are *213* and *087,* respectively. The patient-ids of John in hospital A and hospital B are *037* and *142,* respectively. A medical provider wants to access both the records using a virtualized database.

The physical local databases reside with the respective hospitals as they own the data. Two MPIs are created: one each for John and Mary. Let the MPI for Mary be *5501* and for John be *6702.* The local patient-ids *213* in the hospital A and *087* in the hospital B are mapped to the MPI *5501* (Mary's unique MPI), and the local patient-ids *037* in the hospital A and *142* in the hospital B are mapped to the MPI *6702* (John's unique MPI). The VMR shows two records related to the patients.

Individual hospitals carry their own data, and the corresponding local patient-ids. However, data-marts carry the mappings of patient-ids from individual hospitals to the corresponding MPIs. A healthcare provider makes a request using the corresponding patient-id in his/her organization. The local-id of the patient is mapped to the corresponding *MPI.* Then this MPI value is mapped in the datamart to local patient-ids for the hospitals holding the information about the patient. These local patient-ids are used to retrieve patient-specific data from the owner hospitals and transmitted to the requesting hospital.

4.6.1.1 MPI generation

MPI is created using a combination of unique information associated with patients such as the *social security number* in USA (or *national identification number of other countries*), *date of birth, place of birth, gender* and *mother's maiden name* or *father's name* or *lineage.* There are four major criteria to pick up the information fields that are combined to generate MPI: 1) *relative uniqueness* and 2) *permanence of association with patients*; 3) *acceptability* and 4) *universality.* For example, a current address cannot be used because it is not permanent. The unique and permanent information is joined and transformed using a function to a unique index called *MPI* such that MPI does not reveal any patient-specific information to others.

4.6.1.2 MPI-based operations

MPI involves many operations such as retrieval of data, insertion of data, deletion of data, substitution of data to update the latest conditions and removal of duplicate data. Retrieval of data involves: 1) retrieval of data by a healthcare provider; 2) retrieval of data by the patient himself and 3) retrieval of data by a third party such as insurance agency.

A patient-specific data is inserted after the verification that similar data is already not stored in any local database of the hospital or in a remote database. To check the remote databases, a patient's correct patient-id is essential. However, there are many issues related to patient-ids: 1) a patient may not be registered previously in the hospital, thus there is no mapping between newly issued local patient-id and the MPI in the hospital; 2) the name of the patient may have changed, such as a name change of women after marriage; 3) names misspelled due to phonetic similarity; 4) a misspelled date of birth and/or unique identifying information such as social-security-number (US system) and 5) reversal of the first and last names. In such cases, duplicate records for the same person may be created. These duplicate records need to be removed or minimized using proper grouping of relevant secondary attributes.

MPI verification is also needed to 1) retrieve the patient-specific data correctly; 2) remove duplicated medical data and 3) build correct history of progression of disease for a specific patient. This is difficult due to human-error in collecting patient-related information from appointment forms.

4.6.2 Virtual Medical Record

VMR is a unified view of the patient-data stored in multiple heterogeneous databases in different formats. The purpose of VMR is to provide a uniform, seamless view of patient-data for better care and planning of the patients for *Clinical Decision Systems* (CDS). VMR enables different views of data such as patient–doctor encounter, diagnosis and observations. Types of medical data are: numerical data; spreadsheets; text documents; audio such as doctor's dictation; video such as echogram; images such as X-ray and CAT scan and 3D models such as MRI.

The use of MPI, HL7 and XML plays a major role in transparent unification of patient-centric data stored in different databases. VMR maps across multiple heterogeneous databases, embeds locations of various databases in the VMR schema and uses MPI for cross-referencing the patient data. The data is exchanged using HL7 and built-in adapters at local sites that convert local formats to the common XML-based format and vice versa. The overall schema of VMR is shown in Figure 4.13.

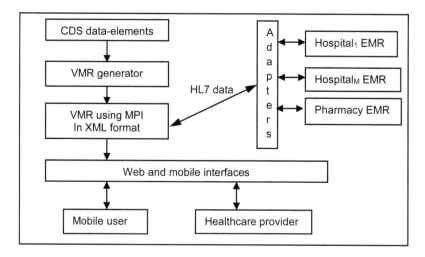

FIGURE 4.13 A simple schema for VMR

Lowest level data-elements used in CDS are: 1) identifying data such as gender, race, age, date of birth, address and care provider; 2) encounter data-elements comprising encounter-identifier, provider type, location type, location, time/interval, encounter status and encounter note; 3) medical procedure data-elements such as procedure-identifier, procedure-code, date-and-time, procedure-site code specifying exact bodypart, procedure modifier code, procedure-status and procedure-note; 4) medication; 5) family history; 6) orders such as prescription, lab results, medications, etc.; 7) lab reports; 8) adverse reactions to treatments and medications; 9) physical findings; 10) CDS context such as user type, information recipient type, information language; 11) patient affiliation and 12) other data-elements such as social history, survey, etc.

4.6.2.1 Retrieval from heterogeneous databases

Search is not limited just to one database. It needs to be searched from other heterogeneous databases. For a patient-specific query, patient-id is used to derive MPI. This MPI is used to identify other heterogeneous databases where patient's records are stored. Using MPI, the corresponding local patient-indices for the heterogeneous databases are picked, and the patient information is retrieved. For similarity-search on attribute-based queries, the attribute-map is derived from the datamarts, and the corresponding databases are searched.

4.6.3 Multimedia Information in Databases

Medical images are of many types: 1) radiology image such as X-ray, CAT-scan, MRI, PET-scan and ultrasound; 2) blood cells, microscope study of biopsy cells, cultured cells; 3) magnified high-resolution images of an internal organ such as a retina; 4) dermatological cells and 5) colored photographs from camera probes such as colorectal cancer, bronchial abnormalities, images of gastrointestinal systems. Multimedia also includes archives of physician's diagnosis dictation, handwritten notes, videos of moving organs such as heart or fetus movement during pregnancy, images of the area being operated on and educational material for training.

Radiology images and ultrasound images are black-and-white and use intensity, texture and shape-based analysis to identify diseases. Multimedia images are also used in remote care to analyze the patient's condition, stress-level including fall detection and postures. Medical images also include the charts used by physicians to mark the disease-affected area of the patient during diagnosis. The use of medical images has been growing rapidly. A remote surgery or history-based diagnosis requires storing, retrieving, editing and transmitting the multimedia objects.

Medical images are archived in a database, transmitted over a distance for the diagnosis by experts, sent for assisted surgery, shown to patients as part of HIPAA requirement and retrieved as evidence in malpractice litigations.

4.6.3.1 Indexing multimedia objects

The multimedia objects are indexed using histograms and other signatures for content-based information retrieval as discussed in Sections 2.5 and 2.6. In addition, the database must support the software packages that perform shape-based similarity matching, caption-based matching and change in the texture and shape of the images to identify abnormalities such as malignant tumors. Images are stored at three levels: 1) actual image for retrieval to be verified by the experts; 2) a histogram of the texture (intensity) of the images for quick similarity-based retrieval of images and 3) a symbolic representation of the image as described in Sections 2.5 and 2.6 for similarity analysis.

4.6.3.2 Characterizing tumors

Medical images such as tumors are characterized by: 1) texture and its uniformity; 2) boundary of the object; 3) shape of the object; 4) intensity contrast and 5) relationship to other objects, including other neighboring tumor growths. Various features such as shape, area, contour and diameter are derived by knowing the contours of the tumor. Tumors do not remain static with time. They evolve and assume new shapes. The evolution results into the change in texture, increase in the size or fusion of two adjacent tumors into one larger mass. A tumor invades, touches an organ, or overshadows an internal organ partially or completely. To understand the progression of tumor, the growth area, spatial area and temporal growth are modeled. The spatial relationships between the objects are characterized by orientations and relative positions.

One problem in representing medical image databases is the huge memory requirement that quickly saturates the available disk space. Image retrievals from the secondary storage and transmission bandwidth are major bottlenecks in transmitting images to another computer.

4.6.4 Temporal Information in Medical Databases

Data conditions of the patients under a surgical procedure, under postsurgical recovery, and in ICU keep changing continuously. These changes are detected and analyzed in near real-time to derive the recovery (or deterioration) rate of the patients' conditions. The analysis is translated into a human comprehensible form to the available nurses for any medical intervention. These necessitate the capture and analysis of a huge amount of sensor data, lab data and the application of relevant intelligent techniques for automated data analytics.

The monitored sensor data includes ECG, blood pressure, drip information, heart rate, respiration rate, hemodynamic condition for kidney patients, blood transfusion information, lab-reports and care-providers' notes. Temporal databases are also needed for logging in the time of patients' admission and discharge from the hospital, provider-encounter time, setting up encounter-time, scheduling nurses' activities.

Many types of queries are asked from these temporal databases such as: 1) What time did an event occur? 2) How long did a condition last? 3) Did the condition occur after a procedure, medical protocol or medication? 4) When did a condition occur before?, etc.

4.6.4.1 Time-stamping

A record in a temporal medical database is time-stamped. The record is represented abstractly as (*<parameter-id, value, context>*, *time-stamp*). A time-stamped record facilitates time-related queries and record comparisons. Time-stamps can be *instant time-stamps,* also called *point time-stamp,* or *interval time-stamps.* Multiple *time-stamps* are summarized to a larger *interval time-stamp* if a monitored condition does not alter during the interval. *Interval time-stamps* have two fields: *start-time* and *stop-time.* An issue in temporal databases is time-granularity. Different data-elements are recorded with different periodicity.

For example, ECG is recorded with a periodicity every few milliseconds, while lab-tests such as blood-sugar analysis of a diabetic patient is done every few hours. The timed facts are valid for certain duration. Beyond that duration, the data is not valid. For example, if a patient had elevated blood-sugar yesterday does not mean that (s)he will have elevated blood-sugar today. The time interval when a timed-fact is valid is called *validity-time.*

Desired features of temporal databases are: 1) compatibility with standard query languages such as SQL-standards; 2) compatibility with nontemporal databases; 3) temporal-joins; 4) fuzzy expression of time such as "around *4* AM," "between *1* and *2* PM," "three days back," etc.; 5) uncertainty-based reasoning to estimate conditions having different time-granularities; 6) automatic summarization of time and 7) statistical summarization using aggregation functions such as mean, median, range, etc. Due to different time-granularities and time-intervals, temporal joining is not straightforward as discussed in Section 2.7.

Automatic summarization of time is needed to compress the past information to reduce memory storage requirements. Statistical summarization is needed to compress and analyze the data.

However, statistical summarization requires a temporal context. For example, a query can be: *what was the average blood-glucose level of a patient last week?* This value will differ completely from the average blood-glucose level today.

4.6.4.2 Temporal inferencing

There are five types of temporal-reasoning used in temporal-inferences from temporal databases: 1) temporal context restriction; 2) vertical temporal inference; 3) horizontal temporal inference; 4) temporal interpolation and 5) temporal-pattern matching. *Temporal context restriction* is necessary for limiting the inference. Context could be related to a specific medical procedure, including surgery for specific medical condition. *Vertical temporal inference* generalizes to or specializes from higher-level concepts. *Horizontal temporal inference* uses similar medical information to derive inference. *Temporal interpolation* joins times-stamps to make bigger intervals for similar data by interpolating that the condition would gradually change between two disjoint intervals or remain unchanged. *Temporal pattern matching* infers based upon complex patterns. For example, one wants to infer the time of a day in a week when the glucose-sugar level goes down for a patient.

Standard relational queries are joined using Boolean operators "AND," "OR" and "NOT." However, time-stamp also necessitates the use of inequality operators to provide a temporal relationship. Two queries Q_1 and Q_2, in a database without time-stamps, can be joined as $(Q_1 \wedge Q_2)$, $(Q_1 \vee Q_2)$, $(\neg Q_1 \wedge Q_2)$, $(\neg Q_1 \vee Q_2)$, $(Q_1 \wedge \neg Q_2)$ and $(Q_1 \vee \neg Q_2)$. However, in the presence of time-stamps, the queries should handle inequality relationship and equality relationship involving time-stamps. The conjunctive query $Q_1 \wedge Q_2$ is further qualified by time $T_1 < T_2$, $T_1 > T_2$, or $T_1 = T_2$.

For example, the query "which patients, who were admitted yesterday, were tested for blood-sugar?" has time T_1 less than time T_2, where T_1 is the time of admission and T_2 is the time of testing blood-sugar. In the query "which patients, who were admitted yesterday, had heart-attack in the past 6 months?" time T_1 is greater than time T_2, where T_1 is the time of admission and T_2 is time of heart-attack. In addition, T_2 is further constrained to last 6 months.

Often, medical providers need to visualize the history of a patient during the diagnosis or the treatment. The time-pattern for data-visualization has been divided into four categories: 1) *snapshot*; 2) *time-interval*; 3) *periodic time-slice*; and 4) their combinations. A *snapshot* is value recorded at a point of time such as a lab-test, time-of-admission, and doctor-patient encounter-time. A *time-interval* is a duration of time when an abnormal condition existed such as: duration of elevated blood-pressure, duration of ventricular arrhythmia – an abnormal heart condition with irregular heart-beats. *Periodic time-patterns* are found in: 1) serial representation of multidimensional data visualization, 2) handling medical conditions that vary periodically such as seasonal allergies, and 3) periodic events such as medications and their effects.

4.7 INFORMATION RETRIEVAL

One of the major aspects of the medical databases is to retrieve the information fast. The information to be retrieved is: 1) patient-id based; 2) attribute-value based or 3) content-based similarity-search as described in Section 2.6.

Patient-id-based information retrieval is an equality-based search and uses standard index-based search retrieved using hashing techniques within the local database. However, when searching heterogeneous database systems, the patient-id is translated to MPI, uses the VMR and finds the MPI mapping to local patient-ids in other databases to retrieve the information.

Attribute-value-based information retrieval involves multiple attribute-value comparisons connected using logical operators. Attribute-values are retrieved using traditional SQL queries. Temporal queries also satisfy time-constraints. Major issues in an attribute-based search are the presence of: 1) typographical errors, 2) homonyms (words with same sound or text), 3) medical terms written differently in different cultures, 4) personal names written differently and 5) same information abbreviated in different forms such as "Dr." and "Doctor," "Ln." and "Lane," "Mister" and "Misseur" etc.

Attribute-based matching also requires the similarity-search that uses a dictionary of subsequence of words with different spellings but sounding the same phonetically such as: 1) "au" or "o"; 2) "e" and "a"; 2) "ee" or "ii" or "i"; 3) "u" and "oo"; 4) "v" or "w"; 5) "c" or "k"; 6) "y" and "ie" or 6) use of "c" to sound like "ch" in many Italian names where one or two characters may get mismatched.

In content-based access, a similarity-based search is performed on the relevant subset of parameters. Images are modeled as histograms, texture-based representation, shape-based representation or strongly connected graphs. Different media use different features for the content-based information retrieval. For example, matching sound requires histograms of frequency patterns.

4.7.1 Matching Similar Records

Records are matched using similarity-based search by: 1) grouping attributes, so they act jointly as a primary key, 2) grouping primary-keys with other attributes such as address or age or last name to reduce the possibility of mismatch due to typographical errors, 3) matching the data values of the records identified by similarity-based search and 4) a combination of first three techniques.

Many matching techniques are used for similarity matching such as: 1) *approximate string matching* using dynamic matrix for strings of unequal sizes; 2) *Hamming distance* to identify the positions where characters mismatch for the same length string; 3) *shift-or algorithm* that returns a binary outcome of match and mismatch: matching positions return *1* and mismatching positions return *0* or *4*) *bigrams* and *trigrams*. *Bigrams*-based matching keeps the count of the consecutive pair of matching characters, and *Trigrams*-based matching keeps the count of three consecutive matching characters.

The advantages of similarity-based search are: 1) removal of duplicated records; 2) identifying the chronological order of tests, visits and diagnosis of a patient; 3) identification of overlap in the information and 4) separating records of different patients.

Example 4.5

Matching the words "Mary" and "Marie" will have a count *1* for the matching trigram "Mar." The trigrams in "Mary" are "Mar" and "ary." The trigrams in "Marie" are "Mar," "ari" and "rie." The matching bigram count for the same words will be *2*. The bigrams for the name "Mary" are "Ma," "ar," "ry," and the bigrams for the name "Marie" are "Ma," "ar," "ri" and "ie."

4.7.1.1 Probabilistic record-linkage

Similarity matches based upon heuristics or data-values of grouped attributes are probabilistic. Probability is higher if many data-values of the grouped attributes match. The probabilities are combined using *Fellegi and Sunter's probabilistic model* of record-linkage. The model uses relative frequency of strings being compared. For example, an attribute-value that occurs with low frequency is more discriminating. The models use statistical *matches* and *unmatches* between the attribute-values to derive the overall weight of an attribute to combine probabilities.

Matches is defined as the fraction of matched-pairs of records in two populations being matched that agree on identifiers. *Unmatches* is defined as the fraction of the matched-pairs of unrelated records in the same population which do not agree on identifiers. *Nonmatches* differ from *unmatches*. *Nonmatches* are defined as the probability of nonmatching records. The agreement between two strings is measured using: 1) *Jaro–Winkler distance* or 2) Levenshtein distance, as described in Section 2.3.

Example 4.6

The *unmatch probability* U of two persons having the same date of birth is greater than $\frac{1}{100} \times \frac{1}{366}$ under the assumption that people live for *100* years or less. The term $\frac{1}{100}$ gives the probability of being born in the same year, and the term *1/366* gives the probability of being born on the same date, including the leap years.

Matching is not perfect leading to many false-positive and false-negative matches. False-positive matches lead to: 1) incorrect information and inference of patient's medical condition, 2) removal of genuine records due to incorrectly being marked as duplicates and 3) merger of genuine record with incorrect records. False-negatives will cause: 1) loss of history or 2) loss of information causing the medical practitioner to reorder lab-tests causing increased cost and increased duplications in the database.

4.7.2 Content-Based Image Retrieval

Content-based image retrieval requires: 1) building the histogram (or image-graph) of the multimedia query using features; 2) converting histogram to an N-dimensional key; 3) searching the N-dimensional space using N-dimensional access trees such as KDB trees, Hilbert R-trees, SS-trees associated with the multimedia database to identify the best matching identifiers; 4) matching the histogram of the query-image with the retrieved histograms of the images and retaining the best matches; 5) performing detailed feature-space matching of query image with the sorted set of best matches and generating similarity scores and 6) sorting the similarity scores to find the best matches.

4.7.2.1 Histogram-based quick retrieval

Histograms are used for the quick retrieval of similar images. A macro-block is a basic unit for image to improve computational efficiency and reduces noise present in individual pixels. The attribute-values in matrix representation of macro-block attributes are summarized to histograms for similarity-based matching of images. A histogram is made from each relevant attribute such as color of the pixels, pixel intensity, texture patterns, distance of the vertices from the centroid for approximating polygons, orientation of the edges in the approximating polygons and their combinations. Each histogram maps to a key that preserves the frequency-pattern in the histogram. The histogram-based similarity-search becomes a similarity-based multidimensional search.

Dissimilarity is calculated from histograms by calculating Euclidean distance of the differences in the frequency patterns in corresponding interval in pixel-intensity space. Mathematically speaking, given two histogram $H_1 = <f_1^1, ..., f_1^N>$ and $H_2 = <f_2^1, ..., f_2^N>$, the dissimilarity-score ds is calculated as $ds = \sqrt{\sum_{i=1}^{i=N}(f_1^i - f_2^i)^2}$, where f_j^i is the frequency in the ith interval of the jth histogram, and there are N such intervals in each histogram. If the dissimilarity-score $ds < threshold$ δ, then the image is selected as a candidate match.

Feature-extraction is used to identify various features in an image. Using the feature-values, a histogram is created. Histogram is processed using an algorithm to create a scalar or N-dimensional index. This index is used to search a K-D tree or R-tree variations to identify M-nearest neighbors. These images are then aligned and matched.

The set of matching images are sorted in the ascending order of dissimilarity scores to rank them. These ranked images are compared using multiple modalities such as texture-based matches, shape-based matches and graph-based matches with the corresponding modes in the query-image. The dissimilarity score from each mode is sorted and combined using weighted multimodal fusion for overall ranking. The overall scheme is given in Figure 4.14.

4.7.2.2 Matching ranked images

The similarity-based matching of images uses a combination of low-level feature-based matching based upon texture, color, shape attributes and spatial relationship with respect to a fixed reference-point, as

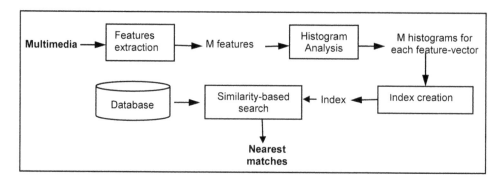

FIGURE 4.14   Content-based retrieval in multimedia database

described in Section 2.5. These techniques cannot identify small variations in normal control images and abnormal images with tumor and other abnormalities. Minute variations, visible to a trained expert's eye, are not detected by the automated techniques. Besides, the textual description of the features and the required thresholds vary based upon the difference in technical vocabulary used by experts having different expertise. To overcome these limitations, multilayer intelligent techniques and semantic interoperability are integrated with the similarity-based matching techniques.

4.7.2.3 Image alignment

An image in a database is aligned with a query image to be compared accurately. The process of alignment is called *image-registration*. The registration process involves finding out a combination of translation, rotation, scaling and intensity transformations. Theoretically, image registration maps points in two images I_1 and I_2 using Equation 4.1.

$$I_2(x, y) = f^I(I_1(f_x(x, y), f_y(x, y))$$ (4.1)

where f^I is the intensity transformation, f_x is the two-dimensional transformation of the x-coordinate and f_y is the two-dimensional coordination of the y-coordinate. If the sensor and the lighting conditions are the same, the transformation f^I becomes an identity transformation, and is ignored.

There are many techniques to align the images correctly such as: 1) correlation-matrix 2) Fourier transforms; 3) matching the same set of externally inserted landmarks in two images and 4) matching of the implicit landmarks in the medical images.

Alignments of radiology images described in Chapter 5 are based upon landmarks. The images are first scaled so the cumulative sum of the Euclidean distances between the adjacent landmarks in the database-image is within a threshold of the distances between the adjacent landmarks in the query-image. The landmarks are then represented as the vertices of the graphs. After achieving scaling invariance, the graph from the query-image is matched with the graph constructed from a recorded image for different combinations of vertex-pairing. The cumulative absolute magnitude of differences of the corresponding edge-weights is calculated for each combination. This cumulative score gives the overall dissimilarity-score. The set of vertex-pairs matchings that provides the minimum dissimilarity-score aligns the graphs. The orientation derived with this minimization corresponds to image alignment. After aligning the images, the regions of interest (ROI) in two images are matched for any change in texture or shape.

4.7.2.4 Invariance

The major problem in the image matching is to ascertain that the information is stored in an invariant form that does not get affected by the image-rotation, image-zooming, noise in the image acquisition under varying ambient conditions. Similarly, the speech-representation should not be affected by change in emotions, sampling rate and background noise. To handle the invariance in images, many techniques are used.

General techniques in providing invariance are to normalize the relevant parameters to minimize the effect the variations under scaling, rotation and noise. If two parameters are equally affected, then their difference provides invariance. If two parameters are affected by the same proportion, then their ratio provides the invariance. Rotational invariance can also be provided by using: 1) relative orientation of the adjacent edges in polygon-based modeling of contours, as described in Section 2.5.2, and 2) normalized length and orientation with respect to the major-axis of an irregular image.

The noise in an image due to uneven lighting is reduced using many heuristic schemes such as: 1) the difference in the intensity values with respect to minimum intensity; 2) dividing the intensity by the ratio of the averages (or median) and 3) using intensity gradient or local binary pattern instead of actual intensity, as described in Section 2.5.1.6.

4.8 INTEROPERABILITY AND STANDARDIZATIONS

Organizations and experts have different ways of encoding the same medical procedure. The same medicinal compound may have different names depending upon the manufacturer. Different experts may describe the disease differently depending upon their domain and audience. Some experts may describe the same abnormal condition as a disease of a subpart of an organ, while others may describe it as a disease of the organ. Some experts may describe the disease to a patient in simple language, while among experts, the disease may be described in technical terms for the use by other healthcare providers.

There are different types of information such as medical images, medical procedures, medications, lab tests that are shared across different healthcare providers. This information needs to be standardized for better comprehension and inferencing. International community has been making study progress in developing standardized codes and international standards such as DICOM, SNOMED, LOINC, ICD and ArdenML to share information.

4.8.1 Knowledge Bases of Medical Terms

There are many ongoing international standardization efforts for the interoperability of different procedures and codes. The major popular ones are DICOM (Digital Image and Communication in Medicine); SNOMED (Systematized Nomenclature of Medicine) and LOINC (Logical Observation of Identifier Names and Codes); ICD (International Statistical Classification of Diseases). Major advantages of a standardization are: 1) improved expressions, 2) reduction in ambiguity, and 3) free electronic flow of data with no modification. The standardized information is shared over the Internet using a standardized XML-based format called HL7.

DICOM is an image standard for patient-specific image communication for disease treatment and management, which has been described in Chapter 6. SNOMED is a structured standard for the exchange of medical procedures' information. It contains terminologies used in medical procedures, observations, histories and medical billings. SNOMED is maintained by IHTSDO (International Health Terminology Standard Development Organization). LOINC is a structured standard for detailed coding of clinical results, lab results and vital signs to facilitate data archival and exchange. ICD is an international structured standard for coding diseases, including external causes of injury, signs and symptoms. It contains a comprehensive list of all the diseases documented by WHO (World Health Organization). ArdenML is a standard XML-based language for representing clinical information in an executable format for clinical decision support systems. It has been integrated with HL7 to transmit embedded executable medical programs and triggerable knowledge-bases over the Internet to different medical providers.

One issue with the standardized codes is that they are still evolving. Often, newer version is incompatible with older version delaying the adoption by the healthcare industry.

4.8.2 SNOMED – Medical Code Standardization

SNOMED is a structured standardized terminology for the exchange of information between heterogeneous databases across different geographical location worldwide. These clinical terms are used by healthcare providers for describing patient care.

SNOMED code describes clinical terms, procedures, clinical histories, physical examination, tests, diagnoses, therapies, their attributes, their connections and relationships. It includes synonyms, variations across subdiscipline, language variations, spelling variations and misspellings by the medical providers. SNOMED is continuously updated to accommodate new clinical terms and technological improvements in drugs and devices.

SNOMED helps in localizing the clinical findings and procedures, specifically to a body-part. It allows for the aggregation and grouping of clinical terms to facilitate: 1) demographic study; 2) minimization of data entry; 3) administrative and financial coding and 4) association with a clinical finding or procedure. It supports many-to-one, one-to-many and many-to-many mappings. Multiple clinical-terms are associated with the same clinical finding. Multiple procedures are associated with the same clinical-term. SNOMED also supports cross-mapping of overlapping clinical terms to other international standards such as ICD-10 and LOINC.

4.8.2.1 SNOMED structure

SNOMED CT consists of a coding scheme, classifications and terminologies: 1) *reference terminology* relating to other concepts and 2) *interface terminology* for the ease of information entry and display. The coding scheme includes: 1) a simple abbreviation of a clinical term; 2) an identity code of the term and 3) simple, concise and consistent meaning of the abbreviation. It allows for: 1) retrieval of information related to the specific or related code using a structured relationship between the concepts; 2) aggregation of information based upon specific requirements; 3) alternative user-friendly synonymous ways of expressing the same term and 4) support for different language groups, clinical disciplines and specialties. It supports multiple dictionaries to express the meaning and synonyms.

There are three building blocks of SNOMED: *concepts*, *descriptions* and *relationships*. Each *concept* is described using a set of *descriptions*, a specified name, a preferred term and multiple synonyms. Several concepts may have the same terms in their descriptions. Concepts are related to other concepts using *relationships*. *Relationships* describe: 1) multiple attributes associated with a concept; 2) relationship between attributes; 3) causative agents and 4) class hierarchy (subtype or supertype) described by "is-a" relationship. "Is-a" relationship facilitates ontology-based reasoning.

Example 4.7

An example of "is-a" relationship is *ventricular arrhythmia* → *arrhythmia* → *cardiac disease*. "Is-a" relationship supports one-to-many mappings. For example, *viral pneumonia* → *infectious pneumonia* → *infectious disease*. It can also map as *viral pneumonia* → *infectious pneumonia* → *pneumonia* → *lung disease*. While the former chain is disease classification, the later chain points out to the location of the disease.

There are around *600,000* concepts, *1,750,000* descriptions and over *2,600,000* relationships. This information is stored in indexed tables. The information is retrieved using clinical-term codes. The identity code of a SNOMED term has four parts: 1) item identifier; 2) partition identifier: "00" for concepts, "01" for description and "02" for relationship; 3) check digit that states the number of characters used in the code and 4) extensions. The extensions are used to allow modifications of the code for international use by healthcare providers while keeping standardized information intact.

4.8.2.2 SNOMED concepts

SNOMED concepts are qualified by different hierarchies/types. There are 41 such *concept-types*. Major concept-types are: 1) a disorder or disease such as myocardial infarction; 2) clinical findings; 2) procedures; 3) an abnormal structure such as a tumor in the brain; 4) an observable entity such as a lump; 5) a substance; 6) an organism such as *Escherichia coli*; 7) a pharmaceutical/biologic product such as *Amoxicillin* and 8) a situation with explicit context such as an accident, genetic history and a patient's medical history. Some other concept-types are: body structure, event, situation, attribute, environment, specimen, cell, assessment, geographical location, cell structure, ethnic group, tumor staging, physical force, religion, lifestyle, racial group, etc.

A name may occur in more than one concept-type. The meaning of a term may change with the context, and represent a different clinical-term. For example, the word "dressing" means: 1) the activity of putting on clothes in the concept-type "observable entity"; 2) wrapping on top of a wound for healing in the concept-type "physical object" or 3) the act of wrapping around the wound in the concept-type "procedure."

Information about concepts is nested shown by the arrow "←" in the following examples. The arrow "←" shows subclass relationship. For example, *organism* ← *microorganism* ← *E. coli* ← *E. coli strain K-12*. A pharmaceutical product may be an *anti-infective agent* ← *antibacterial oral drug* ← *beta lactam antibiotic* ← *penicillin class of antibiotic* ← *amino-penicillin* ← *Amoxicillin* ← *oral form of Amoxicillin* ← *Amoxicillin-dosage*. An observable entity could be *clinical history* ← *cardiovascular observation* ← *ECG* ← *heart rate*. This nested information helps in detailed localization involving different levels.

The major coded components of *clinical findings* are: 1) finding site; 2) associated morphology (structure); 3) associations with site, substance, physical object, physical force, organisms, including microorganisms, pharmacological products, including side-effects of medicines, procedures and events, including trauma causing events; 4) causative agents such as organisms, substances, physical objects, physical force; 5) clinical course – onset and duration; 6) finding methods such as patient-encounter, lab tests, procedures and 7) informer of the findings other than the patient, such as a medical provider, a provider of the history, a performer of the method and a subject of a record.

The major coded components of a procedure are: 1) procedure site; 2) procedure morphology, including abnormal structure; 3) procedure device; 4) procedure method; 5) direct substance. *Procedure-site* is a body structure or anatomical site. *Procedure device* is a device name, an amount of physical force, or an access instrument. *Method* is the standardized and catalogued approach to perform a *procedure*. *Substance* is any biological product needed in a procedure such as anesthetics, an injected substance, etc.

4.8.2.3 SNOMED representation

SNOMED representation uses compositional grammar to combine or refine clinical-terms. The compositional grammar supports: 1) attribute-value pairs describing the meaning of a clinical-term called *expression*; 2) combining two or more expressions connected using logical-AND, denoted by "+"; 3) the refinement of a clinical-term to a combination of simple attribute-value pairs; 4) nesting by defining an attribute-value as an expression; 5) *concrete values* such as integers, strings, floating point numbers and 6) definition status, whether an expression is "equivalent to" for alternate meaning or "nested expression." For example, a case of multiple fractures is expressed as a combination of multiple SNOMED codes one for each fracture. The overall abstract model of the compositional grammar is given in Figure 4.15. Compositional grammar allows SNOMED expressions describing a concept to be transmitted as a string.

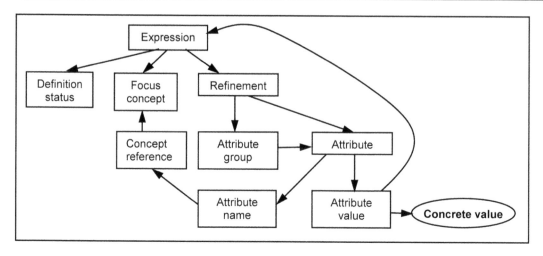

FIGURE 4.15   An abstract model of SNOMED CT compositional grammar

4.8.3 LOINC – Lab and Clinical Data Standardization

LOINC is a standardized code for lab tests, clinical measurements and observations. LOINC database includes: 1) observations reported by clinical laboratories; 2) results of therapeutic drug monitoring; 3) toxicology study of toxic compounds' built-in responses to the administration of therapeutic drugs; 4) serology – study of blood serum; 5) hematology – study of blood-related diseases affecting production and regulation of various components of blood such as blood cells, hemoglobin, blood proteins, bone marrow, platelets and mechanism of coagulations; 6) blood bank study; 7) microbiology, including lab results involving pathogens – disease causing bacteria; 8) cytology – study of cell structure and function including cell culture studies helpful for screening tests related to cancer cell detection using various body-fluids and brush tissues such as a pap-smear; 9) surgical pathology – study of the tissues removed using surgery and 10) fertility studies such as sperm count, ovulation tests, checking of hormone levels. The results include nontest level measurements such as menstrual cycles, drug doses and units of concentrations.

LOINC does not consider various devices, fine details of samples, site of collection such as bedside or clinical lab, verification details, size of samples and priority of tests. However, if the tests involving some lab data are different while measuring the same substance, then the tests are given separate codes. For example, serum-sodium and urine-sodium are treated differently and have different LOINC codes.

4.8.3.1 LOINC code structure

LOINC codes contain information about various properties and units involved in the lab tests. It includes: 1) name of the component; 2) observed property such as density, concentration, mass and volume; 3) timing of the measurement: point-time or time-interval; 4) type of sample; 5) method of measurement; 6) type of scale such as *qualitative, quantitative, ordinal* – values with natural ordering, *nominal* – values with no natural ordering, *multi* – multiple results concatenated as one observation, *document* having some format and *set* and 7) type of method such as coagulation assay, immune blot, immunoassay, complement fixation, DNA probe, etc. The component name may carry an embedded nested structure to take care of class-hierarchy. Each subclass is separated by a period. The code contains abbreviations that have one-to-one mapping with extended terms. For example, "1M" denotes *1* minute and "2H" denotes *2* hours.

Internally, LOINC code uses *short-names* up to *30* characters to represent a *concept*. A *short-name* is unique and maps onto a *long-common-name* that is displayed and used by care-providers and patients. Each LOINC term belongs to a class such as microbiology, blood bank and serology. A result or finding is modeled as a value or a combination of multiple values. Multiple values are associated with a variable. For example, "variable X Findings 1, Findings 2, Findings 3" will associate three findings with the variable X. The use of variable X anywhere in the representation will provide an activity on all three values.

Like SNOMED, multiple expressions are joined using a symbol "+." Different components of an observation are separated by a colon. For example, the components of blood-type-related information are: blood-type of each pack, how much blood was given, time of donation, any adverse reaction to the pack and the pack number.

Components are abbreviated using a lookup table, and the codes are separated by semicolons. There may be more than one pattern of components in an observation type. LOINC identifies the observations. SNOMED is used for other information such as microorganisms, devices and procedures.

Example 4.8

Fully-specified-name (FSN) of the white blood cell count test result is *Leukocytes: NCnc: Pt: CSF: Qn: Manual count*. The long-common-name is *Leukocytes [#/volume] in Cerebral spinal fluid by Manual count*. A corresponding short name is *WBC # CSF Manual*.

4.8.4 ICD (International Classification of Diseases)

ICD is an international standard for coding diseases, and their causes, including external causes of injury, signs and symptoms, complains, and social circumstances. The purpose of ICD is to record and analyze various causes of mortality and morbidity collected in various countries for better health care. ICD is used to translate and archive various textual descriptions of diseases to standardized code. Two versions are being used by medical practitioners ICD-9 (older version) that is transiting to ICD-10 (current version). In the current version, ICD-10, there are over 70,000 codes. There are 21 classifications of various types of diseases under which the codes are distributed.

There may be more than one disease code associated with mortality and morbidity. There are a general principle and three selection rules to derive the cause of mortality and morbidity. The general principle is: If there are more than one condition and the lowest condition can derive all other conditions, then the lowest condition is treated as the cause of mortality or morbidity. In the absence of general principle, the conditions are reported in the order of severity likely to cause mortality. There are three selection rules: 1) If there are more than one cause, then the originating cause is selected; 2) If no reported condition derives the first condition, the first condition is treated as the cause of mortality; 3) in the absence of Rules 1 and 2, if a condition Q is the consequence of another condition P, then the root-condition P is chosen. If a condition reported is a trivial condition unlikely to cause mortality, then more serious condition is chosen.

4.8.4.1 ICD classification

Major classifications are: 1) infectious and parasitic diseases; 2) neoplasm – growth of tumor; 3) blood and immunity-related diseases; 4) endocrine and metabolic diseases such as diabetes; 5) mental and behavior disorders; 6) diseases of the nervous systems; 7) eye diseases; 8) ear and nose diseases; 9) diseases of the circulatory systems; 10) diseases of the respiratory system; 11) disease of the digestive system; 12) diseases of the skin and subcutaneous systems; 13) muscle-related diseases; 14) diseases of the genitals and urinary systems; 15) diseases related to pregnancies and child birth; 16) diseases related to child birth; 17) genetic and cancer-related diseases; 18) symptoms based upon lab findings; 19) external causes of mortality and morbidity; 20) factors influencing health and 21) reserved for other diseases.

Diseases are further grouped into five tables: Tables 1 and 2 are for general mortality; Tables 3 and 4 are for the mortality of children below *4* years and Table 5 is for the morbidity-related diseases. A code is further qualified using additional descriptive terms within a parenthesis to separate between various conditions while retaining the same code for same disease. A parenthesis can also be used to show exclusion of certain term from the disease if preceded by the reserved word "excludes." Square brackets are used to describe synonyms and alternate description of a disease. A colon is used to further qualify a term in the description. Braces denoted by vertical line "|" are used to show mutual dependency of the preceding and following terms. If the causative agent of a disease is unknown then it is marked by the abbreviation "NOS" (Not Otherwise Specified). A "–" as the fourth character means that the subcategory needs to be looked up from the disease table. Cross-references are used to avoid duplication of the terms.

4.8.4.2 ICD code structure

The basic ICD is a list of three alphanumerical characters. The first position uses a letter, and second, third, and fourth positions use a number. The first letter corresponds to the basic *21* classifications. For uniformity, all representations use four characters. Fourth character, a digit, follows a period, and represents one of the ten subcategories. Possible code numbers vary from A00.0 to Z99.9. The letter "U" is not used. The codes "U00" to "U49" are reserved for future diseases of unknown origin.

The top-level three-character classification is based upon frequency, severity or susceptibility to public health intervention. Some of the three-letter categories represent single condition based upon their frequency. If a three-letter category is undivided, then letter "X" is used for the fourth character. A disease is further categorized optionally by diagnosis and manifestation of the disease elsewhere in another symptom.

The primary code for a disease is annotated with a following dagger symbol. The manifestation of a disease is annotated by a following asterisk "*." For example, the manifestation of Parkinson disease elsewhere is *G22**. Its primary code is *A52.1*. The further subdivision allows ICD code to go up to seven characters.

4.8.5 ARDENML – Executable Clinical Information

ArdenML is an official XML-based markup language for representing clinical information in an executable format for CDS. It has been integrated with HL7 to transmit embedded executable medical programs and triggerable knowledge-bases for CDS over the Internet to different medical providers. It incorporates various features suitable for modeling medical information, including point-time and duration units. It is also used for data analytics to predict patient's recovery, mortality and prognosis.

4.8.5.1 Medical logical module

The information is contained in self-contained files called *MLM* (*Medical Logical Modules*). MLMs are activated using explicit calls or time-triggered events. The content of an MLM is divided into three blocks: 1) *maintenance*, 2) *library* and 3) *knowledge*. A *maintenance block* keeps the information related to the author, version number and the creation-date. A *library* block refers to medical information. The *knowledge* block contains actual algorithms, database and decision-making code.

Knowledge is expressed using declarative if-then-else logical rules that support the ease of modification and maintenance. The logical rules combined with procedural codes are embedded in the XML syntax. MLM follows a rigid structure to separate logical rules form procedural codes, including variable declarations, interface with external data-sources, and services to improve maintainability and future enhancement.

Knowledge block is divided into various slots to improve comprehension and transparency. *Knowledge.evoke* slot contains the event-based triggering mechanisms. *Knowledge.logic* contains the logic and procedure for decision making. *Knowledge.action* slot provides various I/O actions such as writing into database and display devices. In future, ArdenML will include fuzzy reasoning. An MLM can trigger another MLM by calling it. MLM can start after a waiting period or periodically.

MLMs created in ArdenML are converted into executable files using XSLT (eXtensible Styles-sheet Language Transformation).

ArdenML syntax provides: 1) portability across different architectures because of XML as middleware, 2) ease of integration with HL7, 3) ease of understandability of MLM code to non-programmers due to the closeness to natural language like syntax, 4) logic independent of any specific programming language and site-specific implementation details, and 5) data-types suitable for clinical terms for medical documentation.

4.9 STANDARD FOR TRANSMITTING MEDICAL INFORMATION

The exchange of clinical data over the Internet is being standardized using an ANSI standard called "Health Level 7" (HL7). There are two versions: 1) older HL7 version 2.x (version 2.1 to 2.5) and 2) HL7 version 3.x. HL7 3.x supports a semantic model, object-model, scope and structure of data representation, message structure for transmitting data and translating data from high level format to low level XML-based protocols.

The foundations of the semantic interoperability model of HL7 are: 1) a common Reference Information Management Model (RIMM); 2) well-defined tool-set specifications for deriving messages; 3) translation of data across various formats; 4) a robust data-type specification; and 5) a formal methodology to link formal concepts to RIMM attributes. Following sections discuss different subcomponents of HL7 3.x and the translation of high level HL7 to low level XML code for transmission over the Internet using SOAP.

4.9.1 Reference Information Management Model (RIMM)

RIMM is a generic object-oriented abstract model that expresses all the information in the healthcare domain. It is independent of any message structure and syntax. It consists of six fundamental classes: *entity, role, act, participation, act-relationship, role-link,* and their specializations. The class *entity* describes persons, organizations, devices, material, place, supply, administered drug, manufacturer, communication, living or nonliving objects that participate in providing clinical healthcare. The class *role* describes the assumed responsibility or part played by an entity in providing the clinical healthcare. The class *act* represents the executed actions, observations and the events occurring during clinical healthcare such as hysterectomy, blood sugar measurement, etc. The class *actRelationship* describes the relationship between two acts such as causality. The class *roleLink* describes the dependency between the roles. The class *participation* describes the context for an action, including the persons involved: patient, medical care provider, date, time and place of action. Inside each box the subclass is described using the relationship *<subclass> → <class>*. For example, *Person → LivingSubject* in Figure 4.16 denotes that *person* is a subclass of *LivingSubject* that is a subclass of the class *entity*.

Entity class has following subclasses: *organization, person, place, container, device, languageCommunication, livingSubject, material, manufacturedMaterial* and *nonPersonLivingSubject. Role* class has following subclasses: *access, patient, licensedEntity* and *employee. Act* class has following subclasses: *account, controlAct, deviceTask, diagnosticImage, diet, financialContract, financialTransaction, invoiceElement, observationParticipation, patientEncounter, procedure, publicHealthCase, substance-Administration, supply* and *workingList. Participation* has a subclass *ManagedParticipation.* The overall structure of these fundamental classes and the corresponding subclasses is illustrated in Figure 4.16.

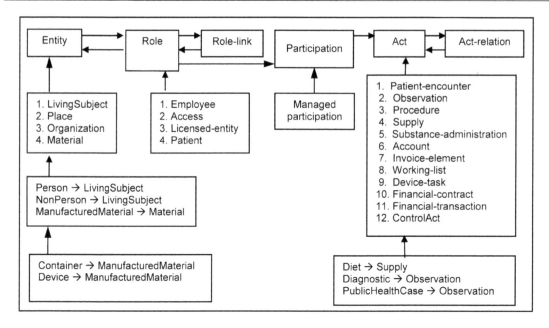

FIGURE 4.16　Major classes in RIM and the corresponding subclass structure

4.9.2　Clinical Document Architecture

Clinical Document Architecture (CDA) is an XML-based markup language for embedding structure and semantics of clinical documents for information exchange. A CDA contains any clinical note such as *continuity of care document* written by physicians during encounter, *diagnostic imaging-report, patient-history, discharge-report, pathology-report, operative note, procedure note, progress note, consultation note* and *daily bedside reports*. It supports *unstructured text* and links to files encoded in .pdf, .doc, .rtf, .jpg and .png. The *structured part* uses various standards such as SNOMED, LOINC and ICD-10. Using CDA many building blocks of documents are built that are archived, accessed and combined to make complex document set for different usage. These building blocks are transmitted electronically for reuse in different formats.

A typical CDA consists of a header and a body like any XML document. Header facilitates: 1) originating organizations, 2) author; 3) legal authenticator; 4) reference to other related documents; 5) reference to transformers needed to compile the document to EMR; 6) language code; 7) confidentiality code and 8) document security.

The body consists of *unstructured* and *structured* text. An *unstructured* text consists of human readable text describing the HL7 document encoded in *structured text*. A *structured text* is written in machine readable version embedded with standard codes such as SNOMED, LOINC and ICD-10 codes describing the clinical conditions of a patient. Body is divided into many *sections*. Each *section* contains one human readable narrative block and variable number of coded *entries*. Each *entry* can describe a medical procedure using a SNOMED code, lab tests using an LOINC code, observations, patient history, care documents using text documents, multimedia objects using a link to the file, administered drug using an ICD-10 code, entity details and their relationships and events. Sections and observations can be nested.

Entries are related to other entries using *entry-relationships. Entry-relationships* are associated with *entry-types*. CDA allows relationships across entries with no rationality check. Some of the entry-relationships are: CAUS (is etiology for), COMP (has component), GEVL (goal evaluation), MFST (is manifestation of), RSON (has reason), SAS (starts after start) and SPRT (has support). "CAUS" shows that source causes target entry.

Depending upon sections, many templates are semi-automatically filled based upon the supplied data. Given a situation, the best matching clinical workflow is chosen. The best matching template for the corresponding clinical workflow is retrieved from the template-database, and is populated by the supplied

```
<clinicalDocument>::  '<ClinicalDocument>' <CDA-header><CDA-body> '<ClinicalDocument>'
<CDA-header> :: '<header>'<language-code><confidentiality-code><organization> <author>
                <creation-date> [<other-information>] '</header>'
<CDA-body>:: '<body>' {<section>}+ '</body>'
<section> :: '<section>' {<entry>}+ '</section>'
<entry> :: '<entry>' <title> [<relationship>] (<unstructured-text> <structured-text>| <entry>) '</entry>'
<title> :: '<title>' <string> '</title>'
<unstructured-text> :: '<text>' {<text-list>}+ | <string> '</text>'
<text-list> :: '<list>' <text-item> '</list>'
<text-item> :: '<item>' <string> '</item>'
<structured-text> :: {<act>}*
<act> :: '<'<actName>'>' [<actRelation>] <SNOMED-text> [LOINC-text] > '<'<actName>'>'
<actName> :: "observation" | "history" | "supply" | "procedure" | "encounter"| ...
<actRelation> :: '<relation>' <relation-code> '</relation>'
```

FIGURE 4.17   A simplistic grammar to illustrate CDA

data. An extended BNF for a simple illustrative grammar for CDA structure is given in Figure 4.17. A CDA document is exchanged using HL7 messages. All the components of HL7, including multimedia objects and needed styling sheets, are exchanged as one unit.

4.9.3 HL7 Based Information Exchange

The standardized transportable format uses XML-based protocol SOAP (Simple Object Access Protocol) as described in Section 2.5. The current popular formats are HL7 2.X and HL7 3.X. "HL" stands for *high level*. This high-level format is translated to low-level XML-based format and transmitted using SOAP. This XML code needs to be enveloped in SOAP format. At the destination, the envelope is removed to get the XML format. XML format is translated back to HL7 format as illustrated in Figure 4.18.

1. A HL7 interface has three components: 1) origin of the message, 2) end-point of the message and 3) the message content. The message is a high-level description translated and transmitted using SOAP protocol. It is recovered into high-level HL7 format at the destination. Some properties for HL7 interface are: 1) flexibility to adapt to different formats, 2) scalability, 3) manageability to monitor and fine-tune the system in a user-friendly way, 4) efficiency and 5) security.

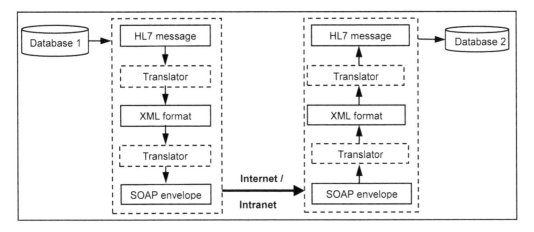

FIGURE 4.18   Transmitting information across Internet between heterogeneous databases

4.9.3.1 HL7 message structure

HL7 message is hierarchical and associated with a *trigger event*. A *trigger event* causes dataflow between the systems such as admission of a patient, surgical procedure, patient-visit to a medical provider, etc. HL7 messages are divided hierarchically into *segments*, *fields* and *components*. A *segment* is a group of fields corresponding to a *datatype*. D*atatype* guides the information structure in the corresponding field, and contains one or more components. There are more than *90* data-types in HL7. *Fields* are separated by a vertical bar "|." Each *field* has multiple components separated by the symbol "^." Consecutive vertical bars denote empty embedded fields. A *component* may have multiple subcomponents separated by "&."

Some of the examples of message segments are: such as "admission/discharge/transfer" abbreviated as "ADT," "admit patient" abbreviated as "A01," a message header abbreviated as "MSH," patient-id denoted by "PID," next-to-kin denoted by "NK1," events abbreviated by "EVN," patient visit abbreviated by "PV1," role abbreviated as "ROL," insurance information abbreviated as "IN1," etc. Each new segment is separated by a blank line.

There are 12 fields in message segment "MSH" as follows: 1) encoding characters, 2) sending application, 3) sending facility, 4) receiving facility, 5) receiving application, 6) date/time stamp, 7) message type, 8) event type, 9) message control ID, 10) processing ID, 11) version ID and 12) sequence number. Similarly, other segments have different templates of fields. Some of the examples of data types associated with individual fields are "AD" (address), "DTM" (date/time), "CQ" (composite quantity with units), "CE" (coded element such as SNOMED code), "CF" (coded element with formatted values), "CN" (composite ID number and name) and "DLD" (discharge location and date).

Example 4.9

The example in Figure 4.19 illustrates the use of various segments, fields and components. The corresponding unstructured message is: "A patient named Mary Shaw with patient-id *pat123* specific to Hillcrest Hospital was admitted in Hillcrest Hospital Room 3 bed 2 on July 1, 2016, at *11:30* AM for surgery. Her next-of-kin is her spouse Peter Shaw. The attending doctor who admitted the patient is Mahesh Seth." The event admits a patient. Other fields and components have been removed for the clarity of illustration.

4.9.3.2 Translating HL7 messages to SOAP

HL7 message is translated to the corresponding XML version, wrapped, and then transmitted using SOAP protocol. At the destination, the reverse process is applied. To translate, a HL7 message is parsed based upon the data-type structure, and appropriate XML-tags are generated. During the translation of a HL7 message, each field of a segment becomes a separate XML-tag in which multiple components are embedded. The general technique is to use the position of the field. For example, the third field of the segment "MSH" will be labeled as "MSH.3".

Wrapping has four layers: 1) HL7 query control wrapper; 2) transmission wrapper; 3) SOAP wrapper and 4) HTTP, SSL layer. *Transmission wrapper* establishes the required interface between the logical sender and the receiver of the message. The root-class of the transmission wrapper is *message class* that

```
MSH|^~\&|ADT1|Hill Crest Hospital| Radiology lab |201607011130| ADT_A01| MSG01...   <CR>
EVN | A01|201607011135|  <CR>
PID ...|PAT123^ADT1^Hillcrest Hospital ...| Shaw^Mary||19900101|F|1234^ Maple Heights^
        OH^ 44242-0001|(330) 555-2004| (330)6720002...  <CR>
NK1 ...|Shaw^Peter|SPO^Spouse|NK^Next-of-kin| ... <CR>
PV1 ...|003^02|0088421^Seth^Mahesh|...|SUR |...|ADM|A0|  ...<CR>
```

FIGURE 4.19 Human readable unstructured message and machine readable structured text

treats a message as an object. A message has following attributes: 1) sender; 2) receiver; 3) respond-to; 4) message-id; 5) creation-time; 6) version-id; 7) interaction-id; 8) processing-code; 9) processing-mode-code; 10) accept-ack code; 11) application-ack-code; 12) security-text; 13) sequence number and 14) attachment texts. Messages can be text or just acknowledgements.

SOAP wrapper is made of five components: 1) the start of SOAP-envelope that identifies the message as a request; 2) the function-name for sending the message; 3) the parameters that include login information and message to be transmitted; 4) function end and 5) SOAP envelope end. After a successful transmission, the SOAP response contains a successful "ACK" message.

4.9.4 WSDL (Web Services Description Language)

WSDL is an XML-based logical description of end-point of communication engaged in the exchange of a message. The description language supports the description of network end-points, declaration of types of messages exchanged, an abstract set of operations supported by one or more end points, binding between the protocol such as SOAP and the data-format specification for supported abstract operations on the end points, and a collection of related endpoints. It also allows the reuse of the abstract definitions.

A WSDL document is a set of abstract definitions. WSDL supports four types of messages: 1) one-way where a message is transmitted to the destination end-point expecting no response; 2) request-response where the source end-point receives a request, and then sends a response to the destination end-point; 3) where the source end-point solicits a response, and receives the corresponding message and 4) a notification where the source end-point receives a message with no request for a response.

4.10 HIPAA – PROTECTING PATIENTS' PRIVACY RIGHTS

HIPAA (Health Information Portability and Accountability Act) was passed by the US Congress in 1996 to protect people against any misuse of their medical information or discrimination for their existing diseases and medical conditions. It has two major components: 1) portability of the information by the patient or for the treatment of patients and 2) privacy of the medical information from being leaked to third parties.

HIPAA limits the knowledge of the disease and medical condition of patients only to the care-giving physicians and medical practitioners. The information and electronic records about patients' medical conditions and the related treatments cannot be disclosed to any other person, including family members, institutions and employer(s) without an expressed written voluntary consent of the patient. This law helps patients not to be discriminated in the society or workplace in the case of serious diseases such as HIV or terminal diseases.

HIPAA makes a person the key decision maker about her medical information. A patient, by law, can see all her medical records and authorize which third party can see her medical information. A patient can also carry her medical records when they change the location of medical caregiver after paying reasonable expenses needed for porting the records. For electronic records, the expenses are nominal. Patients can: 1) inspect and copy their record; 2) obtain a record of past disclosures; 3) request amendments to their records; 4) request restrictions on the uses and disclosures; 5) request for confidential communications and 6) complain about violations to the designated federal agency. In the USA, the designated federal agency is the United States Health and Human Services.

There are exceptions to this privacy law when patient's medical record can be used without explicit consent by the patient such as in the presence of a medical catastrophe that affects the population, or a medical emergency during which a patient is unable to give medical consent. Patients' records are also not protected from their insurance companies, federal government for maintaining public health or law enforcement, health-related benefits, organ donation activities and research purposes, provided the de-identification of data is maintained. Another case when personal information is unprotected when there is a need for national security such as in the case of an outbreak of a rare epidemic such as "Zika virus" that causes brain damage in a fetus during pregnancies.

Different caregivers know the minimal information based upon need to know basis. For example, a patient's primary-care physician or a surgeon performing a major surgery needs to know all medical history, including any genetic disease patient's parents have or had. However, a lab technician, a radiologist, or a physical therapist needs to know only limited information related to the patient's current problem. Similarly, a money-collection agency, hired by the hospital or care provider for nonpayment, need not know the medical condition and the treatment given to a patient. The information held by the insurance agency, hospitals, medical providers cannot be shared (intentionally or through carelessness). There are serious penalties and criminal charges if HIPAA is violated.

The medical decision about how much medical condition and history a healthcare provider needs to know is very tricky. Withholding medical information may cause a medical judgment error. Similarly, hospitals and billing departments used to share the detailed billing information with third parties such as insurers, parents and close relatives paying the bills, and legal advisers are trying to reclaim the disputed bills. Often, the medical information is also divulged to law enforcement agencies to protect or prosecute a person having a medical condition during public court cases. There are also issues when information leaks out accidentally. For example, if two physicians are discussing a case on a phone about a patient in the corridor and the patient's unauthorized relative or other nurses may be listening to the conversation.

Patients' records must be protected against: 1) unintended disclosure; 2) well-intentioned but inappropriate employee behavior; 3) disgruntled employees; 4) self-insured employers; 5) competitors; 6) hackers and 7) data miners. The information must be protected against risks and threats of violation of information breeches. Reasonable security measures and protection mechanisms have to be implemented under specific circumstances to guard patients' personal and disease-related information.

4.10.1 HIPAA Violations and Protection

Some of the common examples of HIPAA violation are unintentional violations. Reasonable effort should be made to avoid these unintentional mistakes to avoid any penalty. Following three examples describe common unintentional mistakes.

Example 4.10

A medical caregiver has a friend whose wife is in a hospital under treatment. The friend asks the medical care giver for suggestions for medical treatment. The medical caregiver is in confusion, whether to look at her chart to find out more background information to give suggestions or not. He is also tempted to ask the nurse-in-charge for the information. However, the correct response is to tell his friend that unless requested by the physician-in-charge or the patient herself, he cannot give any suggestion or look into her disease-related information. However, if the patient (friend's wife) was under a disease state such as a coma and the authority to decide for the patient was with her husband, then he can allow another medical provider to look up the information.

Example 4.11

A surgeon has come out of surgery, and is using his cell phone to discuss the condition with the patient's primary-care physician while he is eating in a cafeteria or walking through a corridor with many people. It is a HIPAA violation because other people can listen to his conversation. However, if he has taken reasonable care to be in a separate private room such as his office, and does not explicitly identify the patient, then he has taken reasonable care, and he is not liable.

Example 4.12

A physician is meeting with a drug company representative promoting an ointment for arthritis pain. The physician gives the representative a list of her patients to send sample tubes of ointment. Although, the intention of the physician is good, it violates HIPAA because the physician has released the medical conditions of patients without their proper authorization to a third party. A correct protocol would be to take samples from the representative and ask the patients privately if they were interested in trying out a sample ointment.

Example 4.13

A person works in the billing department looks into a patient's records because his wife has a similar health condition. It clearly violates HIPAA.

In certain cases, for research data analytics, the health condition of patients is released to the clinical researchers for knowledge discovery, data mining for public health and studying drug efficacy. In such cases, the disease condition and patient responses to various drugs are released with proper de-identification. The administration of release of minimal data for the research purpose lies with Internal Review Board (IRB). IRB is an institution level committee that enforces HIPAA for the clinical research. De-identification makes sure that any information that can be associated with a specific patient's identity is removed from the data used by the clinical researchers. Examples of directly identifying information are: 1) name; 2) geographical sublocation such as address, county, city; 3) elements of major dates associated with patients: date of birth, date of admission, discharge-date from the hospital, date of death, dates indicative of age; 4) communication links such as phone number, fax number, email-id, URLs, IP address, etc.; 5) biometric identifiers such as finger prints, photographs, voice-print, driver's license, employee-id, passport, computer-device id, car license plate, car's vin, etc.; 6) medical identification such as medical record number, bed number, hospital id, hospital location, hospital account number, insurance card number, etc. and) any other identifying characteristic such as voting precinct.

Despite removing direct information, as described in Section 4.10.4, persons can also be identified using meta-analysis. Meta-analysis search the patient's various characteristics without directly knowing them and combines them to prune a large number of possibilities so very few possibilities are left that can be further investigated.

4.10.2 Role Layers and Patient Data Protection

Protection of patient's biomedical data and disease condition is difficult due to unintentional leakage, complex interpersonal connectivity and organizational roles of people in the society. Providing too much control over the patient-data may get a patient little familial and societal support, and providing too much information may hurt his insurability and employment status. There are risks and benefits to different levels of information being provided.

The multiple layers of people's role in the society necessitate different levels of knowledge of patient-related data. The roles are broadly classified in three major layers: 1) core layer; 2) second layer that provides support and cares for the patient and 3) third layer that includes public authorities, fraud detection authorities and friends. The core layer consists of the patient, medical care provider and the payer(s) of the patients' bill. They have complete information about the patients' disease condition and the patients' medical history. The second layer consists of primary-care physician, specialist doctors, administrative staff, public health administrators, law enforcement officials, insurers, clearing houses, claims processors, immediate family and legally authorized representatives who have access to the patients' disease information either through patients' authorizations, patients' immediate connectivity or due to need to know the information. The third layer consists of friends, extended family, community support, internal quality assurance, clinical trial sponsors, national security, bioterrorism detection, business consultants, medical information bureau and fraud detection. Although the information to the third layer is protected from HIPAA, information leak occurs in time of emergencies, need for support from the community and accidental informational leaks and grapevines.

4.10.3 HIPAA and Privacy in EMRs

With the availability of electronic health records that can be easily transmitted over the network, incorporation of HIPAA is very important. The information could be in the form of written reports, charts, X-rays, letters, messages, emails, phone calls, medical meetings, informal conversation, computer records, faxes, voice mails, PDA entries and text messages are the sources of information leaks. In addition, the information can also be leaked when the information is transmitted over the Internet, stored in the insecure clouds or using meta-analysis of patients' database.

The enforcement of HIPAA requires the segregation of data so disease-related records such as lab tests, surgical procedures, patient–physician encounters related to specific disease-related visits, prescription drugs taken by the patients have to be guarded so only relevant persons involved in the patient' disease conditions can have access to the information.

Medical databases should be private, which means no one unrelated to the patient's treatment should have access to the disease-related information and treatment as guided by the HIPAA rules. Security is about preventing an unauthorized person from accessing or corrupting the database. Medical databases are diverse and distributed across various medical organizations. They are transmitted using standardized XML-based protocols such as HL7 over the Internet. It is important that this data remains uncorrupted in the database as well as during transmission.

The types of security issues in distributed databases are: 1) confidentiality of medical record existence, 2) confidentiality of medical record content, 3) identification of a patient by looking data or meta data, 4) lack of linking a database to another database, 5) unauthenticated corruptibility of the database, 6) robust HIPAA-based authentication of users, 7) fine grained HIPAA compatible access and modifiability of database, 8) auditing capability of database access and usage, 9) confidentiality of user access privileges, 10) scalability of security measures, 11) HIPAA compatible access privilege modifiability, 12) long term secure archiving, and 13) security against circumvention.

It is very important to interleave HIPAA-related security at the diagnosis and treatment level for each episode. Criterion #1 ensures that no unauthorized person be even aware of existence of medical data because knowledge of existence of such data motivates circumvention of security. Criterion #2 ensures that no unauthorized person, including doctors unrelated to cases, can access the sensitive medical record. Criterion #3 ensures that no similarity-based search or statistical analysis or meta-analysis on a large data set should be able to identify individual patient's identity. Criterion #4 ensures role-based authorization to care takers and discourages unauthorized access to patients' records. Criteria #5 to #7 are ensured by using fragmentation and encryption.

Medical data is protected according to HIPAA guidelines, and should be guarded by encryption and authentication protocols all the time. This can be achieved by a combination of: 1) encryption of data at the attribute-level with different keys associated with subsets of attributes; 2) fragmenting a record, and joining the fragments to reconstruct the record when needed and 3) keeping different patient-ids with different fragments that can be reconciled only after validating proper authorization. The authorization must be role-based at the record-level to manage the role-based privacy needed in HIPAA. The authority will issue multiple different private keys based upon the role-based access structure. The decryption algorithm allows users to decrypt data based upon the role-based private keys. The attributes are associated using Boolean operator on attributes such as logical-AND, logical-OR and negation.

4.10.4 Data De-Identification

De-identification of patients' information is essential to protect the privacy and security of the patients to follow HIPAA rules. This privacy and security is essential for the patients' financial, social and psychological well-being. De-identification is quite a complex task. It just simply does not mean the removal of the names from the record. Rather, it means removal of any information or a combination of information that will uniquely or near-uniquely identify an individual or a community, so a health-condition can be associated with them. In the wrong hands, this information can hurt an individual. In addition, the information can easily be leaked or passed involuntarily, sometimes with presumably good intention, which can hurt or embarrass the individual or his/her family financially, economically, socially, and psychologically. Some of the strongly protected information-items are: 1) medical history and genetic disorders; 2) care provider information and duration of a care; 3) existing disease conditions and predicted disease conditions and 4) payment and credit history conditions.

A natural question is: what is the minimal set of information that needs to be removed from a data source to de-identify it? Another question is: what are various natures of resources, including archived sources, that can be de-identified? Often, the information is revealed unintentionally by close relatives who may know a patient's condition. Other times, disgruntled employees in the possession of personal information, hackers, unintentional disclosure on social-media, well-intentioned but inappropriate employee-behavior of a medical organization can leak the identifying information.

Indirect information can be derived by data mining reports, health providers' dictations and other meta-level reports. For example, the information available in the public offices such as county offices for registered voters is public and provides a person's name, age, gender and address. A *meta-level query* based upon locality, ethnicity, visit-dates and gender may limit the number of persons to few persons that can be matched with the available public information defeating the de-identification process and HIPAA protection.

The information needed to de-identify are: 1) concrete data and 2) a combination of indirect meta-data. *Concrete data* directly maps to the privacy information, while indirect information uses aggregate data or metadata about the people that does not pertain specifically to an individual. However, the combination of the meta-information can make the corresponding set of individuals so small so that individuals are easily identifiable.

Direct concrete data used to de-identify a person is: 1) *personal identifiers uniquely associated with the person* – name, social security number, date-of-birth, date-of-death, driver license number, passport number, telephone number, fax number, insurance card number, health account number, vehicle serial number, IP address of the computer, biometric identifiers; 2) *associated locations* – home-address, geographic subdivision that is smaller than a state and postal code of residence; 3) *identifiers related to hospital stay* – hospital names, hospital bed-number, date-of-admission, date-of-discharge, medical record number and 4) *other uniquely identifying characteristics* such as employer name, spouse name, children names, personal photographs, race, ethnicity, languages spoken, gender and other specific classifications.

To provide sufficient de-identification, there are two steps: 1) remove the information that uniquely map the fields to a patient and 2) disallow meta-queries in data mining the databases that when combined with public information can associate a person with his past medical history. The first step is taken by suppressing the information fields that uniquely map the patient to his disease conditions and or history.

4.10.4.1 K-anonymity criterion

There are many techniques and algorithms to de-identify a record in the database, Algorithmically, a piece of information becomes meaningful if it can be associated with several records in the database. A commonly used criterion for de-identification is the *k-anonymity criterion* which stipulates, whether K records are associated with a combination of values in one or more fields. For example, if the values of age, gender, and locality returns *20* records in the database, then $K = 20$. If the value of K drops below a certain threshold, then the information can identify a person.

The simplest approach is to remove the identifying information completely. However, often, the information about age, gender and ethnicity information is needed for data-analytics. In such cases, the directly identifying information such as social-security-number, driver's license number, etc. are removed; all the equivalence classes with a K-value less than a predetermined threshold value are suppressed; and other identifying information are expressed in more general form to increase their K-value.

For example, the age *16* may be replaced by an interval group *16–20*. Such replacement must be global. Otherwise, it will be difficult to perform data-analysis due to the incompatibility. During the suppression of information, care must be taken to avoid information-loss needed for data-analytics. For example, if the gender ("male"/"female") is converted to "person," then all gender specific information is lost, and the data becomes useless for gender-related data analysis.

The algorithm to de-identify is based upon creating a lattice of generalization so the leaf node of the lattice satisfies the k-anonymity criterion. A lattice is a directed acyclic graph where the parent node is more general than the descendant node according to a specific asymmetric relation. In a lattice, there need not be any relationship between siblings. When choosing a node in the generalization lattice, only those nodes are picked that do not result into any information loss or least amount of information loss needed for a data-analysis goal.

4.11 INTEGRATING THE HEALTHCARE ENTERPRISE

As the different international coding systems evolve, there is a need to coordinate the efforts for seamless integration and downward compatibility of the overall evolving system. IHE (Integrating the Healthcare Enterprise) has been set up to achieve this goal. Many international organizations are part of this effort. It consists of handling: 1) coordination of information flow: descriptions of cases, actors involved, process-flow and dependencies and 2) transactions and content profile: reference to various international standards, constraints and interactions among the standards.

The overall goal to provide total patient-care coordination, including patient location tracking inside various units in an organization for better care scheduling and mobile access to health documents by healthcare providers and patients. It offers a common language to discuss needs by healthcare providers to integrate various standards.

4.11.1 Patient Care Coordination

Patient-care coordination involves the integration of the information between ambulatory unit and the central repository, paramedics and central repository and information exchange between expert-units in multiple organizations for diagnosis and additional background information.

Multiple patient-related summary documents need to be integrated such as emergency department referrals, emergency department encounters summary, patient plan of care, exchanges of patients' health records, immunization records, labor and delivery records, care summary before and after a child-birth, postsurgery care and emergency medical-system transfer.

The information includes the content-exchange along with the explanation how the information-exchange will affect other summaries. This requires IHE "Cross-enterprise Documents Sharing" (XDS); patient identification (PIX); notification of availability of documents (NAV); IHE "Cross-enterprise media information exchange" (XDM); IHE "Cross-enterprise document reliable interchange and infrastructures to support privacy security," time consistency (CT), audit trails and node identifications (ATNA).

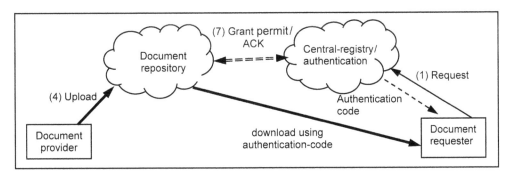

FIGURE 4.20 A schematic of unplanned information exchange between institutions

There are two possibilities: 1) use-cases where the sender knows of the requester based upon the information-flow dependency and 2) unplanned access where the sender does not know about the requester and its need. In the first case, the sender informs the requesters using secure emails, while in the latter case, the requester downloads the information from a *central-document-repository* based upon an authentication-code. The process involves a *requester*; a *data-source*; a *central-registry* that also authenticates the requester; and a *central-document-repository* that retrieves the requested data from the data-source.

The following sequence of actions is taken for data-exchange for the unplanned access based upon a request: 1) a *requesting agency* makes a request to the *central-registry*; 2) the *central-registry* authenticates the identity of the *requester*; 3) the *central-registry* transmits an *authentication-token* to the *central-document-repository*; 4) the *central document-repository* requests the data from the data-source using the *authentication-token*; 5) data is uploaded by the *data-source* in the *central-document-repository*; 6) uploaded data is registered with the *central-registry*; 7) the *central-registry* grants a permit to the data-repository that is acknowledged by the data-repository; 8) *central-registry* sends a *timed authentication-code* for downloading from the document-repository and 9) the *requesting-agency* downloads the data from the *central document-repository* using the *timed authentication-code*. The data is removed from the *central-document-repository*, after the allowed time expires.

The overall scheme for the unplanned version of information exchange is given in Figure 4.20. The bold edges in the figure show the data-link and data transfer, and dashed edges show the authentication links. The dashed double arrow between registry authentication and document repository shows the permission granted by the authentication service to honor the *authentication-code* by the requester. It also carries the information about the privileges and the duration of the privileges.

4.12 SUMMARY

This chapter described issues and solutions in the storage of patient-specific information in a patient-centric manner so that information is private and is safely transmitted across sensors → VMDs → local databases → remote databases using different types of information bus. Each sensor has a virtualized model to exchange information and commands with other sensors and databases. An action on the virtualized model is translated to the sensor action using low-level data and command translators called adapters.

The overall goal is to provide a seamless virtual database even though different sensors and databases use and transmit data in different formats from multiple vendors. The queries from medical databases are complex and require content-based similarity search for matching images besides similarity-based textual search. Besides, databases and queries have temporal information associated with data. The use of time makes the database search more complex due to the temporal relationships and dependencies between events.

Heterogeneity of databases is of two types: 1) different database within the same institution such as pathology database, radiology database, ICU database, etc.; 2) heterogeneous databases connected through the Internet. Local databases within an institution have different characteristics and are connected to each other using an intranet. Geographically separated databases are connected through WAN (Wide Area Network) or wireless network.

The information between the databases is exchanged using HL7 – an XML-based database-independent format. Automated transformation mechanisms transform data from an XML-based format to a database specific format, and vice-versa. Using various translators, HL7 message from the originating institution is translated to low-level XML-based representation and transmitted using SOAP. The information is reconstructed to the HL7-format at the receiver-end, and then added to the database in the remote format using *adapters* – a software format translator.

The information across multiple institutions flows using a common pathway called *HIB* that connects hospitals, emergency clinics, insurance agencies, datamarts, information clearing houses, billing sections, portals, pathology labs, ICU and pharmacies. HL7-format is the common means of data-exchange. Datamarts are smaller functionality-based databases joined to form a bigger data-warehouse. Data-warehouse and datamarts provide better security to archived data.

To maintain the privacy of a patient's data and to avoid its spread beyond curing the patient requires: 1) encryption of data; 2) a dynamic authority based authentication of the care-provider and 3) de-identification of the clinical data. De-identification is removing all the identifying information of patients from the clinical data. For example, name, date-of-birth, address, next-of-kin, employer, car-number, bed-number, combination of (*locality, religion, approximate age* or *gender*) are identifying information.

Complex records are split and archived separately. Meta-queries that can join many individually nonidentifying queries to retrieve information or statistics that can associate a clinical data with a patient are disallowed. One popular technique for the de-identification is K-anonymity criterion. In K-anonymity criterion, only those fields are maintained so search on those fields does not return less than K records. If search leads to a number of records less than the threshold K, then the fields are suppressed from the database-search to maintain anonymity.

EMRs are patient-centric. Different roles, patient, healthcare providers, pharmacy, billing department and insurance agency have different views of the same database depending upon their functions and needs. Each medical institution has its own way of providing patient-id to a patient. These different identifiers need to reconcile with each other to get overall medical information related to a patient and to exchange information regarding the patient with institutions other than the local institution. This problem is resolved using an *MPI* that is a globally unique-id.

The mappings of *local-id to MPI* and *MPI to local-id* are kept in the datamarts. *Local-id to MPI* is a one-to-one mapping. However, *MPI to local-id* is one-to-many mapping. The request for additional information is processed through datamarts holding the table of mappings. The local-id of a patient from the requesting institution is converted to MPI in the datamart, and then this MPI searches all the local-ids in *the MPI to local-ids* table, and a request is generated to those institutions to send the data in the CDA format.

Sometimes, due to data-entry problems or during the registry of a new patient, it may be difficult to have the correct local-id of a patient. In that situation, the information available in multiple identifying fields is combined to map the patient to the corresponding MPI. Even the identifying field may have an error. In that situation, nearest similarity-based search is used to map to the corresponding MPI. The similarity-search uses a dictionary of words that sound the same phonetically, but have different spellings such as "ii" and "ee," "w" and "v," etc.

Medical databases contain text, graphic images, photographs, video clips and sound recordings. A major class of query is to: 1) find out the similar cases and the corresponding diagnosis, prognosis and regimens; 2) derive the growth or remission of abnormal structure caused by a disease; 3) analyze images for any deformity or abnormality. This requires content-based similarity-search. Content-based search can be performed using texture-based histograms or graph-based matching.

In texture-based histograms, a change in intensity pattern is identified using the information about gradient and orientation of pixels in the vicinity of a pixel. One of the popular techniques to model texture-pattern is the *local binary pattern*. The advantage of histograms is that a computational overhead is significantly reduced due to the statistical data representation in histograms. However, multiple images may have very similar histograms causing ambiguities. To perform exact matching, each region-of-interest is modeled as a graph using well-defined criteria. Multiple attributes of graphs, including the *area of the graph, centroid, length, and orientation of edges,* are used for graph-matching. Images are aligned before they are matched. This process is called *image-registration*. Image-registration is done in two ways: 1) external markers placed by the healthcare provider; alternately, 2) identifying the fixed natural organs near the region-of-interests, and treat them as the markers.

Time plays a major role in medical databases as disease conditions and treatments are valid during a time-interval. There are two types of timed-data: 1) interval-based information and 2) snapshot information using point-time. Handling time requires: 1) inequality-based reasoning; 2) reconciling two overlapping time-intervals; 3) reconciling point-time with interval-time and 4) reconciling different units of time such as day, week or year. The use of aggregate function and bigger intervals compresses the time-based information at the cost of losing some detailed information.

Interoperability of databases is achieved by: 1) using standard codes such as SNOMED, LOINC and ICD-10 to code the clinical terms; 2) using HL7 and translated low-level XML format to exchange information between two nodes in the health network and 3) the use of *MPI (Master Patient Index)* to have globally unique identifier.

SNOMED is used for coding medical history, medical procedures, billing and observations. LOINC is used for coding separable lab-tests, observations and toxicological tests. ICD-10 is used for coding disease conditions and the causes. If overlaps occur, there are cross-references to avoid confusion. SNOMED has three building blocks: *concepts, descriptions* and r*elationships*. Concepts are divided into multiple classes called *concept-types*. SNOMED identifies the clinical findings specifically to a location-site of a body. ICD-10 is divided into *21* major classes of diseases. It has four characters with a period after the third character. First character is a letter, and other three characters are digits. ARDENML is an XML-based language for coding information to be exchanged between institutions. All these international standards are continuously evolving.

The current version of HL7 is HL7 3.x, and is a formal object-based model. The model is based upon *Reference Information Module Management* (RIMM) that divides clinical data into multiple classes, which are further divided into subclasses. *CDA* for HL7 has XML-like structure, is based upon RIMM and is used to code the clinical data for the information exchange. CDA-document has a head and a body. *Head* contains message identification information such as originating organization, author, authenticating body, security, language code and confidentiality. Body has two parts: *unstructured-code* for human comprehension and *structured-code* for universal data-exchange across heterogeneous databases.

Unstructured-code refers to image-files archived using DICOM format and clinical notes. The message structure of HL7 is hierarchical: *segments, fields, components* and *subcomponents*. Different fields have been divided into over 90 data-types. High-level HL7 messages go through multiple levels of transformation where transmission protocol level information is wrapped around the message. Finally, the SOAP version of message is transmitted to the destination, where it is transformed back to the high-level HL7 message format to be integrated with the database at the destination.

4.13 ASSESSMENT

4.13.1 Concepts and Definitions

Act; act-relationship; Adapter; ARDENML, asynchronous message; authentication; bigram; clinical decision system (CDS); CDA; clinical interface technology; component; compositional grammar; concept; concept-type; content-based search; cross-enterprise document sharing; cross-referencing; data-element; data semantics; data warehouse; datamart; de-identification; description; destination connector; device communication controller; DICOM; direct concrete information; DIM; dynamic model; EHR; Electronic Health Record; EMR; ESB; entity; Euclidean distance; exact match query; feature-vector; Fellegi and Sunter's probabilistic model; graph matching; graph-based modeling; health claim datamart; HIB; health information exchange (HIX); HL7, HL7 datatype; HSB; heterogeneous database; Hilbert R-tree; HIPAA; histogram; horizontal temporal inference; ICD-9; ICD-10; image registration; inequality-based search, information clearing house; information exchange; information retrieval; IHE; interoperability; interval time; Jaro–Winkler distance; K-anonymity; K-D tree; K-D-B tree; K-dimensional; knowledge block; lattice; Levenshtein distance; local binary pattern; LOINC; MPI; match probability; matches; medical database; medical device data language (MDDL); medical device information bus (MDIB); medical image modeling; MHD; MIB; MLM; meta-data; nonmatches; OPENEMR; PACS; partial match query; participation; patient care coordination; patient location tracking; periodic time-slice; point-time; record linkage; reference information model (RIM); relationship; role-link; segment; similarity search; SNOMED; SOAP; static model; statistical summarization; structured text; synchronous message; temporal; temporal context restriction; temporal interpolation; temporal database; temporal pattern matching; time-consistency; transformation; transformer; trigger event; trigram; unmatch probability; unstructured text; vertical temporal inference; VMD; VMR.

4.13.2 Problem Solving

4.1 Design a relational database of patient-encounter and pharmacy orders such that a doctor orders a medicine over the Internet for a patient after the doctor-patient encounter. The information included for patient-information table, doctor-information table, patient–doctor encounter table, and new relational tables have to be formed optimally to include patient's medications, medications side-effects, medication administration schedule, dosage, duration of the medication, patient's pharmacy name, pharmacy-id, pharmacy address, pharmacy phone number, drug cost, patient's insurance-provider, insurance-id, insurance approval of the medicine brand. Show the relational tables, primary-keys in each relational table and the connectivity.

4.2 Write a program using hash-tables to model patient-doctor encounter using three tables: *patient-table*, *doctor-table*, and *patient-doctor encounter table*. Each relation in patient-table is of the form (*patient-id, social security number, name, date-of-birth, insurance company*). Each relation in the doctor table is (*doctor id, name, affiliation, expertise*). Each relation in the patient-encounter table is of the form (*patient-id, doctor-id, time of appointment, date-of-appointment, type of appointment, diagnostic statement, medications*). Create your own data for ten patients, ten doctors and 20 patient–doctor encounters. The program should be able to answer the following queries: 1) When is the next appointment with which doctor for a patient? 2) Which doctors have specific expertise? 3) What is the diagnosis of a patient? 4) What was the last appointment for a patient? 5) When did a doctor see a patient last?

4.3 Extend the program in Problem 4.2 to display the facial image of a patient. This will require extending one more field in the template that would be the path of the file where the facial image is stored, and displaying the "JPEG" file using a built-in library function to display the image.

4.4 Give an HL7 message showing both structured and unstructured encoding of the following information: "The patient was admitted at *9* AM in the emergency room of Summa Hospital for acute bronchitis by Dr. Nathan Palmrose." This message is being transmitted at *9:03* PM from Summa Hospital to Cleveland Clinic where the patient was finally transferred.

4.5 Read HL7 message structure and give a HL7 representation of the following information about a patient.

> "A patient named John Aston was admitted on June 12, 2018 for pain in the right lower abdomen. He was diagnosed as having enlarged appendix. His surgery is scheduled in 'general surgery' unit of Mount Sinai Hospital at *9* AM on June 14, 2018. His nearest kin for emergency contact is Jane Aston, his wife."

4.6 Give a CDA structure to capture a patient's history that says that "Pamela Painful, a 77-year-old female, has a history of chronic osteoarthritis for the past *10* years. Currently, she cannot walk." Look up the CDA structure in HL7 documents at http://www.hl7.org/implement/standards/product_brief.cfm?product_id=7.

4.7 If a patient has a local-id in Hospital A as "123" and a local-id in Hospital B as "aBC." Assume that his MPI is "178910." Assume that his chest X-ray is in hospital A, and a doctor in Hospital B will like to see his X-ray for possible tuberculosis. Explain how the VMR will link the two heterogeneous databases, and the sequence of operations to transfer the X-ray from Hospital A to B.

4.8 Study various procedures to maintain diabetes II over the Internet and represent them using a dataflow graph similar to Figure 4.4. Some examples of actions are: 1) administer insulin; 2) administer oral medications; 3) reduce the dose; 4) increase the dose; 5) monitor blood sugar; 6) measure A1C periodically; 7) measure lipid panel; 8) monitor blood pressure; 9) albumin test; 10) eye exam for retinopathy; 11) comprehensive foot exam for peripheral neuropathy; 12) dental exam for gingivitis and periodontitis; 13) urine test for sugar and 14) exercises for weight reduction. The workflow should be suitable for senior citizens and should be able to handle different level of diabetes based upon A1C level in different fuzzy ranges.

4.9 For Problem 4.8, write a simple expert system program that acts like a registered nurse, and recommends to the patient what to do, based upon A1C test results, weight increase/decrease and patients existing condition. A nurse can increase the dosage, reduce the dosage, recommend alternate medicine based upon side-effects, and recommend dietary control to manage the sugar level.

4.10 A natural language translator gets confused with ambiguous but equivalent pronunciation of the same words in different language and culture, and has to lookup dictionary before identifying the edit distance errors. Table 4.1 shows such equivalencies among the pronunciations. Table 4.2 shows equivalencies between various equivalent words written differently. For example, "Dr." is pronounced as "doctor"; "Mrs" is pronounced as "Misses," "Ms." Is pronounced as "Miss." "c" may be pronounced as "si," "k" or "ch." Exploiting the information in Tables 4.1 and 4.2 information, match the pronounced phrase with the database in Table 4.3 and calculate the edit distance. The character "_" in Table 4.1 means that the letter could be silent and not pronounced.

The text to speech translator translates the four names and address as: Doctor Johnnee Kokren, Birdy Lane, Loss Anjales; Miss Fenix DyeHaard, Seller Court; Mister Kuku Paije, Eest Noleje Laine; Misses Babler Gaarchio, Kase Drive.

TABLE 4.1 Written letter vs. pronunciations

c	s, k, ch
y	ee, ie, i
a	aa, e, ai
u	u, oo,
p	p, _
k	k, c, _
g	g, j
f	ph
ea	ee

TABLE 4.2 Abbreviations vs. pronunciations

Dr.	Doctor
Mrs.	Misses
Ms.	Miz / Miss
Mr.	Mister
Ln.	Laine
Ct.	Court

TABLE 4.3 A small database of patient title, name and address

PREFIX	NAME	ADDRESS
Dr.	Johnny Cochran	Birdie Ln., Los Angles
Ms.	Phoenix Diehard	Sailor Ct.
Mr.	Cuckoo Page	East Knowledge Ln.
Mrs.	Babler Guercio	Case Dr.

4.11 Give all the bigrams and trigrams of the word "Osteoarthritis." Match the bigrams (trigrams), and give the matching percentage for matching for the strings "Osteoarthritis" and "Osteoporosis" where the matching percentage is defined as (number of matching bigrams or trigrams)/total bigrams (or trigrams) in both the string.

4.12 Study various procedures for maintaining coronary heart disease in senior citizens and represent them using a dataflow graph similar to Figure 4.4. Some actions are: 1) perform bypass surgery; 2) place a stent in the artery; 3) measure CBC (Comprehensive Blood count) periodically; 4) measure electrolyte levels periodically; 5) administer statins orally; 6) administer blood thinners; 7) perform periodic checks; 8) monitor blood pressure; 9) perform brisk walking and exercise; 10) increase the oral medication dose and 11) decrease the oral medication.

4.13 A patient goes into an emergency room complaining of lower abdominal pain that may be due to kidney stones, appendicitis or hernia. Give a workflow diagram for the emergency care that includes: 1) regular vital sign monitoring; 2) urine test; 3) examination by an emergency physician; 4) X-rays of the local region; 5) pain medication; 6) examination by the specialist; 7) admittance to a hospital as an inpatient and 8) release of the patient.

4.14 Read Papers #44 and #45 on HL7 structure, and describe the RIM-based schema for the workflow in Problem 4.12.

4.15 During the postsurgical recovery, a patient's blood pressure level and heart rate were monitored and recorded every *10* minutes. The data was (*60, 90, 56*), (*56, 86, 62*), (*61, 94, 60*), (*62, 98, 68*), (*63, 100, 65*), (*63, 102, 70*), (*64, 105, 70*), (*64, 104, 69*), (*66, 110, 72*), (*67, 112, 70*), (*68, 112, 74*), (*70, 114, 74*), (*72, 116, 74*). Calculate the average blood pressure and pulse rate every *30* minutes for archiving and make a graph if the recovery is smooth and linear from the *30*-minute summarized data. Compare this graph with actual *10*-minute graph.

4.16 A person suffering from diabetes II is monitoring his sugar level three times a day: early morning, just before lunch and just before dinner. His readings for the last week were: (*120, 80, 210*), (*140, 75, 220*), (*150, 80, 120*), (*180, 90, 130*), (*120, 80, 120*), (*120, 82, 135*), (*160, 84, 300*). Discover the range of his blood sugar variation, average blood sugar and patterns coming out of his blood sugar. How should his dosage be adjusted to keeping his blood-sugar level close to *100*? Assume that he is allowed to get small sugar pills, Glipizide tablets – a medication to reduce sugar level and exercise for *30* minutes. Small sugar pills increase the sugar level by *20* units; *30* minutes of brisk walking/exercise decrease his blood-sugar level by *30* units, and every *5* mg of glipizide reduces the blood-sugar level by *20* units. A person cannot take over *15* mg of Glipizide. One can use a combination of Glipizide and exercise. Write a simple expert program that recommends the person the action to take after monitoring the sugar level.

4.17 In a city Dunestown, having a population of *4200*, the percentage of various ethnic communities are: Asian – *2%*, Blacks – *20%*, Hispanics – *15%*, Whites – *60%* and others – *3%*. Assume that the city has 20 localities where people are concentrated, and distributing ethnic communities is uniform. What types of concrete and meta level queries need to be suppressed for a *K*-value of *20*, and why? Explain your answer.

4.18 Repeat this Problem 4.17 with the following age statistics: 1) children below *5: 10%*, 2) youth below *12: 15%*, young adults below *18: 20%*, Adults: *40%*, senior citizens: *5%*.

4.19 Assume that the number of allowed characters in a patient-id in a hospital X are *10*, and a typist makes a mistake with a probability of *4* characters per *1000* characters during typing. Also, assume that there can be no more than one typographical error in each patient-id. Assume that *1,000,000* patients visit the hospital to meet a doctor. Computer the number of errors in patient-id the typist makes? Suggest a strategy to correct these errors in MPI.

4.20 In Figure 4.12, a typist in Hospital A mistypes a patient with patient-id "37" and name "Mary Smith" with a new record "Record #5." Suppose he types patient-id "31" and name "Mury Smit." Give a general-purpose robust scheme to correctly insert this record in the database of Hospital A if there are 10,000 names in each database and there are 400 different patient-ids with the same name. Show how the record will be inserted correctly, and how will it affect the MPI in the figure.

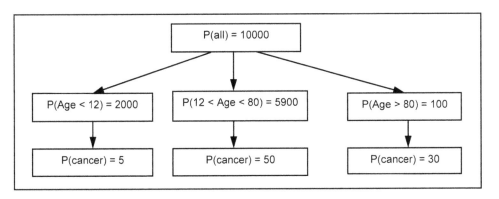

FIGURE 4.21   Hierarchy for de-identification and metaqueries in Problem 4.23

4.21 Look up the following using the Internet, and explain every part of the code using the corresponding code standards:
   a.   ICD code for Arrhythmia, epilepsy, ischemic cardiomyopathy.
   b.   SNOWMED code for CBC (comprehensive blood count), lipid panel.
   c.   LOINC for A1C, Urine protein test, blood-alcohol test.
4.22 For the following natural language text about a lab report recorded by a doctor, use ICD code and SNOWMED code to convert into an XML text. The sentence is:

> "The patient Mary Carrol had a lab test for complete blood count. Her hemoglobin value was 9 gm/dl. She is suffering from anemia." ICD code for Anemia is D64.9. SNOWMED code for CBC concept is 26604007. Use Figure 4.7 and the Internet search to derive the medically relevant phrases and code.

4.23 A city has an overall population of 10,000. Given the following age-related lattice (see Figure 4.21), give possible meta queries that can be answered and that should not be answered for $K = 20$. Note that relationships in a lattice support transitivity, and are anti-symmetric. The function "P" denotes the population.
4.24 Consider these scenarios, describe whether HIPAA is violated or not? Explain your rationale. The scenarios are:

Scenario 1: Two doctors discussing a patient history in a hospital cafeteria.
Scenario 2: A doctor enlisting patients for a rare medical condition with a medical representative who is advertising a new drug for clinical trials.
Scenario 3: Wife of a patient is discussing her husband having a stroke with her mother over the phone.
Scenario 4: A public report showing that seniors in a community suffer from asbestos-related cancer in a community having *20* senior citizens.

4.13.3 Extended Response

4.25 Explain the need for exchanging information across multiple heterogeneous databases using two simple examples.
4.26 Explain the need for content-based search and generic approaches to perform a content-based search.
4.27 Explain the use and need for local binary pattern to model texture-patterns of images.

4.28 Explain the advantages of R+ tree and SS-tree over other similarity-based search schemes.

4.29 Explain the similarity-search tree using Hilbert's curve and their advantages.

4.30 Explain temporality in medical databases and various issues such as handling of time inequality; reconciling different time-interval units and point-time.

4.31 Explain different temporal reasoning used in temporal databases.

4.32 Explain the concept of record linkage and their role in de-identification using a simple example not described in the book.

4.33 Explain the need for patient de-identification, the role of combination of meta-queries in patient identification and K-anonymity criterion.

4.34 Explain the patients' view of the OPENEMR with brief description of each field and record-type used in the database.

4.35 Explain abstract model of HL7 and different components of the abstract model.

4.36 Explain the components of HL7 message structure.

4.37 Explain the RIM Model for HL7.

4.38 Explain the role of adapters in the seamless integration of the heterogeneous databases.

4.39 Explain the role of MPI in seamless integration of heterogeneous databases.

4.40 Explain the role of datamarts and VMR in data sharing.

4.41 Explain various components of ArdenML such as MLM, executable code and event triggering.

4.42 Explain mechanisms to avoid record duplication during a new data entry in the database system.

4.43 Explain the mechanism of identifying whether a patient-entry is a new entry, or an existing entry missed due to wrong name and or DOB.

4.44 Despite de-identification and HIPAA, why is it difficult to maintain the privacy of the patients' disease condition? Explain this with respect to "Role Layers" described in Section 4.7.2. Do literature search.

4.45 Explain the role of match and unmatch probabilities in de-identification.

4.46 Do Internet search and study Fellegi and Sunter's probabilistic model and write a summary in your own words.

FURTHER READING

Medical Data Modeling

Modeling Biosignals

4.1 Hejjel, László, and Roth, Elizabeth, "What is the Adequate Sampling Interval of the ECG Signal for Heart Rate Variability Analysis in the Time Domain," *Physiology Measurements*, 25(6), December 2004, 1405–1411.

Medical Image Compression

4.2 Assche, Steven V., Rycke, Dirk D., Philips, Wilfried, and Lemahieu, Igance, "Exploiting interframe redundancies in the lossless compression of 3D medical images," In *Proceedings of Data Compression Conference*, Snowbird, UT, USA, March 2000, 521–527.

4.3 Ström, Jacob, and Cosman, Pamela C., "Medical Image Compression with lossless Region of Interest," *Signal Processing*, 59(2), June 1997, 155–171.

Automated Data Acquisition and Interfaces

4.4 Gardner, Reed M., Hawley, William L., East, Thomas D., Oniki, Thomas A., and Young, Hsueh-Fen W., "Real Time Data Acquisition: Experience with the Medical Information Bus," In *Proceedings of the 16th Annual Symposium of American Medical Informatics Association*, Baltimore, MD, USA, November 1992, 813–817.

Data Interface Model and Technology

4.5 Franklin, David F., and Ostler, David V., "The P1073 Medical Information Bus," *IEEE Micro*, 9(5), September 1989, 52–60.

Healthcare Data Archival and Interchange

Datamart and Data Warehouse

4.6 Alsqour, Mohammad, Matouk, Kamal, and Owoc, Mieczyslaw, "A Survey of Data Warehouse Architectures – Preliminary Results," In *Proceedings of the Federated Conference on Computer Science and Information Systems*, Wroclaw, Poland, September 2012, 1121–1126.

4.7 Inmon, William H., *Building the Data Warehouse*, 4th edition, John Wiley & Sons, Hoboken, NJ, USA, 2005.

4.8 de Mul, Marleen, Alons, Peter, van der Velde, Peter, Konings, Ilse, Bakker, Jan, and Hazelzet, Jan, "Development of a Clinical Data Warehouse from an Intensive Care Clinical Information System," *Computational Methods and Programs in Biomedicine*, 105(1), January 2012, 22–30, DOI: 10.1016/j.cmpb.2010.07.002.

4.9 Garcelon, Nicolas, Neuraz, Antoine, Salomon, Rémi, Hassan, Faoura, Vincent, Benoita, Arthur, Delapalmea, Arnold, Munnicha, Anita, Burgunb, and Rance, Bastien, "A Clinician Friendly Data Warehouse Oriented Toward Narrative Reports: Dr. Warehouse," *Journal of Biomedical Informatics*, 80, April 2018, 52–63, DOI: 10.1016/j.jbi.2018.02.019.

4.10 Sen, Arun, and Sinha, Atish P., "A Comparison of Data Warehousing Methodologies," *Communications of the ACM*, 48(3), 79–84, March 2005, DOI: 10.1145/1047671.1047673.

Health Information Flow

Health Service Bus

4.11 Menge, Falko, "Enterprise Service Bus," In *Proceedings of the Free and Open Source Conference*, Sankt Augustin, Germany, August 2007, 1–6.

4.12 Ryan, Amanda, and Eklund, Peter W., The Health Service Bus: An Architecture and Case Study in Achieving Interoperability in Healthcare, *MEDINFO*, 160, 2010, 922–926, DOI: 10.3233/978-1-60750-588-4-922.

4.13 Schleifer, Andreas, "Medical Information Bus," In *Seminar Kommukationsstandards in der Medizinetechnik*, Lübeck, Germany, June 2010, 1–12.

Clinical Interface Technology

4.14 Friedman, Carol, Shagina, Lyudmila, Lussier, Yves, and Hripcsak, George, "Automated Encoding of Clinical Documents Based on Natural Language Processing," *Journal of American Medical Informatics Association*, 11(5), September 2004, 392–402.

XML Based Representation

4.15 Saadawi, Gilan M., and Harrison, James H., "Definition of an XML Markup Language for Clinical Laboratory Procedures and Comparison with Generic XML Markup," *Clinical Chemistry*, 52(10), September 2006, 1943–1951.

4.16 *SOAP Version 1.2 specification W3C Recommendation*, World Wide Web Consortium, June 2003, Available at https://www.w3.org/TR/soap/, Accessed June 13, 2018.

Electronic Medical Records

4.17 Jones, Bilan, Yuan, Xiaohong, Nuakoh, Emmanuel, and Ibrahim, Khadija, "Survey of Open Source Health Information Systems," *Health Informatics – An International Journal (HIIJ)*, 3(1), February 2014, 23–31.

4.18 Ratib, Osman, Swiernik, Michael, and McCoy, Michael J., "From PACS to Integrated EMR," *Computerized Medical Imaging and Graphics*, 27(2–3), March–June 2003, 207–215, DOI: 10.1016/S0895-6111(02)00075-7.

Components of Clinical Information Systems

4.19 Rosenbloom, Trent S., Randolph, Miller A., Johnson, Kevin B., Elkin, Peter L., and Brown, Steven H., "Interface Technologies: Facilitating Direct Entry of Clinical Data into Electronic Health Record Systems," *Journal of the American Medical Informatics Association (JAMIA)*, 13(3), May/June 2006, 277–288, DOI: 10.1197/jamia.M1957.

EMR Views

4.20 *OPENEMR Version 5.0.1*, Available at https://www.open-emr.org/, Accessed May 29, 2018.

Heterogeneity in Medical Databases

Master Patient Index

4.21 Schurenberg, Kurt B., Yeager, Robert C., and Johnson, Robin D., "System and Methods for Implementing Global Master Patient Index," *US Patent 20020007284 A1*, January 2002, Available at https://patents.google.com/patent/US20050004895A1/en, Accessed June 2, 2018.

Virtual Medical Record

4.22 Bell, Glen B., and Sethi, Anil, "Matching Records in a National Medical Patient Index," *Communications of the ACM*, 44(9), September 2001, 83–88.
4.23 Marcos, Carlos, Gonza'lez-Ferrer, Arturo, Peleg, Moor, and Cavero, Carlos, "Solving the Interoperability Challenge of a Distributed Complex Patient Guidance System: A Data Integrator Based on HL7's Virtual Medical Record Standard," *Journal of the American Medical Informatics Association (JAMIA)*, 22(3), May 2015, 587–599.

Multimedia Information on Databases

4.24 Acharya, Tinku, and Ray, Ajoy K., *Image Processing – Principles and Applications*, John Wiley and Sons, Hoboken, NJ, USA, 2005.
4.25 Antani, Sameer, Kasturi, Rangachar, and Jain, Ramesh, "A Survey on the Use of Pattern Recognition Methods for Abstraction, Indexing and Retrieval of Images and Video," *Pattern Recognition*, 35(4), April 2002, 945–965.
4.26 Khan, Javed I., and Yun, David Y. Y., "Holographic Image Archive," *Computerized Medical Imaging and Graphics*, 20(4), July 1996, 243–257.

Temporal Information in Medical Databases

4.27 Chu, Wesley W., Hsu, Chi-Cheng, Cárdenas, Alfornso F., and Taira, Rickey K., "Knowledge-Based Image Retrieval with Spatial and Temporal Constraints," *IEEE Transactions on Knowledge and Data Engineering*, 10(6), November/December 1998, 872–888.
4.28 Combi, Carlo, Keravnou-Papilou, Elpida, and Shahar, Yuval, *Temporal Information Systems in Medicine*, Springer Science+Business Media, New York, NY, USA, 2010.
4.29 O'Connor, Martin J., Tu, Samson W., and Musen, Mark A., "The Chronus II Temporal Database Mediator," In *American Medical Informatics Association Annual Symposium*, San Antonio, TX, USA, November 2002, 567–571.

Time Stamping

4.30 Nigrin, Daniel J., and Kohane, Isaac S., "Temporal Expressiveness in Querying a Time-stamp-based Clinical Database," *Journal of the American Medical Informatics Association (JAMIA)*, 7(2), March 2000, 152–163.

Information Retrieval

Matching Similar Records

4.31 Guojun, Lu, "Techniques and Data Structures for Efficient Multimedia Retrieval Based on Similarity," *IEEE Transactions on Multimedia*, 4(3), September 2002, 373–384.

4.32 Hjaltason, Gisli R., and Samet, Hanan, "Index Driven Similarity Search in Metric Spaces," *ACM Transaction on Database Systems*, 28(4), December 2003, 517–580.

4.33 Petrakis, Euripides G.M., and Faloutsos, Christos, "Similarity Searching in Medical Image Databases," *IEEE Transactions on Knowledge and Data Engineering*, 9(3), May 1997, 435–447.

Probabilistic Record Linkage

4.34 Fellegi, Ivan, and Sunter, Alan, "A Theory for Record Linkage," *Journal of the American Statistical Association*, 64(328), December 1969, 1183–1210.

4.35 Winkler, William E., "Using the EM Algorithm for Weight Computation in the Fellegi-Sunter Model of Record Linkage," *Statistical Research Report Series No. RR2000/5*, Statistical Research Division, Methods and Standards Directorate, U.S. Bureau of the Census, Washington DC, October 2000, Available at https://www.census.gov/srd/papers/pdf/rr2000-05.pdf, Accessed June 2, 2018.

Content-Based Image Retrieval

4.36 Alemu, Yihun, Koh, Jong-bin, Ikram, Muhammed, and Kim, Dong-Kyoo, "Image Retrieval in Multimedia Databases: A Survey," In *Proceedings of the Fifth IEEE International Conference on Intelligent Information Hiding and Multimedia Signal Processing*, September 2009, Kyoto, Japan, 681–689.

4.37 Gupta, Amarnath, and Jain, Ramesh, "Visual Information Retrieval," *Communications of the ACM*, 40(5), May 1997, 70–79.

4.38 Kahn (Jr.), Charles E., and Rubin, Daniel L., "Automated Semantic Indexing of Figure Captions to Improve Radiology Image Retrieval," *Journal of the American Medical Informatics Association (JAMIA)*, 16(3), May-June 2009, 380–386, DOI: 10.1197/jamia.M2945.

4.39 Sethi, Ishwar K., and Jain, Ramesh C. (editors), *Proceedings of the SPIE: Storage and Retrieval for Image and Video Databases V*, 3022, January 1997.

4.40 Tang, Lilian H.Y., Hanka, Rudolph, and Ip, Horace H.S., "A Review of Intelligent Content Based Indexing and Browsing of Medical Images," *Health Informatics Journal*, 5(1), March 1999, 40–49.

Image Alignment

4.41 Antoine Maintz, Twan J.B., and Viergever, Max A., "A Survey of Medical Image Registration," *Journal of Medical Image Analysis*, 2(1), March 1998, 1–36.

Interoperability and Standardizations

Knowledge Bases of Medical Terms

4.42 *IEEE Standards Association Healthcare IT Standards*, Available at https://standards.ieee.org/findstds/standard/healthcare_it.html, Accessed June 2, 2018.

4.43 *Introductory Guide MedDRA Version 21.0*, March 2018, Available at https://www.meddra.org/ sites/default/files/guidance/file/intguide_21_0_english.pdf, Accessed June 2, 2018.

SNOMED

4.44 Rosenbloom, Samuel T., Brown, Steven H., Froehling, David, Bauer, Brent A., Wahner-Roedler, Dietland I., Cregg, William M., and Elkin, Peter L., "Using SNOMED CT to Represent Two Interface Terminologies," *Journal of American Medical Informatics Association (JAMIA)*, 16(1), January/February 2009, 81–88.

4.45 IHTSDO SNOMED CT Compositional Grammar Specification and Guide, Version 2.02, May 2015, Available at https://doc.ihtsdo.org/download/doc_CompositionalGrammarSpecificationAndGuide_Current-en-US_INT_20150522.pdf, Accessed June 2, 2018.

ICD

4.46 *ICD-10 – International Statistical Classification of Diseases and Related Health Problem Instruction manual*, 4th edition, World Health Organization, Geneva, Switzerland, Volume 2, 10th revision, 2011, Available at http://www.who.int/classifications/icd/ICD10Volume2_en_2010.pdf, Accessed on June 2, 2018.

LOINC

4.47 Deckard, Jamalynne, McDonald, Clement J., and Vreeman, Daniel J., "Supporting Interoperability of Genetic Data with LOINC," *Journal of Medical Informatics Association (JAMIA)*, 22, 2015, 621–627, DOI: i:10.1093/jamia/ocu01.

4.48 McDonald, Clement J., Huff, Stanley M., Suico, Jeffrey G., Hill, Gilbert, Leavelle, Dennis, Aller, Raymond, et al., "LOINC, A Universal Standard for Identifying Laboratory Observations: A 5-Year Update," *Clinical Chemistry*, 49(4), April 2003, DOI: 10.1373/49.4.624.

4.49 Vreeman, Daniel J., McDonald, Clement J., and Huff, Stanley M., "LOINC®–A Universal Catalog of Individual Clinical Observations and Uniform Representation of Enumerated Collections," *International Journal of Functional Informatics and Personalised Medicine*, 3(4), 2010, 273–291. DOI: 10.1504/IJFIPM.2010.040211.

ArdenML

4.50 Jung, Chai Y., Sward, Katherine A., and Haug, Peter J., "Executing Medical Logic Modules Expressed in ArdenML using Drools," *Journal of American Medical Informatics Association (JAMIA)*, 19(4), 2012, 533–536.

4.51 Samwald, Matthias, Fehre, Karsten, De Bruin, Jeroen, and Adlassnig, Klaus-Peter, "The Arden Syntax Standard for Clinical Decision Support: Experience and Directions," *Journal of Biomedical Informatics*, 45(4), August 2012, 711–718.

4.52 Karadimas, Harry C., Chailloleau, Christophe., Hemery, François, Simonnet, Julien, and Lepage, Eric, "Arden/J: An Architecture for MLM Execution on the Java Platform," *Journal of American Medical Informatics Association (JAMIA)*, 9(4), July/August 2002, 359–68.

4.53 Kim, Sukil, Haug, Peter J., Rocha, Roberto A., and Choi, Inyoung, "Modeling the Arden Syntax for Medical Decisions in XML," *International Journal of Medical Informatics*, 77(10), October 2008, 650–656.

Standards for Transmitting Medical Information

Reference Information Management Model (RIMM)

4.54 Eggebratten, Thomas J., Tenner, Jeffrey W., and Dubbels, Joel C., "A Healthcare Model Based on the HL7 Reference Information Model," *IBM Systems Journal*, 46(1), January/February 2007, 5–18.

4.55 Priyatna, Freddy, Alonso-Calvo, Raul, Parasio-Medina, Sergio, and Corcho, Oscar, "Querying Clinical Data in HL7 RIM Based Relational Model with Morph-RDB, *Journal of Biomedical Semantics*, 8(49), 2017, DOI: 10.1186/s13326-017-0155-8, Available at https://www.ncbi.nlm.nih.gov/pmc/articles/PMC5629785/pdf/13326_2017_Article_155.pdf, Accessed June 13, 2018.

Clinical Document Architecture

4.56 Dolin, Robert H., Alschuler, Liora, Boyer, Sandy, Beebe, Calvin, Behlen, Fred M., Biron, Paul V., and Shabo, Amnon, "HL7 Clinical Document Architecture, Release 2," *Journal of American Medical Informatics Association (JAMIA)*, 13(1), January/February 2006, 30–39.

HL7 Based Information Exchange

4.57 *American National Standard HL7 V3-2015, HL7 Version 3*, Normative Edition, ANSI, 2015, Available at https://www.hl7.org/implement/standards, Accessed July 18, 2018.

4.58 Beeler, George W.Jr., "HL7 Version 3—An Object-Oriented Methodology for Collaborative Standards Development," *International Journal of Medical Informatics*, 48, 1998, 151–161.

HIPAA

4.59 Choi, Young B., Capitan, Kathleen E., Krause, Joshua S., and Streeper, Meredith M., "Challenges Associated with Privacy in Health Care Industry: Implementation of HIPAA and Security Rules," *Journal of Medical Systems*, 30(1), 2006, 57–64.

4.60 *HIPAA Administrative Simplification – Regulation Text, Parts 160, 162 and 164*, March 2013, Available at https://www.hhs.gov/sites/default/files/ocr/privacy/hipaa/administrative/combined/hipaa-simplification-201303.pdf, Accessed July 9, 2019.
4.61 Sengupta, Soumitra, Calman, Neil H., and Hripcska, George, "A Model for Expanded Health Reporting in the Context of HIPAA," *Journal of American Medical Informatics Association (JAMIA)*, 15(5), September/October 2008, 569–574, DOI: 10.1197/jamia.M2207.

HIPPA and Privacy in EMRs

4.62 Wu, Ruoyu, Ahn, Gail-Joon, and Hu, Hongxin, "Towards HIPAA-Compliant Healthcare Systems in Cloud Computing," *International Journal of Computational Models and Algorithms in Medicine (IJCMAM)*, 3(2), April–June 2012, DOI: 10.4018/jcmam.2012040101.

Data De-Identification

4.63 Emam, Khaled E., Dankar, Fide K., Issa, Rome, Jonker, Elizabeth, Amyot, Daniel, et al., "A Globally Optimal K-Anonymity Method for the De-identification of Health Data," *Journal of American Medical Informatics Association*, 16(5), September/October 2009, 670–682.

Integrating the Healthcare Enterprise

4.64 Bortis, Gerald, "Experience with Mirth: An Open Source Health Care Integration Engine," In *Proceedings of the ACM International Conference of Software Engineering*, Leipzig, Germany, 2008, 649–652.

Medical Imaging Informatics

5

BACKGROUND CONCEPTS

Section 1.6.2 Medical Information Exchange; Section 1.6.3 Integration of Electronic Health Records; Section 1.6.7.1 Hidden Markov Model; Section 1.6.8 Machine Learning and Knowledge Discovery; Section 1.6.9 Medical Image Processing and Transmission; Section 2.2.2 Digital Representation of Images; Section 2.4 Statistics and Probability; Section 2.5 Modeling Multimedia Feature Space; Section 2.6 Similarity-Based Search Techniques; Section 2.8.3 Multimedia Databases; Section 2.9 Middleware for Information Exchange; Section 2.10 Human Physiology; Section 3.1 Dimension Reduction; Section 3.4 Machine Learning; Section 3.5 Classification Techniques; Section 3.6.3 Logistic Regression Analysis; Section 3.7 Probabilistic Reasoning over Time; Section 4.1.3 Medical Image Compression; Section 4.6.3 Multimedia Information in Databases; Section 4.7 Information Retrieval; Section 4.9 Standards for Transmitting Medical Information

Medical imaging is a fundamental component of modern health care, used widely for patient diagnosis and treatment applications. As imaging is a key part of the diagnosis of many diseases, automated diagnosis from images has been an active area of research. Computer-aided diagnosis has helped increase diagnostic accuracy and reduce physician time.

An image is a two-dimensional (2D) or three-dimensional (3D) visual representation of a collection of measurements. Medical imaging measurements can be from X-rays (absorption), ultrasound (acoustic amplitude), and MRI (RF amplitude) and other modes. Medical imaging is a crucial component of modern medicine for noninvasive investigation of the internal organs of a human-body.

In 1895, Röentgen discovered that fluorescent screens glowed if exposed to the light emitted from cathode-ray tubes. However, the screens kept glowing even after the tube was placed in a box. Röentgen realized that the tube was also emitting a new invisible radiation that could penetrate objects opaque to the eyes. He called this new radiation X-rays. Röentgen showed that different materials attenuate X-rays differently. By exposing photographic plates (films) to X-ray radiation, internal parts of a human-body were photographed, and medical imaging was born. Modern X-ray systems use digital detectors instead of films.

In the past few decades, several important imaging modalities have been discovered, including nuclear imaging, ultrasound, computed tomography (CT) and magnetic resonance imaging (MRI). These imaging techniques, combined with digital technology and electronic health records, have started a new medical diagnosis field called *imaging informatics*, also sometimes called *radiology informatics* or *medical imaging informatics*. This combination has revolutionized medical diagnosis through information archiving and exchange throughout complex healthcare systems. This subspecialty has gained wide acceptance and is often seen as a mission-critical function in health care.

Virtually every healthcare clinical discipline depends on imaging informatics. Continuous improvements of these imaging modalities have improved their power consumption, time-efficiency and cost-effectiveness. The value is clear; when clinicians have immediate electronic access to all relevant medical images, precious time is saved, allowing for timely medical decisions, reducing unnecessary repetition of examinations and driving costs down.

The science of imaging informatics is the study and application of processes of information and communications technology for the acquisition, manipulation, analysis and distribution of image data. Medical imaging allows acquisition, long-term storage, analysis, retrieval, transmission of medical images and aiding clinical diagnoses. New improved computational intelligence techniques have been employed in various applications of medical imaging. In this chapter, we present some of the latest trends and developments in computational intelligence in medical imaging.

Since the discovery of X-rays, medical imaging has come a long way. New imaging modalities such as CT, MRI, ultrasonic imaging, endoscopy, and nuclear medicine functional imaging techniques such as positron emission tomography (PET) and single-photon emission computed tomography (SPECT) have been developed. Continued improvements in the image quality have made the job of the clinicians easier. The quality of a medical image is generally assessed quantitatively using three parameters, SNR (Signal to Noise Ratio), CNR (Contrast-to-Noise Ratio) and spatial resolution. While all three should be high ideally, instrumental and operational trade-offs should be considered.

This chapter describes various imaging modalities, image formation techniques, image analysis techniques, image archiving and retrieval techniques, image segmentation techniques and the application of these techniques for cancer detection, specifically breast cancer detection.

5.1 RADIOGRAPHY (X-RAY)

X-rays are short-wave electromagnetic radiation in the invisible range with wavelengths in the range of 0.01 to 10 nanometers that can penetrate soft tissues. X-ray-based imaging methods are based on differential absorption. Bones and calcifications absorb X-rays more effectively than soft tissues. The most common example of its applications is to assess bone injury. However, this simple, inexpensive technique is also used extensively to assess lung diseases, kidney stones etc. X-ray propagation is expressed by Equation 5.1.

$$I(x) = I_0 e^{-\alpha x} \tag{5.1}$$

where $I(x)$ is the X-ray intensity after the propagation length x, I_0 is an incident X-ray intensity and α is the linear attenuation coefficient. X-rays are generated using a specialized apparatus called X-ray tubes. An X-ray tube consists of a vacuum tube, an anode and a cathode. A schematic diagram is shown in Figure 5.1.

The cathode current, through thermal excitation, releases electrons at the cathode-end. The voltage difference between the cathode and the anode causes emitted electrons to be accelerated toward the anode. When the electrons hit the anode, their energy is released. Part of this energy is in the form of X-rays.

X-rays are invisible to the human eye and need to be converted to a visual form. This was traditionally done by exposing a photographic silver halide film to the X-ray. However, films are not very sensitive to X-rays. Thus, this technique was partially replaced by "computed radiography" (CR) systems that use a phosphor plate that can be stimulated by light and read with a laser to be converted into a digital image. Improvement in semiconductor technology has led to the development of flat panel semiconductors, capable of detecting X-rays directly, allowing the image data to be read directly off the sensor. X-ray photons are first converted to visible photons using an intensifying screen. Intensifying screens are fluorescent screens, which convert a single X-ray photon to a very large number of visible-light photons, allowing good image quality at lower radiation dosage.

Low-energy (soft) X-rays are completely absorbed by tissues and are not useful in medical diagnostic applications and only increase the radiation-dose. Therefore, the useless lower energy part of the spectrum can be absorbed and filtered out using an "X-ray filter" such as a thin aluminum sheet placed over the X-ray tube window. This process is termed *beam hardening* because it causes the center of the spectrum to correspond to harder (higher energy) X-rays.

A medical X-ray image is called a *radiograph*. For a radiograph, the appropriate body-part of the patient is placed between an X-ray source and an X-ray detector. The source sends a short X-ray burst.

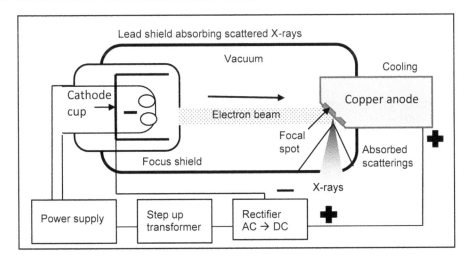

FIGURE 5.1   A schematic diagram of X-ray tube

Different types of tissues absorb X-rays at different rates. For example, calcium in bones has a high absorption rate for X-rays as a result of its relatively high atomic number. Thus, smaller amounts of X-rays reach the X-ray detector making the bones visible in the images.

5.1.1 Components of X-Ray Machine

The basic components of a planar radiography system are: 1) an X-ray tube for the generation of X-rays, 2) a collimator for the reduction of Compton-scattered X-rays and the radiation-dose, 3) an antiscatter grid for a further reduction of scattering X-rays and 4) a detector. Modern X-ray radiography systems use digital detectors. These detectors convert the X-ray energies into the visible spectrum of light. The light is converted into electric voltage using photo-diodes, and the analog volt-value is digitized using an A-to-D (analog to digital) converter.

Figure 5.2 shows the schematic representation of a radiographic imaging chain. A low-energy absorbing filter increases the mean energy of the X-ray beam produced by the source. A collimator limits the area

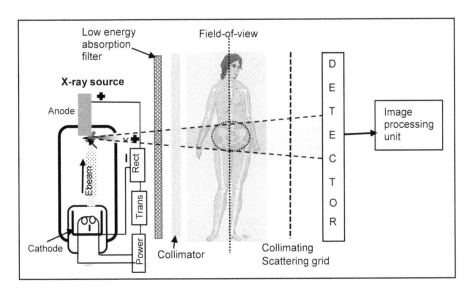

FIGURE 5.2   An illustration of the complete radiographic imaging chain

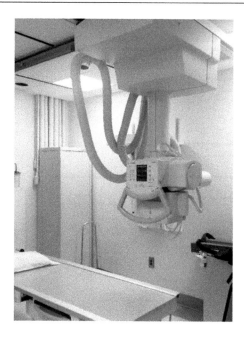

FIGURE 5.3   A modern X-ray radiography system (Image courtesy of Dr. Ashraf Malek, LMA South Center for Liver Disease, Cherry Hill, NJ, USA).

of exposure for the patient. X-ray photons pass through a body, undergoing attenuation (desirable) and scatter (undesirable). The scattered X-ray photons arrive at large angles and add to the image-noise. The collimator stops these scattered photons from arriving at the detector. Figure 5.3 shows a photograph of a modern-day X-ray imaging system.

X-ray radiographs are plain projection images that cannot separate the structures that are on the same line between the X-ray source and detector. Structures that lie on the same line between the source and the detector contribute to the intensity at a single point in the image. Thus, the radiographs lack depth resolution. In many applications, this is not a major problem. Sometimes, additional projection images of the same region are acquired using a near-perpendicular view to visualize the overlapping structures.

5.1.2 Applications of Radiography

Conventional applications of radiography include the detection of a fracture, tissue growth, tissue abnormality, cancers, stenosis, measurement of volume of blood-flow and monitoring of birth-related abnormalities. Clinical applications include all major organs: heart, liver, kidney, lung, muscles, bones, breast, prostate, uterus, teeth-cavities, etc. The current trend is to apply automated image-analysis for various clinical applications.

X-ray imaging is contrast-limited; the difference in attenuation is insignificant for different tissue-types, except bone and teeth, which are distinctly different from soft-tissue types. Thus, traditional X-ray radiographs are useful in imaging the skeletal system and the detection of its pathology, but less useful for soft tissues such as muscle, brain, etc. Although radiographs do not show the differences between different types of tissue, they are useful for detecting specific disease processes in soft tissues. Notable examples include the common chest X-ray, used for the diagnosis of lung-diseases such as lung cancer, pulmonary edema and pneumonia.

Abdominal X-ray is used to assess structures in the abdomen, including stomach, spleen and intestines. KUB (kidneys, ureters, bladder) X-ray depicts the structures in the bladder and kidney. X-rays are also used to detect pathology, such as abdominal masses and fluid buildup, certain types of gallstones, kidney stones, injury to the abdominal tissue, bowel (or intestinal) obstruction, gastrointestinal perforations (from the resultant free air) and ascites (accumulation of fluid in the peritoneal cavity) free fluid. Dentists use dental radiography for the diagnoses of common oral problems, such as cavities, bone-loss, hidden dental structures and benign and malignant masses.

Specialized radiographic techniques are used for different applications. For example, angiography allows radiographic visualization of blood vessels to diagnose vessel constriction and vascular tumors. A nontoxic radiopaque dye is injected into the blood stream before the procedure to improve the visualization of blood vessels.

Mammography is a common imaging technique for the detection of breast-cancer in early stages. It uses low energy X-rays typically of the order of *20* keV to image soft tissue in a breast. To detect microcalcification and trabeculae, spatial resolution better than *0.1* mm is used. Mammography can be performed with or without contrast agents.

The major concern in using the radiation-based modalities is the exposure to the unnecessary radiation and additional cost to get higher resolution. As the technology improves further and the image-quality improves along with a lower-level of radiation required to get high-quality images, radiography will find new applications.

5.2 COMPUTED TOMOGRAPHY

Conventional X-ray methods have many limitations. The images are 2D projections of a 3D structure. Consequently, there is a loss of the depth information. For certain applications, the detection of subtle abnormalities is difficult due to the overlaps of the regions. As mentioned in the Section 5.1, X-ray cannot differentiate well *between* soft tissues. However, radio-opaque dyes can depict certain soft tissues. Additionally, conventional X-rays do not provide quantitative information about tissues.

CT overcomes these limitations by offering excellent sensitivity for differentiating soft tissues. CT produces cross-sectional images of the X-ray attenuation properties of biological tissues. For recording a one-dimensional projection of the cross-section of a patient, a bank of detectors is placed opposite to the bank of X-ray sources. Many X-ray tubes in parallel-beam or a fan-beam geometry are used to cover the field of view. The sources and detectors are rotated through one-half of a complete revolution to complete one set. The second-half is symmetric with the first-half. If this process is repeated at an increment of small angles, the actual attenuation at each point of the scanned slice can be *reconstructed* from the measurements. Image reconstruction process in CT is illustrated in Figure 5.4.

5.2.1 Fundamentals of Computed Tomography

The fundamental principle behind CT is to reconstruct a 2D structure from multiple single-dimensional projections of the structure acquired at different angles. The theoretical basis for CT is *Radon transform* that computes the line integrals from multiple beams in a specific direction. The inverse of the *Radon function* is used to reconstruct the medical images from CT scans. The theory that could be applied for CT was developed by Allan Cormack et al 1963. Godfrey Hounsfield, at EMI in England, built the first practical scanner in 1972. Multiple independent measurements are made at different angles, each giving a different attenuation coefficient α_i. The overall intensity is described by Equation 5.2.

$$I(x) = I_0 e^{-\sum_{i=1}^{i=N} \alpha_i x_i} \tag{5.2}$$

CT is based on the "central slice theorem," which states that the one-dimensional Fourier-transform of a one-dimensional projection of a two dimensional function equals to a slice through the two-dimensional

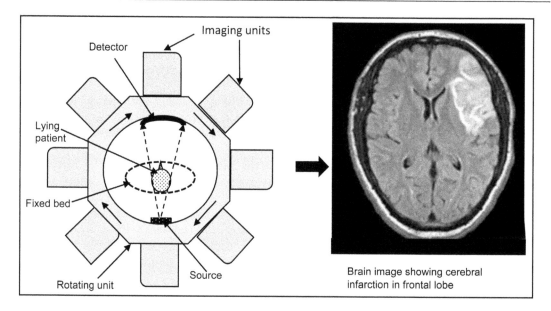

FIGURE 5.4 Reconstructing image from single-dimensional projections (Brain image taken from Wikipedia Commons, website: https://commons.wikimedia.org/wiki/File:FLAIR_MRI_of_cerebral_infarction.png, courtesy: Shazia Mirza and Sankalp Gokhale, *Neuroimaging in Active Stroke*, Available at http://www.smgebooks.com/neuroimaging/chapters/NI-16-06.pdf, original source: Dr. Anvekar, Balaji, Neuroradiology unit, S P Institute of Neurosciences, Solapur, India, permission under Creative Common Attribution 4.0 international license).

Fourier-transform of the function. Thus, it is possible to construct the two-dimensional Fourier-transform of a cross-section from multiple one-dimensional projections, which can estimate the cross-section from its inverse two-dimensional Fourier-transform.

Let $f(x, y)$ be the cross-section being imaged. A projection along the x-axis is given by the Equation 5.3. The two-dimensional Fourier transform of the function $f(x, y)$ is given by the Equation 5.4. The value of the function $F(u, v)$ on $u = 0$ equals the Fourier transform of the projection of $f(x, y)$ along the x-axis ($y = 0$). For $u = 0$, Equation 5.4 reduces to Equation 5.5.

$$g(y) = \int f(x, y)dx \qquad (5.3)$$

$$F(u, v) = \int \int f(x, y)e^{-j2\pi(ux + uy)}dx\ dy \qquad (5.4)$$

$$F(0, v) = \int g(y)e^{-j2\pi vy}dy \qquad (5.5)$$

Equation 5.5 can be generalized for any angle, θ. After deriving the projections at a sufficient number of angles, the function $f(x, y)$ is estimated from the Fourier-transform of the projections using the following approach: 1) Take one-dimensional projections of the object cross-section of interest at many angles; 2) compute the one-dimensional Fourier-transform of each one-dimensional projection; 3) assemble the one-dimensional Fourier-transforms to form the two-dimensional Fourier-transform at corresponding angular coordinates of the projections and 4) reconstruct the cross-section by computing the inverse transform of the 2D Fourier-transform. Because the projections are acquired by rotating the source and the detectors, the Fourier-transforms are acquired using polar-coordinates. For computing the inverse Fourier-transform, the Fourier-transform in the polar-coordinates is interpolated to the Cartesian-grid. However, this interpolation introduces artifacts. Consequently, direct reconstruction is less popular than the filtered backprojection methods, where the slice is reconstructed by backprojecting the acquired projection lines.

A typical CT radiation-dose is significantly higher than in conventional X-rays. The lifetime risk of fatal cancer is approximately one per 2000 CT-scans. The risk of malignancy is age-dependent, being 1 in 10,000 for patients over age 40 and approximately 1 in 500 for neonates. While CT scans saves lives, it could be beneficial for a patient to forgo CT-scans for youth unless the physician considers it necessary.

5.2.2 Application of CT Scans

CT offers good soft-tissue contrast and resolution and is used in orthopedics, including imaging of the protrusion of vertebral discs, complex joints such as the hip or shoulder, etc. CT provides good-quality images of not only bones but also soft tissues, such as lung, brain and liver. CT is useful in depicting the size, location and extent of tumors, especially the large ones. CT images of the cranium and brain show masses, blood clots, blood-vessel defects and enlarged ventricles, etc.

CT also has been used in the interventional applications, such as CT-guided biopsy and minimally invasive surgery. CT is used in radiation therapy planning and monitoring of noninvasive cancer treatments. CT is not overly susceptible to patient motion (such as breathing), because of the relatively short scan-time.

5.3 MAGNETIC RESONANCE IMAGING

MRI essentially shows images of magnetic properties of a tissue. Nuclear magnetic resonance (NMR) was discovered by F. Block and E. Purcell in 1946. NMR was initially used by organic chemists to assess molecular structures. These methods needed to create a homogeneous magnetic field in the sample being tested. Medical use of NMR became a reality when the nuclei in the sample could be spatially mapped by introducing a gradient in the magnetic field. Lauterbur, Mansfield and Damadian pioneered the use of NMR for medical imaging (MRI) in early 1970s. Over 10 million MRI scans are performed every year.

5.3.1 Fundamentals of NMR

NMR is based on the presence of magnetic moments, which requires the nuclei to have an odd number of protons or neutrons. Magnetic moment in a nucleus is like a small magnet. The sum of the magnetic moments in our bodies is zero in the normal state. When the body is placed inside an external magnetic field, magnetic moments align themselves with respect to the external field.

Resonance is defined as a phenomenon of the energy-absorption at a specific "resonant" frequency. For NMR, the absorption of the radiofrequency (RF) energy excites the nuclei of certain atoms (e.g., hydrogen nuclei 1^H in water for biological tissues) to a higher energy state in the presence of an externally applied magnetic field. These nuclei release RF-energy and return to their ground-state after external magnetic field is removed. The emitted RF-energy is the MR-signal, which provides useful information about the molecular environment.

According to quantum physics, a proton has only two allowable quantum states, parallel (spin-up or lower energy) and antiparallel (spin-down or higher energy) states, at a small angle with the direction of the field. Protons, when subjected to an external magnetic field, B_0, align with the field and experience a torque. However, they do not align exactly toward B_0. The spinning protons wobble like a spinning top with a slight tilt from the direction of B_0. This wobbling is known as *precession*. The frequency of this precession around the axis of the field is called *Larmor frequency* f_0 in Equation 5.6.

$$f_0 = \gamma B_0 \tag{5.6}$$

The resonance or *Larmor frequency* is proportional to the applied magnetic field. Here, γ is the *gyromagnetic ratio* – the sensitivity of the nucleus to the MR-signal. It is expressed in MHz/Tesla. Tesla is a unit to measure magnetic field. The *gyromagnetic ratio* is a constant for every isotope. Hydrogen has the highest *gyromagnetic ratio* ($\gamma = 42.58$ MHz/Tesla).

Under an external magnetic field, the parallel (spin-up) state will preferentially be occupied because it has lower energy than the antiparallel (spin-down) state. However, quantum mechanics laws do not allow all spins being in this state. Some spins are in the antiparallel state. The difference in the populations between the two energy levels is given by Equation 5.7.

$$N_{parallel} - N_{antiparallel} = N_{total} \frac{\gamma \hbar B_0}{4\pi kT} \tag{5.7}$$

where $\hbar$ is the Plank's constant (6.63×10^{-34} J-s), k is the Boltzmann's constant (1.38×10^{-23} Joules/Kelvin), T is the temperature measured in Kelvin and N_{total} is the total number of protons. MRI detects only the difference of protons ($N_{parallel} - N_{antiparallel}$), not the total number.

Within a tissue-volume, the aggregation of the spins in the parallel (lower energy) state results in a weak net magnetization-vector $\vec{M}$ that is proportional to the external magnetic-field strength B_0. The magnetic-field $\vec{M}$ is responsible for the emitted MR-signal. A higher magnetic-field increases the MR-signal, which has prompted the use of high-powered magnets. Most of the commercial scanners use a supercomputing magnets capable of generating magnetic-field strength between 0.5 and 3.0 Tesla.

After the fields are turned off, the nuclei relax, and return to their original spin-state. The energy absorbed by the nuclei at Larmor frequency is released into the surroundings. The measured RF-signal returns to zero. There are two types of relaxation-rates associated with nuclei relaxations: 1) T_1 related to thermal-equilibrium and 2) T_2 related to loss of coherence in magnetically aligned protons. It typically requires *four* seconds for water-protons in biological tissues to reach thermal-equilibrium (relaxation-rate T_1). T_1 is an exponential-time constant, which represents the rate at which the magnetization vector returns to its equilibrium by releasing energy. The exponential relaxation-rate T_2 is much faster and will make the RF-signal decay to zero typically within two seconds. Heterogeneity in the external fields also contributes to T_2. The MR-signal weakens due to the exponential dephasing and disappears after the spins return to random phase after a time-interval of approximately $5 \times T_2$. Two relaxations occur concurrently, independent of each other. The relaxation-rate T_2 is more important for MRI because dephasing occurs faster than the release of energy into the molecular environment.

The durations T_1 and T_2 often differ in pathological tissues and surrounding healthy tissues as illustrated in Table 5.1. Information about tissue characteristics is ascertained through the use of pulse-sequences. T_1-weighted and T_2-weighted images are the MR images created from pulse-sequences that emphasize the T_1 and T_2 information.

5.3.2 MRI Resolution

MRI uses hydrogen atoms as it has a high sensitivity to MR-signal and is the most abundant element in biological tissues due to the abundance of water. Furthermore, 99% of all hydrogen isotopes are ^1H, containing one proton and no neutron. The MR-signal depends on the loose-binding of hydrogen nuclei

TABLE 5.1　T_1 and T_2 relaxation times for different types of tissues and water

TISSUE TYPE	T_1 (msec)	T_2 (msec)	TISSUE TYPE	T_1 (msec)	T_2 (msec)
Fat	250	70	Muscle	900	50
Ice	5000	0.0001	Protein	250	0.1–1.0
Gray matter	900	90	Tendon	400	5
Liver	500	40	Water	4000	2000

inside a molecule. Hydrogen atoms are loosely bound in liquids and soft tissues, tilt in the presence of a field and produce a detectable MR-signal. MRI provides an excellent soft-tissue contrast and is nonionizing.

To generate useful MR images, it is essential to localize image pixels/voxels, which is done using magnetic-gradients. Gradients are nonuniform magnetic-fields, superimposed over the static magnetic-field B_0. Magnetic-field gradients create a spatially varying magnetic-field in the tissue. The maximum values of gradients are a small percentage of the static external magnetic-field B_0. After the gradients are superimposed, the effective local magnetic-field becomes $B_0 + G_z z$, which is unique at each location along the gradient, resulting in a unique *Larmor frequency* at each position as shown in Equation 5.8.

$$f_{Larmor} = \gamma B_0 + \gamma(G_z z) \tag{5.8}$$

Application of a gradient allows the spatial location to be encoded into the MR-signal frequency. The first step is to select a plane $z = z_p$ in the tissue-volume by applying a z-gradient in the uniform magnetic-field B_0. At each z-location, the magnetic field is unique, resulting in a unique Larmor frequency. Each z-location is an x-y plane. All nuclei in the plane resonate at the same frequency due to the uniform magnetic-field. MR-signals received at a specific frequency are from one specific plane. This process is called *slice-selection*.

After a slice is selected, additional gradients G_x and G_y are applied during signal measurement, which causes all protons at an angle $\varphi = \arctan \frac{G_y}{G_x}$, to experience an identical magnetic field. The additional gradients reduce the dimensionality from a plane to a line. By changing the gradients, it is possible to acquire data at equally-spaced angular intervals. The measurement using this method is a trajectory of the image to be reconstructed in the Fourier-domain. To reconstruct the image, all values in the Fourier-domain are measured by varying the gradients G_x and G_y. Thus, the straightforward inverse Fourier-transform provides the image of the tissue cross-section without requiring back-projection methods.

5.3.3 MRI System

An MRI system consists of a magnet, three magnetic gradient-coils, an RF-transmitter and a receiver. To create an image, field gradients make the Larmor frequency of each proton to be spatially location-dependent. For imaging, it is essential to spatially localize the origin of a signal. In MRI, well-defined short-duration magnetic-field gradient pulses are superimposed over the main field to "encode" the spatial location of a proton nucleus with a specific resonance frequency and the corresponding phase.

Figure 5.5 illustrates a schematic of an MRI system. The superconducting magnet produces a strong magnetic field (B_0). There are three magnetic gradient coils (only one is shown). Figure 5.5b shows the photograph of an MRI scanner. Figure 5.5c–e show examples of MRI. Figure 5.5c shows the MRI scan of the lateral view of a brain with the brain stem. Figure 5.5d shows the MRI scan of a spine. Figure 5.5e shows the axial view of a brain with lesion caused by vitamin B12 deficiency.

5.3.4 Applications of MRI Analysis

The MR images are detailed, capable of detecting tiny changes inside the body. MR images have excellent resolution and are suitable for diagnosing soft-tissue abnormalities. Major applications of MR images include detection of 1) brain abnormalities, including aneurisms and brain tumors, 2) lung cancer, 3) liver cancer, 4) prostate cancer, 5) renal abnormalities, 6) spinal abnormalities such as herniated discs, 7) cardiological abnormalities, 8) arthritis and 9) muscle and ligament tear, etc. Other applications include the measurements of brain-structure volume, assessment of cartilage injuries, etc. MR images have also been applied for guided computer-aided surgery of a brain.

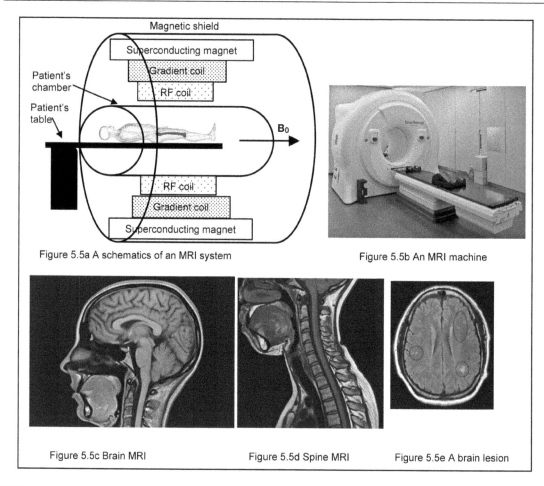

Figure 5.5a A schematics of an MRI system

Figure 5.5b An MRI machine

Figure 5.5c Brain MRI

Figure 5.5d Spine MRI

Figure 5.5e A brain lesion

FIGURE 5.5   A schematic of an MRI system and MRI scans of brain, spine and a brain lesion: (a) Schematics of an MRI system; (b) An MRI machine; (c) Brain MRI; (d) Spine MRI; (e) A brain lesion (Image 5.5b is courtesy Wikipedia Commons, credit: "Digital Cate"/flickr; images of brain MRI (5.5c) and spine MRI (5.5d) are © Siemen's Healthcare, used with written permission; Figure 5.5e is taken from Wikipedia Commons, courtesy Hemlata Bhaskar and Rekha Chaudhary website: https://commons.wikimedia.org/wiki/File:CRIM2010-691563.001.jpg, original source: https://www.ncbi.nlm.nih.gov/pmc/articles/PMC3065218/, Wikipedia Commons images have been used under Creative Commons license attribution 3.0).

5.4 ULTRASOUND

Ultrasonic imaging was developed as an extension to SONAR, which sends out a high-pitched sound pulse into the ocean, and measures the time taken to travel after reflection by an obstructing object. Acoustic waves are longitudinal compressional waves: the direction of the wave propagation is the same as that of particle motion. Wave parameters that describe sound waves include pressure, density, temperature and particle displacement.

Ultrasonic imaging (also known as sonography) is the second most popular medical imaging modality after X-rays. It is the least expensive, the most portable and a safe modality. It has been used in clinical practice for over a half century. Ultrasound is estimated to involve over a quarter of all medical imaging procedures. Among the four main medical imaging modalities, sonography is the only major medical imaging modality that does not employ any electromagnetic radiation.

The first pulse-echo system was described in 1949. Two-dimensional grayscale images were produced during the 1950s, which was followed by the applications of the Doppler technique in 1956. In 1965, Siemens developed an ultrasound scanner that produced first two-dimensional gray-scale image in real time. Image quality saw steady improvement since 1980s. Modern ultrasound scanners can produce very clear images of body-interiors.

5.4.1 Principles of Ultrasound Waves

Ultrasound waves use high-frequency inaudible acoustic waves with a frequency of the order of one to fifteen MHz (wavelengths of *0.1* through *1.5* mm). The frequency f of a wave is related to its period T by the equation $f = 1/T$. For a wave traveling at a speed of c, the wavelength λ is given by c/f. Speed of sound c is given by $\sqrt{\kappa/\rho}$, where κ is the bulk modulus of the tissue and ρ is the density of the medium. Ophthalmic imaging uses very high frequencies in the range *20–50* MHz. Acoustic microscopy uses even higher frequencies up to the *500* MHz range.

Ultrasound imaging is based upon signal reflection from acoustic impedance discontinuities such as a tissue interface. The speed of sound is approximately 1540 meters/second for most biological tissues. The location of tissues generating an echo can be ascertained from the round-trip time of the acoustic wave back to the transducer. Sound waves undergo attenuation during the propagation through tissue, and attenuation increases approximately linearly with the increase in frequency. The lowest diagnostic frequencies (*1* MHz) can propagate through the torso of a large human, whereas the higher frequencies (*20* MHz) attenuate within millimeters. Thus, higher frequencies are more useful for organs such as the eye (low attenuation through vitreous) and skin (low penetration depth). For larger organs or penetration depth (such as liver), lower frequencies are more useful.

The ability to steer the transmitted ultrasonic pulses has allowed the construction of an image from the received echoes by mapping the amplitude of the echoes to brightness. The principle of velocity imaging was based originally on the *Doppler effect*. However, modern tissue-motion estimation methods use more sophisticated techniques.

The phenomena such as scattering, refraction, attenuation, dispersion and diffraction introduce noise in the ultrasound images. The images have *speckles* due to constructive and destructive interferences. *Speckles* are locally correlated intensity variations that especially affect ultrasound images. Speckles reduce the contrast of the image and make edge detection difficult. Besides speckles, ultrasound images also suffer from *acoustics shadowing* that is caused by the absence of signal behind structures that strongly absorb or reflect ultrasonic waves such as lesions and bones. However, images from modern ultrasound scanners can almost rival that of MRI, etc.

5.4.2 Instrumentation

Figure 5.6 shows a block-diagram of the basic ultrasonic imaging system that consists of a pulse-generator, an ultrasound transmitter, a beam former, a receiver, a transmit-receive switch, a processor and a display unit. A piezoelectric transducer comprises a transmitter and a receiver. The transducer transmits and receives the ultrasound signals to a tissue, generally through the skin. The function of the transmit-receive switch is to separate the transmitter and the receiver circuits from each other.

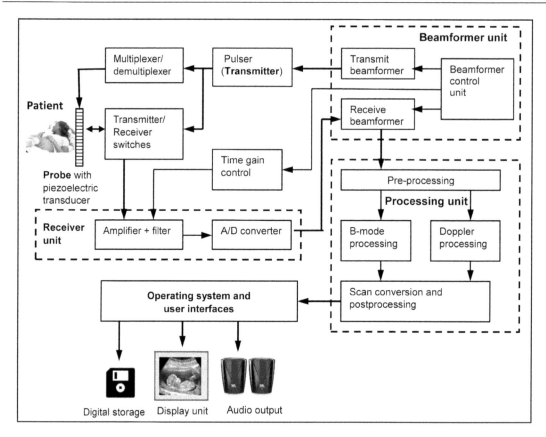

FIGURE 5.6   A block diagram of a medical ultrasound scanner (Patient-image courtesy US FDA, website: www.FDA.gov, public domain; ultrasound image in display taken from Wikipedia Commons, website: https://commons.wikimedia.org/wiki/File:CRL_Crown_rump_length_12_weeks_ecografia_Dr._Wolfgang_Moroder.jpg, © Dr. Wolfgang Moroder, used under GNU Free Documentation License 1.2 and Creative Commons Attribution-share 3.0).

5.4.2.1 Pulse generation

Piezoelectric (or capacitive) transducers are used to generate and receive ultrasound waves. A piezoelectric material generates an AC voltage when subjected to mechanical stress and vibrates when subjected to an AC voltage. Commonly used piezoelectric materials for ultrasound transducers include ceramic, lead zirconate titanate (PZT) and polyvinylidine difluroide (PVDF).

Piezoelectric material vibrates in two ways: 1) by applying a sinusoidal electric-field and 2) by applying a sharp electrical spike across the object. The sharp spike is analogous to hitting a bell with a hammer and causes the material to vibrate at its resonant frequency. The resonant frequency of a piezoelectric slab is determined by its thickness. The thickness equals a half wavelength of sound within the material. To generate ultrasound at *10* MHz using a PZT slab, a *0.2* mm thick PZT is required. The speed of sound in PZT is about *4000* m/s. The wavelength of the ultrasound signal is:

$$\lambda = c/f = (4 \times 10^3 \text{ meters})/(10 \times 10^6) = 0.4 \text{ meters} \times 10^{-3} = 0.4 \text{ mm}$$

5.4.3 Overall Functioning

The pulse-generator sends a periodic short burst of voltage pulses to the transducer. The pulsed signals are amplified and fed via the transmit-receive switch to the beamformer. The beamformer, based upon the desired beam-patterns, provides focusing and beam-steering. The amplified pulses are channeled, via the beamformer, into a transducer-array.

A transducer converts the amplified voltages into the pressure-waves that are transmitted into tissues. The echoes and the backscattered pressure-waves are received using a sensitive receiver at different times, based on the tissue-depth, and are converted back into voltages. The receiving beamformer processes the signal for focusing and steering, as needed. The received signals are amplified using a low-noise amplifier before digitization.

The signal-processing steps include filtering to remove high-frequency noise, envelope-detection, TGC (Time Gain Control, described in Section 5.4.4.5), log compression and scan conversion to convert the signals into a format appropriate for displaying the images. The signals are then stored in the RAM or the hard drive of the scanner. Finally, real-time images are displayed on a monitor.

5.4.4 Image Formation

A simple (single-element) ultrasound transducer only sends and receives ultrasound signals in one direction. The simplest method of building an image using one-dimensional transducer is by laterally moving this transducer to "paint" an image by combining the individual echoes, for each of which the distance and direction is known. We might call this the "lighthouse" image-formation analogy. However, mechanically scanned transducers have complexity and reliability issues, which led to array-based transducers that steer the beams electronically. Modern scanners use linear or curvilinear array transducers. The array transducers comprise many piezoelectric elements arranged on a straight line or a curve. By exciting the elements in a predefined sequence, ultrasonic transmissions can be directed through a predefined pattern.

For many clinical applications, only a small acoustic window is available, for example, heart can be imaged between the ribs. This constraint requires a phased-array transducer to have a small transmission "footprint." Figure 5.7 illustrates the phased-array principle using an array of piezoelectric elements. Piezoelectric elements are excited progressively after a fixed interval δ, creating a linear wavefront. The

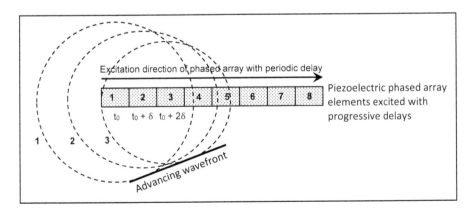

FIGURE 5.7   An illustration of the phased-array principle

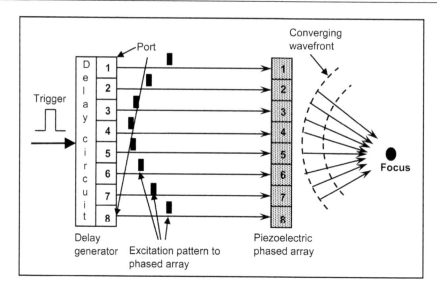

FIGURE 5.8   Focusing using the phased-array principle

figure shows the wavefront created after the 3ʳᵈ piezoelectric element has been excited and before the fourth element is excited.

It uses the allegory of dropping pebbles into the water in a pond. A pebble dropped into water creates ripples that expand in a circular pattern. Multiple pebbles, dropped at once in a straight line, send out a combined ripple wavefront, that becomes straighter as the wave propagates (Huygens' principle). If stones are dropped one by one at a constant interval, the combined waves make an angular straight line.

Similarly, many pebbles can be dropped in a straight line in a sequence so the resultant waves arrive simultaneously at a rock in the pond, focusing toward the rock; the outer pebbles would be dropped first, and the middle one will be dropped last. The waves would reflect off the rock and arrive back at the location of the source pebble after the same time interval it took to reach the rock.

The same principle is used in ultrasound imaging to focus and steer ultrasound beams. The system excites the elements of the transducer in a predefined-pattern to make the wave-fronts arrive at a specific point simultaneously. To obtain the total reflection from the same point, the same phase delays are applied to the echoes at all the transducer elements. In a single transmission, the focus could be only at one point. However, on receive, it is possible to focus on different points by applying appropriate delays (dynamic focusing on receive).

Figure 5.8 illustrates focusing using an array. The edge-elements are excited much earlier compared to the middle element to focus at a near point. Relative delays between elements is much smaller for focusing at a far point.

Figure 5.9 shows wavefront steering. Piezoelectric elements are excited with the first element (leftmost) being activated first. Hence, the wave is steered downwards, in the same direction of the delay-pattern. The wavefront can be steered upwards by applying the delay-pattern first to the last piezoelectric element in the phase-array.

5.4.4.1 Resolution

Resolution is the ability to resolve two separate structures and can be defined as the minimum distance at which two points can be distinguished as separate points. In ultrasound imaging, resolution is proportional to the frequency. Resolution can be improved by decreasing the system pulse-width and increasing the center frequency (decreasing the wavelength).

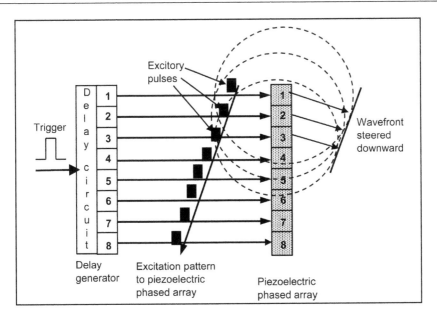

FIGURE 5.9 An illustration of the steering mechanism for the ultrasound waves

Example 5.1: Ultrasound image resolution

Let us assume that a transducer emits *five* periods of a *6* MHz pulse. If the speed of sound through the tissue is *1540* m/s, then the wavelength is $\lambda = c/f = (1.54 \times 10^3)/(6 \times 10^6)$ m $= 0.26$ mm. The length of a five-wavelength pulse is *1.3* mm. We have to remember about the round-trip travel of the pulse and divide the length by *2*. (Why do we divide by *2* and not multiply by *2*?) The resolution would be *0.65* mm as shown in the calculation:

Resolution $= nc/2f = (5 \times 1.54 \times 10^3)/(2 \times 6 \times 10^6)$ m $= 0.65$ mm.

5.4.4.2 Signal attenuation

Intensity of ultrasound decreases exponentially as it propagates. An exponential decay means that per unit length of travel, the same fraction of the energy is removed. Energy lost from an ultrasound wave during the propagation is due to many factors, including reflection and scattering. Further energy is lost due to absorption, and conversion of acoustic energy into heat because of viscosity. Due to the loss of energy, the echoes received from areas deep in the body are much weaker than the echoes from areas close to the surface. Experiments have shown the relationship between attenuation and ultrasound frequency is nearly linear in many tissue-types.

As an ultrasound wave propagates through a medium, its pressure amplitude and intensity decreases due to attenuation, which can be expressed by equations

$$I(z) = I_0 e^{-\mu z} \tag{5.9}$$

$$p(z) = p_0 e^{-\alpha z} \tag{5.10}$$

where $I_0 = I(z = 0)$ and $p_0 = p(z = 0)$. The terms μ and α are the intensity-attenuation and pressure-attenuation coefficients, respectively. The value of μ is twice that of α. Generally, attenuation coefficients are expressed in dB (*20 log_{10}* for pressure and *10 log_{10}* for intensity). Accordingly, μ (dB/cm) $= \alpha$ (dB/cm). The attenuation coefficient can be expressed $f^{1.2}$ for soft tissue and f^2 for water (approximately). A general rule of thumb is that the value of μ is approximately 1 dB/cm/MHz for soft tissues.

Example 5.2: Ultrasound signal attenutation

How much would a 7.5-MHz ultrasound signal attenuate after propagating 6 cm? In soft tissues, attenuation is 1 dB cm^{-1}–MHz^{-1}. For a 7.5-MHz ultrasound signal, the attenuation is 7.5 dB per cm. The attenuation after a propagation of 6 cm is 7.5×6 dB = 45 dB.

With the knowledge about reflection and attenuation, it is possible to calculate the intensity of an ultrasound-pulse returning from the body-parts consisting of multiple tissue-layers. The amplitudes of ultrasound echoes are governed by the reflection-coefficient at each interface, and the attenuation-coefficient of each medium between the boundaries.

5.4.4.3 Reflection and refraction

Echoes are produced due to the reflection of waves. When a wave travels from a medium to another medium with a different impedance property, it is absorbed, refracted or reflected. At every interface, the fraction of energy transmitted through the interface is governed by the difference in the acoustic impedances, which is the square root of the product of the density ρ, and the bulk modulus κ as shown in Equation 5.11.

$$Z = \sqrt{\rho\kappa} \tag{5.11}$$

Since the sound speed can be expressed as $c = \sqrt{\frac{\kappa}{\rho}}$, acoustic impedance can be expressed as the product of the density and sound speed as shown in Equation 5.12.

$$Z = \rho c \tag{5.12}$$

When an ultrasound wave travels through a material with an impedance Z_1 and encounters a material with a different impedance Z_2, only a part of the wave is transmitted through the boundary. The remaining signal is reflected back to the transducer. The fraction of reflected signals depends on the extent of mismatch between the acoustic impedances. Assuming normal incidence, reflection and transmission pressure-coefficients, R_p and T_p, are given in Equations 5.13 and 5.14. The impedances of the first and second mediums are Z_1 and Z_2; and the terms p_r and p_t are the reflected and transmitted pressures.

$$R_p = \frac{p_r}{p_i} = \frac{Z_2 - Z_1}{Z_2 + Z_1} \tag{5.13}$$

$$T_p = \frac{p_t}{p_i} = \frac{2Z_2}{Z_2 + Z_1} \tag{5.14}$$

Reflection and transmission intensity coefficients, R_I and T_I, are given by Equations 5.15 and 5.16, respectively. The impedances of the first and second mediums are Z_1 and Z_2, and the terms I_r and I_t are the reflected and transmitted intensities, respectively.

$$R_I = \frac{I_r}{I_i} = \left(\frac{Z_2 - Z_1}{Z_2 + Z_1} \right)^2 \tag{5.15}$$

$$T_p = \frac{I_t}{I_i} = \frac{4Z_1Z_2}{(Z_2 + Z_1)^2} \tag{5.16}$$

The relationships between reflected and transmitted pressures and reflected and transmitted intensities are given by Equations 5.17 and 5.18, respectively.

$$T_p = R_p + 1 \tag{5.17}$$

$$T_I = 1 - | R_I |^2 \tag{5.18}$$

From the equations, the reflected or backscattered signal is maximized if either impedance Z_1 or Z_2 is zero. If Z_1 is zero, there is no propagation. If Z_2 is zero, the ultrasound beam fails to reach deeper structures in the body. For example, an air-pocket during an imaging of the GI tract will cause a very strong signal from the front of the air-pocket; no signal will be received from behind the air pocket. If Z_1 and Z_2 are equal, no signal is reflected, and the tissue boundary cannot be detected.

For a boundary between muscle ($Z_1 = 1.7 \times 10^5$ g/cm²/s) and fat ($Z_2 = 1.38 \times 10^5$ g/cm²/s), the intensity of the reflected wave is 1.08% of that of the incident wave. At interfaces between most soft tissues, R_I is typically less than 0.1%. It is difficult to image inside a bone due to high acoustic-impedance (7.8×10^5 g/cm²/s).

Example 5.3: Effect of tissue interfaces

It is difficult to perform brain ultrasound due to the presence of multiple interfaces and the presence of high-impedance bone (skull). The various interfaces are: skin (muscle) → skull, skull → cerebral fluid, cerebral fluid → brain tissue. Intensity reflection coefficient between skin (muscle) and skull (bone) is 41%. Only 59% of the ultrasound energy is transmitted. A similar loss occurs between a skull and the cerebral fluid. Less than 36% of the ultrasound energy reaches a brain-tissue. Similar loss occurs during the return.

Figure 5.10 illustrates the reflection and refraction of ultrasound waves at the tissue-layers' boundary. Part of the signal is reflected, and the rest are transmitted. The reflection and refraction coefficients are governed by Equations 5.13 through 5.19 for $\theta_i = \theta_r = \theta_t = 90°$. For nonnormal incidence, the refraction angles are governed by the Snell's law given by Equation 5.19.

$$\frac{\sin\theta_i}{\sin\theta_t} = \frac{c_1}{c_2} \tag{5.19}$$

The terms θ_i and θ_t are the angles of incidence and transmission, respectively. The terms c_1 and c_2 are the sound speeds in the two media.

The direction of the ultrasound wave will change at a boundary if the angle of incidence is not $90°$. This can cause significant issues in imaging. For certain patients, refraction can cause significant image degradations such as blurring and loss of resolution. Furthermore, targets can be incorrectly located.

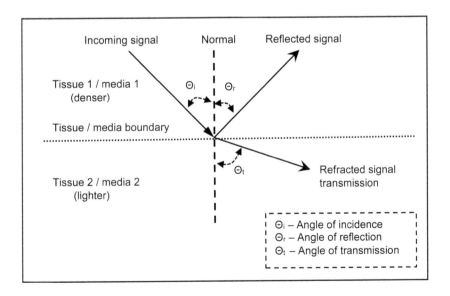

FIGURE 5.10   Reflection and refraction of ultrasound waves at the boundary of two tissue-layers

Example 5.4: Signal refraction at tissue-boundary

Calculate the angle of refraction for ultrasound with an incident angle of $25°$ striking the interface from fat to liver.

$$\frac{\sin 25°}{\sin \theta_t} = \frac{1450}{1570} \Rightarrow \sin \theta_t = \sin 25° \times \frac{1570}{1450} = 0.458$$

$\theta_t = 27.2°$. The angle of refraction is $27.2°$.

5.4.4.4 Probe–skin interface

The ultrasound transducers use material with acoustic-impedance that is significantly greater than that of soft tissues. Used unmodified, this would mean that most of the ultrasound signals would be reflected at the transducer-tissue boundary. To alleviate this, materials with intermediate acoustic impedance ("matching layers") are constructed on top of the transducer elements. Multiple matching layers are used with gradually decreasing impedance. This reduces the reflection at the probe/skin interface and improves the transmission. Air invariably gets trapped between the transducer-face and skin. Because air has a near-zero impedance, impedance mismatch is high. This is overcome by using water-based coupling-gel with acoustic impedance close to that of muscle. These lubricant gels facilitate the manipulation of a probe over the skin-surface.

5.4.4.5 Time-gain compensation (TGC)

B-Mode image is a popular two-dimensional ultrasound image display composed of pixels representing the received ultrasound echoes. The brightness of each pixel is determined by the amplitude of the received signal. The ultrasound signals used to create the B-mode images span a very large dynamic range, sometimes as high as 100 dB. Very strong echo signals can originate at fat–tissue interfaces close to the transducer; very weak signals can come from deeper soft tissue boundaries.

A process termed *time-gain compensation or time-gain control* (*TGC*) is used to compress the dynamic range of the received echoes. TGC increases the amplification as a function of depth. Echoes from structures close to the transducer are amplified less than those farther from the transducer. TGC is controlled by the operator, and generally has a few preset values for clinical imaging applications.

Besides improving the display for the human eye, TGC also helps in the proper functioning of the RF-amplifier that amplifies the received analog signals prior to digitization. RF-amplifiers cannot amplify the signals linearly if the dynamic range is over 40–50 dB. This means that the weaker signals and signals in 100 dB dynamic range will be lost, if TGC is not applied.

5.4.5 Doppler Ultrasound

Noninvasive blood-flow measurements are critical in the diagnosis of many diseases. Ultrasound blood-velocity estimation uses the Doppler-effect: for a moving acoustic source relative to an observer, the observed frequency differs from the transmitted frequency. The pitch is higher than the transmitted frequency when the source is approaching, and lower when the source is receding. For Doppler ultrasound, the governing equation is given by Equation 5.20.

$$\Delta f = 2f_0 \frac{v}{c} \cos \theta \tag{5.20}$$

where Δf is the shift in the frequency f_0, v is the flow-velocity, c is the sound speed and θ is the angle between flow and ultrasound propagation.

For θ = *90°*, the Doppler-shift in the frequency is zero. By measuring Δ*f*, the flow-velocity of the blood can be calculated.

Example 5.5: Doppler effect

If a *10*-MHz transducer makes an angle of *30°* with a vessel through which blood is flowing at a speed of *0.5* meters/second, the Doppler-shift in the original frequency is:

$$\Delta f = 2 \times 10 \times 10^6 \times \frac{0.5}{1540} cos(30°) = 5.6 \text{ kHz}.$$

Pulsed-wave Doppler-effect is used to create flow images. If a transducer transmits a pulse repeatedly in the same direction, the echo from a moving scatter will move from one firing to the next. By estimating the echo movement between firings, the velocity of the scatter could be estimated at all locations on the image. For a duplex image, flow-information is superimposed as color on the regular B-mode image.

5.4.6 Applications of Ultrasonic Image Analysis

Ultrasound waves measure the total kinetic energy of the molecules in biological tissues. It can provide tissue-images in real-time that includes structural and morphological information. Ultrasound imaging has many advantages, including excellent soft-tissue contrast, good resolution, low cost, portability, non-ionizing and almost no biological hazard. Its resolution is the finest in the axial direction.

Clinical applications of ultrasound include cardiac, vascular (including cardiovascular), obstetrics, gynecology, fetal imaging, breast, thyroid, scrotal, prostate, abdominal imaging, etc. It is also useful for visualizing the internal structure of a heart, measuring blood-flow in real-time, identifying cancers and detecting kidney stones. Ultrasonic imaging is also used for guiding biopsy needles and other interventional apparatus.

Ultrasound signal loses its intensity during propagation due to scattering, absorption, attenuation and reflection. Part of the lost energy is converted into heat, raising the tissue-temperature. Physical therapists use ultrasound machines to improve the healing of muscle-injuries by improving the circulation in the affected area. Heating from ultrasound is also used in noninvasive cancer therapy, e.g., therapeutic ultrasound. Furthermore, in radiation therapy, a malignant mass, if heated to *43°*C, can be treated effectively with a much lower dose.

5.5 NUCLEAR MEDICINE

Nuclear medicine involves the diagnostic and therapeutic use of *liquid radionuclides*. The therapeutic use of radiopharmaceuticals began in the early 1930s. Their diagnostic use began in the late 1950s. Nuclear medicine did not gain a quick acceptance because of poor image-quality and resolution. It had difficulty resolving objects less than three millimeters, while X-ray radiography had a resolution of *0.1* mm. However, it is very useful for functional imaging.

Ben Cassen introduced a rectilinear scanner in the early 1950s. The scanner scanned in two dimensions to produce a projection image of the radionuclide concentration in a human-body. Hal Anger developed the scintillation camera that produces a two-dimensional projection image. The projection images, like in X-ray CT, can estimate the underlying spatial radionuclide distribution within a slice or a volume. This tomographic reconstruction was applied to medical applications in the 1970s. This tomographic system is called a SPECT (single-photon emission CT) scanner.

Anger demonstrated that it is possible to detect photon pairs originating after a positron emission by a combination of two scintillation cameras. Positron is the positively charged antiparticle of electron

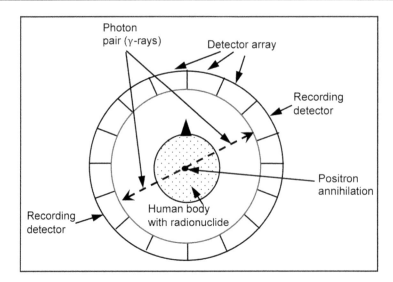

FIGURE 5.11   An illustration of a true coincident event in PET

having the same mass and charge-magnitude. It annihilates after the combination with an electron to produce two *511* KeV (Kilo electron volts) of photons traversing in the opposite directions, which are detected by scintillation cameras. This photon-pair detection is the basis of PET (Positron Emission Tomography). Ter-Pogossian and his team built the first PET system in 1970s. A research led by Phelps and Hoffman led to the development of the first PET scanner (also called PET camera).

PET systems have four major components: 1) a cyclotron that creates unstable radioactive isotopes that produce positrons; 2) produced unstable radiotracer such as 2-deoxy-[18F]-Fluoro-2D-glucose (^{18}F) or 1-methoinine (^{11}C), that is uptaken by the affected sites; 3) a circular ring of detector-pairs to detect the emitted photons after a positron-electron annihilation and 4) scintillation cameras. The size of each detector is between *0.4* and *0.6* centimeter. The arrangement of PET detection is illustrated in Figure 5.11.

The PET scan uses liquid radionuclides such as ^{18}F that contain a fast decay (*110* minutes for ^{18}F) radioactive isotope in a compound instead of oxygen and are absorbed into the cells like glucose. However, due to the absence of oxygen, metabolism does not progress further, and the radioactive isotope is trapped inside the cells until it decays. The radioactive material is traced to image the metabolic activity.

Some of the issues that adversely affect the imaging are: 1) the absorption of the emitted photons before they reach the detectors and 2) Compton scattering – scattering of an emitted photon after the collision with an electron that make the photons come at an angle at the detectors. The Compton scattering contributes to *40%* of the detected radiation and adversely affects the contrast. However, the energy of scattered photon is much less than the unperturbed photons and can be eliminated by placing lead or tungsten (1 mm thick) between the imaging slices.

5.5.1 Applications of PET

PET systems are used routinely to visualize metabolic activities such as blood-flow, oxygen circulation and sugar metabolism in human-body. A PET scan is quite sensitive and is used regularly for detecting cancer, staging of cancer, monitoring the response to oncological medications, heart disease, gastrointestinal, endocrinal and neurological disorders within a human-body. PET systems are capable of identifying the diseases in early stages and have an added advantage of allowing early intervention.

The application of PET images in cancer-detection includes: 1) pulmonary (lung) to separate between benign and malignant tissues; 2) colorectal; 3) malignant tissues in a breast and lymphoma (cancer of lymph nodes); 4) head and neck for lymph node staging; 5) bones to explore suspected metastases and monitoring the response to the treatment; 6) thyroid for metastases; 7) ovarian for metabolically active tumor site and 8) GI (gastrointestinal) tract for metastatic esophageal and gastric cancer. PET systems are being combined with other systems such as CT for better localization of the diseases such as cancer.

5.6 ALTERNATE MODALITIES

MR spectroscopy (*MRS*) is a functional imaging modality. MRS probes tissues for the concentration of various metabolites. The principal of MRS is similar to nuclear magnetic resonance. The basic physics behind MRS is the variation in magnetic-fields in different molecules due to the distribution of orbital electrons in an atom. This variation in magnetic-field causes different resonant frequencies for different molecules, consequently returns different signals.

Optical coherence tomography (*OCT*) is an optical imaging modality, based on low-coherence interferometry. Like ultrasound imaging, OCT is based on signal reflections. The light is split by using two arms. The sample arm contains the item of interest and a reference arm, typically a mirror, providing the reference signal for the interference. The reflected light from the sample and the reference light combines to produce an interference pattern, if both lights traveled the "same" distance (difference between two distances is less than the coherence length). Light outside the short coherence length does not produce an interference pattern. The mirror in the reference arm is scanned to produce a reflectivity profile of the sample. OCT can capture fine-resolution (μm) images from biological tissues, employing near-infrared light. Near-infrared wavelengths allow deeper penetration into the optical scattering media. However, it is limited to imaging the top one to two mm of biological tissues; at larger depths, not enough light is reflected from the sample to be detected.

Hybrid imaging modalities *acousto-optics* and *photo-acoustics* combine optics and ultrasound for medical imaging. In photo-acoustic imaging, short laser pulses are delivered into biological tissues. As the light propagates, part of the energy is converted into heat, which causes a sharp thermal expansion. This creates a mechanical pressure wave that can be received by an ultrasound transducer. By choosing the proper optical wavelength, tissues with certain color, such as oxygenated blood, can be targeted. Acousto-optics measures the change of refractive index due to a propagating sound wave. These methods combine tissue-sensitivity to optics to the spatial resolution of ultrasound.

Elastography is another recent medical imaging modality with the ability to measure the pathological states that affect tissue-elasticity. Elastography is based upon creating a map of tissue-elasticity distributions and quantitative information. Elasticity imaging is divided into two main categories: 1) quasi-static elastography that measures the tissue response to a quasi-static pressure and 2) dynamic elastography that images the tissue response to a dynamic excitation such as Acoustic Radiation Force Impulse (ARFI) imaging and shear-wave imaging. These methods are now implemented in many commercial scanners and aid in the diagnosis and monitoring.

5.7 MEDICAL IMAGE ARCHIVING AND TRANSMISSION

Image archiving and retrieval requires high-quality compression of medical images and videos. Digital imaging, including modalities described in Section 5.2 through 5.6 and three-dimensional images caused an explosion in the number of medical imaging examinations. The staggering volume of images requires compression for archiving and distribution, especially to transmit radiology images, and to access electronic patient records. Image compression decreases the cost of storage and improves transmission efficiency for telemedicine applications.

Effective retrieval of this information can assist physicians, patients, researchers and instructors in improving diagnosis, treatment planning, research and classroom learning. Conventional bibliographic or full-text databases cannot typically exploit figures, illustrations and other visual information contained in online biomedical literature. Figure captions and descriptions cannot fully represent all the visual information needed by human experts.

Image retrieval systems have allowed us to move beyond the traditional text-based searching to searching using a combination of text and visual features. Many medical information retrieval systems are available. These include FigureSearch, BioText, GoldMiner, Yale Image Finder, Yottalook, "Image Retrieval for Medical Applications" (IRMA) and iMedline (from NLM). Each system has certain capabilities and gaps in multimodal (text and image) information retrieval. This is an active area of research, and the capabilities are expected to improve significantly with time.

5.7.1 Medical Image Compression

Digital imaging in medicine emerged in the early 1980s. The number of modalities and acquired images have been increasing with time. Increasingly, large volumes of digital image data brought new challenges in the archiving and management of images. Effective image compression algorithms have played a significant role in image management and transmission. This section briefly reviews the compression algorithms used in medical imaging.

Medical image compression assumes that diagnostically significant data are not discarded. Otherwise, compressed images cannot be used in medical diagnosis. The JPEG is, by far, the most popular image compression standard. JPEG has both lossy and lossless compression standards. Lossless compression techniques include *run-length encoded* (RLE) and the newer JPEG lossless compression standard (JPEG-LS). Lossy compression standards are JPEG standard and JPEG 2000. JPEG stores per pixel of color data in *24* bits. DICOM working group has adopted the wavelet-based JPEG 2000 compression algorithm as a standard besides the existing JPEG compression algorithm. Wavelet-based JPEG 2000 provides higher compression with less distortion over the JPEG standard and is used for image-segments not involved in a diagnostically relevant area.

Studies have shown that no perceptible quality loss is experienced in chest radiographs with JPEG compression ratios of *40:1*. JPEG 2000 provides better performance, with up to *50:1* compressions with no perceptible loss. No difference could be perceived in chest CT scans up to *9:1* compression. The musculoskeletal radiology images allow diagnosis up to *28:1* compressions. Small images can be compressed using JPEG or JPEG 2000 up to *10:1* without the loss of diagnostic information; large images can be compressed up to *20:1*. Teleradiology (transmission of radiology images) applications require compression-ratios in the range of *10:1*. As teleradiology becomes common, higher compression ratios may become necessary to handle archiving and transmission problems.

Video compression is also important for medical applications. MPEG-4 (Moving Picture Experts Group) has become a standard for audio and video coding formats. Further improvements have been achieved in the MPEG-4 compression performances. However, for transmitting medical videos over the Internet, many solutions use proprietary compression and video-formats.

5.7.2 Medical Image Retrieval

The widespread applications of PACS (Picture Archiving and Communication Systems) have resulted in the creation of large digital repositories of digital medical images, to be accessed by clinicians over a network, including online literature databases such as *PubMedCentral, BioMedCentral* and EHRs (Electronic Health Record). These databases contain images from a diverse range of imaging modalities. Descriptions of figures/illustrations in the captions and full-text excerpts cannot effectively represent the semantic information in medical images perceived by human experts through visual examinations.

Physicians assess patients' past images available in the medical image repositories. It also allows interpatient comparisons for image-based diagnosis, treatment planning and monitoring, classroom learning and research. Simultaneously, exam-volumes and numbers of images per study made the tasks of diagnostic radiologists very difficult. Automated assistance with image interpretation will be beneficial. However, PACS search capabilities can only use textual keywords, including patient name, identifiers, imaging equipment, etc.

CBIR (Content Based Image Retrieval) complements the conventional text-based image retrieval of similar images by using quantifiable visual features such as color, texture, shape and spatial relationships between ROIs (Region of Interest). CBIR enables similarity-based indexing and diagnostic support based on the content and the metadata in medical images. CBIR seeks to combine text and visual features in the search queries to retrieve images that are semantically similar for diagnosis, clinical decision support, teaching and research. CBIR techniques offer valuable support to radiologists and save a considerable time for the clinicians. CBIR broadens the scope of an image retrieval system from "image-based search" to "patient" or "case-based" reasoning.

CBIR requires application-specific, user-defined definition of image-features for similarity-based image extraction. The image-features are organized into indices for image-search and fast retrieval from large databases. Feature-extraction and fusion are critical in the design of a CBIR system. A system design requires application-specific features tuned to the problem describing the characteristics unique to the problem domain. General-purpose features may not be appropriate for every application. For example, color has no use for X-ray radiography images. Global features define the overall characteristics of an image. However, visual characteristics that are present only in a small area of the image are not identified by global features. Local features capture the characteristics of a small area comprising a few pixels. Most current approaches focus on using a combination of local and global features.

For a CBIR system to describe the image sufficiently, the selected image-features must minimize the sensory and the semantic gaps. *Sensory gap* is the difference between the recorded image-features and the actual feature of the object. The difference can be caused by low illumination, noise and full or partial occlusion by other objects or even itself. Two-dimensional images of physical 3D objects also exacerbate the sensory gap. The *semantic gap* arises because of the difference between the cognitive information an image contains, and the images retrieved using a limited number of features associated with the images. CBIR systems retrieve images based on image features and not based on image interpretations.

For measuring the similarity of images, weighted Euclidean-distance between the corresponding feature-vectors is used, as described in Section 2.1.2. Since medical images are generally black-and-white, texture and shape-based features are used for comparing two images. Dimension reduction techniques such as PCA (Principal Component Analysis) and LDA (Linear Discriminant Analysis) are used to remove the insignificant features that do not contribute significantly to the image characterization and retrieval. The idea of elastic deformation is used if subtle geometric differences between images need to be captured.

Features such as textures and shapes are matched using multiple techniques, as described in Chapter 2. For quick pruning of images from the database, histogram-based matching is used. For comparing images using a combination of image features and the relationship between objects, graph-based matching is used. The objects are modeled as nodes and the relationships are modeled as edges. Each node may have multiple features modeled as k-tuples where K is the number of features in an object. Statistical classifiers are trained to overcome the semantic gap. Figure 5.12 illustrates a generic framework of a CBIR system.

CBIR algorithms extract features from each image, which are then indexed for searching. For the online query process, the feature-extraction process is repeated on the query image, which are compared to the features of the indexed images in the collection using a predefined similarity criterion, as described in Section 2.6. These similarity results classify the images as "similar" or "not similar." Alternatively, images can be ranked in the order of similarity-scores with the ranking displayed to the user.

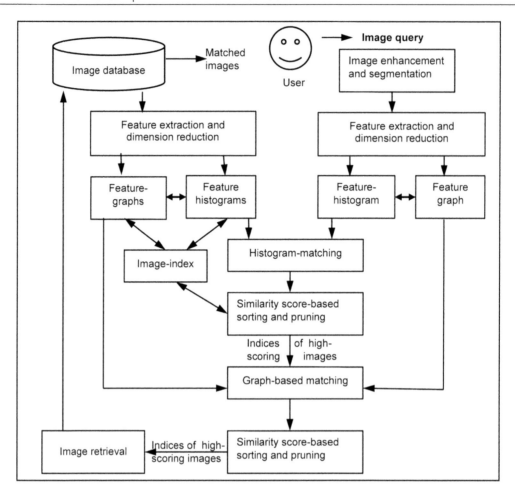

FIGURE 5.12   A generic framework of an image-based CBIR image retrieval system

Two quality measures *precision* and *recall* are used to assess the accuracy of the similarity-based search. *Precision* is the ratio of the relevant retrieved images and the relevant images that a user expected to retrieve. *Recall* is the fraction of similar images in the dataset that the algorithm retrieved. An ideal retrieval system should achieve 100% precision and 100% recall. However, most algorithms cannot find every similar image, and many of the retrieved images are not similar (false-positives).

Medical CBIR techniques have been adopted from standard CBIR shape-based and texture-based techniques, as described in Section 2.5. However, applying these techniques to medical images does not always retrieve the desired images due to continuous changes in disease-states and variabilities among patients. Images of anatomical structures can also vary due to change in viewing angles. Such variability can go up even more for volumetric and multimodality images.

Several online image-retrieval systems are available including FigureSearch, BioText, GoldMiner, Yale Image Finder, Yottalook, Image Retrieval for Medical Applications (IRMA) and NLM's iMedline. All seven systems extract metadata from the full-text about image descriptions and use them to retrieve images in response to a query. Some algorithms allow user-interaction to improve and refine the search until the user is satisfied with the set of retrieved images. Each system has capabilities and gaps related to the combined (text and image) information retrieval. However, only some take advantage of the visual information in the figures and illustrations.

5.7.3 Medical Image Transmission

Teleradiology services, used for expert medical consulting since as early as 1947, are often required due to sparse population, radiologist shortage or a need for a more cost-effective radiological reading. Teleradiology offers an opportunity to improve the image interpretation quality through collaboration of specialist clinicians and radiologists. The Internet and PACS have increased the clinicians' collaboration with their colleagues. Teleradiology now allows small hospitals, clinics, specialty medical practices and urgent care centers to offer coverage not originally possible. Teleradiology is also being adopted for remote monitoring of patients and elderly care from a centralized monitoring station. The computational protocols for the transmission of medical images are explained in Chapter 6.

5.8 MEDICAL IMAGE ANALYSIS

Medical images are analyzed and interpreted by clinicians using qualitative assessment of relevant features within an image. However, many clinical applications do not have standard definitions of the features observed in the images. Human perception and knowledge, even among medical experts, is nonuniform. This can contribute to diagnosis variability. Computer-based intelligent algorithms have been developed that can assist in the automated diagnosis by recognizing and classifying the relevant features in the images. The clinicians use these tools to enumerate and investigate various possibilities and improve their diagnostic accuracy.

Numerous medical-image analysis techniques have been reported in the scientific literature. Common components in these techniques are *image registration, image segmentation, feature extraction/analysis, content-analysis* and *classification. Image registration* is concerned about aligning an image with a control or another image to minimize the distortion due to misalignments of the images. Image alignment is necessary to study the change or progression of a disease. It is another key medical image analysis method that plays vital roles in many medical applications, including in treatment monitoring, disease prediction, wound-care and health-care surveillance. *Image segmentation* identifies various homogeneous regions using multiple techniques, as described in Section 5.8.1. Abnormal tissues differ from healthy tissues in terms of texture, intensity and shape. By identifying the change in the segment attributes, abnormal and healthy tissues can be separated.

5.8.1 Medical Image Segmentation and Feature Extraction

Image segmentation algorithms partition an image into a set of homogeneous regions according to desired characteristics such as texture, color and intensity. Medical image segmentation is inherently modality dependent. A segmentation algorithm that works for MR-images may not be effective for optical-coherence tomography images. Ultrasound images are probably the hardest medical images to segment. However, with the improvement in ultrasound image quality, automated segmentation performance has increased significantly.

Clinicians frequently need to delineate anatomical structures and other areas of interest in medical images for diagnosis. Image segmentation algorithms play a key role in automating radiological tasks in the delineation of the areas of interest in medical images. These algorithms have an important role in clinical applications such as diagnosis, localization, volume quantification, treatment planning and computer-assisted surgery.

Image segmentation methods vary widely for different clinical applications. For example, the requirements for brain-tissue segmentation differ from that of breast-lesion segmentation. No segmentation method exists that can produce acceptable segmentation for every medical image in existence. Although some methods can be applied to a wide variety of images, methods use prior knowledge about specific applications to achieve better performance. This section briefly describes the automated segmentation of anatomical medical images for most common radiological modalities: X-ray radiography, MRI, X-ray CT, ultrasound and lesions in breast ultrasound images.

Segmentation techniques are broadly classified as *hard segmentation techniques* and *soft segmentation techniques*. A pixel is assigned to exactly one class in hard-segmentation techniques. Soft-segmentations allow regions or classes to overlap. Soft-segmentations are important because multiple tissues frequently contribute to a single "blurred" pixel or voxel. In hard-segmentation, a decision is forced whether a pixel is inside or outside a homogeneous region. Both can be useful to the clinicians, depending on the application.

Segmentation techniques can be automated or semi-automated. Semi-automated segmentation techniques allow interaction with an expert during the segmentation and improve accuracy by utilizing a domain-expert's knowledge. However, this is laborious and impractical for a large number of images. Automated segmentation methods require specification of initial parameters. Different initial parameters result into different outcomes. The popular approaches for segmentation are: 1) thresholding; 2) region growing; 3) classifiers; 4) clustering; 5) Markov models; 6) artificial neural networks; 7) deformable models and 8) atlas-guided approaches.

5.8.1.1 Thresholding

Thresholding determines threshold-values to separate an image into the desired homogeneous regions called *segments*. Homogenous regions are identified by extracting values of discriminating features and combining them using a function to derive a scalar value. These scalar values are plotted as a histogram, and all the data-elements belonging to the same range in the histogram belong to the same smaller set. The simplest kind of thresholding converts an image into a binary image: *background segment* and *foreground segment*. The *background segment* is below a threshold, and the *foreground segment* is above a threshold. More than one threshold can generate a gray-level image with a smaller palette. Thresholding is frequently used for medical image segmentation if the images include structures with distinct, contrasting amplitudes. Some of the discriminating features are intensity and texture.

5.8.1.2 Region growing

Region growing derives homogeneous subregions of an image by joining similar pixels in immediate proximity based upon predefined criteria such as similarity of intensity, texture and/or color. The region stops growing when the adjacent pixels in the direction of the growth are no longer similar. Three popular techniques for region growing are: 1) uniform blocking; 2) merge-and-split blocks and 3) merging blocks using some similarity metric such as mean-value, median or max-min difference.

Uniform blocking divides a two-dimensional image into a set of macro-blocks such as 16×16 pixels, and feature-values of the pixels in the macro-blocks are aggregated using statistical values such as mean, median or max-min difference. These aggregated values are attributed to macro-blocks. During the *merge-split phase*, the macro-blocks are merged if they have similar aggregated feature-values.

Example 5.6: Region growing

For merging based upon mean-value, two adjacent macro-blocks having mean intensity value *120* and *124* are merged if the threshold value for dissimilarity is *10*. Similarly, for merging based upon |*max* − *min*|, two blocks are merged if the |*max* − *min*| value is below a predetermined threshold. A bigger region is split into smaller regions if the feature-vector values of two subregions within the same segment are above a threshold-limit. For example, if the absolute value |*max* − *min*| > 40 for a block and the threshold is *30*, then the two-dimensional region is split into four smaller blocks using quad-tree splitting.

In an automatic region-growing approach, first the background macro-blocks are identified using a histogram analysis and are removed from the image to derive the *ROI* (<u>R</u>egion <u>o</u>f <u>I</u>nterest). This *ROI* is divided into uniform blocks of required size. Few pixels are picked as seed-points based upon a heuristics such as mean or median feature-value. Region-growing can also be semi-automatic, requiring a seed-point selected by a domain-expert manually. At the end of region-growing, an image is separated into multiple homogeneous segments.

Like thresholding, region-growing is often used as a part of a sequence of operations. The simplest implementation of region growing uses a manually selected seed-point to grow the region according to a predefined criterion, such as similar intensity-values or texture-values. After the clinician selects a seed -point, the region grows into the similar surrounding areas, while excluding the dissimilar areas.

5.8.1.3 Classifier methods

Classifier methods are supervised pattern recognition methods using some predefined features. Two popular features are: *image intensities* and *texture*. Classifiers require manually segmented training data. *Nearest-neighbor classifier* is a simple classifier, which classifies each pixel or voxel (pixel's analog in three-dimensional space) in the same class as the training set that has the closest Euclidean distance.

Clustering algorithms are fast and require no training data. However, they require initial segmentation. As in the classifier methods, clustering does not incorporate spatial modeling, which makes it susceptible to noise and intensity inhomogeneities. Some popular techniques for medical-clustering are *k-nearest-neighbor (kNN)*, *K-means and expectation-maximization*.

5.8.1.4 Markov random field model

Markov random field (MRF) model is a statistical model for segmentation. It models the spatial relationships between the pixels in a neighborhood. In medical imaging, most pixels are in the same class as the neighboring pixels. Under an MRF assumption, the probability of an anatomical structure consisting of only one pixel is low.

The MRF segmentation is performed by maximizing a posteriori probability of the segmentation by incorporating a prior probability distribution in a Bayesian model. Proper selection of the parameters that control the spatial interactions is difficult. Too high a setting smoothens the segmentation excessively, causing a loss of important structural details. MRF methods are typically computationally intensive. With the availability of increased computational powers, this concern is not as important. MRFs have been used to model segmentation classes and texture-properties.

5.8.1.5 ANN and segmentation

ANNs (*Artificial Neural Network*) have been used for medical image segmentation, including as a classifier where the weights are determined using training data. ANN approaches exploit the learning capability to classify medical images into consistent regions for edge detection and segmentation.

5.8.1.6 Deformable models

Deformable models or *active contours* are model-based techniques for demarcating boundaries. The models use closed parametric curves governed by the internal and external forces. An initial closed curve is placed close to the border that undergoes an iterative process of growth. Internal forces expand the curve outward by identifying the neighboring pixels with similar feature-values. External forces restrict the growth of the segment boundary based upon the dissimilarity of the neighboring pixels. The process is smoothened by removing very small regions of heterogeneity inside a larger homogeneous region.

Example 5.7: Application of active contour

An example of applying an active contour to an MR-image of heart is shown in Figure 5.13. The initial border was a circle that included some areas outside the left-ventricle, which then progressed to the inner boundary of the left-ventricle. Deformable models can generate closed parametric curves based on a smoothness constraint that reduces susceptibility to spurious edges and noise. However, they require manual initialization and the selection of appropriate parameters.

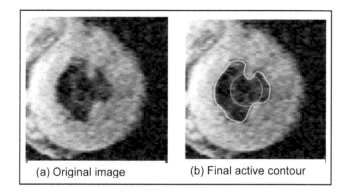

(a) Original image   (b) Final active contour

FIGURE 5.13   Applying active contour to extract the inner wall of the left-ventricle from an MRI: (a) Original image; (b) Final active contour

5.8.1.7 Other segmentation techniques

There are many other approaches such as *atlas-guided*, *model-fitting* and *watershed* algorithms. *Atlas-guided* algorithms use an anatomical atlas to guide the segmentation. Atlases are derived by manually labeling the regions by the domain experts. The manual atlas contours are deformed using various mathematical contour-warping transformations and aligned with the image to identify the segments.

Model-fitting approaches are used to fit well-defined simple geometric structures such as circle, ellipse and parabola that closely match with the location of the extracted features. In a model-fitting approach, matching can take place only after the feature has been extracted. The *watershed algorithm* uses mathematical formulations for edge-detection and morphology for partitioning and generally suffers from over-splitting. The technique uses two-step approaches: 1) partitioning using mathematical formulations and 2) merging the adjacent segmentations based upon feature-similarity.

Intensity inhomogeneity artifacts degrade segmentation of MR-images. Many algorithms have been proposed to handle noise caused by artifacts. Some methods use information gained from the segmentation while estimating the inhomogeneity simultaneously.

5.8.1.8 Multimodal fusion and segmentation

A major problem in image-segmentation is the effect of different types of layered tissues that contribute to the interference in received signal. This interference leads to different amount of attenuation and noise. Segmentation of ultrasound images is challenging due to the presence of artifacts such as *speckles* and *shadows*. As described in Section 5.4, speckles reduce the contrast.

Image fusion has become an important topic in medical imaging, paving the way for multimodal image segmentation. (Multiple views of the same image from the same modality is used also.) With the availability of modern imaging modalities, multimodal image segmentation can use different anatomical information provided by each imaging modality to improve diagnosis. For example, bone imaging is better in CT and soft-tissue imaging is better in MRI. One major issue in multimodal image-fusion is to align the images from different modalities that use different devices, requiring sophisticated image-registration techniques.

5.8.1.9 Applications of image segmentation

Automated segmentation has been used for lesion-detection and demarcation, such as in brain images and mammograms. MRI brain-images, mammograms, herniation of the spinal-cord and tumors in various organs are excellent candidates for automated segmentation. MRI can produce images of soft-tissues with excellent contrast, high resolution (of the order of 1 mm^3 voxels) and a high SNR (Signal to Noise Ratio). Optimal pulse-sequence is important for ensuring quality brain-images. However, variabilities exist between subjects. Optimal pulse-sequences are different for different populations.

Key applications of the automated segmentation in MRI brain-images include: 1) brain volume extraction; 2) identification of gray matter, white matter and cerebral fluid and 3) identification of segment-specific structures such as the hippocampus. Tumor and microcalcification visualization is the major goal of segmentation in digital mammography. Image analysis for the classification and segmentation can be performed concurrently or sequentially. In mammography, a perfect demarcation of masses is difficult. The algorithms are designed and refined. Other areas requiring good automated segmentation include image-assisted surgeries and fetal-image analysis. Image assistance has increased success rates in surgeries while reducing physician-fatigue.

CT is the modality of choice for imaging bone, detecting hairline fractures in bones, and diagnosing bone-tumors due to its cost-effectiveness and resolution that is comparable to MRI. The segmentation of the corresponding CT images is performed using a combination of thresholding, region-growing, Markov random-fields and deformable model techniques.

5.8.2 Image Segmentation for Lesion Identification

A segmentation technique demarcates a lesion by: 1) the localization of the lesion using initial segmentation, 2) placing a seed-point inside the initial segment either manually by a clinician or automatically, 3) using automated region-growing to identify homogeneous regions and 4) automated demarcation of the region boundary by removing very small regions having no consequence. The segmentation is processed in three stages: 1) localization of the lesion, 2) multiple segmentations using different algorithms and 3) fusion of the results derived by different algorithms to enhance accuracy. A recently developed robust breast lesion segmentation method for ultrasound images is illustrated in Figure 5.14.

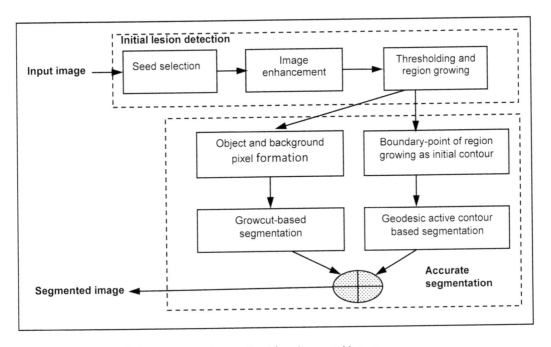

FIGURE 5.14   A breast lesion segmentation method for ultrasound images

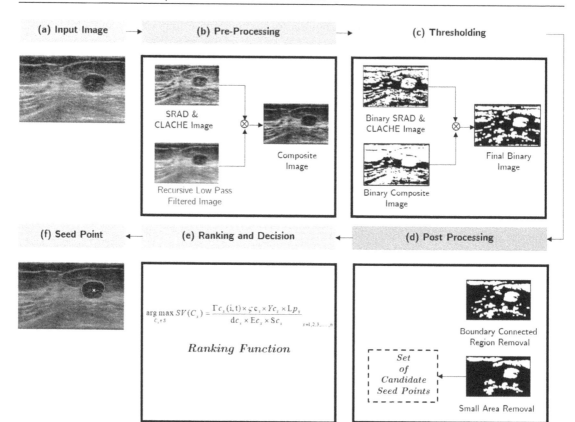

FIGURE 5.15 Localization steps in image segmentation: (a) Input image; (b) Ranking and decision; (c) Thresholding; (d) Post-processing; (e) Ranking and decision; (f) Seed-point

Lesion localization in ultrasound imaging is a four-step process: 1) image-preprocessing for enhancing image quality; 2) thresholding for possible region identification; 3) postprocessing and 4) ranking and decision making to identify the seed-points. Image preprocessing enhances the *hypoechoic regions* (regions not bouncing the ultrasound sufficiently), strengthens the edges and reduces inhomogeneity within the lesions by utilizing different image enhancement methods. SNR can be improved by using multiple images of the same lesion. Thresholding identifies the regions with different features such as intensity and texture. Image is postprocessed to join the similar regions in proximity identified by thresholding; spurious small regions are removed. This decision process may be semi-automated requiring human input. Figure 5.15 illustrates the localization-steps in image segmentation.

5.8.2.1 Enhancing image quality

Ultrasound images suffer from *speckles* degrading the contrast. The image quality can be improved by reducing the speckles, improving the image contrast for better segmentation and edge detection. The image enhancement step should not degrade image contrast.

There are many techniques for speckle reduction including 1) *anisotropic diffusion* such as *SRAD* (Speckle Reducing Anisotropic Diffusion), 2) time-series averaging techniques, 3) wavelet decomposition, and 4) averaging out the local minor variations in the intensity to improve local homogeneity.

SRAD reduces the effect of speckles and improves the image-contrast by unevenly reducing the intensity in different directions while preserving the edges. Image-contrast is also improved using techniques such as CLAHE (Contrast Limited Adaptive Histogram Equalization) and wavelet decomposition. CLAHE is based upon adapting the enhancement of the pixel-intensity based upon the adjacent pixel-values. To reduce the overhead associated with the pixel-level computations, the image is divided into many localized subregions and histogram equalization technique is applied to the subregions.

To reduce the significant intensity variations at the region boundaries, weighted sum of the feature-values of the regions is used to compute the feature-value of a pixel in the region of interest. Wavelet-based decomposition (see Section 7.2.2.1) is used to fuse the multimodal images having different resolutions. Each image is decomposed to a set of coefficients using discrete wavelet-transforms. These coefficients are then combined, and an inverse transform is used to improve the image quality.

5.8.2.2 Binarization

Thresholding methods are used to create binary images. Binary images are processed using multiple filters to enhance various features. The enhanced images are normalized and combined to generate the final binary image. All connected components are identified and labeled with region-numbers in the final binary image. Knowledge-based thresholding identifies the tissue-types of different regions. Regions are characterized by statistically associating the features such as mean intensity-value, texture and region-boundaries to identify different tissue-types. For example, in ultrasound images, white regions in the final binary image correspond to the hypoechoic or anechoic regions of the original B-mode image. Regions having an area less than a user-defined threshold (expressed as a number of pixels) are removed and merged with the neighboring blocks. For example, the threshold for a region is around *1000* pixels for this breast-lesion segmentation technique.

The connected regions that remain after postprocessing are candidate lesions. The collection of the centroids of the candidate lesions forms the set of candidate seed-points. Each centroid is used to generate four additional seeds by shifting the centroid coordinates by *10* pixels right, left, up and down. The next step is to correctly identify the lesion from the set of candidate regions. To achieve the goal, the candidate seed-points are ranked, and the top-ranked seed-points are selected for analysis. The processing steps for segmentation are illustrated in Figure 5.16.

5.8.2.3 Ranking, decision, and segmentation

Potential seed-points are ranked based upon clinical evidence and empirical domain knowledge. A ranking-score R_x is computed for every seed-point candidate. The seed-point with the maximum value of R_x is selected for region-growing. To design the ranking-function, two types of image-features are extracted: 1) echo-pattern descriptors and 2) spatial distribution descriptors. *Echo-pattern descriptors* describe texture information of the regions. *Spatial distribution descriptors* model the probable location of the lesion in the image.

Some of the popular segmentation techniques are: 1) Fuzzy C-means clustering (FMC); 2) Growcut segmentation; 3) Geodesic active contour and 4) Graphcut segmentation. Fuzzy C-means clustering is a technique where every data-point probabilistically belongs to two or more clusters based upon distance from the cluster (see Section 3.5.1).

Growcut segmentation selects a few pixels as seed-points and grows them based upon the strength of the feature-vector. It divides the region in a two-dimensional matrix of homogeneous cells. Unlabeled cells are captured by labeled adjacent cells. An adjacent cell is captured if the neighboring cells are dominating and have higher strength based upon feature-vector values.

Graph cut segmentation is based upon representing the pixels or subregions as nodes of graphs. These nodes are connected by weighted edges so nodes having higher dissimilarity have higher weights. The graph is cut by removing the edges with weights above a threshold. This splits the graph into multiple connected subgraphs. Each of these connected subgraphs represents a homogeneous region. The segmentation outputs from each technique vary somewhat as shown in Figure 5.17. The output of multiple segmentation techniques can be fused to improve lesion segmentation.

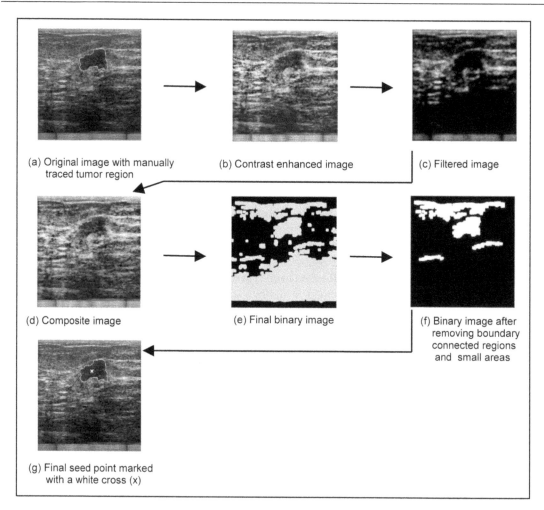

FIGURE 5.16   Processing steps for image segmentation: (a) Original image with manually traced tumor region; (b) Contrast-enhanced image; (c) Filtered image; (d) Composite image; (e) Final binary image; (f) Binary image after removing boundary connected regions and small areas; (g) Final seed-point marked with a white cross (x)

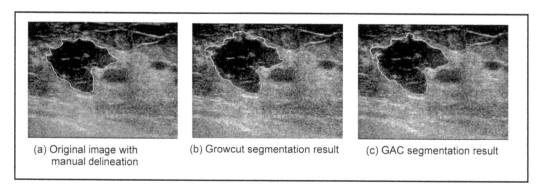

FIGURE 5.17   An illustration of lesion segmentation using multiple techniques: (a) Original image with manual delineation; (b) Growcut segmentation result; (c) GAC segmentation result

5.9 COMPUTER-AIDED DIAGNOSIS

Medical images provide clinicians a great deal of information to be analyzed, evaluated and interpreted comprehensively in a short time. CAD (Computer Aided Detection) and CADx (Computer Aided Diagnosis) systems exploit automated processing of images to facilitate medical image interpretation and diagnosis. One of the major applications of CADx is the classification of masses into malignant tumors or benign masses. The probable tumor-regions are highlighted to assist the radiologists and clinicians. CADx systems also reduce *inter-observer* (different clinicians interpreting the same image) and *intra-observer* (the same clinician interpreting the same image at different times) variability. Radiologists use CADx as a tool to improve the diagnosis and to reduce the interpretation time.

CADx algorithms were first developed to analyze mammograms. The CADx systems identified cancers earlier; however, it also increased false-positives. False-positives cause worry and stress due to a false cancer diagnosis that results into additional hospital visits and pathological tests. However, early diagnosis improves the survival rate of the cancer patients. False-negatives are dangerous because cancers remain undetected and untreated.

5.9.1 Cancer Identification from Medical Images

Early detection of cancer is critical for patient outcome and treatment options. Typically, an incidence of cancer is identified using *histopathology*. *Histopathology* is a technique based on microscopic cell-level analysis of multiple slices of the abnormal tissue-growth, extracted surgically using biopsy. However, biopsy is invasive, painful, expensive and has risks of bruising and infection. Computer-aided diagnosis using automated image-analysis has been an active area of research to diagnose cancer.

Images from different types of cancerous tissues vary and exhibit different features than those from healthy tissues. Some features are identified only by a specific modality. Sections 5.9.1.1–5.9.1.3 describe the automated identification of cancer using various imaging modalities.

5.9.1.1 Identification using CT images

CT is useful in the detection of nodules in many applications, including pulmonary, hepatic and thoracic. CT is more sensitive for the detection of smaller lung nodules than chest X-rays. Lung cancer is one of the most common cancers and accounts for about one out of four cancer-related deaths. Early detection of lung-cancer improves the survival and recovery rates: a substantial fraction of tumors can be treated successfully even before the cancer becomes malignant and invasive.

5.9.1.2 Identification using MR images

MRI is expensive and cannot be used for many patients due to claustrophobia and the use of large magnetic fields. However, it produces good-quality images with high-resolution in all three axes. For some cancers such as prostate cancer, MRI is the only conventional imaging modality that can detect cancerous growth. MRI can depict and identify prostate cancers that are *0.5* mm or larger. CADx systems have shown good performance in identifying cancers from prostate MR-images. CADx systems can analyze MRI-images in identifying gliomas (tumors in brain and spinal-cord).

5.9.1.3 Identification using ultrasound

Breast-cancer is a common cancer, and the second most lethal cancer for women. Doctors use biopsy for a definitive diagnosis of breast-cancer. However, biopsy is an invasive, traumatic and expensive procedure with potential complications. A sensitive and accurate method that reliably identifies benign lesions would reduce unneeded biopsies. However, false-negatives raise important clinical and legal issues and thus, the method should have a very high specificity. Newer, advanced clinical scanners reliably distinguish between cancerous and non-cancerous solid breast tumors. Radiologists frequently use findings from breast ultrasound examinations to recommend periodic follow-up without a biopsy.

Automated analysis of ultrasound images is called quantitative ultrasound (QUS). The goal of QUS is the characterization of the state of tissue such as healthy or abnormal. Sections 5.9.2-5.9.5 discuss a QUS image analysis approach that makes breast-lesion evaluation quantitative, reproducible, and relatively operator-independent.

5.9.2 Quantitative Descriptors

Quantitative descriptors comprising *acoustic-features* and *morphometric features* are used to identify benign and malignant lesions. Quantification facilitates automated identification of cancerous tissues based upon feature-values. Quantitative acoustic-features include *echogenicity, texture-heterogeneity* and *shadowing*. Variations in acoustic-features are caused by *acoustic reflection, dispersion, absorption* and *refraction* of the transmitted pulse from the soft tissues. These acoustic properties vary between healthy tissues, benign lesions and cancerous lesions. *Morphometric features* are concerned with the quantification of lesion's shape, size, and boundary-related features. Morphometric features for benign tissues and cancerous tissues vary significantly.

Noncancerous lesions are typically well textured, homogeneous, have a regular shape with well-defined boundaries, generally have aspect ratio (height divided by width) less than *1.0*, and have no shadow. A cancerous mass has a heterogeneous structure, irregular shape with fuzzy border, with a central shadow. The effects of increasing border roughness are illustrated using four contours in Figure 5.18.

Spiculation is the presence of spoke-like structures in a closed contour and suggests malignancy. Figure 5.18a is a smooth ovoid. Figure 5.18b illustrates a non-spiculated rough border. Figure 5.18c

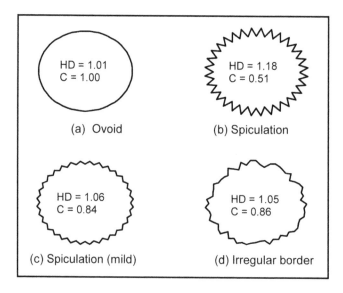

FIGURE 5.18 Illustration of border roughness and its association with malignancy: (a) Ovoid; (b) Spiculation; (c) Spiculation (mild); (d) Irregular border

illustrates a shape with mild spiculation, and Figure 5.18d illustrates moderate spiculation. (*HD* is the hausdorff dimension and *C* is convexity.)

5.9.3 Classifiers

The lesions are classified based on these quantitative features. *Linear discriminant analysis* (LDA) is a popular choice for the classification of the tissue-groups. Logistic regression (see Section 3.6.3) and non-linear approaches such as neural network (see Section 3.5.4) and support vector machines (see Section 3.5.2) are also used. Nonlinear approaches yield better results when quantitative features do not exhibit clear separation between cancerous and noncancerous tissues.

A sample has elements of the class-of-interest and elements not belonging to the class-of-interest. In our case, the class-of-interest is "malignant tumors," and the other class is "benign tumors." Four outcomes are possible after applying the classifiers: 1) malignant tumor identified correctly as the malignant tumor (true-positive); 2) benign tumor incorrectly identified as malignant tumor (false-positive); 3) malignant tumor incorrectly identified as a benign tumor (false-negative) and 4) benign tumor correctly identified as a benign tumor (true-positive). Possibilities one and four are safe. However, the second possibility will lead to invasive biopsy and potentially unnecessary therapy, and the third possibility will allow the cancer to grow and spread. False-positives and false-negatives should be minimized. False-negatives are particularly harmful as it would let a cancer grow.

Classifier performances are measured using true-positive and false-positive fractions. *True-positive fraction* (TPF) is defined as the ratio of samples being correctly identified in the same class. It is expressed by Equation 5.21. *False-positive fraction* (FPF) is defined as the fraction of other classes incorrectly marked as present in a class, as expressed in Equation 5.22. *False-negative fraction* (FNF) is defined as the fraction of the desired class elements identified as belonging to the other class as given by Equation 5.23.

$$TPF, sensitivity = \frac{Number\ of\ correctly\ identified\ malignant\ lesions}{Total\ number\ of\ malignant\ lesions} \tag{5.21}$$

$$FPF = \frac{Number\ of\ benign\ lesions\ incorrectly\ identified\ malignant\ lesions}{Total\ number\ of\ benign\ lesions} \tag{5.22}$$

$$FNF = \frac{Number\ of\ malignant\ lesions\ incorrectly\ identified\ benign\ lesions}{Total\ number\ of\ malignant\ lesions} \tag{5.23}$$

There are two related concepts to measure the performance: *sensitivity* and *specificity*. *Sensitivity* is the same as TPF. *Specificity* is defined as the fraction of the other class accurately identified as the other class as given by Equation 5.24. Specificity is the same as the fourth possibility described above. Specificity and FPF are related: FPF is equal to (*1 – specificity*).

$$specificity = \frac{Number\ of\ benign\ lesions\ correctly\ identified\ benign\ lesions}{Total\ number\ of\ benign\ lesions} \tag{5.24}$$

A curve plotted between TPF and FPF under various threshold is called ROC curve, which is essentially the sensitivity as a function of false alarm. The performance metric in ROC analysis is the Area Under (the ROC) Curve (AUC). The better the classifier, the closer the AUC is to unity. Best methods achieve high sensitivity, high specificity, AUC close to 1, and low FP values.

5.9.4 Machine Learning Approaches

Modern machine learning methods have improved detection-rate and specificity. The performance of machine learning methods relies heavily on proper modeling of tumors by digital features. Digital features are grouped into two categories: *knowledge-based* and *statistical*. In ultrasound, *knowledge-based*

TABLE 5.2   Features of conventional B-mode images associated with malignant and benign lesions (from the BI-RADS criteria). A typical lesion will have only a subset of these identifying features

MALIGNANT LESIONS		BENIGN LESIONS	
MORPHOMETRIC FEATURES	ACOUSTIC FEATURES	MORPHOMETRIC FEATURES	ACOUSTIC FEATURES
Irregular shape/spiculation	Central shadowing	Spherical/ovoid shape	Edge shadowing
Poorly defined margin	Hypoechogenicity	Well-defined margin	Hyper echogenicity
Tall aspect ratio	Heterogeneous texture	Thin capsule	Homogeneous texture
Microlobulation	Calcifications	Gentle bi- or trilobulations	
Architectural distortion		Orientation parallel to tissue plane	

features are derived from the Breast Imaging Reporting and Data System (BI-RADS) lexicon, based on shape, margin, orientation, echo pattern, and acoustic shadowing. *Statistical features* capture the correlation between pixels, which do not necessarily correspond to any observable features. Examples of statistical features are: *auto-covariance coefficients* and *frequency-domain features*.

Digital features are input to machine learning methods that separate benign and worrisome lesions. Machine learning methods popularly employed for the categorization are: *decision tree* (see Section 3.5.3), *artificial neural network* (ANN) and *support vector machine* (SVM). The optimum feature sets might depend on the choice of machine learning methods.

5.9.5 A Case Study

Table 5.2 lists typical characteristics of benign and malignant breast lesions. The malignant lesion has an irregular microlobular shape with a tall aspect-ratio, heterogeneous internal texture and a prominent posterior shadow. The benign lesion exhibits the smooth shape, homogeneous internal texture, central posterior enhancement (sometimes called "anti-shadow") and an aspect ratio of less than one. Some benign lesions may have high aspect ratios. However, the long axis of a benign lesion would "flatten" when a transducer is pressed against the skin for imaging. This is because, unlike malignant lesions, benign lesions are free to move relative to the surrounding tissues. Figure 5.19 illustrates some of these characteristics.

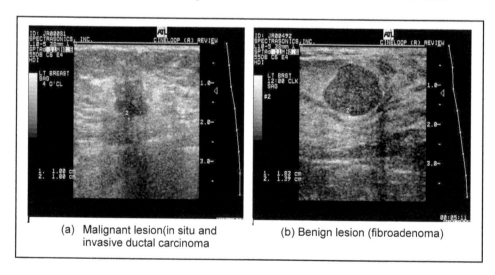

(a) Malignant lesion(in situ and invasive ductal carcinoma

(b) Benign lesion (fibroadenoma)

FIGURE 5.19   Comparison between malignant lesion and benign lesion: (a) Malignant lesion (in situ and invasive ductal carcinoma; (b) Benign lesion (fibroadenoma)

B-mode ultrasound is routinely used to separate benign masses (such as cysts and fiboradenomas) from cancerous masses. Breast-lesion segmentation algorithms can distinguish between the textured appearance and geometry of a lesion, normal tissue and artifacts, shadowing and the poorly defined cancerous lesion borders.

5.10 SUMMARY

Medical image is a two-dimensional or reconstructed three-dimensional image that depicts the internal structure of the body. It employs different measurements based upon intensity changes of original signals due to different physical phenomena corresponding to different modalities. Advances in medical imaging have accelerated since 1990s due to the availability of better technology, minimally invasive nature of imaging, low cost of diagnosis and better disease management. There are two major classes of medical imaging: 1) radiation-based technology and 2) acoustics-based technology. These technologies are based upon the measurement of change in energy while traversing through the tissues either through absorption or reflection from the boundaries of different tissue layers.

Radiation based technologies is classified as: 1) technologies based upon light and 2) technology based upon magnetic resonance. X-rays are invisible rays beyond ultraviolet wavelengths. X-rays are emitted in an X-ray tube when electrons, emitted by cathode, strike anodes. X-ray imaging is based on absorption and good for imaging bones, but not soft tissues. X-ray technology cannot separate the objects that lie in the same straight line between source and detector.

Combinations of X-rays and computer-based analysis of multiple images taken from different angles are combined to generate 3D-image of an organ. In CT, both X-rays and computer processing of multiple two-dimensional images are combined to derive a 3D image of the region-of-interest. CT has been used to localize the abnormalities in three-dimensional space. Image reconstruction techniques in CT technology use inverse Fourier 2D-transform that takes combination of multiple Fourier-transforms of one-dimensional projections to build a 2D cross-sectional image. One of the major concerns of radiation-based technology is the radiation overdose that has the potential to cause life-threatening malignancies.

MRI is based upon the variation of emitted RF by excited protons in water molecules within soft-tissues that return to the normal states after the removal of the external magnetic field. The protons in water molecules are excited when an external magnetic field is applied, and emits RF when external magnetic field is removed. There are two types of relaxations: 1) water molecules returning to thermal equilibrium and 2) loss of coherence of magnetic alignment of protons. The relaxations decay exponentially. The second relaxation-time is much smaller than the first relaxation time, and plays a major role in the measurement. Since the relaxation-times of healthy tissues and abnormal tissue vary significantly, the measured difference is used to identify abnormal tissues. To localize a specific point in three-dimensional space in a soft tissue, linear magnetic field gradients are applied dynamically in addition to the static magnetic-field. This allows the position to be uniquely coded by the resonance frequency (frequency encoding). MRI has high resolution, and has been used to record three dimensional details of human organs such as brain, heart, kidney and liver.

Ultrasound imaging is based upon the scattering of the emitted acoustic pulses in 1–20 MHz range while propagating through the tissue. It provides good image quality and resolution without a risk of radiation exposure. Ultrasound waves are reflected differently from different tissue layers due to different acoustic impedances of tissue layers. The measurement of the received echo is used to identify various soft-tissue layers. An ultrasound equipment consists of a transmitter and a receiver (typically within the same transducer). Ultrasound is not good for the tissues enclosed inside bones such as brain tissue because most of the signal gets reflected from the bones, and there is insufficient intensity of the waves reaching the soft tissues embedded inside the bones.

An ultrasound transducer is made of piezoelectric materials, which vibrate when a voltage is applied and generate a voltage when subjected to vibration. The vibration property is used in the transmission of acoustic waves, and the generated voltage is used by the receiver to acquire the reflected acoustic waves. An array transducer is used to focus and steer ultrasound beams and form images. An array can be of different formats such as linear-array, curvilinear-array, and sector arrays for imaging different organs. Doppler ultrasound is used to measure the blood-flow.

Ultrasound equipment consists of pulse-generator, beam-former, transducer-array, amplifier, processor and a display unit. The transducer transmits the signals and receives the echoes. The signal is then amplified and processed to form the image. A process called time-gain compensation is used to compensate for attenuation at varying depth.

Other modalities include nuclear medicine (such as PET scan), MR spectroscopy, optical coherence tomography (OCT), photo-acoustics, acousto-optics, and elastography. PET scan uses liquid radionucleotides that contain a fast-decay radioactive isotope in a compound instead of oxygen and are absorbed into the cells like glucose. However, due to the absence of oxygen, metabolism does not progress further, and the radioactive isotope is trapped inside the cells until it decays. The radioactive material is traced to image the metabolic activity. PET has been used for detecting cancer, heart-disease, kidney-disease, liver-disease, dementia, Alzheimer's, other neurological disorders and bacterial infections. MRS is a functional-imaging variant of MRI that probes tissues for various metabolites. It can detect the concentrations of different molecules from the signals returned by them due to their slightly different resonant frequencies. OCT, photoacoustics, and acousto-optics are based on light. OCT is based on interference. Source light-beam is split into two: one beam is reflected from the sample. The other beam provides the reference signal for the interference. Both photoacoustics, and acousto-optics combine optics and ultrasound in different ways to image biological tissues. Elastography is based upon making the map of elasticity distribution of tissues in response to a pressure or dynamic excitation applied to tissues.

Medical imaging has become the mainstay of noninvasive or minimally invasive diagnosis, treatment monitoring and surgery guidance. Imaging data needs to be stored, retrieved and transmitted over the Internet. The image-databases need to be searched using image-based similarities. This requires two special requirements: 1) image compression for efficient transmission of data and 2) content-based image-retrieval. Image compression requires that region of interest should be compressed with no significant information loss that would make diagnosis difficult. Other regions can be compressed more if they do not contribute to the diagnosis. JPEG and its variations have been established as standard for image compression.

Content-based image-retrieval requires summarization of the image-features to reduce the available information and matching this summarized information. Information can be summarized using histograms and their refinements. Features extracted in medical images are texture, shape and intensity. The histograms can be made globally. However, they lose spatial information. To incorporate spatial arrangement, the image is derived into local regions, and histogram is taken of local regions. The arrangement of these local regions is mapped as graphs with adjacent regions corresponding to an edge in the graph. The local regions may have a fixed-size as in macro-blocks, or they could be homogeneous segments derived using segmentation techniques. This refined histogram information along with the meta-level information about the image has been used for efficient content-based image-retrieval. Meta level information consists of: 1) disease specific information; 2) de-identified patient profile such as age, gender and weight; 3) medical history; 4) lab results and 5) external symptoms. CBIR algorithms work on the principle of matching the extracted feature-vector with the feature-vector of the images in the database using similarity scores or tree-based similarity search such as SS-tree, R-tree and KD-tree, as described in Section 2.6.

Automated segmentation and classification are key areas for diagnosis, treatment planning and treatment. Many areas are already mature and used regularly in the clinic. Segmentation can use many approaches. In one approach, certain features such as texture and intensity can identify different homogeneous regions each having a different feature-value. The classification of these segmented regions is used to detect and localize malignancies. For example, the texture between the inside and outside of the lesion can differentiate between malignant and benign lesions.

5.11 ASSESSMENT

5.11.1 Concepts and Definitions

2D image; 3D image; absorption; acoustic impedance; acoustic shadow, acoustic wave; acousto-optics; active contour; antiscatter grid; artificial neural networks; atlas-guided models; attenuation; attenuation coefficient; B-mode image; beam hardening; beamformer; biological tissues; bulk modulus; cancer; capacitive material; central slice theorem; classification; classifiers; CNR; Computer Aided Detection; Computer Aided Diagnosis; CT; CR; CT-image; content-based image-retrieval; contrast; clustering; deformable models; delay generator; density; DICOM; diffraction; digitization; Doppler ultrasound; dynamic range; echoes; echogenicity; elastography; envelope; feature extraction; focusing; FMC segmentation; FPF; Fourier transform; frequency; fuzzy border; gradient coils; gradient field; graphcut segmentation; growcut segmentation; gyromagnetic ratio; Hamming window; histogram; Huygen's principle; image archiving; image artifacts; image binarization; image compression; image enhancement; image fusion; image guided surgery; image reconstruction; image registration; image segmentation; image transmission; imaging informatics; incidence; JPEG; JPEG 2000; K-means clustering; Larmor frequency; LDA; lesion borders; lesion; lesion identification, lesion localization; linear array; liquid radionuclides; lossless compression; lossy compression; magnetic gradient; magnetic moments; magnetization vector; mammography; Markov random field models; medical image compression; medical image transmission; medical imaging; meta information; microcalcification; modalities; model fitting; MPEG; MRI; MRS; multimodal fusion; multimodal segmentation; noise; nuclear magnetic resonance; nuclear medicine; OCT; PACS; PCA; PET; phased array; photoacoustics; piezoelectric material; pixel; precession; propagation; pulse generator; pulse-sequences; quantum states; radiation toxicity; radiography; radiology; Radon transform; ranking; receiver; reflection; refraction; region growing; region of interest; relaxation time; RF signals; resolution; seed-point; sensitivity; shadow; SNR; soft tissues; sonography; SONOR; sound speed; specificity; speckle reduction; speckles; spiculation; support vector machine; teleradiology; texture; therapeutic ultrasound; thresholding; time-domain correlation methods; time-gain compensation; TPF; transducer; transmission; transmitter; treatment monitoring; tumor; two-dimensional projection; ultrasound imaging; voxel; watershed algorithm; wavelength; Wavelets; X-ray; X-ray tube; Young's modulus.

5.11.2 Problem Solving

5.1 The equation for 140 keV X-ray in fat tissue is given by $I = I0 \, e^{-0.140 \, x}$, where x is the thickness of the tissue in centimeters. Compute when *50%* of the X-ray would be absorbed in the tissue. Repeat this calculation for bone with the equation $I = I0 \, e^{-0.28 \, x}$. Plot the two curves on the same graph and compare the decline in the intensity.

5.2 Which element produces the MRI signals? Given that gyromagnetic ratio, $\gamma = 42.58$ MHz/T protons. What is the Larmor frequency if the main magnetic field is *3*T (hint: use Lamour equation)?

5.3 If the gyromagnetic ratio for electron is *27,204* MHz/Tesla, what will be its precession frequency in a magnetic field of *3.0* Tesla.

5.4 A person has a broken finger. Model his finger as a bone layer embedded between two fat layers, if each fat layer is one cm thick, and the bone is two cm thick. The incident energy is *140* keV, and the attenuation factors are *0.16* cm^{-1} for fat and *0.27* cm^{-1} for bone.

5.5 Calculate the distance *3*-MHz and *7.5*-MHz ultrasound beams have to propagate for the intensity to drop *50%* in air (*45* dB-cm^{-1}-MHz^{-1}), liver (*0.47* dB-cm^{-1}-MHz^{-1}), and bone (*8.7* dB-cm^{-1}-MHz^{-1}).

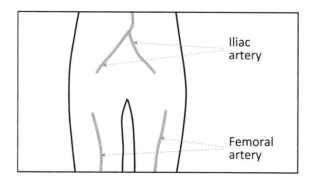

FIGURE 5.20   Locations of iliac and femoral arteries for Problem 5.9

5.6   Ultrasound signals are typically generated using PZT transducers. Calculate the thickness of a PZT slab for generating a 4-MHz ultrasound signal assuming that the speed of sound is *4000* m/s inside the PZT crystal. Calculate the wavelength in the PZT material and in tissue where the speed drops to *1540* m/s.

The intensity of the *4*-MHz ultrasound beam entering tissue is *40* mW/cm^2 and the soft-tissue attenuation is *1* dB-cm^{-1}-MHz^{-1}, calculate the intensity at a depth of *5* cm.

If a transducer emits *10* periods of a *15*-MHz pulse and the speed of sound through the tissue is *1,540* m/s, what is the resolution for the received ultrasound echo signals?

5.7   An ultrasound image is being used to identify cysts in a liver. The ultrasound passes through gel → fat → muscle → liver. Calculate the angle of refraction at the fat-muscle interface and the muscle-liver interface, if ultrasound strikes the gel-fat interface at an incident angle of *20°*. Also, calculate the reflection and transmission pressure coefficients and reflection and transmission intensity coefficients if the ultrasound beam is exactly perpendicular to the interface and for an incident angle of *20°*.

Use the following speeds of sound: gel: *1450* m/s, fat: *1450* m/s, muscle: *1590* m/s, and liver: *1570* m/s. Acoustic impedance: gel: 1.48×10^5 g-cm^{-2}-s^{-1}, fat: 1.38×10^5 gm-cm^{-2}-s^{-1}, muscle: 1.7×10^5 g-cm^{-2}-s^{-1}, and liver: 1.65×10^5 g-cm^{-2}-s^{-1}. Show the calculations.

5.8   Ultrasound is being used to measure the thickness of abdominal fat layer. If the time of flight from the fat-muscle interface is *t* milliseconds, write the equation for the thickness of the fat layer. The speed of sound in the fat is *1450* m/s.

5.9   An sonographer tests for poor blood circulation in the legs of an elderly patient and measures the Doppler-shift for the blood flowing in iliac and femoral arteries (see Figure 5.20). If the Doppler-shift is *9.8* kHz in iliac artery and *8.2* kHz in femoral artery for a beam angle of *30°* and a center frequency of *12* MHz, does the patient have poor circulation? Use a sound speed of *1540* m/s. Normal peak velocity in the iliac artery is *100–150* centimeters/second. Normal peak velocity in the femoral artery is *80–110* centimeters/second. (hint: use Section 5.4.5 on Doppler Ultrasound).

5.10 A small part of an image is shown in the *8 × 8* matrix in Table 5.3, which contains the intensity of a radiographic image where *255* is white and toward *0* is black. If the threshold value is *63*, binarize the image. Binarization maps the values less than threshold to "*0*" and values greater than or equal to threshold to "*1*." Identify the edges between "*1*" and "*0*," and mark the boundary.

5.11 Write an algorithm for binarization that takes as input a matrix as in the above Problem 5.10, converts into a binarized matrix, identifies the cells where *0* and *1* are in neighboring cells and displays the cell index.

5.12 Table 5.4 shows an *8 × 8* matrix with *1* and *0*. The clusters of "*1*" show tumors. Write an algorithm that identifies these clusters of "*1*" in an *8 × 8* matrix. The algorithm should merge the segments if the minimum distance between two segments is just one cell.

5.13 Chain-code is a technique to describe the boundary of an image-region using *8* directions: N, NE, E, SE, S, SW, W, NW. Each direction is modeled as one bit in an *8*-bit byte. Another byte

TABLE 5.3   A matrix of intensity for Problem 5.10

48	45	58	62	64	70	120	145
47	49	55	67	72	120	124	150
37	49	59	64	72	145	134	170
42	41	52	69	75	130	122	160
47	49	55	67	72	120	124	150
25	39	45	57	62	80	94	110
35	29	45	67	62	89	99	104
25	34	55	57	59	70	91	110

TABLE 5.4   A binarized matrix for segmentation in Problem 5.12

0	0	0	0	0	0	0	0
0	1	1	1	0	0	0	0
1	1	1	1	0	0	0	0
0	1	1	0	0	0	1	1
0	0	0	0	0	0	1	1
0	0	0	0	1	0	0	1
1	1	0	1	1	0	0	1
1	1	0	1	1	0	0	1

is used to describe the magnitude of the travel in a direction. The traverse is started consistently from one direction, e.g., the leftmost corner and continues counterclockwise. Chain coding compresses the image significantly. Take Table 5.4 and assume that a tumor is modeled as (coordinates of the start cell, sequence of chain-code). Each coordinate of the start cell is modeled as one byte. Each chain-code direction is modeled as 2-bytes. Give a chain-code for the tumors.

5.14 Using a classification technique on a dataset, *34* malignant tumors are identified correctly and *9* malignant tumors are identified as benign tumors. *12* benign tumors are incorrectly identified as malignant tumors and *20* benign tumors are identified correctly. Compute the sensitivity and specificity of the technique and compute false-negative and false-positive. Show your computation using the basic equations.

5.15 Write a program that identifies the areas (in a binary image) containing *1s* above a threshold limit (say *98%* of pixels), and then substitutes the *0s* within that area with *1s*. The minimum number of pixels for an area should be *50*. Use random number generator to create a *30 × 30* matrix and test your program.

5.16 Extend the program in Problem 5.15 to grow the regions of *1s* as described in Problem 5.15, and then identify the centroids of the regions of *1s*. Derive the distances between the centroids, and construct a weighted graph with distances as weights, and display it as a weighted matrix.

5.17 Given the original grayscale image in Table 5.5, binarize the image using a threshold (e.g., *128*). Form the initial clusters of *1s*. Average out the actual values of the cells that have "*1*" in the binarized matrix. Replace the *1s* by the average value. Retain the *0s* as they are. Make a graph of the clusters with the weight of the edges as the difference in the average intensity value. Group the nodes if the edge-weight is less than sixteen. Repeat the process and show the final graph.

5.18 Write an algorithm for solving Problem 5.13 (chain-code).

5.19 An MRI scan of a brain comprises *21* images of *1690 × 1744* pixels. Calculate the amount of data to be stored and transmitted. Assume that *10%* of the image is the region-of-interest to detect a lesion, and is not compressed. Remaining *90%* image is compressed to *10%* of the original size. What is the new memory requirement of the transformed image to be transmitted? What is the effective compression percentage? Show the computation.

TABLE 5.5   A matrix of intensity showing embedded tumor

24	23	22	22	21	11	17	18
34	189	200	190	24	25	33	22
172	192	193	194	25	43	42	42
38	210	220	44	43	50	234	244
42	45	40	50	55	56	213	235
51	52	53	55	198	54	50	240
198	178	59	212	220	53	34	239
167	200	59	220	224	52	10	215

TABLE 5.6   Region of macro-blocks for Problem 5.21

212	208	185	195
204	202	188	205
180	167	45	33
120	170	29	16

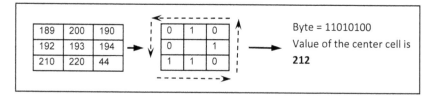

FIGURE 5.21   An illustration of local binary pattern for Problem 5.15

5.20 A linear binary pattern (LBP) is used to express the texture of an image. LBP is calculated using a 3×3 template and moving it across the image matrix. The center pixel value in this 3×3 block is compared against its adjacent eight neighbors. If the value (center-pixel) $\geq$ value (neighboring-cell) then that cell is mapped to *1*. Otherwise, the neighboring cell value is mapped to zero. After computing this 3×3 template with *1s* and *0s*, the neighboring cells are traversed starting from the left-most bottom-most corner in a counter-clockwise manner, and the bit-sequence is arranged into the form of a byte. The integer value of this byte is computed and placed as the center cell-value as shown in Figure 5.21. Assume that the end row and column cells are extended to an additional row and column with duplicated values for computing LBP. Write a program that generates a 10×10 matrix, and calculates the LBP for every cell. Print out the original matrix and the matrix generated using LBP. The four corners get the average value of their immediate neighbors.

5.21 In a region growing scheme, two adjacent heterogeneous macro-blocks are joined into one bigger homogeneous macro-block, if the average intensity of two macro-blocks is below a threshold. The overall intensity of the bigger macro-block is the average of the average intensities of the individual macro-blocks that are gathered together. Let us take a 4×4 matrix of macro-blocks as shown in Table 5.6. If there are *16* regions, use the above technique to join the regions. Write an algorithm to grow regions using this scheme. Assume that the threshold is *20*. Merge only horizontally or vertically adjacent regions.

5.11.3 Extended Response

5.22 Explain the causes of weakening of intensity of the received signals in ultrasound imaging. Explain how it is corrected using time-gain control for human viewing.

5.23 Explain the principle of X-ray imaging.

5.24 What is "beam hardening" in X-ray radiography? Why and how is it done?

5.25 Explain the role of histograms in image matching and fast retrieval. Also, explain the limitations of the histograms.

5.26 Explain the central-slice theorem, which is basis of image reconstruction in Computed Tomography. Is this theorem used in practice for the reconstruction?

5.27 Explain the principle of forming a magnetic resonance image.

5.28 Explain the difference between focusing and steering using a linear-array ultrasound transducer.

5.29 What are the advantages and disadvantages of X-rays, CT, MRI, ultrasound, nuclear medicine, and the modalities based on optics?

5.30 Read the review articles and briefly explain the following segmentation techniques: 1) Atlas-guided segmentation; 2) graph cut segmentation; 3) growcut segmentation; 4) Markov random field model and 5) geodesic-contour segmentation.

5.31 Explain the need for content-based image-retrieval for medical images.

5.32 Explain various quantifiable features used in cancer detection.

5.33 Explain the steps used in region-growing for image segmentation.

5.34 Read a few review articles and discuss at least five applications of different modalities.

5.35 Explain various steps used in cancer analysis using MR-images and ultrasound.

5.36 Read the articles 5.52–5.60 in the section "Further Reading," and write a two-page summary of how cancer is diagnosed using various imaging techniques.

5.37 Read two review articles on JPEG 2000 format and explain in a page how JPEG 2000 works.

5.38 Read two survey articles on automated MRI-segmentation of brain and provide a two-page summary.

5.39 Explain the proton relaxation mechanism in NMR and compare two types of relaxations.

5.40 Explain why a burst of pulses are transmitted in ultrasound systems. Also, explain the role of beamformers.

5.41 Do a literature search on multimodal scheme and discuss three combinations of multimodal schemes used. Explain various advantages and cost aspect of multimodal schemes.

5.42 For the following scenario, which type of radiology scheme will you use and why? Explain your rationale.

(a) Detecting a brain tumor and its location

(b) Hair-line fracture on the skull

(c) Increase in the thickness of heart

(d) Heart enlargement

(e) Cancer of the liver

(f) Coronary artery diseases

(g) Blood clots in a lung

(h) Fetus movement and growth

FURTHER READING

Medical Imaging Fundamentals

5.1 Bushberg, Jerold T., Seibert, Anthony J., Leidholdt Jr., Edwin M., and Boone, John M., *The Essential Physics of Medical Imaging*, 3rd edition, Lippincott Williams and Wilkins, Philadelphia, PA, USA, 2011.

5.2 Paragios, Nikos, Duncan, James, and Ayache, Nicholas, (editors), *Handbook of Biomedical Imaging – Methodologies of Clinical Research*, Springer, New York, NY, USA, 2015.

5.3 Prince, Jerry L., and Links, Jonathan M., *Medical Imaging: Signals and Systems*, 2nd edition, Pearson Education, Upper Saddle River, NJ, USA, 2015.

5.4 Smith N, Webb A, *Introduction to Medical Imaging: Physics, Engineering and Clinical Applications*, Cambridge University Press, Cambridge, UK, 2011.

5.5 Suetens, Paul, *Fundamentals of Medical Imaging*, 2nd edition, Cambridge University Press, Cambridge, UK, 2009.

5.6 Birkfellner, Wolfgang, *Applied Medical Image Processing: A Basic Course*, CRC Press, Boca Raton, FL, USA, March 2014.

5.7 Huang, Hai K. (Bernie), *PACS and Imaging Informatics: Basic Principles and Applications*, Wiley-Blackwell, New York, NY, USA, 2004.

5.8 Bransetter, Barton F., Rubin, Daniel L., Griffin, Scott, and Weiss, David L., *Practical Imaging Informatics: Foundations and Applications for PACS Professional*, Society for Imaging Informatics, Springer Science+Business Media, New York, NY, USA, 2013.

Radiology (X-ray)

5.9 Carter, Christi, and Veale, Beth, *Digital Radiography and PACS*, 2nd edition, Elsevier, Matfield Heights, MO, USA, 2014.

5.10 Herring, William, *Learning Radiology: Recognizing the Basics*, 3rd edition, Elsevier Health Sciences, Philadelphia, PA, USA, 2015.

5.11 Martz, Harry E., Logan, Clint M., Schneberk, Daniel J., and Skull, Peter J., *X-ray Imaging: Fundamentals, Industrial Techniques and Applications*, Chapman and Hall/CRC Press, Boca Raton, FL, USA, 2016.

Computed Tomography

5.12 Buzug, Thornsten M., *Computed Tomography: From Photon Statistics to Modern Cone-Beam CT*, Springer, Berlin, Germany, 2008.

5.13 Herman, Gabor T., *Fundamentals of Computerized Tomography: Image Reconstruction from Projections*, 2nd edition, Springer-Verlag, London, UK, 2009.

MRI and NMR

5.14 Katti, Girish, Ara, Syeda A., and Shireen, Ayesha, "Magnetic Resonance Imaging (MRI) – A Review," *International Journal of Dental Clinics*, 3(1), January-March 2011, 65–70.

5.15 Liange, Zhi -P., and Lauter, Paul C., *Principles of Magnetic Resonance Imaging: A Signal Processing Perspective*, IEEE Press, New York, NY, USA, 2000.

5.16 Rinck, Peter A., *Magnetic Resonance in Medicine: A Critical Introduction*, 12th edition, 2nd printing, The Round Table Foundation/Book on Demand, Norderstedt, Germany, 2018.

5.17 Wang, Yi, *Principles of Magnetic Resonance Imaging: Physics Concepts, Pulse Sequences, & Biomedical Applications*, ISBN: 9781479350414, Createspace Independent Publishing Platform, 2012.

5.18 De Graaf, Robin A. *In Vivo NMR Spectroscopy: Principles and Techniques*, John Wiley & Sons, Chichester, West Sussex, England, 2007.

Ultrasound

5.19 Hughes SW, "Medical ultrasound imaging," *Phys Educ* 36(6):468, 2001

5.20 Szabo, Thomas L., *Diagnostic Ultrasound Imaging: Inside Out*, 2nd edition, Elsevier, Amsterdam, The Netherlands, 2014.

Nuclear Medicine and PET

5.21 Dixon-Brown, Ann, and Soper, Nigel DW, "Nuclear medicine: diagnosis and therapy", *Phys Educ, Vol. 31, Number 2,* 1996, pp 96–100.

5.22 Shukla, Arvind K., and Kumar, Utham, "Positron Emission Tomography: An Overview," *J Med Phys,* 31(1), January 2006, 13–21.

Elastography

5.23 Ophir, J, Alam, SK, Garra, BS, Kallel, F, Konofagou, E, Krouskop, T, and Varghese, T, "Elastography: ultrasonic estimation and imaging of the elastic properties of tissues," *Proc Inst Mech Eng H: J Eng Med,* vol. 213, no. 3, pp. 203–233, 1999.

5.24 Gennisson JL, Deffieux T, Fink M, Tanter M, Ultrasound elastography: principles and techniques. *Diagn Interv Imaging*. 2013 May; 94(5):487–495.

5.25 Barr RG, Elastography in clinical practice. *Radiol Clin North Am*. 2014 Nov; 52(6):1145–1162.

5.26 Garra, Brian S., "Elastography: History, Principles, and Technique Comparison," *Abdom Imaging*, 40(4), April 2015, 680–697.

5.27 Sigrist, Rosa MS., Liau, Joy, Kaffas, Ahmed E., Chammas, Maria C., and Willmann, Juergen K., "Ultrasound Elastography: Review of Techniques and Clinical Applications," *Theranostics,* 7(5), 2017, 1303–1329.

Alternate Modalities

5.28 Hrynchak, Patricia, and Simpson, Trefford, "Optical Coherence Tomography: An Introduction to the Technique and its Use," *Optom Vis Sci,* 77(7), July 2000, 347–356.

5.29 Su, Jimmy L., Wang, Bo, Wilson, Katheryne E., Bayer, Carolyn L., Chen, Yun-Sheng, Kim, Seungsoo, Homan, Kimberly A., and Emelianov, Stanislav Y., "Advances in Clinical and Biomedical Applications of Photoacoustic Imaging," *Expert Opin Med Diagn,* 4(6), November 2010, 497–510.

Medical Image Archiving and Transmission

Medical Image Compression

5.30 Kalyanpur, Arjun, Neklesa, Vladimir P., Taylor, Caroline R., Daftary, Aditya R., and Brink, James A., "Evaluation of JPEG and wavelet compression of body CT images for Direct Digital Teleradiologic Transmission," *Radiology*, 217(3), December 2000, 772–779.

5.31 Koff, David A., and Shulman, Harry, "An Overview of Digital Compression of Medical Images: Can We Use Lossy Image Compression in Radiology?" *Can Assoc Radiol J,* 57(4), October 2006, 211–217.

Medical Image Retrieval

5.32 Akgül, Ceyhun B., Rubin, Daniel L., Napel, Sandy, Beaulieu, Christopher F., Greenspan, Hayit, and Acar, Burak, "Content-Based Image Retrieval in Radiology: Current Status and Future Directions," *J Digit Imaging*, 24(2), April 2011, 208–222.

5.33 Castelli, Vittorio, and Bergman, Lawrence D., *Image Databases: Search and Retrieval of Digital Imagery*, John Wiley and Sons, New York, NY, USA, 2002.

5.34 Kumar, Ashnil, Kim, Jinman, Cai, Weidong, Fulham, Michael, and Feng, Dagan, "Content-based Medical Image Retrieval: A Survey of Applications to Multidimensional and Multimodality Data," *J Digit Imaging*, 26(6), December 2013, 1025–1039.

5.35 Muramatsu C. Overview on subjective similarity of images for content-based medical image retrieval. *Radiol Phys Technol*. 2018 Jun;11(2):109–124.

5.36 Wei S, Liao L, Li J, Zheng Q, Yang F, Zhao Y. "Saliency Inside: Learning Attentive CNNs for Content-based Image Retrieval." *IEEE Trans Image Process*. 2019 May 2.

Medical Image Transmission – See Chapter 6

Medical Image Analysis

5.37 Birkfellner, Wolfgang, *Applied Medical Image Processing – A Basic Course*, 2nd edition, Chapman and Hall/CRC Press, Boca Raton, FL, USA, 2014.

5.38 Dhavan, Atam P., *Medical Image Analysis*, 2nd edition, IEEE Press Series in Biomedical Engineering, Series editor: Akay, Metin, John Wiley and Sons, Singapore, 2011.

5.39 Toennies, Klaus D., *Guide to Medical Image Analysis: Methods and Algorithms*, Springer-Verlag, London, UK, 2017.

Medical Image Enhancement

5.40 Stark JA, "Adaptive Image Contrast Enhancement Using Generalizations of Histogram. Equalizations," *IEEE Trans Image Process,* 9(5), May 2000, 889–896.

5.41 Yu, Y, Acton ST, "Speckle Reducing Anisotropic Diffusion," *IEEE Trans Image Process,* 11(11), 2002, 1260–1270.

Medical Image Segmentation and Feature Extraction

5.42 El-Baz A, Jiang X, Suri JS, *Biomedical Image Segmentation – Advances and Trends*, Chapman and Hall/CRC Press, Boca Raton, FL, USA, 2016.

5.43 Noble JA, Boukerroui D. Ultrasound image segmentation: a survey. *IEEE Trans Med Imag,* 2006 Aug; 25(8):987–1010.

5.44 Pham DL, Xu C, Prince JL, "Current methods in medical image segmentation," *Annu Rev Biomed Eng.* 2000;2:315–37

ANN and Segmentation

5.45 Jiang J, Trundle PR, Ren J, "Medical Imaging Analysis with Neural Networks," *Comput Med Imaging Graph,* 34(8), 2010, 617–631.

Deformable Models

5.46 McInerney T, Terzopoulos D, "Deformable Models in Medical Image Analysis: A Survey," *Med Image Anal,* 1(2), June 1996, 91–108.

Applications of Image Segmentation

5.47 Middleton I, Damper RI, "Segmentation of Magnetic Resonance Images Using a Combination of Neural Networks and Active Contour Models," *Med Eng Phys,* 26(1), January 2004, 71–86.

5.48 Baka N, Leenstra S, van Walsum T, "Random Forest-Based Bone Segmentation in Ultrasound," *Ultrasound Med Biol.* 2017 Oct;43(10):2426–2437.

Ultrasound Lesion Segmentation

5.49 Madabhushi A, Metaxas DN, "Combining Low-, High-Level and Empirical Domain Knowledge for Automated Segmentation of Ultrasonic Breast Lesions," *IEEE Trans Med Imaging,* 22(2), February 2003, 155–169.

5.50 Shan J, Cheng HD, Wang Y., "Completely Automated Segmentation Approach for Breast Ultrasound Images Using Multiple-domain Features," *Ultrasound Med Biol,* 38(2), February 2012, 262–275.

5.51 Huang Q, Luo Y, Zhang Q. Breast ultrasound image segmentation: a survey. *Int J Comput Assist Radiol Surg.* 2017 Mar;12(3):493–507.

Computer-Aided Diagnosis

Cancer Identification from Medical Images

5.52 Chang RF, Wu WJ, Moon WK, Chen DR, "Improvement in Breast Tumor Discrimination by Support Vector Machines and Speckle-emphasis Texture Analysis," *Ultrasound Med Biol,* 2003 May;29(5), 679–686.

5.53 Chen DR, Chang RF, Kuo WJ, Chen MC, Huang YL, "Diagnosis of Breast Tumors with Sonographic Texture Analysis Using Wavelet Transform and Neural Networks," *Ultrasound Med Biol,* 2002 October;28(10), 1301–1310.

5.54 Brem RF, Lenihan MJ, Lieberman J, Torrente J. Screening breast ultrasound: past, present, and future. AJR Am J Roentgenol. 2015 Feb;204(2):234–240.

5.55 Guo R., Lu G., Qin B., Fei B., Ultrasound Imaging Technologies for Breast Cancer Detection and Management: A Review. *Ultrasound Med Biol.* 2018 Jan;44(1):37–70.

5.56 Hasan MK, Hussain MA, Ara SR, Lee SY, Alam SK, "Using Nearest Neighbors for Accurate Estimation of Ultrasonic Attenuation in the Spectral Domain," *IEEE Trans Ultrason Ferroelectr Freq Control,* 60(6), June 2013, 1098–1114.

5.57 Alam SK, Feleppa EJ, Rondeau M, Kalisz A, Garra BS, "Ultrasonic Multi-feature Analysis Procedure for Computer-aided Diagnosis of Solid Breast Lesions," *Ultrason Imaging,* 33(1), 2011, 17–38.

5.58 Oelze ML, Mamou J., "Review of Quantitative Ultrasound: Envelope Statistics and Backscatter Coefficient Imaging and Contributions to Diagnostic Ultrasound." *IEEE Trans Ultrason Ferroelectr Freq Control,* 2016 Feb;63(2):336–51.

5.59 Mamou J, Oelze ML, *Quantitative Ultrasound in Soft Tissues*, Springer, New York 2013.

5.60 Mallidi, Srivalleesha, Luke, Geoffrey P., and Emelianov, Stanislav, "Photoacoustic Imaging in Cancer Detection, Diagnosis, and Treatment Guidance," *Trends Biotechnol,* 29(5), May 2011, 213–221.

DICOM – Medical Image Communication

6

BACKGROUND CONCEPTS

Section 1.6.2 Medical Information Exchange; Section 1.6.3 Integration of Electronic Health Records; Section 1.6.9 Medical Image Processing and Transmission; Section 1.6.3.2 Heterogeneity and Interoperability; Section 2.2.3 Image Compression; Section 2.8.3 Multimedia Databases; Section 2.9 Middleware for Information Exchange; Section 4.1 Modeling Medical Data; Section 4.4 Health Information Flow; Section 4.7 Information Retrieval; Section 4.9 Standards for Medical Image Transmission; Section 4.10 HIPAA - Protecting Patients' Privacy Rights; Section 5.7 Medical Image Archiving and Transmission

This chapter discusses a framework and technical standards for medical image management and transmission over the Internet. There has been a vast amount of fascinating developments in this field over the past three decades. Communication between two application-entities residing anywhere around the world is a sophisticated technological feat. It requires many technological pieces to come together. Computer networking has greatly advanced the services surrounding medical imaging. These include patient image management, study management, archiving of patient images and automation of medical imaging processes in the hospitals. Appropriate use of digital communication can improve the quality, reliability and efficiency of these services.

Attempts to use computer communication to improve digital image sharing started as early as the 1970s. Until the mid-1980s, most devices stored images in a proprietary format and transferred files of these proprietary formats over a network or on removable storage media. Attempts to develop interchangeable formats and protocols emerged around 1993 through an image standardization effort called *DICOM* (Digital Imaging and Communications in Medicine). It focused on defining the imaging information model and key services around their exchange that can be built over TCP/IP. Technology like DICOM is critical to achieve compatibility and to improve workflow efficiency between imaging systems and other information systems in healthcare environments worldwide.

6.1 OVERVIEW OF NETWORK AND APPLICATION LAYERS

The communication system responsible for today's feat is normally organized into multiple communication layers. DICOM lies near the top in the application-ended layers. This chapter discusses DICOM. However, an overview of other layers concerned with the network layer message-transfer are discussed in Sections 6.1.1–6.1.2. Each layer solves a subset of network problems focusing on an aspect of this complex communication puzzle, and adds a specific set of capabilities. This divide-and-conquer engineering principle is reflected in the layers, as illustrated in Figure 6.1.

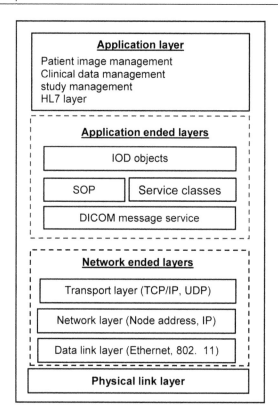

FIGURE 6.1   Layers for medical imaging communications

The *physical-link layer* uses network hardware or medium to transmit data encoded as signals. The *network-ended layers* are above the physical-link layers and ensure the transmission of data-packets from the source to the destination through a complex network. The *application-ended layers* are above the *network-ended layers* and translate the information being sent from an application software to the data packets used by the *network-ended layer*. The application layer is the top-level layer and is above the *application-ended layer*. The *application layer* interfaces with the application-level software. Application software run at various locations and are linked to each other through a communication network.

6.1.1 Physical-Link Layer

The *physical-link layer* is the fundamental layer. This layer transmits signals between two end-points of a physical medium. The *physical-link layer* ensures that a *signal* from one end is carried to the other connected-end reliably. The layer can be: 1) a copper wire; 2) an optical fiber; 3) a wireless medium; 4) infrared light; 5) direct laser light or 6) microwave. Signals are sent via empty space as electromagnetic radio wave, infrared light, visible light or their combinations. Each medium has different physical properties and constraints to be overcome.

6.1.2 Network-Ended Layers

The data-link layer (DLL) is immediately above the *physical-link layer*. It is ensured of signal communication using the *physical-link layer*. It focuses on how the signals carry a *data-packet* between the two end-points of a directly connected link. A *data-packet* is a collection of binary bits, size-information, and parity-check bits for error detection and correction. Bits are also used to represent various types and forms of information that are used by the upper layers.

All end-points are not directly connected in a communication network. A network is built of many intermediate points. The *network layer* focuses on how a packet can be delivered between any two reachable end-points without the restriction of direct connection. This requires complex capabilities such as modeling network as a graph, assigning unique addresses to each node, locating a destination, finding optimum routes, forwarding and buffering packets, etc. The *network layer* ensures that a data-packet with a destination-address can reliably and efficiently reach the destination-end-point in a graph-like communication network.

The *network layer packets* are size-limited. Some applications use shorter data, but many need to send a large amount of data, more than the maximum allowable packet-size of the network layer packet. Many applications often send data as bit-streams, arriving irregularly and intermittently; Entire data may not be available at a time. In the real world, despite the diligence of the lower-level layers, data-packets are often randomly and probabilistically lost. Such packet-losses need to be detected and corrected.

Multiple applications share the same communication network. The correct mapping between a data-packet and the applications needs to be established at the end-points to deliver it to the right receiver. The *transport layer* focuses on solving these issues so two applications residing at different end-points of a complex network can exchange data-streams and data-units of varying size. It uses the underlying packet service of the *network layer.*

The *network-ended layers* ensure that a data-stream is sent over a network from one application end-point to another, despite the network configuration and connectivity. However, it cannot convert high-level information into a data-stream. But, if a community of applications would like to use the distributed network to exchange information, coordinate each other's activities and work together as a coherent system, the communication subsystem must deal with high-level information, and use a common language for communication.

Simple and private applications (such as a file transfer system or email system), with a sender IP-address and receiver IP-address, custom-made for a sender, need only network-ended layers and a simple underlying language for the data-exchange by each sender-receiver pair.

6.1.3 Application-Ended Layers

This section discusses the *application-ended layers* for medical-image communication. A medical imaging communication system is complex and requires: 1) a sophisticated digital communication language; 2) translation of the information from HL7 into the communication language and 3) translation of complex communication language to a simple language used by network-ended layers.

A distributed information system needs to identify the types of data-elements along with their key-attributes. Attributes are coded after a key-entity is identified. The medical image communication world involves patients, doctors, healthcare providers, radiology labs, surgical units, expert radiologists, researchers involved in clinical analysis, medical boards, data warehouses, ICUs, patient-visits, and equipment-modalities besides the actual images. Each entity is well-defined. Information pertaining to these entities is stored in *IODs* (Information Object Definitions).

The next application-ended layer is the *service layer*. In this layer, objects are created, stored, retrieved, archived for a longer period to maintain the history, moved from one database or URL to another, deleted, and updated. Objects are represented using many data-formats. Moving different objects can differ vastly, depending upon the object-types. End-users often print or display certain type of objects such as images. In the medical imaging world, each object-class requires a specific set of plausible well-defined operations. The combinations of specific services along with the corresponding objects are called *SOPs* (Service Object Pairs).

The applications encode the domain-knowledge in the form of *IODs* and *SOPs*. Encoding these entities and operations requires a language. *IOD and SOP layer* in Figure 6.1, provides a group of library utilities to create, read, express, and manipulate these entities.

Information transfer between two applications requires a series of coordinated service-requests and transportation of the associated data-elements between two physically-separated application components. The *messaging service layer* provides message-packet structures, and coordination protocols to help the application to pack the *SOPs* and *IODs* as network data-streams. At the receiving end, the messaging service layer receives the transmitted data-streams and converts the received data-streams back into the application-specific *SOPs* and *IODs*.

These three application-ended layers translate and create the format and "meaning" of the binary data transported by the combination of network-ended layer and physical link layer. The meaning of the binary data varies with the application domains.

6.2 MODELING MEDICAL IMAGING INFORMATION USING DICOM

The following sections describe the major design principles and protocols used in the application-ended-layers for medical image communication, and the information-objects.

6.2.1 Entity-Relationship Model

Medical entities include patient, doctors, hospitals, radiology lab, pathology lab, medical equipment, patients' close relatives and images. Medical images are interpreted in a context. Medical image information communication involves the transmission of medical images, along with the corresponding contexts that includes the entities and their relationships.

Medical entities are connected to each other through a *relationship*. DICOM models medical image information using this *entity-relationship* model. A relationship can be *symmetric* or *asymmetric*. *Symmetric relationships* are bidirectional, and *asymmetric relationships* are unidirectional. Each entity-class has certain attributes. An entity-instance has different attribute-values and relationships to other entities.

In Figure 6.2, an entity is depicted as a rectangular box. The diamond shows a relationship between two entities. Relationships could be one-to-one, one-to-many, many-to-one, or many-to-many. E-R diagrams also express the cardinality of the relationship between two entities.

Example 6.1

Labels "*a*" and "*b*" show the source and destination cardinalities of the relationship respectively. For example, $a = 1, b = 1$ means that one source entity is related to only one destination entity. Similarly, $a = 1$, $b = n*$ means one source entity is related to at most n (including zero) destination entities. Source cardinality can also be greater than 1. For example, $a = n+, b = 1$ means n source entities ($n \geq 1$) are related to one destination entity. Similarly, $a = n+, b = m+$ means n ($n \geq 1$) source entities map to m ($m \geq 1$) destination entities.

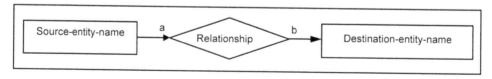

FIGURE 6.2   An illustration of entity-relationship diagram

6.2.2 DICOM's Entity Relationship Model

DICOM world is patient-centric. A *patient* is normally an individual. However, in DICOM, the term "patient" can also refer to a group of humans or even animals being treated or studied simultaneously as one unit. An example of a group situation is a mother and fetus(es) during an obstetric ultrasound. Another example is multiple specimens in a single tissue microarray or a group of small animals being studied together.

A patient undergoes various *studies* as an outcome of patient–doctor encounters. Each encounter with a doctor starts a study to diagnose and monitor a disease. Certain standardized guidelines of the study are provided by the doctor. After a visit, a *study-request* is generated. Each *study-request* results in one or more *image-series* in one or more steps. Each *image-series* contains multiple images using different modalities, such as X-ray, CT-scan, MRI, ultrasound and microarray analysis (as described in Section 10.4.2). Medical devices create various types of complex imageries such as two-dimensional, three-dimensional stack or videos.

A study involves one or more diagnostics organizations. Each imaging-session might be divided into multiple days. Some prescribed procedures are refined based on the equipment used in a facility. Some orders may not even be fulfilled for patient conditions. The workflow for various types of medical procedures and diagnostics can vary.

A *study* contains one or more *procedure-steps*. Each *procedure-step* includes multiple series of image-related data. Each *series* involves multiple modalities and devices that generate multiple images and other associated information. The use of an equipment creates multiple *series*. *Series* are also defined by optional *frames-of-reference*. Often, a study may not have an associated *frame-of-reference*. However, if a *frame-of-reference* is present, the series must contain the *registration information* which will specifically identify the orientation of the data-element captured in the series with respect to the *frame of-reference*. With the above view of the world, a baby DICOM information system can be created. For each element of the objects, a data structure containing the parameters is defined.

An object can be: 1) a presentation state; 2) a registration for aligning images; 3) an MR image; 4) a fiducial; 5) an X-ray image; 6) a CT scan; 7) an ultrasound image; 8) an image using other modalities; 9) raw data such as vital sign monitoring data or lab results; 10) an encapsulated document; 11) an SR document; 12) a stereoscopic relation; 13) a waveform such as ECG, EEG and EMG; 14) a tractography result; 15) a measurement; 16) a real-world mapping and 17) a surface.

Figure 6.3 illustrates an E-R model with various visits and studies. The notation $n*$ denotes zero or more; the notation $n+$ denotes one or more and the notation [1] denotes "optional" one. The words describing the entities are placed in bold within rectangles, and the words describing the relationships are placed in a diamond. The relationships can be one-to-one or one-to-many. Some occurrences can be optional.

6.2.3 Medical Imaging

To interpret medical images correctly, other important data along with the basic image are needed. For example, for a two-dimensional X-ray, it is important to know the geometric *frame-of-reference* from which the picture is taken. To specify the *frame-of-reference*, the coordinate system of the capturing equipment is specified.

There are other medical image objects that are far more complex than X-ray and MR images, such as radiotherapy (RT) objects. RT-objects require encoding of information related to the patient's anatomy. These entities are typically identified on devices such as CT-scanners, physical or virtual simulation workstations, treatment planning systems and RT-plan containing geometry and dosimetry data – data related to the absorption of ionizing radiation.

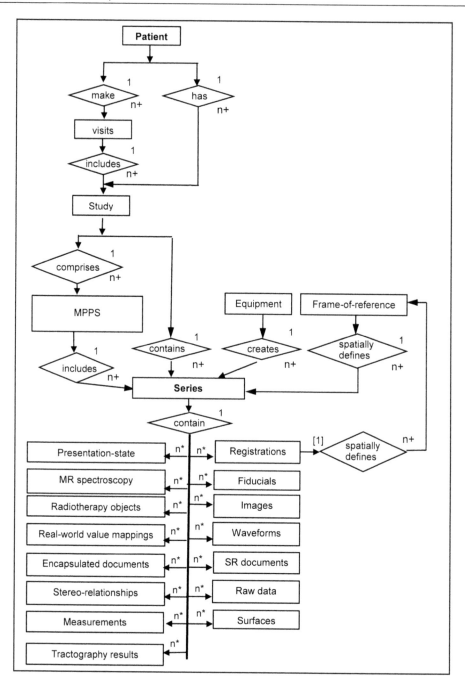

FIGURE 6.3   An entity relationship model for patient-services

All the data in a medical image object may not be added only by the initial capturing equipment. The RT-plan entity may be created by a simulation workstation, and enriched by a treatment planning system before being passed on to a "record and verify system" or treatment device. An instance of the RT-object usually references an RT-structure set to define a coordinate-system and a set of patient-structures. The actual RT-images are obtained on a conical imaging geometry such as those found on conventional simulators and portal imaging devices. These can also generate new calculated images such as digitally reconstructed radiographs (DRRs).

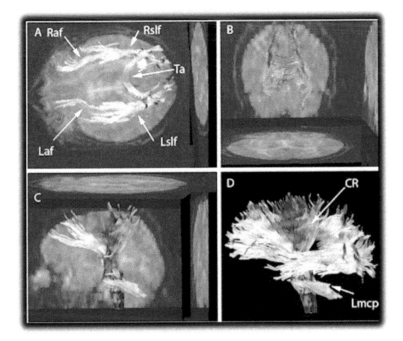

FIGURE 6.4 An illustration of tractography – a brain tractographic image set (Images adopted from English Wikipedia, Courtesy Aaron G. Filler, Md, PhD, website: https://commons.wikimedia.org/wiki/File:DTI_Brain_Tractographic_Image_Set.jpg, distributed under GNU Free Documentation License and Creative Commons Attribution-Share Alike 3.0 (https://creativecommons.org/licenses/by-sa/3.0/)).

Example 6.2

An example of a complex secondary product imagery is a *tractograph* which shows neural tracts. Neural tracts are not normally identifiable by direct exam, CT or MRI scans. These are generated by image analysis on a primary set of images of MRI or other forms. The results are presented as two-dimensional and three-dimensional images. Figure 6.4 illustrates different views of neural images in three-dimensions constructed using MRI images and computer-aided analysis. The four views are showing the same area of interest. Different views help a doctor get a better understanding of the subject in three-dimensional space.

In medical imaging, often, a group of images, including duplicate shots, shots taken from multiple views, or more ordered sequence such as slices are used. A primary unit is an *image-series* that also includes a single image. An *image-series* also contains the associated context data-elements such as parameters' values of the images, patient-identifying data, stereoscopic relationships between images, description of objects such as fiducials (reference objects/markers to help the measurements) and SR-documents (Structured Reporting documents) intended for human reading. The leaves in the E-R diagrams show a variety of these context elements.

6.2.4 Life Cycle of Medical Imaging

A medical image starts in a *service episode* and goes through a life-cycle. The stages in a life-cycle are: 1) request for a procedure, 2) specification of various requirements, 3) scheduling the steps as per valid protocols and 4) performing the procedures. To manage and archive medical imaging, the process-flow context of the image generation is illustrated in Figure 6.5.

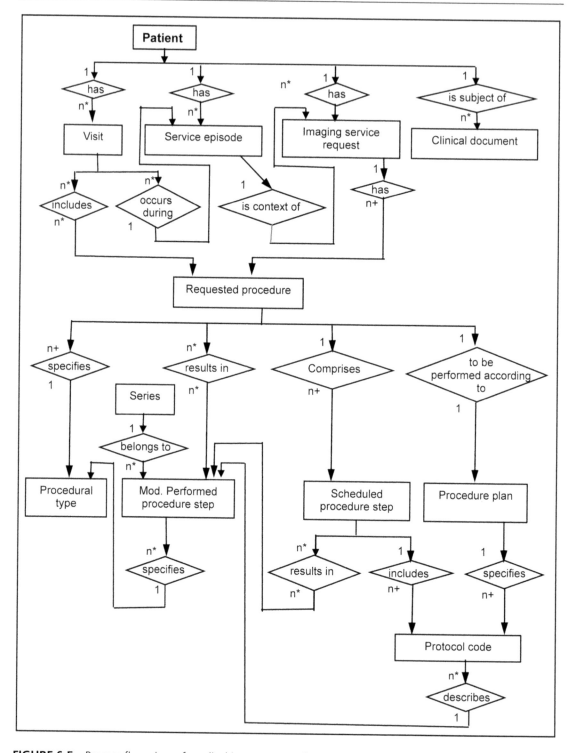

FIGURE 6.5   Process flow view of medical image generation

6.2.4.1 Service-episode

A medical image originates in a *service-episode*, which refers to all the treatments to manage one incident of medical conditions. The definition of the start-time, stop-time and included events of a *service-episode* are flexible. It may include a single outpatient visit or hospitalization, or a treatment for an extended period such as the duration of a pregnancy, an oncology treatment regimen, a cardiac episode from infarction through rehabilitation, or an extended illness such as diabetes lasting multiple years. A service episode may involve one or more healthcare organizations.

A *visit* is a subset of a *service-episode*. A *service-episode* describes several administrative aspects of health care, while a *visit* is limited to the description of one patient-visit to a facility. It can be a collection of events. However, a *visit* in our ER-model falls under the accountability of a specific healthcare organization in a single facility. A *visit* may be associated with one or more physical locations (e.g., different rooms, departments, or buildings) within a healthcare facility, with admission and discharge diagnoses, and with the time-range of the visit. In the modality work-list, *SOP-class* and the attributes of a *service-episode* are defined in the corresponding visit modules.

6.2.4.2 Imaging service request

An *imaging service request* is a set of one or more requested procedures selected from a list of *procedure-types*. A *procedure-type* identifies a class of procedures.

In imaging services, a *procedure-type* is an item in a catalog of imaging procedures that can be requested by a physician during a visit and reported in an imaging-service facility. An instance of a procedure-type typically has a name and one or more other identifiers. Specification of a procedure from a procedure-type does not prescribe protocols to be followed.

A *procedure-type* is associated with one or more pre-designed *procedure-plans*. A *procedure-plan* means execution of one or more specific protocols. A *requested procedure* must be done following one *procedure-plan,* and the protocols specified in the plan.

An *imaging-service request* is submitted by an authorized imaging-service requester to an authorized imaging-service provider in a *service-episode*. An *imaging-service request* is associated with one or more visits, occurring within the same *service-episode*. An *imaging-service request* results in one or more *imaging-service reports* distributed to one or more destinations.

6.2.4.3 Requested-procedure

A *requested-procedure* is the smallest unit of service that can be requested, reported, coded and billed. Performance of an instance of a *requested-procedure* is specified by exactly one *procedure-plan* (see Section 6.2.4.5). The *procedure-plan* is defined by the imaging-service provider based on the *procedure-plan templates* associated with the procedure-types needed for the study. A *requested-procedure* leads to one or more scheduled procedure-steps involving protocols specified in the *procedure-plan*. A *requested-procedure* may be associated with many imaging-service requests, and involve many imaging devices and modalities.

6.2.4.4 Modality scheduled procedure-step

A *modality scheduled procedure-step* (*MSPS*) is a scheduled unit of service specified by the *procedure-plan* (see Section 6.2.4.5) of a *requested procedure*. It prescribes a *protocol* identified by one or more protocol codes. It involves a set of medical devices and modalities as described in Chapter 5, anesthesia equipment, surgical equipment, vital signs monitoring equipment and even transportation equipment. It may also involve the engagement of human resources, specify consumable supplies and indicate specific location and time: start-time, stop-time, duration for the requested procedure.

In *imaging services*, the performance of an instance of an *MSPS* may include only a general designation of imaging modality. However, it may involve multiple images, which could be satisfied by multiple instances of the same equipment-type. A *procedure-step entity* is provided to support management of the logistical aspects of procedures (e.g., material-management, human resources and scheduling). The full description of the contents of *procedure-steps* and *protocols* is implementation dependent.

An *MSPS* may contribute to more than one *requested-procedure*. For example, an intravenous iodine contrast injection might be shared by an intravenous pyelogram and a CT-examination. However, for billing, an instance of an *MSPS* is typically considered a part of *single requested-procedure*.

6.2.4.5 Procedure-plan

A *procedure-plan* is a specification that defines a set of *protocols* to perform the *scheduled procedure-steps* of a *requested-procedure*. Each *scheduled procedure-step* is performed according to a single *protocol* identified by one or more *protocol-codes* described in a *defined procedure-protocol*. The actual protocols performed during a *procedure-step* may be recorded in a *performed procedure-protocol* and may differ from those prescribed in the related *procedure-plan*. It is important to record them separately. Comparison of *actual protocols* versus the *protocols* in a *prescribed procedure-plan* is an important element of quality-control audit.

6.2.4.6 Protocol

A *protocol* is a specification of actions to perform a *scheduled procedure-step*. A *scheduled procedure-step* contains only one *protocol* identified by one or more *protocol-codes*. A *protocol* may be specified by a *defined procedure-protocol*. A *protocol* is documented as the *performed procedure-protocol* after a *procedure-step* has been performed.

A *defined procedure-protocol* describes a set of parameters, corresponding values, constraints associated with the parameters (such as acceptable ranges) and other details associated with the prescribed action. A *defined procedure-protocol* is generic and is not associated with any patient or *scheduled procedure-step*. However, it may contain parameters specific to a model, device-version, or generic parameters common to multiple device models. A *defined procedure-protocol* also includes the information such as the clinical purpose, indications and appropriate device models.

A *performed procedure-protocol* encodes the parameter values used. It is always associated with a specific patient and the *performed procedure-step*. The *performed procedure-protocol* may reference the *defined procedure-protocol* on which it is based. However, it does not record the original constraints and their satisfaction.

6.2.4.7 Performed procedure-step

A *requested-procedure* maps into multiple *scheduled procedure-steps*. A *scheduled procedure-step* has many-to-many mappings to *performed procedure-steps*. A *performed procedure-step* (*PPS*) is a unit of performed service. It may not be the same as the *scheduled procedure-step*. Logically, a *scheduled procedure-step* maps to a *performed procedure-step*. However, there may not be one-to-one correspondence between the two. Actual conditions may cause substitutions, alterations, splitting a *scheduled procedure-step* into multiple *performed procedure-steps*, or fusion of two or more *scheduled procedure-steps* into one *performed procedure-step*.

For example, two or more *scheduled procedure-steps*, generated by different referring physicians may be satisfied by a single *PPS* at the discretion of an operator. A single scheduled procedure-step may be satisfied by multiple performed procedure steps on different types of medical devices, due to clinical need or the availability of equipment.

A *PPS* also contains information describing the *performed procedure*. This information is represented by the *performed protocol* defined by one or more *protocol-codes* and the corresponding parameters. The *PPS* also contains information about its state: in progress, discontinued or completed.

A *modality performed procedure-step* (*MPPS*) is a *PPS* that results from an activity such as the acquiring images of a patient (or other imaging subject) using an imaging modality, as described in Chapter 5. It contains information describing the performance of a step of an imaging procedure, including data about the performance of the procedure itself, and data for billing and material-management. The *MPPS* refers to zero or more *image-series* and other composite *SOP* instances created as part of the *performed procedure-step*.

While an *MPPS* represents a unit of service within a workflow, the specification of the workflow is beyond the scope of the standard. An MPPS does not identify or control any subsequent activities to be performed. For example, a modality may create both "for processing" images for automated analysis and "for presentation" images for human review from the same acquisition. The standard does not specify whether the production of these is a single unit of service, or two. A single *MPPS* instance could list both the "for processing" images and the "for presentation" images, regardless of storage semantics. Alternatively, the modality may treat these two sets of images as two separate units of service, and send two separate *MPPS instances.*

Similarly, A *radiation dose SR* from the irradiation events of an acquisition could be referenced in the same *MPPS-instance* as that of the acquired images, again irrespective of where such a radiation dose SR might be sent, if at all. Alternatively, the modality may treat the production of the *radiation dose SR* as a separate unit of service, and report it in a distinct *MPPS.*

Acquisition MPPS cannot be modified once computed. After an acquisition MMPS is completed, the subsequent instances will necessarily be referenced in a new *MPPS* instance. For example, consider the case of thin-and-thick slice CT-images acquired from the same acquisition (raw) data. When the reconstruction of both sets of images is prospectively defined and automatically initiated by the protocol selection, then both sets are referenced from a single MPPS instance. However, if the reconstruction of one or the other set is performed retrospectively by manual intervention after the acquisition *MPPS* had been completed, a new MPPS has to be used.

The completion of an *MPPS* is a significant event that triggers or enables downstream activity. However, MPPS does not require the modality to be configured to manage such activity. The *units of service* that the modality describes in an *MPPS,* and how the modality relates those *performed procedure-steps* to *scheduled procedure-steps,* are implementation decisions beyond the scope of the standard. The *IHE radiology scheduled workflow* profile guides the implementation.

An *MPPS* may describe acquired instances that may never be stored. For example, a modality may have the capability of storing a CT acquisition as multiple single-frame CT-image storage *SOP* instances, as a single multiframe enhanced CT-image storage *SOP* instance, or as several enhanced CT-image storage *SOP* instances that together comprise a *concatenation.* An *MPPS* may describe all three possibilities, even though only one mode is used, depending on the negotiated capabilities of the storage recipient. Alternatively, separate *MPPS* instances could be used for different storage *SOP* classes. An *MPPS* contains only the instances that a modality creates, not the instances converted and created subsequently in response to a query (e.g., during legacy conversion). The *MPPS* is not a substitute for, nor is equivalent to, a storage commitment request, nor an instance availability notification.

6.2.4.8 Clinical document

A *clinical document* contains clinical observations and services and is part of a patient's medical record. *Clinical documents* are a subclass of *healthcare structured documents.* Structured documents are not necessarily related to a specific patient. *Structured documents* may be used for imaging procedure operational instructions such as product labeling, procedure-plans and patient-care plans. The format and semantics of *structured documents* are outside the scope of the DICOM standard and are generally part of HL7. *Clinical documents* may be implicitly associated with *service-episodes, service-requests, requested-procedures* or other entities related to a patient's treatment. Such associations are not explicitly modeled.

The core text in a clinical document has several important characteristics: 1) persistence – it must remain unaltered for a period defined by the local and regulatory requirements; 2) stewardship – it is maintained by an authorized organization entrusted with its care; 3) authentication mechanism – it is legally authenticated; 4) context – it must establish default context for its contents; 5) wholeness – authentication should apply to the whole document and 6) human readability – it must contain additional information to facilitate human comprehension. *Clinical documents* provide significant context for performing imaging and related procedures. The context information about a patient is: 1) clinical history, 2) lab-test results and 3) advance medical directives.

6.2.5 Representation of Information Entities

The identification of the entities in an E-R diagram is the first step to define a data abstraction for recording, and organizing the information needed for the elements. The second step of the abstraction process is to identify various attributes of these entities. The *attributes* have specific semantics. The attributes-value pairs define an information object – an instance of an *IOD*. Many information objects share similar types of attributes. For example, both doctor and patient are humans, and should have many common attributes associated with humans such as name, contact information, and address, etc.

DICOM dictionary defines over five-thousand attributes to capture information for the delivery of *medical imaging services* in a modern healthcare system. These attributes are logically grouped into *attribute-modules*. An *attribute-module* has two major fields: 1) attribute-name and 2) tag – a pair of information-fields. The first field shows encoding for the corresponding attribute-type. The second field encodes the specific attribute.

Example 6.3: Information Associated with a Patient

Tables 6.1–6.3 show the attribute-modules about a patient's identification, patient's profile, patient's medical attributes and attributes related to patient's visit to healthcare organizations. Table 6.1 illustrates the patient identification module – group of the attributes such as patient's name, patient's id, patient's birth-name, information about his parents, and photograph. Table 6.2 illustrates the grouping of attributes for patient's profile module. This module focuses on attributes related to the patient's profile during admission such as: 1) birth-date, 2) birth-time, 3) gender, 4) weight, 5) size, 6) occupation, 7) military service, 8) ethnicity, 9) religious preference, 10) country of residence, etc. Table 6.3 groups attributes for the patient medical information module. The module focuses on the medical knowledge of the patient such as: 1) medical alerts to inform about drugs' side-effect, contagious conditions and any significant evocable disorder; 2) allergies; 3) smoking status; 4) pregnancy status; 5) last menstrual date and issues; 6) body-mass index (BMI); 7) other special needs such as wheelchair, oxygen, translator, etc. and 8) the patient's current state such as comatose, disoriented or vision impaired.

The key difference between Tables 6.1 and 6.2 is the permanence of the attributes in Table 6.1. Some attribute-values in Table 6.2 change with time. For animals, additional attributes related to the species description (i.e., genus, subgenus, species or subspecies) and breed, are added to the module in Table 6.2. The grouping of attributes is based upon functionality, and is flexible. Same attributes (such

TABLE 6.1   Patient's identification module

ATTRIBUTE NAME	TAG
Patient's name	(0010,0010)
Patient-ID	(0010,0020)
Other patients' IDs	(0010,1000)
Type of patient-ID	(0010,0022)
Patient's birth-name	(0010,1005)
Patient's mother's birth-name	(0010,1060)
Patient photo	(0010,1100)
......	....

TABLE 6.2 Patient's profile module

ATTRIBUTE NAME	TAG
Patient's age	(0010,1010)
Occupation	(0010,2180)
Confidentiality constraint	(0040,3001)
Patient's birth-date	(0010,0030)
Patient's birth-time	(0010,0032)
Patient's gender	(0010,0040)
Patient's language(s)	(0010,0101)
Patient's size	(0010,1020)
Patient's weight	(0010,1030)
Military rank	(0010,1080)
Country of residence	(0010,2150)
Ethnic group	(0010,2160)
Patient's religious preference	(0010,21F0)
Responsible person	(0010,2297)
Patient species description	(0010,2201)
Patient breed description	(0010,2292)
Strain description	(0010,0212)
…	…

TABLE 6.3 Patient medical module

ATTRIBUTE NAME	TAG
Medical alerts	(0010,2000)
Allergies	(0010,2110)
Smoking status	(0010,21A0)
Pregnancy status	(0010,21C0)
Last menstrual date	(0010,21D0)
Body mass index	(0010,1022)
Special needs	(0038,0050)
Patient state	(0038,0500)
…	…

as name and patient-id) can be present into multiple modules. The attributes in multiple patient-related modules can be joined to create an inclusive module containing the union of attributes.

Example 6.4: Information Associated with a Visit

Consider modules for the entity *visit*. Table 6.4 illustrates the grouping of attributes for the module *visit identification* module. This module contains the information about patient's schedule to visit a health-care organization. However, he is yet to go through the imaging-service. The module has attributes such as institution-name, institution-code, referring physician-id, service-id and service-description. There is a separate module to track the progress of a visit. Table 6.5 shows the attributes for the *visit status module*. This module groups attributes such as status-id, patient current-location, patient's institution and visit comments. The attribute *status-id* has values such as *not yet scheduled*, *not yet admitted*, *admitted* or *discharged*, etc. The attribute patient's current-location has values such as: 1) residence ward; 2) floor, room, bed and 3) outdoor.

TABLE 6.4   Visit identification module

ATTRIBUTE NAME	TAG
Institution name	(0008,0080)
Institution-id	(0008,0082)
Referring physician-id	(0038,0010)
Service-id	(0038,0060)
Service-description	(0038,0062)
...	...

TABLE 6.5   Visit status module

ATTRIBUTE NAME	TAG
Visit status-id	(0038,0008)
Patient's current-location	(0038,0300)
Patient's institution	(0038,0400)
Visit comments	(0038,4000)
...	...

Example 6.5

After a patient is admitted, the information pertaining to the admission is grouped into a *visit admission module* as illustrated in Table 6.6. The module has attributes such as referring physician's name, referring physician's id, referring physician's address, referring physician's phone, consulting physician's name, admitting diagnosis' description, admitting diagnosis' codes, admitting date and time, etc.

Example 6.6

Table 6.7 shows a subset of grouped attributes for a *general equipment module*. The attributes are: 1) manufacturer, 2) administrative institution name (hospital, clinic etc.), 3) station name, 4) model name, 5) device serial-number, 6) software versions and 7) date and time of the last calibration, etc. *Specific equipment modules* store many additional attributes specific to the modality and the type of instrument. After collecting raw image-data, many equipment-related attributes are needed for interpreting the data.

Example 6.7

Tables 6.8 and 6.9 illustrate *specific equipment modules* for a cathode-ray tube (CRT) and a computed tomography (CT) equipment, respectively. Table 6.8 has *computer radiography (CR) series module* attributes: body-part examined, view-position, filter-type, collimator/grid-name, focal spot(s), plate-type, phosphor-type, etc. Table 6.9 describes *CT image module* attributes: image-type, samples-per-pixel, allocated-bits, stored-bits, most significant bit, data collection diameter, reconstruction diameter, source-to-detector distance, source-to-patient distance, gantry/detector tilt, table-height, rotation-direction, exposure-time, etc. Both tables mostly have equipment-related attributes describing parameter-settings in a specific equipment.

TABLE 6.6   Visit admission module

ATTRIBUTE NAME	TAG
Referring physician's name	(0008,0090)
Referring physician's phone	(0008,0094)
Consulting physician's name	(0008,009C)
Admitting diagnoses codes	(0008,1084)
Admission date	(0038,0020)
...	...

TABLE 6.7   General equipment module

ATTRIBUTE NAME	TAG
Manufacturer	(0008,0070)
Administering institution name	(0008,0080)
Station name	(0008,1010)
Manufacturer's model name	(0008,1090)
Device serial number	(0018,1000)
Software versions	(0018,1020)
Gantry id	(0018,1008)
Spatial resolution	(0018,1050)
Date of last calibration	(0018,1200)
Time of last calibration	(0018,1201)
…	…

TABLE 6.8   CRT series module

ATTRIBUTE NAME	TAG
Body-part examined	(0018,0015)
View position	(0018,5101)
Filter type	(0018,1160)
Collimator/grid name	(0018,1180)
Focal spot(s)	(0018,1190)
Plate type	(0018,1260)
Phosphor type	(0018,1261)
…	…

TABLE 6.9   CT image module

ATTRIBUTE NAME	TAG
Image type	(0008,0008)
Samples per pixel	(0028,0002)
Bits allocated	(0028,0100)
Bits stored	(0028,0101)
Most significant bit	(0028,0102)
Data collection diameter	(0018,0090)
Reconstruction diameter	(0018,1100)
Source to detector distance	(0018,1110)
Source to patient distance	(0018,1111)
…	…

6.2.6 Information Objects

An *information-object (IO)* is a higher-level abstraction and the basic unit of information transfer. Each *information-object definition (IOD)* is a logical collection of modules, which are transmitted or received under a *medical-service context*. *Medical-service* stores a medical image from the capturing device to a PAC, searches patient-specific information in a PAC, and prints an image.

TABLE 6.10 Radio therapy (RT) plan *IOD*

INFORMATION ELEMENT	MODULE	USAGE
Patient	Patient	M
	Clinical-trial subject	U
Study	General study	M
	Patient study	U
	Clinical-trial study	U
Series	RT-series	M
	Clinical trial series	U
Frame-of-reference	Frame of reference	U
Equipment	General equipment	M
Plan	RT general plan	M
	RT prescription	U
	RT tolerance tables	U
	RT patient setup	U
	RT fraction scheme	U
	RT beams	C
	RT brachy application setups	C
	Approval	U
	General reference	U
	SOP common	M
	Common instance reference	U

DICOM has over *100* IODs. Table 6.10 shows a standard IOD. The first column shows the information elements. The second column shows the modules in the information elements. It has 20 modules connecting six information elements: patients, study, series, frame-of-reference, equipment and plan.

All modules are not mandatory under a service context. The third column shows the mandatory or optional nature of these modules in a transmitted IOD. Mandatory type (M) modules must be present in an IOD. Conditional (C) modules are present if certain condition is met. User-defined (U) modules are optional.

6.2.6.1 Concatenation

IODs can be very large, and it may not be practical to transport them as single object. To maintain practical limits on the maximum size of an individual SOP-instance, large IODs such as content of a multi-frame images are split into more than one SOP-instance. These SOP-instances together form a concatenation, which is a group of SOP-instances within a series. The value of the concatenation UID is the same for all parts. The receiver-end can reassemble them as one object by looking at the concatenation UID-value.

6.2.6.2 Dimension organization

Objects are often grouped naturally. For example, a volumetric representation can be made by a series of CT-slices. Each series represents a two-dimensional planer image. In such a case, only the z-depth attribute changes among these objects. *Dimension organization* allows a way to order grouped objects. It is defined as a set of ordered attributes that change on a per-frame basis in a predetermined manner before an image is acquired. Dimension organization is set by the capturing-application, which generates the IOD. The IOD contains a list of attributes designated as dimension-item and is intended for end-application. When there are multiple dimension attributes, the first item of a dimension-index sequence shall be the slowest varying index. Other attributes among these groups of objects may also change on a per-frame basis, but if they are not listed in the dimension organization, they are not considered significant for the organizational purpose.

6.2.6.3 Types of data elements

The IODs and modules are extensive and list almost an exhaustive set of attributes. The modules and the attributes support a wide variety of current and conceivable future services, applications, and scenarios. Every application may not require values for all the attributes. However, an application requires a minimum number of values for the attribute-fields. It is important to specify requirement constraints on the data-elements. Medical imaging communication defines three major types of data-elements: 1) type-1 data-element, 2) type-2 data-element and 3) type-3 data-element.

The value-field of a *type-1 data-element* must contain a valid data. The length of the value-field cannot be zero. The absence of a valid value in a *type-1 data-element* is considered a protocol violation. For compound data-elements such as sequence, multiple values are allowed. The presence of a single value or item satisfies the *value* requirement, unless specified otherwise in the attribute description. Other value-fields may be empty, unless specified otherwise by the IOD. The presence of one or more delimiter alone, without any value, cannot satisfy the *type-1* requirement. It is a protocol violation if the specified conditions are met and the *type-1 C data-element* is not included.

Type-2 data-elements are also mandatory. The value-field must contain the known value as defined by the elements VR (value representation) and VM (value multiplicity). However, it is permissible to encode an unknown value with zero length, and no entry in the value-field. The absence of value is treated as a protocol violation. The intent of *type-2 data-elements* is to allow a zero length to be conveyed when the operator or application does not know its value. A *type-2 sequence data-element* will contain zero or more items, as defined by the IOD. *Type-2C data-elements* (*type-2 conditional data-elements*) have the same requirements as *type-2 data-elements* if and only if specified conditions are true. It is a protocol violation if the specified conditions are true, and the data element is not included.

Type-3 data-elements are optional. The absence of a *type-3 data-element* from a data set conveys no significance and is not a protocol violation. *Type-3 data-elements* may also be encoded with zero length and no value. The meaning of a zero-length *type-3 data-element* is the same as that element being absent from the data-set.

6.3 NETWORK ENCODING AND COMMUNICATION

Each *IOD* represents a meaningful object or process in the medical service world. These *IODs* and the corresponding information objects are eventually represented as binary strings, and packed as a message. During a communication, the sending-end must encode the data using a standardized scheme. The receiving-end uses a standardized scheme to decode the binary string.

6.3.1 Network Encoding Levels and Formats

The objective of network encoding is to convert the IODs and its modules into a binary bit-stream. Network encoding requires three sets of techniques: 1) schema for basic encoding; 2) schema for compound encoding and 3) schema for semantic encoding.

The first level is the base encoding techniques. It encodes basic symbols such as integers, floating-point numbers, alphabet sets from world languages and special characters into binary representations. The three key design considerations at this level are: 1) the order of bits in a computer memory; 2) length of a code and 3) a symbol-table for the lookup of symbols.

The second encoding level defines the composite objects such as sequence, tables, multidimensional matrices and nested data-elements. Examples of sequences are: alphanumerical strings and lists of numbers. The three key design considerations at this level are: 1) order of symbols; 2) variable length of a code 3) and compression (encoding) of large data-items. DICOM supports all well-known image formats developed and accepted by medical imaging community. Medical images and videos are compressed using *JPEG, JPEG2000, MPEG2, MPEG7*, etc. New compression standards will be progressively adopted by DICOM.

The third encoding level highlights the semantic components of an attribute-value. The attribute-values should accommodate different schemes used by the equipment vendors, hospitals and cultures. For example, a personal name is encoded differently by different cultures. There are many conventions of specifying names around the world. In the medical imaging world, a personal name is encoded as a character string using five parts: 1) family name, 2) given name, 3) middle name, 4) name prefix and 5) name suffix. In encoding these, the order should be exact, and adjacent parts should be separated by a delimiter "^" (05EH). Any of the five components may be a null string. Multiple entries are permitted in each component and are encoded as natural text strings, in the format preferred by the named person.

For naming animals brought for veterinary services, the same convention is reused. However, the first two of the five components in their order of occurrence are: 1) responsible party's family name or responsible organization's name; and 2) patient name. The remaining components are not used, and are omitted.

Example 6.8

The name "Dr. Javed Iqbal Khan, Professor, Computer Science" is encoded as "Khan^Javed^Iqbal^Dr. Professor, Computer Science," indicating one family name "Khan"; one given name "Javed"; one middle name "Iqbal"; one prefix "Dr."; and two suffixes: "Professor" and "Computer Science."

Example 6.9

A dog named coco belonging to "Bansal" family will be encoded as "Bansal^Coco." A horse name "Jet Runner" belonging to the organization "Kaisar Farms" will be encoded as the encoding "Kaisar Farms^Jet Runner."

Generally, a name is written ideographically for printing purpose only. Many names are pronounced differently and have a different phonetic representation. To incorporate phonetic representation of names, the name representation has been extended to include three components: 1) an alphabetic representation; 2) an ideographic representation and 3) a phonetic representation. Each component is separated by the delimiter "=" (3DH). Any component group may be absent, including the first component group.

A date is represented as a string of characters of the format YYYYMMDD. It has three components: 1) YYYY contains four digits of a year; 2) MM contains two digits of a month, and DD contains two digits for a date in the Gregorian calendar system. For example, the string "20170822" represents the date August 22, 2017.

Time is represented as a concatenated string in the format: YYYYMMDDHHMMSS. FFFFFF&ZZXX. It has four types of information: date; time within a date, including a fraction of a second up to one-millionth; and coordinated universal time. *Coordinated universal time* is a common 24-hour clock across the world, and is maintained using an atomic clock. UTC offsets are calculated as "local time minus UTC." The offset for a date time value in UTC shall be *+0000*.

The components of the time-string, from left to right, are: 1) YYYYMMDD describing Gregorian calendar date; 2) HH – two digits of the hour (range *00–23*); 3) MM – two digits of the minute (range *00–59*); 4) SS – two digits of the second (range *00–59*); 5) FFFFFF – fractional part of a second up to one-millionth of a second (range *000000–999999*) and 6) & ZZXX – an optional suffix for offset from the *coordinated universal time* (UTC), where & = "+" or "–" ZZ = hours of offset and XX = minutes of offset.

A 24-hour clock is used. Midnight is denoted as *0000*. The fractional second component contains one to six digits. The string is padded with trailing SPACE characters. Leading and embedded spaces are not allowed. A component omitted from the string is termed a null component. Trailing null components of date indicate that the value is imprecise. The YYYY component shall not be null. Nontrailing null components are prohibited. The optional suffix is not considered as a component. A time-value without the optional suffix is interpreted to be in the local time-zone of the application creating the data-element, unless explicitly specified by the time-zone offset from UTC (*0008, 0201*).

Age is encoded as a string of characters with one of these formats: *nnnD, nnnW, nnnM, nnnY*. Here *nnn* contain the number of days for D, weeks for W, months for M or years for Y, for example: "018M" would represent an age of 18 months. Medical imaging needs to be precise about time, and data can move across different time-zones.

6.3.2 IOD Encoding

Medical imaging systems communicate a sequence of *information objects*. An *information object* is represented as a *data-set* of an ordered sequence of *data-elements*. Each *data-element* contains the encoded value of an *attribute* of the corresponding information object. Each data-element has four fields: *tag*, *VR-field*, *value-length* and *value-field*.

A *tag* is a number that uniquely identifies a data-element whose semantics is defined by the dictionary of attributes. The *VR-field* identifies the representation scheme to encode the value-bits. The field *value-length* identifies the length of the encoded *value bit-stream* (excluding the tag, VR-field and value-length field) in bytes. The *value-field* is the bit-stream with the actual value of the attribute. Figure 6.6 illustrates the overall organization of a data-set. There are many types of data-elements.

The *data-elements* in a data-set are normally packed in an ascending sorted order of tag numbers. *Data-elements* can be nested. The VR-type *SQ* provides a flexible encoding scheme for representing simple structures of repeating sets of data-elements. The *VR-type SQ* (sequence) represents zero or more items. *SQ* data-elements can also be used recursively to contain multilevel nested structures. A subset of data types is illustrated in Table 6.11.

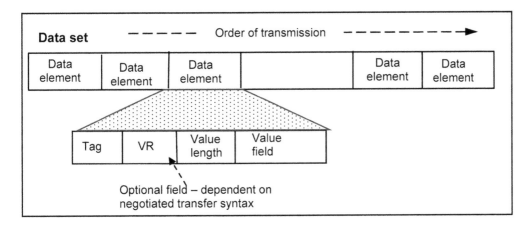

FIGURE 6.6   Organization of data-elements

TABLE 6.11   Datatypes of the data-elements in VR-field

ABBREVIATION	DESCRIPTION	ABBREVIATION	DESCRIPTION
AE	Application entity	OD	Other double
AS	Age string	OF	Other float
AT	Attribute tag	OL	Other long
CS	Code string	OW	Other word
DA	Date	PN	Person name
DS	Decimal string	SH	Short string
DT	Datetime	SL	Signed long
FL	Floating point single	SQ	Sequence of data-items
FD	Floating point double	SS	Signed short
IS	Integer string	ST	Short text
LO	Long string	TM	Time
LT	Long text	UC	Unlimited characters
OB	Other byte	UI	Unique identifier

6.3.3 Communication Services at Application Level

Medical applications can be very complex. The medical applications can use their local data-models, internal database technology and design to keep and process the medical information as they want. However, the data must be converted to the standardized common format before two such medical applications can exchange information using DICOM communication services. Communication service in DICOM requires information-exchange framework besides data-formatting and packet definitions to exchange IODs between two application entities (AEs) reliably and efficiently.

Figure 6.7 illustrates an information-exchange mechanism between two AEs: AE_1 and AE_2. These AEs are called *peer service-users*. An application entity can play either of the two roles: 1) an *invoking* entity and 2) a *performing* entity with respect to a data-exchange service request. Each information-exchange service requires many micro-steps called *primitives*. Except for very few services classes (such as ABORT) all application-level services must go through the full cycle of request-indication-response-confirmation steps.

There are also two corresponding gatekeeping agents SP_1 and SP_2 corresponding to these application service endpoints (AE_1 and AE_2). In Figure 6.7, the gatekeeper SP_1 is on the side of the invoking application entity AE_1, and the gatekeeper SP_2 is on the side of the performing application entity AE_2.

To complete a typical service, the following four sequential micro-steps occur in sequence: i) request, ii) indication, iii) response and iv) confirmation. The invoking service-user AE_1 issues a *request primitive*

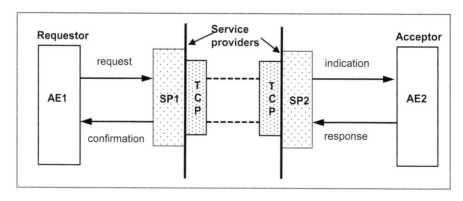

FIGURE 6.7   A schematic for association establishment

to the service-provider SP_1, intended for the performing service-user AE_2. The service-provider SP_1 receives the *request primitive*, and issues an *indication primitive* to the service-provider SP_2, intended for the performing service-user AE_2. The application-entity AE_2 receives the *indication primitive* from the service-provider SP_2 and performs the requested-service. The performing application entity AE_2 issues a *response primitive* to the service-provider SP_2, intended for the service-requester AE_1. The invoking side service-provider SP_1 receives the *response primitive* from the performing-side service-provider SP_2, and issues a *confirmation primitive* to the invoking service-user AE_1. The invoking service-user AE_1 receives the *confirmation primitive* from the service-provider SP_1 completing the *DIMSE service* (DICOM Message Service Element). A typical *application-level service* (ALS) goes through all these four stages without fault.

An authentication service may result in an acceptance or the denial of authentication. Even the denial (just like the acceptance) must be communicated by a *response primitive* and delivered back to AE_1 via a *confirmation primitive*. There are two types of services: 1) confirmed type and 2) unconfirmed type. In *confirmed type* services, the requesting service-user AE_1 waits for the confirmation from the performing service-user AE_2. In an *unconfirmed type service*, the requesting service-user AE_1 does not wait to receive a confirmation from the performing service-user AE_2 after a service request.

DICOM services are divided in five different classes: 1) Association Service - connects two application entities; 2) Data transmission services - performs information-exchange between two applications; 3) search-and-retrieve service - matches a query with the data-elements in the storage and retrieves the data elements; 4) storage services - an application requests to store the data-elements for some duration; and 5) verification service - verifies the identity of the requesting application. These services use DICOM Messaging Services (DIMSE) that has a kernel set of commands to search, store, move, transmit and verify data-elements.

Unlike lower-level communication services such as UDP or TCP, DIMSE framework supports communication suitable for deeper semantics that can be requested by one medical application entity to another medical application entity.

Even, before performing information-exchange between two applications, two applications (service-requester and service-provider) must be associated with each other as described in Section 6.3.5.

6.3.4 Types of DIMSE (DICOM Messaging Service)

There are two types of DIMSE services: 1) DIMSE-N for normalized SOPs and 2) DIMSE-C for composite SOPs. Normalized IODs are single entities, while composite entities include various attributes and contexts as described in Section 6.2. Both DIMSE-N and DIMSE-C are confirmed services, and confirmation from the performing service-user is required. There are 11 services as listed in Table 6.12. Five services are for *composite data-elements* and six are for *normalized data-elements*.

Five services for composite data-elements are: 1) storing (*C-STORE*) composite IOD objects, 2) retrieving (*C-GET*) composite IOD objects, 3) moving composite IOD objects from a service-user to a third-party (*C-MOVE*), 4) finding attribute-values by matching against the attributes (*C-FIND*) in IOD objects and 5) verifying the information (*C-ECHO*).

TABLE 6.12   DIMSE services

DIMSE-C GROUP		DIMSE-N GROUP	
NAME	TYPE	NAME	TYPE
A-ASSOCIATE	meta	N-EVENT-REPORT	notification
C-STORE	operation	N-SET	operation
C-GET	operation	N-GET	operation
C-MOVE	operation	N-ACTION	operation
C-FIND	operation	N-CREATE	operation
C-ECHO	operation	N-DELETE	operation

Six services for single entities are: 1) modification of an information in one of the peer service-users (*N-SET*), 2) retrieving information (*N-GET*), 3) invoking an action (*N-INVOKE*), 4) creating a *normalized IOD object* (*N-CREATE*), 5) deleting an *IOD-object* (*N-DELETE*) and 6) reporting a *SOP-event* (*N-EVENT-REPORT*).

6.3.5 Association Service

A connection should be established between the two AEs before they can invoke any service for information-exchange. Establishment of this connection is called *association service*. The process of association achieves three goals: 1) it introduces and identifies peers, including support for mutual authentication at the application level; 2) it helps exchanging information about the supported commands and medical objects by each party and 3) it helps AEs to decide the specific encoding of the objects.

An *association* is analogous to "transport" (such as UDP or TCP) between two end-points. However, it is specific to a group of standard applications. A typical list of association commands includes: 1) associate; 2) release; 3) abort; 4) A-P-abort and 5) process data. After starting a TCP/IP connection, applications negotiate the association-parameters to agree what can be done during the association, send the DICOM objects, close the association, and then close the TCP/IP connection.

The establishment of an *association* is performed by an *A-ASSOCIATE* service that uses four steps: *request, indication, response* and *confirmation*. The initiator of the service is called a *requestor*, and the service-user that receives the *A-ASSOCIATE* indication is called the *acceptor*.

6.3.5.1 DICOM upper layer protocol

An information-exchange between *AEs* is done using common message formats called *PDUs* (Protocol Data Units). It consists of protocol control information and user data. *PDUs* have two types of fields: 1) mandatory fixed fields, and 2) variable fields containing the data-items. The DICOM upper layer has seven PDUs: 1) *A-ASSOCIATE-RQ*; 2) *A-ASSOCIATE-AC-PDU*; 3) *A-ASSOCIATE-RJ PDU*; 4) *P-DATA PDU*; 5) *A-RELEASE-RQ PDU*; 6) *A-RELEASE-RP PDU* and 7) *A-ABORT PDU*. There are three A-ASSOCIATE services to establish an association between two AEs, two A-RELEASE services to release the association after the successful completion of information-exchange or action, and one A-ABORT service to abort a data-transfer.

An *association* is negotiated by a series of *A-ASSOCIATE* commands, each carrying a specific type of information to help the negotiation. After an association is established, the *P-DATA* service is used to perform the medical information related services by exchanging the *SOP commands* and related *IOD* objects. After the information-exchanges are completed, *A-RELEASE* service is used by the requesting application (*AE₁*) to end the association gracefully, and close the TCP-connection. Only the requestor *AE₁* initiates the *A-RELEASE* request after receiving *confirm primitive* for all the invoked operations. The requestor *AE₁* issues no further primitives other than an *A-ABORT request primitive* until it receives an *A-RELEASE confirmation primitive*.

An association is aborted if some error is detected by the *AEs*. An error can be unrecognized, unsupported encoding, or sudden loss of application-level resources. An *A-ABORT* command is used to abort the association. The command carries only one parameter identifying the source originating the request. It is an *unconfirmed service* and the *AE* receiving the *A-ABORT indication* need not issue the response. Similarly, *AE* invoking the request does not wait to receive the *confirmation*. Unexpected problems can also occur at lower levels. The service provider SP_1 or SP_2 issues *A-P-ABORT* primitive, if fatal errors are detected by the service-providers. The primitives originate at the service-providers' level, and propagate in two opposite directions as shown in Figure 6.8.

The *A-P-ABORT* message carries a reason parameter to indicate problems such as an unrecognized or unexpected *PDU* (Protocol Data Unit), an unexpected session-service primitive, unrecognized or invalid *PDU* parameters, etc. An *A-P-ABORT* message can also have user defined reasons to reflect errors that caused the abort, and originated in the transport-network, data-link, or physical layers. *A-P-ABORT indication primitives* are issued towards both the *AEs* after a service provider (SP_1 or SP_2) detects an internal error. After that, neither *AE* issues any additional primitive for the association.

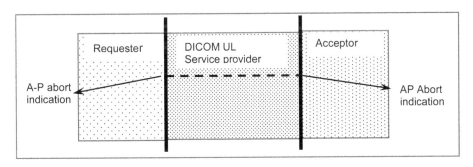

FIGURE 6.8 Termination of an association

6.3.5.2 Application identity

At the beginning of a request, the application entity AE_1 declares the operational context under which it intends to communicate during the session. For example, our two DICOM 3.0 conformant applications declare their *application context name* parameter as "DICOM 3.0." This is done with a UID of DICOM 3.0 "1.2.840.10008." The value *1* in the first field identifies ISO (International Standard Organization); the value *2* in the second field identifies the ISO-member bodies' branch; the value *840* identifies ANSI as the ISO member-body representing the USA; and the value *10008* identifies NEMA. The identified organization accepts the responsibility to properly register these suffixes to ensure uniqueness.

The *requesting application-entity AE₁* specifies its common-name in the field *calling AE title*. It also expresses the common-name of the called application in the field *called AE title*. The application entity AE_2 checks the titles, and sends back its name in *AE-title field* in the response message. The called and responding AE titles should match. The requesting application entity AE_1 (or the corresponding service provider SP_1) also specifies its network address (IP address and TCP port number) in the field *calling presentation address*. The responding application-entity AE_2 (or the corresponding service provider SP_2) provides the *responding presentation-address* field.

A single *application-entity title* can be associated with multiple network-addresses assigned to a single system. A single *application-entity title* can be associated with multiple TCP-ports using the same or different IP-addresses. A single network access-point (IP-address and TCP-port) can support multiple application entity titles. A medical imaging application on a network may support several applications identified by different DICOM *application-entity titles*.

The sub-items' *implementation class UID* and *implementation version-name* enable applications to identify each-other uniquely. Well-established software packages register with the standard's organization and declare their conformance with various profiles managed by the organization. Each registered software gets an *implementation class UIDs* and optionally a version number. It is useful to distinguish between various implementation environments, and maintain interoperability within complex evolving environments. Equipment of the same type or product-line having different serial numbers can use the same *implementation class UID* if they share the same software-environment. Key parameters of association services are described in Table 6.13.

User-identity negotiation is used to notify the association-acceptor about the user-identity of the association-requestor. The association-requestor may also optionally request that the association-acceptor also respond with the server-identity. If this sub-item is not present in the *A-ASSOCIATE request*, then *A-ASSOCIATE response* shall not contain a user-identity in the response.

The association-requester conveys in the *A-ASSOCIATE request* either a username, username with passcode, a time-bound user-authentication based upon standard security protocols and an indication, whether a positive server-response is requested. There are two popular security protocols: *Kerberos* for Windows environment and *SAML* (Security Assertion Markup Language) for the Internet-based usages. A security

TABLE 6.13   Key association service parameters

A-ASSOCIATION PARAMETER NAME	REQUEST	INDICATION	RESPONSE	CONFIRMATION
mode	UF	MF(=)		
Application context name	M	M(=)	M	M(=)
Calling AE title	M	M(=)	M	M(=)
Called AE title	M	M(=)	M	M(=)
User-information(subitems listed)	M	M(=)	M	M(=)
Maximum PDU length (51H)	M			
Implementation class UID	M			
Implementation version name (55H)	optional			
Asynchronous operation window (53H)	optional			
SCP/SCU role selection	optional			
User identity negotiation (58H)	optional			
SOP class extended (56H)	C			
Common SOP class extended (57H)	C			
Result			M	M(=)
Result source				M
Diagnostic			U	C(=)
Calling presentation address	M	M(=)		
Called presentation address	M	M(=)		
Presentation context definition list	M	M(=)		
Presentation context definition list result			M	M(=)
Responding AE title			MF	MF(=)
Responding presentation address			MF	MF(=)
Presentation and session requirement	U	UF(=)	U	UF(=)

Abbreviations: M = mandatory; U = user-defined; C = required if error; MF = mandatory with fixed value; UF = user-defined with fixed value.

protocol has three components: 1) a requestor, 2) an acceptor and 3) a third party both the requestor and the acceptor trust. The third party is used by the requestor to assert the genuineness of its identity-claim.

The association-acceptor does not provide an *A-ASSOCIATE response* unless a positive response is requested, and user-authentication succeeds. An *A-ASSOCIATE response* shall contain a user-identity sub-item. If a *Kerberos ticket* is used for the user-identity authentication, then the response includes a *Kerberos server-ticket*. If the *association-acceptor* does not support user-identification, it will accept the association without making a positive response.

The *association-acceptor* may utilize the *user-identity information* during an association negotiation to: 1) populate the user information fields in the audit-trail messages; 2) perform authorization controls during the performance of other DIMSE transactions on the same association and 3) modify the performance of DIMSE transactions for other purposes, such as workflow optimizations. The *user-identity SOP* conveys user-identity to support usages such as authorization controls and audit controls. It does not specify their behavior.

User-identity authorization controls may be simple "allow/disallow" rules, or they can be more complex scoping rules. For example, a query could be constrained to apply only to return information about patients associated with the identified user. The issues surrounding authorization controls is complex, especially with HIPAA constraints.

The option to include a passcode along with the user-identity enables many non-Kerberos secure interfaces. Sending passwords in the clear is insecure, but there are single use password systems such as *RFC-2289* and the smart-tokens that do not require protection. The password might also be protected by *TLS* (Transport Layer Security) or other mechanisms.

6.3.6 Negotiation

6.3.6.1 Capacity negotiation

The *maximum PDU length*, a sub-item within the user-information item, defined by each communicating *AE*, limits the size of the data for each *P-DATA* (Presentation Data) payload (*P-DATA* indication). Different maximum lengths can be specified for each direction of dataflow on an association. This notification is required.

The item *presentation context definition list* declares the type of required-services and is needed to negotiate the encoding for the objects supported by both sides. Each definition in this list can be a sub-channel within an association to handle one type of *SOP-object*. An association may manipulate multiple types of *SOP-objects*. There should be one presentation context definition for each anticipated *SOP-object*.

Each definition in this list has three components: 1) a *presentation context identification*; 2) an *abstract syntax name* and 3) a list of one or more *transfer syntax names*. The first component is a serial-number assigned by AE_1 to label the subchannel. The *abstract syntax name* is the *UID* of the *SOP-class objects* manipulated by the subchannel. *Transfer syntax* is used to describe the file format and the network-transfer methods. Three main variables in the *transfer syntax* are: 1) implicit or explicit *VR* (Value Representation); 2) *endianism* to represent a data-format: little-endian or big-endian and 3) *pixel-data compression standard*. For explicit VR, the data type code of every element and the byte-order of multi-byte data-types should be ordered.

Example 6.10

The corresponding *VR* for an element with *unsigned short data-type* is *US*. It has two bytes, and reading order of bytes to be inserted in the buffer is specified. Compressed pixel-data transfer syntax is always explicit *VR* little endian.

Table 6.14 lists the most common DICOM *data-transfer syntaxes*. The imaging formats appeared long before DICOM. The medical imaging equipment uses a variety of image-compression standards, such as JPEG, MPEG and TIFF.

TABLE 6.14 Some common transfer syntax

COMPRESSION CLASS	DATA TRANSFER SYNTAX
Uncompressed	1.1 Implicit VR little-endian
	1.2 Explicit VR little-endian
	1.3 Explicit VR big-endian
Lossless compressed	2.1 JPEG lossless
	2.2 JPEG lossless first order
	2.3 RLE lossless
	2.4 JPEG 2000 (lossless)
	2.5 JPEG lossless
Lossy compressed	3.1 JPEG baseline
	3.2 JPEG extended
	3.3 JPEG 2000 (lossy)
	3.4 JPEG – LS (lossy)
MPEG transfer-syntax	4.1 MPEG-2
	4.2 MPEG-4
Special transfer systems	5.1 Deflate
	5.2 JPIP
	5.3 JPIP + deflate

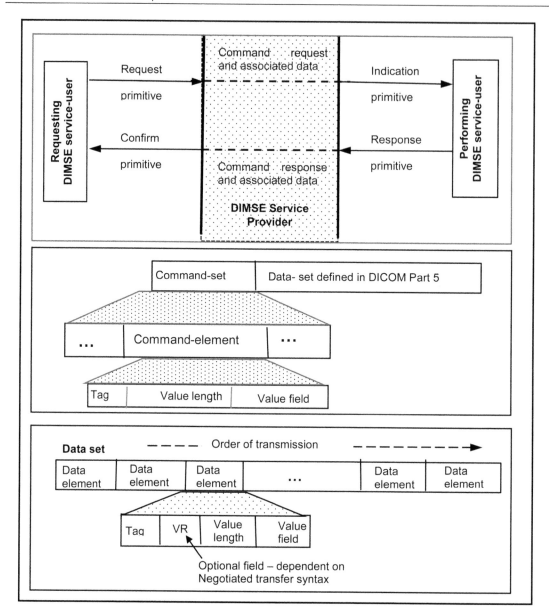

FIGURE 6.9   A schematic for association set up and information exchange

The requestor AE_1 can specify multiple transfer-syntaxes for each *abstract syntax-name*. The acceptor AE_2 accepts a syntax if it supports the named SOP and at least one of the encoding-types. It selects only one *transfer-syntax* for each *abstract syntax*. It returns the accepted value on the presentation context definition list, and the requestor must use that *transfer-syntax*. If the acceptor cannot support the named SOP-class or the listed *transfer-syntax,* it rejects the proposed context. An overall association set-up is described in Figure 6.9.

6.3.6.2 Role selection negotiation

Each of the SOP-class has a specific role for *AEs*. An *AE* plays either the role of a service-user (*SCU*) or a service-provider (*SCP*). By default, the invoking AE_1 plays the role of *SCU*, and the acceptor AE_2 plays the role of *SCP*. However, the roles are negotiable by the peers during association establishment for each SOP service-type. For each *abstract syntax name*, the *association-requester* can optionally propose one of the three roles for itself: 1) *SCU* only, 2) *SCP* only or 3) both *SCU* and *SCP*. The *association-acceptor* either accepts the *association-requester's* proposal by returning the same value or turns down the proposal by returning the value *0* in the *A-ASSOCIATE* response.

6.3.6.3 *Communication window negotiation*

Many services can be requested in one association. These services are performed either in *synchronous mode* or *asynchronous mode*. In synchronous mode, these services are performed in a fixed order. AE_1 in *SCU* role must wait for a response from the *SCP AE_2* before it can invoke any other operation (or suboperation/notification if AE_1 is *SCP* and AE_2 is *SCU*). The overall synchronous mode of communication is illustrated in Figure 6.10.

The arrow denotes the sequential activities: *request for SCP-action 1 → receive SCU-action 1 request by SCP → local SCP-actions for SCU-request → response by SCP for SCU-action request 1 → SCU receives SCP-response → local SCU-actions₁*. The instruction-set *local SCU-actions₁* waits for receiving the response by *SCP*.

In the asynchronous mode, the restriction of serial processing is relaxed. The invoking AE_1, on an established association, may continue to invoke further operations or notifications for AE_2 without waiting for a response. In multiprocessing systems, asynchronous mode speeds up the service. The number of asynchronous operations depends on the available resources.

Figure 6.11 illustrates the asynchronous mode of processing. The *SCU (AE_1)* sends a request to *SCP* for an action. The sequence of actions is: *request for SCP-action 1 → receive SCU-action 1 request by SCP → local SCP-actions for SCU-request → response by SCP for SCU-action request 1 → SCU receives SCP-response*. The instruction set *concurrent local SCU-actions₁* does not wait for receiving the response by *SCP* and does not depend on the sequential set of actions between *SCU (AE_1)* and AE_2.

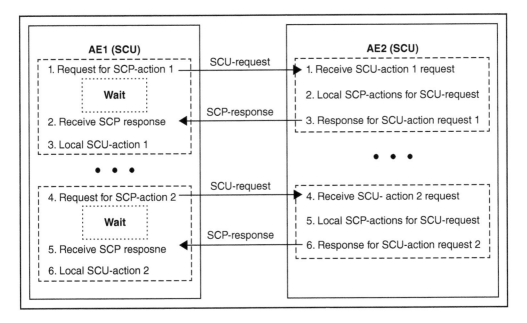

FIGURE 6.10 Synchronous mode of communication

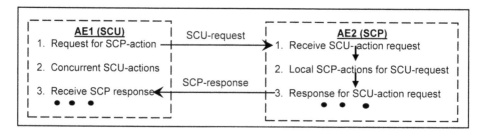

FIGURE 6.11   Asynchronous mode of communication

An association supports asynchronous-operations window to identify a maximum number of supported asynchronous operations. This is done by sending two optional values in *A-ASSOCIATE* to negotiate the maximum number of outstanding operation (for the *SCU* role) and suboperation requests or notifications (for the *SCP* role). The invoking *IE* proposes two values. The *association-acceptor* replies with two values in the *A-ASSOCIATE* response. The values in the response are equal, if it accepts the proposed values. Returned values are reduced if the acceptor has fewer resources. In the absence of a negotiation, the windows default to synchronous mode and returns the value "1." The *association-requester* indicates unlimited operations by sending a value "0." The association-acceptor agrees to unlimited operations by returning a value "0." It negotiates the parameter by conveying a value other than "0."

6.4 VERIFICATION SERVICE

The intention of *verification service* is to occasionally verify the availability of communication between two application end-points in an association. It is similar to the echo mechanism used by other communication protocols. This is implemented using a command *C-Echo*. *Verification* service can be initiated by *SCU* or *SCP*. After receiving a *C-Echo request primitive*, the DIMSE-C protocol machine constructs a message using the *P-DATA* request-service and issues a *C-Echo indication primitive* to the DIMSE-service-performer. The DIMSE-service-performer replies back.

After receiving a *C-Echo response primitive*, issued by the DIMSE-service-performer, the DIMSE-C protocol machine constructs a message conveying the *C-Echo-RSP* and sends the message using the *P-Data* response service. After receiving a message conveying a *C-Echo-RSP* the DIMSE protocol machine issues a *C-Echo* confirmation primitive to the invoking DIMSE-service-user. This completes the *C-ECHO* procedure. Figure 6.12 illustrates the overall verification service using *C-Echo* command. Although it uses P-DATA messages, these messages carry no payload.

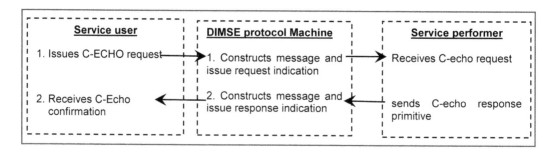

FIGURE 6.12   A schematic of the verification process using C-ECHO DIMSE service

6.5 STORAGE SERVICES

The objective of the storage services is to facilitate the transfer of information instances. It allows an *AE* to send images, waveforms, reports, etc. to another *AE*. Two *AEs* implement a *storage service SOP-class* with one serving in the *SCU* role and the other serving the *SCP* role. *Storage SOP-classes* are used in transferring images 1) from various modalities to workstations or archives; 2) from archives to workstations and 3) from workstations to archives. Each *IOD* must include at least the patient, study and series information entities.

For each major type of image or *IOD*, there is a definition of storage service. Separate storage services are defined for transferring different IOD-objects such as digital X-ray image storage SOP-classes, digital mammography X-ray image storage SOP-classes, digital intra-oral X-ray image storage SOP-classes, soft-copy presentation state storage SOP-class, structured reporting storage SOP-class, enhanced MR-image storage class, enhanced CT-image storage SOP-class, enhanced MR color image storage SOP-class, etc.

6.5.1 Storage Services Implementation

Storage service SOP-classes are implemented using the *C-STORE DIMSE-C* service. A peer may not have the ability to store all possible image or *IOD types*. A successful completion of the *C-STORE* means both ends (*SCU* and *SCP*) support the type of information to be stored. An information is stored in some medium for some time, and the information may be accessed.

The duration of the storage is implementation dependent and is exchanged in the conformance statement of the *SCP*. Because of the variety of equipment used, support for a specific *storage service SOP-class* does not imply support for a query-response (*QR SOP-classes*). Information access method is implementation dependent and may require an implementation dependent operation at the *SCP*.

The *SCU* invokes a *C-STORE DIMSE service* with a *SOP-instance* that meets the requirements of the corresponding *IOD*. The *SCU* recognizes the status of the *C-STORE* service and acts based upon the success or failure of the service. Table 6.15 illustrates various status codes in a *C-STORE* response.

6.5.2 Levels of Service

A provider may not store all the attributes of the IOD sent to it. Normally during association negotiation, the provider sends a conformance statement indicating one of these levels:

1. *Level-0* (local) *conformance* indicates that a user-defined subset of the image-attributes will be stored, and all others will be discarded. This subset is defined in the conformance statement of the implementer.

TABLE 6.15   SCP response for storage request

SERVICE STATUS	DESCRIPTION
Failure	Refused: out of resources
	Error: data set does not match SOP class
	Error: cannot understand
Warning	Coercion of data elements
	Data set does not match SOP class
	Elements discarded
Success	Attributes matched

2. *Level-1* (base) *conformance* indicates that all type-1 and type-2 attributes defined in the *IOD* associated with the *SOP-class* will be stored, and may be accessed. All other elements may be discarded. The *SCP* may validate that the attributes of the *SOP-instance* meet the requirements of the *IOD*.

3. *Level-2* (full) *conformance* indicates that all type-1, type-2 and type-3 attributes defined in the IOD associated with the *SOP-class*, and any standard extended attributes (including private attributes), included in the instance of *SOP-class,* will be stored and may be accessed. The *SCP* may, but need not validate that the attributes of the instance meet the IOD. A *level-2 SCP* may discard without storing type-3 attributes that are empty (zero length and no value), since the meaning of an empty type-3 attribute is the same as the absence of an attribute-value. A *SCP* that claims conformance to *level-2* (full) support of the *storage service class* may accept any *presentation context negotiation* of a *SOP-class* that specifies the *storage service class* during the negotiation, without asserting conformance to that *SOP-class* in its conformance statement.

6.5.3 Coercion

In theory, an *SCP* of a *storage service class* may modify the values of certain corresponding attributes such as *patient-id*, *study-instance UID* and *series-instance UID* to integrate the information with the records of other databases carrying the information about the same patient. This process is called *coercion*. The *SCP* may also modify the deprecated values of *code sequence attributes* to convert to valid values.

If an SCP modifies the attribute-values, it shall return a C-STORE response with a status of "warning." Modification of the attributes may be necessary if the SCP also processes the query/retrieve SOP-classes. For example, an MR scanner may be implemented to generate study-instance UIDs for images generated by the MR. The mechanism to perform coercion is implementation dependent.

Modification of attributes that may reference a *SOP-instance* by another *SOP-instance* (such as *study-instance UID* and *series-instance UID*) will make that reference invalid. Modification of attributes may affect digital signatures referencing the content of the SOP-instance. Three levels of digital signature support are defined for an *SCP*. Level-1 support means that the *SCP* may not preserve digital signatures and does not replace them. Level-2 support means *SCP* does not preserve the integrity of incoming digital signatures, but validates the signatures of *SOP-instances* being stored, takes implementation-specific measures to ensure the integrity of data stored and will add replacement digital signatures before sending *SOP-instances* elsewhere. Level-3 support means that the *SCP* preserves the integrity of incoming digital signatures.

6.6 STORAGE COMMITMENT SERVICE

<u>S</u>torage <u>C</u>ommitment <u>S</u>ervice (SCS) allows a *SCU* to transmit images to another *SCP*. The *SCP* does not explicitly take responsibility for the safekeeping of data. In theory, the *SCP* can delete any archived item anytime depending upon its resource availability. However, reliable storage is a major part of medical image communication. The *SCS-class* enables an *SCU* to request an *SCP* to commit for the safekeeping of the *SOP-instances* such that a transferred object will be kept for a specific time and can be retrieved later.

The request for safekeeping is conveyed through the *N-ACTION-RQ* message. After accepting *SCU's* request by returning an *N-ACTION-RSP* message, the *SCP* will then asynchronously notify the client about its commitment to store and safekeep the requested set of DICOM objects. A single *N-ACTION-RQ* message may request storage commitment for multiple objects. The *SCP* upon receiving the message assesses its resources and capabilities. It then sends the response back. The notification of storage commitment by the *SCP* succeeds only if the *SCP* commits to safekeep all the DICOM objects requested by the *SCU* in its last request. The *SCP* returns the *N-ACTION-RSP* with a code "unsuccessful" if it cannot commit.

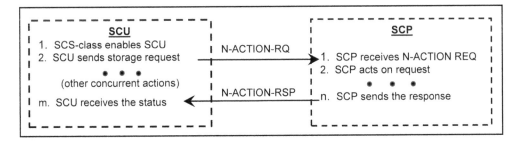

FIGURE 6.13 An illustration of storage commitment service

The notification of storage commitment by an *SCP* may happen also on a distinct association with respect to an earlier *storage commitment request* by the *SCU*. This notification is conveyed through the *N-EVENT-REPORT* message by the *SCP* later in a separate association. *N-EVENT-REPORT* message contains the *transaction-ID* of the earlier request to reference back to the previous *N-ACTION-RQ* message. The *N-EVENT-REPORT* can convey that it has accepted the commitment request of a previous request fully (code 1) or partially (code-2). For the latter case, the message contains the list of objects for which the commitment has been made.

A medical imaging device acquires many DICOM studies in a day, stores them in its internal storage and sends them to a PACS server for long-term archiving using the storage service. The device may delete one or more images from its internal storage after its internal storage becomes full. However, before deletion, for the safety purpose, it sends a *storage commitment request* to the PACS-server listing all the images selected for the deletion.

Figure 6.13 illustrates an instance of *SCS* between two AEs acting as *SCU* and *SCP*. In this scenario, the *SCU* sends an asynchronous request to the *SCP*. The *SCP* acts on the request asynchronously and checks its resources. The *SCU* keeps working concurrently on other tasks. The *SCP* sends the final response after some time. The final response could be "success," "partial commitment" or "failure." The three dots in the figure denote other tasks being performed by the *SCU* and *SCP* asynchronously, during message transfer and storage services.

6.7 QUERY/RETRIEVAL SERVICE

The *query/retrieve (QR) service class* defines an application-level class of service to search-and-retrieve *IODs*. This service is more complex than an SQL query/retrieval mechanism. A *QR-service* requires implicit context for the operations to be performed. Search into medical information normally extends into the multiple related IODs. DICOM *QR service* requires an information model that extends beyond single *IOD*.

6.7.1 QR Information Model

QR is a good example of performing a semantic search in an unstructured information-space. DICOM *QR-information-model* allows two types of searches: 1) conventional unrestricted relational query using any combination of keys and 2) baseline search in the hierarchical information-space. Semantically meaningful searches involving ontology require hierarchical information-space found in medical data.

All searchable attributes are in a four levels patient-tree hierarchy in the order of: 1) *patient*; 2) *study*; 3) *series* and 4) *composite object-instance* as shown in Figure 6.3. A patient can have one or more studies. Each study can have one or more series, and each series can have one or more *IOD-instances*. Table 6.16 lists the attributes and the corresponding tags (identifying the attributes) in this

TABLE 6.16 Attributes for patient root QR information model

LEVEL-1: PATIENT LEVEL ATTRIBUTES

Attribute Name	Tag	Type
Patient's Name	(0010,0010)	R
Patient ID	(0010,0020)	U
Issuer of Patient ID	(0010,0021)	O
Referenced Patient Sequence*	(0008,1120)	O
>Referenced SOP Class UID	(0008,1150)	O
>Referenced SOP Instance UID	(0008,1155)	O
Patient's Birth Date	(0010,0030)	O
Patient's Birth Time	(0010,0032)	O
Patient's Sex	(0010,0040)	O
Other Patient IDs	(0010,1000)	O
Other Patient Names	(0010,1001)	O
Ethnic Group	(0010,2160)	O
Patient Comments	(0010,4000)	O
Number of Patient Related Studies*	(0020,1200)	O
Number of Patient Related Series*	(0020,1202)	O
Number of Patient Related Instances*	(0020,1204)	O
All other Attributes at Patient Level		O

LEVEL-2: STUDY LEVEL ATTRIBUTES

Attribute Name	Tag	Type
Study Date	(0008,0020)	R
Study Time	(0008,0030)	R
Accession Number	(0008,0050)	R
Study ID	(0020,0010)	R
Study Instance UID	(0020,000D)	U
Modalities in Study**	(0008,0061)	O
SOP Classes in***	(0008,0062)	O
Referring Physician's Name	(0008,0090)	O
Study Description	(0008,1030)	O
Procedure Code Sequence	(0008,1032)	O
>Code Value	(0008,0100)	O
>Coding Scheme Designator	(0008,0102)	O
>Coding Scheme Version	(0008,0103)	O
>Code Meaning	(0008,0104)	O
Name of Physician(s) Reading Study	(0008,1060)	O
Admitting Diagnoses Description	(0008,1080)	O
Referenced Study Sequence	(0008,1110)	O
>Referenced SOP Class UID	(0008,1150)	O
>Referenced SOP Instance UID	(0008,1155)	O
Patient's Age	(0010,1010)	O
Patient's Size	(0010,1020)	O
Patient's Weight	(0010,1030)	O
Occupation	(0010,2180)	O
Additional Patient History	(0010,21B0)	O

TABLE 6.16 *(Continued)* Attributes for patient root QR information model

LEVEL-3: SERIES LEVEL ATTRIBUTES

Attribute Name	Tag	Type
Modality	(0008,0060)	R
Series Number	(0020,0011)	R
Series Instance UID	(0020,000E)	U
Number of Series Related Instances****	(0020,1209)	O
All Other Attributes at Series Level		O

LEVEL-4: IOD LEVEL ATTRIBUTES

Attribute Name	Tag	Type
Instance Number	(0020,0013)	R
SOP Instance UID	(0008,0018)	U
SOP Class UID	(0008,0016)	O
Alternate Representation Sequence*	(0008,3001)	O
>Series Instance UID	(0020,000E)	O
>SOP Class UID	(0008,1150)	O
>SOP Instance UID	(0008,1155)	O
>Purpose of Reference Code Sequence	(0040,A170)	O
>>Code Value	(0008,0100)	O
>>Coding Scheme Designator	(0008,0102)	O
>>Coding Scheme Version	(0008,0103)	O
>>Code Meaning	(0008,0104)	O
Related General SOP Class UID	(0008,001A)	O
Concept Name Code Sequence	(0040,A043)	O
>Code Value	(0008,0100)	O
>Coding Scheme Designator	(0008,0102)	O
>Coding Scheme Version	(0008,0103)	O
>Code Meaning	(0008,0104)	O
Content Template Sequence	(0040,A504)	O
>Template Identifier	(0040,DB00)	O
>Mapping Resource	(0008,0105)	O
Container Identifier	(0040,0512)	O
Specimen Description Sequence	(0040,0560)	O
>Specimen Identifier	(0040,0551)	O
>Specimen UID	(0040,0554)	O
All Other Attributes at COD Instance Level		O

Additional Attributes for QR Search

* These numbers are related to result of the QR-Search.

** The number of distinct modalities found in the study.

*** SOP classes found in the study.

**** Alternate encodings found in the search.

information model for each hierarchy. These attributes can be: 1) *required* denoted by *"R"*; 2) *unique* denoted by *"U"* and 3) *optional* denoted by *"O."* Multiple entities may have the same value for the required keys. Two entities at the same hierarchical-level cannot have the same unique key.

The patient-level QR-service contains QR-attributes associated with a *patient Information Entity* (*patient IE*) of the *composite IODs. Patient IEs* are modality-independent. *Patient's name* is a required attribute. *Patient-id* is a unique attribute. All other patient-related attributes are optional.

A single patient may have multiple studies. The study-level attributes are associated with the *study IE* of the composite *IODs. Study IEs* are modality-independent. Date, time, accession-number, *study-ID* and *study-IOD* are required attributes. S*tudy-instance UID* is a unique attribute.

A single study may have multiple series. The series-level attributes are associated with a *series IE, frame-of-reference* and *equipment IEs* of the composite IODs. A *series IE is* modality-dependent, has modality and series-related attributes. The *series instance ID* is a required attribute. All other attributes are optional.

The lowest level is the *composite object-instance level.* The level contains attributes associated with the *composite object IE* of the *composite IODs.* A *series* may contain multiple *composite object-instances. Composite object IEs* are modality-dependent. The set of optional keys at the *composite object-instance level* includes all attributes defined for any composite IOD.

The topmost level is the patient-level, when one can query and retrieve all the image or IOD-instances searching by attributes of the patients. Alternately, QR-service can be performed rooted at a level down focusing on *study attributes* that belong to a single patient. Similarly, QR-service can start rooted at the lowest level of a composite object (that belongs to a series belonging to a single study belonging to a single patient).

Though the QR models primarily use attributes assigned to various *IOD entities*, the QR-search can use special information elements (like attributes) which are meaningful only in a QR-search, such as: 1) how many matches have been found? and 2) how many alternate encodings have been found? A list of the additional search attributes is described in Table 6.17.

TABLE 6.17   Additional QR attributes

ATTRIBUTE NAME	TAG	ATTRIBUTE DESCRIPTION
Number of patient-related studies	(0020,1200)	The number of studies that match the patient level QR search criteria.
Number of patient-related series	(0020,1202)	The number of series that match the patient level QR search criteria.
Number of patient-related instances	(0020,1204)	The number of *composite object instances* that match the patient level QR search criteria.
Number of study-related series	(0020,1206)	The number of series that match the study level QR search criteria.
Number of series-related instances	(0020,1209)	The number of *composite object instances* in a series that match the series level QR search criteria.
Number of study-related instances	(0020,1208)	The number of *composite object instances* that match the study level QR search criteria.
Modalities in a study	(0008,0061)	All distinct values used for modality (0008,0060) in a study-series.
SOP-classes in a study	(0008,0062)	The *SOP-classes* in a study.
	(0008,3001)	A sequence of Items, each identifying an alternate encoding of an image that matches the instance level QR search criteria.

6.7.2 Commands for QR Service

SCU and *SCP* use commands *C-FIND* followed by *C-GET* or *C-MOVE* to implement the query-and-retrieval service. Each of the three operation at QR-level is a distinct *SOP* service in DICOM. There are nine unique *SOP-class UIDs*.

The command C-FIND is performed to search objects with specific criteria. The search-keys are encoded in the field called an *identifier* in a *C-FIND* command. It contains specific values to be matched against the attribute-values of the entities in the SCP using the hierarchical model. A *SCU* requests a SCP to perform a match of all the keys specified in the *identifier* of the request against the information possessed by the SCP at the *level* specified in the request. The SCP generates a *C-FIND* response for each match with a *return identifier* containing the values of all key-fields and all the known requested attributes. After the attributes match, the SCU (requesting *AE*) can retrieve the matched *composite object-instances* from the SCP using *C-GET* primitive. Alternately, the SCU invoking the C-FIND can direct a storage to transfer to another *AE* (instead of itself receiving the object) using the *C-MOVE* primitive.

A DICOM search can take long time on an *SCP*. Responses by an *SCP* contain a status "pending" indicating that matching is incomplete. After matching is completed, a *C-FIND* response is sent with a status of "success." A search request may also be refused or fail. In such cases, the *SCP* may send a status "refused" or "failed." The *SCU* may also cancel the *C-FIND* service by issuing a *C-FIND-CANCEL* request during the processing of the *C-FIND* service. The *SCP* will interrupt the process of matching, and return a response with the status "cancelled."

For all three operations, there are baseline and extended behaviors. Baseline behavior specifies a minimum level of conformance required to facilitate interoperability. Extended behavior enhances the baseline behavior to provide additional features. Extended *QR-service* behaviors are negotiated independently at association establishment time.

Figure 6.14 illustrates a typical successful search and retrieve operation using two *AEs* acting as *SCU* and *SCP*. There are six steps to *QR-service*: 1) association establishment; 2) negotiation during association establishment; 3) asynchronous *C-FIND* request along with UID and attributes to search the media-set holding the entities; 4) asynchronous *C-FIND* response returning the status along with the information about the media-set containing the requested information-objects; 5) *SCU* sending *C-GET* request along with the identifiers and the requested attributes of the information-objects and 6) *SCP* transferring the objects asynchronously during the *C-GET* response.

An *SCU* sends an asynchronous *C-FIND* request to the corresponding *SCP*. After receiving the request, The *SCP* starts level-wise search for every attribute. While it is searching, it sends a response to the SCU with a status "pending." After the search is complete, the *SCP* sends the final status to the *SCU*. Under the assumption that the search succeeds, the *SCU* sends a *C-GET* request with identifier values along with the requested attributes. The *SCP* returns the data-elements in the *C-GET* response at a later time asynchronously.

6.7.3 Tree Maintenance Requirements

All entities managed by *C-FIND*, *C-MOVE* and *C-GET SCPs* shall have a specific nonzero length unique key-value; two entities at the same level cannot have the same unique key-value. A unique-key must be contained in the identifier of *C-MOVE* or *C-GET* requests. *SCPs* for the operations *C-FIND*, *C-MOVE* and *C-GET* must support the existence and matching of all the unique keys.

C-FIND SCPs shall support existence and matching of all required keys defined by a *QR information model*. *Required keys* are contained optionally in the identifier of a *C-FIND* request. Required keys shall not be contained in the identifier of *C-MOVE* and *C-GET* requests. If a *C-FIND SCP* manages an entity with a *required key* of zero-length, the value is considered unknown. A matching against the zero-length *required key* shall be considered a successful match.

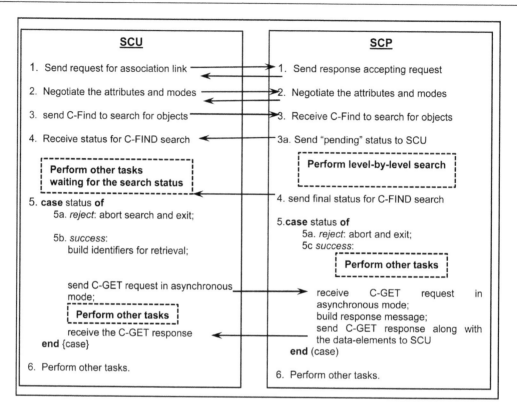

FIGURE 6.14 *QR-service* steps between *SCU* and *SCP* to retrieve desired objects

Optional keys may be optionally contained in the identifier of a *C-FIND* request. However, optional-keys are not contained in the identifier of *C-MOVE* and *C-GET* requests. The conformance statement of the *C-FIND SCP* lists the supported optional-keys. *Optional keys* in the identifier of a *C-FIND* request may have three types of behavior: 1) if the *C-FIND SCP* does not support the existence of an *optional key*, then the attribute-value is not returned in *C-FIND responses*; 2) if the SCP supports only the existence check of the *optional key*, then the *optional key* is processed like a *zero-length required key*: the value specified to be matched for the optional-key is ignored but a value may be returned by the SCP for the *optional key* and 3) if the *C-FIND SCP* supports the existence of both check and matching of the *optional key*, then the optional key is processed like a *required key.*

6.7.4 Types of Matching

A *C-FIND* process supports several types of attribute-matches: 1) *single value match*; 2) *list of UID match*; 3) *universal match*; 4) *wild-card match*; 5) *range match* and 6) *sequence match*. The actual search depends on the semantics of actual attributes.

The simplest type of match is the *single value*. It is successful if two strings match perfectly at the binary-code level. A single value case-sensitive match is performed, if the value specified for a key attribute has nonzero length. Additional constraints for a single value match are: 1) the single value should not be a *personal name*, a *date, time, datetime* and 2) the single value should not contain any wild-card character.

For attributes with a *PN (Patient Name) value representation*, an application may perform case-sensitive literal matching or that is insensitive to case, position, accent and other character encoding variants based upon an extended negotiation. The two sides may also negotiate fuzzy semantic matching rather than literal matching of *PN value representation* during the negotiation. Two sides may also negotiate to support phonetic matching.

Example 6.11

A query for "Swain" may also phonetically match "Swayne." A query for "Smith^Mary" may also match "Mary^Smith" or "Mary Smith" or "Smith Mary."

Fuzzy semantic matching may separate single-byte, ideographic and phonetic name component groups. *SCU* applications need to be careful, if it has negotiated fuzzy semantic matching of *PN attributes* as it may cause unexpected responses. Additional filtering of responses is performed by the *SCU* when it processes the responses obtained from fuzzy matching.

In the absence of an extended negotiation, a date, time or datetime is matched using exact literal match, even when the time-zones are specified. However, if the time-zone query adjustment is negotiated, time and dates are converted appropriately before matching. When a time-zone query is negotiated, the sides also agree to handle the missing time-zone offset specification.

A *list of UIDs* is encoded by using the backslash ("\") as a delimiter between UIDs. Each item in the list shall contain a single *UID* value. Each *UID* in the list contained in the identifier of the SCU-request may generate a match.

All entities match the attribute having a zero-length value. An attribute that contains a *universal match* specification in a *C-FIND* request provides a mechanism to return the corresponding attribute-value in the response by the *SCP*.

In a nonstandard datatype attribute (see Table 6.11), wild character "*" matches any sequence of characters, including a zero-length value; the character "?" matches any single character.

6.7.4.1 *Temporal range matching*

Range matching involves lower bound and upper bound of dates and times in DICOM. In the absence of an extended negotiation, a string of the form "$\tau^L - \tau^U$" ($\tau^L < \tau^U$) matches anytime τ such that $\tau^L \leq \tau \leq \tau^U$. A string of the form "$-\tau^U$" matches all the occurrences of date or time prior to and including τ^U. A string of the form "τ^L-" match all the occurrences of τ^L and subsequent dates or times. The offset from the *universal coordinated time*, if present in an attribute-value, is also used for the match. When a key-attribute has multiple values, then all the matched values are returned.

Figure 6.15 shows different scenarios of a temporal-match. The temporal-match is broadly classified into five categories: 1) point-match; 2) range-match; 3) range-inclusion match; 4) partial-range match and 5) no overlap. In a point-match, the point-time τ is included in the range $\tau^{L-}\tau^U$, satisfying the condition

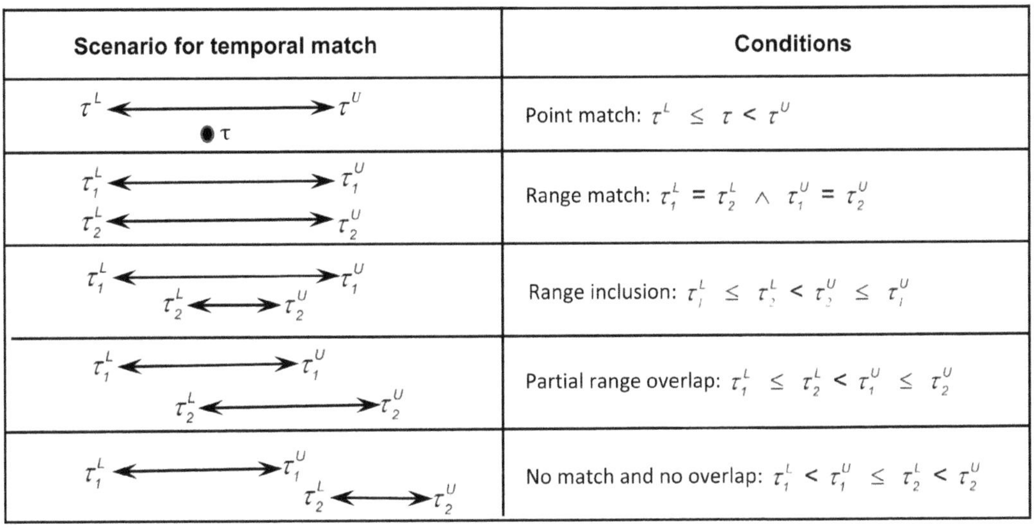

Scenario for temporal match	Conditions
$\tau^L \longleftrightarrow \tau^U$ $\bullet\,\tau$	Point match: $\tau^L \leq \tau < \tau^U$
$\tau_1^L \longleftrightarrow \tau_1^U$ $\tau_2^L \longleftrightarrow \tau_2^U$	Range match: $\tau_1^L = \tau_2^L \wedge \tau_1^U = \tau_2^U$
$\tau_1^L \longleftrightarrow \tau_1^U$ $\tau_2^L \longleftrightarrow \tau_2^U$	Range inclusion: $\tau_1^L \leq \tau_2^L < \tau_2^U \leq \tau_1^U$
$\tau_1^L \longleftrightarrow \tau_1^U$ $\tau_2^L \longleftrightarrow \tau_2^U$	Partial range overlap: $\tau_1^L \leq \tau_2^L < \tau_1^U \leq \tau_2^U$
$\tau_1^L \longleftrightarrow \tau_1^U$ $\tau_2^L \longleftrightarrow \tau_2^U$	No match and no overlap: $\tau_1^L < \tau_1^U \leq \tau_2^L < \tau_2^U$

FIGURE 6.15 An illustration of different scenarios for temporal range matching

$\tau^L \leq \tau \leq \tau^U$. The corresponding lower and upper bounds of the two ranges match in the exact range-match, and the condition $\tau_1^L = \tau_2^L \wedge \tau_1^U = \tau_2^U$ is satisfied. In a range-inclusion match, one range is a subrange of another range, and the condition $\tau_1^L \leq \tau_2^L < \tau_2^U \leq \tau_1^U$ is satisfied. In a partial-range match, two ranges overlap, and the condition $\tau_1^L \leq \tau_2^L < \tau_1^U \leq \tau_2^U$ is satisfied. In no match, two ranges do not overlap, and the condition $\tau_1^L < \tau_1^U \leq \tau_2^L < \tau_2^U$ is satisfied.

6.7.4.2 Sequence matching

If a key-attribute in the identifier of a *C-FIND* request is matched against an attribute structured as a sequence of data-elements, the key-attribute shall be structured as an item with a sequence of subitems. The matching is performed iteratively with one subitem at a time. If all the subitem key-attributes match, a successful match is generated. A sequence of matching items containing only the requested attributes is returned in the corresponding *C-FIND* responses. If the key-attribute in the identifier of a *C-FIND* request contains zero-length item tag, then all the matching entities are returned.

6.8 SEARCH AND RETRIEVAL PROCESS

6.8.1 Search Process

The *C-FIND* search primitive uses the parameter *SOP-class UID* to identify the *QR information model* against which the search is to be performed. An *SCP* can also specify an optional priority attribute that defines a requested priority of the *C-FIND* operation with respect to other DIMSE operations being performed by the same *SCP*. The support for the priority processing is stated in the conformance statement of the *SCP*.

The structure of the *identifier* parameter in *C-FIND* request contains: 1) key-attributes (including optional keys) along with the values to be matched against the values of storage *SOP-instances* managed by the *SCP* and 2) the *QR-level*. However, it does not contain the attributes absent in the request.

6.8.1.1 Actions on response to search

The *SCP* action in response to *C-FIND* request supports: 1) *retrieval of AE title data-element* and 2) retrieval of the *storage media file set UID data-elements*. The *storage media file-set UID* uniquely identifies the storage media on which the *composite object-instance(s)* resides. The *C-FIND SCP* may also optionally support the *instance availability* of a data-element. *Instance availability* defines how rapidly *composite object-instance*(s) becomes available for transmission after a *C-MOVE* or *C-GET* request. These can be: 1) "ONLINE" resulting into immediate availability; 2) "NEARLINE" meaning availability on slower secondary medium such as optical disk, or requiring time-consuming format conversion; 3) "OFFLINE" meaning the instances need to be retrieved manually and 4) "UNAVAILABLE" meaning the instances cannot be retrieved.

The availability of the least readily available instance is returned. If this data element is not returned, the availability is "unknown" or "unspecified." A null value (data element of zero-length) is not permitted. Finally, it returns one of the status codes: "failure," "canceled," "success" or "pending." A status of "success" indicates that a response has been sent for each match known to the *SCP*.

6.8.1.2 Hierarchical search

The *SCP* generates a *C-FIND* response for each match using the *hierarchical search method*. All such responses contain an *identifier* whose attributes contain values from a *single match*. Starting at the top level in the *QR information model*, the *SCP* continues until the level specified in the *C-FIND* request is reached. After reaching the specified level, all key match-strings contained in the *identifier* of the *C-FIND request* are matched against the values of the key-attributes for each entity at the current level. For each entity, for which the attributes match the specified strings, corresponding *identifiers* are constructed. Each *identifier* shall contain the unique keys at higher levels and all the values of the matching attributes. The *SCP* returns a response for each such *identifier*. If there are no matching keys, the *SCP* returns a "success" response with no identifier.

C-FIND can support typical open relational-queries. The *C-FIND* service with full relational queries allows any combination of keys at any level in the hierarchy. The *unique key-attribute* associated with the QR level shall be contained in the *C-FIND* request and may specify *single value matching, universal value matching* or *list of UID matching*.

Often, *C-FIND* is used to retrieve information related to a single entity such as all the information about one patient, or all the information from a single study. During the establishment of an association, an *SCU* can negotiate few extended options such as support for enhanced multiframe image conversion or full relational queries.

6.8.2 Retrieval Process

A *C-MOVE* or *C-GET* request is performed to retrieve or move the matched-objects at multiple composite-objects instance levels such as patient-level, study-level, series-level and image-level. The retrieval and transfer of stored *SOP instances* occurs from the same storage level as of *composite object instance*. For example, A *C-FIND, C-MOVE* or *C-GET* command at the *patient-level* indicates that all *composite object-instances* related to a patient are being retrieved or transferred. More than one entity may be retrieved if the *QR level* is *image, series* or *study* level, using *List of UID matching*. However, only single matching value is specified for a patient-ID (0010, 0020).

The *SCU* of *C-GET* supplies *unique key-value* to identify an entity. The *SCP* then generates *C-STORE* sub-operations for the corresponding storage *SOP-instances* by matching the *unique key-value*. These *C-STORE* suboperations occur on the same association as the *C-GET* service, and the *SCU/SCP* roles are reversed for the *C-STORE*. The *SCP* optionally generates responses to the *C-GET* with status equal to "pending" during the processing of the *C-STORE* suboperations. The *C-GET* responses indicate the number of remaining *C-STORE* suboperations. After the number of remaining *C-STORE* suboperations reaches zero, the *SCP* generates a final response with the last status. A *UID-list* is returned if the status of a *C-STORE* suboperation is "failed."

C-MOVE is an alternate operation after a successful match. After the responses of the *C-FIND* have been received by the querying *SCU* from the SCP, *SCU* may request to move the stored objects to another application-entity AE_3 that may not reside on the same system. This requires a complex set of operations requiring additional associations to be established between AE_2 and the AE_3. The *SCP* role of the *QR SOP-class* and the *SCU* role of the *storage SOP-class* may be performed by different AEs that may or may not reside on the same system.

6.9 MEDICAL IMAGE SECURITY

Medical image security is very important due to: 1) the enforcement of HIPAA; 2) the transmission of medical images to multiple geographically separated experts and institutions; 3) increasing multidoctor collaboration to handle a medical episode; 4) transmission of medical images for added expert opinion(s) and 5) storage of medical images in a digital format for a longer period. During this period, authorized personnel may change.

There are three types of transmissions of medical images: 1) within a hospital network covered by a firewall; 2) across the public networks requiring added encryption and security and 3) a user's access to medical images over the web not covered by any firewall. The medical images can be attacked and corrupted during the storage, altered during the transmission and illegally shared between unauthorized persons, including intruders.

Security is needed for: 1) maintaining the integrity of the image data; 2) enforcing HIPAA privacy and confidentiality rules and 3) supporting the authenticity of the medical images. *Authenticity* is defined as the correctness of the ownership and approval authority of data and has legal consequences. *Integrity* is defined as the absence of tampering of the original data. *Confidentiality* is defined as an assertion that the data is unavailable to unauthorized personnel and to an authorized personnel for unintended use.

DICOM image files embed sensitive information about patients in their headers. The header information includes a patient's name, date-of-birth, gender, ethnicity, other identification details and disease-related information. The images can be altered without any suspicion and lead to wrong diagnosis and treatment. Avoiding this problem requires strong encryption and verifiable digital signature for authentication. Authentication is important for the legality of the document.

6.9.1 DICOM Standards and Image Security

DICOM is a standard only for the information exchange of medical images. While security is very important during image transmission, NEMA specifications do not endorse any specific security standard. Rather, they consider security as an implementation issue. The data is transmitted and protected during transmission using an asymmetric encryption mechanism involving two keys: *private key* and *public key*. Private key and public key are mathematically related. However, the knowledge of public key does not enable the knowledge of private key.

DICOM has adopted the use of *Cryptographic Message Syntax* (CMS) for encryption and protecting the data movement over the Internet. The mechanism is based upon: 1) encrypting the header information selectively that contains the patient's identifying information, medical information and associated demographics information and 2) creating a private-and-public key-based system. CMS syntax allows for more than one digital signature by multiple persons approving the document.

DICOM specification protects sensitive portions of DICOM objects by specifying the portions to be encrypted within the DICOM objects. Since the header of the image contains the patient-identifying data, only the header-part is encrypted in DICOM specifications; bulk raw-image data is not covered to avoid the computational overhead during an information exchange.

The combination of a sender's private key and the patient-identifying part of the DICOM document are hashed using a *cryptographic hash function* to generate a *digital signature*. This digital signature shows the authenticity of the sender. Receiver's public key is used to encrypt the concatenation of the signature and the DICOM's patient-identifying image-header information. Encryption mechanism employs well-known encryption algorithms. At the receiver's end, receiver's private key is used to decrypt the incoming message. After decryption, the transmitted decrypted copy of the signature and the original signature are compared to verify any tampering of the signature or encrypted transmission. After successful signature matching, sender's public-key is used to authenticate the sender in the embedded digital signature of the concatenated document. Since the data can only be decrypted by the receiver's guarded private-key, there is no possibility of breaking the encoded message during transmission. After the authentication of a valid sender, the header information containing patient-specific information is extracted.

DICOM has changed object definitions to facilitate encryption of the sensitive part of the objects. DICOM specifies: 1) how encrypted DICOM objects can be read at the receiver's end; 2) how to use a TLS (transport layer security) connection in DIMSE protocol and 3) and how to use an encrypted HTTPS connection. A scheme for secure DICOM transmission based upon digital signature is illustrated in Figure 6.16.

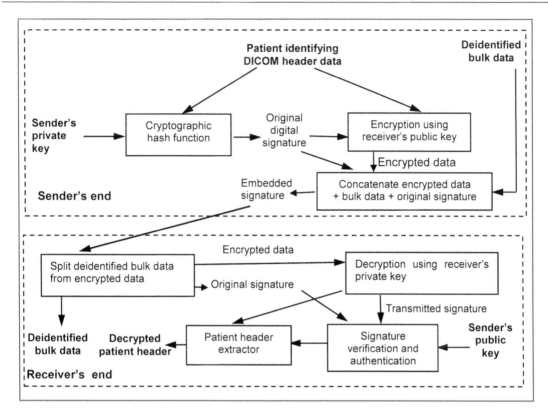

FIGURE 6.16 An illustration of secure DICOM transmission

DICOM does not mandate the encryption as the healthcare providers will provide privacy and confidentiality. The rationale for the approach is that security is also related to safe storage of data, policies related to access of the data, audit trails to verify the absence of tempering the data at the source and physical protection of storage units.

6.9.2 Watermarking

Many techniques have been developed by data-providers to securely transmit data and to authenticate the data. The techniques include *variations of digital signature* and *watermarking. Watermarking* is a technique to embed a piece of information such as digital signature or the tracking information of the last users in an encrypted form inside the image (or any media file to be protected) in such a way that it cannot be tempered with.

Watermarking is being used for integrity protection and authentication of medical images. In recent years, watermarking has become important to trace the illegal transmission of medical images for data-mining purposes. In a collaborative environment involving multiple doctors, medical images are shared, and it is difficult to keep track of image transmission. Watermarking based upon tracking the information travel as part of the watermarking can be used for finding out illegal transmission of medical images.

Inside a DICOM image, there are two parts: 1) uncompressed region of interest and 2) compressed background part. Watermarking embeds the identifying information such as digital signature in the background part. The information is hidden using *steganographic methods*. Watermark may contain additional visual information displaying the identity of the institution where the information is stored. Given a medical image of the form *B (background part)* + *ROI* (region of interest) and a digital signature *S*, the encoding algorithm will produce B^S + *ROI*, where B^S has embedded signature in it. The decoding algorithm takes the watermarked image and extracts the background *B* and the signature *S*. For example, a simple watermarking scheme is to periodically change a pixel value that includes a byte from the signature. The change will not be perceptible, and it does not affect the region of interest. Another technique is to modify the least significant bit of pixels by a bit of the information to be added. The key idea is to embed the information without making the change perceptible. If kth pixel of the original image is substituted by a byte of the watermarked information, the peak signal to noise ratio (PSNR) is computed by Equation 6.1.

$$ PSNR = 10 \; log \left(\frac{\sum_k (2^b - 1)^2}{\sum_k (x_k - x_k')^2} \right) \tag{6.1} $$

where *b* is bit-size of the image, x_k and x_k' are the values of the original image and the substituted image, respectively, at the kth pixel. Due to nonlinear nature of the visual-perception, exact effect of the noise on the perception cannot be measured accurately. Other indices have been proposed to take care of the nonuniform nature of the perception.

The presence of watermark inside a media is derived by extracting a test watermark and computing its correlation with a reference watermark. Strong correlation between the test and reference watermark shows the presence of the reference watermark in the media.

6.10 SUMMARY

This chapter describes the basics of DICOM, including entity-relationship information model and the *IOD*s, encoding of patient information and associated workflow information, and application level image transfer protocols. It also describes core application services, association and capability negotiation, storage, search, and retrieval.

There are multiple layers of network transmission. The transmission task is divided into five types of layers: 1) physical-link layer, 2) data-link layer, 3) network layer, 4) message layer, 5) application-ended layers and 6) application layer. The physical link layer is concerned with transmission using a physical medium. Data-link layers are concerned about splitting and fusing information into fixed size data-packets. The network layer models a complex network as a graph, and labels the nodes with an identifier (*UID*) and automates routing of the packets from one *UID* to another *UID* through a network. The application-ended layers convert a high-level data structure specified using the HL7 format into a data-stream, and use DIMSE service to associate AEs to each other for information transfer. At this level, images and associated information are modeled as *objects*. These objects are transferred between AEs using DIMSE service operations. A message service layer is used to pack objects into the form of messages transmitted using network-ended layers to the destination.

Each patient makes multiple visits. Each visit may result into multiple studies. Each study comprises a series (or procedure-steps) of images that are taken using one or more modalities and medical imaging devices. Each modality may have different types of imaging devices with different specifications. The final parameters of the images taken are guided by the specifications of the equipment. To serve correct diagnostics and infer the abnormalities, medical images require many associated parameters such as equipment-setup, patient-orientation, etc. along with the raw image. This whole process can be modeled using a patient-centric workflow model.

DICOM requires an entity-relationship model to abstract the workflow to generate and refer to patient-centric images generated using multiple modalities during a study. The entity-relationship model has three components: 1) entity; 2) relationships connecting the entities and 3) mapping between two entities. Mapping can be one-to-one, one-to-many or many-to-many.

Each service episode is authorized by a service provider before performing the services. Each procedure has a procedure-plan that consists of one or more procedure-steps. Each procedure-step can be satisfied by one or more modalities. Each procedure-step follows a desired protocol for a patient. The specification of procedure-step does not fix the performed procedure and modality. The final procedure-steps are decided by the available modalities at the equipment-level at the organization performing the prescribed procedure.

Each service episode generates multiple images. Clinical documents provide an important context to these images. The images along with the contextual data such as structured reports (SR), raw vital sign data and waveforms are archived, retrieved and transferred across applications residing on different computers over the network. The patient-centric information is structured in a hierarchical model having four hierarchy levels: 1) patient-level; 2) study-level; 3) visit-level and 4) composite data-level.

Patient-level information consists of multiple study-level objects. Study-level information consists of multiple visit-level objects. Visit-level information consists of many *composite data-elements*. A *composite data-element* is made of many *information-objects*. These *information-objects* are grouped based upon *IODs*.

Each *information-object* consists of many attribute-value pairs, including the information about attribute-type and the corresponding encoded name. Attributes are grouped into multiple modules that can share attributes. These modules represent the patient-centric information and abstract services provided by medical facilities such as patient-profile, patient-history, medical-visit and hospital-admission.

Information objects are the basic unit of information-exchange between two application-level entities. DIMSE (DICOM Medical SErvices) primitives are used that set up the association between to AEs (application entities) before information-exchange can take place. There are many DIMSE primitives to perform a search, store, retrieve, verify and transfer data-elements between two application entities. Each service employs an association between two AEs to exchange data-elements or perform requested actions.

Four basic actions are involved in most of the DIMSE operations for information-exchange: 1) request by SCU (service user) to SCP (service performer) through service-providers; 2) DICOM service-provider informing SCP about the request; 3) SCP responding to SCU for its request through DICOM service-provider and 4) SCU receiving the response from the DICOM service-provider. The association is released after an A-RELEASE command. The association can be aborted anytime if a major error occurs during data-transfer or association link-up.

For network level data-transfer, data-items are encoded at three levels: 1) basic data encoding; 2) composite data encoding and 3) semantic encoding. Basic data encoding contains the data-types of single data elements. Composite data encoding includes nested structures, sequence, tables, multidimensional matrices and image formats. Semantic data includes the semantic component of an attribute-value. Time representation is very important for data analytics and history management. There are multiple formats to manage date and time including temporal range. While comparing a point in time or a subrange with a range, a match occurs if the point or subrange is included in the range.

There are many types of searches using C-FIND. The search is made level-wise by attribute-matching. If the search results into success, then the information is retrieved by the *SCU* using C-GET (or N-GET) DIMSE operation or moved from the storage media to another storage media using C-MOVE DIMSE operation. The search and information-exchange can be done both in the synchronous or asynchronous mode. Asynchronous mode is more efficient because the AEs acting as SCU or SCP do not wait for the response. However, resources are committed for the search and QR in advance in both synchronous and asynchronous modes. In the absence of sufficient resources, SCP rejects the request by an SCU. A DICOM search may be long, and SCP may send an initial response of "pending" and keep working until the final response is ready. Often, due to the incompatibility of databases and the modalities, data stored in one format may have to be coerced to another format to match the values.

Attributes being matched during the search operations can be unique keys, required keys or optional attributes specific to modalities and equipment. Unique keys have unique value for an entity. Required keys may have more than one value. Attribute-matchings are of many types such as simple single value match, universal match, range match, wild-card match and sequence match. The semantics of the values also plays a role in matching. Fuzzy match relaxes the conditions of a match and required filters at the SCU end to remove incompatible values. Search operation can be canceled any time either by SCU/SCP, using C-CANCEL DIMSE operation.

DICOM is an evolving protocol. Every major medical imaging vendor has incorporated the standards into their product design. Most vendors are actively participating in the enhancement of the standards. DICOM will soon be used by virtually every medical professional, medical organization and vendor, that utilize images within the healthcare industry. These include cardiology, dentistry, endoscopy, mammography, ophthalmology, orthopedics, pathology, pediatrics, radiation therapy, radiology, surgery, etc. DICOM is also used in veterinary medical imaging applications.

Encryption and authentication are needed to enforce HIPAA during transmission and to protect tempering of the disease-related data in the DICOM image. DICOM specification facilitates encryption of the patient identifying area only in the header of the information objects and images; the remaining data is de-identified, and is not encrypted to save huge computer overhead.

DICOM specifies the encryption for the patient identifying areas in a patient identifying header. This information is encrypted using well-known encryption algorithms using asymmetric key-pairs that are mathematically connected. However, knowledge of one does not enable the knowledge of the other. The sender information is authenticated using digital signature created using an encrypted hashing function and detects the application of the private key in the signature using public key of the sender at the receiver-end. The authentication information also uses watermarking in the nonsignificant compressed region of DICOM images. DICOM facilitates encryption. However, it does not enforce encryption.

6.11 ASSESSMENT

6.11.1 Concepts and Definitions

A-ABORT service; A-ASSOCIATE service; A-P-ABORT service; A-RELEASE service; abstract syntax name; acceptor; application-entity (AE); application layer; application level service; application-ended layers; association; association acceptor; association entity; association attribute-module; association service; attribute matching; attribute type; basic encoding level; C-ECHO; C-FIND; C-GET; C-MOVE; C-STORE; capacity negotiation; clinical document; coercion; communication mode; compound object; compound encoding level; compound matching; concatenation; confirmation primitive; CRT series module; data-element; data type; defined procedure protocol; destination entity; DICOM; DICOM message service; dimension organization; endianism; entity-relationship model; frame-of-reference; general equipment module; healthcare structured document; imaging service; imaging service request; implementation class; indication primitive; information objects; IOD; IOD encoding; Kerberos protocol; key attribute; messaging service layer; modality; modality scheduled procedure-step; QR-service; module; N-ACTION; N-CREATE; N-DELETE; N-EVENT-REPORT; N-GET; N-SET; negotiation;

network encoding; network encoding levels; network-ended layer; optional key; P-data; patient identification module; patient information entity; performed procedure-step; physical link layer; presentation context definition; presentation state; procedure-plan; procedure-step; protocol; protocol data-unit; QR information model; radio therapy; range matching; request primitive; requesting application entity; required key; retrieval service; RT-objects; RT-plan entity; RT-structure set; SAML protocol; scheduled procedure-step; semantic component; service; service episode; service levels; service-object pairs, SOP-class; SOP-object; source-entity; storage commitment service; storage service; structured reporting; study; tractography; transfer-syntax name; type-1 data element; type-2 data element; type-3 data element; unique key; universal match; user-identity negotiation; value field; value representation; verification service; video-compression algorithm; visit; visit admission module; visit identification module; visit status module; watermarking.

6.11.2 Problem Solving

6.1 A Dentist's office has wireless network connected devices such as two DICOM conformant X-ray machines, four radiological workstations to investigate the images, one generic color printer, one server to store the medical images, two billing computers, one wireless access points to interconnect the servers, several Bluetooth connected measurement devices such as blood pressure, temperature, EEG and weight machines that wirelessly report to a patient data collection device, which reports to patient data collection software in the server, one firewall, a broadband modem to connect to the Internet and external hospital information systems. Explain from the network layer perspective which network-layers are required (and possibly implemented) in which devices.

6.2 A composite IOD comprises various modules. Investigate the IOD (Information Object Definition) for Computed Tomography (Section A3.3 and Table 3.1), CT Image IOD modules in DICOM standard 5.3 about IODs). Check the attribute description of the modules. List all the mandatory attributes that are part of this IOD.

6.3 Describe a pseudo-code to implement the exact match for date-time attribute with GMT conversion in an SCP for a C-FIND command.

6.4 A Radiology workstation would like to retrieve all the radiological IODs available for a patient named "Bob Jones" from a PAC. Explain with a diagram the DIMSE steps at the command level executed between the workstation as SCU and PAC as the SCP, including the association setup stage.

6.5 Consult DICOM standard references (mostly Part 4). A radiological CT machine wants to start a new MPPS object (to save captured image series for a new patient) in a workstation. It will use N-CREATE messaging services. Create a minimum case. Construct a sample N-CREATE request message showing all the required (Type 1 and Type 2) attributes.

6.6 Explain the difference of Type 1 and 1C attribute elements. Give five examples of each with their Tag numbers.

6.7 Construct three examples of identifiers corresponding to DICOM queries rooted at i) patient level, ii) study level, and iii) IOD level.

6.8 Patient Emily is seeing Dr. Cureall for the first time. Dr. Cureall orders a CT-image to be taken in Akron General Hospital Radiology Department. Identify the DIMSE messaging involved in placing this order. Step-by-step explain the DIMSE services those will flow between the two AEs when the doctor's office initiates the order from start to complete.

6.9 Patient Emily's film needs to be electronically transferred back to Dr. Cureall's office after the filming. Identify the DIMSE messaging involved in this transfer. Step-by-step explain the DIMSE services those will flow between the two AEs when the doctor's office initiates the transfer from start to complete.

6.10 Dr. Cureall's office wants Akron General to transfer all of patient Emily's CT-images taken during one study to be electronically sent to a PAC server for archiving. Identify the DIMSE messaging involved in this request and transfer. Step-by-step explain the DIMSE services those will flow between the two AEs when the doctor's office initiates the transfer from start to complete.

6.11 An elderly patient is complaining of pain in her hip-joint. She goes to her primary care physician. The physician recommends her to an orthopedist. The orthopedist examines her, orders X-ray slides of the area covering the hip-joint and studies the hip-joint. He suspects that the person has osteoarthritis and orders an MRI for her. After the MRI, he recommends her for a hip reconstructive surgery. Model this scenario as an entity-relationship diagram similar to Figures 6.3 and 6.5. Discuss your diagram.

6.12 Describe the complete procedure plan for the Problem 6.11. Look up the SNOWMED CT and ICD-10, identify all the procedure-steps and diseases recommended by the specialist doctor. Describe various SNOWMED and ICD-10 codes for the identified procedure-steps and ICD-10.

6.13 Using DICOM library and the Internet, get all the SOP Id (Service Object Pair) and SOP names associated with the procedure-steps in Problem 6.11. A SOP is a pair of the form (IOD – Information Object Definition, DIMSE – DICOM Service Element) as described in Section 6.3.4. Examples of IOD are CT images, MR images, ultrasound, schedule list, print queues (see Sections 6.2.5 and 6.3.2). Examples of DIMSE are *store, get, find, move* (see Section 6.3.4). For the detailed DICOM information, consult the website (http://dicom.nema.org/medical/-dicom/current/output/chtml/part03/PS3.3.html).

6.14 To retrieve images, the stored timed and request times should match through the relationship inclusion as discussed in Section 6.7.4. The matching is also required to avoid duplication in the storage. Consider the following pair of storage and request of retrieval. Explain the common time between storage and requested time and how many records will be retrieved.

Scenario 1: *stored data:* radiology report of March 1, 2000 – March 30, 2000.
Requested data; first half of 2000.

Scenario 2: *stored data: Yearly* Cat-scan data from 1997 – 1999.
Requested data: last CAT scan data

Scenario 3: *Stored data:* Monthly average summary report of blood pressure for January 2017 – March 2018
Requested data: Blood pressure in May 2016.

6.15 A coronary disease specialist, Dr. Cardio, will need to see the list of MR images of heart for the last 5 years to check for cardiomyopathy. Cardiomyopathy means increase in the thickness in the heart. Generate a request for images, including patient's details and details about the patient–doctor encounter. Create the corresponding dataset, including the tags and data elements, using information in Tables 6.1–6.4 and 6.11.

6.16 An image file in DICOM has patient information as a header followed by the dataset containing the equipment details of the image as given in Tables 6.6–6.9. Create a dataset using the proper tags for the information in Tables 6.6–6.9 to be attached as a header to the image.

6.17 A woman suffering from breast cancer goes for mammogram testing every year. The mammogram testing includes: 1) manual testing for lumps, 2) creation of a work order by the physician, 3) approval of the work order by the radiologist, 4) getting a new mammogram, 5) automated comparison of the old mammograms with the new mammogram, 6) manual reading of the automated results by the radiologists, 7) final report by the radiologist, 8) approval by the oncologist and 9) final diagnostic report and treatment advice by the oncologist. Create a process-flow view of the above service episode marking visits, events along with various time ranges.

6.18 A veterinarian is treating a dog named "Bobo" that belongs to "Robin Smith." The dog is a male, lives at 123 Elm Drive, Akron, Ohio, USA, and was born on July 4, 2017. Please code this information suitable for DICOM transmission (hint: read various coding schemes in Section 6.3.1).

6.19 A neurologist reads the 42 slides of MRI about a person with severely herniated vertebral discs on January 5, 2017. He recommends neurosurgery by fusing two vertebrae and asks the slides to be stored for the next 6 months starting January 5, 2017. There are 42 slides to be stored. Give the sequence of storage commitment services for the request and commitment.

6.20 Given a dataset in a DICOM file, write a simple abstract algorithm, that reads one data-element at a time, looks up the tag and translates the information into a human readable form. Try this algorithm on Problems 6.15 and 6.16.

6.11.3 Extended Response

6.21 Describe different classes of network layers used in the transmission of images from the source-end to the destination-end.

6.22 Explain the difference between application-ended layers and network-ended layers. Describe each layer briefly.

6.23 Explain the overall service flow model for a patient–doctor encounter in terms of medical images using a simple figure.

6.24 Explain the key function of DICOM association.

6.25 Explain how DICOM helps user-level authentication during association establishment.

6.26 Explain SOP (Service-Object-Pair)? Consult DICOM standards and list the UID of three associated SOP-services when one application tries to find an object in another DICOM application entity.

6.27 Explain the process of creation of unique identifiers in DICOM.

6.28 Explain the importance of unique attribute at each level to support the hierarchical search in DICOM?

6.29 Explain coercion within DICOM data search and match.

6.30 Explain the role of DIMSE operations during an information-exchange between two application entities.

6.31 Discuss the correspondence between real-world medical entities and the IODs. Why do real-world medical entities differ from IODs?

6.32 Discuss the advantages and disadvantages of asynchronous IOD transfer.

6.33 Explain the C-Find search for composite data-objects. Clearly explain different levels of search and main search parameters.

6.34 Explain different modes of search operations, including synchronous negotiation for the asynchronous search modes.

6.35 Explain the retrieval of the composite data-elements, including images in DICOM.

6.36 Explain various JPEG standards for storing images using lossy and lossless compression. Use the Internet and journal articles for literature search.

6.37 Explain the concept of storage commitment service using a simple figure.

6.38 Explain the concept of storage services using simple figures.

6.39 Explain the integrated QR search and retrieve operation that uses diverse types of attribute matches using simple figures.

6.40 Explain the mechanism of secure transmission over the Internet for DICOM objects.

6.41 Read the articles on CMS syntax and algorithms by Housley and the DICOM security specification document (see Reference 6.38), and write an extended summary of DICOM security specification in two single-spaced pages.

FURTHER READING

Overview of Network and Application Layers

6.1 Bonaventure, Olivier, *Computer Networking: Principles, Protocols and Practice*, December 2017, Available at http://cnp3book.info.ucl.ac.be/2nd/cnp3bis.pdf, Accessed June 6, 2018.

Modeling Medical Imaging Information Using DICOM

6.2 Kagadis, George C, and Langer, Steve G., (editors), *Informatics in Medical Imaging*, CRC Press, Boca Raton, FL, USA, 2012.

6.3 Kahn, Charles E., Langlotz, Curtis P., Channin, David S., and Rubin, Daniel L., "Informatics in Radiology: An Information Model of the DICOM Standard," *RadioGraphics*, 31, 2011, 295–304.

6.4 Tirado-Ramos, Alfredo, Hu, Jingkun, and Lee, K. P., "Information Object Definition-based Unified Modeling Language Representation of DICOM Structured Reporting: A Case Study of Transcoding DICOM to XML," *Journal of American Medical Association (JAMIA)*, 9(1), January/February 2002, 63–71.

PACS

6.5 Bellon, Erwin, Feron, Michel, Deprez, Tom, Reynders, Reinoud, and Bosch, Bart Van den, "Trend in PACS Architecture," *European Journal of Radiology*, 78(2), May 2011, 199–204.

6.6 Bransetter, Barton F., Rubin, Daniel L., Griffin, Scott, and Weiss, David L., *Practical Imaging Informatics: Foundations and Applications for PACS Professional*, Society for Imaging Informatics, Springer Science+Business Media, New York, NY, USA, 2013.

6.7 Bueno, Josiane M., Chino, Fabio, Traina, Agma J. M., and Traina, Caetano, "How to Add Content-based Image Retrieval Capability in a PACS," In *Proceedings of the 15th IEEE Symposium on Computer Based Medical Systems (CBMS 2002)*, Maribor, Slovenia, June 2002, DOI: 0.1109/CBMS.2002. 1011397

6.8 Huang, Hai K., *PACS and Imaging Informatics – Basic Principles and Applications*, 2nd edition, Wiley-Blackwell, Hoboken, NJ, USA, 2010.

DICOM Fundamentals

6.9 Clunie, David A., "DICOM Structured Reporting: An Object Model as An Implementation Boundary," *Proceedings SPIE International Society for Optical Engineering*, 4323, 2001, pp. 207–215.

6.10 DICOM Working Group 9, *Digital Imaging and Communications in Medicine (DICOM) Supplemental 130: Ophthalmic Refractive Measurements Storage and SR SOP Classes*, Rosslyn, Virginia, USA, January 2008, Accessed June 6, 2018.

6.11 Flanders, Adam E., and Carrino, John A., "Understanding DICOM and IHE," *Seminars in Roentgenology*, 38(3), July 2003, 270–281.

6.12 Graham, Richard N. J., Perriss, Richard W., and Scarsbrook, Andrew F., "DICOM Demystified: A Review of Digital File Formats, and their Use in Radiological Practice," *Clinical Radiology*, 60(11), November 2005, 1133–1140.

6.13 Güld, Mark O., Kohnen, Michael, Keysers, Daniels, Schubert, Henning, Wein, Berthold B., Bredno, Jörg, and Lehmann, Thomas M., "Quality of DICOM Header Information for Image Categorization," In *Proceedings of SPIE Medical Imaging 2002: PACS and Integrated Medical Information Systems*, (editors: Siegel, Eliot L., and Huang, Hai K.), 4685, May 2002, 280–287.

6.14 Horiil, Steven C., Prior, Fred W., Bidgood, Jr., W. Dean, Parisot, Charles, Claeys, Geert, *DICOM: An Introduction to the Standard*, 1994, Available at: http://www.csd.uoc.gr/~hy544/mini_projects/Project8/DICOM %20(Paper-Parisot).Doc, Accessed June 6, 2018.

6.15 Madema, Jeroen, Horn, Robert, and Tarbox, Lawrence, *Digital Imaging and Communication in Medicine (DICOM): Introduction and Overview*, National Electrical Manufactures Association, Roslyn, VA, USA, 2018, Available at http://dicom.nema.org/medical/dicom/current/output/html/part01.html, Accessed June 6, 2018.

6.16 Madema, Jeroen, Horn, Robert, and Tarbox, Lawrence, *Digital Imaging and Communications in Medicine (DICOM) standards*, NEMA PS3.1/ISO 12052, National Electrical Manufacturers Association, Roslyn, VA, USA, 2018, Available at https://www.dicomstandard.org/current/, Accessed on June 6, 2018.

6.17 Mildenberger, Peter, Eichelberg, Marco, and Martin, Eric, "Introduction to DICOM Standard," *European Radiology*, 12(4), April 2002, 920–927, DOI: 10.1007/s003300101100.

6.18 Mustra, Mario, Delac, Kresimir, and Grgic, Mislav, "Overview of the DICOM Standard," In *Proceedings of the 50th International Symposium (ELMAR-2008)*, Zadar, Croatia, September 2008, 39–44.

6.19 Noumeir, Rita, "Benefits of the DICOM Structured Report," *Journal of Digital Imaging*, 19(4), December 2006, 295–306.

6.20 Pianykh, Oleg S., *Digital Imaging and Communications in Medicine (DICOM): A Practical Introduction and Survival Guide*, Springer-Verlag, Berlin, Germany, 2012.

Verification

6.21 Kobayashi, Luiz O. M., Furuie, Sergio S., and Barreto, Paulo S. L. M., "Providing Integrity and Authenticity in DICOM Images: A Novel Approach," *IEEE Transactions on Information Technology in Biomedicine*, 13(4), July 2009, 582–589.

Storage Services

6.22 Huang, Hai K., Zhang, Aifeng, Liu, Brent, Zhou, Zheng, Documet, Jorge, King, Nelson, and Chan, Lawrence W. C., "Data Grid for Large-Scale Medical Image Archive and Analysis," In *Proceedings of the 13th Annual ACM International Conference in Multimedia*, Singapore, November 2005, 1005–1013.

6.23 Liu, Brent J., Cao, Fei, Zhou, Meng-Zhi, Mogel, Greg T., and Documet, Jorge, "Trends in PACS Storage and Archive," *Computational Medical Image and Graphics*, 27(2), March–June 2003, 165–174.

Image Distribution

6.24 Hai K. Huang, "Enterprise PACS and Image Distribution," *Computerized Medical Imaging and Graphics*, 27(2–3), March–June 2003, 241–253.

6.25 Zhang, Jianguo, Sun, Jianyong, and Stahl, Johannes N., "PACS and Web-based Image Distribution and Display," *Computerized Medical Image and Graphics*, 27(2–3), March–June 2003, 197–206.

Search and Retrieval Process

6.26 Traina Jr, Caetano, Traina, Agma J.M., Araújo, Myrian R. B., Bueno, Josiane M., Chino, Fabio J. T., Razente, Humberto and Azevedo-Marques, Paulo M., "Using an Image-Extended Relational Database to Support Content-based Image Retrieval in a PACS," *Computer Methods and Programs in Biomedicine*, 8(Suppl), December 2005, pp. S71–S83.

6.27 Costa, Carlos, Freitas, Filipe, Pereira, Marco, Silva, Augusto, and Oliveira, José L. "Indexing and Retrieving DICOM Data in Disperse and Unstructured Archives," *International Journal of Computer Assisted Radiology and Surgery*, 4(1), January 2009, 71–77.

DICOM Image Visualization

6.28 Escott, Edward J., and Rubinstein, David, "Free DICOM Image Viewing and Processing Software for Your Desktop Computer: What's Available and What it Can Do for You," *RadioGraphics*, 23(5), September 2003, 1341–1345.

6.29 Kroll, Michael, Melzer, Kay, and Lipinski, Hans-Gerd., "Accessing DICOM 2D/3D-Image and Waveform Data on Mobile Devices," In *Proceedings of the 2nd Conference on Mobile Computing in Medicine*, Heidelberg, Germany, 2002, 81–86.

Standards Interface and Image Transmission

6.30 Bojan Blazona and Miroslav Koncar, "HL7 and DICOM Based Integration of Radiology Departments with Healthcare Enterprise Information Systems," *International Journal of Medical Informatics*, 76(3), December 2007, S425–S432.

6.31 Fatehi, Mansoor, Safdari, Reza, Ghazisaeidi, Marjan, Jebraeily, Mohamad, and Habibi-Koolaee, Mahdi, "Data Standards in Tele-radiology," *Acta Informatica in Medicine*, 23(3), June 2015, 165–168.

6.32 Jimenez-Rodriguez, Leandro, Auli-Llinas, Francesc, and Marcellin, Michael W., "Visually Lossless Strategies to Decode and Transmit JPEG2000 Imagery," *IEEE Signal Processing Letters*, 21(1), January 2014, 35–38.

6.33 König, Helmut, "Access to Persistent Health Information Objects: Exchange of Image and Document Data by the use of DICOM and HL7 Standards," *International Congress Series*, 1281, March 2005, 932–937.

Medical Image Security

HIPAA Related Security

6.34 Cao, Fei, Huang, Hai K., and Zhou, Xiaoqun., "Medical Image Security in a HIPAA Mandated PACS Environment," *Computerized Medical Imaging and Graphics*, 27(2–3), March–June 2003, 185–196.

6.35 Kallepalli, Vijay N. V., Ehikioya, Sylvanus A., Camorlinga, Sergio, and Rueda, Jose A., "Security Middleware Infrastructure for DICOM Images in Health Information Systems," *Journal of Digital Imaging*, 16(4), December 2003, 356–364.

6.36 Lee, Chien-ding, Ho, Kevin I.-J., and Lee, Wei-Bin, "A Novel Key Management Solution for Reinforcing Compliance with HIPAA Privacy/Security Regulations," *IEEE Transactions on Information Technology in Biomedicine*, 15(4), July 2011, 550–556.

DICOM Standards and Image Security

6.37 *DICOM PS3.15 2013 – Security and System Management Profiles*, DICOM Standards Committee, 2013, Available at http://dicom.nema.org/dicom/2013/output/chtml/part15/PS3.15.html, Accessed June 18, 2018.

6.38 Housley, Rick, *Cryptographic Message Syntax*, Network Working Group, The Internet Society, 2002, Available at http://www.ietf.org/rfc/rfc3369.txt, Accessed June 18, 2018.

6.39 Housley, Rick, *Cryptographic Message Syntax (CMS) Algorithms*, Network Working Group, The Internet Society, August 2002, Available at https://www.rfc-editor.org/pdfrfc/rfc3370.txt.pdf, Accessed June 18, 2018.

6.40 Lou, Der-Chuyan, Hu, Ming-Chiang, and Liu, Jiang-Lung, "Multiple Layer Data Hiding Scheme for Medical Images," *Computer Standards and Interfaces*, 31(2), February 2009, 329–335.

Watermarking

6.41 Li, Mingyan, Poovendran, Radha, and Narayanan, Sreeram, "Protecting Patient Privacy Against Unauthorized Release of Medical Images in a Group Communication Environment," *Computerized Medical Images and Graphics*, 29, 2005, 367–383.

6.42 Tan, Chaun K., Ng, Jason C., Xu, Xiaotian, Poh, Chueh L., Guamn, Yong L., and Shea, Kenneth, "Security Protection of DICOM Medical Images Using Dual-Layer Reversible Watermarking with Tamper Detection Capability," *Journal of Digital Imaging*, 24(3), 2011, 528–540.

6.43 Zain, Jasni M., Baldwin, Lynne P, and Clarke, Malcolm, "Reversible Watermarking for Authentication of DICOM Images," In *Proceedings 26th Annual International Conference on Engineering in Medicine and Biology Society (EMBC 2004)*, Vol. 2, San Francisco, CA, USA, 2004, 3237–3240.

Web Based Security

6.44 Choe, Jun, and Yoo, Sun K., "Web-based Secure Access From Multiple Patient Repositories," *International Journal of Medical Informatics*, 77(4), April 2008, 242–248.

Bioelectric and Biomagnetic Signal Analysis

7

BACKGROUND CONCEPTS

Section 1.6.10 Biosignal Processing; Section 2.4 Statistics and Probability; Section 2.5.1.8 Wavelets; Section 2.10 Human Physiology; Section 3.5 Classification Techniques; Section 3.7 Probabilistic Reasoning over Time; Section 4.1 Modeling Medical Data; Section 4.2 Automated Data Acquisition and Interfaces; Section 4.7.2 Content-Based Image Retrieval

A human body emits many types of signals based upon electrical and magnetic activities caused by the circulation of an electric charge in the body. As discussed in Section 2.4.3, the contraction and relaxation of the heart allowing it to pump blood to the body is based upon electrical activity in the heart. This electrical activity is due to the periodic charge imbalance between the interior and exterior of the heart-cells, as explained in Section 7.1. The brain generates many types of waves called brain-waves in different cognition-states, emotional states and disease-states. Muscle movement generates signals. By knowing the combinations of these electrical activities in different parts of the body, different activities and abnormalities in these parts can be derived probabilistically. The advantages of bioelectrical signal analysis are that we can identify or predict many abnormalities without performing invasive surgeries. Using signal analysis, we can study the growth of the diseases noninvasively, study the remission of the diseases noninvasively and monitor the signals to detect the state of the organ and the body post-surgery.

Four major types of signals have been used to study disease states: ECG (Electrocardiogram) – surface level recording due to electrical activities in heart; 2) EEG (Electroencephalogram) – surface level recording due to electrical activities in a brain; 3) MEG (Magnetoencephalogram) – surface level recording of magnetic fields due to electrical activities in a brain and 4) EMG (Electromyogram) related to electrical activities needed for muscle excitation. ECG has been studied extensively and has been used for analyzing heart-related diseases such as conduction abnormalities in tissues, blood-flow problems, oxygen content in the blood, acidity level in the blood, nutrition content in the blood and other types of cells related to body repair and immune system response.

Besides these signals, there are subclasses of the bioelectrical signals used for monitoring health of various organs in the body. Three other major subclasses are: *electroneurogram* (ENG), *electrogastrogram* (EGG) and *phonocardiogram* (PNG) that are used for various diagnoses. ENG measures the nerve potential activity when a muscle nerve is activated. It differs from the EMG. EMG is a surface level potential change in response to electrical signal by motor neurons while ENG is in response to externally applied stimulus. ENG is used to test the conduction capability of nerve fibers. EGG is the measurement of the muscle activity of the gastric system when the patient is lying motionless on the back. PNG is the sound signal recording of the overall heart vibration. It is recorded by putting a microphone at the thorax – upper part of the human body between neck and the abdomen near the center of the rib that contains both heart and lungs. In this chapter, we will discuss four major bioelectrical signals: ECG, EEG, MEG and EMG.

Heart-related diseases are caused due to three types of disorders: 1) electrical; 2) blood circulation and 3) structural. Often these three defects are related. For example, electrical signal problem in a heart can cause the heart-contraction problem. The heart contraction problem can result into the blood circulation problem. Electrical problem causes different types of irregular heartbeats. Irregular beats in atria can cause blood clots that can lead to brain strokes and blockages. Irregular beats in ventricles can cause *sudden cardiac death* (SCD). Irregular beat can also be caused due to the lack or excess of different cells, minerals and oxygen carried by blood. Low oxygen from lung can cause heart to beat faster because an adequate amount of oxygen needs to be supplied to the cells. The blood-circulation problem is also caused when cholesterol or blood-clot blocks the blood flow.

Two major diseases are caused by the blockages of blood circulation: 1) *ischemia* (blockage in the arteries inside a heart) leading to oxygen deprivation to nearby cells and 2) *myocardial infarction*—death of the heart cells due to the lack of the blood supply or prolonged ischemia resulting into deletion of glycogen. Structural problems are: thickening of the heart-muscles and valve problems resulting into leakage of blood between heart-chambers. Thickening of the muscle or valve problem causes a heart to pump low volume of blood that can cause fatigue, less blood supply to brain and other parts of a body, causing blockages and blood clots.

EEG waves are surface level recordings of electrical activities generated in the brain that can be further classified to different waveform patterns. These brain waves have frequency less than 100 Hz and have amplitude in the order of microvolts. These electrical signals vary with mental activities, mental relaxation, emotions, stress level, sleep, eye-motion, brain abnormalities such as epilepsy, migraines, brain-tumors, Alzheimer's disease, Parkinson's disease and brain injuries. Continuous EEG monitoring is used to identify the epileptic seizure in advance to provide timely help to patients suffering from epilepsy. One surface level electrode is affected by many electrical activities inside the brain. Hence, EEG itself is not sufficient for *localization* of lesions and tumors inside the brain. A combination of EEG and MEG analysis has been used successfully to localize a brain-lesion.

MEG is the study of the magnetic fields inside the brain due to electrical activities in the neuronal cells. At the cellular level, due to the flow of charged ion, there is a slow ionic current. The flow of current causes magnetism. The overall effect of millions of neuronal activities causes a small measurable magnetic field that is measured using powerful superconductors called SQUID (Superconducting Quantum Interference Device) that amplifies and detects the magnetic fields. MEG provides the required spatial localization with sub-millisecond precision of major neuronal activities in the brain. A combination of MEG and functional MRI (fMRI) allows localization of abnormalities inside the brain. Functional MRI detects the area of signficant blood activities in a brain.

Muscles comprise many fibers intertwined in a complex way for motor activities of the body. Muscle movement and coordination between muscle-movements are essential for the balanced locomotion, gait-cycle and other coordinated activities. Lack of muscle coordination causes many diseases related to body balance, back pain, hip motion, shoulder problems, eyes-coordination problem and a lack of coordination in muscle activities.

Muscles are controlled by electrical activities of the nerves generated by the motor-neurons in the brain. Varying lengths and branches of nerve-fibers cause variations in these electrical activities. A measurement of these electrical activities at the surface level is called EMG (Electromyography), can help us in diagnosis of muscle-health and diseases born of muscle dysfunction and the lack of muscle coordination. EMG analysis is becoming important to help in the management and rehabilitation of people with motor disabilities.

This chapter describes four major biomedical signals: ECG, EEG, MEG and EMG and techniques to analyze these signals and the application of the analysis of these biosignals in identifying many disease-states related to the corresponding organs such as heart, brain and muscles.

7.1 ELECTROCARDIOGRAPH (ECG) FUNDAMENTALS

An electrocardiograph is a surface level measurement of the cumulative effect of ionic electrical activities in the heart cells that cause a heart to contract and relax. The two activities are called *depolarization* and *repolarization*. *Depolarization* causes cell-contraction, and *repolarization* brings the cells back to the original relaxed form. Depolarization is caused by the migration of excessive positive ions: Na^+ and Ca^{++} inside the cell, and repolarization is caused by the removal of excess positive ions from the cells.

The measured electrical signal at the surface level is in the order of millivolts. The measured signal is amplified for automated recording. An ECG recording is a repeated sequence of P-QRS-T waves. P-wave corresponds to depolarization of the atria – the upper chambers, and QRS-wave complex corresponds to the depolarization of the ventricles – the lower chambers. After the depolarization of atria, oxygenated blood coming from lungs moves from the left-atrium to the left-ventricle, and deoxygenated blood coming from the body moves from the right-atrium to the right-ventricle. After the depolarization of the ventricles (corresponding to QRS wave-complex), the oxygenated blood is pumped into the body from the left-ventricle, and deoxygenated blood moves from the right-ventricle to the lungs for reoxygenation. After the depolarization of ventricles, cells in the ventricles get repolarized. A T-wave corresponds to the repolarization of the ventricles. The overall repeat-cycle is described in Figures 7.1 and 7.2. There is also a weak U-wave that corresponds to repolarization of Purkinje fibers—a network of conduction cells for spreading the electrical activity in the ventricles.

These signals are measured at the surface level. Due to the available impedance between the origin of electrical activities and the surface, the signals are weakened. Raw signals have several types of noise caused by power-line interference, EMG interference due to muscles' movements, baseline wanders and the interference caused by involuntary movements of the body. These noises are removed from the ECG signals to get a corrected P-QRS-T signal. Multiple techniques have been applied on the filtered electrical signals to identify P-QRS-T waves.

Some popular techniques for the analysis of ECG signals are: *morphological analysis*; *wavelet transforms*; *Fourier transforms*; *artificial neural network analysis*; *Markov processes* and *HMM* (hidden

Step 1	SA node fires, and depolarization of the atrium starts. P-wave starts
Step 2	Atrium depolarizes and contracts. The oxygenated blood flows from left atria to left ventricle. Deoxygenated blood goes from right atrium to right ventricle. Deoxygenated blood enters from body to right atrium. P-wave terminates to Isoelectric base line.
Step 3	The electrical signal travels to AV-node (atrioventricular node) that separates atria from ventricles. Gets delayed for few milliseconds in AV-junction giving time to valves between the atria and ventricles to shut down.
Step 4	Depolarization of the ventricle starts starting from Purkinje fibers. Right ventricle starts first generating a small Q wave. Left ventricle electrical activity starts a little later but takes over. The depolarization wave travels from endocardium (inner layer) to epicardium (outer layer) generating R-wave. The contraction of the ventricles takes place causing S-wave. Due to ventricular contraction, oxygenated blood flows from left ventricle to the body, and deoxygenated blood flow from the right ventricle to the lung.
Step 5	Ventricles repolarize generating T waves.

FIGURE 7.1   Depolarization and repolarization cycle of a heart

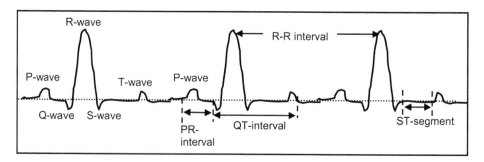

FIGURE 7.2 P-QRS-T wave

Markov model). Morphological analysis is based upon measuring the spread, slope and amplitude of different waveforms. These measurements are correlated with healthy-state and various disease-states. Wavelet analysis (see Section 2.5.1.8) is based on applying mathematical function that splits the complex waveforms into different frequency distributions and uses both time-domain and frequency-domain. The waveforms with the highest amplitude are easy to pick. The peaks with the highest amplitudes correspond to the R-waves peak and become the anchor-points for identifying other waveforms.

7.1.1 Polarization and Electrical Conductance

The cells depolarize and repolarize based upon the difference between the ion concentrations of Ca^{++} (calcium ion), NA^+ (sodium ion) and K^+ (potassium ion) across the cell walls. During the resting phase, the cell potential is around -90 mV due to outward leakage of K^+ (potassium ion) ions from the cell. During resting phase Ca^{++} (calcium ion) and Na^+ (sodium ion) channels for the inflow of ions into the cells are closed. Before a depolarization starts, Na^+ (sodium ion) concentration outside the cell is higher than inside the cell. Electrical activity starts from the SA-node (Sino-atrial node) located on the top right corner of the right-atrium, and then it spreads through the atrium.

The SA-node excites the right-atrium, and sends an electrical signal to the left-atrium through a bundle of fibers called *Bachmann bundle*. The excitation by the SA-node initiated electric pulse selectively, opens time-dependent sodium ion-channels for Na^+ ions to enter the cell. The voltage inside the cell quickly reaches above -40 mV. At this time, Ca^{++} (calcium ion) ion-channels open; the time-dependent Na^+ channels automatically close to avoid the influx of excessive charge in the cell. The Ca^{++} ion-channels remain open to take voltage above 0 mV for the contraction to occur. At this point, some K^+ channels open briefly. K^+ leaks out of the cell down its *concentration gradient*. The outward flow of K^+ returns the trans-membrane potential to 0 mV. Ca^{++} channels are still open, and there is a small inward current of Ca^{++}. These two counter-currents are balanced, and trans-membrane potential is maintained at 0 mV. Ca^{++} channels are gradually inactivated. Persistent outflow of K^+ ions brings the potential back to -90 mV (the resting potential) to prepare the cell for new cycle. The voltage change inside the cell is illustrated in Figure 7.3.

After the atria are depolarized, the electrical signal reaches the AV-node (atrioventricular node) and gets delayed. The electrical signal travels from the AV-node to ventricles through the *bundle of His* that traverses between the left and right-ventricles and splits into two parts: 1) left bundle going into the left-ventricle and 2) right bundle going into the right-ventricle. From there, it further splits into small fibers called *Purkinje fibers,* which go around the ventricles in three dimensions exciting cardiac muscles very quickly.

The left-ventricle is larger than the right-ventricle. First, the right-ventricle gets excited creating a Q-wave, which is quickly overcome by a larger electrical activity in the left-ventricle generating an R-wave. Each cell's activity is modeled as a vector that can be added to give an overall vector. The measured signal is the strongest if the lead-placement matches with the direction of the vector (or *180°* exactly in the opposite direction giving maximum negative value). Q-wave is negative because the vector corresponding to the electrical activity in the right-ventricle is away from the overall vector.

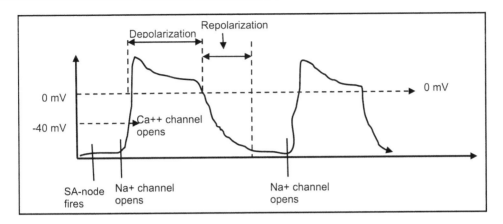

FIGURE 7.3   Cell depolarization and repolarization cycle

Cardiac muscle cells conduct the electric pulse slowly, while cells in the conduction pathway, including *Bachman bundle, bundle-of-his* and *Purkinje fibers* conduct the electrical activity faster for quicker depolarization. The depolarization–repolarization cycle in a healthy SA-node excitation is around *60–100* beats/minute. In the absence of SA-node excitation, AV-node fires at the rate of *40–60* times/minute. In the absence of SA-node and AV-node, ectopic-nodes (group of fibrous tissues) in one or more chambers excite asynchronously at a rate of *20–40* times/minute.

7.1.2 Measurement Planes and Leads Placement

Electrical activities of the heart are divided into two planes: *vertical plane* and *horizontal plane*. Multiple leads are placed on the chest, shoulder and legs to study the electrical activities. Three leads, AVL (left shoulder), AVR (right shoulder) and AVF (left leg) measure the electrical activities in the vertical plane. Six leads are put on the chest near the heart between the 4th and 6th rib-cage to measure the electrical activities along the horizontal plane. The overall lead placement is illustrated in Figure 7.4. There is an additional lead put on the right leg. However, it is not used in ECG measurement.

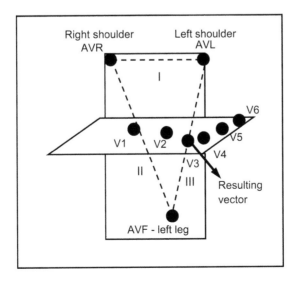

FIGURE 7.4   Measurement plane and lead placements

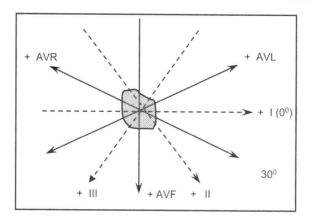

FIGURE 7.5   All six leads in vertical plane

V1 is placed just left of the trachea on the 4th rib-cage; V2 is placed just right of trachea on the 4th rib-cage; V3 and V4 are placed further right on the 5th rib-cage and V5 and V6 are placed laterally on the 6th rib-cage. This type of lead placement is necessary to measure electrical activities from multiple angles because multiple leads better resolve different waveforms.

Although only ten leads (nine measurement leads) are placed on a body, the popular measurement technique is called 12-lead placement because three additional virtual positions and the corresponding waveforms, I, II and III that are computed using AVL, AVR and AVF as illustrated in Figure 7.5. The direction of the virtual lead I is from the lead AVR to the lead AVL. The direction of the virtual lead II is from the lead AVR to the lead AVF, and the direction of the lead III is from the lead AVF to the lead AVL. The output is reported as *12* waveforms: V1 to V6 in the horizontal plane; and AVL, AVR, AVF, I, II and III on the vertical plane.

In Figure 7.5, the solid lines show the directions of AVF, AVL and AVR. Dashed-lines show the virtual leads I, II and III. The arrows show the positive directions of the leads. Lead II is *60°* apart from the lead I, lead III is *60°* apart from lead II and *120°* part from lead I. AVF makes *120°* angle with AVR and AVL, and AVR makes *120°* angle with AVL. The lead I makes *30°* angles with AVL and *150°*. Lead II makes *30°* angles with AVF and *150°* angle from AVR. Lead III makes *30°* angle with AVF and *150°* angle from AVF.

7.1.3 Polarization Vectors

The overall polarization vector is the vector-sum of all small electrical-vectors generated by each cell in a heart. A vector-sum is described in Figure 7.6. The maximum depolarization-vector in a healthy heart is aligned around V4 lead in horizontal plane and lead II in the vertical plane and shows maximum amplitude for the QRS-complex.

The orientation of the actual resulting vector varies from heart to heart depending on the heart shape, heart-size, ischemic or injured cells and conduction activities in a heart. For example, in the case of cell-death in left-ventricle or conduction block in left-ventricles, the resulting depolarization vector will move right toward the lead V2. Similarly, in the case of an electrical conduction problem in the right-ventricle, the resulting depolarization-vector will move left toward the lead V5.

The depolarization-vector changes with time and the orientation of multiple measurement-leads. If the angle between the resulting depolarization-vector and a measurement-lead is θ, then the electric potential seen at the lead is $V \times cos(\theta)$. That means that the measured value (through a lead) changes based on the direction between the angle made by the measuring-lead and the resulting potential-vector. The measured amplitude decreases from *0°* to *90°* and eventually becomes zero at *90°*. The magnitude starts increasing in the negative direction between *90°* and *180°*.

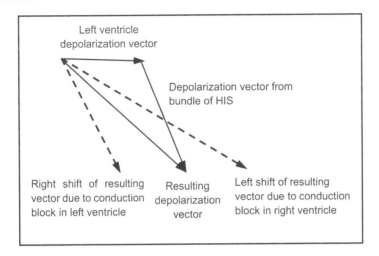

FIGURE 7.6 Resulting polarization vector and shift

All other leads show smaller amplitudes because of the misalignment with the resulting depolarization-vector. If the angle becomes $90°$ (with respect to lead II aligned with the resulting vector) as in the case of AVL, the amplitude of QRS-complex in AVL will become 0. All twelve leads need to be studied to derive the maximum amplitude of different waveforms.

7.1.4 Variations in P-QRS-T Waveforms

P-wave is a superimposition of P-wave from the right-atrium and P-wave from the left-atrium. The normal size of P-wave is less than 0.25 millivolts (2.5 mm of the ECG graph paper) with a duration of less than 0.12 seconds. The two P-waves are superimposed because the delay between P_{left} and P_{right} is quite small. However, if the left-atrium enlarges, P_{left} will be delayed causing P-wave to spread. If the right-atrium dilates, then both will superimpose causing the P-wave peak to become more pronounced and the spread of P-wave shortens.

In the case the SA-node does not fire, then there are three possibilities: 1) ectopic-node(s) in the atria provides initial electrical activity; 2) AV-junction provides initial activity or 3) ectopic-node(s) in ventricles provides the initial electrical activity. The firing of ectopic-nodes is irregular, the electric signal generated is quite weak, and due to the nonuse of proper electric-conductance fibers, it takes longer to travel.

In the first case, P-waves are spread and have less amplitude than normal P-waves. Since ectopic-nodes fire irregularly, P-wave may not occur at regular intervals. In the second case, first electric signal appears in AV-junction. P-wave is negative in V3 and V4 because the wave travels away from the lead-directions and often gets embedded in QRS-complex. In the third case, the electrical signal is weak. Depolarization of the ventricles is irregular and weak. Some P-waves may be missing due to the lack of depolarization in atria, and others may get embedded in QRS-complex or T-waves.

7.1.5 ECG Metrics

The measurement paper is mapped to a special format for the ease of reading ECG. The X-axis of the graph paper shows time, and Y-axis shows the amplitude of the ECG waves in millivolts. Each centimeter on Y-axis corresponds to 1.0 mV. Each centimeter on X-axis corresponds to 0.4 seconds; each millimeter corresponds to 0.04 seconds. The graph paper is divided into blocks of 5 mm (0.20 seconds) $\times 5$ mm (0.5 mV).

The whole graph paper is around *30* blocks or *6* seconds. For a healthy man having 60 heartbeats/second, each P-QRS-T wave is around one second (*5* blocks), and there are approximately six P-QRS-T waves in a graph paper.

In a healthy person, the PR-interval (the interval between P-wave and R-wave) is between *0.12* seconds (*3.0* mm) and *0.20* seconds (*5.0* mm). The amplitude of a regular P-wave is around *0.25* mV (~*2.5* mm). The QRS-complex's duration is around *0.10* seconds (~*2.5* mm), and the amplitude varies for different people. It depends upon the size of the chamber and the amount of muscles in the ventricles.

The major time-segments that carry meaning in the ECG are *RR-interval, PR-interval, QT-interval, QT^C-interval* (corrected *QT-interval*) and *ST-segment. RR-interval* measures the interval of one heartbeat. It is chosen as a standard because R-wave is the tallest and the easiest to detect. *PR-interval* is the time-interval between the start of a P-wave to the start of the following QRS-complex. PR-segment shows the total time taken for the depolarization in atria until the start of the depolarization of the left-ventricle. Shorter PR-segment shows shorter depolarization in atria, and longer PR interval shows -additional delay in AV junction or conduction-block in the *bundle-of-His.*

QT-interval is the duration between the start of Q-wave to the end of the T-wave. A *QT-segment* shows the duration of electric depolarization and repolarization of ventricles. *QT-intervals* shorten with faster heart-rates. *QT-intervals* are measured in lead II. Due to varying heart-rate, *corrected QT-interval* called *QT^C-interval* shows the approximate *QT-interval* at the heart-rate of 60 beats/minute.

Some of the popular formulae to compute QT^C are: 1) *Bazett's formula* (see Equation 7.1); 2) *Fridericia's formula* (see Equation 7.2); 3) *Framingham's formula* (see Equation 7.3) and 4) *Hodge's formula* (see Equation 7.4). The time in the formulae is given in milliseconds. Different formulae show better accuracy at different heart-rates. Bazett's formula is the most popular and is used to correct heart-rate between *60* and *100* beats/minute. Fridericia's and Framingham's formulae are used outside the range of *60–100* BPM. All four formulae use two independent variables to derive QT^C: *QT-interval* and *RR-interval* (or heart-rate). Normal QT^C is around *400* milliseconds. QT^C value less than *350* milliseconds and greater than *440* milliseconds are abnormal and indicate irregular heartbeats.

$$QT^C = QT / \sqrt{RR\ interval} \tag{7.1}$$

$$QT^C = QT / RR^{1/3} \tag{7.2}$$

$$QT^C = QT + 0.154\,(1,000 - RR\ interval) \tag{7.3}$$

$$QT^C = QT + 1.75\,(heart\ rate - 60) \tag{7.4}$$

ST-segment is the interval between the end of S-wave and the start of T-wave. Ideally, it should be flat isoelectric baseline. However, in the presence of ischemia (lack of oxygen) and myocardial infarction, ST-segment becomes elevated.

7.2 ECG ANALYSIS

Recorded raw ECG is a low amplitude signal with noise. It needs to be preprocessed to: 1) remove noise and 2) improve the signal-to-noise ratio (SNR). Most common types of noise in ECG are: power-line interference; muscle-related electrical signals (EMG); motion-related artifacts; baseline-drift due to the respiration and poor electrode-contacts. Baseline-drift is a very low frequency wave, while muscle-related electrical signal range from *0* to *500* Hz with a maximum energy median around 100–150 Hz.

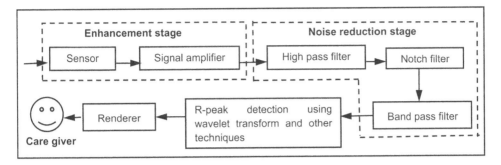

FIGURE 7.7   An overall schema to derive P-QRS-T waves from raw signal

Three steps to process raw signals before ECG analysis are: 1) filtering out the noise; 2) amplification of ECG signal without distorting the waveforms and 3) identification of P-QRS-T waveforms. In these subsections, we discuss various techniques to: 1) remove noise; 2) enhance ECG waves and 3) identify P-QRS-T waveforms. The overall scheme for P-QRS-T detection is shown in Figure 7.7.

After the localization of the R-peaks, P-QRS-T waveforms are identified. These P-QRS-T waveforms are further analyzed for aberrations from the average waveforms of healthy persons. These deviant patterns have been studied using both statistical techniques and morphological analysis by heart specialists to correlate deviations with different heart-conditions. In recent times, automated ECG-analysis has been developed to assist cardiologists, paramedics and emergency physicians to monitor the heart-conditions regularly during office-visits, postsurgery and in the intensive-care units.

7.2.1 Noise Removal

Popular filtering techniques are: 1) a high-pass filter to remove low-frequency baseline-drift between *0.05* and *0.67* Hz; 2) a notch filter (a type of narrow band-pass filter) to remove *50–60* Hz frequency noise introduced by the power lines; 3) an adaptive band-pass filters to remove frequencies caused by the muscle-movement frequencies; 4) Savitzky–Golay digital-filter to smoothen the signals and enhance the SNR without distorting the signals; 5) a Gaussian filter to smoothen the waveforms by suppressing high-frequency noise and 7) statistical corrections.

Statistical techniques use principal component analysis and independent component analysis in the statistical domain to derive the correlations between the signal peaks. Care is taken to avoid the attenuation of the original ECG-signal during the noise-filtering. After the noise removal, wavelet-transform and other techniques are used to localize the different peaks starting with R-peak that is the most pronounced peak in P-QRS-T waveforms.

The sensor recording is modeled as $x(t) = s(t) + n(t)$, where $x(t)$ is the recording of the signal, $s(t)$ is the actual information content and $n(t)$ is the noise content. Signal-to-noise ratio (SNR) is defined as $s(t)/n(t)$. SNR should be high. The denoising algorithms improve the SNR (signal-to-noise ratio) while preserving the sharpness of the actual P-QRS-T waves. ECG enhancement techniques should not introduce any false peak or attenuate P-QRS-T waveform significantly. For example, low-pass filters can attenuate R-peak of an ECG, and high-pass filters can distort ST-segment of ECG.

Wavelet-transforms decompose analog ECG signals into a combination of mathematical functions by dilation and translation of wavelet function $\psi(t)$. After applying the wavelet-transforms, low-frequency components are retained; high-frequency noise is filtered out. The inverse-transform recovers the original analog ECG-signal with much reduced noise. Summing up of different components causes phase-distortions due to disproportionate phase-shifts with respect to the change in frequency. A linear phase-filter is used to reduce the phase-distortions. A linear phase-filter maintains the phase relationship between the different frequency components of ECG waves. Other techniques used to recover ECG signals are: 1) an artificial neural network (ANN)-based approach and 2) energy-thresholding and Gaussian distribution based approach.

Removal of EMG signals from the ECG waves is tricky because removal of any signal above *35* cycles/minute, and below *400* cycles/minute distorts the ECG signal. ECG signals are around *60–100* beats/minute for a healthy individual, lower than *60* beats/minute for persons suffering from bradycardia (low heartbeat rate) and greater than *100* beats/minute for tachycardia (high heartbeat rate). To circumvent this problem, a carefully tuned bandpass filter with layered attenuation factor is used that attenuates the signal below *35* cycles/minute and above *200* cycles/minute. It has been experimentally observed that even the careful use of low-pass filter for *35* cycles/minute attenuates S-wave. However, noise due to EMG is suppressed.

7.2.2 P-QRS-T Waveforms Detection

P-QRS-T waves have been identified using four major techniques: 1) wavelet-transforms; 2) model-based approach using Gaussian mixture; 3) ANN and 4) HMM. Model-based approach is based upon template matching. Gaussian mixture is a mixture of two or more Gaussian curves. Many other algorithms use nonlinear transforms to accurately identify QRS-complex such as: 1) MOBD algorithm (Multiplication of Backward Difference); 2) Okada algorithm and 3) Hamilton–Tompkins algorithm. These algorithms are based upon squaring the backward difference (signal-sample(t) – signal-sample($t − 1$))^2. Since the QRS complex is big, the difference between the QRS-complex and other signals gets further amplified. This amplification helps in the detection of the peaks.

To differentiate healthy heartbeats from abnormal heartbeats, following factors are significant: 1) the predictable pattern of the presence/absence of a waveform; 2) the amplitude and the spread of waveforms; 3) variations in the segments within P-QRS-T waveforms; 4) repeating patterns of specific waveforms and 5) the rate of heartbeats. It is important to localize the waveforms and identify the amplitudes and intervals.

7.2.2.1 Wavelet approach

The technique based on a wavelet-spectra uses wavelet-transform to identify P-QRS-T waveform. After applying a wavelet-transform, ECG waves are decomposed into smaller wavelets. Wavelet-transforms use the scaling (dilation) to pick up waveforms with different frequencies. Eight levels of scales are used in analyzing ECG waves. QRS-complex corresponds to 2nd, 3rd and 4th scales. Signals are separated by sliding-scale windows.

Daubechies 6 (Db6) wavelet closely resembles QRS-complex. DB6 is used to match decomposed waveforms as illustrated in Figure 7.8 and extract the wavelet coefficients for QRS-complex. The inverse-transform is used to rebuild ECG-wave. Since the R-wave-peak has the highest magnitude, it is easier to localize R-peak

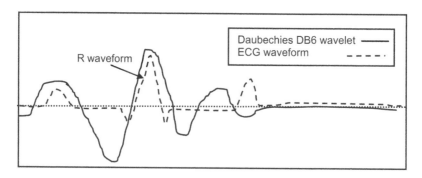

FIGURE 7.8   Matching peak of Daubechies DB6 wavelet and R-peak

while attenuating P and T waveforms. The distance between two successive R-peaks gives the time-interval of an ECG waveform, and time-average of RR-intervals is used to derive the heartbeat/minute.

After the R-peaks are identified; Q-peaks and S-peaks are derived by: 1) identifying the intersection of the ECG signals with the baseline and 2) identifying other points of zero inflections around R-peak after the baseline crossing. These inflection-points give Q-peak and S-peak. T-peak is identified by the next maxima occurring after S-peak. T-onset and T-offset are identified as the minimum potential on either side of the T-peak within a reasonable window justified by the Gaussian distribution of T-wave interval. Average P-R interval is used to localize the P-waveform. Further analysis is done to derive the actual interval, amplitude and slope of the waveforms using windows of expected time-intervals around the waveform-peaks.

7.2.2.2 Model-based approach

Model-based approach simulates the template of a realistic waveform and matches the template with the actual ECG to localize the position of different waveforms. The technique uses Gaussian mixture models where many overlapping Gaussian-waveforms are mixed to generate a combined Gaussian wave that looks like an ECG waveform. The idea is based upon the statistical depolarization of cells in the atria and ventricles. Gaussian-waveform is used to model the population statistics. It is also suitable for modeling collective electric-impulse measurement of heart-cells. The peak of the matching Gaussian-waveforms provides the approximate peak of the waveforms. These waveforms are then analyzed to identify actual peak points.

7.2.2.3 HMM-based approach

HMM-based model assumes that each waveform-peak is a state, and the baseline between PQ, ST and TP segments is also states. In a healthy person, transition is in the sequential order $P \rightarrow PQ\text{-}segment \rightarrow Q \rightarrow R \rightarrow S \rightarrow ST\text{-}segment \rightarrow T \rightarrow TP\text{-}segment \rightarrow P$, where PQ-state corresponds to the isoelectric-line between P-waveform and Q-waveform and represents delay in the AV-node. ST-state corresponds to isoelectric-line between S and T waveform and represents the additional time taken for ventricle-contraction to complete before repolarization starts. The state TP corresponds to the isoelectric-line between the current T-waveform and next P-waveform. After a R-peak is derived, the remaining waveforms are derived using relative order of occurrence of other waveforms, derivatives and their morphological characteristics.

7.2.2.4 Artificial neural network approach

ANN-based approach has been used for the detection of QRS-complex. It uses adaptive filtering to model lower frequencies of ECG. As mentioned in Section 7.2.1, adaptive filtering removes time-varying nonlinear noise from the ECG-waves. Linear adaptive filters provide protection against varying noise in ECG signals. The noise in ECG-signals is correlated where normal filters do not work. A preprocessing step is used to remove the correlation between noises. This improved filter is called *whitening filter*. ECG-signals are inherently nonlinear that can be handled by ANN. In this approach, multiple hidden layers are added in a neural-network to handle nonlinearity in noise. The resulting approach is called *neural-network-based nonlinear whitening filter*.

7.2.2.5 Identification of P and T waveforms

Analysis and identification of P and T waveforms is challenging in the presence of noise and the low magnitudes of P-peaks and T-peaks. Many other probabilistic techniques such as Bayesian decision network and Monte-Carlo method have been applied to detect low amplitude P and T waves. Bayesian decision network is based upon calculating the probability of occurrence of events using past evidence and expected probabilities of the next event given the occurrence of the current event. Monte-Carlo simulation allows for possible outcomes with expected probabilities for events with inherent uncertainties. It computes the probabilities based upon the randomness of samples.

The identification of P-wave is also challenging in the case of arrhythmias where the electrical activities may originate either in AV-node, ectopic-nodes in atria or ectopic-nodes in ventricles. In such cases, the sequence P-QRS-T is not followed. Often, P-wave gets embedded in QRS-complex or preceding T-waves; P-waves may be missing; P-wave may occur after QRS-complex; multiple P-waves may occur before a QRS-complex or T-waves may be inverted.

7.3 HEART DISEASES AND ECG

ECG is one of the core vital signs that is monitored before a surgery, during a surgery and after a surgery. ECG is also measured and analyzed to derive abnormalities in a heart. The ECG waves are analyzed for *morphological changes*, *duration changes* and *rhythm-pattern changes*. The changes are caused by heart abnormalities such as *ischemia*—lack of oxygen in heart or a part of a body due to the lack of blood or lack of oxygen in the blood; deposits in the arteries, including inside a heart, causing cells to starve of nutrients and oxygen; cardiomyopathy—thickening of heart-muscles; hypertrophy—enlargement of heart; calcification of the arteries; conduction blockages in left or right ventricles causing heart not to contract completely; lack or excess of ions such as Na^+ (sodium ion), K^+ (potassium ion), Ca^{++} (calcium ion) needed for depolarization and repolarization of cells; *myocardium infarction* - death of heart-cells possibly due to ischemia. Irregular beats of heart may lead to blood clot or sudden cardiac death. Blood clots may travel to brain resulting into brain stroke and paralysis.

Heart disease is also caused by various valves in the heart that may become defective either due to: 1) elongation of valves caused by *cardiomyopathy*; 2) calcification causing hardening of the tissues; 3) *stenosis* — narrowing of a valve and 4) *bacterial endocarditis* — bacterial infection of the valve-leaves causing them not to close properly. Abnormality in a valve causes continuous leakage of blood, and the appropriate pressure needed for the blood flow does not build up.

There are two types of heart activities: 1) initiation and conduction of electrical activity in the cells causing a heart to depolarize and pump blood regularly, and 2) the regulation of the amount of blood and its flow. A healthy heart follows a regular sequence, magnitude and interval of P-QRS-T waveform. Abnormal electrical activities in the heart are caused by: 1) irregular origination of electrical activities causing heart to beat irregularly; 2) irregular conduction of electrical signals from *His bundle → [Right/ left bundle] → Purkinje fibers*; 3) enlargement of heart-muscles causing the spread of the corresponding waveform due to longer time taken by the electrical signal to travel and 4) starvation/death of the muscle cells hindering the depolarization and repolarization of the heart.

The result of these abnormalities in a heart causes: 1) improper sequence in *{P, QRS, T}* waveforms; 2) irregular beat-patterns; 3) change in amplitude and interval of P-wave, QRS -wave and T-waves; 4) inversion of T-waves and 5) a combination of regular and irregular beats. Depending upon the origin, intensity and randomness of the originating electrical activity, order could be shuffled randomly. A P-wave may be embedded in the corresponding QRS-wave, may come after the corresponding QRS-complex but before T-wave or could be embedded in the corresponding T-wave. Different combinations of these ECG-waves indicate different irregularities in the heart. Many conditions lead to brain-stroke, sudden cardiac death, improper flow of blood, increase in blood pressure and lack of oxygen circulation to the cells in the body causing multiple diseases. Variations in ECG may also be caused by the imbalance in the electrolyte levels of Na^+, K^+ and Ca^{++} ions.

Following sections discuss six types of heart diseases: 1) cardiomyopathy, 2) arrhythmia, 3) ischemia, 4) myocardial infarction, 5) left/right bundle block and 6) deficiency of electrolytes.

7.3.1 Cardiomyopathy

Cardiomyopathy is characterized by thickening of one or more muscles in the four chambers of the heart. The thickening of the heart-muscles causes the heart chamber(s) to be less flexible and incapable of pumping blood. If the muscle thickens without the enlargement of heart, then the volume inside the chamber decreases, and less blood accumulates and pumped from the chamber. As heart pumps a lesser amount of blood, heart-beat increases to pump required blood needed by the body. An enlargement of heart causes the heart-muscles to get weaker and over stressed and may lead to heart failure.

There are four subclasses of cardiomyopathy: 1) hypertrophic cardiomyopathy; 2) dilated cardiomyopathy; 3) restrictive cardiomyopathy and 4) unclassified cardiomyopathy. In *hypertrophic cardiomyopathy*, the enlargement of the heart is associated with the thickening of the heart-muscles. This causes less blood to flow in the body causing fluid to build-up in different parts of the body. Thickening of the muscles also deforms the valves and makes them inflexible. This deformity in the valve(s) causes blood to leak continuously.

In *dilated cardiomyopathy*, one of the four chambers gets enlarged. Ventricular enlargement is common. The enlargement causes heart-muscles to get weaker and cause heart failure or heart-valve problems. In *restrictive cardiomyopathy,* the muscles neither thicken nor enlarge, but they lose flexibility causing the heart to exert more pressure on muscles to pump the blood. Since stiff muscles do not relax, it pumps less blood to flow in the body. The fibrillation of heart-muscles stiffens them, and the corresponding tissues develop ectopic-nodes. Fibrillation in atria is called *supraventricular fibrillation*, and fibrillation in a ventricle is *ventricular fibrillation*. This fibrillation leads to irregular and low amplitude heartbeats. Ventricular fibrillation is very serious and may lead to *sudden cardiac death.*

7.3.1.1 Heart enlargement

Since the number of cells increases, electric vectors increases and amplitude of the P-wave and QRS-wave increases. If a chamber gets enlarged, then more time is taken to traverse the chamber spreading the corresponding waveform. P-wave comprises: 1) depolarization-wave due to the left-atrium and 2) depolarization-wave due to the right-atrium. Normally, the difference between the peaks of P^{left} and P^{right} is small. Hence, the superimposition gives a perception of a single waveform as shown in Figure 7.9a.

With the enlargement of the left-atrium, P^{left} shifts right, and the resulting superimposed P-wave becomes like a saddle with a higher spread (*>120* ms) as shown in Figure 7.9b. The P-wave axis shifts in counterclockwise direction. With enlargement of the right-atrium, P^{right} shifts slightly toward P^{left} causing the amplitude to rise as shown in Figure 7.9c. The P-axis vector shifts in clockwise direction.

The enlargement of a right-ventricle causes the Q-wave (related to right ventricle depolarization) to shift right and get partially occluded inside the corresponding R-wave. The enlargement of a left ventricle

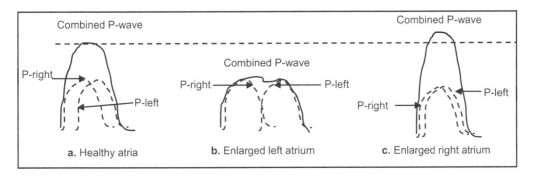

FIGURE 7.9   Effect of atrium enlargement on combined P-wave: (a) Healthy atria; (b) Enlarged left atrium; (c) Enlarged right atrium

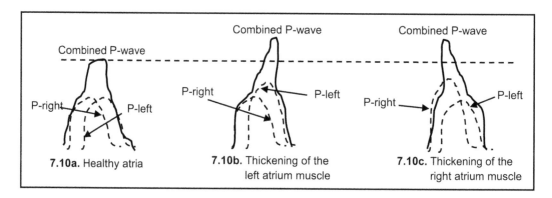

FIGURE 7.10   Effect of atrium hypertrophy on P-wave: (a) Healthy atria; (b) Thickening of the left-atrium muscle; (c) Thickening of the right-atrium muscle

causes R-waves to spread causing the QRS-complex to spread. The thickening of heart-muscles increases the number of cells increasing electrical activity; the corresponding waveform gets more amplitude and spread.

The thickening of the left-atrium muscles causes P^{left} to get larger. The combined P-wave gets bigger and wider, and somewhat asymmetric with peak shifting to the left. The thickening of the right-atrium causes P^{right} to become larger and more spread. The combined P-wave becomes more spread with a higher peak and somewhat asymmetric with the peak shifting to the right. The atrium hypertrophy is illustrated in Figure 7.10.

The thickening of the right-ventricle in a heart causes Q-wave to become pronounced. Since part of the depolarization wave is embedded inside the R-wave in a negative direction (vector points away in lead V4), the amplitude of R-wave in lead V6 becomes smaller. In the case of left ventricular hypertrophy, the amplitude of QRS-complex becomes very pronounced and spike-like (>25 mm) in the lead V5 or V6, and the conduction is delayed causing QRS interval to become bigger (> 0.12 ms). The sum of S-wave amplitude in V2 and R-wave amplitude in V5 increases beyond 35 mm. Figure 7.11 illustrates the effect of ventricular hypertrophy on QRS and T waveforms.

In the case of left ventricular hypertrophy (LVH) or left atrium enlargement, J-point depression and large T-wave inversion (>3 mm) are common. The magnitude of the inverted T-wave-peak in the lead V6 is greater than the inverted T-wave-peak in the lead V3. T-wave is asymmetric. The differences between ECG waveforms for ischemia (see Section 7.3.4) and LVH are complex and they look similar.

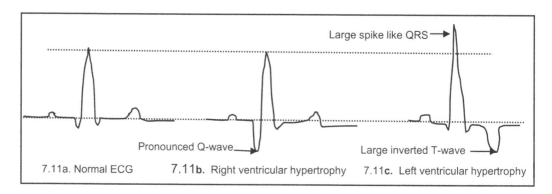

FIGURE 7.11   ECG waves for ventricular hypertrophy: (a) Normal ECG; (b) Right ventricular hypertrophy; (c) Left ventricular hypertrophy

7.3.2 Ectopic-Nodes and Arrhythmia

Arrhythmia is a common problem in old age. The heartbeat becomes irregular in arrhythmia. Heartbeat can either slow down or speed up. *Bradycardia* is a subclass of arrhythmia when the rate of heartbeats slows down. *Tachycardia* is another subclass of arrhythmia when the heartbeat speeds up. *Tachycardia* is divided into two subclasses: 1) *supraventricular arrhythmia* caused by abnormalities in the atria and 2) *ventricular arrhythmia* caused by abnormalities in the ventricles. Three major problems cause irregular heartbeats: 1) abnormality in ion-channels of heart cells that disturb the depolarization cycle of the SA-node and/or heart-cells; 2) change in the origin of the electrical activity in the heart and 3) fibrillation of heart-muscles. Fibrillation causes: 1) random alignment in heart-muscles disorienting the regular electrical pattern and 2) fibrous tissues that become an alternate source of irregular electrical activities in addition to the SA-node.

Normally, the electrical activity starts from the SA-node, goes through some delay in the AV-node, passes through *His bundle*, bifurcates to the right and left bundle branch and is followed by conduction through Purkinje fibers. In the case of excessive firing of the SA-node, only a smaller number of the signals can pass to the ventricles causing a circulating charge in atria. If the SA-node stops originating the electrical activity, then ectopic-nodes or AV-nodes become the originator of electrical activities, the sequence of the P-QRS-T waveforms is disturbed as the atria and ventricle are polarized concurrently and/or asynchronously. Ectopic-nodes are slow to conduct, have low amplitude and fire asynchronously causing irregular waveforms.

7.3.2.1 Supraventricular arrhythmia

There are six other causes of supraventricular arrhythmia: 1) rapid electrical firing from SA-node; 2) atrial fibrillation—arrhythmia with rapid unsynchronized electrical activity starting in the ectopic-nodes of the atria; 3) junctional tachycardia—electrical activities starting from the AV-node, His-bundle or immediately surrounding area ; 4) atrial flutter—a periodic reentrant circulating current formed due to the travel of electrical pulse to the left-atrium to depolarize at the rate to *250–300* beats/minute; 5) AVNRT—the depolarization wave from the slow path in AV-node that cancels out the repolarization wave from the fast path in the AV-node and (6) AVRT—atria-ventricular reentry tachycardia caused by additional asynchronous depolarization due to the additional conduction path between ventricles and atria. The simplest tachycardia is sinus tachycardia where SA-node fires rapidly (>*100* impulses/minute). The passage of an electrical signal is the same as the normal rhythm except that it happens at a faster pace. Different types of atrial fibrillations and the corresponding P-QRS-T waveforms are illustrated in Figure 7.12.

In atrial fibrillation (AFib), the waves starting from the atria may be from one or more ectopic-nodes in the pulmonary veins (vein carrying oxygenated blood from lungs to the left-atrium) or fibrillated atrial muscles. If the ectopic-nodes are further away from the AV-node, then the P-wave will be spread because it takes much longer for an electric impulse to travel slowly through the slow-conducting heart-muscles. In addition, the amplitude of the P-wave will be smaller.

If there are too many ectopic-nodes firing asynchronously, they interfere with each other causing P-wave interval, location and amplitude to change randomly. The techniques to identify the P-QRS-T waveform may even not be able to recognize the P-waves from the noise. AV-node does not pass all the electrical activities generated in the atria to the ventricles. There are more P-waves than QRS-T wave complexes causing multiple P-waves to be grouped followed by one QRST-T wave complex. Ventricles also depolarize at a faster rate. Atrial contraction is often incomplete due to the low magnitude electrical activities in atria.

Blood starts pooling in the atria due to insufficient atria contraction causing blood-clots to form. These clots occasionally dislodge from the left-atria, move with blood to the left-ventricle and move through the artery to the body. These blood-clots may travel to the brain and get stuck in smaller capillaries in the brain causing paralyzing brain-stroke. Blood clots can get stuck in other parts of the body also causing ischemia due to the lack of blood-flow. Like atria, ventricles do not get enough time to compress fully. Hence, enough blood does not go to body causing fatigue.

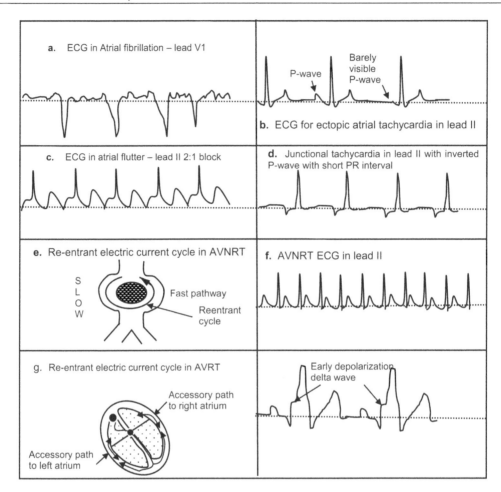

FIGURE 7.12   Cases of supraventricular arrhythmia and the corresponding ECG: (a) ECG in atrial fibrillation—lead V1; (b) ECG for ectopic atrial tachycardia in lead II; (c) ECG in atrial flutter—lead II 2:1 block; (d) Junctional tachycardia in lead II with inverted P-wave with short PR interval; (e) Re-entrant electric current cycle in AVNRT; (f) AVNRT ECG in lead II; (g) Re-entrant electric current cycle in AVRT

The waveform for atrial fibrillation is shown in Figure 7.12a. P-waves are mostly replaced by rapid low amplitude electrical activities called *F-waves* (Fibrillation waves). The peak of the F-wave is around *50* microvolts. The atrial excitation pattern is irregular. AF varies from *250* to *500* beats/minute. The RR-interval becomes irregular. Leads II and V1 have the largest ventricular and atrial waves' amplitudes and are used for the analysis of atrial fibrillation.

In ectopic atrial tachycardia (EAT), there are few discrete ectopic-nodes in the atria firing at a faster rate than SA-node. The electrical activity of the SA-node is weak, and it is interfered by the electrical activities of the ectopic-nodes. The axis of P-waves keeps changing, and the compression of the atria is erratic. The amplitude traveling through the AV-node is weak causing a low amplitude QRS-complex. Symptoms include shortness of breath, dizziness and palpitations. Due to extremely fast conduction activities in the AV-node, a heart may go through hemodynamic derangement and cardiac-arrest. The waveform for EAT is shown in Figure 7.12b.

In *Atrial Flutter (AFlu)*, the electric impulses circulate in the atria causing a circulating current that causes depolarization at the rate of *250–300* beats/minute. It occurs due to uneven delay in conduction in different parts of the heart-muscles caused by the inactivation of sodium ion-channels. The discharge of depolarizing current from the reentrant loops causes negative P-waves *90%* of the time, and P-wave may be missed around 10% of the time. Sometimes negative P-waves get superimposed with T-waves showing up as a missing T-wave or a T-wave with reduced magnitude. The waveforms for atrial flutter in lead II are given in Figure 7.12c. Atrial flutter occurs in patients suffering from cardiomyopathy, hypertension and diabetes.

In *Junctional tachycardia*, electrical activities arise around the *AV-node* and above the *bundle of His*. Atria and ventricles get depolarized concurrently, making P-wave in inferior leads (leads II, III and AVF) negative. PR-interval is negligible, and many times P-waves are embedded in or trail QRS-complex. The QRS-complex is usually narrow. Markov model-based analysis shows that negative P-wave with short PR-interval occurs *86%* of the times, and *14%* of the times P-wave is embedded in the QRS-complex or occurs after QRS-complex. The waveform for junctional tachycardia is shown in Figure 7.12d.

In *Atrio-Ventricular Nodal Reentry Tachycardia (AVNRT)*, AV-Junction has two paths: a slow path and a fast path based upon the conduction fibers present in the AV-junction. The fast pathway repolarizes slowly, and the slow pathway repolarizes quickly. This causes many possibilities: 1) wave entering the slow pathway and getting suspended for a longer period and 2) slow pathway repolarizing quickly and the electric signal traveling to both the fast pathway and slow pathway. This asynchronous behavior causes a reentrant cycle as illustrated in Figure 7.12e. The depolarization in atria and ventricles occurs concurrently causing P-waves to be negative. There are two major possibilities: 1) negative P-waves are embedded in the QRS-complex or T-waves *60%* of the times and 2) negative P-waves occur after QRS-complex and before T-waves *22%* of the time.

Normally, the depolarization waveform travels from *SA-node → AV-node → His bundle → Purkinje fibers* and stops. The depolarization is followed by the repolarization and waits for the next depolarizing electric pulse. However, in *Atrio Ventricular Reentry Tachycardia (AVRT)*, there is an accessory pathway through the external muscle-walls between the *Purkinje fibers* and the atria, which carries the depolarizing wave back to atria before the SA-node fires again. This causes atria to depolarize even before regular depolarization resulting into frequent atria contractions (*250–300* beats/minute) and is felt as heart palpitations. The pathway is illustrated in Figure 7.12g, and the corresponding waveform is shown in Figure 7.12f.

Wolff–Parkinson–White's (WPW) syndrome is a variation of AVRT. Like AVRT, tachycardia is caused due *to SA-node → atria → AV-node → His bundle → Purkinje fiber → extra-pathway → atria* causing atria to depolarize before regular SA-node fires. In addition, the conduction capability of the fast-conducting extra-pathway and the slow conducting AV-node causes a time-differential. The electric *impulse SA-node → atria → extra-pathways → ventricles* start depolarizing the ventricles before the regular pathway involving *AV-node → His bundle*. QRS-complex has an initial bulge caused by the early depolarization. This bulge is called *delta-wave*. The delta-wave is followed by the regular QRS-complex.

7.3.2.2 *Ventricular arrhythmia*

Ventricular arrhythmias occur when the electrical activity originates in multiple ectopic-nodes in ventricles instead of the SA-node, AV-node or ectopic-nodes in atria. Since the ectopic-nodes occur in ventricles, the P-waves are invariably embedded in QRS-complex and are of negative amplitude. Depending upon the locations of the ectopic-nodes, ventricular tachycardia has been classified into: 1) *Premature ventricular complex*; 2) *ventricular tachycardia*; 3) *ventricular flutter* and 4) *ventricular fibrillation*.

Premature ventricular complex (PVC): The ectopic-nodes in the ventricles fire causing an irregular long-interval QRS-complex. A negative P-wave is embedded inside the corresponding QRS-complex. T-waves are in the opposite direction to the QRS-complex. Often, PVC is combined with regular P-QRS-T wave due to irregular firings of SA-node. This causes strange mixture of regular waveforms and PVC-waveforms. Two most common subclasses are *bigeminy* and *trigemini*. In *bigeminy*, a regular P-QRS-T wave arising from the SA-node is followed by an irregular PVC-waveform. In *trigemini*, two regular P-QRS-T waveforms arising from the SA-node are followed by an irregular PVC-waveform. This class of combined regular P-QRS-T-waveform and periodic fixed-number PVC-waveforms is called *regularly irregular waveforms*.

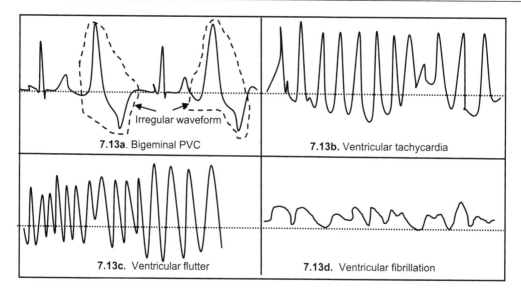

FIGURE 7.13 Ventricular arrhythmia: (a) Bigeminal PVC; (b) Ventricular tachycardia; (c) Ventricular flutter; (d) Ventricular fibrillation

An example of a PVC wave and bigeminy is shown in Figure 7.13a. In an irregular PVC waveform, the electric impulse conduction through the heart-muscles is slow spreading the QRS-complex. T-waves are inverted and have large duration, showing large repolarization time. In ventricular tachycardia, the origin of the electric impulse is the region just below the region of *His bundle*. Both atria and ventricles depolarize in a disassociated manner. The PR interval is short, or the negative P-wave is embedded inside the QRS-complex. Ventricular tachycardia starts suddenly and stops suddenly with no pattern. Heartbeats are over *100* beats/minute. Figure 7.13b illustrates the ECG for ventricular tachycardia.

Ventricular flutter is a fast tachycardia with *250–300* beats/minute. The waveform is more like a sinusoidal wave with no clear P-wave, QRS-complex or T-wave. It happens due to severe ischemia. Ventricular flutter, if left alone, turns quickly into ventricular fibrillation. The waveform for ventricular flutter is shown in Figure 7.13c.

Ventricular fibrillation is characterized by rapid, asynchronous, low amplitude electrical activities insufficient for contractions. The blood stops flowing. It is characterized by nonidentifiable rapidly changing (*350–500* beats/minute) irregular beats with no identifiable waveform. Often, it is irreversible. It leads to *sudden cardiac death*, if untreated immediately. A waveform-pattern for the ventricular fibrillation is illustrated in Figure 7.13d.

7.3.3 Ischemia and Myocardial Infarction

Ischemia occurs when there is lack of oxygen in the blood or there is lack of blood supply to the cells due to a shortage of blood-flow. Blood-flow is obstructed due to the blockage in the artery caused by an excessive deposit of cholesterol or the presence of a blood-clot. The blockage of an artery causes the lack of oxygen going to the cells around that area and causes death of the neighboring cells in the heart-muscles. Dead cells do not depolarize or repolarize. However, cells die in a random manner with no pattern causing a mix of dead and alive cells that becomes the source of ectopic-nodes. Currents around the ischemic cells go in bizarre cyclic patterns. Untreated ischemia leads to myocardial infarction—death of the muscle cells. The formation of ectopic-nodes leads to arrhythmia as described in the Section 7.3.3.

Ischemia is characterized in ECG waves by symmetrical and sometimes inverted T-waves in the lead V5. The inverted T-wave is less than 3 mm in the lead V6, and the magnitude of inverted T-wave-peak in

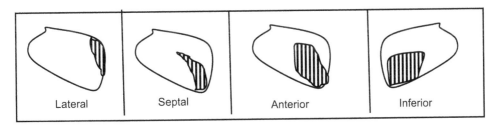

FIGURE 7.14 Regions of myocardial infarction depending upon blocked arteries

the lead V6 is less than or equal to magnitude of the T-wave-peak in the lead V3. Ischemia also depresses ST-segment that can be identified by the J-point—the junction point between the end of the S-wave and the start of isoelectric line. Ischemia has also been identified using frequency-analysis of ST-segment. Ischemic beats contain lower frequency compared to normal beats. Many techniques such as wavelet transforms, neural networks, PCA analysis and combinations of these techniques have been used to derive J-points and ST-segment depression. One issue is the presence of low-frequency noise that makes the identification of depression difficult. Another issue is that ST-segment depression is also caused by oxygen deficiency due to pulmonary embolism—blockage in the lung artery caused by blood clots traveling from other parts of a body.

Complete myocardial infarction is characterized by an elevated J-point and elevated ST-segment due to incomplete depolarization caused by the dead cells in the heart-muscles and incomplete repolarization resulting in abnormal T-waves for the same reason. Depending upon the region of the cell-death, different leads will show varying deviations of the ST-segment and J-point. For example, dead tissues in the lateral wall (side wall) will affect leads I, AVL, V5 and V6; dead tissues in the inferior wall will affect the leads II, III and AVF; dead tissues in the anterior wall (front wall) will affect the leads V3 and V4; and dead tissues in the septal wall (wall partitioning left and right chambers) will affect the leads V1 and/or V2. This information is used to localize the region of infarction and corresponding blocked arteries causing infarction. The different types of infarction are shown in Figure 7.14.

Complete myocardial infarction is also known as STEMI (ST-segment Elevated Myocardial Infarction) and is associated with intense pain in the chest that often radiates to neck, jaw, shoulder and arm. Certain enzyme biomarkers are also elevated. STEMI can cause arrhythmia due to fibrillation of the heart-muscles that forms ectopic-nodes. Often, the artery is not completely blocked causing random death of some cells; the limited blood supply keeps other cells alive. In that case, ST-segment does not get elevated and is called NSTEMI. Figure 7.15a shows a typical ECG with ischemia without myocardial infarction, and Figure 7.15b shows a typical ECG with myocardial infarction.

If a blockage occurs in the AV-node, then the time taken by the electric impulse from atria to the ventricles increases elongating the PR-interval. If the blockage only causes delay in the electric impulse, then the heart beats at a slower rate, and blood is pumped to the body at the slower rate not showing any serious outward symptom. However, occasional and random blockage of electric impulses in the AV-node

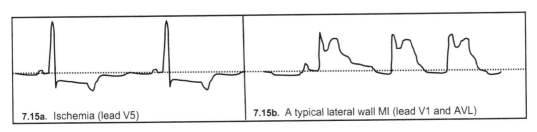

FIGURE 7.15 A typical ECG for ischemia and myocardial infarction: (a) Ischemia (lead V5); (b) A typical lateral wall MI (lead V1 and AVL)

due to scarred tissues causes random skipped contractions of ventricles resulting in lack of oxygen and nutrients in the body-cells. A person with skipped heartbeats feels light-headed and dizzy. This form of block is called *second degree block*. This can be corrected using pacemakers.

The last type of the blockage is the *third degree block* where electrical impulses are excessively blocked by the AV-node. In the case of complete block, electrical pulse is not conducted from atria to ventricles. The electric depolarization impulse may not travel to the ventricles often resulting into many P-waves not followed by the corresponding QRS-complex as ventricles do not depolarize. The effective heart-rate drops significantly to *30–40* beats/minute. The QRS-complex in the ventricles is generated by the ectopic-nodes in the ventricles having no synchronization with contractions in atria. QRS-complex becomes large around *360* milliseconds.

7.3.4 Electrolyte Imbalance

The imbalance in sodium, potassium and calcium can cause an imbalance of electrolyte levels in the heart-cell affecting the depolarization and repolarization of the heart-cells. Na^+ and Ca^{++} ions intake plays a key role in cell depolarization, and K^+ leaks out of the cells in the resting phase to maintain the resting potential. For out of range ion-concentrations, the ECG changes and the person is treated for electrolyte imbalance. A normal potassium level is between *3.6* and *5.2* millimoles/liter.

Hyperkalemia (excessive potassium) or hypokalemia (low potassium) cause abnormal heart rhythms. *Hyperkalemia* suppresses the depolarization of heart suspending the heart activity thus causing heart-failure. Hyperkalemia can also affect other organs also such as kidney failure or muscle paralysis. Hyperkalemia results into bradycardia in the form of AV-blocks, causes slow depolarization resulting into QRS-complexes of longer duration. In *hyperkalemia*, T-wave is symmetrical with high amplitude (>*2.5* mm), QT-interval is shortened and the amplitude of P-wave decreases. In the case of *hypokalemia* (less potassium), the changes in ECG are caused by the delayed ventricular repolarization. The T-wave can either get flattened or get inverted with a prominent U-wave and the depression of ST-segment. QRS-complex may get widened too. QU-interval is elongated. In addition, PR-interval is elongated from normal *0.2–0.32* second.

Low concentration of sodium in blood is called *hyponatremia*. A low concentration of Na^+ ion causes the lack of cell-depolarization resulting into lack of contraction of atria and ventricles reducing the blood-flow in the body. The body-cells get deoxygenated and lack necessary nutrients. Deoxygenation of body-cells leads to confusion, nausea and headaches. The PQ-interval and QRS-complex get elongated showing slow depolarization caused by lack of sodium concentration outside cells during depolarization. Severe case of hyponatremia can result into first-degree of AV-block and later turn into second-degree and third-degree AV-block.

In case of *hypercalcemia* (excessive calcium), the QT-segment and ST-segment shorten showing that it takes less time for depolarization. The T-waveform flattens and has longer duration showing that it takes more time to repolarize due to longer duration taken to flush Ca^{++} ions from the cells. If *hypocalcemia* occurs (less calcium), the QT segment and ST segment lengthen showing that it take more time to depolarize. The duration of the T-waveform shortens showing that it takes less time flush out Ca^{++} ions from the cells.

7.4 COMPUTATIONAL TECHNIQUES FOR DETECTING HEART DISEASES

Based upon the disease states, the changes in ECG waves are classified as: 1) change in durations of P-waveform, QRS-complex and T-waveform; 2) interval changes such as PR-interval change, QT-interval change, ST-interval change and RR-interval changes; 3) change in slope such as ST-segment depression or ST-segment elevation; 4) change in the amplitude and slope of such as P-wave, R-wave and T-wave; 5) inversion of the T-waves; 6) specific waveform pattern as in atrial flutter, ventricular flutter and ventricular fibrillation; 7) splitting of P-waves into P^{left} and P^{right} during heart enlargement; 8) changes in the order of occurrence of P-waves in the case of ectopic-nodes in atria, AV-node and ventricles and 9) various combinations of changes specified in (1) to (8).

The entire analysis is classified into three categories: 1) morphology analysis about shape, slope and duration of different waveforms; 2) interval analysis such as QT-interval, QT^C interval, PR-interval and ST-segment and 3) probabilistic temporal analysis of transition between P-QRS-T waveforms. Morphology analysis identifies the diseases based upon magnitude changes, T-wave inversion, elongation or compression of time-intervals and shape of the waveforms.

Probabilistic temporal analysis is good for the analysis and the identification of the different subclasses of arrhythmia where atria and ventricular depolarization can occur asynchronously due to the irregular excitation of ectopic-nodes. Probabilistic temporal analysis models the transition between P-QRS-T waveform statistically and makes a Markov model for each disease using a large sample of P-QRS-T waveforms. These Markov models are stored in a library. A patient's ECG is analyzed to extract the P-QRS-T waveforms. A *Probabilistic Transition Graph (PTG)* is made out of these P-QRS-T waveforms acquired in a time-window in real-time. A *Probabilistic Transition Graph (PTG)* is like a Markov model with a limited sample-size. The PTG is matched with the Markov models in the library to identify the best-match. The best-matching Markov-model identifies the specific subtype of arrhythmia.

7.4.1 Morphological Analysis

Morphological analysis of ECG is based upon measuring various segment-intervals, slope of the waveforms, peaks and inversion of the waveforms in different leads and connecting the conditions using logical operators. These measurements are compared against the statistical analysis of the waveforms of healthy individuals. Different classes of diseases give rise to different types of deviations that can be derived using clustering and verified using common-sense reasoning of the underlying physiological process as described in Section 7.3. Various morphological features are: 1) P-wave duration; 2) slope of the P-wave; 3) amplitude of the P-wave; 4) presence of a saddle in P-wave; 5) PQ interval; 6) PR interval;7) P-wave inversion; 8) missing P-waves; 9) magnitude of Q-wave-peak; 10) magnitude of R-wave-peak; 11) QRS interval; 12) slope of the QRS-complex;13) magnitude of S-wave-peak; 14) QT interval and QT^C interval; 15) ST segment elevation; 16) J-point location; 17) T-wave interval; 18) magnitude of T-wave peak; 19) T-wave inversion and 20) RR-interval.

Morphology-based detection is suitable for those diseases where the shape and intervals of the waveforms are altered. However, the sequence of P-QRS-T waveforms should be maintained in ECG. The examples of such abnormalities are: ischemia, myocardial infarction and electrolyte imbalance. RR-interval analysis can also reason about frequency variations of the ECG waveforms and can identify bradycardia (slow beats) and tachycardia (fast beats). Morphology analysis can also help in a limited way in heart enlargement and myopathy.

7.4.1.1 Limitations of morphology analysis

Morphology analysis fails to identify different subclasses of arrhythmia such as different types of supraventricular arrhythmia or ventricular arrhythmia that require analysis and inferencing in the presence of irregular beat-patterns. Since the treatment for different subclasses is different, it is imperative that computational techniques should identify an exact subclass of arrhythmia.

Further subclassifications of arrhythmia are not usually possible by simple morphology analysis due to: 1) multiple disease-states that interfere with each other and 2) analysis of transitory temporal variation of P-QRS-T waveform is not possible. Ectopic-nodes causing ventricular arrhythmia or junctional arrhythmia alter the order of P-waves randomly: P-wave can get embedded in QRS-complex; come after QRS-complex; or get embedded in T-waves. In the case of conduction block, multiple P-waves may occur before a regular QRS-complex occurs. Analysis of such waveforms requires statistical analysis. Markov model is one such technique to perform statistical analysis to study the pattern variations of P-QRS-T waveforms.

7.4.2 Applying Markov Models

P-QRS-T waveforms have been derived using Markov Model with eight states: 1) one each for five waveforms and 2) three states for PQ, ST and TP segments. Markov models have been used to find subclassifications of different types of arrhythmia in real-time because it can model missing waveforms, embedded waveforms and change in the sequence of P-QRS-T waveforms.

The technique first trains from a curated ECG database to make a Markov graph for each subclass of arrhythmia. The patient's ECG is collected at run-time using a sliding window. A probabilistic transition graph is derived using this limited number of P-QRS-T waves in the window, and the resulting graph is matched with the Markov graphs of the subclasses of arrhythmia to derive the exact disease label.

Example 7.1

Figure 7.16 illustrates Markov model for atrial fibrillation – one of the supraventricular diseases caused by irregular rhythm. Since the electrical activity originates in the atrium, P-wave from the left-atrium and P-wave from the right-atrium are split to give four states P_{11}, P_{12}, P_{21} and P_{22} instead of just one state. P_{11} corresponds to the rising edge of the P-wave from the left-atrium; P_{12} corresponds to the falling edge of the P-wave from the left-atrium; P_{21} corresponds to the rising edge of the P-wave from the right-atrium and P_{22} corresponds to the falling edge of the right-atrium. The splitting of P-wave gives a lot more information about statistical variations of the P-wave morphology changes.

This Markov model holds all the information for the characterization of *atrial fibrillation*. The transitions are: $Iso_3 \rightarrow P_{11}$ (P-waves occurrence with low probability); $Iso_3 \rightarrow Iso_1$ (P-wave missing with high probability 0.9); $P_{11} \rightarrow P_{22}$ (conditional probability *0.31*); $P_{11} \rightarrow P_{12} \rightarrow P_{21} \rightarrow P_{22}$ (P-wave splitting, conditional probability 0.69); $P_{22} \rightarrow Iso_1$ (P-wave to isoelectric line always if P-wave is present); $Iso_1 \rightarrow Q$ (Q-wave follows PR-segment with high probability 0.91); $Iso_1 \rightarrow R$ (Q-wave missing with low probability 0.09 possibly due to overlap with P-wave); $Q \rightarrow R$ (R wave always follows Q-wave); $R \rightarrow S \rightarrow Iso_2$ (S-wave follows R-wave with higher probability *0.67*); $R \rightarrow Iso_2$ (S-wave missing with probability *0.33*); $Iso_2 \rightarrow T \rightarrow Iso_3$ (T-wave present with high probability *0.87*); and $Iso_2 \rightarrow Iso_3$ (T-wave missing with low probability *0.13*). In atrial fibrillation, multiple ectopic sites depolarize asynchronously. Low amplitude action-potentials fire rapidly. Thus, P-waves may not be seen on the ECG as illustrated by the transition $Iso_3 \rightarrow Iso_1$.

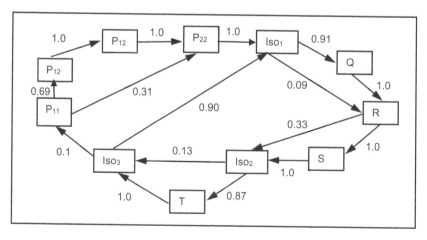

FIGURE 7.16   A Markov model for atrial fibrillation

7.4.2.1 Matching PTG with Markov models

PTG has a similar graph as the Markov model. It has the same number of nodes. However, some transitions may be missing or may have different probability (edge-weights). The graph-matching algorithm works in two steps: 1) makes sure that all the edges present in the PTG are also present in the Markov model and 2) finding dissimilarity-score between PTG and all the related Markov models that pass the step one. The disease corresponding to the Markov model with the least dissimilarity score is used to label the patient's condition.

A dissimilarity score between the PTG and a Markov model is calculated by summing up the absolute difference between the edges of the most probable path in PTG and the corresponding edges in the Markov model as given in the Equation 7.5. The most probable path is calculated using probabilistic forward reasoning with expectation maximization.

$$Dissimilarity\ score = \sum_{e_i \in MPP} \left| \left(weight\left(e_i^{PTG} \right) - weight\left(e_i^{Markov\ model} \right) \right) \right| \tag{7.5}$$

Example 7.2

Consider a PTG from a patient in Figure 7.17. It contains all the corresponding edges as in the Markov model. Assume that the threshold value is *0.1*. Hence, the first test is met. The most probable path in the PTG is $Iso_3 \rightarrow Iso_1 \rightarrow Q \rightarrow R \rightarrow S \rightarrow Iso_2 \rightarrow T \rightarrow Iso_3$. The cumulative differences probability of the path is given by $|0.94 - 0.90| + |0.73 - 0.91| + 0 + 0 + 0 + |1.0 - 0.87| + 0 = 0.35$. The comparisons with all other Markov models should yield dissimilarity scores > 0.35 if this PTG qualifies for *atrial fibrillation*.

7.4.3 Neural Network-Based Analysis

Neural network has been applied as a classifier on various extracted features to identify various disease states such as ventricular hypertrophy, ischemia and myocardial infarction. The features are different variations and deviations in duration and amplitude. The accuracy of the neural network approach varies from *63% to 80%*. Similarly, time-domain analysis of duration of the waveforms, segment durations and RR-interval has been applied to identify ventricular fibrillation and atrial fibrillation to an accuracy of *95%*. However, ANN-based schemes are somewhat limited in identification of arrhythmia due to the lack of reasoning of probabilistic transition between various states and the embedding of P-waveforms when the electrical activities originate from ectopic-nodes near the junction area or in the ventricles.

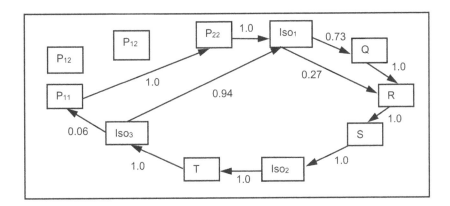

FIGURE 7.17   An illustration of the probabilistic transition graph of a patient

7.4.4 Other Techniques

Other AI techniques have also been used with success on a subset of heart diseases with moderate success. The techniques include: 1) the use of HMM for arrhythmia analysis; 2) use of the support vector machines to identify supraventricular arrhythmia and ventricular arrhythmia using morphological features of QRS-complex and RR-interval variations; 3) clustering using GMM (Gaussian Mixture Model) to identify arrhythmia using R-wave amplitude and FFT (Fast Fourier Transform) coefficients and 4) clustering based techniques using morphological features of different waveforms and RR-interval.

7.5 ELECTROENCEPHALOGRAPHY (EEG)

EEG is the surface level recording of various electrical activities in a brain. Unlike ECG where a repeat pattern is well understood in terms of the underlying physiological process, the physiology of the EEG waves is not fully understood. EEG waves comprise many types of waveforms with different frequencies. EEG signal is a combination of *alpha waves, beta waves, gamma waves, delta waves, mu waves* and *theta waves*.

When an electrical activity occurs in the brain, a faint magnetic activity is generated in accordance with Maxwell's equations of electromagnetism. MEG is a surface recording of these magnetic fields generated due to neuronal electrical activities. The major advantage of MEG is that it is not affected by noise, and is localized to the same surface area where the neuronal activity occurs. This property of MEG makes them suitable for localizing the origin of abnormal neuronal activities. A combination of MEG with functional MRI (fMRI) has been used to localize tumors, aneurysm of the arteries and cerebrovascular lesions.

7.5.1 Lead Placements for EEG Measurement

Leads are placed on the scalp using 10–20 scheme that uses 23 electrodes placed in a two-dimensional plane as shown in Figure 7.18. The arrangement is called 10–20 system because the skull perimeter is

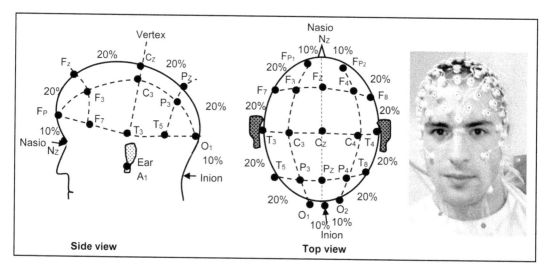

FIGURE 7.18 Placement of electrodes using 10–20 scheme on a human scalp (EEG cap placement image is courtesy Douglas Myers, website: https://commons.wikimedia.org/wiki/File:EEG_cap.jpg, public domain).

evenly divided into *10%* on the transverse and 20% on the longitudinal lines. The nomenclature of the electrodes is based upon the placement location.

The electrodes are aligned with: 1) ear lobe denoted as $A_{i(1 \leq i \leq 2)}$; 2) central – midline of the scalp denoted as $C_{i(i>0)}$; 3) nasopharyngeal – aligned with the nose and denoted as F_P; 4) parietal (towards back of the skull) denoted by P_i; 5) frontal polar denoted as $F_{i(i>0)}$; 6) temporal-lobe (side of the skull near ear) denoted by $T_{i(i>0)}$ and 7) occipital (back of the skull) denoted by $O_{i(i>0)}$. The reference points are the bottom of an earlobe, two eyes, nasion: top mid-point of a nose, and inion: mid-point of the back of the skull as shown in Figure 7.18. This causes 23 points as illustrated in Figure 7.18. Even numbered electrodes are placed on the right hemisphere, and odd numbered are placed on the left hemisphere. The subscript "z" denotes the central longitudinal line joining *nasion* and *inion*.

Electrodes are placed on each of these terminals and intersection points. The nomenclatures for the electrodes are: A_1 (left earlobe), F_{P1}, F_{P2}, F_7, F_3, F_Z, F_4, F_8, T_3, C_3, C_Z, C_4, T_4, A_2 (right earlobe), T_5, P_3, P_Z, P_4, T_8, O_1, O_2, I_Z and N_Z. There are other configurations that divide a skull with a 10-10 configuration using extra electrodes for better resolution. In 10-10 configuration transverse and longitudinal lines are drawn at every *10%* of the perimeter. Recent arrangements include as big as *128* or *256* electrodes on the scalp for much higher resolution.

7.5.2 EEG Recordings

There are two types of recordings: *monopolar* and *bipolar*. In *monopolar recording*, one electrode is put in the interest area on the scalp, and the other electrode is put away from the scalp near an ear-lobe. In bipolar recording, both the electrodes are put on the scalp and measure the difference in electrical activities between two electrodes on the scalp. A typical EEG signal has an amplitude of *10–100* microvolts. EEG signals are dynamic and nonlinear in nature.

Medications are discontinued during EEG recordings. EEG is usually recorded based upon the Nyquist criterion of sampling at the twice the frequency of the highest frequency component. Hence, it is sampled around *100–200* Hz. EEG signals are recorded in two modes: 1) spontaneous relaxed mode in a resting condition with eyes closed to reduce the artifacts caused by eye-blinking and 2) in response to a stimulus such as light flash or solving a task. Figure 7.19 shows a near *10* second region-wise recording of an epoch for a healthy 24-year-old patient. The channels are grouped into regions. The first group of four channels shows cerebral activity. The second group shows left parasagittal (plane parallel to the bisecting line of the head) region. The third group shows right parasagittal region. Groups four and five show the signals in left and right temporal-lobe region respectively. By associating these regions with the functional regions of the brain, different waveform under abnormal conditions can be studied in different functional regions.

7.5.3 EEG Artifacts and Noise Removal

EEG rhythm is affected by many things such as sleeping mode, diseases, sweating, eye-motion, motion of other muscles, heart-beats, electrode saturation, power-line, medications and stress. The signal is also distorted due to excessive amplification required to process the signal. The noise caused by stress, sleep deprivation and eye-motion is filtered out to study the pattern changes in the EEG waves.

Artifacts due to eye-blinking are monitored by putting the electrodes next to the eyes at the top and bottom. Muscle activity is characterized by high-power activity around *15–20* Hz, while EEG is low-power activity usually above *25* Hz. Noise due to muscles interferes with β-waves. Filters are used to reduce the effect of noise due to muscles. However, the use of filters also distorts the EEG waveforms. Sharp amplitude changes due to sweating are checked by setting up cutoff amplitude or significant deviation of the amplitude. EEG at rest is identified by taking the epoch (significant period) with minimal power in the band *1.5–7.5* Hz.

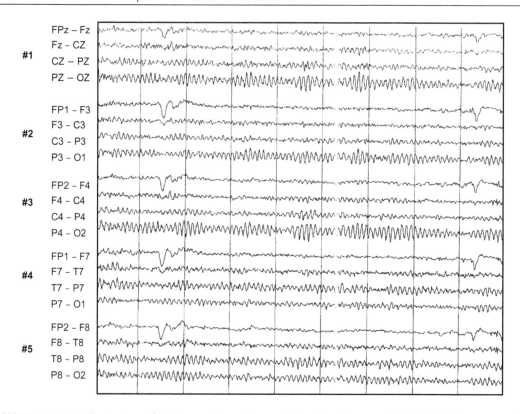

FIGURE 7.19 An illustration of EEG recording of normal healthy 24-year-old individual (Image courtesy St. Louis, Erik K, and Frey, Lauren C. (editors), *Electroencephalography – An Introductory Text and Atlas of Normal and Abnormal Findings in Adults, Children, and Infants,* Chicago, IL, American Epilepsy Society (www.AESNET.org), 2016, available at https://www.ncbi.nlm.nih.gov/books/NBK390354/pdf/Bookshelf_NBK390354.pdf, DOI: 10.5698/978-0-9979756, Figure 1, permitted under e Creative Commons Attribution-NonCommercial-ShareAlike 4.0 International Public License (BY-NC-SA: http://creativecommons.org/licenses/by-nc-sa/4.0/legalcode)).

Noise removal in EEG waves is difficult because artifacts have similar frequency range as EEG waveforms. Noise removal is done in two ways: 1) selecting the good epochs (usually 20–60 seconds) with little noise and 2) cleaning up the artifacts from the EEG signals. Artifacts are identified by: 1) setting cutoff amplitudes; 2) checking spectral power in certain frequency bands; 3) deriving the variance and checking for the significant deviations and 4) monitoring for signals or electric potentials that cause the noise such as eye-motion and muscle-signals. Shorter epochs around *20* seconds give better results, especially for high-frequency components. The most popular techniques to separate EEG waveforms are ICA (Independent Component Analysis described in subsection), CWT (Continuous Wavelet Transforms) and the combinations. The preferred mother wavelets used in CWT are: Daubechies (Db), coiflets (coif) and dmey. Different mother wavelets are shown in Figure 7.20.

7.5.4 EEG Waveforms

There are six types of waveforms in an EEG recording: alpha-wave, beta-wave, gamma-wave, theta-wave, delta-wave and mu-waves. Figure 7.21 illustrates alpha, beta, gamma, theta and delta waves.

Alpha-waves range from *8* to *13 Hz*, is found on the posterior (back), parietal and occipital region of the brain and has an amplitude around *20–200* microvolts. Alpha-waves appear only during the time a

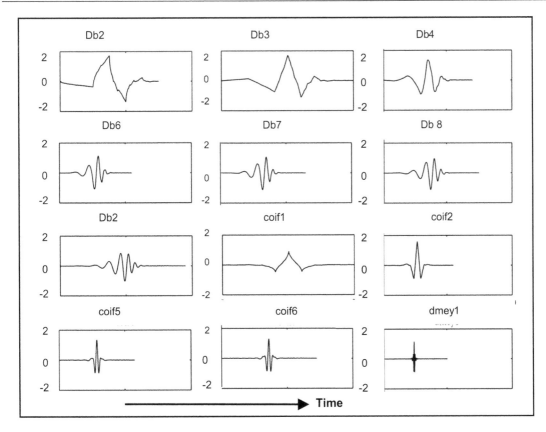

FIGURE 7.20 Different mother wavelets used in EEG waveform extraction (Image courtesy Hindawi Publishing Corporation; Al-Qazzaz, Noor K., Bin, Sawal, H., Ahmad, Siti A., Chellappan, Kalaivani, Islam, Mohammed S., and Escudero, Javier, "Role of EEG as Biomarker in the Early Detection and Classification of Dementia," *The Scientific World Journal*, Vol. 2014, Article Id 906038, 2014, DOI: 10.1155/2014/906038, Figure 8, permitted under Creative Commons Attribution License).

person is awake, relaxed and mentally alert with eyes closed without any engagement in activities. Alpha-waves disappear during sleeping and are replaced by asynchronous waves when a person is involved in intense activities.

Beta-waves range from *13* to *30* Hz and are found in the frontal electrodes. The amplitude of beta waves varies from *5* to *10* microvolts. Beta waves appear in response to the excitation of the central nervous system. Beta waves are associated with anxiety, arousal state or intense mental activities. Maximum amplitude of beta waves is around *20* microvolts. The frequency of beta-waves increases during anxiety and stress. Beta waves replace alpha waves during cognitive impairment.

Gamma-waves are in the range of *30–100* Hz and are associated with cognitive functions. The waveform is recorded in the somatosensory cortex of a brain. It increases during intense mental activities such as perception or enhanced mental tasks. They are associated with sensory processing, object-recognition, sound-recognition and tactile sensation.

Delta-waves have frequency below *3.5* Hz and are found in young infants. Its amplitude ranges from *20* to *200* microvolts. Delta-waves are associated with deep non-REM sleep. Delta-waves are stronger in the right hemisphere and originate in thalamus part of the brain. Delta-waves play a major role in the formation and consolidation of episodic memory.

Theta-waves range from *4* to *8* Hz and are found during sleeping, emotional stress, drowsiness and meditation. Its amplitude ranges from *5* to *10* microvolts. It often occurs among infants and young

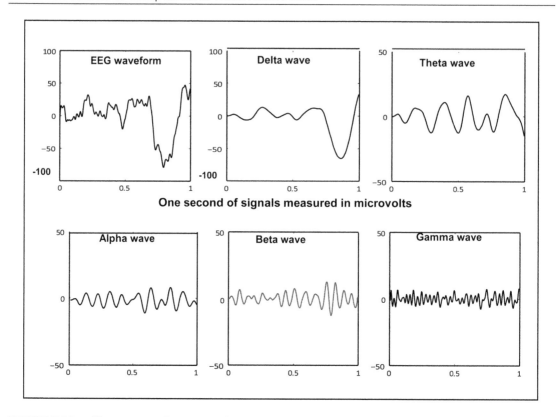

FIGURE 7.21   Different types of EEG waves (Image courtesy Hindawi Publishing Corporation, Al-Qazzaz, Noor K., Bin, Sawal, H., Ahmad, Siti A., Chellappan, Kalaivani, Islam, Mohammed S., and Escudero, Javier, "Role of EEG as Biomarker in the Early Detection and Classification of Dementia," *The Scientific World Journal*, Vol. 2014, Article Id 906038, 2014, DOI: 10.1155/2014/906038, Figure 8, permitted under Creative Commons Attribution License).

children. It is also associated with arousal in older children and adults, emotional stress and idling. This wave is associated with focusing, attention, mental effort and stimulation processing.

Mu-waves range from *7.5* Hz to *12.5* Hz and are associated with motor-activity. Mu-waves are detected in the motor-cortex and are most prominent when a body is in the resting phase. Mu waveforms originate as synchronized activity of neurons. However, when a body moves, that adds to desynchronization, and mu-waves diminish. Mu-waves also diminish when a brain observes movements of another person.

7.5.5 EEG Analysis

EEG signals are multivariate time-series data associated with some stochastic process. It also has to be compared against MEG, MRI and PET data. After preprocessing the EEG waves, features are extracted in the time-domain and in the frequency-domain. Feature extraction involves identifying the presence or absence of different types of waves, their mean-amplitude and the variance in the amplitude. The presence and absence of different types of waveforms (*alpha, beta, gamma*, etc.) are ascertained using *Fast Fourier Transform* that separates the waveforms in the frequency-domain.

After the feature extraction, various AI techniques such as neural network, HMM and cluster analysis are used: 1) to classify the waveform to different types of EEG waves and 2) to derive the deviation from the normal EEG waves. The prediction of epilepsy is done using techniques that detect the lack of random nature of EEG before epileptic attack as described in the Section 7.7.3. Dementia-related diseases are predicted using the change in the composition of different frequencies of waveforms using

Fourier transforms that analyze in the frequency-domain or CWT (Continuous Wavelet Transform) that analyzes EEG both in the frequency-domain and time-domain as described in Section 7.7.2.

7.6 MAGNETOENCEPHLOGRAPHY (MEG)

The basic principle of electromagnetism states that flow of electric current generates electro-magnetism around it. The neuronal electrical activities are due to: 1) axon potential generating fast discharge of current and 2) potential difference between basal (base) and apical (end) dendrites or cell-nucleus during cell excitation or inhibition. Individually, the current is small. However, when cells fire in a synchronized manner, then the electric and the corresponding magnetic activity can be detected.

7.6.1 MEG Recording

MEG sensing technology includes magnetometers – a pickup coil to measure magnetism. The coils are put on the scalp. The magnetic wave has two components: vertical going away from the source of electrical activity and the scalp position and parallel to the scalp. The vertical component of the magnetic field generates an electric current in the coil, which is magnified using superconductors cooled by liquid helium. Since the magnetic activity is very weak, noise-filtering and boosting the signal are done in a magnetically shielded room. Signal-to-noise ratio is very important. The magnetic activity is sampled at *5,000* Hz. Magnetic activity is expressed in *femto-Tesla (10^{-15} Tesla)*.

7.6.2 Localization of Neural Activities

Lesion or tumor in the brain is localized before performing any surgery. The deviation from regular EEG or MEG can be exploited to localize the change in electrical or magnetic activities caused due to tumor. EEG measures the surface-potential differences at multiple electrodes, and any electrical activity deep in the brain affects all the electrodes due to radial spread of electrical activities. In addition, 1) random noise is present due to EMG, eye-blinking, emotional stress and 2) sources of electrical activities interfere with each other at the surface level electrodes. MEG has advantage over EEG for localization because: 1) MEG signal does not deteriorate while traveling from origin to the surface and 2) measurements are independent of any reference.

There are two approaches to localizing the source of original electrical activities: 1) forward modeling and 2) inverse modeling. *Forward modeling* is a simulation technique that calculates the value of the electric and magnetic potentials at surface electrodes for a single electrical activity in the brain using Maxwell's equation. The technique is based upon these principles: 1) The curl of the magnetic field at a location r (in spherical coordinates) depends upon the density and 2) divergence of the magnetic field is zero. The model uses a sphere or multiple-layer sphere to calculate the electric potential. The model uses a *beam-forming approach* to calculate the surface potential and filter the data from other sources that cancel out the electric potential due to other sources.

Beam-former constructs the electric and magnetic potential at the surface from one source of electrical activities in the brain. There are two limitations of forward modeling approach: 1) number of sources of intense electrical activities should be known in advance and 2) it works only for a small number of sources inside the brain. The inverse model calculates the source of origin based upon the surface electric potential using nonlinear numerical methods.

MEG can identify temporal characteristics of a localized neuronal electrical activity in the range of milliseconds. However, the issues with MEG are: 1) low signal to noise ratio and 2) nonrepeatability of the signals make it difficult for any classifier based on pattern analysis to identify a repeat-pattern.

7.7 IDENTIFYING BRAIN ABNORMALITIES

Pattern changes in EEG, based upon change in mental states, are not well understood. However, EEG has been successfully used to identify *epilepsy* before it occurs and other forms of dementia. In recent years, techniques such as *short-term Fourier transform* (SFT) and *continuous wavelet transform* (CWT) have been used to analyze deviations in EEG waves to detect neurodegenerative diseases such as dementia, including Alzheimer's disease and Parkinson's disease, before the onset of visual symptoms with moderate accuracy. EEG can also predict the stages of dementia to a limited extent.

7.7.1 Predicting Dementia

Dementia includes Alzheimer's disease, Pick's disease, vascular dementia, frontotemporal dementia and Parkinson's disease. Dementia exhibits the loss of cognitive function, memory, attention and the lack of executive function. Dementia results from the progressive death of brain-cells. Dementia increases beyond the age of sixty. Current estimate is that around fifty million people suffer from dementia. Half of the people above the age of 85 have Alzheimer's disease. People above the age of 65 also suffer from MCI (Mild Cognitive Impairment) that results in the loss of cognitive functions and impairment without dementia. People with MCI have high risk of developing dementia. Early detection of Alzheimer's and other forms of dementia will help in delaying the onset of the diseases and gives more useful years to the patients. EEG-based detection is a noninvasive biomarker.

Alzheimer's disease is caused by the amyloid plaque formation in the glial-cells and fibrillation in different parts of the brain. Vascular dementia is caused by the ischemia of the brain-cells due to the lack of oxygenated blood-supply. The glial-cells in the brain are supporting cells with multiple functions: holding neurons in place, separating a neuron from another, supplying nutrients and oxygen to neurons, removing dead neurons. The patients with the early stages of Alzheimer's and vascular dementia show increase in lower frequency-bands: delta waves and theta waves and decrease in higher frequency-bands: beta waves and gamma waves. The patients with MCI show variations in the subrhythms of alpha waves.

7.7.2 Localizing Brain Tumors and Lesions

MEG is being utilized for better spatial localization of brain tumors and ischemic lesions because magnetic signals do not spread out radially like electric signals. MEG is also being utilized for preoperative evaluation of epileptic patients because it can spatially localize the brain-section where abnormal neuronal activity is taking place. With brain-tumors, variations in MEG have been used to identify the damaged part of a brain, especially caused by *meningioma* – a slow-growing brain-tumor causing serious damage by compression, by localizing the source of intense delta-band activities. In recent years, MEG has been combined with the functional MRI images to localize accurately many neurological problems.

7.7.3 Predicting Epilepsy and Epileptic Attacks

Epilepsy is a chronic neurological disorder characterized by recurrent unprovoked seizures. These seizures are caused by abnormal neuronal activity in the brain. Approximately, fifty million people suffer from epilepsy around the world. The frequency of occurrence is greater in youth and the elderly.

Normally, EEG waveforms are more random. This randomness is less before the onset of epilepsy due to regular synchronous neuronal activities in patients suffering from epilepsy and becomes excessive during an epilepsy attack. Epileptic EEG shows sudden spikes, and the amplitude of the signals becomes excessive

TABLE 7.1   EEG-parameter–based separation of different states in epilepsy patients

PARAMETERS	NORMAL EEG	PRE-ICTAL EEG	ICTAL EEG
Approximate entropy (ApEn)	Highest	Most reduced	Reduced
Correlation dimension (CD)	Highest	Slightly reduced	Most reduced
Fractal dimension	Smallest	Most elevated	Elevated
Hurst exponent	Least	Most elevated	Elevated
Largest Lyapunov exponent	Smallest	Elevated	Most elevated

during seizure. In addition, the amplitude of the waveforms from certain areas of the brain increases many folds that can be detected. This lack of randomness is measured using multiple statistical techniques to derive the onset of epilepsy attack.

There are four classes of EEG waveforms: 1) normal EEG; 2) interictal (EEG waveforms during seizure-free intervals for epileptic patients); 3) preictal (background EEG waveforms in epileptic patients before epileptic attacks) and 4) ictal (EEG waveforms during the epileptic attack). To predict epileptic attack several minutes before it happens, *preictal* waveform has to be studied and separated from *ictal* waveforms. However, the duration and the actual event of epileptic attack after the onset of preictal EEG waveform are uncertain and may not happen. Similarly, to identify a tendency of epilepsy in a brain, the lack of randomness in the EEG has to be detected. Three types of analysis are done on EEG signals: 1) frequency-domain analysis; 2) time-domain analysis and 3) nonlinearity analysis.

Major quantitative parameters used to measure the deviations from normal EEG are: 1) approximate entropy (ApEn) approach is a log-likelihood measure to identify the regularity and closeness of the patterns; 2) correlation dimension (CD) predicts; 3) fractal dimension is used to find a common pattern; 4) Hurst component measures the self-similarity and correlation properties and 5) largest Lyapunov exponent measures chaos and predictability. In addition, deep convolution neural networks have also been used to detect the onset of epilepsy. Lower value of the *ApEn* shows increased regularity of the waveform observed in epileptic patient's EEG. The value of *ApEn* further decreases in preictal EEG. Table 7.1 shows the qualitative increase or decrease of the parameters in normal EEG, epileptic seizure and preictal period. Gaussian mixture model and support vector machines are used to investigate the performance of the above measures.

7.8 ELECROMYOGRAPHY (EMG)

Electromyograph is recording of the electrical activities of the muscle. Muscles are controlled by brain using electric signal from motor-neurons $\rightarrow$ muscle-fibers that contract using depolarization using a mechanism involving Na^+, Ca^{++} and K^+ ions similar to depolarization and repolarization of heart-muscles. In the normal mode, an ionic equilibrium is maintained in the muscle cells around -80 to -90 millivolts. After an electric signal is transmitted from motor-neurons through axon-interfaces to muscle-fibers (see Figure 7.22), sodium channel is activated, and the NA^+ ions enter the muscle cells followed by the release of Ca^{++} ions depolarizing the muscle-cells.

The depolarization causes muscles to contract based upon electromechanical coupling involving multiple proteins: *actin, troponin, myosin* and *tropomyosin*. Calcium ions cause the triggering of the contraction. Depolarization is followed by repolarization when Na^+ ions exit out of the cells, and K^+ ion maintains an ion-equilibrium. Action potential threshold for depolarization is around -40 mV. During the depolarization, the action potential shoots up to $+30$ mV before dropping back to -80 mV during relaxation. The range of surface voltage measured for EMG ranges from 0 to 10 mV.

This depolarization-repolarization cycle causes a waveform of active potential that can be measured at the surface like ECG. Unlike ECG wave-pattern, EMG signals are a superimposition of multiple *Motor*

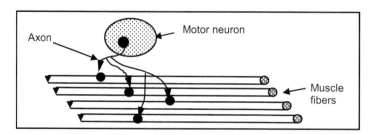

FIGURE 7.22 Motor neuron and muscle fiber interface

Unit Action Potentials (*MUAP*) due to asynchronous firings of different motor units having different amplitudes and frequencies. These superimposed MUAPs cause an inference pattern measured at the surface and is called EMG or surface EMG (SEMG). EMG shows up as a sequence of waveform-bursts followed by no activity periods as shown in Figure 7.23.

7.8.1 EMG Measurement and Noise Removal

EMG can be measured using two types of leads: 1) surface electrodes and 2) needle electrodes. Surface electrodes are noninvasive. However, their frequency-bandwidth is limited and has more noise. Needle-based recording uses a disposable concentric needle inserted into a local section of muscle that contains around 100 muscle fibers within *1* mm cross-section of the needle. In this region, there are generally *4–6* fiber units/motor-unit. The recording picks up activity from one or more motor units. Initially, when the needles are inserted, motor units give *6–10* spikes/second.

The voluntary firing of the muscle spikes can be reduced by relaxing the muscles and changing the posture of the corresponding limb to more relaxed form. The recorded raw EMG signal also contains white Gaussian noise. An EMG signal is modeled by the Equation 7.6, where $v(n)$ is the modeled signal, $e(n)$ represents the firing impulse, $h(r)$ represents the MUAP, $w(n)$ denotes zero-mean white Gaussian noise and *N* is the number of motor unit firings. A typical raw EMG is shown in Figure 7.24.

EMG signal detection consists of three major steps: *denoising*, *spike detection* and *spike separation*. EMG has many types of noises such as ECG, powerline interference in the range of *50–60* Hz, baseline wanders and noise due to electrode-connection. EMG varies from *1* to *500* Hz. Powerline interference

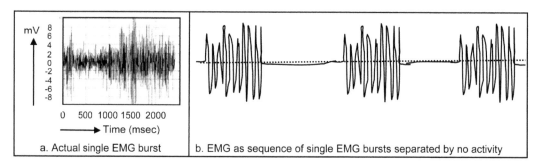

FIGURE 7.23 EMG wave bursts: (a) Actual single EMG burst; (b) EMG as sequence of single EMG bursts separated by no activity (Image in Figure 7.23a courtesy Reaz, Mamun B. I., Hussain M. Sazzad, and Mohd-Yasin Faisal, Techniques of EMG signal analysis: detection, processing, classification and applications. *Biological Procedures Online* 2006; 8(1): 11–35, DOI: 10.1251/bpo124, Figure 5, Permitted under Creative Common License Attribution 4.0 or Creative Commons 1.0 Public Domain Dedication Waiver as part of BMC license agreement).

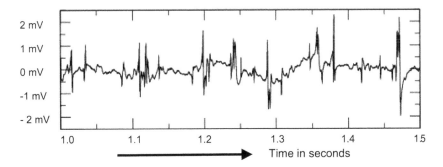

FIGURE 7.24 A raw recording of EMG waves (Image courtesy Physiobank, website: https://physionet.org/physiobank/database/emgdb/courtesy of Seward Rutkove, MD, Department of Neurology, Beth Israel Deaconess Medical Center/Harvard Medical School, Boston, MA, USA, permitted under Physionet's ODC Public Domain Dedication and License v1.0 available at https://opendatacommons.org/licenses/pddl/1.0/).

is removed using an adaptive notch filter that filters out *50–60* Hz. The electric potential of motor units for a healthy muscle is around 2 millivolts with a duration of 5–10 milliseconds. ECG is removed using wavelet-transforms and adaptive filters. Multiple mathematical techniques are used to extract and measure EMG.

$$v(n) = \sum_{r=0}^{r=N-1} h(r)e(n-r) + w(n) \tag{7.6}$$

Some of the major techniques are: *wavelet-transforms, Fourier transforms* and *statistical techniques*. Wavelet-transforms are used to decompose multiple superimposed EMG-waves due to multiple motor units. Like ECG, wavelet-transforms are used to map the waveforms in wavelet domain and match with different wavelet functions. *Mexican hat* wavelet is a typical shape used to match the MUAP waveforms. During a muscle contraction, the myoelectric signals are compressed toward the lower frequencies. Fuzzy clustering and neural network have also been used to separate EMG waveforms into different MUAPs.

7.8.2 EMG Analysis and Classification

EMG is analyzed using *autoregressive moving average* (ARMA) model and other AI techniques. *Autoregressive model* is used to statistically analyze randomly time-varying processes, where the value of the output variable depends upon its own previous values and a stochastic term as described in Equation 7.7. The signal X_t depends upon model-parameters φ_i and a previous value X_{t-i}, a white-noise ε_t that varies with time and a constant c. *Moving average model* describes a time-series data modeled using an average μ, model-parameter θ_i, and the white-noise ε_{t-i} associated with different moving average parameters. Moving average model is described by Equation 7.8. The equation of *autoregressive moving average model* (ARMA) is an integration of *autoregressive model* and *moving average model* and is given in Equation 7.9. The value of the parameters is decided by a correlation-function between the current value and the lagging value.

$$X_t = \sum_{i=1}^{i=p} \varphi_i X_{t-i} + \varepsilon_t + c \tag{7.7}$$

$$X_t = \mu + \varepsilon_t + \sum_{i=1}^{i=q} \theta_i \varepsilon_{t-i} \tag{7.8}$$

$$X_t = \mu + \varepsilon_t + \sum_{i=1}^{i=q} \theta_i \varepsilon_{t-i} + \sum_{i=1}^{i=p} \varphi_i X_{t-i} + c \tag{7.9}$$

EMG signals have also been analyzed using a combination of FFT (Fast Fourier Transform) on short intervals for feature-extraction and fuzzy clustering or variants of neural networks for classification of the signals. The extracted features in an EMG signal are: number of active motor units, the MUAP waveform (Motor Unit Action Potential) and various frequency bands (both low and high) in MUAPs from different muscles and amplitudes and durations of the extracted waveforms. Fourier transforms for short interval is called SFT (Short-term Fourier Transform). During the short-interval employed by SFT, the frequency characteristic of a muscle during the interval is nearly stationary. The advantages of fuzzy clustering and fuzzy logic are that variations in the features can be tolerated.

One technique to separate the signals from different muscles is called *blind source separation. Blind source separation* uses neural network approach to classify the frequencies coming out of different muscles. A variation of neural network where the activation function of each node is a wavelet-function integrates wavelet-based analysis and neural networks. The output becomes a weighted sum of multiple wavelet functions.

There are two major problems with EMG classification and mapping to specific muscles: 1) multiple signals from different MUs may share the overlapping frequency and duration making linear techniques to separate difficult and 2) frequency composition of the EMG even from the same muscle keeps changing with time. Slow nonstationary frequencies are due to the accumulation of many metabolites that causes muscle fatigue.

7.8.3 EMG Application and Muscular Abnormalities

EMG has been used to identify the level of activities in muscles. EMG measurement has been to diagnose muscle fibrillation, lesion, inactive muscles, muscle fatigue and many neuromuscular diseases such as *periodic paralysis* due to hypokalemia (lack of potassium), spinal injury, carpal tunnel syndrome (CTS) related to wrist injury and psychophysiological stress. During muscle fatigue, the muscle slows down progressively after initial excitation and the frequency of the waveform decreases. CTS is caused due to the nerve-compression causing numbness, tingling or burning sensation. EMG signal has also been used for man–machine interface by interpreting the EMG signals to control the machine without the explicit use of a hand.

7.9 SUMMARY

This chapter describes four types of biosignals to derive abnormalities in heart, brain and muscles. The corresponding waveforms are called ECG (electrocardiogram), EEG (electroencephalogram), MEG (Magnetoencephalogram) and EMG (electromyogram). MEG is the measurement of changes in magnetic fields caused by electrical activities in the brain. The basic principle in ECG and EMG is a sequence of periodic depolarization and repolarization caused by change in Na^+ and Ca^{++} concentration inside the cell in response to an electric signal.

Heart consists of two types of chambers: atria and ventricles. Atria are further divided into left-atrium and right-atrium. Ventricles are further divided into left-ventricle and right-ventricle. Oxygenated blood comes from lungs to the left-atrium. The deoxygenated blood comes from the body to the right-atrium. The oxygenated blood flows from the left-ventricle to the body during the contraction of ventricles. Deoxygenated blood flows out from the right-ventricle to lungs during the compaction of ventricles. Proper contraction of atria and ventricles is essential for the continuous and sufficient flow of the blood within the heart–lung–body complex. Insufficient contraction of atria will result into blood-clots that may travel through the ventricles to any part of the body choking the blood-flow. A blood-clot in a brain causes that part of the brain-cells to die causing a stroke. Improper contractions of ventricles reduce the supply of the oxygenated blood to the body causing fatigue and cell-death. Improper contraction in ventricles also causes sudden cardiac death.

Electrical activities in heart are very important for the contractions of atria and ventricles. In a healthy heart, electrical activities are facilitated by a special class of muscle-fibers that conduct ionic activities faster than rest of the muscle-cells. The conduction channel consists of SA-node (origin of electrical activities), AV-node that separates atria from ventricles, His-bundle—main conduction channel in ventricles that gets bifurcated in left and right bundles and Purkinje fibers that supply electric-excitation to different parts of ventricles. Electrical activity starts with depolarization of atria resulting into contraction of atria, followed by the depolarization of ventricles resulting into the contraction of ventricles. In a healthy heart, the sequence of electrical signal conduction is *SA-node → atria → AV-node → His bundle → left and right bundles → Purkinje fibers*. In a healthy heart, there is no return conduction path between ventricles and atria.

ECG is the recording of the electrical activities in the heart-muscles. ECG wave recording is done using electrodes on the surface of the body. The body is divided into two planes: vertical plane and horizontal (lateral) plane. The leads AVL, AVF and AVR define the vertical plane. The leads V1, V2, V3, V4, V5 and V6 make the horizontal plane. In addition, there are three virtual leads, leads I, II and III in the vertical plane calculated using the recordings of AVL, AVR and AVF. Effectively, there are twelve leads: nine actual and three derived. The voltage in the ECG recordings is in the order of millivolts.

Each active cell of the heart muscle depolarizes and produces a vector of electric signal. The overall direction of electrical activity in the heart is the vector-sum of the electrical activities. Normally, it is along the direction of lead II and between V3 and V4. However, the vector changes if there is a deviation in the electrical activities of the heart cells. The deviation can be caused by many reasons: 1) death of heart cells; 2) lack of regular electrical activity in the conduction path causing ectopic-nodes to fire asynchronously and 3) blockage in the conduction pathway. For ECG recording purpose, the positive direction during depolarization is coming toward the lead II and V3 (or V4), and negative direction is going away from it. During repolarization, the polarity reverses: negative direction is coming toward the lead II, and positive direction is going away from lead II.

Electrical activity measurements are governed by the equation measured potential $= |E| \, cos(\theta)$, where $|E|$ is the maximum magnitude of the electrical activity and θ is the angle that the vector makes with the direction of the lead. The same electric potential will show different values in different leads due to different directions of the lead. Since $cos(\theta)$ varies from -1 to 1 and $cos(\theta)$ is maximum for $\theta = 0°$ or $\theta = 180°$. The lead showing the maximum magnitude shows the resultant electrical vector in the positive or negative direction for the current electrical activity. Different leads may show maximum magnitude due to heart abnormalities.

ECG consists of six waveforms: P-QRS-T-U. P waveform is associated with the depolarization of atria. QRS-complex is associated with the depolarization of ventricles. T-waveform is associated with the repolarization of ventricles. U-waveform is mostly unobservable and is associated with the repolarization of Purkinje fibers.

The ECG waveform comes with multiple types of noises such as power line, muscle signals, signals due to body-movement, sensor-noise and low-frequency baseline-wander. The noise needs to be filtered out for the accurate identification of P-QRS-T waveforms. Multiple types of filters are used to remove the noise. High-pass filter that removes waveforms less than 30 Hz is used to remove baseline-wander and muscle-wave noises; adaptive notch filter is used to remove power line interference. Band-pass filter is used to remove both low-frequency noise and high-frequency noise.

Even after various attenuations during the filtering process, wavelet-transform is used to identify P-QRS-T waveforms by using functions that simulate QRS waveform and match with the similar pattern in the transformed signal. The transformed signal is inverse-transformed to get the QRS waveform. Q and S waveform peak is derived by derivative analysis of the waveform near the corresponding R-peak. T-wave can be recognized because it follows R-peak and has the second largest amplitude. P-wave is identified using a window near QRS-complex.

There are different types of abnormalities both in atria and ventricles. Some of the common abnormalities are: 1) thickening of the heart-muscles; 2) enlargement of the heart; 3) blockage in the conduction system; 4) fibrillation of the heart-muscles resulting into ectopic-nodes; 5) blockage in the arteries causing oxygen starvation of the heart-muscles and 6) electrolyte imbalance. Many diseases also cause blood pressure imbalance and mechanical problems in the heart valves. Thickening of the heart muscle increases the amplitude of the ECG waves due to additional cells. Conduction blockages are due to: 1) lack of firing or misfiring of SA-node; 2) firing of asynchronous ectopic-nodes; 3) electrical activities starting from AV-node instead of SA-node and 4) blockages or delays in His bundle: left-bundle or the right-bundle. The disease may also have origin in the lack of oxygen coming from the lung that may speed up the heartbeat.

Abnormality in the atria causes P-wave to get deformed or deviated from its usual position. Abnormality in ventricles affects mainly QRS-complex and sometimes T-waves. Due to this abnormality, the amplitude and durations of the waveforms vary. Slower depolarization results into longer duration of the corresponding waveform, and faster depolarization results into the shortening of the durations of the corresponding waveforms. The excess of NA^+ or Ca^{++} ions causes faster depolarization and slower repolarization, and vice versa.

ECG analysis is concerned with the analysis of morphology, occurrence order of the P-QRS-T waveforms and the superimposition of the waveforms. The analysis of P-QRS-T waveform is done using: 1) morphology analysis and 2) statistical techniques based upon HMM or variations of the Markov model. Morphology analysis is based upon: 1) measuring intervals between the waveforms; 2) magnitude of the waveform-peaks; 3) variation in the spread of the waveforms; 4) change in orientation of vector; 5) slope of the waveforms and 6) the interval-segments and their correlation with depolarization and repolarization of the heart-muscles.

EEG waveforms are a surface-level recording of multiple waveforms having different frequencies. Major waveforms are: *alpha, beta, gamma, delta, theta* and *mu*. The EEG waveforms are in the range of microvolts and are generated inside the brain due to neuronal excitation. However, we can only measure the electrical activities at the surface of the scalp. The measuring electrodes are traditionally placed in a 10-20 configuration or a 10-10 configuration. 10-20 configuration divides the perimeter at regular intervals of 20 degrees laterally or 10 degrees longitudinally. There are even higher resolution EEG recordings with large number of electrodes.

EEG has been used to identify epilepsy before it occurs, and many types of dementia by studying the change in the randomness and change in frequency content of the EEG waveforms. Before an epileptic attack, the randomness in EEG decreases. Three types of analyses are done on EEG; frequency-domain analysis, time-domain analysis and nonlinearity analysis. Major parameters employed to identify EEG waveform changes before epileptic attacks are: approximate entropy (ApEn), correlation dimension (CD), Hurst component, fractal dimension and largest Lyapunov exponent. Dementia is characterized by the increase in the lower frequency bands and decrease in the upper frequency bands.

MEG waveforms are the recordings of magnetic activities measured at the surface of the scalp. Magnetic activities are caused by electrical activity at the neuron level. However, they do not move radially outward causing them less susceptible to noise and more favorable for localizing the neuronal activities. Localization of neuronal activities inside the brain uses two types of techniques: *forward modeling* and *inverse modeling*. In forward modeling, the simulation of electrical activity is propagated to surface and matched with the surface potential using spherical coordinates and Maxwell's equations. In inverse modeling, the values from multiple electrodes are used to localize the site of neural activity using numerical modeling techniques. Combined with fMRI, MEG has also been used to localize brain-tumors.

EMG is the recording of the waves generated due to electrical activities in the muscle. Electrical activities are generated in the neurons that are passed through axon–muscle interfaces to muscle-fibers. Inside the muscle-cells, depolarization occurs due to the influx of NA^+ ions followed by CA^{++} ions. This depolarization combined with actions by many proteins involved in muscle-cells causes muscles to contract. Muscles repolarize after the depolarization balancing the ion-level inside and outside the cells. EMG-waveform is a superimposition of multiple motor-waveforms. It has noise due to heartbeats, powerline using interference, baseline-wanders and electrode-connections. Different filters, wavelet-transform and Mexican hat wavelet have been used to isolate specific EMG. EMG is used to diagnose and study many neuromuscular diseases, motor inactivity and spinal injuries.

7.10 ASSESSMENT

7.10.1 Concepts and Definitions

10-10 system; 10-20 system; 12-lead ECG; 1st-degree block; 2nd-degree block; 3rd-degree block; adaptive filter; alpha wave; Alzheimer's disease; approximate entropy; ARMA; arrhythmia; atrial fibrillation; atrium; autoregressive model; autoregressive moving average model; AV-node; AVNRT; AVRT; Bachman bundle; backward difference; band-pass filter; Bazett's formula; beta wave; bradycardia; cardiomyopathy; continuous wavelet transform (CWT); Daubechies wavelet; delta wave; dementia; depolarization; difference squaring; dilated cardiomyopathy; dilation; dipolar recording; ECG wave; ectopic-nodes; EEG; EGG; EMG; ENG; epilepsy; F-wave; Framingham's formula; fibrillation; forward modeling; Fridericia's formula; gamma wave; Hamilton-Tomkins algorithm; high-pass filter; His-bundle; Hodge's formula; hypercalcemia; hyperkalemia; hypertrophic cardiomyopathy; hypocalcemia; hypokalemia; hyponatremia; inverted P-wave; inverse modeling; inverted T-wave; ischemia; J-point; junctional tachycardia; lead placement; left atrium enlargement; left bundle; left bundle branch block; left ventricular hypertrophy; linear adaptive filter; localization; low pass filter; magnetometer; Maxwell's equation; measurement plane; MOBD algorithm; monopolar recording; morphology analysis; motor neuron; moving average model; mu wave; myocardial infarction; noise reduction; notch filter; NSTEMI; Okada algorithm; P-QRS-T waveform; P-wave; PCG; powerline interference; PR-interval; Purkinje fibers; Q-wave; QRS-complex; QT interval; QT^C interval; repolarization; restrictive cardiomyopathy; right atrium enlargement; right bundle; right bundle branch block; R-peak; R-peak detection; right ventricular hypertrophy; RR-interval; SA-node; Savitzky–Golay filters; scaling; squaring difference; SQUID; ST-segment; STEMI; T-wave; tachycardia; theta wave; translation; vascular dementia; ventricle; ventricular fibrillation; ventricular flutter; ventricular tachycardia; wavelet spectra; wavelet-transform; whitening filter; Wolff–Parkinson–White's (WPW) syndrome.

7.10.2 Problem Solving

7.1 Given two vectors, *4i + 8j* and *3i–2j*, given the resulting vector and its angle with respect x-axis, y-axis, first vector and the second vector.

7.2 Given a cumulative electric potential vector is aligned with lead II on the vertical axis and the lead V5 on the horizontal axis. The peak R-value along the horizontal lead V5 is *2.4* and *1.8* mV along lead II in vertical axis. Give the value of R-peak along all other leads in 12-lead ECG using the trigonometric function $V_i = V_{max} \times cos(\theta)$.

7.3 A variation of wavelet-transforms called "Mexican hat" is used to extract the QRS-waveform. A specific property of this transform is that it looks like an upside-down conical Mexican hat and falls quickly due to low variance. The equation for the Mexican-hat transform is:

$$\psi(t) = \frac{2}{\pi^{1/4}\sqrt{3\sigma}}\left(\frac{t^2}{\sigma^2} - 1\right)\exp^{\left(-\frac{t^2}{\sigma^2}\right)},$$ where σ is the standard deviation in the Gaussian function.

Plot the function for $\sigma = 1$ and $\sigma = 2$ and study the properties of the function.

7.4 Model the P-QRS-T waveform using a Markov model where each waveform (P, Q, R, S and T) is a vertex of a directed graph, and the edge-weight is the frequency of transition between the waveforms. The Markov graph is nontrivial because of missing or embedded waveforms in different disease such as arrhythmia. Try this model on a sample size of *100* beats, where P wave occurs *70* times, QRS-complex occur *95* times and T-waveform occurs *80* times. Both T and P waves are missing together *10* times. When P wave misses, then the transition occurs from T to Q, and when T wave is missing, then the transition occurs between S node and P node. The transition occurs between S node and Q node if both T and P waveforms are missing. Show the probability of transition as the edge-weights.

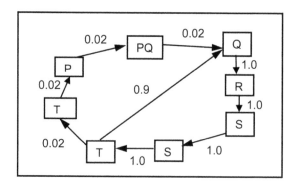

FIGURE 7.25   A Markov model graph for Problem 7.6

7.5 Extend the model in Problem 7.4 to an 8-vertex Markov model where the baselines PQ, ST and TP are also treated as vertices besides five waveforms. Besides the above transition, assume that P wave is missing *30* times. Out of which, ten times it is embedded inside QRS-complex. Hence, PQ-segment is missing. Ten times both P and T waves are missing, and P-wave is cancelling out the inverted T-waves. Hence ST-segment and TP-segment are missing. The rest of the time transitions are regular. Show the graph for the Markov model with weighted edges as the probability of transition.

7.6 Read the transition graph for 8-vertex Markov model for the graph in Figure 7.25 of a type of arrhythmia, give its characteristics and identify the possible arrhythmia. Explain your answer.

7.7 Assume that ECG waveforms are given as a sequence of sampled amplitudes with no noise. Write an algorithm to find out the maximum peak and its index in the sequence. Based upon the peak indices, extend your algorithm to find out RR-interval as an average of the adjacent indices.

7.8 Extend your algorithm to detect Q and S waves. Assume that Q and S waveforms have negative magnitude, Q-waveform occurs at most *15* milliseconds before the corresponding R-peak and S-waveform occurs at most *15* milliseconds after the corresponding R-peak.

7.9 Extend your algorithm to detect T-waveforms under the assumption it is the second largest peak occurs between S-waveform and P-waveform or S-waveform and R-waveform. Explain how your algorithm separates detection of T-waveform from the P-waveform. (Hint: use the magnitude and distance from R-wave and read literature.)

7.10 Modify your algorithm to detect elevated ST-segment where S wave is missing, and ST-segment starts before the downward R-waveform touches the baseline. This is a sign of myocardial infarction as shown in Figure 7.15b.

7.11 Modify the algorithm in Problem 5 to identify skipped beat. A skipped beat is a random blockage in AV node that causes ventricles to skip depolarization as described in Section 7.3.5. A skipped beat can be identified using analysis of the RR-interval and comparing individual interval against average RR-interval. Derive mean and variance of the RR-interval population and find out the outliers with duration twice the average RR interval.

7.12 Give the ECG waveform for left atrium enlargement and right atrium enlargement and compare the two ECG waveforms.

7.13 If SA-node becomes inactive and is substituted by AV-node as the start of electrical activities, then give the possible ECG waveforms and explain.

7.14 Give the ECG waveform for ischemia and myocardial infarction and explain.

7.15 Give the ECG waveform when multiple ectopic-nodes in ventricles fire irregularly and SA-node is not working. Explain the waveform.

7.16 Give the waveform for hyperkalemia and explain.

7.17 Matching two EEG signals is complex. However, multiple techniques can be used to match similarity of two EEGs. Distribution patterns of intervals between zero crossing and amplitudes of the waveforms can match the two EEG signals. Histogram-based dissimilarity can prune out a large class of mismatching EEG waveforms. The dissimilarity between histograms is measured as $\sum_{i=1}^{i=N} \sqrt{\left(f_i^1 - f_i^2\right)}$, where f_i^1 is the frequency in the ith bucket of the first histogram, and f_i^2 is the corresponding frequency in the ith bucket of the second histogram. The amplitude can also be negative. Two sequences of 30 amplitude-values (in microvolts) are as follows:

Sequence 1: *1, 8, 4, –3, 12, 5, –2, 1, 11, 15, 2, –4, 3, 4, –5, 6, 2, 7, – 2, 3, 12, –5, 3, – 3, 1, 7, 5, 4, 9, –1*
Sequence 2: *1, 9, 3, –2, 5, 3, 11, 7, –4, 3, 8, –3, 5, 4, 6, –4, 3, –6, 8, 12, –1, 11, 4, – 2, 9, 4, 7, – 2, 8, 12*

Derive the histogram and the dissimilarity for the two sequences.

7.18 Identify the heart abnormality in the ECG waveforms given in Figure 7.26 and explain your reasoning. The answers are in the random order: 1) ischemia; 2) bradycardia; 3) arrhythmia; 4) myocardial infarction; 5) ventricular tachycardia; 6) atrial flutter; 7) left bundle block; 8) right bundle block and 8) atrial fibrillation. A sample may have one or more of these conditions.

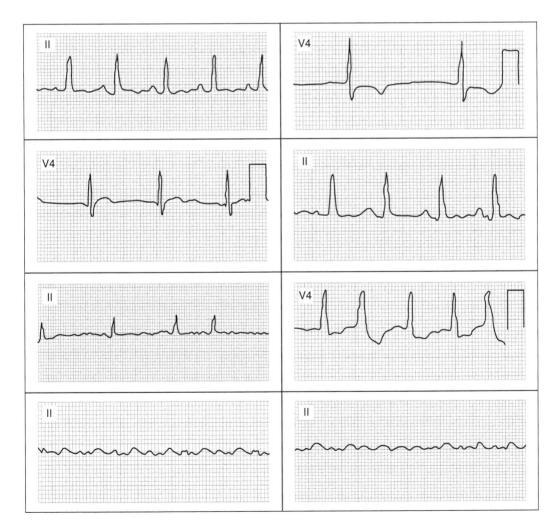

FIGURE 7.26   De-identified samples of abnormal ECGS for Problem 7.18

7.19 Another technique to perform EEG analysis is to make a histogram of polarity changes for a fixed time-interval and match the histogram. To calculate polarity changes, the values x(t) > 0 is mapped to +1, and the value *x(t) < 0* is mapped to *−1*. The whole sequence of values is mapped to a sequence of {+1, −1}. After that the sequence is mapped to a sequence of numbers so each number represents the number of discrete time-intervals when the polarity was same. For the sequence in Problem 7.17, make the polarity change based histogram and calculate the dissimilarity.

7.20 Write a program that uses sine waves of different frequencies to simulate different (*alpha, beta, theta, delta*) EEG waves with an interval slot of *100* milliseconds and displays the combination of amplitude for each interval. The interval slot, amplitude and frequency should be parameterized so that it can be easily altered based upon user interaction.

7.21 Apply the program developed in Problem 7.20 to simulate the waves during: 1) awake but relaxed state; 2) eyes closed and relaxed state; 3) deep sleep state; 4) alert state and 5) performing mental arithmetic. These states correspond to different combinations of various EEG wave-types and amplitudes. Do literature search, using the Internet and research literature, to identify the combinations.

7.10.3 Extended Response

7.22 Explain the principle of depolarization and repolarization of cells.

7.23 How is the resulting vector formed at the surface in ECG? Explain.

7.24 Explain the placement of leads and their orientation for recording ECG.

7.25 Explain the reason for different leads having different readings for the same ECG waveform.

7.26 Explain the techniques to extract P-QRS-T waveform from raw electric potential fluctuation readings in ECG.

7.27 Explain the distinct types of filters used to remove noise from the raw ECG signals.

7.28 Describe different components of heart and their roles in the blood-flow in the body.

7.29 Explain wavelet-transform and its role in extracting P-QRS-T waveforms in ECG.

7.30 Explain arrhythmia and describe different subclasses of arrhythmia with the corresponding waveforms.

7.31 Describe how the enlargement of heart-muscles will affect ECG.

7.32 Describe the change in waveforms caused by thickening of heart-muscles.

7.33 Explain ischemia and myocardial infarction, and how do they affect the ECG of a patient suffering from ischemia and/or myocardial infarction.

7.34 Explain the role of ectopic-nodes in ventricular arrhythmia and sudden cardiac death.

7.35 Explain the origin of several types of circulating current in a heart, and how do they affect the ECG.

7.36 Describe diverse types of waveforms that constitute EEG and their roles in mental activities.

7.37 Describe the fundamentals of MEG and their relationship to neuronal electrical activities.

7.38 Describe 10-20 lead placement for EEG.

7.39 Explain the techniques used in identifying dementia.

7.40 Describe the localization using fMRI and MEG. Why is spatial localization better in MEG?

7.41 Explain the principle of muscle contraction using various ions.

7.42 Explain the various techniques to remove noise from EMG signals.

7.43 Explain the techniques for the classification of EMG signals to identify the originating muscles and the problems associated with the signal classification.

7.44 Explain the application of EMG in various diseases.

FURTHER READING

Biosignal Processing and Medical Image Processing Fundamentals

7.1 Semmlow, John L., and Griffel, Benjamin, *Biosignal and Medical Image Processing*, 3rd edition, CRC Press, Boca Raton, FL, USA, February 2014.
7.2 Sörnmo, Leif, and Laguna, Pablo, *Bioelectrical Signal Processing*, Elsevier Academic Press, Burlington, MA, USA, 2005.

ECG Fundamentals

7.3 Bers, Donald M., "Cardiac Excitation-Contraction Coupling," *Nature*, 415, January 2002, 198–205.
7.4 Garcia, Tomas B., and Holtz, Neil E., *Introduction to 12-Lead ECG: The Art of Interpretation*, Jones and Bartlett Publishers, Sudbury, MA, USA, 2003.
7.5 Gregg, Richard E., Zhou, Sophia H., Lindauer, James M., Helfenbein, Eric D., and Giuliano, Karen K., "What is Inside the Electrocardiograph," *Journal of Electrocardiology*, 41(1), January/February 2008, 8–14.

ECG Analysis

7.6 Kligfield Paul, Gettes Leonard S., Bailey James J., et al., "Recommendations for the Standardization and Interpretation of the Electrocardiogram: Part I: the Electrocardiogram and its Technology," *Journal of American College of Cardiology*, 49, March 2007, 1109–1127.
7.7 Schläpfer, Jürg, and Wellens, Hein J., "Computer-Interpreted Electrocardiograms: Benefits and Limitations," *Journal of American College of Cardiology*, 70(9), August 2017, 1183–1192.

Noise Removal

7.8 Addison, Paul S., "Waveform Transforms and the ECG: A Review," *Physiology Measurements*, 26(5), October 2005, R155–R199.
7.9 Luo, Shen, and Johnson, Paul, "A Review of Electrocardiogram Filtering," *Journal of Electrocardiology*, 43(6), November/December 2010, 486–496.
7.10 Sameni, Reza, Shamsollahi, Mohammad. B., Jutten, Christian, and Clifford, Gari D., "A Nonlinear Bayesian Filtering Framework for ECG Denoising," *IEEE Transactions on Biomedical Engineering*, 54(12), December 2007, 2172–2185.
7.11 Sayadi, Omid, Sameni, Reza, and Shamsollahi, Mohammad B., "ECG Denoising Using Parameters of ECG Dynamical Model as the States of an Extended Kalman Filter," In Proceedings of the International Conference on Engineering in Medicine and Biology (EMBC), August 2007, Lyon, France, 2548–2551.

P-QRS-T Waveforms Detection

Wavelet and HMM-Based Approaches

7.12 Crouse, Matthew S., Nowak, Robert D., and Baraniuk, Richard G., "Wavelet-Based Statistical Signal Processing Using Hidden Markov Models," *IEEE Transactions on Signal Processing*, 46(4), April 1998, 886–902.
7.13 Koski, Antti, "Modeling ECG Signals with Hidden Markov Models," *Artificial Intelligence in Medicine*, 8(5), October 1996, 453–471.
7.14 Martínez, Juan P., Almeida, Rute, Olmos, Salvador, Rocha, Ana P., and Laguna, Pablo, "A Wavelet Based ECG Delineator: Evaluation on Standard Databases," *IEEE Transactions on Biomedical Engineering*, 51(4), April 2004, 570–581.
7.15 Martis, Roshan J., Acharya, Rajendra U. and Min, Lim C., "ECG Beat Classification Using PCA, LDA, ICA and Discrete Wavelet Transform," *Biomedical Signal Processing and Control*, 8(5), September 2013, 437–444.
7.16 Andreão, Rodrigo V., and Boudy, Jérôme P., "Combining Wavelet Transform and Hidden Markov Models for ECG Segmentation," *EURASIP Journal on Advances in Signal Processing*, Article ID 56215, December 2006, DOI: 10.1155/2007/56215.

Model-Based Approach

7.17 Sayadi, Omid, and Shamsollahi, Mohammad B., "A Model-based Bayesian Framework for ECG Beat Segmentation," *Physiological Measurements*, 30(3), February 2009, 335–352.

Artificial Neural Network Approach

7.18 Xue, Qiuzhen, Hu, Yu-Wen and Tompkins, Willis J., "Neural-Network-Based Adaptive Matched Filtering for QRS Detection," *IEEE Transactions on Biomedical Engineering*, 39(4), April 1992, 317–329.

Identification of P and T Waveforms

7.19 Castells, Francisco, Laguna, Pablo, Sörnmo, Leif, Bollmann, Andreas, and Roig, José M., "Principal Component Analysis in ECG Signal Processing," *EURASIP Journal on Advances in Signal Processing*, Article ID 74580, Hindawi Publishing Corporation, February 2007, DOI: 10.1155/2007/74580.

7.20 Chen, Po-Ching, Lee, Steven, and Kuo, Cheng-Deng, "Delineation of T-wave in ECG by Wavelet Transform Using Multiscale Differential Operator," *IEEE Transactions on Biomedical Engineering*, 53(7), July 2006, 1429–1433.

7.21 Lin, Chao, Kail, Georg, Giremus, Audrey, Mailhes, Corinne, Tourneret, Jean-Yves, and Hlawatsch, Franz, "Sequential Beat-to-Beat P and T Wave Delineation and Waveform Estimation in ECG Signals: Block Gibbs Sampler and Marginalized Particle Filter," *Signal Processing*, 104, November 2014, 174–187.

7.22 Sagie, Alex, Larson, Martin G., Goldberg, Robert J., Bengtson, James R., and Levy, Daniel, "An Improved Method for Adjusting the QT Interval for Heart Rate (the Framingham heart study)," *American Journal of Cardiology*, 70(7), September 1992, 797–801.

Heart Diseases and ECG

7.23 Surawicz, Borys, Childers, Rory, Deal, Barbara J., Gettes, Leonard S., "AHA/ACCF/HRS Recommendations for the Standardization and Interpretation of the Electrocardiogram: Part III: Intraventricular Conduction Disturbances," *Journal of the American College of Cardiology*, 53(11), 2009, 976–981.

7.24 Go, Alan S., Mozaffarian, Dariush, Roger, Véronique L., Benjamin, Emelia J., Berry, Jarett D., Borden, William B., Bravata, Dawn M. et al., "AHA Statistical Update, Heart Disease and Stroke Statistics—2013 update," *Circulation*, 127(1), June 2013, e6–e245.

7.25 Lilly, Leonard S., *Pathophysiology of Heart Diseases*, 6th edition, Wolters Kluwer, September 2015.

7.26 Rautaharju, Pentti M., Surawicz, Borys, Gettes, Leonard S., "AHA/ACCF/HRS Recommendations for the Standardization and Interpretation of the Electrocardiogram Part IV," *Journal of American College of Cardiology*, 53(11), March 2009, 982–991.

7.27 Zipes, Douglas P., Libby, Peter, Bonow, Robert O., Mann, Douglas L., Tomaselli, Gordon F., and Braunwald, Eugene, *Braunwald's Heart Disease – A Textbook of Cardiovascular Medicine*, 11th edition, Elsevier Health Sciences, Philadelphia, PA, USA, 2018.

Cardiomyopathy

7.28 Ogura, Riyo, Hiasa, Yoshikazu, Takahashi, Takefumi, Yamaguchi, Koji, Fujiwara, Kensuke, Ohara, Yoshikazu et al., "Specific Findings of the Standard 12-Lead ECG in Patients with 'Takotsubo' Cardiomyopathy – Comparison with the Findings of Acute Anterior Myocardial Infarction," *Circulation Journal*, 67(8), 2003, 687–690, DOI: 10.1253/circj.67.687.

7.29 Ommen, Steve R., "Hypertrophic Cardiomyopathy," *Current Problems in Cardiology*, 36(11), November 2011, 409–453, DOI: 10.1016/j.cpcardiol.2011.06.001.

7.30 Birnbaum, Yochai, Drew, Barbara J., "The Electrocardiogram in ST Elevation Acute Myocardial Infarction: Correlation with Coronary Anatomy and Prognosis," *Postgraduate Medical Journal*, 79(935), September 2003, 490–504, DOI: 10.1136/pmj.79.935.490.

Ectopic-Nodes and Arrhythmia

7.31 Adams, Mary G., and Pelter, Michele M., "Ventricular Escape Rhythms," *American Journal of Critical Care*, 12(5), September 2003, 477–478.

7.32 Garcia, Tomas B., and Miller, G. T., *Arrhythmia Recognition: The Art of Interpretation*, Jones and Bartlett, Sudbury, MA, USA, 2013.

7.33 Graffigna, Angelo, Vigano, Mario, Pagani, Francesco, and Salerno, Giorgio, "Surgical Treatment of Ectopic Atrial Tachycardia," *The Annals of Thoracic Surgery*, 54(2), August 1992, 338–343.

7.34 Petrutiu, Simona, Ng, Jason, Nijm, Grace M., Al-Angari, Haitham, and Sahakian, Alan V., "Atrial Fibrillation and Waveform Characterization," *IEEE Engineering in Medicine and Biology Magazine*, 25(6), November/December 2006, 24–30.

Ischemia and Myocardial Infarction

7.35 Amit, Guy, Granot, Yair Abboud, Shimon, "Quantifying QRS Changes During Myocardial Ischemia: Insights from High Frequency Electrocardiography," *Journal of Electrocardiography*, 47(4), July/August 2014, 505–511.

7.36 Zizzo, Claudio, Hassani, Aimen, and Turner, Delphine, "Automatic Detection and Imaging of Ischemic changes during Electrocardiogram Monitoring," *IEEE Transaction on Biomedical Engineering*, 55(3), March 2008, 1343–1348.

Electrolyte Imbalance

7.37 Diercks, Deborah B., Shumaik, George M., Harrigan, Richard A., Brady, William J., and Chan, Theodore C., "Electrocardiographic Manifestations: Electrolyte Abnormalities," *The Journal of Emergency Medicine*, 27(2), August 2004, 153–160.

Computational Techniques for Heart Diseases Diagnosis

Morphological Analysis

7.38 García, Constantino A., Otero, Abraham, Vila, Xosé, and Márquez, David G., "A New Algorithm for Wavelet-based Heart Rate Variability Analysis," *Biomedical Signal Processing and Control*, 8(6), 2013, 542–550.

Applying Markov Models

7.39 Chang, Pei-Chann, Lin, Jyun-Jie, Hsieh, Jui-Chien, Weng, Julia, "Myocardial Infarction Classification with Multi-lead ECG Using Hidden Markov Models and Gaussian Mixture Models, *Applied Soft Computing*, 12(10), October 2012, 3165–3175.

7.40 Andreão, Rodrigo V., Dorizzi, Bernadette, and Boudy, Jérôme P., "ECG Signal Analysis Through Hidden Markov Models," *IEEE Transactions on Biomedical Engineering*, 53(8), July 2006, 1541–1549.

7.41 Coast, Douglas A., and Stern, Richard M., "An Approach to Cardiac Arrhythmia Analysis Using Hidden Markov Models," *IEEE Transactions on Biomedical Engineering*, 37(9), September 1990, 826–836.

7.42 Gawde, Purva R., Bansal, Arvind K., and Nielson, Jeffrey A., "ECG Analysis for Automated Diagnosis of Subclasses of Supraventricular Arrhythmia," In Proceedings of the International Conference on Health Informatics and Medical Systems *(HIMS)*, Las Vegas, NV, USA, July 2015, 10–16.

7.43 Gawde, Purva R., Bansal, Arvind K., and Nielson, Jeffrey A., "Integrating Markov Model and Morphology Analysis for Automated Finer Classification of Ventricular Arrhythmia in Real Time," In Proceedings of the IEEE International Conference in Biomedical and Health Informatics, Orlando, Florida, USA, February 2017, 409–412.

7.44 Jastrzebski, Marek, Kukla, Piotr, Czarnecka, Danuta, and Kawecka-Jaszcz, Kalina, "Comparison of Five Electrocardiographic Methods for Differentiation of Wide QRS-Complex Tachycardias," *Europace*, 14(8), 2012, 1165–1171.

7.45 Lee, Jinseok, Nam, Yunyoung, McManus, David D., and Chon, Ki H., "Time-varying Coherence Function for Atrial Fibrillation Detection," *IEEE Transactions on Biomedical Engineering*, 60(10), October 2013, 2783–2793.

7.46 Martis, Roshan Joy, Acharya, Rajendra U., and Min, Lim Choo, "ECG Beat Classification Using PCA, LDA, ICA and Discrete Wavelet Transform," *Biomedical Signal Processing and Control*, 8(5), September 2013, 437–448.

Neural Network-Based Analysis

7.47 Telemachos Stamkopoulos, Konstantino Diamantaras, Nicos Maglaveras and Michael Strintzis, "ECG Analysis Using Nonlinear PCA Neural Networks for Ischemia Detection," *IEEE Transactions on Signal Processing*, 46(11), November 1998, 3058–3067.

Other Techniques

7.48 Faust, Oliver, Acharya, Rajendra U., Sudarshan, Vidya K., Tan, Ru San, Yeong, Chai Hong, Molinari, Filipo, Ng, Kwan Hoong, "Computer Aided Diagnosis of Coronary Artery Disease, Myocardial Infarction and Carotid Atherosclerosis Using Ultrasound Images: A Review," *Physica Medica*, 33, 2017, 1–15.

7.49 Park, Jinho, Pedrycz, Witold, and Jeon, Moongu, "Ischemia Episode Detection in ECG Using Kernel Density Estimation, Support Vector Machine and Feature Selection," *Biomedical Engineering Online*, 11:30, June 2012, DOI: 10.1186/1475-925X-11-30, Available at http://www.biomedical-engineering-online.com/content/11/1/30, Accessed June 8, 2018.

Electroencephalography (EEG)

7.50 Fisch, Bruce J., *EEG Primer – Basic Principles of Digital and Analog EEG*, 3rd edition, Elsevier, 1999.

7.51 St. Louis, Erik K, and Frey, Lauren C. (editors), Electroencephalography – An Introductory Text and Atlas of Normal and Abnormal Findings in Adults, Children, and Infants, Chicago, IL, American Epilepsy Society (www.AESNET.org), 2016, Available at https://www.ncbi.nlm.nih.gov/books/NBK390354/pdf/Bookshelf_NBK390354.pdf, DOI: 10.5698/978-0-9979756.

EEG Recordings and Noise Removal

7.52 Diez, Pablo F., Mut, Vincent, Laciar, Eric, and Avila, Enrique, "A Comparison of Monopolar and Bipolar EEG Recordings for SSVEP Detection," In Proceedings of 32nd Annual International Conference of the IEEE EMBS, Buenos Aires, Argentina, August/September 2010, 5803–5806.

7.53 Teplan, Michal, "Fundamentals of EEG measurement," *Measurement Scientific Review*, 2(2), 2002, 1–11.

EEG Waveforms and Analysis

7.54 Gasser, Theo, "The Analysis of the EEG," *Statistical Methods in Medical Research*, 5, 1996, 67–99.

7.55 Lotte, Fabien, "A Tutorial on EEG Signal Processing Techniques for Mental State Recognition in Brain-Computer Interfaces," Chapter 7, *Guide to Brain-Computer Music Interfacing* (editors: Miranda, Eduardo Reck, and Castet, Julien), Springer, London, 2014.

7.56 Subha, Puthankattil D., Joseph, Paul K., Acharya, Rajendra, and Lim, Choo Min, "EEG Signal Analysis: A Survey," *Journal of Medical Systems*, 34(2), April 2010, 195–212.

Magnetoencephalography (MEG)

7.57 Hari, Riita, and Puce, Aina, *MEG-EEG Primer*, Oxford University Press, New York, NY, USA, 2017.

7.58 Braeutigam, Sven, "Magnetoencephalography: Fundamentals and Established and Emerging Clinical Applications in Radiology," *ISRN Radiology*, July 2013, Article Id 529463, 18 pages, DOI: 10.5402/2013/529463, Available at https://www.hindawi.com/journals/isrn/2013/529463/, Accessed June 8, 2018.

Identifying Brain Abnormalities

Predicting Dementia

7.59 Al-Qazzaz, Noor K., Bin, Sawal, H., Ahmad, Siti A., Chellappan, Kalaivani, Islam, Mohammed S., and Escudero, Javier, "Role of EEG as Biomarker in the Early Detection and Classification of Dementia," *The Scientific World Journal*, Article Id 906038, 2014, DOI: 10.1155/2014/906038.

Localizing Brain Tumors and Lesions

7.60 Oshino, Satoru, Kato, Amami, Wakayama, Akatsuki, Taniguchi, Masaki, Hirata, Masayuki, and Yoshimine, Toshiki, "Magnetoencephalographic Analysis of Cortical Oscillatory Activity in Patients with Brain Tumors: Synthetic Aperture Magnetometry (SAM) Functional Imaging of Delta Band Activity," *NeuroImage*, 34(3), February 2007, 957–964.

Predicting Epilepsy and Epileptic Attacks

7.61 Lehnertz, Klaus, Andrzejak, Ralph G., Arnold, Jochen, Kreutz, Thomas, Mormann, Florian, Rieke, Christoph, Widman, Guido, and Elger, Christian, "Nonlinear EEG Analysis in Epilepsy," *Journal of Clinical Neurophysiology*, 18(3), 2001, 209–222.

7.62 Wong, Stephen, Gardner, Andrew B., Krieger, Abba M., and Litt, Brian, "Stochastic Framework for Evaluating Seizure Predicting Algorithms Using Hidden Markov Models" *Journal of the Neurophysiology*, 97(3), March 2007, 2525–2532.

Electromyography (EMG)

EMG Fundamentals

7.63 Konrad, Peter, *The ABC of EMG: A Practical Introduction to Kinesiological Electromyography*, Version 1.0, April 2005, Available at https://hermanwallace.com/download/The_ABC_of_EMG_by_Peter_Konrad.pdf, Accessed June 8, 2018.
7.64 Merletti, Roberto, and Parker, Philip, *Electromyography Physiology, Engineering, and Noninvasive Applications*, IEEE Press, Hoboken, NJ, USA, 2004.
7.65 Mills, Kerry R., "The Basics of Electromyography," *Journal of Neurology, Neurosurgery & Psychiatry*, 76 (sppl II), June 2005, ii32–ii35, Available at http://jnnp.bmj.com/content/jnnp/76/suppl_2/ii32.full.pdf, Accessed June 8, 2018.

EMG Measurement and Noise Removal

7.66 Abbaspour, Sara, Fallah, Ali, Lindén, Maria, Gholamhosseini, Hamid, "A Novel Approach for Removing ECG Interferences from Surface EMG Signals Using a Combined ANFIS and Wavelet," *Journal of Electromyography and Kinesiology*, 26, February 2016, 52–59.
7.67 Lu, Guohua, Brittain, John-Stuart, Holland, Peter, Yianni, John, Green, Alexander L., Stein, John F., Aziz, Tipu Z., and Wang, Shouyan, "Removing ECG Noise from Surface EMG Signals Using Adaptive Filtering," *Neuroscience Letters*, 462(1), September 2009, 14–19.
7.68 Zhang, Xu and Zhou, Ping, "Filtering of Surface EMG Using Ensemble Empirical Mode Decomposition," *Medical Engineering and Physics*, 35(4), April 2013, 537–542.

EMG Analysis and Classification

7.69 Phinyomark, Angkoon, Limsakul, Chusak, and Phukpattaranont, Pornchai, "A Novel Feature Extraction for Robust EMG Pattern Recognition," *Journal of Computing*, 1(1), December 2009, 71–80.
7.70 Raez, M. B. I, Mohammad-Yasin, F., "Techniques of EMG Signal Analysis: Detection, Processing, Classification and Applications," *Biological Procedures Online*, 8(1), March 2006, 11–35, DOI: 10.1251/bpo115.

EMG Application and Neuromuscular Abnormalities

7.71 Kimura J., *Electrodiagnosis in Diseases of Nerve and Muscle: Principles and Practice*, 3rd Edition. Oxford University Press, New York, 2001.
7.72 Reaz, Mamun B. I., Hussain, M. Sazzad, and Mohd-Yasin, Faisal, Techniques of EMG Signal Analysis: Detection, Processing, Classification and Applications. *Biological Procedures Online*, 8(1), 2006, 11–35, DOI: 10.1251/bpo124.
7.73 Sparto, Patrick J., Parnianpour, Mohamad, Baria, Enrique A., and Jagadeesh, Jogikal M., "Wavelet and Fourier Transform Analysis of Electromyography for Detection of Back Muscle Fatigue," *IEEE Transactions on Rehabilitation Engineering*, 8(3), September 2000, 433–436.
7.74 Subasi, Abdulhamit, "Classification of EMG Signals Using PSO Optimized SVM for the Diagnosis of Neuromuscular Disorders," *Computers in Biology and Medicine*, 43(5), June 2013, 576–586.

Clinical Data Analytics

8

BACKGROUND CONCEPTS

Section 1.2 Modeling Healthcare Information; Section 1.6.1 Medical Databases; Section 1.6.7 Intelligent Modeling and Data Analysis; Section 1.6.8 Machine Learning and Knowledge Discovery; Section 1.6.11 Clinical Data Analytics; Section 2.1 Data Modeling; Section 2.4 Statistics and Probability; Section 2.5.1 Texture Modeling; Section 2.7 Temporal Abstraction and Inference; Section 2.8 Types of Databases; Section 2.10 Human Physiology; Section 2.11 Genomics and Proteomics; Section 3.4 Machine Learning; Section 3.5 Classification Techniques; Section 3.6 Regression Analysis; Section 3.7 Probabilistic Reasoning over Time; Section 3.8 Data Mining; Section 4.7 Information Retrieval; Section 5.8 Medical Image Analysis; Section 5.9 Computer-Aided Diagnosis

A large amount of clinical and lab data specific to various diseases and abnormalities are being gathered in large data warehouses due to the availability of electronic data archival and retrieval systems and inexpensive data-storage facilities. The data is analyzed using statistical and artificial intelligent techniques to derive: 1) disease biomarkers; 2) disease features; 3) new classifications of diseases and other abnormal conditions; 4) develop new drugs and study their efficacy and toxicity; 5) risk analysis of various life-threatening diseases at different stages and 6) optimizing the clinical processes.

One of the major tasks of data analytics is to identify and discover temporal progression of the diseases from large-scale data. The intelligent data analytics has been used to: 1) provide missing data in the management of chronic diseases for better management and 2) to predict many classes of diseases such as heart diseases, cancer, hepatitis, diabetes and other long-term diseases. Other advantages of clinical data analytics are: 1) better treatment at reduced cost; 2) identification of dosage for faster recovery; 3) reduced health-maintenance cost for the populace by predicting the outbreaks of epidemics and 4) enhanced medical planning for the community.

An important aspect of clinical data analysis is the preparation of data. Data is extracted from different components of a body with minimal invasion. The components can be: 1) tissue from the localized diseased part to analyze the changes in affected cells; 2) body-fluid such as urine, spinal fluid and blood to study biochemical and biophysical changes in the biomarkers; 3) changes in the electrolyte concentration in the body; 4) radiology images and report that describe the texture, size, thickness of a vital organ or abnormal tissues; 5) density changes of the affected mass; 6) MRI report and 7) monitored bioelectrical signals. Patient-specific clinical and lab data needs to be de-identified to maintain the HIPAA mandated privacy and confidentiality. A study may be quite broad and may include multiple fields, including clinical trials in multiple phases involving animals and human volunteers.

Intelligent data analytics requires the application of automated machine learning techniques, biostatistical techniques, decision support systems and artificial intelligent techniques to predict, study or improve: 1) new knowledge and associations: 1) disease-related information; 2) biomarkers for chronic and life-threatening diseases such as cancer; 3) drug-development-related information; 4) toxicity-related information; 5) meaningful patterns and trends in large-scale data, including time-series data for abnormalities; 6) diseases that affect large class of demographics in various localities or 6) regimen.

AI techniques in clinical data analytics are: 1) cluster analysis, support vector machines and artificial neural networks for automated classification; 2) predicting the trend and correlation using regression analysis; 3) Gaussian mixture based analysis; 4) fuzzy analysis; 5) associative rule mining; 6) clustering of time-series of vital signs and 7) clustering of texture information of radiology images.

These AI techniques are exploited for the automated analysis of: 1) echograms for pregnancy-related data; 2) spectral analysis of the intravascular ultrasound to study blood-flow in various parts of the body for the identification of oxygen starvation; 3) analysis of MRI and PET-scan data of different organs such as liver, brain, kidney and heart; 4) chronic disease management; 5) disease diagnosis and discovery and 6) biomarker identification. *Microarray data clustering* and *Genome Wide Association Studies* have also been applied to predict the associated genes, pathways and biomarkers for the genetic diseases.

Decision support systems are the semi-automated software systems that assist human experts in the decision-making process. These systems analyze a large amount of data to derive association-rules, generalized if-then-else rules and develops reports for the human-experts to facilitate the decision. Interactive decision support systems also assist physicians and experts by providing multiple diagnosis-choices and probable actions to fix disease-conditions.

The role of human experts in a decision support system is important for the following reasons: 1) a computational intelligent system uses an approximate model of disease with a small number of parameters to reduce the computational overhead; 2) all the parameters modeling the effect of an abnormality are not available; 3) generalized rules in a decision support system are unable to handle the exceptions due to the oversimplification of the underlying mathematical model and 4) the liability of the decision cannot lie with a program if the outcome is incorrect.

Randomized trial and the *lack of bias* are very important features of clinical data analysis. The patients who are being administered for medical intervention or experimental medication should be picked randomly. There are multiple ways a procedure or medication can be administered to the volunteers: 1) *unblinded study* – both the medication administrators and the participants know each other; 2) *single-blind study* – participants are unaware of medication being administered, while providers have full knowledge of medication and 3) *double-blinded study* – both the drug administrator and the participants are unaware of the drug being administered.

8.1 CLINICAL DATA CLASSIFICATION

Clinical data classification facilitates: 1) the identification of the factors that increase the risk of getting disease or side-effect while other parameters are unchanged and 2) the identification of the medication effectiveness for recovery from a disease state. For example, for the same age group, gender, ethnicity and same disease-state, we will like to find the optimum drug-dosage for quickest recovery with minimum toxicity. In such cases, different test-groups are given different dosage, and their recovery outcome is studied using clustering techniques. Another example is to study the effect of an ingested compound as a daily diet on heart attack, cancer or Alzheimer's disease. Another example is to study the risk of brain-stroke caused by hypertension.

It is not always possible to keep all the parameters constant; there may be more than one factor that may be classified as *risk-factors*. The effects of individual risk-factors are separated by modeling the combinations of the individual effects. Nonlinear combinations are difficult to handle; most of the times, the modeling uses linear approximations.

Statistical data analysis also studies the correlation between seemingly different independent variables. Correlation analysis reduces the number of seemingly independent variables by identifying strong correlation between two or more variables. Regression analysis between independent variables and dependent variables needs to be backed up by statistical significance analysis. In case of multiple independent variables, correlation analysis involves linear or nonlinear combination in the form $Y = f_1(X_1) + f_2(X_2) + \ldots + f_n(X_n)$, where Y is a dependent variable; $f_1, ..., f_n$ are the functions; and $X_1, ..., X_n$ are uncorrelated independent variables. The simplest form of multivariate combination is *multivariate linear combination* of the form $Y = A_1X_1 + A_2X_2 + ... + A_n X_n + B$, where all X_i ($1 \leq i \leq n$) are the independent variables.

Classification study is also done on patients with deteriorating health to study their *survivability*. Their survival-rate is based upon curve-fitting on data of the patients with similar disease. There are many types of curves that are used to model survivability. Most common curve-fitting models are exponential curves using multivariate analysis. For example, in the case of gene-aberration causing cancer, a patient's survivability depends upon age, cancer-staging and the type of cancer. Some types of cancer are slow growing, while others are more aggressive. The predicted outcome is validated using statistical techniques such as *chi-square test* for valid confidence-interval. Popular confidence-intervals for statistical significance are *95%, 97.5%* or *99.5%*.

8.1.1 Clustering of Clinical Data

A classification of data requires: 1) careful planning to collect the relevant data so that the output states of the patients are very similar and 2) identification of the appropriate set of features for clustering. The feature-set is sent to the experts for consensus. The parameters include: 1) historical finding; 2) lab results such as protein-level changes, biochemical and biophysical property changes of tissues and bodily fluids and 3) changes in radiological properties such as texture, shape and size. Biochemical property involves the formation of new compounds and changes in color. Biophysical properties involve changes in texture, shape and size or the amount of eluting fluid.

Clustering derives: 1) the effectiveness of a medication; 2) the optimal dosage of the medication for the effective recovery; 3) the dosage toxicity-level that would hurt other bodily functions and 4) the side-effects that occur due to the dosage-levels of the administered medications. Many times, a disease is sub-classified into different stages based upon: 1) evidence-based analysis; 2) spread of the disease and 3) the patient's prognosis for the survivability. These terms are interrelated. The analysis involves clustering of the time-series data of monitoring a patient's vital signs and other symptoms over a period. The identified clusters are labeled based upon variations of the associated parameters.

K-means clustering has been used in the disease identification and comparison of the two groups with different parameters such as age, gender and ethnicity among other attributes. Diverse sample groups are clustered and compared for consistent patterns and variations. Consistent patterns in diverse sample groups give credence to features characterizing a disease. Other popular clustering techniques are: 1) mixed Gaussian method involving a combination of Gaussian curves to model the data-distribution; 2) a combination of mixed Gaussian method at lower level and hierarchical clustering at higher level; 3) model-based clustering and 4) nearest-neighbor clustering.

Model-based clustering is based upon probabilistic distribution of a point to different clusters. The cluster models can be of different types such as: 1) spherical; 2) constant variance; 3) equal volume and 4) varying volume model. It is useful in the cases where the clusters are irregular (nonspherical) and can be modeled as a mixture of multiple small numbers of clusters.

8.2 BIOSTATISTICS IN CLINICAL RESEARCH

Biostatistics is used to derive and validate certain hypothesis related to the nature of specific diseases, biomarkers, spread of diseases, recovery in response to a medication and the dosages of a medication for the appropriate response. The hypothesis is based upon the statistical analysis of feature-values of interest in an observed population. Measurements can be explicit or conceptual that can be visualized.

Measured features are referred to as independent or dependent variables. It is assumed that independent variables are not correlated to each other. A query is made to study the relationships between independent and dependent variables. Independent variables could be disease stage, recovery state. Dependent variable could be the concentration of biomarkers. The population assumes a shape of distribution. These shapes of distribution are analyzed for mean, median, dispersion (variance) and bias (lopsidedness of the curve) and outliers. *Outliers* are the measurements outside the scope of the data-distribution.

For a generalized study of data-distribution, various models such as Gaussian distribution and mixed Gaussian distribution are used. Using statistical analysis, we measure the probability to a high level of confidence that a measured data-point is part of the data-distribution described by the corresponding model.

8.2.1 Sensitivity, Specificity and Accuracy

The metrics used to measure the probability of a clinical event to occur are: *sensitivity, specificity, positive predictive value, negative predictive value* and *overall accuracy*. *Sensitivity* is defined as the probability of identifying a condition (or abnormality) based upon a prescribed diagnostic test or the change in biomarker provided the condition is truly present. *Specificity* is described as the probability that the absence of a disease and abnormal condition will be clearly recognized by a diagnostic test or a biomarker. *Positive predictive value* is defined as the probability that a person who have a disease is identified by the diagnostic tests or biomarkers to have the disease. *Negative predictive value* is the probability that person who has tested negative does not have the disease. O*verall accuracy* is defined as the probability that a randomly selected person is diagnosed correctly of his/her health-state based upon a diagnostic-test or biomarkers. Consider a positive indication based upon a diagnostic-test or a biomarker as $\wp+$, a negative indication based upon a diagnostic-test or a biomarker as $\wp-$, presence of a disease as $Q+$ and the absence of the disease as $Q-$. Then, the sensitivity is denoted as conditional-probability $prob(\wp+ \mid Q+)$; specificity is denoted as the condition-probability $prob(\wp- \mid Q-)$; positive predictive value is denoted as condition-probability $prob(Q+ \mid \wp+)$, and negative predictive value is denoted as $prob(Q- \mid \wp-)$.

There are four more related terms associated with diagnostic tests: *true-positive, true-negatives, false-positive* and *false-negative*. *True-positive* is defined as the number of subjects who have a disease and have been identified correctly as having the disease by a diagnostic-test or a biomarker. *True-negatives* are the number of subjects that do not have a disease and have been correctly identified by a diagnostic test not to have the disease. *False-positives* are the number of subjects who do not have a disease, yet have been detected by the diagnostic tests or biomarkers as having the disease. *False-negatives* are the number of subjects who have a disease, yet have not been diagnosed by the test or the biomarker as not having the disease. Specificity and sensitivity are closely related to the concept of *false-positive* and *false-negative*. Sensitivity is defined by Equation 8.1; specificity is defined by Equation 8.2; *positive predictive value* is defined by Equation 8.3 and the *negative predictive value* is defined by Equation 8.4.

$$Sensitivity = \frac{truePositive}{truePositive + falseNegative} \tag{8.1}$$

$$Specificity = \frac{trueNegative}{trueNegative + falsePositive} \tag{8.2}$$

$$Positive\ predictive\ value = \frac{truePositive}{truePositive + falsePositive} \tag{8.3}$$

$$Negative\ predictive\ value = \frac{trueNegative}{trueNegative + falseNegative} \tag{8.4}$$

Example 8.1

Consider a biomarker test for predicting diabetes II – an insulin-based disease that causes sugar imbalance in a person. A biomarker used for diabetes II is A1C test that checks the average sugar level for the past three months. Out of *100* persons tested, *40* persons have diabetes, and the remaining *60* persons do not exhibit the disease. Out of these 40 persons, only *34* people have elevated A1C level; the remaining six persons do not show sugar elevations. Out of the remaining *60* persons not having the disease, eight people show elevated A1C, and remaining *52* persons do not show an elevated-level. In this case, true-positive is *34*, false-positive is *8*, false-negative is *6* and true-negative is *52*. *Sensitivity* is $34/(34 + 6) = 0.85$; *specificity* is $52/(52 + 8) = 0.87$; *positive predictive-value* is $34/(34 + 8) = 0.81$ and *negative predictive-value* is $52/(52 + 6) = 0.90$.

There are two types of statistical studies in a clinical analysis: *observational* and *experimental*. In an observational study, values of various variables are observed, without any human control, in response to the natural changes. In an experimental study, certain independent variables are altered in a regulated manner while keeping other independent variables constant to see their effect on the outcome (dependent variables). For example, occurrence of various types of cancer, heart abnormalities, brain abnormalities and their correlation with different biochemical will be labeled as observational study. However, checking the efficacy of a medicine and dosage of a medication needed for the recovery of the patients in a control-group in response to a medication is an experimental study.

8.2.2 Application

Biostatistics uses many data-distribution models to reason about the confidence-factor of a new data being part of an event, verification models to validate the confidence-factor and probabilistic decision-network models to infer the probability of the outcome and overall advantage of taking an approach. There are many clinical activities that require data-modeling using various distribution models. Some of these activities are: 1) study of disease occurrence such as tumor growth models using radiology data analysis; 2) clinical trials to study the efficacy, toxicity and other side-effects of a medication; 3) dose-response models; 4) machine learning models to learn about prognosis of a patient and the action to take; 5) toxicity analysis using animal and/or human models and 6) risk analysis models to assess the adverse outcome of a clinical treatment such as surgery, administering anesthesia, medication or a combination of medications interfering with each other. The data-distribution models involve many types of statistical curves such as Gaussian and its variations, Poisson and its variations and binomial distribution.

8.3 RANDOMIZED CLINICAL TRIALS

Before clustering, relevant clinical data are collected and compared. One of the issues in clustering is to pick up a right sample. General comparison measures the effect of some treatment, surgical procedure, use of medical devices, medication or dosage. In such cases, the patient-space is divided into two pools: 1) those who are administered an experimental regimen and 2) those who do not go through the experimental regimen. The people should be chosen randomly in two pools to avoid any bias in clinical clustering.

8.3.1 Hidden Inherent Clusters and Bias

In real life, there are many types of clusters that do not represent randomized groups as follows: 1) person may be going to the same doctor or clinic; 2) people may be from the same ethnic group, age group, income group or gender preferring to go to the same clinic and 3) hospitals or clinics may be following similar guidelines for treating the patients affecting the outcome of the randomized study. The patients within inherently hidden groups are correlated in some way based upon the similar attributes such as age, ethnicity, treatment and guidelines.

The ideal way is to randomly pick patients who do not belong to just one of these inherently hidden clusters. Both *control-group* (patients not taking the medication) and *intervention-group* should be present randomly in these inherently hidden clusters. Each of the hidden clusters should be analyzed for the correlation-differences between the control-group and intervention-group, and the result should be averaged out across these inherent groups for accuracy. Ignoring the effect of inherently hidden clusters within the same cluster (control or intervention) leads to large standard deviations and higher false-negative results. Ignoring the clusters in randomized trial study biases the outcome because inherently hidden clusters will have uneven distribution in control and intervention groups, and the variance due to differences between control and intervention group will not be properly reflected in the outcome. The overall variance will be underestimated.

8.3.2 Cluster Contamination

Another problem in the randomized trial is *contamination*. *Contamination* occurs when the intervention-group or control-group members are treated differently and in the worst case, change the role: a control-group member may be treated as an intervention-group member. *Contamination* attenuates the dissimilarity between the control-group and intervention-group causing reduced knowledge of the actual effect. In order to reduce contamination, control-group and intervention-group are clearly separated from each other. Ideally, both the groups are carefully monitored for the administration of a regimen. The sample-size is increased by a factor *1/(1 − contamination)²* to reduce the attenuation-effect caused by contamination where *contamination* is defined as the proportion that has been attenuated.

Two types of models are used to reason about correlation in clusters. In the *fixed model,* the effect of clusters considered is fixed on the outcome under the assumption that the sample-size does not affect, and the population is uniform. In the *random model*, the sample-size is considered randomly chosen from a large population, and its effect on the outcome varies with the sample-size and the sample itself. Random models are more general and realistic.

8.3.3 Correlation

The sample-size for clinical trial is adjusted to take care of intracluster and intercluster variations in the randomized trials. Standard sample-size is smaller and underestimated. The corrected sample-size for randomized trails is given by *standard-sample-size* $\times$ *(1 + (m − 1)ρ)*, where *m* is the average cluster size, and ρ is the correlation-coefficient of the clusters within and across each other. The correlation-coefficient is calculated using Equation 8.5.

$$\rho = \frac{\sigma_b^2}{\sigma_b^2 + \sigma_w^2} \tag{8.5}$$

The factor σ_b^2 is the variance between the clusters, and σ_w^2 is the variance within a cluster. For the *correlation-coefficient = 0*, the variance σ_b^2 is zero that means variance between the clusters is zero. The *correlation-coefficient = 1* means that σ_w^2 is equal to *0*, and there is no variability within the clusters. Correlation-coefficients are derived from experimental data or previous studies.

8.3.4 Meta-Analysis

Meta-analysis uses statistical techniques to combine the findings of similar test-studies. For example, drug-efficacy may have been studied in multiple hospitals on different samples. Individually, they may not give a definitive picture. However, when the results are combined, a clear picture emerges. Meta-analysis reduces the errors of individual studies and strengthens the evidence. Meta-analysis is used in studying the efficacy of the drugs, dosage analysis and effects of a regimen.

There are two approaches to the meta-analysis: 1) using aggregate data (AD) of an individual study, and merging the effect of individual studies; 2) performing statistical analysis after pooling the raw data of *individual participants' data* (IPD). The goal of the meta-analysis is to study the overall result. Both the approaches yield similar results if the population pool is homogeneous. IPD approach is used for standardizing statistical analysis, summarizing the results, examining interactions, assessing participant-level effects and checking the model. However, AD based approach is easier as they can combine the result of individual studies.

8.3.4.1 Control-group vs. intervention-group

In every sample, there are two types of data: 1) data related to the *control-group* and 2) data related to the *intervention-group*. The *control-group* is not administered any medication (or therapeutic procedure). The corresponding *intervention-group* is administered the medication or therapeutic procedure. Much metrics for meta-analysis can be modeled for the ith sample using the number of patients n_i^c in the control-group, number of the patients n_i^{in} in the intervention-group, the ratio of the patients with some event as R_i^c for the control-group, the ratio of the patients with some event R_i^{in} for the intervention-group, the event-probability P_i^c for the control-group, and the event-probability P_i^{in} for the intervention-group. Meta-analysis studies the true treatment effect β_i, mean μ of all the studies, variance σ_i^2 of individual studies from the combined mean, risk-difference $(P_i^{in} - P_i^c)$, relative-risks (P_i^{in} / P_i^c) and the relative-odds. *Relative-odds* is the ratio of the odds of an event happening in an intervention-group against the odds of an event happening to the control-group. Mathematically, *relative-odds* is defined as $\frac{P_i^c/(1-P_i^c)}{P_i^{in}/(1-P_i^{in})}$. Relative-odds is a popular metrics.

8.3.4.2 Fixed vs. random effect

The general assumption is that a pool of samples is homogeneous, and the different sample-studies carry some errors; the overall effect is fixed. However, in reality, due to variations in the characteristics of the pools or biased control-groups and intervention-groups, the studies may have random-effects varying the overall result. The fixed-effect model can be modeled easily using Equation 8.6 where γ_i is the estimated outcome, θ is the fixed-effect, and ε_i is the random sampling error. The random-effect model assumes that each sample θ_i varies randomly and can be modeled as a normal Gaussian distribution with mean θ. The random-effect model is represented by Equation 8.7 where μ_i represents the study variance σ^2.

$$\text{Fixed-effect model}: \gamma_i = \theta + \varepsilon_i \tag{8.6}$$

$$\text{Random-effect model}: \gamma_i = \theta + \mu_i + \varepsilon_i \tag{8.7}$$

In the case, random-effect meta-analysis is modeled as meta-regression analysis where each study is plotted as a point in the two-dimensional plane. Equation 8.7 is substituted by Equation 8.8 such that the term θ is substituted by the term βx.

$$\text{Meta-regression analysis model}: \gamma_i = \beta x + \mu_i + \varepsilon_i \tag{8.8}$$

8.4 SURVIVABILITY AND RISK ANALYSIS

An important concept in clinical data analysis is to model patient's survivability after a life-threatening incident such as cancer surgery, cardiac bypass surgery, kidney transplant, heart transplant or brain tumor surgery. *Survivability* means the probability of a patient to live after a hazardous incident occurs under a fixed treatment plan up to sometime in the future. Another related concept is *hazard*. *Survivability* is related with the probability that a patient can live for a specific number of years without the relapse of the same or correlated incident or death. *Hazard* is defined as the probability of the relapse of the next incidence after the previous occurrence of the event. Both the concepts are related, yet different. Survivability is related to the incident not occurring in the future, while hazard is related to the relapse. The increase in relapse increases the hazard and reduces the survivability at the same time.

Reasoning about survivability requires statistical analysis of the patients with a similar condition. The statistical analysis is not easy because the study is done on the real patients over a longer period. During that period, multiple factors hinder the study as follows: 1) a patient's age keeps increasing triggering other abnormal conditions; 2) the treatment for a patient may change due to the availability of a new therapy; 3) a patient may leave the study group; 4) a patient may die before the next phase of the study; 5) a patient may suffer from an additional hazard affecting the survivability and 6) censor—missing information of an incidence before and after the monitoring period. There are many models to estimate survivability: 1) Kaplan–Meier survival estimate; 2) log-rank test for comparing survivability and hazard in different study groups; 3) non-parametric test based upon Chi-square distribution function to compare different survival-group studies; 4) combining multivariate regression analysis with the exponential model of survivability function or hazard function.

A person may join the survivability analysis after the occurrence of an incidence. Another possibility is the lack of knowledge about the occurrence of an incidence before monitoring starts. This affects the survivability information and is called *left-censor*. Similarly, a patient may be lost during the survivability study, but may be still alive. In this case, the information about them is missing and is called *right-censor*. Both left-censor and right-sensor adversely affect the survivability analysis.

8.4.1 Nonparametric Estimates

Kaplan–Meier estimate is based on the notion that the probability of survival in the future is directly related to the probability of survival in the immediate past and is given by Equation 8.9. The time space is discretized based upon the follow-up visits of the patient. The term $S(t_i)$ denotes the survivability at time t_i, $S(t_{i-1})$ is the survivability at discrete time-step $t_{(i-1)}$, n_i is the number of surviving patients at time just before time t_i, and d_i is the number of cumulative events such as death or another relapse right before the time t_i. The initial survivability at time t_0 is assumed to be *1*. Using Equation 8.9, a survivability curve is drawn to plot survivability against time.

$$S(t_i) = S(t_{i-1}) \times \left(1 - \frac{d_i}{n_i}\right) \qquad (8.9)$$

Log-rank test is a mechanism to relate hazard with survivability. Under the assumption that survivability can be modeled using an exponential curve, Equation 8.10 defines the hazard function $h(t)$ as the change of rate of the log of the survivability. The cumulative hazard function $H(t)$ is defined as the integral of hazard function $h(t)$ and is equal to $-ln(S(t))$. The equation can be used to calculate the hazard from the survivability curve and vice-versa.

$$h(t) = -\frac{d}{dt}(\ln S(t)) \qquad (8.10)$$

Survivability in two or more groups can be compared using non-parametric chi-square test that estimates the number of events from the previous event under the assumption that two groups are very similar. The chi-square distribution is given by the cumulative sum of the square of the difference between the observed events and expected events in each group.

Kaplan–Meier estimate is suitable for univariate analysis when there is only one risk-factor that affects the survivability of a patient-group. However, more than one factor called *covariates* effect the survivability. This combined effect of covariates is modeled using regression analysis. The simplest form is linear multivariate regression analysis. Some of the examples of covariates are: 1) different types of incidences such as tumors, age, kidney failure, fibrillation in heart-muscles; 2) age and 3) gender. The covariates could be continuous like tumor or various types of cancer or Boolean like gender. The effect of regression analysis is combined with the exponential model of the hazard function using Cox's PH-model (proportional hazard model) described in Equation 8.11.

$$h(t) = h_0(t) \times e^{(\alpha_1 x_1 + \alpha_2 x_2 + ... + \alpha_n x_n + \beta)} \tag{8.11}$$

where $h(t)$ is the hazard function at a discrete-time t; $h_0(t)$ is the baseline hazard function; $x_1, ..., x_n$ are the covariates; and α_i, β are coefficients for combining the covariates. As any one of the factor increases, the corresponding hazard function increases lowering the survivability as shown by Equation 8.11.

Another variation of the PH-model is to link the multiple regression analysis with an exponential model of survivability as shown in Equation 8.12.

$$S(t) = S_0(t) \times e^{(\alpha_1 x_1 + \alpha_2 x_2 + ... + \alpha_n x_n + \beta)} \tag{8.12}$$

where $S(t)$ is the survivability function at a discrete-time t; $S_0(t)$ is the baseline survivability function; $x_1, ..., x_n$ are the covariates; and α_i, β are coefficients for combining the covariates. This model is used to predict survivability in different stages of the disease by changing the corresponding coefficient values.

8.4.2 Parametric Estimates

There are variations of the exponential Ph-model such as *Weibull model, Gompertz model* and *AFT model* based upon different probability distributions for the failure-rate. *Weibull probability-density function* has three parameters: t, λ and γ. The Weibull probability-density function is given in Equation 8.13. The factor γ is the *shape parameter*, and the factor λ is the *scale parameter*. By changing the shape and scale parameters, the shape and size of the distribution-function can be altered, respectively. A value of $\gamma > 1$ denotes that survivability decreases faster with time, and the hazard rate increases with time.

$$f(t; \lambda, \gamma) = \begin{cases} \lambda\gamma(\lambda t)^{\gamma-1}e^{-(\lambda t)^{\gamma}} & \text{if } t \geq 0 \text{ and } \lambda, \gamma > 0 \\ 0 & \text{otherwise} \end{cases} \tag{8.13}$$

The survivability function $S(t)$ is given by Equation 8.14, and the corresponding hazard function is given by Equation 8.15.

$$S(t) = \int_t^{\infty} f(t)dt = e^{-(\lambda t)^{\gamma}} \tag{8.14}$$

$$h(t) = -\frac{d}{dt}(ln\, S(t)) = \lambda\gamma \times (\lambda t)^{(\gamma-1)} \tag{8.15}$$

Gompertz probability function is given by Equation 8.16. It also has the shape parameter and the scale parameter. The factor γ is the scale parameter, and the factor λ is the shape parameter.

$$f(t;\gamma, \lambda) = e^{\frac{\lambda + \gamma t - 1}{\gamma(e^{\lambda + \gamma t} - e^{\lambda})}} \qquad (8.16)$$

The corresponding survivability-function and the hazard function for the *Gompertz probability* distribution are given by Equations 8.17 and 8.18, respectively.

$$S(t) = e^{\frac{-e^{\lambda}}{\gamma(e^{\gamma t} - 1)}} \qquad (8.17)$$

$$h(t) = e^{(\lambda + \gamma t)} \qquad (8.18)$$

The expected value of life μ is given by multiplying time t with the distribution-function $f(t)$ and integrating over the time from 0 to ∞ as shown in Equation 8.19. The equation reduces to Equation 8.20 in terms of survivability function $S(t)$.

$$\mu = \int_{t=0}^{t=\infty} t\, f(t)dt \qquad (8.19)$$

$$\mu = \int_{t=0}^{t=\infty} S(t)dt \qquad (8.20)$$

A major problem in survivability analysis is to model the people leaving the group due to additional information (or misinformation) about the medication such as side-effects and toxicity. This phenomenon is called *informative censoring,* and biases the survivability study. However, if the percentage of informative censoring is small, then it does not affect the study significantly.

8.4.3 Risk Analysis

Risk is defined as the probability of the occurrence of an undesirable event. In clinical risk analysis, an event is the single occurrence of a serious disease. Disease risk requires surveillance of a considerable number of cases. Clinical risk analysis and management are an important topic to handle the chronic and life-threatening diseases such as diabetes, cancer, brain-related dementia, cognitive impairment diseases, age-related macular degeneration, kidney failure and heart diseases. In risk analysis, probability of occurrence of the disease is derived for better management of the disease. There are five levels in risk analysis and management: 1) establish the context; 2) identifying the risk; 3) risk analysis; 4) risk evaluation and 5) treating the risks.

Risk-factors are the variables that affect the disease. Risk-factors are monitored and regulated for risk-management. *Risk analysis* involves the comparison of the probabilities of a risk of the persons having the risk-factor against the probability of the persons without the risk-factor to get a disease. This ratio is called *relative risk.*

Example 8.2

For example, in a study relating exercise to diabetes, there are two groups: 1) people who exercise and 2) people who do not exercise. The risk-factor for diabetes is "not doing exercise". Let us assume that the probability of people getting diabetes without doing exercise is P_1, and the probability of the people getting diabetes while doing exercise is P_2. In that case, the relative-risk of not doing exercise is given by the ratio P_1/P_2.

There are other metrics to calculate risks such as *hazard-ratio* and *odds*. *Hazard-ratio* is defined as the ratio of a hazardous event to occur in a fixed time for persons having the risk factor and the hazardous event to occur for the persons not having the risk-factor. *Hazard ratio* is computed using nonlinear equations. *Odds* are defined as the ratio of the probability of an event to occur and the probability of an event not to occur.

8.5 CLINICAL DECISION SUPPORT SYSTEM

A *clinical decision support system* (CDSS) is an intelligent system that generates pertinent information based upon inputted patient data to suggest a course of actions. The action could be the diagnosis of an ailment, choice of a medication, a possible surgery, their combinations and the patient-care. CDSS is a special type of decision support system (DSS) that integrates knowledge management, intelligent inferencing, machine learning, data warehousing and electronic health records. The intelligent inferencing and machine learning techniques include cluster analysis, Bayesian decision network, uncertainty-based reasoning, data mining and content-based information retrieval from EHR. CDSS is needed to help the medical practitioners make a quick informed decision that involves handling larger amount of data and deduction using rule-based systems that can handle uncertainties and probability-based reasoning.

CDSS has been used for antibiotic prescription, simple patient-ailment diagnostics, drug-dosing and patient-care. The main advantage of CDSS is the integration of human reasoning and intelligent computational analysis involving a large amount of data. In the cases of significant uncertainty, an expert in the loop navigates through the solution-space using fast computational and artificial intelligent techniques. Humans have vast knowledge-base and complex reasoning power, and a rule-based system integrated with EHR can process the data intelligently and efficiently.

There are four major components of a decision support system: 1) input data, 2) triggers, 3) offered choices and 4) intervention. *Triggers* are the events that invoke a rule in the decision support system. *Offered choices* are multiple possible actions that a human in the loop can take. The execution of rules invokes an action. The action is called *intervention*.

The issues in the design of CDSS are: 1) better interface that is patient-friendly and clinician-friendly that also autocorrects the omissions and errors; 2) better summarization technique that can help medical practitioner visualize individual patient's data such as pertinent medical history, physiological parameters and current treatments clearly and intuitively; 3) prioritize the recommendation based upon prognosis, mortality reduction, side-effect reduction, patient's preference and lifestyle and type of insurance; 4) combine the effects of multiple disease conditions and the effects of medications, including drug-interference; 5) support natural language text to extract information from the knowledge-bases and databases; 6) use data-mining of existing databases; 7) create an architecture that supports sharing of the standard modules and 8) creation of the Internet accessible decision support repositories that can be easily accessed and used in CDSS. The NLP system should support alerting, monitoring, natural data input and reminding.

8.5.1 Clinical Knowledge-Based Systems

Applications of clinical knowledge-based systems include: 1) protocol design and maintenance tools, 2) automated preparation of reports from laboratory tests, 3) drug advisory systems that check for multiple drug-interactions, 4) optimum visualization of data stored in EHR,) image analysis to identify abnormalities, 5) automated training systems to educate clinical practitioners and patients, 6) clinical reminders systems and 7) temporal abstraction systems derived from time-series databases.

Huge amount of data is collected over a large period through lab testing, patient monitoring and data mining. Time is a major dimension. Clinical data is both quantitative and qualitative. If all the data is stored in a raw-format, then huge amount of storage-space will be required. The data needs to be compressed using interval-logic in such a way that: 1) facilitates the visualization of data in a clinically meaningful way for medical practitioners to facilitate a diagnosis and 2) to show trends in the data.

The whole process requires *time-abstraction*. The inputs to the time-abstraction are: 1) parameter values, 2) events and 3) abstraction goals. The outputs from the time-abstraction process are: *intervals, interval based parametric values* and *trends within a context*. The parameter values have different data-types, different units of measurements, may arrive in random order and may have different temporal properties. The time-abstraction task requires five concurrent methods: 1) *context formation* by *domain-restriction* for accurate analysis; 2) abstraction using *vertical inference*; 3) temporal *horizontal inference*; 4) *temporal interpolation* for interval alignment of data and 5) *temporal pattern matching* over diverse interval representations. An example of the restricting the domain for context formation is to study the effect of a drug on a disease state.

The *context formation* requires interval analysis of multiple events and tasks. *Vertical inference* involves forming of higher-level abstractions by the events and parameter values at the lower level. *Horizontal inference* joins the inference from two different events at the different time intervals at the same level to create a new abstraction. *Temporal interpretation* bridges the gap between two semantically similar but temporally disjoint episodes to form a bigger episode that is represented using a truth-value. *Temporal pattern matching* connects the disjoint intervals by matching the patterns of several types of parameter values. Connected with ontology, the time-abstraction mechanism defines a patient's disease-state.

Different intervals may have different units and need to be aligned by standardizing the *used units*. The intervals are connected using *Allen's synchronization primitives* that temporally relate two intervals. There are 13 such primitives that relate the start and end of two intervals and delay between them as described in Section 2.7.

Example 8.3

A diabetic patient shows both qualitative and quantitative symptoms. Qualitative symptoms are based on the visual observation of the patient's condition, and quantitative symptoms are based on the lab tests, glucose-level monitoring and drug dosage sampled over a period. A patient's state is modeled as (*drug-type, dosage, glucose-state, physical symptoms*). The *glucose-state* is inferred by multiple glucose tests using blood and urine samples. The visual symptoms could be the presence of yeast-infection on genitals, reported light headedness, and reported frequency of urination. The events could be measurements before and after the meals, before and after exercise, before and after insulin and other diabetes drug intake. The various intervals are included in the measurements. Together, these parameters are combined to derive a patient's state.

8.5.2 Neural Network-Based Diagnosis

Neural networks have been applied to diagnose different types of diseases. Some examples of the applications are: 1) image analysis of different tissue samples for tumor detection, 2) clinical decision support systems, 3) heart disease prediction, 4) epilepsy detection, 5) tuberculosis detection, 6) detection of diabetic retinopathy, 7) colorectal cancer detection, 8) colon cancer detection, 9) gynecological disease detection and management, 10) early diabetes detection from blood samples, 11) survival prediction from terminal diseases such as cancer or renal failure and 12) blood and urine sample analysis to derive diverse biomarkers.

The development of a neural network based system involves: 1) feature-selection based on the evaluation of pathological data and clinician-guided parameter extraction; 2) building a database of extracted feature-set; 3) training and verification of the ANN; 4) testing ANN on the medical cases. The database used for training should contain enough reliable cases to enhance the reliability of the system.

Backpropagation algorithm is used to train the ANN. False-positives and false-negatives are handled by using the discriminative power of the clinicians.

The major advantages of a neural network based system are: 1) evidence-based identification; 2) use of large amount of data during training phase that significantly improves accuracy and robustness and 3) reduced time for diagnosis.

8.5.3 Clinical Data Mining

Data mining in clinical databases is called *clinical data mining*. Clinical data mining is important because a massive amount of data is collected in EHR from various medical institutions. Data mining can derive: 1) appropriate medication and dosage for different population groups; 2) biomarkers; 3) trends of medical changes in a population area; 4) predictions of patients' conditions for a drug and the corresponding dosages; 5) risk-factor for abnormal conditions; 6) survivability analysis; 7) optimal practice and techniques to improve the clinical care; 8) diagnosis of diseases from patients' symptoms and lab-tests.

Examples of clinical data mining systems are: 1) computer-aided diagnosis of cancer from radiology images; 2) monitoring tumor response to chemotherapy; 3) screening of different types of cancer such as breast, colon and prostate cancer episodes; 4) identification of metabolic diseases; 5) mining healthcare literature; 6) diagnosis of dementia and Alzheimer's disease; 7) diagnosis and monitoring of neuromuscular diseases and 8) molecular profiling of different types of cancers.

The knowledge discovery process in databases requires: 1) dataset selection using data-quality assessment techniques; 2) preprocessing to de-identify and cleaning the data; 3) choice of techniques such as clustering, Bayesian classification, regression analysis and temporal abstraction; 4) selection of an appropriate algorithm for pattern extraction; 5) pattern extraction along with hypothesis testing to validate the pattern; 6) interpretation of the pattern using known physiological process and common-sense reasoning in the specific domain and 7) application of the patterns for better clinical care and timely intervention. *Common sense reasoning* is a rule-based reasoning technique that represents knowledge as a graph, and traverses the knowledge-graph using the relationship between the various entities.

The data may include lab tests, monitored data, and patient survey. The derived rules are represented in the form of if-then-else rules. The data may include noise, contradictory data and missing data. The data across heterogeneous databases may have different formats, including raw text due to the prevailing lack of standardizations among the vendors. The data may be missing because different medical practitioners give different weights to the parameters responsible for the same medical condition, and less-weighted parameters are ignored in computational systems due to time and resource burden.

The data is represented as a two-dimensional matrix where rows represent multiple cases or events, and the columns represent the parameters. Each cell represents the corresponding parameter value for a specific case. Certain pattern analysis cases require tables with a large number of parameters. In such cases, an existing hypothesis or knowledge of existing patterns is used as the starting point. The use of the prior knowledge prunes the number of parameters used in analyzing the patterns.

The data in multiple formats and unstructured data are converted into a standard format before processing the data. Natural language processing is used to convert raw text into a structured format. The variable values are normalized to the proper unit, and the weights are readjusted to reduce the bias of specific parameters due to large values. Many times, the same concept is expressed by multiple parameters. In such cases, the related parameters are combined and represented as a single parameter to avoid redundancies. Time-series data are transformed to time-interval data using temporal-abstraction. Missing values are handled using following three techniques: 1) a parameters with large number of missing values are removed, if not associated with rare conditions; 2) Data is used as it is or 3) Records with missing values are removed from the data mining set. The parameters are reduced and prioritized using: 1) an automated dimension-reduction technique such as PCA or LDA; 2) using domain-experts who can pick important parameters and 3) entropy-based information gain measures with each associated parameter to priotize the parameters and adjust their weights accordingly.

The data mining methods use clustering, associative rule formation, regression analysis, summarization and dependency modeling to study the correlation between the variables. The techniques are selected based upon the overall goal of data mining. For example, clustering is used to classify different classes of tumors. These classes are associated with different regimens. Regression analysis is used to correlate: 1) biomarker concentration with disease stages or 2) drug-dosage with the efficacy and/or toxicity. Summarization and associative rule mining are used to derive the association-rules. The popular algorithms used for data mining include K-means clustering, multivariate linear regression analysis, decision trees, support vector machines, neural network and Bayesian decision rules.

8.5.4 Effect on Clinical Practices

Healthcare organizations are increasingly using CDSS to provide patient-care, including reminder systems to reduce the medication errors and to provide preventive care. The use of CDSS has significantly improved the intervention procedures resulting into improved health care. Most of the intervention successes happen when the suggestion by the CDSS is automatic, and the clinicians are prompted with multiple choices along with the reasoning and certainty factors. It is found that clinicians directed solution generations are limited and are not exploited by the clinicians. It has also been found that CDSS is more effective when they are applied to individual cases at the time of decision making.

Important parameters that improve the acceptance of CDSS are user-friendly interface, workflow integration, upgradability, compatibility with the legacy systems and clinical effectiveness that shows improvement in patient care.

8.6 CLINICAL PROCESS MINING

Clinical processing-mining is the analysis and optimization of the patterns of activities and event logs required for diagnosing and treating a patient for a specific disease. Process-mining is also used for improving the efficiency of the various processes involving lab activities, doctors, patients and nurses to reduce patients' waiting time, reducing the response time, improving the resource productivity and improving transparency of the processes. The improvement is based upon actual evidence and existing treatment activities. Most hospitals use clinical pathways for various interventions based on clinical trials and patients' responses after monitoring many patients for the similar conditions. Clinical pathways are based upon the consensus among the physicians based on the analysis of the clinical data.

Evidence-based medicine is quite useful in maintaining chronic diseases such as diabetes or improving hazard conditions in terminal and life-threatening diseases such as post-cancer surgery, major organ-transplant where a new drug or regimen is introduced, and their administration are continuously adjusted based upon the analysis of the administration and the observed improvements. Recently, the availability of large-scale EHRs has become a natural resource of data for automated clinical process-mining using multiple AI techniques.

Given a database of activities recorded in the hospital databases, event-logs are created from the administration of the medication, treatment and the observation of patients' conditions. The sequences of event-logs are analyzed to improve the efficiency, and remove bottlenecks as illustrated in Figure 8.1. The events could be a patient-doctor encounter, a new diagnosis, significant change in physiological condition or value of a biomarker, a medical procedure, change in medication. The technique finds a common sequence of data that takes a patient from one observable outcome to another. The events are ordered based upon the temporal order.

There are multiple techniques for process mining, including clustering, probability-based techniques such as LDA (Latent Dirichlet Allocation) and HMM (Hidden Markov Models).

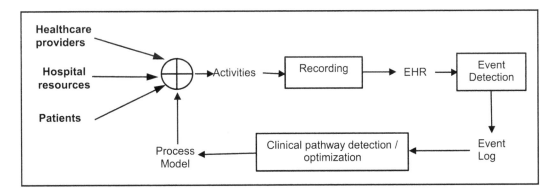

FIGURE 8.1   An illustration of process detection and enhancement model

8.6.1 LDA-Based Derivation

Event-logs are derived using multiple features such as length of stay, bed-utilization, medical service level, etc. The event-logs are quite complex, and event-log analysis to derive the process model derives a probabilistic model. The probabilistic model is derived using Latent Dirichlet Allocation (LDA). As described in Section 3.10, LDA is an unsupervised probabilistic clustering technique to derive the common topics from the text-documents by identifying cooccurrences of the words in proximity in a text-document. Clinical pathways are the patterns of the clusters derived using LDA.

One of the major differences in using LDA for text documents and LDA to derive clinical pathway patterns is the presence of time-stamps in event-logs in clinical records that plays a major role in deriving events and pathways. An event e_i is defined as a pair (*action, time-interval of the action*). A *trace of activities* is a sequence of activities related to the same treatment and patient. A *clinical event-log* is a set of trace of activities for all the patients. The LDA scheme generates first the treatment activities, and then time-stamps for these treatment activities to derive the sequential pattern of the treatment activities.

For a particular patient-trace of activities τ, LDA picks up the CP (Clinical Pathway) pattern θ, using a multinomial probability distribution $prob(\theta|\tau)$. The importance of each clinical activity α is modeled using a multinomial probability distribution $prob(\alpha|\theta)$. The overall probability of an activity α in a patient-trace is given by Equation 8.21, where M is the number of clinical pathway patterns in a patient's activities. Each patient-trace is a mixture of multiple CP patterns. By heuristically assigning an activity to a pattern, the overall probability maximizing combination is derived.

$$prob(\alpha \mid \tau) = \sum_{\theta=1}^{\theta=M} prob(\alpha \mid \theta) \times prob(\theta \mid \tau) \tag{8.21}$$

8.6.2 HMM-Based Modeling

HMM-based modeling is based upon identifying the states and the transition between the states using multiple patients' records of treatments of similar disease conditions. A state is derived using clustering of the activities where each state is characterized by multiple feature-values such as lab-tests, doctor's diagnosis and radiology tests. Multiple activities are grouped in the same state using clustering if the features-values are quite similar. A state may have transition to itself if the condition of the patient remains similar to the previous visit or action. A treatment-trace is a sequence of states and treated as a pathway of the derived HMM. The observations are emitted signals of the HMM.

To identify the similarity between the traces of two patients, multiple distance measures are used. One such heuristics is the notion of *longest common subsequence (LCS)* that is defined as the longest subsequence $S(x, y)$ in the traces that is common in two patient-traces x and y (see Equation 8.22). The distance between traces x and y is given by Equation 8.23. Another scheme can be to use dynamic programming as described in Section 2.3.2. This notion of distance between traces can be used to identify similar traces statistically and can be used to derive census clinical pathways.

$$LCS(x, y) = max\{|u| : u \in S(x, y)\} \qquad (8.22)$$

$$distance(x, y) = |x| + |y| - 2 \times LCS(x, y) \qquad (8.23)$$

8.7 DISEASE MANAGEMENT AND IDENTIFICATION

For chronic diseases, the data collected from patient is quite large, and temporal analysis of data is needed to monitor the progression of chronic disease. Cluster analysis has also been used to identify *spatial clusters* – clusters that identify regions that are frequently affected by a prevailing disease. Spatial clusters have been used to identify specific diseases such as malignant cancers among children. Spatial cluster analysis is based upon analyzing the clinical data of the body-areas where the concentration of the occurrence of disease is much higher than the statistical mean.

Another major problem in the management of the chronic diseases such as diabetes and coronary diseases is that patient forgets to follow the regimen correctly. S(he) may miss taking the regular glucose readings, blood-pressure readings, forget to show up for lab tests or expensive MRI and forget to take the medications regularly resulting into missing or erroneous data. The data is discrete and suitable for *Dynamic Bayesian Network* (DBN) analysis. As described in Section 4.7.4, *dynamic Bayesian network* is an iterative version of a Bayesian network that is repeated in the next unit of time. The clinical data collected at time t is probabilistically related to the clinical data produced at time $t + 1$. Hidden Markov model (HMM) is a special case of DBN. First, a model for the disease is made for the specific chronic disease, and the parameters are populated using time-series data. After populating the parameters with data, the model is used to predict the missing data using forward-backward reasoning. *Forward-backward reasoning* is a technique in DBN to probabistically reason about missing data in the past or predict the data in future using existing evidence. DBN can also be used to predict the progression of disease.

The acquired data may be nonlinear and/or fuzzy. Multiple approaches have been taken to analyze and label the data. The popular approaches are variations of K-means clustering, associative-rule mining, Levenberg–Marquardt algorithm based upon minimization of the least-square error during curve fitting, fuzzy expert systems, naïve Bayesian classifier, neural networks .and decision trees.

The hospitals acquire a huge amount of data due to lab tests and accumulation of vital signs of patients being monitored. The distributed raw data in EHR is collected and automatically analyzed using data-mining techniques to derive association between multiple attributes associated with the disease. This has major advantages in the development of evidence-based medicine and improves healthcare management by identifying correct correlation between multiple attributes.

There are variations of two types of clustering: *K-means clustering* and *K-nearest neighbors*. Both techniques have been used to identify diseases from large data samples. The techniques have been applied to identify different types of diseases such as dementia, Parkinson's disease, coronary diseases and pulmonary disease. Cluster analysis has also been used to derive the gene-clusters related to genetic diseases. The parameters for clustering are carefully analyzed based upon the correlation with the outcome.

Example 8.4

Heart diseases are correlated with various parameters such as BMI index, waist size, high-density lipoproteins level, cholesterol level, triglyceride level, glucose level, insulin level and blood-pressure level. By periodically measuring these parameter values, the progression or the probability of worsening the condition of existing heart-condition is predicted. The data is further divided based upon age and gender. The BMI and waist size are age adjusted using linear regression analysis.

8.7.1 Genetic Disease Identification

Genetic diseases are caused due to gene insertion/deletion and gene mutation caused by nucleotide substitution, insertion and deletion. Gene's mutation leads to variations in translated proteins and their functionality: nucleotide substitution, deletion and insertion cause a different chain of amino-acids during the translation process changing the effectiveness of the corresponding proteins and/or enzymes in the biochemical reactions in metabolic and signaling pathways.

There are two types of genetic diseases: 1) those caused by mutation in a single gene and 2) those caused by mutations of multiple genes possibly in the same signaling or metabolic pathway because many times functionally related genes do not work in isolation. The diseases are identified by macro-level symptoms called *phenotypes*, while genetic mutations are called *genotypes*.

Linking genotype aberrations to phenotypes associated with specific diseases is quite a challenge. It has been attempted in two ways: 1) hypothesis formation based upon wet lab analysis and testing the gene mutations and change in functionality based upon the hypothesis and 2) genome-wide mapping of genetic mutations to the corresponding genetic disease using statistical study of patients' data and probabilistic reasoning technique such as Bayesian network and conditional Bayesian formula. Major genetic techniques to derive genetic diseases are: 1) GWAS (Genome Wide Association Studies); 2) SNP (Single Nucleotide Polymorphism) studies; 3) microarray analysis of gene-expression; 4) SAGE (Serial Analysis of Gene Expressions) and 5) CAGE (Cap Analysis of Gene Expressions). These techniques are described in Chapter 10.

The *genetic techniques* associate mutations in gene (or gene clusters), or insertion/deletion of associated genes with specific diseases. The popular computational techniques used for the analysis are: 1) logical regression analysis; 2) Bayesian association mappings using Bayesian network; 3) multifactor dimensionality reduction by identifying independent variables that do not contribute to disease factors; 4) Bayesian model selection and 5) polymorphism interaction analysis.

Logical regression analysis uses Boolean combinations (Boolean-AND and Boolean-OR) of independent variables as predictors. Logical regression analysis has been used in predicting different types of cancer and heart diseases. There are multiple variations of logic regression analysis that have been used in identification of diseases. The popular ones use *simulated annealing, genetic programming, Monte-Carlo Logic regression* and *Bayesian logic regression*. Monte-Carlo method is a simulation technique that correlates the actual randomly sampled data with a deterministic outcome.

Bayesian association mapping is based upon high correlation of the SNPs in vicinity of each other. This vicinity is called a *block*. The correlation weakens beyond the block-boundaries. Bayesian association mapping identifies blocks where SNPs are locally highly correlated using Monte-Carlo method. After identifying the blocks, the logical combination of multiple blocks is associated with a specific disease. The technique is based upon partitioning the set of SNPs in a gene such that there is a strong correlation between the SNPs in the same block. Using Bayesian formula, the probability of blocks mapping to disease is calculated. The technique has been used to calculate many genetic diseases such as autism, macular degeneration and Alzheimer's disease.

8.7.2 Identifying Biomarkers

Biomarkers detection has focused on the identification of specific molecules, proteins or some chemical end-products that deviate from the healthy human range. There are four conditions for a biomarker to be called an effective biomarker: 1) the change in quantity of the biomarker should be verifiably highly correlated with the biological process being measured; 2) the mapping should be one-to-one meaning other biological processes should not alter the quantity of the biomarker; 3) the correlation should be such that a trend can be measured to identify the degree of the biological process or the disease-state and 4) it should be possible to measure the biomarkers accurately. It has been established that biomarkers start showing up in the body long before the medical symptoms appear. Thus, biomarkers can identify diseases long before disease becomes unmanageable. Some of the examples of these diseases are Alzheimer's disease,

Parkinson's disease, different types of cancer, HIV, heart diseases, kidney-related diseases, liver-related diseases and so on.

Biomarkers may not be definitive all the time. Rather, they are predictive. Once a disease state is suspected using biomarkers, then more definitive tests including invasive tests such as biopsy are used to ascertain the disease-state. Biomarkers have also been used to measure the long-term side-effects of many drugs for internal organs such as heart, kidney and liver that have long-term side-effects on a certain percentage of patients. Based upon the biomarkers, the drug may be substituted by another drug, or may be discontinued. For example, many drugs to treat heart-conditions caused by excessive cholesterol level or high blood-pressure level may also adversely affect the kidney and the liver. Thus, a biomarker that measures the kidney abnormalities can facilitate the choice of the right drug for cardiovascular disease.

A major requirement in identifying biomarkers for diseases is deriving the specificity to a disease. The gene-expression (described in Chapter 10) of various types of proteins involved in the growth of tissue, enzymes, transportation of the protein, trans-membrane receptors interacting with foreign bodies and transcription regulators are targets for biomarkers. The known pathways associated with the corresponding diseases are studied to identify any deviation from the healthy reference. The techniques used to identify biomarkers use microarray-based gene-expression analysis and clustering to identify the gene-clusters that are coexpressed and show a proportionate ratio of an increase in the intensity of the disease-state. The cluster analysis is done in N-dimensional space using the attributes such as physical and chemical properties, amino-acid sequence and structural features.

Different diseases correspond to different pathways. For example, in cancer, 13 pathways are upregulated (gene-expression enhances), and 22 pathways are downregulated (gene-expression slows down). The upregulated pathways include cell-communication, cell-cycle, cell-adhesion and downregulated pathways include fatty-acid metabolism and insulin signaling pathway. The expression of genes varies with age and gender, and care should be taken during clustering to derive proper groups. Different groups of genes show the different amount of gene-expression as the disease progresses. By careful analysis of the extent of gene-expression increase or decrease the staging of the disease can be approximated.

8.8 APPLICATION OF AI TECHNIQUES

Clinical data mining techniques have been used in disease detection such as cancer detection, diabetes detection, dementia detection, heart disease analysis, fatty liver analysis, biomarker discovery, prescription extraction, drug-efficacy analysis, trauma severity analysis, prognosis modeling and so on. Multiple classification techniques such as ANN, clustering, regression analysis, support vector machines, Bayesian network analysis, decision tree analysis, hidden Markov model, temporal abstraction and time-series indexing have been applied. This section describes three specific applications and associated techniques: 1) cancer detection using serum proteomic profiling; 2) multiple organ-failures and 3) fatty liver disease classification.

8.8.1 Case Study I: Cancer Detection

The technique is based upon proteome analysis of the serum. Proteome is the entire set of proteins found in the human genome. The technique uses a combination of wet-lab techniques to identify amino-acid sequences, bioinformatics and data-mining techniques to identify probabilistically disease-associated proteins. Laser absorption-based mass-spectrometer based analysis identifies various proteins in the serum. Mass spectrometry can be applied a patient's tissue cells, blood, serum or other body-fluids for a similar analysis. To uncover the differences in mass-spectral pattern of proteins, data-mining techniques are used. The goal is to extract a protein pattern that corresponds to the presence of the disease based upon sensitivity and specificity analysis. The technique is useful for the detection of different types of cancer such as ovarian cancer, prostate cancer, breast cancer, liver cancer and colon cancer. The AI techniques used are decision trees, neural networks, clustering and statistical methods.

After extracting the data-points using mass-spectrometers, feature-extraction is used. Feature selection is classified into two categories: 1) filter approach and 2) wrapper approach. *Filter approach* uses Euclidean distance as similarity measure, information measures, dependency measure and consistency measure and is more efficient but less accurate. The *wrapper approach* uses classifier error rate and genetic algorithm to find the individuals with the highest feature-values. The data-points are grouped into intervals, and means and standard deviations are calculated. Each group of intensity-values acts as a potential feature. However, only few most significant features are selected.

Given the two groups: cancer and noncancer, the filter approach uses Equation 8.24 to compute the distance between the two groups. The equation uses the ratio of the absolute difference of the mean of the spectrometer intensities for cancer and noncancer groups divided by the cumulative deviation $\sqrt{\sigma_1^2 + \sigma_2^2}$. Based upon this distance comparison for each potential feature, around ten most significant features are selected for further classification. After identifying the distance between the points, SVM (support vector machine) or other classification technique is used to separate the classes.

$$distance\ (cancer\ group,\ noncancer\ group) = \frac{|\mu_1 - \mu_2|}{\sqrt{\sigma_1^2 + \sigma_2^2}} \qquad (8.24)$$

8.8.2 Case Study II: Dynamic Organ-Failure in ICU

Patients in intensive-care unit are continuously monitored for multiple organ-failures in response to deteriorating physiological conditions. The deteriorating multiple organ-conditions are measured using many scoring systems such as <u>L</u>ogistic <u>O</u>rgan <u>D</u>ysfunction <u>S</u>ystem (LODS) and the <u>S</u>equential <u>O</u>rgan <u>F</u>ailure <u>A</u>ssessment (SOFA).

A SOFA score depends upon the combination of six different scores: respiratory (lung), hepatic (liver), cardiovascular (heart), coagulation (blood), renal (kidney) and neurological (nervous system and brain). *Respiratory-score* is measured using oxygen-content in the blood; *Neurological-score* is measured using *Glasgow coma-scale*; *Cardiovascular-score* is measured using arterial pressure; *Hepatic-score* is measured using the concentration of bilirubin in liver; *Coagulation-score* is measured using the density of platelets in the serum. These individual scores vary from *0* (no organ-failure) to *+4* (complete organ-failure). The overall SOFA score varies from *0* (no organ-failure) to *3* (multiorgan-failure).

These scores are continuously monitored to assess the seriousness of the patients' conditions in ICU until the patient is discharged from the ICU. The temporal aspects of the measurements are summarized using horizontal temporal abstraction. As described in Sections 8.5.1 and 2.7, *horizontal temporal abstraction* joins the inference from two different events at the different time-intervals at the same level to create a new abstraction. The time-unit is dependent upon the severity of a patient's condition, may vary from hours to a single day. Markov models are used to model the states and transitions. Using the Markov model analysis, following questions are answered: 1) the effect of an organ-failure on the stay-duration in ICU; 2) the increase in probability of death after an organ-failure assuming organ-failures are independent of each other and 3) effect of an organ-failure on the increase in probability of another organ-failure. There are $2^6 = 64$ possible Markov states under the assumption that each organ can have two possible states: nonfailed or failed.

The model is simplified using regression analysis and hierarchical dynamic Bayesian network. Regression analysis is used to estimate the progression of severity of organ condition. Hierarchical dynamic Bayesian network is used because a variable representing an organ's condition can be further split into multiple variable-value pairs that are combined to give overall severity.

8.8.3 Case Study III: Detecting Fatty Liver Disease

Fatty liver disease is caused by the accumulation of triglyceride fats in the liver cells due to excessive alcohol consumption, obesity due to insulin resistance and metabolic syndrome. It leads to liver inflammation (steatohepatitis), cirrhosis – a chronic liver disease where healthy cells are replaced by scar tissues and liver cancer. The disease can be reversed if found earlier. There are two ways to detect it: biopsy and radiology images such as ultrasound, MRI and CT scan. Ultrasound images are most inexpensive and have high specificity and sensitivity. Biopsy is invasive, uncomfortable, requires extra healing-time and is expensive. In either case, the images need to be automatically analyzed for the detection of fat-accumulation. An example of fatty liver disease is shown in Figure 8.2. The left-hand side columns show a normal ultrasound of healthy liver, and the right-hand side column shows the ultrasound images of fat accumulation.

The accumulation of triglycerides is identified using higher-order spectra-based features, texture-based features and analysis of the features using classification techniques such as decision trees, fuzzy classifiers and statistical analysis. Higher-order spectra-features are derived using Radon-transforms and Fourier-transforms. Texture-based features are derived using run-length matrices or other techniques as described in Section 2.5.1. The lack of homogeneity indicates the presence of fatty tissues.

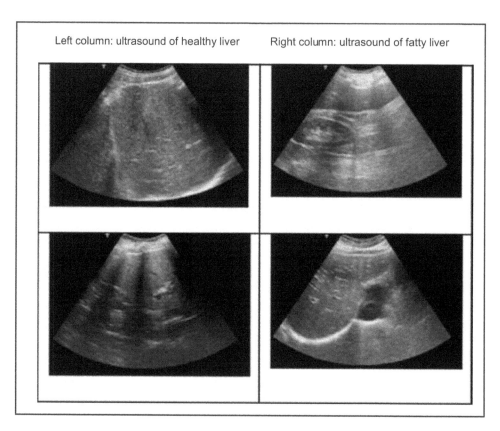

FIGURE 8.2 Ultrasound images of normal liver (left column) and fatty liver (right column) (Image courtesy John Wiley and Sons publishers, Acharya, U. Rajendra, Ribeiro, Ricardo, Krishnamurthi, Ganapathy, Marinho, Rui T., Sanches, João, and Suri, Jasjit S., "Data Mining Framework for Fatty Liver Disease Classification in Ultrasound: A Hybrid Feature Extraction Paradigm, *Medical Physics*, 39(7-1), July 2012, 4255–4264, DOI: 10.1118/1.4725759, Figure 2, published with permission).

8.9 SUMMARY

Intelligent clinical data analytics is concerned with the discovery of new diseases, disease diagnosis, identification of new biomarkers, dosage for better recovery and reduced toxicity, development of new vaccines, better patient care, study of patients' survivability and risk factor using artificial intelligence techniques and statistical analysis. The major artificial intelligence techniques used are: variations of clustering techniques, artificial neural network techniques, variations of regression analysis, support vector machine, associative data mining, Bayesian network and Bayesian conditional formula for probabilistic reasoning.

In these studies, the corresponding feature-sets characterizing the disease are derived, and AI techniques are applied on these techniques to derive the outcome. The correlation between the seemingly independent variables is studied to reduce the dimensionality in regression analysis and clustering. Popular clustering techniques are K-means, K-nearest neighbor and mixed Gaussian. Clustering has been used to derived efficacy and dosage of the medicine, spread of the disease, patients' prognosis and survivability.

Biostatistics is used to derive and validate the hypothesis when the amount of data is quite large. Statistical study involves the derivation of mean, median, variance, standard deviation, correlation between variables, specificity, sensitivity, false-negatives, false-positive and overall accuracy. The data can be collected either through a controlled experiment or through observation.

Randomized clinical trials are used to study the efficacy of a medicine. It requires two types of groups: control-group and intervention-group. Control-group is not given any dose of experimental medicine; intervention-group is given regular dose of medication. An important aspect of randomized trial is to keep the sample unbiased. However, it is not always possible due to many reasons such as people belonging to same ethnicity or same age group or same gender going to the same clinic. Complete randomized studies are also not possible due to the contaminations caused by change in technology or people leaving due to death or side-effect. Ideally, control-group and intervention-group should be completely separated. However, it is not always possible and medications can get mixed. Sample-size is carefully selected, and the hypotheses are validated using Z-score. Gaussian distribution has two tails. However, we study the confidence-level using just the half of the curve either left or right of the mean-value.

Risk analysis is concerned with the probability of an untoward event specially during chronic diseases or terminal diseases. Risk analysis is also concerned with the ratio of person getting the disease and persons not getting the disease. Meta-analysis is concerned with combining the results of the similar studies and studying the statistical metrics of studies such as mean, median and variance.

Survivability analysis is concerned about the predicting the probability of a person surviving for a number of years after a life-threatening disease. It is closely related to the hazard-function that is defined as the probability of life-threatening disease occurring again. Reasoning about survivability requires statistical study of patients under similar conditions. However, such statistical studies are adversely affected due to many persons leaving the study due to the availability of a new treatment, leaving the treatment due to the knowledge of new side-effects and/or fatalities. The study has two popular estimates: 1) Kaplan–Meier estimate and 2) log-rank test. The Kaplan–Meier estimate is based upon the principle that survivability in the future is related to survivability in the past and is suitable for single variable dependence. Log-rank test relates survivability to the hazard-function.

Survivability in two or more groups is compared using nonparametric chi-square test. The combined effect of multiple risk-factors on survivability is achieved using multivariate regression analysis. Both hazard-function and survivability are modeled as exponential functions where the exponential-factor is decided by the multivariate regression analysis. There are other models to derive the exponential function such as *Weibull probability-density function* and *Gompertz distribution*.

Clinical decision support systems are semi-autonomous intelligent system to help a provider make the decision in data-intensive evidence-based environment. The major components of a clinical decision support system are: input data, triggers, intervention and offered choices. CDS systems should have user-friendly interface, intelligent summarization capability and sufficient NLP support to interact with patients and medical providers.

Clinical knowledge-based systems are intelligent rule-based and data-mining tools to analyze large databases consisting of EHR, pathological-lab data, radiology images and pharmacology data. They are used to study abnormalities, educate patients and medical practitioners, perform clinical reminders and create temporal-abstraction from data. Temporal-abstraction requires context formation, domain restriction, vertical and horizontal inference and temporal-alignment of data.

CDSS is being used increasingly in the healthcare organizations as it has been found that medical practitioners make better decisions if they are provided with the choices that they may have otherwise overlooked. Artificial neural network systems have been used to analyze the patterns from the known systems based upon the training of the ANN. Pattern analysis has been used to diagnose patients' abnormalities based upon symptoms, observations, images and lab data.

Clinical data mining is about the knowledge and pattern discovery using machine learning techniques. Multiple machine learning techniques are used in deriving knowledge from data such as clustering, regression analysis, decision trees, summarization and dependency modeling. It has been used in identifying a correlation between various attributes and identifying chronic and terminal diseases, efficacy information for drugs and staging of terminal diseases. Before mining, clinical data should be cleaned, and temporal data should be abstracted to match the units. The derived rules after data-mining are of the form of probabilistic if-then-else rules. Noise such as missing values is corrected by ignoring the attribute.

Genetic diseases are of two types: 1) those that are caused by mutations in a single gene and 2) those that are caused by mutations in multiple genes in a pathway. Genetic diseases are identified using two major techniques: 1) genome-wide association of SNPS and 2) microarray-data analysis of pathways based upon some hypothesis formed by experimental wet-lab data. Genome-wide association studies find out the probability of association of highly correlated SNP blocks to the phenotypes. Microarray-data analysis uses clustering to identify those genes that are coexpressed. These genes may be upregulated or downregulated based upon the pathway. Multiple computational techniques such as logical regression analysis and Bayes association mapping are used to derive gene–gene interactions. Biomarkers are identified using cluster analysis on gene-expression data.

8.10 ASSESSMENT

8.10.1 Concepts and Definitions

AFT model; Allen's synchronization primitives; Bayesian association mapping; Bayesian classification; block; Bayesian model selection; biomarker; binomial distribution; biostatistics; CAGE; Chi-square model; clinical decision support system (CDSS); clinical knowledge-based system; contamination; context formation; control-group; control sample; correlated variables; covariates; Cox-model hazard function; decision support system; dependency modeling; domain restriction; double-blinded study; experimental study; false-negative; false-positive; fixed-model; Gaussian curve, gene-expression; gene–gene interaction; genome-wide association studies; genotype; Gompertz distribution model; hazard; hazard-ratio; horizontal inference; hypothesis verification; informative censoring; inter-cluster variation; intervention-group; intra-cluster variation; K-means clustering; Kaplan–Meier survival estimate; Levenberg–Marquardt algorithm; linear multivariate analysis; logical regression analysis; log-rank test; margin-of-error; meta-analysis; mixed Gaussian model; model based clustering; multivariate analysis; negative predictive value; non-parametric test; observational study; odds; overall accuracy; phenotype; population; population mean; population-size; positive predictive value; process-mining; random model; randomized clinical trial; relative odds; relative risk; risk analysis; risk difference; SAGE; sample mean; sample-size; sensitivity; single-blinded study; SNP; spatial cluster; specificity; summarization; survivability;

temporal abstraction; temporal interpretation; temporal pattern matching; time-abstraction; triggers; true-negative; true-positive; unblinded study; vertical inference; Weibull model; Z-score.

8.10.2 Problem Solving

8.1 For the following data, calculate mean, median, mode, standard deviation and covariance. Use curve-fitting equation to derive the straight-line equation for regression analysis. Assume that the first coordinate(x-coordinate) is an independent variable, and the second coordinate (y-coordinate) is a dependent variable. Use the equation for small samples to compute the variance. Show the calculations. The data gives two-dimensional coordinates as follows: (2, 3), (3, 4), (11, 12), (17, 21), (14, 16), (20, 24).

8.2 A cluster is modeled as a tuple *(centroid's x-coordinate, centroid's y-coordinate, radius, cluster-population)*. *K-nearest neighbor* places a point in the nearest cluster using a distance criterion. The centroid's x-coordinate gets altered as centroid's *((x-coordinate × cluster-population + new-point's x-coordinate)/cluster-population + 1)*. A similar calculation is done for centroid's y-coordinate using y-coordinate of the new point. The radius gets altered, within a reasonable limit, to the *maximum(old-radius, distance of the new-point from the new centroid)*. Given the set of clusters as {(3.4, 4.5, 4, 21), (11.2, 14.0, 3, 34)} and the new points' coordinates as (3, 2), (2, 5), (5, 4), (11, 12), (12, 13), calculate new centroids and the corresponding radii, after inserting the points in the cluster one-by-one. Show the calculations.

8.3 Given that a biomarker B for a disease is correlated to two risk-factors X_1 and X_2. The regression equation for the correlation is given as $B = 0.1\,X_1 - 0.2\,X_2$. Explain how the two risk-factors may change without affecting the biomarker.

8.4 A clinical trial of a home-based pregnancy test was performed on a sample of *1,000* women. In the population, *100* women are truly pregnant, and the remaining *900* women are not pregnant. The test finds that *80* women out of *100* women are pregnant. It also finds that *20* women of the remaining *900* women are pregnant. What can you infer about sensitivity and specificity? Explain.

8.5 Given a sample-size of *1,000*, there are *453* true-positives, *151* false-positives, *340* are true-negatives and *56* are false-negatives. Calculate the sensitivity, specificity, positive predictive value, negative predictive value and overall accuracy from the given data. Show your calculations.

8.6 In a randomized clinical trial, sample-size of *1,000* patients was divided into two groups: *500* in the control-group and *500* in the intervention-group. However, medications got mixed up, and *50* people in the control-group were given the experimental medicine. Another *60* persons from the intervention-group did not get the experimental medicine. Calculate the contamination-factor. Then, how much the sample-size should be increased to counter the effect of the contamination.

8.7 Given a Gaussian distribution for the occurrence of a sexually transmitted disease vs. age of the population, the mean age of maximum occurrence is *34.3*, the standard deviation is *10* years. Assume that the sample-size is *12,000*. Calculate the Z-score and margin of error for *95%* confidence-level. Show the calculation.

8.8 In a randomized trial of a medicine in multiple hospitals, using a total sample-size of *943* volunteers of age *60* and above, fourteen patients withdrew from the study due to serious side-effects. Each patient was given *10* mg orally until s(he) recovered. The collected data from different hospitals is given as a triple *(number of patients in the cluster, average recovery period in days, variance in the recovery period)*. The data is as follows: *(30, 20, 7.2)*, *(40, 23, 8.4)*, *(36, 25, 2.1)*, *(66, 21, 5.3)* *(43, 19, 4.7)*, *(100, 24, 3.2)*, *(72, 18, 9.3)*, *(26, 20, 1.8)*, *(56, 23, 4.5)*, *(23, 22, 4.9)*, *(18, 25, 2.8)*, *(34, 24, 3.9)*, *(29, 19, 4.5)*, *(88, 22, 2.3)*, *(73, 20, 3.1)*, *(53, 21, 4.0)*, *(11, 18, 2.4)*, *(61, 20, 1.5)*, *(31, 23, 3.9)*, *(39, 20, 2.6)*. Calculate the variance σ_b^2 between the clusters. The variance σ_b can be calculated by treating each mean-value as a data point, and finding variance of these *20* points. Calculate the correlation-coefficient between the clusters and the corrected sample-size after calculating average sample-size of the cluster.

8.9 For a controlled random trial study to test a cancer drug for preventing the progression of stage II cancer, the number of patients in the control-group are *112* (stage II cancer); the number of patients in the intervention-group are *114* (stage II cancer). The patients in the control-group who progress to stage III in two years are *34*, and the patients in the intervention-group that progress to the stage III are *15*. Calculate the associated probabilities of progressing to stage III in both the groups, the difference in the probabilities, relative-risk and relative-odds.

8.10 In a medical study to reduce lung cancer, a pharmaceutical company spawns a volunteers-based study in ten hospitals across the nation. The sample-size for each hospital over a three-year study was *100*: *50* patients were in the control-group, and another *50* were given the medication. The data for maintaining the cancer level to the same stage was collected as the pair (*patients in the control-group not progressing to the next stage of cancer, patients in the intervention-group not progressing to the next stage of cancer*). Using meta-analysis (see Section 8.5), calculate the individual and overall associated probabilities, risk-difference, relative-risks and relative-odds. Assume that the data is correlated. The data is as follows: (*3, 10*), (*4, 12*), (*5, 14*), (*4, 13*), (*2, 11*), (*6, 13*), (*2, 17*), (*5, 16*), (*3, 18*), (*1, 11*).

8.11 The survivability of a patient *P* in the beginning is *50%* for the duration of one year. He is given a medication that improves the survivability by starting the remission. The statistical analysis shows that *30%* patients suffering from the disease die after one year even after taking the medication. What is the survivability of patient *P* after one year? Plot the graph for the next five years, find the slope of the curve and calculate the cumulative hazard-function $H(t)$ and the hazard-function $h(t)$.

8.12 The survivability of a patient suffering from cancer is estimated by the function $S(t) = 5 \times e^{-t}$. The patient starts taking a cancer drug and starts exercising daily. The regression curve for improving the survival is given by $y = 1.2\ m + 0.4\ e$, where *m* is the medication regularity and *e* is the exercise regularity. Give the new survivability function and hazard function $h(t)$ and draw a table that shows a survival-rate with *20%* medication regularity, *40%* medication regularity, *80%* regularity of medication and *30%*, *60%* and *90%* regularity of exercise.

8.13 Given the probability distribution is given by Weibull function with parameters $\lambda = 3$ and $p = 1.5$, plot the curve for the survivability function and the hazard function.

8.14 Write a program for calculating the survival-function and hazard-function based upon Kaplan–Meier's estimate using parameterized regression-analysis as given by Equations 8.9 and 8.10.

8.15 A patient is being monitored after surgery. His postsurgical data would be collected for further analysis. Give a generic model of the state of the patient that changes every two hours after he is administered medication.

8.16 Given three parameters with different units and ranges such that first parameter has a weight of *0.6*, second parameter has a weight of *0.3* and the third parameter has a weight of *0.1*, give overall steps to calculate how the parameters will be normalized and combined using the corresponding weights to map on one combined scalar value.

8.17 Express the probability of a disease in terms of a subset of conditional probabilities $prob(T|D)$, $prob(\neg T|D)$, $prob(T|\neg D)$ and $prob(\neg T|\neg D)$. The symbol '$\neg$' denotes logical negation. Repeat this exercise for finding out the probability of not having a disease. (Hint: Use Bayesian theorem to connect condition probability with absolute probability.) Try your equation on the following data: $prob(T|D) = 0.9$, $prob(\neg T|D) = 0.1$, $prob(T|\neg D) = 0.02$, $prob(\neg T|\neg D) = 0.98$. Explain your answer.

8.18 A biomarker shows *97%* effectiveness about predicting cervical cancer. It also produces *2%* false-positives. The local population has *1%* probability of having cervical cancer between the age of *20* and *25*. A young woman takes the test and tests positive. What is the probability that she has the cervical cancer? Show your calculations and explain your answers.

8.19 After making some changes in a drug, a pharmaceutical company makes a better non-drowsy cough syrup and runs a clinical trial about dosage and the increase in the amount of sleep. It uses the regression analysis results from the previous version as the prior. The prior slope and intercept for the linear curve fit were: *increase-in-sleep-hours* = $m \times dosage + b$, where *dosage* is in milliliters and *increase-in-sleep-hours* is given in hours. The intercept is given in hours. The slope m and intercept b have Gaussian distributions and change with different ethnicity, climate conditions and age. The Gaussian distribution for slope has mean *0.02* and standard deviation of *0.02*. The Gaussian distribution for the intercept has mean as *1.0* and the standard deviation of *1.0*. The new drug was tested in ten different communities. New regression analysis results yielded the following (*slope, intercept*) pair. What can you say about the claim of the company? Make reasonable assumptions. Explain your answer (hint: find out the correlation between the various slope-values and intercept-values. Also, check if the Gaussian distribution of the prior slope includes the mean value from other studies. You can also do meta-analysis with some sample-size assumption and compare the mean and variance results with the prior).

(*0.03, 1.1*), (*0.15, 2*), (*0.02, 0.5*), (*0.1, 2.0*), (*0.1, 1.0*), (*0.1, 1.4*), (*0.13, 1.2*), (*0.12, 1.8*), (*0.14, 2.1*), (*0.13, 1.7*).

8.20 Write a C++/Python/Java program that takes two sequences comprised of characters "A", "G", "C" and "T" of equal length and identifies the position where the characters are different. It saves the position and the different character-set in a list (or equivalent data structure to store a sequence) and returns the list in the end. Such a program can be used for SNP analysis.

8.10.3 Extended Response

8.21 Explain the randomized clinical trial and the issues faced in a randomized clinical trial.

8.22 Explain the problem of contamination in clinical random trials.

8.23 Explain Z-score and its relationship to confidence-interval, margin-of-error and sample-size. Explain all the related equations.

8.24 Explain Gaussian distribution and Poisson distribution and compare the two distributions in terms of their usage to different clinical scenarios.

8.25 Explain the correlation-coefficient between multiple clusters in the clinical random trials

8.26 Explain the notion of meta-analysis, its relationship to clinical random trials and its advantages.

8.27 Explain survivability analysis and various techniques to derive survivability specially when multiple independent variables affect survivability.

8.28 Explain hazard-function and various techniques to derive hazard-function. Explain how hazard-function is related to survivability.

8.29 Explain the components and issues in the development of clinical decision support systems.

8.30 Explain temporal-abstraction and its role in the development of clinical knowledge-based systems.

8.31 Read one review article on time-interval matching and synchronization and discuss interval-based matching and time-abstraction.

8.32 Read three articles on the application of ANN for tumor-detection and summarize the overall approach in one single spaced page.

8.33 Read three articles on clinical data-mining and discuss various issues in clinical data-mining.

8.34 Explain various metrics and their relationships to each other to measure the probabilities of a clinical event to occur.

8.35 Explain various advantages and issues in clinical data classification.

8.36 Explain the differences between K-means clustering and K-nearest neighbor clustering.

8.37 Explain the various relationships and their significance between false-positive, true-positive, false-negative and true-negative.

8.38 Explain the techniques to derive genetic diseases and biomarkers.

8.39 Read articles on the application of Bayesian association mapping and discuss how it can be applied for identifying genetic diseases using SNPs.

FURTHER READING

Fundamentals of Clinical Data Analytics

8.1 Portney, Leslie G., and Watkins, Mary P., *Foundations of Clinical Research*, 3rd edition, Pearson/Prentice Hall, Upper Saddle River, NJ, USA, 2008.

8.2 Reddy, Chandan K., and Aggarwal, Charu C., *Healthcare Data Analytics*, Chapman and Hall/CRC Press, Boca Raton, FL, USA, 2015.

Clinical Data Classification

8.3 Gage, Brian F., Waterman, Amy D., Shannon, William, Boechler, Michael, Rich, Michael W., and Radford, Martha J., "Validation of Clinical Classification Schemes for Predicting Stroke: Results from the National Registry of Atrial Fibrillation," *Journal of American Medical Association (JAMA)*, 285(22), June 2001, 2864–2870.

Clustering of Clinical Data

8.4 Katzmarzyk, Peter T., Srinivasan, Sathanur R., Chen, Wei, Malina, Robert M., Bouchard, Claude, and Berenson, Gerald S., "Body Mass Index, Waist Circumference and Clustering of Cardiovascular Disease Risk Factors in a Biracial Sample of Children and Adolescent," *Pediatrics*, 114(2), May 2013, e198–e205.

8.5 Van Rooden, Stephanie M., Heiser, Willem J., Kok, Joost N., Verbaan, Dagmar, and Van Hilten, Jacobus J., "The Identification of Parkinson's Disease Subtypes Using Cluster Analysis," *Movement Disorders*, 25(8), June 2010, 969–978.

Biostatistics in Clinical Research

8.6 Altman, Douglas G., *Practical Statistics for Medical Research*, Chapman and Hall/CRC Press, London, UK, 2003.

8.7 Bland, Martin, *An Introduction to Medical Statistics*, 4th edition, Oxford University Press, Cambridge, UK, 2015.

8.8 DeShea, Lise, and Toothaker, Larry E., *Introductory Statistics for the Health Sciences*, 1st edition, Chapman and Hall/CRC Press, Boca Raton, FL, USA, 2015.

8.9 Muth, James E.D., "Overview of Biostatistics Used in Clinical Research," *American Journal of Health-Systems in Pharmacy*, 66(1), January 2009, 70–80.

8.10 Pagano, Marcello, and Gauvreau, Kimberlee, *Principle of Biostatistics*, 2nd edition, Chapman and Hall/CRC Press, Boca Raton, FL, USA, 2018.

8.11 Winner, Larry, *Introduction to Biostatistics*, Department of Statistics, University of Florida, Gainesville, Florida, USA, 2004, Available at http://web.stat.ufl.edu/~winner/sta6934/st4170_int.pdf, Accessed June 8, 2018.

Randomized Clinical Trials

8.12 Chuang, Jen-Hsiang, Hripcsak, George, and Heitjan, Daniel E., "Design and Analysis of Controlled Trials in Naturally Clustered Environment," *Journal of American Medical Association (JAMA)*, 9(3), May/June 2002, 230–238.

8.13 Friedman, Lawrence M., Furberg, Curt D., DeMets, David L., Reboussin, David M., and Gragner, Christopher B., *Fundamentals of Clinical Trials*, 5th Edition, Springer International Publishing, Cham, Switzerland, 2015.

Cluster Contamination

8.14 Borm, George F., Melis, René J.F., Teerenstra, Steven, Peer, Petronella G., "Pseudo Cluster Randomization: A Treatment Allocation Method to Minimize Contamination and Selection Bias," *Statistics in Medicine*, 24(23), December 2005, 3535–3547, DOI: 10.1002/sim.2200.

8.15 Hahn, Seokyung, Puffer, Suezann, Torgerson, David J., and Watson, Judith, "Methodological Bias in Cluster Randomized Trials," *BMC Medical Research Methodology*, 5(10), March 2005, DOI: 10.1186/1471-2288-5-1.

Correlation

8.16 Hung, Man, Bounsanga, Jerry, Voss, Maren W., "Interpretations of Correlations in Clinical Research," *Postgrad Medical Journal*, 129(8), November 2017, 902–906, DOI: 10.1080/00325481.2017.1383820.

8.17 Miot, Hélio A., "Correlation Analysis in Clinical and Experimental Results," *Journal Vascular Brasileiro*, 17(4), October–December 2018, 275–279, DOI: 10.1590/1677-5449.17411.

Meta-analysis

8.18 DerSimonian, Rebecca, and Laird, Nan, "Meta-analysis in Clinical Trials," *Controlled Clinical Trials*, 7(3), September 1986, 177–188.

8.19 DerSimonian, Rebecca, and Laird, Nan, "Meta-analysis in Clinical Trials Revisited," *Contemporary Clinical Trials*, 45(Pt A), November 2015, 139–145.

8.20 Irwig, Les, Macaskill, Petra, Glasziou, Paul, and Fahey, Michael, "Meta-analytic Methods for Diagnostic Test Accuracy," *Journal of Clinical Epidemiology*, 48(1), 1995, 131–132.

8.21 Kelley, George A., and Kelley, Kristi S., "Statistical Models for Meta-analysis: A Brief Tutorial," *World Journal of Methodology*, 2(4), August 2012, 27–32, Available from: http://www.wjgnet.com/2222-0682/full/v2/i4/27.htm.

8.22 Lipsey, Mark W, and Wilson, David B., *Practical Meta-Analysis, Applied Social Research Method Series*, 49, Sage Publications, Thousand Oaks, London, UK, 2001.

8.23 Thompson, Simon J., and Higgins, Julian P.T., "How Should Meta-Regression Analyses be Undertaken and Interpreted?" *Statistics in Medicine*, 21(11), June 2002, 1559–1573.

Survivability and Risk Analysis

8.24 Bradburn, Michael J., Clark, Taane G., Love, Sharon B., and Altman, Douglas G., "Survival Analysis Part II: Multivariate Data Analysis – An Introduction to Concepts and Methods," *British Journal of Cancer*, 89(3), August 2003, 431–436.

8.25 Bradburn, Michael J., Clark, Taane G., Love, Sharon B., and Altman, Douglas G., "Survival Analysis Part III: Multivariate Data Analysis – Choosing a Model and Assessing its Adequacy and Fit," *British Journal of Cancer*, 89(4), August 2003, 605–611.

8.26 Clark, Taane G., Bradburn, Michael J., Love, Sharon B., and Altman, Douglas G., "Survival Analysis Part I: Basic Concepts and Analyses," *British Journal of Cancer*, 89(2), July 2003, 232–238.

8.27 Clark, Taane G., Bradburn, Michael J., Love, Sharon B., and Altman, Douglas G., "Survival Analysis Part IV: Further Concepts and Methods in Survival Analysis," *British Journal of Cancer*, 89(5), September 2003, 781–786.

8.28 Lee, Elisa T., *Statistical Methods for Survival Data Analysis*, 4th edition, John Wiley and Sons, London, UK, 2013.

8.29 Lee, Elisa T., and Go, Oscar T., "Survival Analysis in Public Health Research," *Annual Reviews in Public Health*, 18, 1997, 83–104.

8.30 Massonnet, Goele, Janssen, Paul, and Burykowski, Tomasz, "Fitting Conditional Survival Models to Meta-analytic Data by Using a Transformation Towards Mixed Effect Models," *Biometrics: Journal of the International Biometric Society*, 64(3), September 2008, 834–842.

8.31 Stevenson, Mark, *An Introduction to Survival Analysis*, December 2007, Available at http://www.biecek.pl/statystykaMedyczna/Stevenson_survival_analysis_195.721.pdf, Accessed on June 8, 2018.

Nonparametric Estimates

8.32 Läuter, Henning, and Liero, Hannelore, "Nonparametric Estimation and Testing in Survival Models," In *Probability, Statistics and Modeling in Public Health* (editors: Nikulin Mikhail, Commenges Daniel, Huber Catherine), Springer, Boston, MA, USA, 2006, 319–331.

8.33 Rodríguez, Germán, *Non-Parametric Estimation in Survival Models*, Available at https://data.princeton.edu/pop509/NonParametricSurvival.pdf, Accessed date July 11, 2019.

Parametric Estimates

8.34 Cai Tianxi, Wei Lee-Jen, and Liero, HanneloreWilcox M., "Semiparametric Regression Analysis for Clustered Failure Time Data," *Biometrika*, 87(4), December 2000, 867–878.

Risk Analysis

8.35 Suo, Qiuling, Xue, Hongfei, and Gao, Jing, "Risk Factor Analysis Based on Deep Learning Models," *ACM Conference on Bioinformatics, Computational Biology, and Health Informatics*, Seattle, WA, USA, October 2016, 394–403.

Clinical Decision Support System

8.36 Burner, Eta S. (editor), *Clinical Decision Support System – Theory and Practice*, 3rd edition, Health Informatics Series (editors: Hannah, Kathryn J., and Ball, Marion J.), Springer International Publishing, Switzerland, 2016.

8.37 Sittig, Dean F., Wright, Adam, Osheroff, Jerome A., Middleton, Blackford, Teich, Jonatahn M., Ash, John S., Campbell, Emily, and Bates, David W., "Grand Challenges in Clinical Decision Support," *Journal of Biomedical Informatics*, 41(2), April 2008, 387–392.

8.38 Valkenhoef, Gert V., Ternoven, Tommi, Zwinkles, Tijs, Brock, Bert D., and Hillege, Hans, "ADDIS: A Decision Support System for Evidence-Based Medicine," *Decision Support Systems*, 55, 2013, 459–475.

Neural Network-Based Diagnosis

8.39 Amato, Filippo, Lôpez, Alberto, Peña-Méndez, María E., Vaňhara, Petr, and Hampl, Aleš, "Artificial Neural Networks in Medical Diagnosis," *Journal of Applied Biomedicine*, 11(2), January 2013, 47–58.

8.40 Lisboa, Paulo J., and Taktak, Azzam F. G., "The Use of Artificial Neural Networks in Decision Support in Cancer: A Systematics Review," *Neural Networks*, 19(4), May 2006, 408–415.

Clinical Data Mining

8.41 Bellazi, Riccardo, and Zupan, Blaz, "Predictive Data Mining in Clinical Medicine: Current Issues and Guidelines," *International Journal of Medical Informatics*, 77(2), 2008, 81–97.

8.42 Iavindrasana, Jimison, Cohen, Gilles, Depeursinge, Adrien, Müller, Henning, Meyer, R., Geissbuhler, Antoine, "Clinical Data Mining: A Review," *IMIA Yearbook Medical Informatics*, 18(1), January 2009, 121–133.

8.43 Jensen, Peter B., Jensen, Lars J., and Brunak, Søren, "Mining Electronic Health Records: Toward Better Research Applications and Clinical Care," *Nature Reviews/Genetics*, 13, June 2012, 395–405.

8.44 Palaniappan, Sellappan, and Awang, Rafiah, "Intelligent Heart Disease Prediction System Using Data Mining Techniques," *IEEE/ACS International Conference on Computer Systems and Applications (AICCSA)*, Doha, Qatar, March/April 2008, 108–115.

Effect on Clinical Practices

8.45 Garg, Amit X., Adhikari, Neil K., McDoanld, Heather, Rosas-Arellano, M. Patricia, Devereaux, Philip J., Beyene, Joseph, and Haynes, R. Brian, "Effects of Computerized Clinical Support Systems on Practitioner Performance and Patient Outcomes: A Systematic Review," *Journal of American Medical Association (JAMA)*, 293(10), March 2005, 1223–1238.

8.46 Kawamoto, Kensaku, Houlihan, Caitlin A., Balas, Andrew E., and Lobach, David F., "Improving Clinical Practice Using Clinical Decision Support Systems: A Systematic Review of Trials to Identify Features Critical to Success," *BMJ*, March 2005, DOI: 10.1136/bmj.38398.500764.8F.

Clinical Process Mining

8.47 Huang, Zhengxing, Dong, Wi, Ji, Lei, Gan, Chenxi, Lu, Xudong, and Duan, Huilong, "Discovery of Clinical Pathway Patterns From Event Logs Using Probabilistic Topic Models," *Journal of Biomedical Informatics*, 47, February 2014, 39–57.

8.48 Rojas, Eric, Munoz-Gama, Jorge, Sepúlveda, Marcos, Capurro, Daniel, "Process Mining in Healthcare: A Literature Review," *Journal of Biomedical Informatics*, 61, June 2016, pp. 224–236, DOI: 10.1016/j.jbi.2016.04.007.

8.49 Yang, W, Hwang, S., "A Process-Mining Framework for the Detection of Healthcare Fraud and Abuse," *Expert System Applications*, 31(1), January 2006, 56–68.

8.50 Uzark, K., "Clinical Pathways for Monitoring and Advancing Congenital Heart Disease Care," *Program Pediatrics Cardiology*, 18, 2003, 131–139.

8.51 Zhang, Yiye, Padman, Rema, and Patel, Nirav, "Paving the COWpath: Learning and Visualizing Clinical Pathways from Electronic Health Record Data," *Journal of Biomedical Informatics*, 58, December 2015, 186–197, DOI: 10.1016/j.jbl.1015.09.009.

Disease Management and Identification

Genetic Disease Identification

8.52 *Genes and Diseases*, National Center for Biotechnology Information, Available at https://www.ncbi.nlm.nih.gov/books/NBK22183/pdf/Bookshelf_NBK22183.pdf, Accessed June 8, 2018.

8.53 Han, Bing, Chen, Xue-wen, Talebizadeh, Zohreh, and Xu, Hua, "Genetic Studies of Complex Diseases Characterizing SNP-disease Associations Using Bayesian Network," *BMC Systems Biology*, 2012, 6(suppl. 3), S14, Available at http://www.biomedcentral.com/1752-0509/6/S3/S14.

8.54 Hirschhorn, Joel N., and Daly, Mark J., "Genome-wide Association Studies for Common Diseases and Complex Traits," *Nature Reviews Genetics*, 6(2), February 2005, 95–108.

8.55 Sreekumar, Kodangattil R, Aravind, Laxminarayanan, and Koonin, Eugene V., "Computational Analysis of Human-Disease Associated Genes and their Protein Products," *Current Opinion in Genetics and Development*, 11(3), June 2001, 247–257.

8.56 Sun, Peng Gang, Gao, Lin, and Han, Shan, "Prediction of Human Disease-Related Gene Clusters using Cluster Analysis," *International Journal of Biological Sciences*, 7(1), January 2011, 61–73.

8.57 Szymczak, Silke, Biernacka, Joanna M., Codell, Heather J., González-Recio, Oscar, and König, Inke R., "Machine Learning in Genome Wide Association Studies," *Genetic Epidemiology*, 33 (suppl 1), November 2009, S51–S57.

8.58 Yu, Zhang, Zhang, Jing, and Liu, Jun S., "Block-Based Bayesian Epistasis Association Mapping with Application to WTCCC Type 1 Diabetes Data." *The Annals of Applied Statistics*, 5(3), 2011, 2052–2077.

Identifying Biomarkers

8.59 Cui, Juan, Chen, Yunbo, Chou, Wen-chi, Sun, Liankun, Chen, Li, Suo, Jian et al., "An Integrated Transcriptomic and Computational Analysis for Biomarker Identification in Gastric Cancer," *Nucleic Acids Research*, 39(4), March 2011, 1197–1207.

Applications of AI Techniques

Case Study I: Cancer Detection

8.60 Li, Lithua, Tang, Hong, Wu, Zuobao, Gong, Jianli, Gruidi, Michael, Zou, Jun, Tockman, Melvyn, Clark, Robert A., "Data Mining Techniques for Cancer Detection using Serum Proteomic Profiling," *Artificial Intelligence in Medicine*, 34(2), October 2004, 71–83, DOI: 10.1016/j.artmed.2004.03.006.

8.61 Van der Gaag, Linda C., Renooij, Silja, Witteman, Cilia L. M., Aleman, Berthe M. P., Tall, Babs G., "Probabilities for Probabilistic Network: A Case Study in Oesophageal Cancer," *Artificial Intelligence in Medicine*, 25(2), June 2002, 123–148.

Case Study II: Multiple Organ Failure in ICU

8.62 Peelen, Linda, Keizer, Nicolette F. D., Jonge, Evert D., Bosman, Robert-Jan, Abu-Hanna, Ameen, Peek, Niels, "Using Hierarchical Dynamic Bayesian Networks to Investigate Dynamics of Organ Failure in Patients in the Intensive Care Unit," *Journal of Biomedical Informatics*, 43(2), April 2010, 273–286.

8.63 Sandri, Micol, Berchialla, Paola, Baldi, Ileana, Gregori, Dario, Blasi, Roberto A. D., "Dynamic Bayesian Networks to Predict Sequences of Organ Failures in Patients Admitted to ICU," *Journal of Biomedical Informatics*, 48, April 2014, 106–113.

Case Study III: Detecting Fatty Liver Disease

8.64 Acharya, Rajendra U., Ribeiro, Ricardo, Krishnamurthi, Ganapathy, Marinho, Rui T., Sanches, João, and Suri, Jasjit S., "Data Mining Framework for Fatty Liver Disease Classification in Ultrasound: A Hybrid Feature Extraction Paradigm," *Medical Physics*, 39(7), July 2012, 4255–4264, DOI: 10.1118/1.4725759.

Pervasive Health and Remote Care

9

BACKGROUND CONCEPTS

Section 1.3 Medical Informatics; Section 1.6 Overview of Computational Health Informatics; Section 2.1 Data Modeling; Section 2.2 Digitization of Sensor Data; Section 2.7 Temporal Abstraction and Inference; Section 2.8 Types of Databases; Section 2.10 Human Physiology; Section 3.4 Machine Learning; Section 3.5 Classification Techniques; Section 3.7 Probabilistic Reasoning over Time; Section 4.1 Modeling Medical Data; Section 4.2 Automated Data Acquisition and Interfaces; Section 4.4 Health Information Flow; Section 4.5.1 Components of Clinical Information System; Section 4.6 Heterogeneity in Medical Databases; Section 4.9 Standards for Transmitting Medical Information; Section 4.10 HIPAA – Protecting Patients' Privacy Rights; Section 4.11 Integrating the Healthcare Enterprise; Section 6.1 Overview of Network and Application Layers; Section 6.9 Medical Image Security

Pervasive care is an integration of mobile computing, widespread Internet-connectivity and miniaturized medical devices employing intelligent techniques to provide remote monitoring of the elderly and chronic patients. Pervasive care is also needed to reach out to remote areas that are beyond the reach of the clinical infrastructure connectivity such as: 1) distant villages and 2) localities isolated due to natural disasters such as earthquakes floods, snowstorms or war.

Remote care is about taking care of patients in their own homes, nursing homes where the patients are residing or remote clinics that lack sufficient clinical infrastructure that patients can access without physically meeting medical experts. During remote care, portable intelligent medical monitoring devices are used to collect vital signs' data and transmit it electronically over a communication link to a centralized medical center where the medical experts analyze the data and prescribe remedies.

Remote care has become a social and medical necessity. Life-expectancy has improved more quickly than medical infrastructures, and the number and ratio of the elderly population needing care are continuously increasing. Hence, the use of resources such as hospitals, hospital-beds, nursing staff and healthcare providers needs to be optimized. In addition, the cost of providing health care in hospitals and clinics is constantly increasing.

Maintaining the health of elderly patients serves to: 1) cut down the cost of expensive hospital care; 2) avoid hazardous transportation and 3) allow psychologically more satisfying care from relatives and friends. In addition, the rural community will suffer more as the growth of care-providers and clinical infrastructure lags behind the growth in the elderly population.

Mobile health monitoring is required due to the increasingly mobile nature of modern society. With increasing life-span, more people are actively involved in work and excursions at an advanced age, and in need of help when symptoms of ill health occur. Mobile health monitoring and communication are also needed for emergency cases when paramedics are transporting patients with life-threatening conditions. The paramedics may not have the required expertise nor the sophisticated instruments for diagnostics and treatment, and may be unable to perform lab-tests needed to diagnose the patients' conditions. However, they can act based upon the physician's advice in an emergency room. Mobile health monitoring is also needed when seriously ill patients are transported between different units in the same hospital.

Telemedicine and teleradiology are emerging areas, which can improve the quality of care in remote, under-served areas, reducing socioeconomic disparities. This requires the integration of intelligent analysis of field-data and real-time transmission of multimedia data: radiology-images, video conversation, heart-beats, the voice of the patient or doctor and textual form input. This integration requires combining techniques from diverse fields such as electrical engineering, biomedical engineering, computer science and medicine.

Wearable medical devices are being used to monitor vital signs such as heart conditions, oxygen levels, sugar levels and biochemical composition of sweat during perspiration. Intelligent wearable devices also analyze the collected data using embedded software; suspected emergencies and abnormalities are shared with the hospital. Medical experts check the data periodically or intervene when an emergency occurs.

This chapter describes various technologies for remotely monitoring the patients' conditions.

9.1 PERVASIVE HEALTHCARE TAXONOMY

To provide pervasive healthcare systems, one has to consider: 1) the type of illness; 2) body-functions to measure; 3) environmental factors; 4) personal factors, including psychological factors associated with elderly patients and the availability of caregivers; 5) available communication infrastructure such as wireless connectivity and network infrastructure such as a bandwidth; 6) available sensor capability, the medical device's functionality and computing capacity and 7) data collection and analysis capability of the medical facility where a patient is located.

A loss of body-functions includes: 1) mental functions such as dementia; 2) sensory functions such as blindness, 3) voice and speech functions; 4) functions of the vital organs such as the cardiovascular system, respiratory system, hematological system and immunological system and 5) movement-related functions such as Parkinson's disease and bone fractures. The knowledge of body-functions and the associated environmental conditions provide the context for the impairment a patient has.

A patient and the healthcare provider treating the patient should be able to retrieve the authorized healthcare information about the patient, based upon the current location of the patient. A patient should also be able to authorize the use of his(her) medical record to a healthcare provider using a secure channel from anywhere and should be able to retrieve the information on her/his mobile-unit after identifying himself/herself. Such data include prescription medicines, past diagnosis, lab-tests results and dental records. In these scenarios, the wireless network plays a major role. A wireless network for authentication and health-related data-transfer should be secure and reliable with sufficient fault-tolerance; Public Switched Telephone Networks (PSTN) are insufficient for ubiquitous health care.

A location-tracking network should have high accuracy for acquiring immediate medical help. RFID (Radio Frequency Identifier) tags are utilized to track, localize and register a person when s(he) enters a zone where the RFID tag can be read automatically.

Some direct usages of pervasive healthcare technologies are: 1) fall management of elderly persons; 2) effective rehabilitation of patients; 3) providing assistance for daily needs of the people with neuromuscular diseases and 4) tracking people with Alzheimer's, dementia, epilepsy, apnea and people prone to sudden cardiac death due to ventricular fibrillation and myocardial infarction.

9.1.1 Context-Aware Systems

An important aspect of pervasive health care is the timely response by caregivers in an emergency. This requires interpreting the adverse vital signs in the context. For example, falling down of a youth differs from falling down of an elderly person. An abnormal reading of ventricular arrhythmia for an elderly person with a history of ventricular fibrillation requires an emergency response with defibrillators. The knowledge of context also helps in separating noise from the actual data.

Context information is based upon: 1) knowing the medical history of the person being monitored; 2) knowing the age and gender of the person; 3) knowing the environment in which a person is moving or residing and 4) knowing the time, day and location of the person being monitored. For example, detecting the posture of a person lying on a bed during sleep (with a regular breathing pattern and a normal ECG and pulse rate) causes no alarm. However, a person lying down while walking in a park causes concern and shows possible accident requiring an emergency response.

Context-aware systems use: 1) RFID to tag the monitored person; 2) storing sufficient daily-behavior-pattern of the person in an accessible computer system; 3) fusion of the data from multiple sensors to reduce false interpretation; 4) intelligent analysis of the data based upon statistical analysis and logical inferences of behavior patterns 5) wireless network and mobile phones to inform an emergency response system and 6) a two-way communication system, including video-conferencing for the care-provider and the emergency response system to assess the patient's condition.

At the application level, a context-aware system should enable a healthcare provider and the person being monitored to enter sufficient patient-related relevant information, behavior patterns, phone numbers of the nearest care-providers and emergency-response system for an accurate interpretation and quick response. The interface should be simple, graphical, user-friendly and menu-driven for the ease of adoption by patients and caregivers. Figure 9.1 shows a schematic of a context-aware healthcare network.

The context framework management comprises multiple components: 1) context provider; 2) context aggregator; 3) a knowledge-base; 4) an inference engine; 5) a query engine and 6) a context discoverer. A *context provider* collects raw data from environment sensors, video sensors and wearable body sensors to discover the exact location and condition of a patient. A *context aggregator* joins individual context information provided by context providers. *Context knowledge-base* is the centralized time-stamped storage unit where the context-related information is stored.

The information is used for a logical inference using an *inference engine*. The outcome of the inference engine is stored in a *knowledge-base*. The information stored in the *knowledge-base* is retrieved by the authorized caregivers and healthcare professionals using a *query engine*. A schematic of the context framework for health management is given in Figure 9.2.

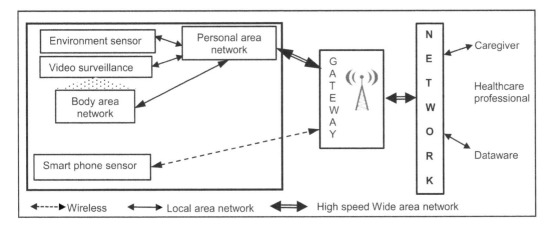

FIGURE 9.1   A schematic of a context aware remote health management network

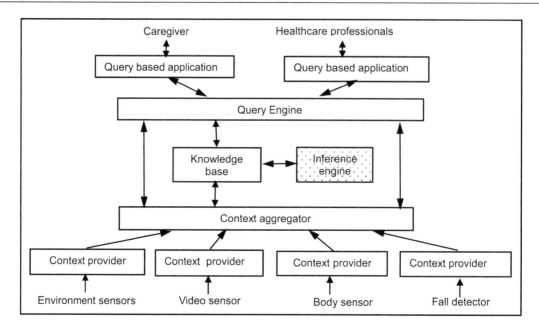

FIGURE 9.2   A schematic of a framework of context-aware system

9.1.2 Remote Monitoring

As the age till which people remain active has increased, there are many more medical conditions that need to be managed. These medical conditions include diabetes, dementia, heart abnormalities such as ischemia and arrhythmias, muscle abnormalities and neuromuscular diseases such as Parkinson's disease and pulmonary diseases that may cause seizures, asthma and hypertension. The outcome becomes quite serious if a medical facility is not readily available, as in the case of rural areas and remote locations.

Remote care also applies to elderly patients with limited cognitive capability as they tend to: 1) forget the location where they are in; 2) forget to eat; 3) forget to take medication; 4) overdose on the medication; 5) be unable to perform daily activities due to sudden worsening of conditions and 6) be prone to sudden falls. Despite being in an assisted-living scenario, such patients are largely unattended and are responsible for their own daily needs. Remote care also applies to post-surgical patients who are recovering in a residential environment. They may not have sufficient strength for all their daily needs, may fall, or need assistance with medication and home-based tests such as blood-sugar level, blood-pressure monitoring and ECG monitoring.

Remote care also applies to babies and toddlers because they cannot take care of their daily needs such as feeding and protection against accidental injury. However, babies and toddlers cannot wear multiple wearable devices because of the lack of available body-surface area and possible injury the device can cause. They have to be monitored using ambient motion-sensors and sound-sensors. Remote care also applies to active pregnant women who can lose consciousness suddenly, may get affected by sudden abnormal hormonal changes or may suffer from sudden falls.

With an increase in Internet connectivity and the miniaturization of wearable devices, it has become possible to wear the devices that capture and analyze the biosignals and transmit the abnormalities directly to the hospitals and emergency-response systems. The wearable devices may transmit data to the Internet directly, or through a mobile phone. Remote monitoring supports the people in different age-groups, with a spectrum of cognitive and physical abilities, to reduce morbidity and fatality by providing timely intervention.

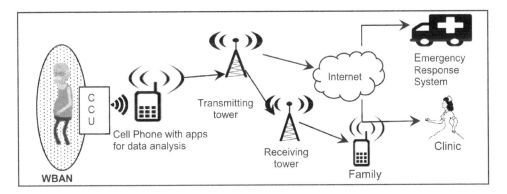

FIGURE 9.3 A Schematic of remote care using WBAN and integrated Internet-based system

Remote care reduces the lack of connectivity during a medical event even when the caregiver is not in the immediate vicinity. Combined with the built-in location finding capabilities available in GPS-based devices and cell phones, remote care enables quick emergency-response. While remote care is not a substitute for the continuous monitoring by caregivers needed for acute-care patients, it facilitates lifesaving by summoning medical care quickly. Most of the emergency situations, including heart-attacks and sudden falls, allow for few minutes of response time.

A remote care network requires: 1) miniaturized wearable devices often embedded in a physical unit, 2) *wireless body area network* (WBAN) connecting various wearable devices and RFID tags to a *central control unit*, 3) adapter that transforms raw data to an abstract analyzable format; 4) data analysis software for analyzing data, 5) a *wireless personal area network* (WPAN) to integrate ambient sensor readings for providing context-awareness and 6) a software interface and a gateway to an available wide area network (WAN). WAN could be a wireless network as used by mobile phones, a high-speed cable network or a satellite system as illustrated in Figure 9.3.

9.1.3 Automated Data Collection

Automated data collection is needed for: 1) a patient in an intensive-care unit or post-surgical recovery unit requiring continuous monitoring of vital signs; 2) remote-care patients in an in-home care system and 3) chronic disease patients going through long-term rehabilitation for monitoring disease specific data periodically but less frequently. The data transmission is classified into four major categories: 1) continuous raw data; 2) periodic raw data; 3) event-based analyzed data and 4) periodic analyzed data. In addition, the system sends an alert in case the analyzed data identifies or predicts an unusual emergency condition that requires urgent intervention by the care-provider or physician.

Home-care devices analyze raw data and send the outcomes to save the bandwidth requirement. The sensor units in ICU send raw data to a central unit that analyzes the data. The advantages of the central control-unit in the hospital are: 1) it is continuously monitored by one or more nurses or emergency doctors who can intervene immediately and 2) it analyzes the data of multiple patients. Hence, a single nurse or doctor can monitor multiple patients simultaneously.

9.1.4 Telemedicine and Teleconsultation

Telemedicine is defined as an interaction between a patient and a care-provider remotely for the diagnosis or sharing of knowledge about an abnormal condition with the use of fast Internet-based technology and applications, including videoconferencing. During teleconsultation, images of patient(s), radiology images, images of infected areas and physiological data are exchanged with the care-provider.

The intent of telemedicine and teleconsultation is to: 1) amplify the presence of a care-provider to remote areas where the density of the caregivers is low; 2) provide the presence of a caregiver for smaller emergencies at odd hours where a patient cannot traverse to a clinic or hospital; 3) provide consultation to paramedics in an emergency; 4) provide consultation to patients if a natural disaster occurs and 5) provide expert opinion to a general physician at a distance where in-person collaboration may not be cost-effective. Telemedicine has also been used by paramedics to send a patient's medical status to an emergency unit of a hospital to facilitate the unit's readiness when the patient arrives. Telemedicine has found applications in radiology interpretation, ECG interpretation, psychiatric consultation, dermatology, emergency health care, chronic disease management and pathology reports consultation.

Remote satellite clinics have a telemedicine booth with the required sensor-equipment. Equipment transmit data through the Internet to stations monitored by nurses, physicians and/or radiologists. The telemedicine booth may have a nurse measuring various symptoms of an elderly patient or a patient under rehabilitation. The raw data or images are analyzed, and the outcome of the analysis is accurately displayed to a physician (or a radiologist). The physician (or the radiologist) collaborate/consult with each other and transmit the diagnosis/treatment over the Internet to the patient, as illustrated in Figure 9.4.

With the availability of 5G wireless technologies that would soon replace 4G wireless technologies, the data-transfer rate for image-transmission and video-conferencing will no longer be a limitation. Many hospitals have started a smaller satellite hospital in remote areas with sufficient self-diagnostic equipment, including visual inspection, a pulse-meter, a glucometer and an oximeter. A patient can transmit data instantaneously using a high-speed data-transfer link and can talk to an attending physician or registered nurse for the diagnosis.

The communication between a patient and a care-provider is done in one of the three modes: 1) instant data transmission and video-conferencing; 2) store-and-forward data collected by a patient that is analyzed by a healthcare-provider asynchronously before video-conferencing; 3) bidirectional store-and-forward data (and diagnostics) and the receipt of a response from the physician based upon her/his availability. In store-and-forward modes, the data is stored onto a secure website looked up by authorized care-providers using a secure mechanism.

The major issue in telemedicine is to transmit the physiological and radiological data using patient-friendly equipment (or software application running on the patient's PDA) by a distressed patient over the Internet with high accuracy and security. The transmitted data for the Internet-based video-conferencing is compressed, and may not be always suitable for sending information about a patient's condition accurately. However, this limitation is absent in satellite hospital units, which are connected to the main campus through high-speed data-links.

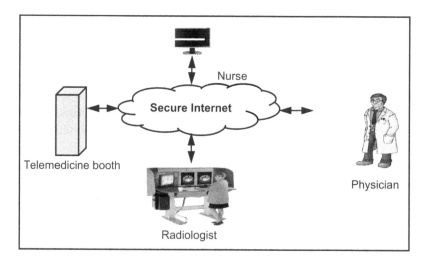

FIGURE 9.4   A schematic showing telemedicine and teleconsultation

Example 9.1

An endocrinology clinic has maintained the diabetes history of its patients aggressively to help them with their dietary control. It provided them a wireless connected glucometer that sends blood-sugar readings to a secure website by pushing a button. The patients measure the glucose-level every day before the meal and four hours after the meal. The glucometer sends the latest glucometer reading, seven-day average, and time and date over the website. The nurse discusses the patients' history weekly and helps them decide on their diet for the following week.

Example 9.2

A diabetic patient has developed a skin wound below the knee. He calls a skin-specialist and uploads a magnified high-quality JPEG image of the wound on a secure website. The doctor calls him later in the day and discusses his condition. Based upon the patient history available in the secure website, she prescribes an ointment and antibiotic to him. The prescription is automatically transmitted to the patient's pharmacist. The patient picks up the medicine from the pharmacist later in the day.

Example 9.3

A hospital starts a weekly rehabilitation program where a patient participates in a group-exercise-activity with other post-surgical patients, and an exercise-training specialist. During the exercise-session, a patient can talk to other members of the group and the trainer. During this session, her ECG, respiration rate and blood pressure are continuously monitored and stored in a secure website. Later, a nurse discusses her vital-signs variations. The cardiologist discusses her case weekly and adjusts the medicine dosages.

Telemedicine implementation requires: 1) user-friendly interface for the doctor, nurse and patient to see the information from web-based EHR; 2) a web-based cooperation system for collaboration among multiple parties and 3) an adaptive intelligent interface for ease of uploading information. Telemedicine has many advantages during post-surgical rehabilitation, minor cases of dermatology, pregnancy-related consultations and chronic disease management. It has been shown in many studies that telemedicine for cardiac consultation, diabetic consultation and dermatology has been effective. However, its adoption has been slow.

9.1.5 Home Health Care

Home health care is used for the maintenance of chronic diseases, elderly care, post-surgical recovery or conditions where a patient is confined to a bed for a longer period. Such conditions include patients suffering from an advanced stage terminal cancer, injured patients, patients with a pulmonary embolism (blood-clot in an artery in a lung) and post-surgical patients after bypass surgery or paralysis strokes. Such patients are characterized by frequent monitoring of vital signs in a residential setting. The major requirements are: 1) frequent monitoring of vital signs; 2) analysis of the vital sign data for any emergency; 3) store-and-forward mode of data-transmission for the nonemergency vital-signs data; 4) immediate connectivity to the emergency-response system and clinical physician in an emergency and 5) telemedicine capability for periodic conversations between the patient and the physician or homecare-provider, including nurse.

The situation differs from the mobile or ubiquitous health care due to the availability of the reliable connectivity and higher bandwidth. Compared to a hospital system, there is less reliance on RFID tags in a home-healthcare-system. Rather, many ambient sensors are used. Besides wearable-body-sensors, the homecare-systems use many ambient-sensors such as video-cameras, vibration-sensors, pressure-sensors, door-sensors, motion-sensors, light-sensors, fall-detectors, etc.

Video cameras are used to image a patient. The image is analyzed using software residing on the desktop computer to: 1) identify the posture and motion of the patient; 2) recognize activities and 3) recognize the expression of emotion or pain. Privacy is maintained. Pressure-sensors are used to discover a patient's presence on various furniture such as beds, chairs and toilet seats. Vibration-sensors are used to detect the liveliness of the patient on the bed. Context is derived from the collected data from various sensors. Light-sensors, door-sensors and motion-sensors detect the presence, entry or exit from a subunit such as a bedroom, kitchen, bathroom, etc.

There are subtle differences between a homecare patient and a monitored patient in a hospital:

1. A homecare patient is more mobile compared to a monitored patient in a hospital. She can move between various subunits and go for a walk while needing mobile care when she is outside the dwelling area;
2. A homecare patient is more independent and does much of the daily routine work herself; and
3. A homecare patient requires less frequent collection of vital signs data, and the data is preferably analyzed on a local desktop. The analyzed data is sent periodically or when an adverse event occurs.

The entry to the front-door contains a door-sensor to sense moving in and out of the house or rooms. The house-space is monitored by video cameras to detect activities and falls. Falls are also detected using a neck-worn sensor that has an embedded accelerometer for a sudden change in height and velocity. The places that require privacy such as bedrooms and bathrooms use motion-sensors instead of a camera. Kitchens and other places where major activities take place also contain vibration-sensors embedded in the furniture. The bed contains a pressure-sensor and a vibration-sensor. The bedroom also contains a light-sensor. In addition, the house-lighting is automated based upon motion-sensing. The details of the localization, activity monitoring, fall detection and tracking for home-health care are given in Section 9.5.

9.1.6 Acute Care Technologies

Acute care is defined as the heath-related services and care-delivery platforms used to treat sudden urgent or emergency episodes of injury and illness that can lead to death or disability without immediate intervention. The term *acute care* includes emergency care, trauma care, post-surgical recovery, immediate surgery to relieve a life-threatening condition, sudden cardiac arrest and short-term inpatient stabilization. Acute care is given to prevent sudden death or disability in an unexpected medical condition. During the acute care, it is assumed that the patient is not in a condition to take care of herself, has an excessive pain-condition, and possibly needs a life-support system. It requires continuous monitoring of vital-signs.

Examples of acute care are sudden high fever, loss of consciousness, brain stroke leading to full or partial paralysis, sudden loss of blood due to an accident, a serious side-effect of a medication, a sudden stroke, an epileptic seizure or migraine attack, food-poisoning and sudden cardiac arrest. The key factor in acute care is the timely intervention to stabilize the patient's condition. It may require sudden additional allocation of resources, including nursing resources, which are usually unavailable to a patient.

9.2 WEARABLE HEALTH-MONITORING SYSTEMS

Wearable health-monitoring systems are suitable for monitoring vital signs of patients suffering from chronic diseases, elderly people for any emergency need such as an accidental fall, postoperative patients during rehabilitation, persons taking part in extreme excursions or training and persons with special abilities. Wearable sensors measure ECG, EEG, EMG, heartbeats, blood-pressure, body-temperature, oxygen-saturation and respiration.

The major advantage of wearable devices is the automated monitoring of a patient's condition and an immediate alert system that can: 1) inform care-providers of fluctuations in the patient's vital signs; 2) provide automated reading and intelligent analysis and 3) save lives by an alert system that facilitates quick intervention. In addition, it helps elderly people to keep an active healthy life by performing statistical analysis of varying vital sign data on a daily and weekly basis.

Wearable sensors are classified as: 1) sensors that detect biosignals such as ECG, EEG, EMG, level of oxygen, respiration rate, heartbeats per minute, etc.; 2) sensors that detect ambient conditions such as gaseous composition and toxicity, external temperature and radiation level; 3) sensors that detect the biochemical composition of the body fluids; 4) safety-related sensors such as fall detectors and 5) tracking devices suitable for persons with a neurological disorder, macular disorder and post-surgical recovery. In addition, capsule size cameras are used to image abnormalities (such as cysts, inflammation or ulcer) in the gastrointestinal system. These cameras transmit high-resolution images using wireless transmission.

9.2.1 Wearable Sensors Placement

The popular ways to attach wearable devices are: 1) sticking it on the body such as ECG leads; 2) fastening it on the wrist, arms or chest; 3) implanting it as with pace-makers or renal implants; and 4) inserting it in the special fabric. With the recent advancement in e-fabric technology, many sensors can be embedded as part of the e-fabric.

Figures 9.5 and 9.6 illustrate wearable sensors connected to the WBAN of individual patients. Due to limited space on a body, multiple wearable sensors are packed as one unit. The fastened system is worn either on the wrist, legs, chest or ear (for females). Examples of the wearable sensors on a wrist are: pulse rate, pulsed-oximeter (photo-plethysmograph [PPG]), blood-pressure and skin-temperature. Similarly, a hidden chest-belt monitors ECG, blood-pressure and respiration.

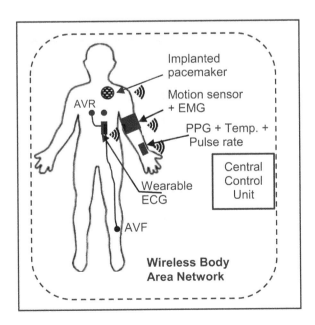

FIGURE 9.5 Wearable and implanted units

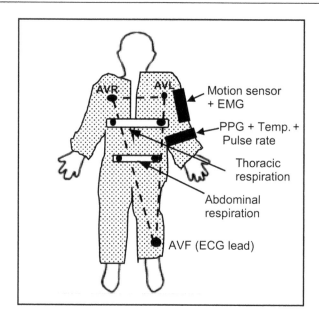

FIGURE 9.6 Wearable sensor fabric with sensors

9.2.2 Wearable Sensors Design

Wearable sensors should be aesthetically acceptable, or hidden and comfortably placed on a patient. It requires miniaturization and innovative design. Many sensors are used as ornaments such as a ring, while others can be placed on the neck or ear. A ring-based sensor has been designed to measure the oxygen-saturation level (miniaturized oximeter). It can also diagnose high blood-pressure and congestive heart-failure.

A blood-monitor that can be worn on the arm has been designed. It utilizes two MEMS (Micro Electro Mechanical System) accelerometer that measure the hydrostatic pressure of the blood using a PPG sensor. PPG sensor illuminates the skin and measures the absorption of light. It can measure the pressure pulse when blood is pumped out of the heart by measuring the distension in the artery. The amplitude of the PPG is proportional to the distension in the artery. By measuring the amplitude and frequency of the PPG, cardiac cycle and the blood pressure can be measured. By measuring the PPG at neck, the respiration cycle s measured. PPG can be used for long-term wearing as an ear-ring with no discomfort.

The rate of breathing has also been measured by placing a miniaturized microphone that picks up the acoustics each time a person breathes. The environmental noise is filtered using a band-pass filter. The signal provides the rate of breathing with an accuracy of *90%*.

Wearable devices include the measurement of the blood-sugar using biochemical sensors. A drug-delivery system has been developed that measures the blood-sugar using the minimal invasion, and then delivers the required amount of insulin using micro-level perfusion. The device uses Bluetooth wireless communication for data-transmission. Using biochemical-sensors and miniaturized spectrometer, one can analyze the chemical composition of any protein that acts as a biomarker for a disease.

9.2.3 Wireless Connectivity and Data Collection

Wireless connection of the wearable devices has to be safe and convenient for patients. The communication devices have to be robust to enable continuous data-communication. Four types of networks are used to transmit data to the care-provider: 1) WBAN, 2) WPAN, 3) Wireless Local Area Network (WLAN) and 4) WAN to transfer the data over the Internet to a care-provider or a healthcare professional.

WBAN supports short-range communications with a large amount of data-transfer for: 1) cardiac diseases such as myocardial infarction, arrhythmias, hypertension; 2) gastrointestinal disorders; 3) pulmonary diseases such as asthma and 4) use in neurological disorders such as epilepsy detection or long-term behavior modeling. It can also help people with disabilities such as patients with retinal transplants. It also supports nonmedical applications such as data-file transfer, gaming, social networking applications and monitoring forgotten things.

Power consumption in WBAN is very low. The data-transfer rate in WBAN is much higher to support in the excess of 250 kbps. WBAN is connected wirelessly to a personal computer or a mobile phone, which are readily connected through to WPAN. WPAN and other communication networks are discussed in Section 9.5. As the wireless technologies become more robust and 5G technologies are implemented, the wearable technologies, centralized repository, and emergency response system would become better integrated.

Data collected from each sensor are passed to a computer that tags the contextual information (*time, date, biosensor unit, type of the information*), structures the information into a more abstract form suitable for human interpretation and sends the information to the communication unit for transmitting to the wide area wireless network.

9.2.4 Issues in Wearable Devices

There are many issues with wearable technologies such as unpredictable battery drainage, sudden connectivity loss with the Internet, false-positives and false-negatives in analyzing data for emergency conditions. Other important issues are their adaptability, robustness, aesthetics, safety issues in a home environment, miniaturization for a low form-factor to fit on a human-body, chemical inertness, ergonomics, user-friendly controls, exposing a patient to minimal microwave-radiation during data-transmission and privacy. There are cultural and psychological barriers where a patient will like to wear minimal devices in an aesthetically pleasant way during social interaction periods.

9.2.5 Sensor Data Fusion

Homecare-based systems do not have as much accuracy as hospital equipment. However, this shortcoming is reduced by combining the information from two or more sensors. Fusion of biosensors is needed due to: 1) data-loss because of improper functioning or failure of a sensor; 2) limited spatial coverage of a sensor in the body-part; 3) imprecision present in low-cost homecare-based sensors and 4) the uncertainty present in the sensor-attributes.

A sensor is placed only on a specific part of the body and cannot cover all the readings. For example, reading from just one ECG-lead is insufficient for the diagnosis of ischemia or impending myocardial infarction. Another example is the neuromuscular problems. An EMG biosensor on one arm is insufficient to derive the extent of a neuromuscular problem.

The fusion of information from multiple sensors: 1) improves the SNR (Signal-to-Noise Ratio); 2) reduces inherent uncertainty caused by the limited number of features available in one signal; 3) improves reliability and robustness; 4) provides fault-tolerance in case of failure and 5) improves resolution and precision. Fusion of the information results in more accurate diagnosis and improves efficiency.

Information fusion is classified in two ways: 1) collaboration-type and 2) data-abstraction type. There are three types of collaboration: *competitive*; *complementary*; *cooperative*. In terms of data-abstraction, the data-fusion is done at multiple levels: 1) at the *raw-data level*; 2) after *feature-extraction*; 3) at similarity score level after *matching various features* and 4) at the *decision-level*.

Competitive fusion means multiple types of sensors are used to derive the same (or similar) information due to the differences in calibration and resolution. The use of multiple sensors provides fault-tolerance. However, most of the fusion in a body sensor-network is either *complementary* or *cooperative* where different sensors provide different information.

The data-fusion at the raw-data level is simplest. However, it is suitable for homogeneous devices: two devices measuring the same feature-values. Data from homogeneous devices is easily combined using average readings or maximum value (as in the amplitude of P-QRS-T from different leads), or deleting an imprecise reading. In wearable body-sensors, this scheme measures pulse-rate from multiple devices. Most devices are heterogeneous and measure different features.

Data-level fusion is also used to derive different classes of activities. For example, during exercise, the inertial sensors such as accelerometer, gyroscope and magnetometers are used. The same set of sensors is also used in fall detection. *Accelerometer* measures the nongravitational acceleration and vibration of an object. *Gyroscope* measures rotation and angular position in space with respect to an axis, and *magnetometer* measures the direction of a movement.

The fusion after feature extraction: 1) concatenates various feature-values; 2) removes features that contribute a little to the diagnostics of an abnormal condition; 3) normalizes all the feature-values to a common scale and 4) provides higher weights to more important features. Reduction of the feature-set is done using a combination of experts' knowledge, dimension reduction techniques such as PCA (Principal Component Analysis), LDA (Linear Discriminant Analysis), and machine learning techniques such as decision trees (see Section 3.5).

Fusion is an important mechanism to combine diverse information needed to diagnose a disease that requires inferencing of a complex pathway requiring conditions of two or more organs. For example, cardiopulmonary diseases require cardiac feature-values and pulmonary feature-values: both respiration rate and pattern and ECG are measured. There are many cardiopulmonary diseases such as COPD (Chronic Obstructive Pulmonary Disease), chronic bronchitis, asthma, hypertension, and stroke.

Two sets of feature-values are normalized using statistical techniques and concatenated to perform fusion. In the minmax technique, the normalized value is derived by (*actual value − minimum value*)/(*maximum value − minimum value*). The normalized scores are weighted based upon the importance of specific features. The features with smaller weights are removed without losing accuracy to ensure computational efficiency.

Example 9.4

Two feature-vectors are given by $<v_{11}, v_{12}, ..., v_{1N}>$ and $<v_{21}, v_{22}, ..., v_{2N}>$. The corresponding minimum and maximum values for feature-set-1 are $<min_{11}, min_{12}, ..., min_{1N}>$ and $<max_{11}, max_{12}, ..., max_{1N}>$. Similarly, the corresponding minimum and maximum values for feature-set-2 are $<min_{21}, ..., min_{2N}>$ and $<max_{21}, max_{22}, ..., max_{2N}>$. The normalized values for the *feature-vector-1* is given as $\left\langle \frac{v_{11}-min_{11}}{max_{11}-min_{11}}, \frac{v_{12}-min_{12}}{max_{12}-min_{12}}, ..., \frac{v_{1N}-min_{1N}}{max_{1N}-min_{1N}} \right\rangle$. The normalized values for the *feature-vector-2* are given as $\left\langle \frac{v_{21}-min_{21}}{max_{21}-min_{21}}, \frac{v_{22}-min_{22}}{max_{22}-min_{22}}, ..., \frac{v_{2N}-min_{2N}}{max_{2N}-min_{2N}} \right\rangle$. After the concatenation of the two feature-vectors, the fused feature-vector becomes $\left\langle \frac{v_{11}-min_{11}}{max_{11}-min_{11}}, ..., \frac{v_{1N}-min_{1N}}{max_{1N}-min_{1N}}, \frac{v_{21}-min_{21}}{max_{21}-min_{21}}, ..., \frac{v_{2N}-min_{2N}}{max_{2N}-min_{2N}} \right\rangle$

9.2.6 Privacy and Confidentiality

Privacy and security issues are important because data from the sensor networks are collected and transmitted continuously (or frequently) from the homes to the hospitals and data centers. Given the distributed nature of network, data has to be protected against malicious attacks that capture and misuse the personal information and medical data. There is sufficient controversy on ownership of collected data and information after the analysis. It is uncertain whether hospitals, patients or insurance companies own the data. The role of third parties in accessing the patients' data is also uncertain.

Different users have different privileges over the data. For example, attending doctors and nurses can modify or insert new data based upon the analysis and diagnosis. Insurance companies and billing departments cannot alter medical data. Rather, they can read limited amount of data for medical billing purposes. In an emergency, the data may also be disclosed to others such as close relatives, caretakers and law enforcement agencies, even without the explicit consent of the patient.

Data is de-identified and anonymized as described in Section 4.10.4 to maintain a patient's privacy. However, data sent from remote centers and home-care must have sufficient hospital-issued tags to identify patient-related data. To provide data and information security, both encryption and authentication mechanisms have to be provided as described in Sections 4.10 and 6.9.

9.3 PATIENT CARE COORDINATION

A patient is identified in the hospital using a bar-code along with multiple bar-code readers at different units. However, this requires human-effort and is not very convenient for automated tracking of patients in real-time. Another popular alternative is the use of RFID. A typical RFID system has one reader that can read multiple UHF (Ultra High Frequency) RFID tags.

9.3.1 Radio Frequency Identification (RFID)

RFID tags have two major advantages over bar-code readers: 1) they do not require manual reading and 2) they can be used for monitoring a patient's vital signs and movements. The frequency range of operation of a typical RFID is between *865* and *920* MHz. An RFID tag can be *active* or *passive*. *Active tags* have their own power source having an ultra-small form factor. A *passive tag* uses the radio energy received from the readers, and backscatters the signal. Since RFID tags have ultra-low energy requirement, there is no adverse long-term impact on a body. RFID readers have two interfaces: 1) to read the ID and the monitored signal values and 2) communication interface with the server.

Besides identifying a person or medical instrument, RFID can also monitor the operated part of a body to track the healing. For example, chest movement monitoring indicates patient breathing. Unexpected movement can be studied to detect epilepsy during the recovery and rehabilitation period. RFID tags can also monitor the body-temperature. Hence, it is very useful for remote monitoring where the patient is incapable of interacting such as monitoring infants, monitoring comatose patients and monitoring patients in ICU (Intensive Care Units). Many RFID tags can be placed on a body.

RFID tag comprises three major components: 1) analog front-end that contains a flat antenna, power-source and a clock, 2) a digital module, including memory and 3) a printed sensor with an ultra-small form-factor. The sensor could be a temperature-sensor, a pressure-sensor or an accelerometer. The sensor is placed in the digital-module. The front-end supplies power and clock to the digital-module, modulates and demodulates the sensor-readings and transmits signals to readers. The digital-module converts signal-values to a standard digital format and transmits to the front-end. The sensor-module measures the signal and transmits the measured values to the digital-module.

There are two types of RFID antennas: 1) wristband type and 2) fabric-worn antenna. Wristband antennas are easily replaceable and give more stable readings. Fabric-worn antennas are easily hidden and have more space. Antenna should be very flexible to withstand a lot of strain that occurs due to bending of the fabric. The size of a typical patch-like antenna required for transmitting the signal is approximately *49* mm × *60* mm × *4* mm. The wavelength of *865* MHz waveform is *34.5* cm that is in the same range as human limb, facilitating the study of human motion.

9.3.1.1 RFID collision

RFID suffers from three types of interference: 1) responses of multiple RFID tags communicating at the same time to a single reader for the same query; 2) multiple readers querying to a single RFID tag and 3) low power weak responses of the tags compared to strong signals from neighboring other readers. The first interference causes delay, while the second and third type interference affects the locational accuracy of the tags. As a consequence, second and third types of interferences affect the estimation of the positional and movement-related accuracies of the entity they are tagged on. Figure 9.7 illustrates types of collisions.

There are two types of ranges: *reading-range* and *interference-range*. Reading-range is much smaller than the interference-range because: 1) reading-range involves the traversal of the transmitted signal twice the reading-range and 2) the signal must be above a threshold to be acceptable to the reader to maintain higher SNR.

Figure 9.7 shows two types of RFID tags: 1) those who are affected by the interfering readers and 2) those who are unaffected by readers. The RFID tags affected due to interference are shaded as dark, and unaffected tags have a pattern. The RFID tags are marked as T_{ij} where the subscript i is the index of the corresponding reader, and the subscript j is the tag-index within a specific reader circle. The readers are marked as R_i where the subscript i is the reader-index. The bold circles with solid lines show the reading-range with the center as the corresponding reader, and the dashed circles show the corresponding interference-range.

Example 9.5

Figure 9.7 illustrates all three types of collisions. The interference circles of the readers R_1 and R_2 include each other. Hence, R_1 and R_2 get the signal directly from each other showing *type-3 collision*. The reader R_1 has four associated RFID tags: T_{11}, T_{12}, T_{13} and T_{14}. All four tags almost simultaneously scatter back the received signal to R_1 causing *type-1 collision*. Similarly, the reader R_2 has three associated RFID tags: T_{21}, T_{22} and T_{23}. All three tags almost simultaneously reflect back the signal to R_2 causing *type-1 collision*. Four RFID tags T_{13}, T_{14}, T_{21} and T_{22} interfere with the readers R_1 and R_2: T_{13} and T_{14} are being interfered by R_2, and T_{21} and T_{22} are being interfered by R_1. The interfered RFID tags show *type-2 collision*: the reader R_1 gets inaccurate readings from T_{13} and T_{14}; the reader R_2 gets inaccurate readings from the transmitted signal received from R_1. Both R_1 and R_2 have multiple tags that are associated with the same reader showing *type-1 collision*.

Careful planning of the RFID readers can avoid the collision so that the signal received by other readers is weak enough not to cross the reading threshold of the readers.

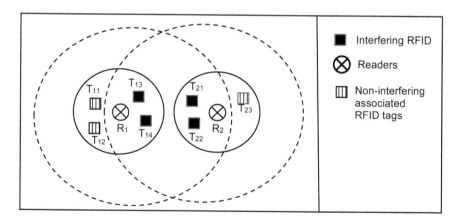

FIGURE 9.7   An illustration of three types of collision in RFID-based sensor system

9.3.2 RFID-Based Localization and Tracking

Localization problem is defined as identifying the exact position of an object within a specific region. Localization can be outdoor or indoor. An example of *outdoor localization* is finding the location of: 1) an elderly person suffering from dementia and 2) a patient in rehabilitation going for a morning walk. Examples of *indoor localization* are: 1) a patient going from one unit of a hospital to another unit of the hospital and 2) an elderly patient under in-home care going from one room to another room. The two systems are quite different. Outdoor localization requires the use of GPS; indoor localization requires RFID, ambient sensors, wearable sensors, and cameras. The two systems need to be integrated for patients under rehabilitation or elderly home-care persons to cover their indoor and outdoor movements.

This section describes hospital-based indoor localization using RFID tags and readers. The indoor localization using a combination of ambient sensors, camera-based scene analysis and sound analysis is described in Section 9.4.

Localization techniques use various parameters to identify the position such as the *angle of arrival*, *time of arrival*, the *time-difference between the transmitted signal and the corresponding received signal from the readers to the RFID tags with known locations*. Localization uses three types of techniques: 1) *triangulation*; 2) *scene analysis* and 3) *proximity (distance from the reader)*.

Triangulation uses three or more RFID readers and the estimated distance from each reader to identify the location of the entity. Triangulation is further divided into *multilateration* and *angulation*. *Multilateration* uses *time of arrival*, *time-difference of transmitted and received signal*, and *the reduction in received signal strength (RSS) by the reader from a passive RFID with known location*. In angulation technique, the *angle of arrival* is used to estimate the relative direction of the reader-tag. Together, these two techniques can accurately estimate the position of the target RFID.

Indoor localization includes stochastic movement of persons and other objects. In addition, different surfaces cause uneven absorption and reflection of radio waves. Hence, the distance calculation is an estimate and varies even if the object's position does not alter. Another problem with *multilateration* is the assumption that each reader reads the signal uniformly from all directions. However, the reading capability changes with the change in direction. This irregularity has to be considered during triangulation.

Scene analysis uses statistical analysis of the *RSS* of the objects within a specific area. The signal strength is used to create a radio map of the region being monitored. The problem of localization reduces to matching the RSS with the signal strength stored in the database, and retrieving the corresponding location from the radio-map of the locations. Multiple readers with different power-levels are used to read the scattered signals from RFID tags with established locations.

One technique is to select K-reference tags with a signal strength closest to the target tag's signal strength. The average of the positions of the K-nearest reference tags is used to estimate the position of the target RFID tag (patient's RFID). After localizing the target, the movement of the target is predicted using Kalman-filters under the assumption that the movement is uniform. The reference RFID tags are arranged on the floor in a well-defined patterns such as repeating square-patterns or repeating triangular-patterns. There are two problems with this approach: 1) multiple locations may map to the similar signal strength causing ambiguity and 2) stochastic movements of people and object interfere and disturb the RSS. Multiple machine learning techniques such as SVM (Support Vector Machine,), and Bayesian approach with forward-backward reasoning to maximize the posterior probability have been used to improve the accuracy of location prediction.

9.3.2.1 Modeling RFID-based localization

The whole area is divided into $m \times n$ cells. Each cell contains passive RFID tags as shown in Figure 9.8. RFID readers are placed in the center at the ceiling. The patient or the patient's ambulatory cart carries an RFID tag. RFID-readers know the exact locations of all $m \times n$ reference tags. At any point, it receives $m \times n + 1$ RSS readings. Based upon the *RSS* reading, *K*-nearest readings are identified. The average of the locations of the identified *K* reference tags estimates the location of the patient's RFID tag.

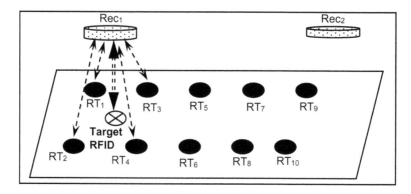

FIGURE 9.8 Estimating the location of the target RFID using reference-tags and their locations

Mathematically, let the original strength of the reader transmitted signal be S_T. Let the attenuation of the power be S_A. The attenuation is given by $S_T - RSS(RFID_i)$, where $RSS(RFID_i)$ is the RSS from the ith RFID. Let the distance between a reference tag and reader be d_i. The amplitude gain of the reader during transmission and the receiving are G_T^{reader} and G_R^{reader}; the amplitude gain of the reference-tag is given by G_T^{tag} and G_R^{tag} for transmission and receipt, respectively; the reflection coefficient of the tag is ρ and the wavelength of the radio wave is λ. Then using the scattering formula for the electromagnetic waves, the attenuated energy for each reference tag is given as $G_T^{reader} \times G_R^{tag} \times \rho \times G_T^{tag} \times \frac{\lambda^4}{(4\pi d_i)^4}$ where the last term is due to scattering of the electromagnetic waves under the assumption that there is no collision. The reader knows the distance d_i for every reference-tag. Hence, it can isolate the energy attenuation of every reference-tag. Since transmitted energy is invariant, the attenuated energy affects the RSS uniformly. Thus, by sorting on RSS values, the K-nearest readings are identified. The locations of the corresponding reference RFIDs are also identified.

Figure 9.8 illustrates a schematic to locate the target RFID using multiple readers and multiple rows of reference RFID tags. The readers are denoted as Rec_i, and the reference tags are denoted as RT_j. Each reader takes the reading from the target RFID within its reading-range. However, only the nearest reader is used based upon the RSS. Once the reader is identified, the signal strength is sorted, and the K readings nearest to the signal strength of the target RFID are measured. Their locations are looked up based upon previous training and similarity-based search.

Example 9.6

In Figure 9.8, both Rec_1 and Rec_2 read attenuated received signals. The RSS $RSS(Rec_1) > RSS(Rec_2)$. Hence, Rec_1 is chosen. The reference tags RT_1 to RT_6 are in the reading-range of the reader Rec_1. The sorted RSS-values show that $RSS(RT_1)$, $RSS(RT_2)$, $RSS(RT_3)$, and $RSS(RT_4)$ are nearest to $RSS(target\ RFID)$. The $location(RT_1)$… $location(RT_4)$ are looked up from the database using similarity-based search using their corresponding values. Location($target\ RFID$) is modeled as the centroid of the $location(RT_1)$, $location(RT_2)$, $location(RT_3)$, and $location(RT_4)$.

9.4 HOME-BASED MONITORING

With the growth of the elderly population, the need for providing elder-care is increasing. However, more elderly patients are deferring their movement to an assisted residential living due to the cost-factor and the loss of independence. Elderly persons have age-specific issues such as dementia, neuromuscular diseases and other medical conditions. Unlike patients recovering from surgery and chronic diseases, elderly persons prefer to use minimal wearable devices necessary for their well-being despite being connected to the world through the Internet and wireless link.

Some common age-specific problems among the elderly need to be addressed for their living. Elderly persons statistically have more falls caused by balancing problems due to: 1) neuromuscular problems; 2) fluid buildup in the brain; 3) sudden loss of consciousness; 4) accidental missteps; 5) weak vision that may cause to misjudge stairs; 6) dementia and deterioration of vision processing in the brain and 7) problems in the vestibular complex.

Besides falls, elderly persons are prone to age-specific diseases such as sudden cardiac arrest, pulmonary embolism (a clot in the lung-artery causing breathing problem), sudden fatigue, stroke caused by a blood-clot in the brain, dementia and neuromuscular problems. Elderly patients in early states of dementia and Alzheimer's disease tend to live independently. However, they have episodes of suddenly forgetting the context or becoming unaware of their whereabouts. They need help in doing daily activities. Thus, they need to be tracked voluntarily in their surroundings. The tracking has to be interpreted in a specific context to identify an abnormal behavior.

Localization and tracking are also important when a person is going through recovery or rehabilitation to provide better intervention in the case their vital signs become irregular. Localization and tracking devices should be well-connected to the emergency response system and to the caregivers so that they can respond in time.

This section describes available techniques and technologies for detecting fall and tracking elderly patients in residential surroundings.

9.4.1 Activity Recognition

Monitoring a person's activity in the context of location and time of the day can derive unexpected behavior and medical conditions. It can be used during rehabilitation, cognitive diseases, musculoskeletal diseases and fall and imbalance assessment. Elderly people have a greater tendency of a delayed response and sudden falls due to the lack of motor-control. Associated with ECG waveforms and oxygen saturation readings, the movement activity can infer the cardiopulmonary emergencies.

There are many devices to monitor activities such as home-based cameras, body-worn accelerometers, gyroscopes and magnetometers. The body sensors are worn on ankles, arms and the upper body to assess different motions. The body-worn sensors measure the linear and angular motion and acceleration. These raw values are analyzed to suggest the irregular motion and falls. The overall system for activity recognition is given in Figure 9.9.

There are three phases to the activity detection: 1) training phase; 2) adaptation phase and 3) recognition phase. The training phase uses a large number of patients having similar conditions to identify the combination of feature-vectors and a sequence of feature-vectors that correspond to unusual activities. Each individual has a somewhat different combination of diseases and feature-vectors for the same unusual activities. In the adaptation phase, the system adapts to the condition of the specific patient. After the adaptation phase, the information about unusual activities is stored in a feature-database.

During a daily activity, body-worn sensors continuously send raw data to a central device that could be an app in the mobile phone or a desktop. The software algorithms remove noise from the signals, enhance the signals, normalize the values and combine them to form feature-vectors. Classification and recognition algorithms are used to look up the feature-database and match them with the feature-vectors to derive various activities such as walking-motions, joint-monitoring, unusual motions and change in height.

The walking-motions are standing, walking, jogging, running, bending, ascending or descending a slope, ascending or descending stairs. Joint-monitoring includes measuring joint angles and flexing joints. Joint angles can be combined into feature-vectors. These feature-vectors are statistically analyzed and classified using variations of Markov models and clustering such as GMM (Gaussian Mixture Models) and K-means clustering. The recognition algorithm uses nearest-neighbor match.

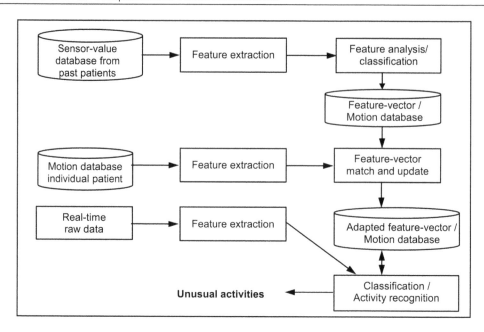

FIGURE 9.9   An illustration of unusual activity recognition system

9.4.2 Fall Detection

For elderly care, fall-detection is very important to provide immediate help. It is anticipated that *30%–35%* elderly persons above the age of *70* fall due to weak motor-activity, muscle-strength and sensor-problems. A fall-detection system is an activity recognition system or posture recognition system that can identify a person in a fallen state. A fall-detection system, linked to an emergency response system, can effectively reduce the injury-time, and provide much needed medical care in time.

In the last few years, many technologies have been developed to detect falls that include: 1) analysis of sudden posture changes; 2) the use of accelerometers to check unusual motion and 3) the use of gyroscope to measure imbalance. The analysis includes data from: 1) ambient sensors such as floor-sensors, infrared-sensors, pressure-sensors; 2) images from cameras; 3) wearable sensors such as accelerometer and gyroscope and 4) a combination of camera and sensors. Ambient sensors are generally fixed and well placed in the locations where the fall is predicted. Wearable sensors are mobile and can be used anywhere along with the motion of the patients or elderly persons.

Camera-based sensors use computer vision techniques and perform silhouette analysis and match acquired silhouette with a database of silhouettes associated with the fallen states. Similarly, ambient sensors-based and wearable sensor-based systems are analyzed to classify the sensor-readings between the normal activities and fallen states. Video analysis techniques are popular, and use multiple metrics on silhouette analysis such as: 1) change in the ratio of the major-axis and minor-axis of the silhouette; 2) change in the orientation of the silhouette; 3) change in illumination and 4) change in the centroid coordinate of the silhouette. The major-axis and minor-axis can also be estimated using height and weight ratio of the person.

The fall states can be: 1) lying down in a stretched position; 2) lying down in a tucked position and 3) sideways, backward or forward fall. The intelligent technique used in the fall-detection are: 1) Kalman-filter-based motion tracking; 2) Bayesian segmentation; 3) shape analysis; 4) Mel-frequency cepstral coefficient for fall-related sound analysis; 5) K-means clustering for different body-posture analysis; 6) Markov model analysis having two or more states such as walking, falling, fallen and getting up; 7) Gaussian mixture model for analyzing shape-deformation during falling;

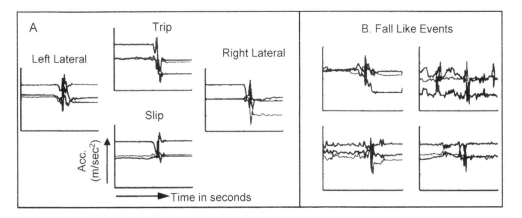

FIGURE 9.10   An illustration of simulated accelerometer readings for falls and fall-like events (Courtesy PLoS One (www.plosone.org), a creative common license journal, Publication: Ibert, Mark V., Kording, Konrad, Herrmann, Megan, Jayraman, Arun, "Fall Classification by Machine Learning Using Mobile Phones," *PLoS One*, 7(5), May 2012, e6556, DOI: 10.1371/journal.pone.0036556, Figure 2, permitted under Journal's Creative Common License Attribution).

8) histogram-based representation of various features extracted from silhouette shape and 9) support vector machine based classification.

Most of the wearable devices based on fall-detection use tri-axial accelerometers and gyroscopes. Gyroscopes are used to derive the position of the fallen person, and accelerometers measure the rate of change of motion. Techniques used to measure the fall detection are: 1) threshold-based techniques and 2) machine learning-based techniques. Machine learning-based methods include the use of: 1) support vector machine; 2) GMM; 3) K-means clustering; 4) decision-trees; 5) neural network; 6) meta-heuristic searches such as particle swarm optimization; 7) Bayesian reasoning system and 8) Sparse Multinomial Logistic Regression (SMLR). Based upon the accelerometer readings, the analysis can identify a soft fall and a hard fall. Combined with the camera readings, a fall location can be identified such as a fall on the stairs, and falling during walking. Figure 9.10 shows tri-axial accelerometer readings for a fall-like event. The experimental results show that SVM classifiers, SMLR classifiers, decision-trees and neural-network classifiers exhibit better accuracy in fall detection. With the continuous technological improvement in smart phones, many sensors such as accelerometer are being built-in smart phones to provide inexpensive computational analysis capabilities.

9.4.3 Localization and Tracking

Localization and tracking are very important for elderly care for many reasons: 1) they may encounter hazardous conditions in the house due to sudden accidents and forgetfulness about gas leaks and running water taps; 2) a sudden fall or loss of consciousness and 3) sudden fluctuation of vital signs leading to a serious health condition such as a stroke. For this reason, the patients have to be monitored and tracked continuously.

Outdoor localization and tracking are also important for patients suffering from dementia, Alzheimer's disease, autism and Down syndrome. These patients are in a confused state and wander off without any memory of how to return and where to return. Many wandering patients are killed. With the advancement of GPS devices and wearable devices, such persons can be tracked, and caregivers are alerted if they leave the home parameters. The miniature GPS devices can be attached to the body or the wearable clothes, and the tracked GPS can be compared with the latitude and longitude values of the perimeter.

9.4.3.1 Indoor tracking

The indoor monitoring requires a combination of sensors such as a vibration sensor for feeling the activity in the bed, motion sensors to sense the motion in bathrooms and toilets, pressure sensors in various furniture, including beds, light sensors in bedrooms, cameras in a kitchen and living rooms to check any abnormal motion and wearable fall detectors and vital sign monitors. In addition, the ambient sensors such as a temperature sensor, carbon-dioxide sensor and humidity sensor are also used to assess the environmental conditions.

These devices are interconnected with the computational device and gateways using wireless communication. The computational device also aggregates the data, builds up the context before transmitting the alert information to the caregivers, medical emergency systems, and healthcare professionals. The context could be "walking," "taking medication," "cooking," "sleeping," "getting up," "sitting," etc. Daily intake of water is also monitored to derive the causes of unusual activities.

Tracking a person requires multiple machine learning techniques that use parameters such as amount of light, temperature gradients across the rooms, motion pattern inside an enclosed space, time of the day, and availability of the wireless signal. These signals are collected during the training phase from various locations in the house using wearable sensors such as 3D magnetometer, fluorescent light detector and temperature sensor and mobile phone with wireless signal strength application software.

The data acquisition is performed periodically. Knowing the map of the dwelling and the time of the day, activity and location are predicted. It is further improved by statistical analysis and learning the activity patterns of the elderly persons (or recovering patients).

9.4.3.2 Outdoor tracking

Human activities and mobility are quite random. Different people have different preferences, and daily activities change based upon their plan and needs. However, the activities of the elderly persons are limited and fixed such as morning walks at specific times and daily activities indoors. Hence, by analyzing the spatio-temporal patterns, their activity can be tracked. Outside the house, mobile phones and wearable GPS devices are used to track them.

An elderly person suffering from dementia may walk outside the residential-premises in a confused disoriented state. By comparing the history of his walks and mapping his movement on grid-cells of a digital map, his regular activities is subtracted from the daily walk, and the disoriented walks are recognized. The walking activity is modeled as a graph where the nodes are the places and landmarks frequently visited by the elderly person. For example, a graph vertex can be community store, drug store, home, grocery store, frequently-visited friends or kin's home. The edges of the graph carry the information about frequency-of-visit, duration-of-walk, measure of irregularity of a walk between two vertices and the order of traversal.

The measure of irregularity of a walk is tracked by measuring the time taken to traverse between the two vertices, a change in the order of vertices, traversal to new parts of the map not traversed regularly and so on. A person with dementia attack may walk completely randomly, or stay at a place for a longer period. Random activities are determined using spatio-temporal analysis. An illustration of an overall model is given in Figure 9.11.

9.5 HEALTHCARE NETWORK AND INTERFACES

An architecture of healthcare network for automated data collection requires: 1) WBAN for continuous data collection from implanted and wearable sensors and 2) WPAN including Zigbee, MICS and Bluetooth from other sensor devices embedded in the wearable fabric and WBAN central control unit;

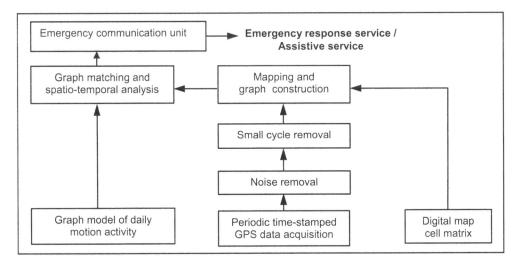

FIGURE 9.11   An illustration of outdoor tracking and localization analysis system

3) WLAN from WPAN to the remote PC and 4) gateway interface to the Internet. A schematic of the wireless connectivity is illustrated in Figure 9.12.

9.5.1 WBAN (Wireless Body Area Network)

WBAN is a short-range ultra-low power network that connects medical devices for remote medical monitoring using multiple-hop technology to extend the range. Central Control Unit (CCU) of WBAN connects to a Personal Area Network (PAN) or a gateway device such as desktop or mobile phone depending upon the location of the person.

The network controller should have an ultra-small form factor (*shape, size and weight*), and power consumption should be very small. Small power consumption helps adoption of the system because it resides on the human body or the sensor fabric, and is changed sparingly. A typical network controller is few millimeters in size and consumes power in the order of few milliwatts. Due to the ultra-small form

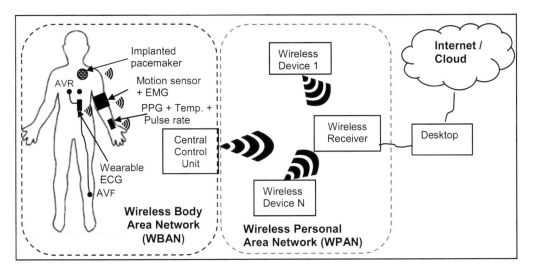

FIGURE 9.12   Healthcare network and interfaces

factor, battery weight has to be kept low. Battery changing is not easy for the patients and elderly persons. Mobility, portability, real-time data-transfer and reliability of the network controller are very important. Mobility and portability is important because the network controller is placed on a human body for a long period even during walking, running and other activities.

To ensure real-time data-transfer, the monitoring delay should be small, and message delivery should be quick and reliable. To ensure reliable message-delivery, multicast routing scheme is used. A multicast routing scheme sends a message to multiple destinations to ensure robustness against message loss and delay. The system should have multi-hop capability to extend the range of the system. A multi-hop ad hoc wireless network is a collection of wireless devices that connect to each other to form a network. A message can jump multiple devices to increase the range. Multi-hop is needed in the case of WBAN due to the small range of individual wearable body sensors. To secure the messages, key-based encryption and authentication is used.

WBAN uses mainly NB (narrow-band), UWB (ultra-wide band) and MICS (protocols for wearable and implanted devices) to detect signals such as ECG, EEG, EMG, PPG, pulse rate, blood-pressure and skin-temperature.

9.5.2 WPAN (Wireless Personal Area Network)

PAN is related to other ambient devices that are used for the localization of the patient in a home-care unit, devices with RFID tags, PDA that communicates with the medical devices involved in WBAN. The devices include video cameras, RFID readers, temperature-sensors, pressure-sensors, humidity-sensors, GPS sensing devices, vibration-sensors, light-sensors, motion-sensors and door-sensors. These sensors are used to track the location and posture of a patient in a home setting or during walking. Many times, the devices used by a patient have an embedded RFID tag, which can be wireless sensitive, and can be tracked. Simple bar-coded RFID tags should be visible to a camera to improve resolution for the identification.

Power consumption of nonmedical tracking devices should be kept low since most of the time no specific event occurs. Additional resolution requires more data-transfer and faster sampling rate that increase the data transfer rate, and increase the power consumption. The energy efficiency of the MAC (Medium Access Control) layer is an important issue in a multi-hop ad hoc network. Hence, in the regular mode, the devices work at low-resolution and low data-transfer mode to save power consumption. The resolution and data-transfer is increased after the detection of an event such as the entry of the person being monitored in the sensing-range and detection of an abnormal behavior or pattern.

Example 9.7

A camera works in a low resolution and low sampling rate in a nonvideo mode unless a moving object is identified. After the moving object is identified, it is switched to a high-resolution video mode to monitor every action of the patient. The action of a human is much slower than *30* frames/second video; lower number of frames is sufficient for human activity analysis.

Example 9.8

ECG analysis is done periodically with low SNR and is analyzed by a local device such as the cell phone or laptop using an intelligent software application. In the case of abnormality or in response to a command from the emergency response system, the application sends a command to increase the resolution and SNR for high-quality ECG data that can be analyzed accurately.

Example 9.9

A node in an adhoc network has two data-transmission modes: low data-transmission and high data-transmission. Assume that during low data-transmission, a video camera transmits one image/second and remains idle for the remaining time. In the high data-transmission mode, the video camera transmits five high-resolution image per second. Assume that the node transfers the data at the rate of 4 Mbps, and low-resolution image size is *128* pixels × *128* pixels, and the high-resolution image is *512* pixels × *512* pixels. The image is compressed to *20%* of the actual size.

The initial image size is *16,384* bytes. After the compression, the image-size becomes *3276* bytes. Assuming that the packet length is *1500* bytes, and the cumulative size of headers (preamble + packet length header + MAC header + IP header) is *576* bits, the total time taken to transmit one packet is $(1500 \times 8 + 576)/(4 \times 10^6) = 3.1$ ms, and the number of packets in a low-resolution image are $\lceil 3276/1500 \rceil = 3$ regular packets. The time estimate to send one low-resolution image is $3.1 \times 3 = 9.3$ ms. Using similar calculation, the time taken to transmit one high-resolution image is *108.5* ms. Thus, five images/second will take 5×108.5 ms $= 542.5$ ms. The power consumption increases proportionately to increase data-transmission time by a factor of $542.5/9.3 = 58.3$.

9.5.3 Gateway to Wide Area Network

The gateway during a mobile-activity is a cell phone running the required apps or a personal computer inside the smart-home running the corresponding data analysis software. The gateway sends raw data, or the processed information through a WAN to an emergency response system or a healthcare provider. The gateway is prone to the Internet data-congestion and unsecured lines. The integrity of the data needs to be protected by encrypting the data using a session-based key.

9.5.4 Application Software Interface

An application software is an intelligent signal processing software with sufficient user-friendly graphical user-interface. The interface changes with the demographics and the physical ability of the patients. For example, the interface for elderly persons should be more tolerant of typing errors, should help in correcting the spellings, should magnify smaller font characters and read it out to support vision impairment. The signal-processing part should denoise the signal before any interpretation of the data. The data should be critically analyzed in real-time using the context and past statistics of the patient to identify any unusual event, and trigger appropriate alert response.

9.6 STANDARDS FOR DATA COMMUNICATION

The most common technologies used in home devices or body-worn devices are Zigbee (IEEE standard 802.15.4), Bluetooth (IEEE standard 802.15.1), IrDA (Infrared), MICS (Medical Implant Communication Service), WBAN, WMTS (Wireless Medical Telemetry System) and UWB.

In order to standardize WBAN and to provide secure transmission, an IEEE standard 802.15.6 has been developed. The standards support short-range large amount of data transfer for: 1) cardiac disease such as myocardial infarction, arrhythmias, hypertension; 2) gastrointestinal disorders; 3) pulmonary diseases such as asthma and 4) use in neurological disorders such as epilepsy detection or long-term behavior modeling. It helps people with disabilities such as patients with retinal transplants. It also supports non-medical applications such as a data-file transfer, gaming, social networking applications and monitoring forgotten things.

IEEE standard 802.15.6 supports *UWB* and *Human Body Communication* layer. It supports four modes of data-transfer: 1) unsecured; 2) authentication only; 3) encrypted and 4) authenticated transmission of medical data. In the *unsecured mode,* there is no mechanism to protect the data-integrity and the confidentiality of a patient. However, it is faster and useful for local raw data-transfer where security is not an issue. The *authentication only mode* authenticates a senders' identity. However, the data is not encrypted, and it does not protect the confidentiality of patients. The *authentication and encryption mode*, the data is both authenticated and encrypted.

9.6.1 Zigbee

Zigbee (IEEE Standard 802.15.4) is relatively new, wireless standard protocol. It has low power requirement and supports two-way communication. It uses *2.4* GHz (*868–915* MHz frequency band in Europe and USA), has a bandwidth of *5* MHz with a data-rate of *250* kbps. The transmitted power is less than 100 mw (milliwatt), and its range is ten meters. Zigbee chipset consumes around 1 mw of power. It is used in multiple consumer electronics, smart homes, IoT devices, industrial controls, and many medical applications. One of the problems with Zigbee is interference with other *2.4* GHz-based devices.

The major advantages of Zigbee are: 1) interoperability between manufacturers; 2) low cost; 3) open standards and 4) the ease of deployment. It supports a large number of nodes, a wide variety of topologies including mesh networking as shown in Figure 9.13, and 128-bit encryption. Mesh network has three types of nodes: *coordinator, router* and *end device*. The *coordinator* ensures communication and distributes addresses to newly connected devices. A *router* finds a device in the network.

Each Zigbee device consists of a compatible RF-chip, including memory and device interface, microcontroller unit, sensors or actuators, battery for DC power and an energy harvester unit that collects the energy from radio waves, and charges the DC battery. The networking uses: 1) application layer, 2) network layer, 3) MAC layer and 4) physical layer.

The application layer consists of an application sublayer, Zigbee Device Object (ZDO) and application framework consisting of multiple application objects. *ZDO* defines the role of the device in the network as coordinator, router or end-device.

9.6.2 MICS (Medical Implant Communication Service)

MICS is a multi-hop wireless network used by WBAN to collect data from various physiological sensors implanted in the body. MICS uses the frequency band of *402–405* MHz, has a bandwidth of *3* MHz and supports a data transfer rate of *76* kbps. The transmitted power is much lower around *25* micro watt, and it has a range of less than *10* meters. Limited power requirement means that implanted devices communicate using use short-duration messages within a communication session. MICS communication session is initiated in response to programmer, control transmitter or a medical event. A medical event is an emergency situation that requires the medical-data to be transmitted immediately. Examples of MICS applications are cardiac pacemakers, implantable cardiac defibrillators and neuro-simulators.

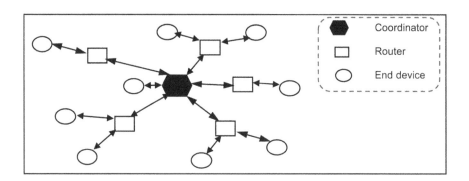

FIGURE 9.13   A Zigbee-based mesh network for wireless connectivity of wearable devices

9.6.3 WMTS (Wireless Medical Telemetry Service)

WMTS uses multiple frequency bands: *608–614* MHz, *1395–1400* MHz, *1429–1432* MHz; has a bandwidth of *6* MHz with a data transfer rate of *six Mbps*. The power transmission of WMTS is relatively large to support the range of *100* meters. WMTS is used by intermediate nodes for longer wireless communication inside a medical facility.

9.6.4 UWB (Ultrawide Bandwidth)

UWB (IEEE Standard 802.15.6) is an ultra-low power, ultra-low range (*1.2* meters), higher data transfer rate (*850* kbps–*20* Mbps) and higher bandwidth (>*500* MHz) protocol to connect various wearable and implanted devices in the WBAN. The major advantage of using UWB is to support continuous wireless data-transfer from multiple sources without the risk of signal interference.

9.6.5 Clouds in Health care

Cloud computing includes both the Internet-based applications, hardware and system resources available remotely by a provider. Cloud services are also refereed as SaaS (Software as a Service), IaaS (Infrastructure as a Service) and PaaS (Platform as a Service). Cloud can be private or public. Public clouds are available upon Pay-as-you-go basis.

Some advantages of the cloud-based services are: 1) freedom from technology updates; 2) freedom from in-house security issues; 3) availability of flexible resources based upon the requirement; 4) no upfront infrastructural cost and 5) freedom from system crashes. Some issues are: 1) security during data-transfer; 2) data-transfer bottlenecks and delays and 3) performance unpredictability due to the use of virtual machines and sharing of the secondary storage devices. Cloud-based monitoring services are especially advantageous for remote homecare as the monitored data can be continuously stored asynchronously in the cloud, and can be analyzed periodically.

9.7 ACCEPTANCE AND ADOPTION

Multiple types of technologies have been implemented for pervasive health care: 1) RFID technology, 2) wearable sensors, 3) implantable devices, 4) ambient and environmental sensors, 5) camera, 6) fall detection sensors such as an accelerometer, 7) wireless communication devices and standards with different bandwidth and data transfer rates, 8) cloud-based technology for automated storage of sensor data, 9) intelligent software for temporal and pattern analysis for identifying unusual activities and 10) miniaturized system-on-chip implementation. The use of RFIDs has improved patients' safety and the utilization of medical assets and staff. The increased pervasive care technology is useful for: paramedics, elderly care, patient's recovery in ICU, home-based patient recovery and utilization of medical staff and assets thus improving the workflows.

The use of these technologies is making health care more cost-effective. However, the key to the acceptance and adoption is the perception of medical benefit, inconvenience of putting on numerous wearable devices, perception of losing privacy, forgetfulness of elderly persons in putting on the devices, and lack of faith of the patients and caregivers to get timely medical help in emergency situations. Another limitation is the inconvenience for the patient and caregivers to learn the use of new technology.

Another major problem is the nonavailability of home-based equipment and skilled operators for: 1) radiological images of abnormalities inside the body and 2) major pathological tests such as antibacterial tests, blood tests and urine related tests. Current technology in wearable sensors is not intelligent enough to draw the sample and analyze.

Another major problem in the adoption is the cost of setting up of the overall infrastructure. Wearable devices, RFID, associated wireless hardware and software are still expensive to set up and maintain for elderly persons and hospitals. In addition, the overhead of the training caregivers and hospital staff to use the new technology is significant. As a consequence, the adoption-rate by the hospitals and the homecare facilities has been low.

9.8 SUMMARY

The growth of medical infrastructure and staff is lagging behind the population-growth of elderly persons. As a result, more elderly patients are staying back in the comfort of home. This trend, augmented with improvement in miniaturization of wearable medical devices and wireless connectivity, has fueled the growth of pervasive health care. Pervasive health care includes wireless monitoring, remote care, home-healthcare using wearable devices that monitor the vital signs and the use of RFID tags to track the patients and elderly persons.

The use of pervasive health care has amplified the medical infrastructure, improved the process flow and optimized the use of resources. The major challenge in the pervasive health care is to miniaturize (with ultra-low power consumption) the devices to be placed aesthetically on the limited surface area on a human body.

The use of pervasive health care is facilitated by the context-aware systems that make use of the environmental condition, ambient condition, scene analysis, patient's condition, past history and demographic data such as age and gender. The context-aware systems continuously read and analyze the environmental and ambient conditions along with the patient's vital signs to understand the current context under which patient's medical data are interpreted.

In addition to the miniaturized wearable devices, multiple ambient sensors and camera are also used to assess the patients' actual activity and condition. One of the common problems of wearable devices is standardization of the networks used to connect wearable devices to each other and to the central data collection and analysis unit. Multiple standards have been developed for this purpose such as Zigbee, MICS, WMTS, and UWB.

The requirement of wireless standards near a body is to have low power consumption, high data-transfer rate and ease of deployment. In addition, they should have a capability to connect to multiple devices. There are two types of network standards: 1) that support short-distance near body and 2) standards that support long-range connectivity. Short-range standards consume less power, while long-range standards such as WMTS support higher bandwidth but send short messages to consume less power.

RFID tags are used for tracking a large number of patients, medical devices and medical staff simultaneously. The use of RFID improves the tracking and localization of many patients, medical devices and medical staff to improve their utilization and optimize the workflow.

There are two types of RFIDs: 1) passive RFIDs that use and scatter the energy of incoming radio waves and 2) active RFIDs with their own power source. RFID tag consumes very little power and has three major components: 1) a flat antenna with small form-factor for communication; 2) digital module with memory and 3) a sensor. RFIDs go through collision problem that interferes with localization and tracking. Localization uses three techniques: 1) triangulation; 2) scene analysis and 3) proximity from the tag reader. Proximity is determined by the strength of the transmitted signal and the attenuation of the received signal. Triangulation uses two techniques: 1) multilateration and 2) angulation.

Home-based monitoring is increasing. In addition to vital signs to detect emergency conditions, activity recognition and fall-detection are quite important. Localization and tracking in home-based care are done using a context-aware system that combines information from wearable devices, environmental sensors, ambient devices and camera-images for the scene analysis. Wearable devices include the use of accelerometers and gyroscopes. The signals are initially trained on a general population, and then fine-tuned on an individual using past-data, and then deployed to separate between regular and unusual activities. Fall-detection uses silhouette-analysis acquired from camera-images, and the use of accelerometers. Multiple machine-learning schemes are used for activity-classification such as decision-trees, SVM, K-means clustering, neural network and Bayesian network.

Patient-tracking can be indoor tracking or outdoor tracking. Indoor tracking is used for identifying unusual activities and fall-detection. Outdoor tracking is used to provide immediate emergency care, especially for elderly people suffering from dementia. To track dementia, a graph-based model is used to mark their usual activity, and the time taken to traverse between the well-identified location. Spatio-temporal analysis and graph-matching are used to derive the deviations from the regular activities.

While the technology is continuously improving, the placement of wearable devices, their aesthetics, the inconvenience of handling and maintaining the devices and the violation of personal privacy are still making the adoption of the pervasive health care slow. As described in Chapter 11, pervasive health care also has to be sensitive to the emotional needs and independence of the patients and caregivers for them to be more extensively adopted.

9.9 ASSESSMENT

9.9.1 Concepts and Definitions

Acceptance and adoption, activity recognition, active RFID, acute care technology, ambient sensors, angulation, automated data collection, BAN, cloud-based storage, collision, collision avoidance, context aware system, elderly care, fall-detection, ECG, EEG, gyroscope, home health care, IEEE standard 15.4.4, IEEE standard 15.4.6, intelligent interfaces, IrDA, localization, magnetometer, MEMS, MICS, mobile health care, multilateration, passive RFID, patient care coordination, PAN, pervasive health care, PPG, real-time data collection, remote care, RFID, RFID reader, RFID tag, RFID-based localization, scene analysis, tele-consultation, telemedicine, tracking, triangulation, ubiquitous health care, UWB, wearable fabric, wearable sensors, wearable health monitoring system, WAN, WBAN, wireless monitoring, Zigbee.

9.9.2 Problem Solving

9.1 A wearable glucometer measures a patient's glucose level three times a day. The glucometer is connected to the doctor's clinic through the secure Internet. The thresholds are: 1) if the glucometer reading goes above *150* mg/dl, then a local alarm is set that warns patient to stop eating/drinking sugary material. If the reading goes above *200* mg/dl, then it stores the time-stamped value in the database, and alerts the patient. If the reading goes above *300* mg/dl, then it immediately informs the nurse station for immediate advice and additional medication. If the reading goes up above *400* mg/dl, it dials for an emergency service. He also gives weekly advice from a nurse about the dosage and maintenance of the sugar level. The patient's reading for *8* days is given as follows. Plot the readings on a graph along with threshold values, and mark the events and actions taken. The readings are taken at *7* AM, *4* PM and *11* PM, respectively.

Day 1: (*90, 120, 180*); day 2: (*95, 155, 215*); day 3: (*100, 170, 220*), day 4: (*80, 100, 140*); day 5: (*72, 120, 135*); day 6: (*83, 140, 180*); day 7: (*130, 220, 240*); day 8: (*200, 180, 270*).

9.2 A person is recovering after cardiac surgery in his two-bedroom smart home. During cardiac surgery, a ventricular pacemaker was implanted in his heart. He has started going for his daily routines and is independent. However, he does not go to the kitchen and does not cook. Provide the various wearable devices, ambient devices, including pressure-sensors, motion-sensors, speech recognition systems and camera to monitor his vital signs and camera without violating his privacy. Provide the network connectivity using proper standards, and the required data transfer rate. The wearable devices should also include devices to detect activities and fall.

9.3 In an intensive care unit, an ECG is transmitted every minute to a central station, along with the oxygen reading. Blood-pressure is being transmitted every *10* minutes. The breathing pattern is measured every *15* seconds. The breathing pattern measures the volume of air taken in and blown out. Assume that the ECG is built by sampling every *10* milliseconds. Each sample is represented by *24* bits, and each ECG monitors heart for *10* seconds. The oxygen level takes one byte to transmit; blood-pressure and pulse-rate take three bytes. The breathing pattern measurement takes two bytes every *15* seconds. Calculate the overall data-transfer rate required. Will Zigbee be able to handle the data-transfer rate? Explain your answer and show the calculations.

9.4 A child regularly breathes *20* times in a minute with a heart rate of *75* beats/minute and oxygen saturation level of *97%–99%*. However, when he gets an asthma attack, his breathing rate increases in the range of *25–30*; his heart rate shoots up in the range of *100–120* and his oxygen saturation level drops to *92%–94%*. The boy is resting on the bed with the wearable devices to measure these three readings. The breathing pattern is measured continuously and sampled every second. It measures the amount of air inhaled and exhaled and time-stamps the measurement. The heartbeat is measured every *30* seconds for *10* seconds, and time-stamped; the oxygen-level is also measured every *15* seconds and time-stamped. Write a program to generate the data using a random number generator for *10* minutes and analyze the data for a possible asthma attack. An asthma attack is considered when the condition remains there for at least 1 continuous minute. Plot the data. Use a linear plot.

9.5 The range for the heartbeats for a patient is *60–120*; the breathing rate is *16–30*; the oxygen saturation level is *90–99*. Three different sensors are reading the data as shown below. The data has to be normalized and fused to form a feature-vector for each reading. Assume that all the readings are taken concurrently. For the following reading set, normalize the readings using the max-min technique and create the feature-vector.

Heart-beats: *65, 72, 69, 70, 74, 71, 68, 73, 66, 68, 74, 80, 83, 88, 90, 95*
Breathing rate: *20, 21, 20, 20, 21, 21, 20, 21, 19, 20, 22, 23, 24, 24, 25, 25*
Oxygen saturation: *97, 97, 98, 97, 96, 96, 97, 95, 96, 97, 96, 95, 95, 95, 96, 95*

9.6 In a semi-private room inside a hospital, there are two readers. Assume that the reading circle of the reader is *10* feet, and the interference circle is *20* feet. Each patient has four RFID that identify the patients, visitors to the patient, nurse and the food-tray. The RFID readers have to be placed such that RFID tags of patients, and his food tray do not interfere. Assume that the nurse and visitors will be at most *10* feet away from their respective patients. Design the room layout such as the room-size is minimal without any RFID collision. Calculate the dimensions of the room, and the distance between the RFID readers.

9.7 A patient with an active tag is located in a room. There are three receivers (*R1, R2* and *R3*) in three corners of a room as shown in Figure 9.14. The room measures *12 × 14* feet. Each receiver emits a short burst of a signal that is received by the moving RFID tag on the patient's chair, and transmitted back. Assume that the receivers are synchronized and send the signal one after another in quick succession. The ratios of the time taken for *R1, R2* and *R3* are *0.4, 0.5* and *0.8*. Calculate the possible location of the patient in the room.

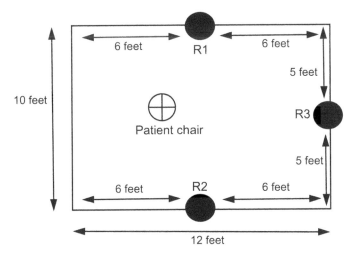

FIGURE 9.14   Localization using triangulation for Problem 9.7.

9.8 In a hospital corridor of the size *10 × 100* feet, rows of RFID readers are placed as shown in Figure 9.8. Each RFID tag on the floor is *10* feet away. The ceiling is *10* feet high, and a reader (receiver) can receive from *40* feet away. Give a layout of the readers so that they can cover the whole corridor with minimal interference and maximum coverage. Identify the locations of the readers. Calculate the energy attenuation received by reader-1 and reader-2 by the target-RFID as the patient moves from the reader-1 end to the reader-2 end, assuming that the maximum energy received by the reader-1 is *1.0* when the RFID tag is just below the reader 1. Assume that the attenuated energy is given according to the formula $G_T^{reader} \times G_R^{tag} \times \rho \times G_T^{tag} \times \frac{\lambda^4}{(4\pi d_i)^4}$ as given in Section 9.3.4.

9.9 Discuss the features that can be extracted from the silhouette of a fallen person (see Figure 9.15). Use these features to derive the feature-vectors of the following images. Show how your feature-vector can separate a walking person from a fallen person. Your feature-vector should also separate persons lying on the floor in the sleeping position and fallen position. Images are not to the scale.

9.10 In Figure 9.8, assume that the passageway is divided into 2×5 cells: RT_1 is in cell (*1, 1*); RT_2 is in cell (*2, 1*), and so on. The RSS-value from RT_1 to RT_{10} for Rec_1 is given as: (RT_1: 0.5; RT_2: 0.5; RT_3: 0.5; RT_4: 0.5; RT_5: 0.7; RT_6: 0.7; RT_7: 0.9; RT_8: 0.9; RT_9: 1.0; RT_{10}: 1.0). The RSS value for the target RFID is *0.4*. Use K-means clustering to identify the possible location of the patient's chair to the nearest cell.

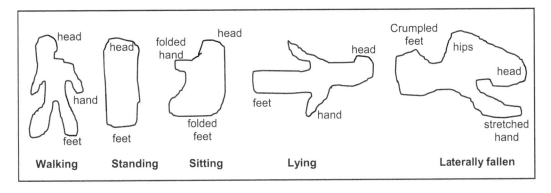

FIGURE 9.15   Silhouette shapes for fallen person analysis in Problem 9.9

9.11 Given the following motion of a person suffering from dementia, and the given map with her regular visiting location marked, remove the noise and the small cycles, and draw the graph. Assume that the GPS readings are taken every minute. The visiting locations are *100–500* feet apart, and the person walks *50–100* feet every minute on a regular day. The location is given with respect to the map-cells that are marked as *20* feet × *20* feet for each cell. The map has *50* feet × *50* feet cells. Table 9.1 gives the readings for *24* minutes of a regular walk when she is sober. Calculate the graph and mark the edge-weights as triples of the form (Euclidean distance between cell centroids, time taken, wait time spent at the location). Identify the small cycles, and subtract the cycle time from the time taken to traverse between two locations.

Initial location: (*24, 20*).
Daily visited locations (not in visiting order): (*10, 15*); (*18, 17*); (*25, 21*); (*30, 19*);

9.12 The patient in Problem 9.11 wanders off on another day due to her dementia. The new measurement table being developed in real time is given in Table 9.2. Describe the unusual patterns. An unusual pattern could be when she breaks the sequence of travel or wanders off into a new direction not in her regular graph, or goes repeatedly in a local area of *60* feet for more than 5 minutes. Describe at what time the assistive services should be informed. Assuming that assistive service takes around ten minutes to arrive there, calculate the location where the patient would be found.

9.13 Give the data structure for ISAX-based indexing and TPR tree for the data in Tables 9.1 and 9.2.

9.14 Write a program to analyze a database table as given in Table 9.1, and develop a graph with the information needed in Problem 9.11.

9.15 Write a program that will take the table in Problem 9.12 and a graph constructed from Problem 9.14 as input, identify the cycles and other unusual activities not following the graph, and print out the unusual activities.

9.16 Build an HMM graph for a person walking with Parkinson's disease. Assume that there are the following states: 1) left feet on ground; 2) left feet lifted up; 3) left feet forward in the air; 4) right feet on the ground; 5) right feet lifted up; 6) right feet forward in the air; 7) head near the hip; 8) a person lying down and 9) standing. Assume that the person falls *10%* of the time, and he falls laterally to the left when he lifts the left feet. The remaining time, he walks regularly. The initial state is "standing." While standing, there is a *40%* chance that he will keep standing, a *20%* chance that he would sit down, a *10%* probability that he would lift the left foot for walking, and *30%* probability he would lift his right feet. After he sits down, there is a *90%* probability that he would keep sitting down, and a *10%* probability that he would stand up.

TABLE 9.1   Walking pattern of a dementia patient for Problem 9.11

TIME	CELL	TIME	CELL	TIME	CELL	TIME	CELL	TIME	CELL	TIME	CELL
1	**25,21**	5	30,18	9	27,16	13	14,16	17	10,15	21	25,16
2	28,21	6	30,19	10	22,16	14	15,16	18	14,16	22	24,16
3	30,20	7	29,19	11	18,16	15	14,16	19	19,16	23	24,18
4	**30,19**	8	30,19	12	**18,17**	16	**10,15**	20	24,16	24	**24,20**

TABLE 9.2   Dementia patient's travel for Problem 9.12

TIME	CELL	TIME	CELL	TIME	CELL	TIME	CELL	TIME	CELL	TIME	CELL
1	**25,21**	6	30,17	11	20,14	16	14,8	21	16,11	21	15,12
2	28,21	7	30,19	12	18,12	17	15,9	22	16,10	22	15,11
3	30,20	8	29,19	13	18,10	18	16,9	23	15,11	23	15,10
4	**30,19**	9	30,19	14	16,10	19	16,10	24	15,11	24	16,10
5	30,18	10	29,19	15	15,9	20	16,10	25	15,11	25	16,10

9.17 Extend your program in Problem 9.14 to analyze multiple tables under the assumption that the person has mild dementia and can take different routes and visit the places in a different order and build an HMM with transition probability matrix.

9.18 Write a program for developing shape-based histogram for various silhouettes, as explained in Chapter 2.

9.19 Compress the time-series data in Tables 9.1 and 9.2 using techniques described in Section 2.1.7.1 so that the maximum relevant information such as wandering and reaching a desired location is retained. Show the graph using mean data (see Figure 2.17) as well as the linear curve fitting (Figure 2.18). To retain the data, you must retain the actual temporal reading of reaching the desired location.

9.20 Write a program that calls the graph generating program developed in Problem 9.14 and the takes Table 9.1 as input and generates the information generated in Problem 9.19.

9.21 Assume that a camera in a living room takes two images per second while the person is walking. The original image is of the size 64×64 pixels. The image is compressed to 10% of the original, and transmitted over the Internet. However, as soon as the accelerometer data-analysis predicts a potential fall, a 512×512 pixel image is sent with only 50% compression. The regular bandwidth is 8 Mb/second, and the overhead of packaging in each frame is 576 bits. Compute the time taken to transfer each image before and after the suspected fall.

9.9.3 Extended Response

9.22 Give at least five reasons with rationales for the use of pervasive health care.

9.23 Read the original IEEE standards for WBAN and write an extended summary in two pages.

9.24 Explain the difference between WPAN and WBAN.

9.25 Explain the various components of a context-aware system for remote care.

9.26 Describe various wearable sensors and the underlying scientific principle such as the use of MEMS and PPG.

9.27 For the following situations, you have to design a framework that includes the minimum number of wearable sensors and the corresponding network. Draw a schematic of the network connectivity in each case:

1. Pervasive care for a pregnant executive whose assignment includes weekly travel.
2. Pervasive care of an adult going on a mountain excursion.
3. Pervasive care of a youth going to a summer camp that involves trekking in a forest and water reservoirs.
4. Pervasive care of an elderly person with an episode of myocardial infarction going for a brisk walk.
5. Pervasive care for a person with Parkinson's disease inside the dwelling area.
6. Pervasive care for a toddler just learning to walk.
7. Pervasive care for an elderly lady who is suffering from a vision-problem.
8. Pervasive care for a post-surgical heart-patient who just had a triple bypass surgery.
9. Pervasive care for an elderly diabetic patient who suffers from dementia, and is dependent upon insulin to regulate his sugar level.
10. Pervasive care for a person suffering from ventricular fibrillation going on a long tour involving air travel.

9.28 Discuss the tracking mechanism in a hospital setting using RFID and ambient sensors.

9.29 Read three recent research articles, and discuss the characteristics of a user-friendly interface for a context-aware pervasive care system.

9.30 Explain RFID collision and the resolution using a simple example.

9.31 Explain the graph-based system for identifying wandering of dementia patients using a simple example.

9.32 Explain the reason for slow acceptance and adoption of the pervasive healthcare system.

9.33 Explain the privacy and confidential concerns in the pervasive healthcare system, and how they are resolved.

9.34 Compare the Zigbee standard with the MICS system in terms of power requirement, frequency bandwidth, and range. Explain the rationale for the choices made.

9.35 Describe the various components of an active RFID tag.

9.36 Explain the various components of a Zigbee-based device.

9.37 Describe different types of falls, and how will they be projected in the motions of X, Y and Z-axes. Explain the need for tri-axial accelerometer in modeling these motions and fall.

9.38 Read research articles on five different types of wearable sensors and summarize their implementation, size, power requirement and data-transfer rate in one page.

9.39 Explain the difference between the pervasive care requirements in an intensive care unit and an elderly home-care unit.

9.40 Explain the need for telemedicine and a teleconsultation system in a modern healthcare system.

9.41 Explain the data-transfer requirement and the rationale for a teleconsultation system.

FURTHER READING

Pervasive Healthcare Taxonomy

9.1 Varshney, Upkar, "Pervasive Healthcare: Applications, Challenges and Wireless Solutions," *Communications of the Association for Information Systems*, 16(3), July 2005, 55–72.

9.2 Orwat, Carsten, Graefe, Andreas, and Faulwasser, Timm, "Towards Pervasive Computing in Health Care – A Literature Review," *BMC Medical Informatics and Decision Making*, 8(26), DOI: 10.1186/1472-6947-8-2., Available at https://www.ncbi.nlm.nih.gov/pmc/articles/PMC2467411/pdf/1472-6947-8-26.pdf.

Context-Aware Systems

9.3 Paganelli, Federica, Spinicci, Emilio, and Giuli, Dino, "ERMHAN: A Context-aware Service Platform to Support Continuous Care Networks for Home-Based Assistance," *International Journal of Telemedicine Applications*, Article Id 867639, 2008, DOI: 10.1155/2008/867639.

9.4 Wood, Anthony D., Stankovic, John A., Virone, Gilles, Selāvo, Leo, He, Zhimin, Cao, Qiuhua, Doan, Thao, Wu, Yafeng, Fang, Lei, and Stoleru, Radu, "Context-aware Wireless Sensor Networks for Assisted Living and Residential Monitoring," *IEEE Network*, 22(4), July 2008, 26–33.

9.5 Zhang, Daqing, Yu, Zhiwen, and Chin, Chung-Yau, "Context-Aware Infrastructure for Personalized Healthcare," In *Personalized Healthcare Systems* (editors: Nugent, Chris D., McCullagh, Paul J., and McAdams, Eric T.,), *Studies in Health Technology and Informatics Series*, Vol. 117, IOS Press, Amsterdam, Netherlands, 2005, 154–163.

Remote Monitoring

9.6 Ren, Hongliang, and Meng, Max Q.H., "Bioeffects Control in Wireless Biomedical Sensor Networks," *Proceedings of the 3rd Annual Conference of IEEE Communications Society on Sensor and Ad Hoc Communications and Networks (SECON'06)*, vol. 3, Reston, VA, USA, September 2006, 896–904.

9.7 Triantafyllidis, Andreas K., Koutkias, Vassilis G., Chouvarda, Ioanna, Adami, Ilia, Kouroubali, Angelina, and Maglaveras, Nicos, "Framework of Sensor-Based Monitoring for Pervasive Patient Care," *Healthcare Technology Letters*, 3(3), August 2016, 153–158, DOI: 0.1049/htl.2016.001.

Telemedicine and Teleconsultation

9.8 Chakroborty, Chinmay, Gupta, Bharat, and Gosh, Soumya K., "A Review on Telemedicine-Based WBAN Framework for Patient Monitoring," *Telemedicine and E-Health*, 19(8), August 2013, 619–626.

9.9 Pattichis, Kyriacou C. E., Voskarides, S., Pattichis, M., Istepanian, R., and Schizas, C., "Wireless Telemedicine Systems: An Overview," *IEEE Antennas and Propagation Magazine*, 44(2), April 2002, 143–153.

9.10 Perednia, Douglas A., "Telemedicine Technology and Clinical Applications," *Journal of Medical Association (JAMA)*, 273(6), February 1995, 483–488.

9.11 Raza, Tasleem, Joshi, Manish, Schapira, Ralph M., and Agha, Zia, "Pulmonary Telemedicine – A Model to Access the Subspecialist in Underserved Rural Areas," *International Journal of Medical Informatics*, 78(1), January 2009, 53–59.

Home Health Care

9.12 Aziz, Omar, Atallah, Louis, Lo, Benny, ElHelw, Mohamed, Wang, Lei, Yang, Guang-Zhong, and Darzi, Ara, "A Pervasive Body Sensor Network for Measuring Postoperative Recovery at Home," *Surgical Innovation*, 14(2), July 2007, 83–90.

9.13 Fensli, Rune, Gunnarson, Einar, and Gundersen, Torstein, "A Wearable ECG-recording System for Continuous Arrhythmia Monitoring in a Wireless Tele-Homecare Situation," In *Proceedings of the 18th IEEE Symposium of the Computer-Based Medical Systems*, Dublin, Ireland, June 2005, 407–412.

9.14 Lee, Hak J., Lee Sun H., Ha Kyoo-Seob, Jang Hak C., Chung, Woo-Young, Kim, Ju Y., Chang, Yoon-Seok, and Yoo Dong H., "Ubiquitous Healthcare Service Using Zigbee and Mobile Phone for Elderly Patients," *International Journal of Medical Informatics*, 78, 2009, 193–198.

9.15 Lee, Ren-Guey., Chen, Kuei-Chien, Hsiao, Chun-Chieh, and Tseng, Chwan-Lu, "A Mobile Care System with Alert Mechanism," *IEEE Transactions on Information Technology in Biomedicine*, 11(5), November 2007, 507–517.

9.16 Scanaill, Clidohna N., Carew, Sheila, Barralon, Pierre, Noury, Norbert, Lyons, Declan, and Lyons, Gerard M., "A Review of Approaches to Mobile Telemonitoring of the Elderly in Their Living Environment," *Annals of Biomedical Engineering*, 34(4), April 2006, 547–563.

Acute Care Technologies

9.17 Frost, Megan C., and Meyerhoff, Mark E., "Real-time Monitoring of Critical Care Analytes in the Bloodstream with Chemical Sensors: Progress and Challenges," *Annual Review of Analytical Chemistry*, 8, July 2015, 171–192, DOI: 10.1146/annurev-anchem-071114-040443.

9.18 Torjman, Marc C., Dala, Niti, and Goldeberg, Michael E., "Glucose Monitoring in Acute Care: Technologies on the Horizon," *Journal of Diabetes Science and Technology*, 2(2), March 2008, 178–181.

Wearable Health-Monitoring Systems

9.19 Kang, Dong-Oh, Lee, Hyung-Jik, Ko Eun-Jung, Kang, Kyuchaing, and Lee, Jeunwoo, "A Wearable Context Aware System for Ubiquitous Healthcare," In *Proceedings of the 28th Annual International Conference of the IEEE Engineering in Medicine and Biology*, New York, NY USA, August 2006, 5192–5195.

9.20 Nugent, Chris D., McCullah, Paul J., McAdams, Eric T., and Lymberis, Andreas, *Personalized Health Management Systems: The Integration of Innovative Sensing, Textile, Information and Communication Technologies*, IOS Press, Amsterdam, Netherlands, 2005.

9.21 Pantelopoulos, Alexandros and Bourbakis, Nikolaos G., "A Survey of Wearable Sensor Based Systems for Health Monitoring and Prognosis," *IEEE Transactions on Systems, Man and Cybernetics – Part C: Applications and Reviews*, 40(1), January 2010, 1–12.

Wearable Sensors Placement

9.22 Atallah, Louis, Lo, Benny, King, Rachel, and Yang, Guang-Zhong, "Sensor Placement for Activity Detection Using Wearable Accelerometers," *International Conference on Body Sensor Networks*, 2010, 24–29.

9.23 Majumder, Sumit, Mondal, Tapas, and Deen, M. Jamal, "Wearable Sensors for Remote Health Monitoring," *Sensors*, 17(1), January 2017, DOI: 10.3390/s17010130.

Wearable Sensors Design

9.24 Patel, Shyamal, Park, Hyung, Bonato, Paolo, Chan, Leighton, and Ridgers, Mary, "A Review of Wearable Sensors and Systems with Application in Rehabilitation," *Journal of Neuroengineering and Rehabilitation*, 9(21), April 2012, DOI: 10.1186/1743-0003-9-21.

9.25 Asada, Haruhiko H., Shaltis, Philip, Reisner, Andrew, Rhee Sokwoo, and Hutchinson, Reginald C., "Mobile Monitoring with Wearable Photoplethysmographic Biosensors," *IEEE Engineering in Medicine and Biology Magazine*, 22(3), May–June 2003, 28–40.

9.26 Bates, Andrew, Ling, Martin J., Mann, Janek, and Arvind, Damal K., "Respiratory Rate and Flow Waveform Estimation from Tri-axial Accelerometer Data," In *Proceedings of the International Conference of Body Sensor Networks*, June 2010, 144–150.

9.27 Barbaro, Massimo, Caboni, Alessandra, Cosseddu, Piero, Mattana, Giorgio, and Bonfiglio, Annalisa, "Active Devices Based on Organic Semiconductors for Wearable Applications," *IEEE Transactions on Information Technology in Biomedicine*, 14(3), May 2010, 758–766, DOI: 10.1109/TITB.2010.2044798.

9.28 Corbishley, Phil, and Rodriguez-Villegas, Esther, "Breathing Detection: Towards a Miniaturized, Wearable, Battery-Operated Monitoring System," *IEEE Transactions on Biomedical Engineering*, 55(1), January 2008, 196–204, DOI: 10.1109/TBME.2007.910679.

9.29 Dudde, Ralf, Thomas, Vering, Piechotta, Gundula, and Hintsche, Rainer, "Computer-Aided Continuous Drug Infusion: Setup and Test of a Mobile Closed-Loop System for the Continuous Automated Infusion of Insulin," *IEEE Transactions on Information Technology in Biomedicine*, 10(2), May 2006, 395–402, DOI: 10.1109/TITB.2006.864477.

9.30 Lam Po Tang, Sharon, "Recent Developments in Flexible Wearable Electronics for Monitoring Applications," *Transaction of the Institute of the Measurement and Control*, 29 (3/4), 2007, 283–300, DOI: 10.1177/0142331207070389.

9.31 Milici, Stefano, Lázaro, Antonio, Villarino, Ramon, Girbau, David, and Magnarosa, Marco, "Wireless Wearable Magnetometer-based Sensor for Sleep Quality Monitoring," *IEEE Sensors Journal*, 18(5), March 2018, 2145–2152.

9.32 Morris, Deirdre, Schazmann, Benjamin, Wu, Yangzhe, Coyle, Shirley, Brady, Sarah, Fay, Cormac, Hayes, Jer, Lau, King T., Wallace, Gordon, and Diamond, Dermot, "Wearable Technology for Bio-Chemical Analysis of Body Fluids During Exercise," In *Proceedings of the 30th Annual International Conference of the IEEE Engineering in Medicine and Biology Society*, Vancouver, BC, Canada, August 2008, 5741–5744, DOI: 10.1109/IEMBS.2008.4650518.

9.33 Patterson, James A. C., McIlwraith Douglas G., and Guang-Zhong Yang, "A Flexible, Low Noise Reflective PPG Sensor Platform for Ear-Worn Heart Rate Monitoring," In *Proceedings of the Sixth International Workshop on Wearable and Implantable Body Sensor Networks*, June 2009, Berkley, CA, USA, 286–291, DOI: 10.1109/BSN.2009.16.

9.34 Shaltis, Philip A., Reisner, Andrew, and Asada, Harry H., "Wearable, Cuff-less PPG-based Blood Pressure Monitor with Novel Height Sensor," In *Proceedings of the IEEE Engineering in Medicine and Biology Society*, Vol. 1, August/September 2006, New York, NY, USA, 908–911.

9.35 Yong-Sheng Yan, Yuan-Ting Zhang, "An Efficient Motion-Resistant Method for Wearable Pulse Oximeter," *IEEE Transactions on Information Technology in Biomedicine*, 12(3), May 2008, 399–405.

9.36 Han, Seungyong, Kim, Min Ku, Wang, Bo, Wie, Dae Seung, Wang, Shuodao, and Lee, Chi Hwan, "Mechanically Reinforced Skin-Electronics with Networked Nanocomposite Elastomer," *Advanced Materials*, 28(46), December 2016, 10257–10265, DOI: 10.1002/adma.201603878.

Wireless Connectivity and Data Collection

9.37 Catrysse, Michael, Puers, Robert, Hertleer, Carla, and Lagenhove, Lieva V., "Towards the Integration of Textile Sensors in a Wireless Monitor Suit," *Sensors and Actuators A Physical*, 114(2), September 2004, 302–311, DOI: 10.1016/j.sna.2003.10.071.

9.38 Chung, Wan-Young, Lee, Young-Dong, and Jung, Sang-Joong, "A Wireless Sensor Network Compatible Wearable U-Healthcare Monitoring System Using Integrated ECG, Accelerometer and SpO2," In *Proceedings of the 30th Annual International Conference of the IEEE Engineering in Medicine and Biology*, Vancouver, BC, Canada, August 2008, 1529–1532.

Issues in Wearable Devices

9.39 Bergmann, Jeroen H. M., Chandaria, Vikesh, and McGregor, Alison, "Wearable and Implantable Sensors: The Patient's Perspective," *Sensors*, 12(12), December 2012, 16695–16709, DOI: 10.3390/s121216695.

9.40 Gaff, Brian M., "Legal Issues with Wearable Technology," *IEEE Computer*, 48, September 2015, 10–12, DOI: 10.1109/MC.2015.280.

9.41 Saleem, Kasif, Shahzad, Basit, Orgun, Mehmet A., Al-Muhtadi, Jalal, Rodrigues, Joel J. P. C., and Zakariah, Mohammed, "Design and Deployment Challenges in Immersive and Wearable Technologies," *Behaviour and Information Technology*, 36(7), 2017, 687–698, DOI: 10.1080/0144929X.2016.1275808.

Sensor Data Fusion

9.42 Lee, Hyun, Park, Kyungseo, Lee, Byongyong, Choi, Jaesung, and Elmasri, Ramez, "Issues in Data Fusion for Healthcare Monitoring," In *Proceedings of the 1st international conference on Pervasive Technologies Related to Assistive Environments (Petra'08),* Athens, Greece, July 2008, Article No. 3.

9.43 Medjahed, Hamid, and Boudy, Jerome, "A Pervasive Multi-sensor Data Fusion for Smart Home Healthcare Monitoring," In *Proceedings of the IEEE International Conference on Fuzzy Systems,* Taipei, Taiwan, June 2011, 1466–1473.

Privacy and Confidentiality

9.44 Al Ameen, Moshaddique, Liu, Jingwei, and Kwak, Kyungsup, "Security and Privacy Issues in Wireless Sensor Networks for Healthcare Applications," *Journal of Medical Systems,* 36(1), February 2012, 93–101, DOI: 10.1007/s10916-010-9449-4.

9.45 Baranchuk, Adrian, Refaat, Marwan M., Patton, Kristen K., Chung, Mina K., Krishnan, Kousik, Kutyifa, Valentina et al., "Cybersecurity for Cardiac Implantable Electronic Devices: What Should you Know," *Journal of American College of Cardiology,* 71(11), March 2018, 1284–1288, DOI: 10.1016/j.jacc.2018.01.023.

9.46 Camara, Carmen, Peris-Lopez, Pedro, and Tapiador, Juan E., "Security and Privacy Issues in Implantable Medical Devices: A Comprehensive Survey," *Journal of Biomedical Informatics,* 55, June 2015, 272–289, DOI: 10.1016/j.jbi.2015.04.007.

9.47 Meingast, M., Roosta, T., and Sastry, S., "Security and Privacy Issues with Health Care Information Technology," In *Proceedings of the 28th Annual International Conference of the IEEE Engineering in Medicine and Biology Society (EMBS '06),* New York, NY, USA, August–September 2006, 5453–5458.

Patient Care Coordination

Radio Frequency Identification (RFID)

RFID Collision

9.48 Gandino, Fillipo, Ferrero, Renato, Montrucchio, Bartolomeo, and Rebaudengo, Maurizio, "Probabilistic DCS: An RFID Reader-to-Reader Anti-Collision Protocol," *Journal of Network and Computer Applications,* 34(3), May 2011, 821–832.

9.49 Occhiuzzi, Cecilia, and Marraco, Gaetano, "Human Body Sensing: A Pervasive Approach by Implanted RFID Tags," In *Proceedings of the 3rd International Symposium on Applied Sciences in Biomedical and Communication Technologies (ISABEL),* Rome, Italy, November 2010, DOI: 10.1109/ISABEL.2010.5702812.

RFID-Based Localization and Tracking

9.50 Bai, Yijian, Wang, Fusheng, and Liu, Peiya, "Efficiently Filtering RFID Data Streams," In *Proceedings of the First International Workshop on Clean databases,* September 2006, Seoul, Korea.

9.51 Ku, Wei-Shinn, Sakai, Kazuya, and Sun, Min-Te, "The Optimal K-Covering Tag Deployment for RFID-Based Localization," *Journal of Network and Computer Applications,* 34(3), May 2011, 914–924.

9.52 Ma, Haishu, Wang, Yi, and Wang, Kesheng, "Automatic Detection of False Positive RFID Readings Using Machine Learning Algorithms," *Expert Systems with Applications,* 91, January 2018, 442–451, DOI: 10.1016/j.eswa.2017.09.021.

9.53 Najera, Pablo, Lopez, Javier, and Roman, Rodrigo, "Real-Time Location and Inpatient Care System Based on Passive RFID," *Journal of Network and Computer Applications,* 34(3), May 2011, 980–989.

9.54 Ni, Lonel M., Zhang, Dian, and Souryal, Michael R., "RFID-Based Localization and Tracking Technologies," *IEEE Wireless Communications,* April 2011, 45–51.

9.55 Papapostolou, Apostila, and Chaouchi, Hakima, "RFID-Assisted Indoor Localization and Impact of Interference on its Performance," *Journal of Network and Computer Applications,* 34(3), May 2011, 902–913.

9.56 Golding, Andrew, and Lesh, Neal, "Indoor Navigation Using a Diverse Set of Cheap, Wearable Sensors," In *Proceedings of the 3rd IEEE International Symposium of the Wearable Computers,* Washington, DC, USA, October 1999, 29–36.

9.57 Wang, Fusheng, and Liu, Peiya, "Temporal Management of RFID Data," In *Proceedings of the 31st International Conference on Very Large Databases,* Trondheim, Norway, August/September 2005, 1128–1139.

Home-Based Monitoring

Activity Recognition

9.58 Barman, Joydip, Usweatte, Gitendra, Sarkar, Nilanjan, Ghaffari, Touraj, Sokal, Brad, "Sensor-Enabled RFID System for Monitoring Arm Acitvity in Daily Life," In *Proceedings of the 33rd Annual International of IEEE EMBS,* Boston, MA, USA, August–September 2011, 5219–5223.

9.59 Pärkkä, Juha, Ermes, Miikka, Korpipää, Panu, Mäntyjärvi, Jani, Peltola, Johannes, and Korhonen, Ilkka, "Activity Classification Using Realistic Data from Wearable Sensors," *IEEE Transactions of Information Technology in Biomedicine,* 10(1), January 2006, 119–128.

Fall Detection

9.60 lbert, Mark V., Kording, Konrad, Herrmann, Megan, and Jayraman, Arun, "Fall Classification by Machine Learning Using Mobile Phones," *PLoS One,* 7(5), May 2012, e6556, DOI: 10.1371/journal.pone.0036556.

9.61 Igual, Raul, Medarno, Carlos, and Plaza, Inmaculada, "Challenges, Issues and Trends in Fall Detection Systems," *Biomedical Engineering Online,* 66, 12, DOI: 10.1186/1475-925X-12-66.

9.62 Lee, RYW, and Carlisle, AJ, "Detection of Falls Using Accelerometers and Mobile Phone Technology," *Age and Aging,* 40, 2011, 690–696.

Localization and Tracking

9.63 Lin, Qiang, Zhang, Daqing, Connelly, Kay, Ni, Hongbo, Yu, Zhiwen, and Zhou, Xingshe, "Disorientation Detection by Mining GPS Trajectories for Cognitively-Impaired Elders," *Pervasive and Mobile Computing,* 19, 2015, 71–85.

9.64 Teipel, Stefan, Babiloni, Claudio, Hoey, Jesse, Kaye, Jeffrey, Kirste, Thomas, and Burmeister, Oliver K., "Information and Communication Technology Solutions for Outdoor Navigation in Dementia," *Alzheimer's and Dementia,* 12(6), June 2016, 695–707, DOI: 10.1016/j.jalz.2015.11.003.

Healthcare Network and Interfaces

9.65 Alemdar, Hande, and Ersoy, Cem, "Wireless Sensor Networks for Healthcare: A Survey," *Computer Networks,* 54(15), October 2010, 2688–2710, DOI: 10.1016/j.comnet.2010.05.003.

9.66 Stankovic, John A., Cao, Quihia, Doan, Tam T., Fang, Lei, He, Zhijun, Lin, Shan, Son, Sang H., Stoleru, Radu, Wood, Anthony D. et al., "Wireless Sensor Networks for In-home Healthcare: Potential and Challenges," In *Proceedings of the High Confidence Medical Device Software and Systems (HCMDSS) Workshop,* Philadelphia, PA, USA, June 2005, Available at https://rtg.cis.upenn.edu/hcmdss/papers.php3, Accessed December 22, 2018.

WBAN (Wireless Body Area Network)

9.67 Gravina, Raffaele, Alinia, Parastoo, Ghasemzadeh, Hassan, and Fortino, Giancarlo, "Multi-sensor Fusion in Body Sensor Networks: State-of-the-Art and Research Challenges," *Information Fusion,* 35, May 2017, 68–80, DOI: 10.1016/j.inffus.2016.09.005.

9.68 Hao, Yang, and Foster, Robert, "Wireless Body Sensor Networks for Health Monitoring Applications," *Physiological Measurement,* 29(11), November 2008, R27–R56.

9.69 Farella, Elisabbetta, Pieracci, Augusto, Benini, Luca, Rocchi, Laura, and Acquaviva, Andrea, "Interfacing Human and Computer with Wireless Body Area Sensor Networks: The WiMoCa Solution," *Multimedia Tools and Applications,* 38(3), July 2008, 337–363.

9.70 Hayanjeh, Thaier, Almashaqbeh, Ghada, Ullah, Sanah, and Vasilokas, Athansios V., "A Survey of Technologies Coexistence in WBAN: Analysis and Open Research Issues," *Journal of Wireless Networks,* 20(8), November 2014, 2165–2199.

9.71 Ullah, Sana, Higgins, Henry, Bream, Bart, Latre, Benoit, Blondia, Chris, Moerman, Ingrid, et al., "A Comprehensive Survey of Wireless Body Area Networks: On PHY, MAC, and Network Layers Solutions," *Journal of Medical Systems,* 36(3), August 2012, 1065–1094, DOI: 10.1007/s10916-010-9571-3.

WPAN (Wireless Personal Area Network)

9.72 Akyildiz, Ian F., Su, Wei, Sankarasubramaniam, Yogesh, and Çayirci, Erdal, "Wireless Sensor Networks: A Survey," *Computer Networks,* 38(4), March 2002, 393–422, DOI: 10.1016/S1389-1286(01)00302-4.

9.73 Anastasi, Giuseppe, Conti, Marco, Francesco, Mario D., and Passarella, Andrea, "Energy Conservation in Wireless Sensor Networks: A Survey," *Ad Hoc Networks*, 7(3), May 2009, 537–568, DOI: 10.1016/j. adhoc.2008.06.003.
9.74 Fariborzi, Hossein, and Moghavvemi, Mahmoud, "Architecture of a Wireless Sensor Network for Vital Signs Transmission in Hospital Setting," In *Proceedings of the International Conference on Convergence Information Technology (ICCIT 2007)*, Gyeongju, Korea, November 2007, 745–749.

Standards for Data Communication

9.75 *IEEE Standard 802.15.6-2017 – IEEE Standard for Local and Metropolitan Area Networks – Part 15.6: Wireless Body Area Networks*, February 2012, Available at https://standards.ieee.org/standard/8802-15-6-2017. html, Accessed December 22, 2018.
9.76 Kwak, Kyung S., Ullah, Sana, and Ullah, Niamat, "An Overview of IEEE 802.15.6 Standard," In *Proceedings of the 3rd IEEE International Symposium on Applied Sciences in Biomedical and Communication Technologies (ISABEL)*, November 2010, DOI: 10.1109/ISABEL.2010.5702867.

Zigbee

9.77 Tran, Thang V., and Chung, Wan-Young, "IEEE-802.15.4-based Low-Power Body Sensor Node with RF Energy Harvester," *Biomedical Materials and Engineering*, 24(6), November/December 2014, 3503–3510, DOI: 10.3233/BME-141176.
9.78 Varchola, Michal, and Drutarovský, Milos, "ZIGBEE Based Home Automation Wireless Sensor Network," *Acta Electrotechnica et Informatica*, 7(4), 2007, 1–8, Available at http://www.aei. tuke.sk/papers/2007/4/11_ Varchola.pdf, Accessed December 23, 2018.

MICS (Medical Implant Communication Service)

9.79 Islam, Mohd N., and Yuce, Mehmet R., "Review of Medical Implant Communication System (MICS) Band and Network," *ICT Express*, September 2016, 188–194, DOI: 10.1016/j.icte.2016.08.010.
9.80 Yuce, Mehmet R., "Implementation of Body Area Networks for Healthcare Systems," *Sensors and Actuators*, 162(1), July 2010, 116–129, DOI: 10.1016/j.sna.2010.06.004.
9.81 Savci, Huseyin S., Sula, Ahmet, Wang, Zheng, Dogan, Numan S., and Arvas, Ercument, "MICS Transceivers: Regulatory Standards and Applications," In *Proceedings of the IEEE SoutheastCon*, 2005, Chicago, IL, USA, 179–182.

UWB (Ultra Bandwidth)

9.82 Zhang, Jinyun, Orlik, Philip V., Sahinoglu, Zafer, Molisch, Andreas F., and Kinney, Patrick, "UWB Systems for Wireless Sensor Networks," *Proceedings of the IEEE*, 97(2), March 2009, 313–331, DOI: 10.1109/ JPROC.2008.2008786.

Clouds in Health Care

9.83 Yeh, Lo-Yao, Chiang, Pei-Yu, Tsai, Yi-Lang, and Huang, Jiun-Long, "Cloud-Based Fine-Grained Health Information Access Control Framework for Lightweight IoT Devices and Dynamic Auditing and Attribute Revocation," *IEEE Transaction of Cloud Computing*, 6(2), April–June 2018, 532–543.

Acceptance and Adoption

9.84 Katz, James E., and Rice, Ronald E., "Public Views of Mobile Medical Devices and Services: A US National Survey of Consumer Sentiments towards RFID Healthcare Technology," *International Journal of Medical Informatics*, 78(2), February 2009, 104–114, DOI: 10.1016/j.ijmedinf.2008.06.001.
9.85 Nasir, Suphan, and Yurder, Yigit, "Consumers' and Physicians' Perceptions about High Tech Wearable Health Products," *Procedia – Social and Behavioral Sciences*, 195, July 2015, 1261–1267, DOI: 10.1016/j. sbspro.2015.06.279.
9.86 Yazici, Hulya J., "An Exploratory Analysis of Hospital Perspectives on Real-Time Information Requirements and Perceived Benefits of RFID Technology for Future Adoption," *International Journal of Information Management*, 34(5), October 2014, 603–621.

Disease Prediction and Drug Development

10

Section 1.3.2.2 Pharmacoinformatics; Section 1.6.13 Bioinformatics for Disease and Drug Discovery; Section 1.6.14 Pharmacokinetics and Drug Efficacy; Section 2.3 Approximate String Matching; Section 2.4 Statistics and Probability; Section 2.10 Human Physiology; Section 2.11 Genomics and Proteomics; Section 3.2.5 Nature-Inspired Metaheuristics; Section 3.4 Machine Learning; Section 3.5 Classification Techniques; Section 3.6 Regression Analysis; Section 3.7 Probabilistic Reasoning over Time; Section 3.8 Data Mining; Section 8.1 Clinical Data Classification; Section 8.7 Disease Management and Identification

During the last 20 years, human genome sequencing and microbial genome sequencing have given us new insight into the structure and function of genes and their collective behavior. A gene is a double-stranded sequence of four types of biomolecules: "A" (adenine), "T" (thymine), "G" (guanine) and "C" (cytosine). An adenine ("A") biomolecule in one strand is bound to thymine ("T") in the complementary strand through hydrogen bonds. A guanine ('G') biomolecule in one strand is bound to cytosine ('C') in the complementary strand through hydrogen bonds. At the micro-level, genes are responsible for every small function needed by the body to transform the food we eat and to protect our body against microbial imbalance and foreign-body invasion. The chain of reactions involving genes is called a *pathway*. There are two types of pathways: 1) the metabolic pathway involved in the transformation of biomolecules and 2) the signaling pathway involved in the immune system to protect our body against imbalance and foreign invasion.

Metabolic pathways use a cascade of reactions to produce and transform different biomolecules, chemical substances and energy. A special class of proteins called *enzymes* act as catalysts for the reaction. Their job is to magnify or attenuate the rate of reaction in biochemical pathways. Besides *enzymes*, some chemicals act as *coenzymes*. *Coenzymes* enhance the power of catalysts.

A foreign-body invades and changes the balance of the signaling pathway using *receptors* – a part of the surface-proteins in a signaling pathway that interact with other biomolecules. The receptor acts as a docking area for external molecules. The attachment of another biomolecule(s) at the receptor changes the three-dimensional configuration of the protein molecule altering its functionality. This change in the functionality alters the corresponding signaling pathway and the reaction rate.

A cell is a complex and dynamic feedback-based structure comprising genes, biochemicals and proteins (including metal-ions) that are continuously interacting and producing (or reproducing) new material needed for the reproduction, functioning, protection and repair of a cell. Any abnormality in these functions causes an aberration in the cell-function that is corrected using a feedback system that may also trigger a latent pathway. If the aberration is not correctable, it results in observed abnormal behavior at a higher level.

The abnormality generation is caused by: 1) regular mutation of genes that skip the repair (or repaired incorrectly); 2) attack by a foreign-cell or genetic material that disrupts the regular working by gene–protein or protein–protein interactions in the pathways and 3) change in environmental conditions such as PH-value, pressure, temperature, radiation, chemicals such as salt, metal-ions and so on that affect the functioning of pathways. A cell's defense system uses latent pathways that get triggered when cell-function becomes erratic.

465

The foreign biomolecule that triggers the correction or aberration in the cell-function is called an *antigen*. These antigens are proteins that interact with the receptors in the cell to alter the function of the cell, or allow the entry of foreign biomolecules inside the cell causing aberrations. The body has a mechanism to protect against such attacks by identifying and tagging foreign-bodies and killing these foreign-bodies. The biomolecules that neutralize the *antigens* are called *antibodies*. The internal system that identifies, destroys and cleans the foreign-bodies is called *immune-system*. Immune-system gets activated when it senses a signature of a biomolecule or a protein not being produced by the body.

Diseases are caused because of the following: 1) an aberration in the gene-production causing a slightly deviant gene that does not function properly; 2) a fault in gene-repair mechanism; 3) a lack of proper functioning of the certain proteins called *antibodies* that are responsible to protect body's control system by binding and/or destroying the foreign bodies called *antigens* that affect our immune system adversely; 4) presence of an excess of a microbe that consumes or produces certain biochemicals in an imbalanced way affecting the chemical composition of the body adversely and 5) loss of a regulation mechanism that controls the production of genes and proteins causing an imbalance in the gene/protein system.

Our body has many portals that interact with outside world: 1) nose breathing in foreign-bodies from surrounding air; 2) wound that gets infected due to exposure to a foreign substance; 3) mouth that ingests foreign-bodies along with food; 4) urinary tract and 5) skin (including eyes) that is continuously in touch with foreign-bodies. In addition, any fluid injected through syringes can also infect human-body. The human-gut is a complex biosystem that contains millions of different strains of bacteria, some useful, some harmful, some opportunistic and some neutral to the surrounding. Opportunistic bacteria are generally benign. However, they get pathogenic (disease causing) under certain conditions.

Bacteria are a self-reproducing complex feedback system with mechanisms to protect themselves against adverse condition such as changed PH-value, changed pressure, changed temperature and the amounts of nutrients and water. Under altered circumstances, opportunistic bacteria can also become harmful to a human-body. Bacteria evolve using one or more mechanisms: 1) borrowing gene(s) from other bacteria or host to become more robust and survive in the environment; 2) invoking latent pathways to survive in an adverse environment and 3) mutating and changing the gene functions to undo the effect of agents that harm them such as antibiotics. The result is the evolving into more resistant strain of bacteria that can grow much faster and are resistant to antibiotics.

To protect against a foreign-body invasion, three approaches are used: 1) administer the biomolecules externally to block the actions of antigens; 2) boost the immune system to fight against the foreign bodies and 3) the growth of the infecting bacteria is inhibited using biomolecules that interfere with bacterial growth and/or kill them. The proteins that block the actions of antigens are called *antibodies*. Antibodies are natural candidates for drugs. The cells or set of molecules that boost the immune-system by faking an active foreign-body attack are called *vaccines*. The drugs that affect the growth of undesired bacteria are generically called *antibiotics*.

Many antibiotics attack the reproduction-system or the cell-wall formation of the foreign-bodies. However, the action has to be specific for it not to harm beneficial bacteria that help a human-body. Unfortunately, specific drugs are difficult to develop; it may not work on variable strains of a bacterium. In the absence of a correct diagnosis of a bacterium, specific antibiotics may not work.

The main goal of the genomics and proteomics is to facilitate the discovery of the cause of the diseases, including genetic diseases and find the corresponding drugs and vaccines. The key process in the drug-development is to identify the cause of aberrations in the pathway. The aberration can be caused by: 1) a single mutation in a gene that changes its functionality such as in sickle-cell anemia; 2) an insertion or deletion of foreign or mutated genes/proteins that change the pathway; 3) changes in the concentration of enzymes that affects the reaction rate; 4) change in the three-dimensional configuration of the proteins due to protein–protein interaction between a receptor and an antigen and 5) environmental changes that either speed up gene mutation or reduce the capacity of gene-repair.

After recognizing the cause of aberration, specific biomolecules are developed to alleviate the cause of aberrations. For antigens, antibodies are developed. For many diseases, including viral diseases, bacterial diseases and cancer, immune system has to be enhanced that will stop the growth of the cause of diseases. For bacterial diseases, chemical molecules are identified that would disrupt, destroy or weaken the stability or growth of harmful bacteria.

Bioinformatics has a major role in: 1) identifying the causes of the diseases; 2) improving the efficacy of drugs by 3D modeling of the proteins for studying protein–protein interaction between the antigens and antibodies and 3) searching the library of similar chemical compound with improved efficacy and lower ill effect (toxicity) after a candidate compound has been identified that disrupts the causes of a disease.

To identify the causes of the diseases, deriving the role of genes in macroscale function (phenotype) and the function of the proteins involved in a biochemical pathway is very important. Similarity-based search and sequence-alignment has been used to approximate the function of the proteins and genes in the newly sequenced genomes. The original functions of genes and proteins are derived using wet-lab experiments because protein is a 3D dynamic structure with potentially multiple functions.

Sequence-based matches do not give a lot of information due to: 1) 3D distance between different amino-acids in the structure for affinity-based interaction; 2) the presence of functionally relevant domains; 3) currently active domains; 4) purpose of the protein and 5) some mutations are more important than others; all mutations are not equally weighted. Comparing 3D structure of proteins is limited due to: 1) the lack of availability of a big library of actual 3D-structures and 2) dynamic variation in 3D-structure based on the interaction with the environment and the foreign-bodies. Most of the 3D-structures are modeled using approximation techniques based upon energy-minimization assumption.

Despite the limitations, bioinformatics has contributed to the genome sequencing, identification of gene-boundaries and genes, identification of similar genes using similarity-based search like BLAST and its variants, three-dimensional modeling of the proteins, affinity analysis of amino acids in protein–protein interaction, identification of gene-groups involved in common pathways, graph-based modeling of metabolic pathways, gene-expression analysis and detection of gene silencing mRNA (messenger RNA) and their impact on various genetic diseases such as cancer. Bioinformatics is only used for prediction and pruning the pool of candidates. The actual verification and development come from the wet-lab through actual biochemical experiments in chemical laboratories.

This chapter discusses various aspects of bioinformatics that facilitates the prediction of the causes of diseases and the drug-discovery.

10.1 GENOMICS

Genomics is related to the study of structure and function of genes and proteins. It is also about gene-to-protein translation. A gene is a sequence of four types of biomolecules called *nucleotides:* "A" (adenosine), "G" (guanine), "C" (cytosine) and "T" (thymine). Genes have a double-helical strand. Both strands are complementary to each other: The nucleotide "A" binds to the nucleotide "T" using two hydrogen bonds; the nucleotide "G" binds to the nucleotide "C" using three hydrogen bonds. A hydrogen bond is an electrostatic attraction between a hydrogen ion (proton) and a negatively charged ion or molecule such as hydroxyl ion (OH−). A lot of energy is required to break these bonds. In higher organisms, genes are packed into a chromosome, and many chromosomes make a genome.

There are two types of nucleic molecules: 1) DNA and 2) RNA. A *DNA* (deoxyribonucleic acid) molecule is made of nucleotides "A," "C," "G" and "T." An RNA (Ribonucleic Acid) molecule is made of nucleotides "A," "G," "C" and U (uracil). DNA molecules are used for the formation of genes. RNA molecules are used for the translation of genes to proteins. The nucleotide "U" substitutes the nucleotide "T" when DNA is transformed to RNA during gene-to-protein translation process. A gene has two ends: 3′ end and 5′ end. A 3′ end terminates in a hydroxyl group. A 5′ end terminates in a phosphate group.

| Double stranded DNA (complement strands A ↔ T and G ↔ C | (5' end) ... A T T G C C G T T A G G C A T T A C ... (3' end) \| \| \| \| \| \| \| \| \| \| \| \| \| \| \| \| \| \| \| \| \| (3' end) ... T A A C G G C A A T C C G T A A T G ... (5' end) |
| Messenger RNA Corresponding to top DNA strand | ... A U U G C C G U U A G G C A U U A C ... |

FIGURE 10.1   An illustration of nucleotides in DNA strand and messenger RNA

Example 10.1: DNA-RNA correspondence

Figure 10.1 describes a double-stranded subsequence of DNA, and the corresponding mRNA created for translating the top strand of DNA to mRNA. All the occurrences of "T" are replaced by "U."

Proteins are the building blocks in cells. Proteins are made of a sequence of 20 amino-acids. Nucleotide patterns form the template for the generation of 20 amino-acids combinations in a protein. A *codon* is a group of three consecutive nucleotides, and maps uniquely to an amino-acid during the gene-to-protein translation. There are $4^3 = 64$ such combinations of nucleotides corresponding to a codon. However, only the first two nucleotides in a codon cause variations in the amino-acid formation. Mostly, the third nucleotide in a codon does not alter the amino-acid mapping. Thus, 64 possible combinations map to 20 amino acids as illustrated in Figure 10.2. There is a *stop-codon* to mark the end of the coding-region of a gene. Methionine ("M") also acts as a *start-codon* to mark the start of a coding-region.

Example 10.2: Amino acid sequence

Figures 10.2 and 10.3 illustrate an example of the nucleotide-sequence of a gene that translates to produce the amino-acid sequence of an enzyme *IDS* shown in Figure 10.3. *IDS* is an enzyme involved in a metabolic pathway *Chondroitin sulfate and dermatan sulfate metabolism*. The lack of *IDS* causes Hunter's syndrome – a disease associated with multiple severe symptoms and an early death. The symptoms include coarseness of facial features, thickening of heart valves. Immobility of joints, enlargement of spleen and liver and carpel-tunnel syndrome.

```
atgccgccaccccggaccggccgaggccttctctggctgggtctggttctgagctccgtctgcgtcgccctcggatccgaaa
cgcaggccaactcgaccacagatgctctgaacgttcttctcatcatcgtggatgacctgcgcccctccctgggctgttatggg
gataagctggtgaggtccccaaatattgaccaactggcatcccacagcctcctcttccagaatgcctttgcgcagcaagcag
tgtgcgccccgagccgcgtttctttcctcactggcaggagacctgacaccacccgcctgtacgacttcaactcctactggagg
gtgcacgctggaaacttctccaccatccccccagtacttcaaggagaatggctatgtgaccatgtcggtgggaaaagtctttca
ccctgggatatcttctaaccataccgatgattctccgtatagctggtcttttccaccttatcatccttcctctgagaagtatgaaaa
cactaagacatgtcgagggccagatggagaactccatgccaacctgctttgccctgtggatgtgctggatgttcccgagggc
accttgcctgacaaacagagcactgagcaagccatacagttgttggaaaagatgaaaacgtcagccagtcctttcttcctgg
ccgttgggtatcataagccacacatcccccttcagataccccaaggaatttcagaagttgtatcccttggagaacatcaccctg
gcccccgatcccgaggtccctgatggcctacccctgtggcctacaaccccctggatggacatcaggcaacgggaagacgt
ccaagccttaaacatcagtgtgccgtatggtccaattcctgtggactttcagcggaaaatccgccagagctactttgcctctgt
gtcatatttggatacacaggtcggccgcctcttgagtgcttggacgatcttcagctggccaacagcaccatcattgcatttacc
tcggatcatgggtgggctctaggtgaacatggagaatgggccaaatacagcaattttgatgttgctacccatgttcccctgata
ttctatgttcctggaaggacggcttcacttccggaggcaggcgagaagcttttcccttacctcgaccctttgattccgcctcac
agttgatggagccaggcaggcaatccatggaccttgtggaacttgtgtctcttttttcccacgctggctggacttgcaggactgc
aggttccacctcgctgcccgttccttcatttcacgttgagctgtgcagagaaggcaagaaccttctgaagcattttcgattccg
tgacttggaagaggatccgtacctccctggtaatccccgtgaactgattgcctagccagtatccccggccttcagacatcc
ctcagtggaattctgacaagccgagtttaaaagatataaagatcatgggctattccatacgcaccatagactataggtatact
gtgtgggttggcttcaatcctgatgaatttctagctaactttttctgacatccatgcaggggaactgtatttgtggattctgaccca
ttgcaggatcacaatatgtataatgattcccaaggtggagatctttccagttgttgatgccttga
```

FIGURE 10.2   An illustration of the nucleotide sequence of a gene corresponding to *IDS* enzyme

```
MPPPRTGRGLLWLGLVLSSVCVALGSETQANSTTDALNVLLIIVDDLRPSLGCYGDKLVRSPNIDQLASHSLLFQ
NAFAQQAVCAPSRVSFLTGRRPDTTRLYDFNSYWRVHAGNFSTIPQYFKENGYVTMSVGKVFHPGISSNHTDDSP
YSWSFPPYHPSSEKYENTKTCRGPDGELHANLLCPVDVLDVPEGTLPDKQSTEQAIQLLEKMKTSASPFFLAVGY
HKPHIPFRYPKEFQKLYPLENITLAPDPEVPDGLPPVAYNPWMDIRQREDVQALNISVPYGPIPVDFQRKIRQSY
FASVSYLDTQVGRLLSALDDLQLANSTIIAFTSDHGWALGEHGEWAKYSNFDVATHVPLIFYVPGRTASLPEAGE
KLFPYLDPFDSASQLMEPGRQSMDLVELVSLFPTLAGLAGLQVPPRCPVPSFHVELCREGKNLLKHFRFRDLEED
PYLPGNPRELIAYSQYPRPSDIPQWNSDKPSLKDIKIMGYSIRTIDYRYTVWVGFNPDEFLANFSDIHAGELYFV
DSDPLQDHNMYNDSQGGDLFQLLMP
```

FIGURE 10.3   An illustration of amino-acid sequence of the enzyme *IDS* involved in Hunter's syndrome

Example 10.3: Codon to amino-acid translation

Consider the messenger-RNA in Example 10.1. If the first nucleotide "A" is the start of a codon, the mRNA pattern shows six codons: *<AUU, GCC, GUU, AGG, CAU, UAC>*. Looking up the Table 10.1, each codon correspond to an amino acid. The corresponding amino acid subsequence coming out of ribosomes is "… *isoleucine alanine valine arginine histidine tyrosine*…"

Proteins have multiple functions such as interaction with other proteins, acting as catalysts in metabolic-pathway to regulate the biochemical reactions, interacting with the environment around the cells, carrying different nutrients from the environment inside the cells, translating a DNA-strand to messenger-RNA strand during gene-to-protein translation, gene-repair and so on.

TABLE 10.1   Codon and amino-acid correspondence

		SECOND NUCLEOTIDE			
		U	*C*	*A*	*G*
F I R S T	U	UUU → F (Phe) UUC → F (Phe) UUG → L (Leu)	UCU → S (Ser) UCC → S (Ser) UCA → S (Ser) UCG → S (Ser)	UAU → Y (Tyr) UAC → Y (Tyr) UAA → **Stop** UAG → **Stop**	UGU → C (Cys) UGC → C (Cys) UGA → **Stop** UGG → W (Trp)
N U C L E O T I D E	C	CUU → L (Leu) CUC → L (Leu) CUA → L (Leu) CUG → L (Leu)	CCU → P (Pro) CCC → P (Pro) CCA → P (Pro) CCG → P (Pro)	CAU → H (His) CAC → H (His) CAA → Q (Gln) CAG → Q (Gln)	CGU → R (Arg) CGC → R (Arg) CGA → R (Arg) CGG → R (Arg)
	A	AUU → I (Ile) AUC → I (Ile) AUA → I (Ile) AUG → M (Met)/**Start**	ACU → T (Thr) ACC → T (Thr) ACA → T (Thr) ACG → T (Thr)	AAU → N (Asn) AAC → N (Asn) AAA → K (Lys) AAG → K (Lys)	AGU → S (Ser) AGC → S (Ser) AGA → R (Arg) AGG → R (Arg)
	G	GUU → V (Val) GUC → V (Val) GUA → V (Val) GUG → V (Val)	GCU → A (Ala) GCC → A (Ala) GCA → A (Ala) GCG → A (Ala)	GAU → D (Asp) GAC → D (Asp) GAA → E (Glu) GAG → E (Glu)	GGU → G (Gly) GGC → G (Gly) GGA → G (Gly) GGG → G (Gly)

Three-letter abbreviations ≡ amino-acid names

1. Ala ≡ Alanine	2. Arg ≡ Arginine	3. Asn ≡ Asparagine	4. Asp ≡ Aspartic acid
5. Cys ≡ Cysteine	6. Gln ≡ Glutamine	7. Glu ≡ Glutamic acid	
9. His ≡ Histidine	10. Ile ≡ Isoleucine	11. Leu ≡ Leucine	8. Gly ≡ Glycine
13. Met ≡ Methionine	14. Phe ≡ Phenylalanine	15. Pro ≡ Proline	12. Lys ≡ Lysine
17. Thr ≡ Threonine	18. Trp ≡ Tryptophan	19. Tyr ≡ Tyrosine	16. Ser ≡ Serine
			20. Val ≡ Valine

A ribosome is made of *ribosomal-RNAs*. Ribosome is the unit that chains different amino-acid molecules together based upon the codon being scanned. Different types of RNA have different yet complementary functions in gene-to-protein translation. There are five types of RNA: 1) mRNA (messenger RNA); 2) t-RNA (transfer RNA); 3) rRNA (ribosomal RNA); 4) siRNA (Small Interfering RNA / silencer RNA) and 5) miRNA (micro-RNA). *mRNA* carries the message about gene-template to ribosomes (a unit that converts mRNA to proteins). *tRNA* transports amino-acids to ribosomes to chain the amino-acid molecules together during the protein-translation. *siRNA* is a double-stranded RNA molecule that down-regulates the mRNA-translation to protein by degrading the mRNA. *miRNA* is a small noncoding RNA that inhibits the gene-to-protein translation for a class of mRNAs. miRNA destabilize the mRNA after transcription. The difference between siRNA and miRNA is that siRNA is specific to downregulating one mRNA, while miRNA downregulates multiple mRNAs simultaneously.

10.1.1 Genome Structure

The human genome and microbial genome differ in many respects. The microbial genome is a mostly single chromosome organism, and human genome comprises multiple chromosomes. The organization of microbial genomes is that each microbial-cell is a set of genes, proteins and multiple chains of reactions.

The human-body contains around *75–100* trillion (*10^{12}*) cells. Each cell contains two copies of each *chromosome*. Human genome comprises *46* chromosomes. The chromosomes are divided into two sets of *23* chromosomes each. Each chromosome is a packed set of *genes*. Thirty-third chromosome is the sex-chromosome: females have two copies of X-chromosomes, and males have one copy of *X-chromosome* and a copy of *Y-chromosome*. Around *50,000* genes have been annotated in a human genome.

Genes become defective due to the alterations in a nucleotide-subsequence. A nucleotide-subsequence is altered either by deletion, insertion or modification of one or more nucleotides. This alteration causes the codon sequence to morph that results into production of a defective protein that may have a different configuration and function. Most of these alterations are repaired by DNA-repair proteins. However, unrepaired alterations become the cause of a genetic defect.

Proteins in the final form have a 3-D folding pattern that can be altered dynamically due to the presence of another protein or another small biomolecule that binds to a *receptor* on the surface of the protein. This *protein–protein interaction* or protein interaction with other molecules alters the original functionality of a protein. There are many types of receptors on the surface of a protein. Some receptors bind to specific classes of proteins, while others may be generic. *Antigens* bind to these receptors to change the behavior of the proteins. In response to the docking of the external biomolecule, a protein 1) may change 3D-configuration that alters its functionality; 2) split in two parts changing the functionality and 3) change the reaction-rate.

There are three major processes in every cell: 1) mitosis; 2) meiosis and 3) transcription and translation for protein formation. The process of cell replication is called *mitosis*. During mitosis, all the genes are replicated precisely with an equal number of chromosomes. In meiosis, a cell divides into four daughter-cells, each with half the number of chromosomes from the parent-cell.

10.1.1.1 Mitosis

Mitosis is a multiphase process. The main stages are: 1) prophase; 2) prometaphase; 3) metaphase; 4) anaphase and 5) telophase. In the *prophase*, the chromosome-strands called *chromatids* separate. However, the chromatids remain connected at the constricted region joining the chromatids. After that, chromatids undergo a condensation process on a bipolar fibrous material within a cell. This fibrous material is called *mitotic spindle*. During *prometaphase*, the nuclear envelope within a cell is divided into multiple fluid-filled fragments called a *vesicle*. These vesicles are divided eventually in daughter-cells. In the *anaphase*, the sister-chromatids separate abruptly. The *spindle-poles* separate using *motor proteins*. Mitosis ends with *telophase* when the separated chromosomes reach the poles of the spindle, and the chromosomes begin to decondense from the *spindle*. *Telophase* is followed by *cytokinesis* – the division of the cytoplasm into identical cells: cells having identical genetic composition.

10.1.1.2 Meiosis

Meiosis involves a two-stage process: 1) Meiosis I and 2) Meiosis II. Before meiosis begins, DNA of each chromosome in a cell is replicated to generate identical sister-chromatids held together by cohesion. During meiosis I, homologous (similar) chromosomes exchange genetic information using chromosomal crossover. During meiosis II, the two daughter-cells created after meiosis I again divide keeping half of the chromosomes of their parent-cells. Meiosis differs from mitosis because it retains only half of the chromosomes of the original cell, and the chromatids in chromosomes go through crossover. Thus, the chromosomes are altered. Meiosis is involved in reproduction.

10.1.2 Gene Structure

Genes are classified into two major types: 1) bacterial genes and 2) eukaryotic genes in higher-order organisms, including yeast, flies, birds, plants and mammals. A bacterial gene has two parts: 1) regulatory region (promoter) and 2) coding-region. A eukaryotic gene such as a human-gene or a plant-gene has three parts: 1) regulatory region (promoter), 2) noncoding-region called *intron* and 3) coding region called *exon*. The promoter-region is used by the transcriptase (a specific protein) to latch on the gene to create the mRNA-template corresponding to a DNA strand. Bacteria do not have a noncoding-region. The role of the noncoding region is to protect the coding-region from degradation. In addition, there are many genes responsible for the repair of genes if undesired mutation occurs or damage occurs due to environmental factors such as radiation, extreme temperature or pH-value.

The major differences between the genes in prokaryotes and genes in eukaryotes are: 1) prokaryotic genes do not have *introns*, while eukaryotes have introns and 2) the transcription enhancers (promoters) are upstream (coming before the coding region) in prokaryotes, while they can be upstream or downstream (coming after the coding-region) in eukaryotes. Introns are very important for gene-expression and translation of gene-to-protein in eukaryotes. Introns increase the gene-expressions and protein formation by many folds. The difference between prokaryotic gene-structure and eukaryotic gene-structure is shown in Figure 10.4. Prokaryotes and bacteria have been used interchangeably in this discussion despite some finer subclassifications.

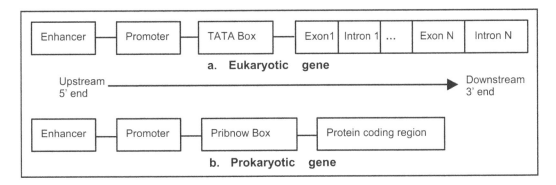

FIGURE 10.4   Gene-structure in eukaryotes and prokaryotes: (a) Eukaryotic gene; (b) Prokaryotic gene

As shown in Figure 10.4, both eukaryotic gene and prokaryotic gene have a regulatory-region consisting of an *enhancer, a promoter* and a *TATA-box* that acts as a marker that separates the regulatory-region from the coding-region. The sequence of *TATA-box* is conserved, and is of the form "TATAAA" in Eukaryotes. The corresponding region in prokaryotes is called "TATAAT," and is called *Pribnow box.*

Enhancers promote the binding of transcription-factors to the promoters to start transcription. With eukaryotes, the corresponding enhancer could be both upstream (occurring before) or downstream (occurring after) the coding-region. However, in a bacterial gene, the corresponding enhancer always occurs upstream. The coding-region of the eukaryotic genes is a repeated sequence of (*exon, intron*) pairs, and there can be one or more such pairs.

10.1.3 Gene-to-Protein Translation

Gene-to-protein translation involves the use of a specific protein that transcribes the DNA to mRNA that acts as a template for the translation of DNA to the corresponding protein. The gene translation process involves: 1) transcription of DNA → mRNA, 2) mRNA moving through a ribosome, 3) polymerase creating a chain of amino acids brought by tRNA into ribosomes based upon the codons in mRNA and so on 4) maturation of a chain of amino-acids to protein. During transcription, *enhancers* (or *repressors*) play a major role. Enhancers increase the transcription-rate. Conversely, repressors suppresses the transcription-rate.

During the transcription process, a *transcriptase* binds to the promoter-region, strips the hydrogen bonds between the A≡T and G≡C pairs and separates the strand. In the next phase, a thymine molecule ("T") is substituted by an uracil ("U") molecule. The transcription of a bacterial gene is straightforward due to the absence of introns. However, in eukaryotes, the primary transcript after the transcription process also includes the corresponding intron version. The primary transcript goes through an additional process of *splicing* that removes the intron-part from the primary transcript and joins the exons to create the corresponding mRNA. The process of transcription of DNA to mRNA for eukaryotic gene is illustrated in Figure 10.5.

tRNA has a tri-axial structure as shown in Figure 10.6. The center axis contains an *anti-codon.* An anti-codon is a complementary subsequence that binds to a codon. For example, anti-codon for the codon "AAU" is "UUA." This anti-codon binds to the codon being scanned in a mRNA-strand. After the binding

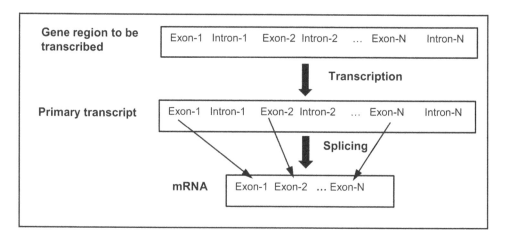

FIGURE 10.5 A schematic of DNA transcription to mRNA

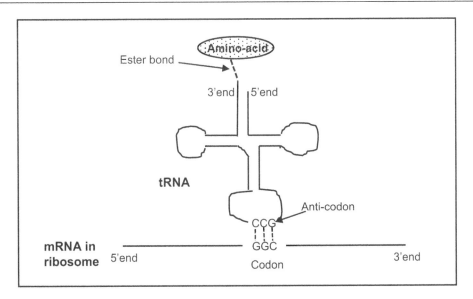

FIGURE 10.6   An illustration of role of tRNA in building a chain of amino acid sequence

of codon and anti-codon, tRNA drops the amino-acid that is added to the existing amino-acid chain. Ribosome traverses the whole length of mRNA generating the corresponding chain of amino-acids. The resulting amino-acid chain is still not robust and has to go through a protein-maturation process.

In eukaryotes, a ribosome comprises two ribosomal units: *60S* and *40S*. Each ribosomal component contains multiple proteins involved in mRNA-translation. The *40S* unit comprises *18S* rRNA (*1900* nucleotides) and *33* proteins, and *60S* subunit comprises *5S* rRNA (*160* nucleotides), *28S* rRNA (*4700* nucleotides), *5.8S* rRNA (*160* nucleotides) and *46* proteins. The symbol "S" stands for the *Svedberg sedimentation coefficient* – a measure on the size of a particle. Many tRNAs are attached by a ribosome concurrently during a translation.

Many ribosomes work on the same mRNA-strand concurrently to build the chain of amino-acids. Thus, one mRNA-strand generates multiple amino-acid chains. It takes around one to two minutes for all the codons to be scanned by a ribosome. A special protein within ribosomes called *eRF* (<u>R</u>elease <u>F</u>actor) identifies the stop-codon in an mRNA-string, stops the synthesis of the amino-acid chain and the mRNA-string is released by the ribosome. Protein-synthesis consumes energy provided by ATP (*Adenosine Triphosphate*) and GTP (*Guanosine Triphosphate*). Using a biochemical reaction, GTP is converted to ATP.

A chain of amino-acids is called a *polypeptide*. It has to go through a maturation phase to become a stable 3D-protein. There are many chaperon-proteins in *cytosol* (intra-cellular fluid) that facilitate the maturation of the polypeptides. Many structure-holding bonds such as covalent disulfide bonds ($-S-S-$) are added. The process also involves removing *peptidases* (proteins responsible for clipping peptides), addition of phosphates, sugar or lipid to specific amino-acids and *glycosylation*. *Glycosylation* is a complex process involving carbohydrate–protein bonds changing the structure and function of the proteins. Defective glycosylation causes many diseases.

Protein-folding structure depends upon the property of amino acids' physical and biochemical properties. Amino-acids such as glycine are very flexible, while tryptophan and phenylalanine are rigid. Amino-acids such as leucine, isoleucine, valine, methionine and phenylalanine are hydrophobic (water avoiding). Amino-acids such as glutamine, asparagine, histidine, serine, threonine, tyrosine and cysteine are hydrophilic (affinity for water).

Every amino-acid contains a common chain $H_3N-CHR-CO=O$. H_3N is the amine-group and is also denoted as "R." "R" could be as simple as an H^+ ion as in glycine or a complex structure such as $CH_3-CH-CH_3$ as in valine. The biophysical and biochemical properties of "R" contribute to the corresponding properties of amino-acids.

TABLE 10.2   Amino acid properties

AMINO ACID	PROPERTIES	AMINO ACID	PROPERTIES
Alanine (A)	non-polar; hydrophobic	Leucine (L)	non-polar; hydrophobic
Arginine (R)	+ve charged; polar; hydrophilic	Lysine (K)	+ve charged; polar hydrophilic
Asparagine (N)	no charge; polar; hydrophilic	Methoinine (M)	non-polar; hydrophobic
Aspartate (D)	–ve charged; polar; hydrophilic	Phenylalanine (F)	non-polar; hydrophobic
Cysteine (C)	no charge; nonpolar; hydrophilic	Proline (P)	non-polar; hydrophobic
Glutamate (E)	–ve charged; polar; hydrophilic	Serine (S)	no charge; polar; hydrophilic
Glutamine (Q)	no charge; polar; hydrophilic	Threonine (T)	no charge; polar; hydrophilic
Glycine (G)	no charge; non-polar; hydrophilic	Tryptophan (W)	no charge; non-polar; hydrophobic
Histidine (H)	+ve charged; polar; hydrophilic	Tyrosine (Y)	no charge; polar; hydrophilic
Isoleucine (I)	nonpolar; hydrophobic	Valine (V)	non-polar; hydrophobic

Depending upon the stereochemistry of the R-group, an amino-acid can be flexible or rigid. Depending upon the R-group, an amino-acid can be nonpolar, polar but neutral, negatively charged or positively charged. Nonpolar group (except cysteine and glycine) is generally hydrophobic. Polar group is hydrophilic. Charged amino-acids are hydrophilic. The biochemical properties of the amino-acids are described in Table 10.2. The hydrophobic amino-acids curl up and make a pocket in water. This pocket acts as a storage for the specific molecule or metal-ion crucial for the protein functionality. Hydrophobic amino-acids are also found on the cell-surface to create a barrier against water.

10.1.4 Protein–Protein Interactions

For the drug development, understanding protein–protein interaction is important. Protein–protein interaction is involved in host–pathogen interaction such as human-cell and bacterial–cell interaction, antigen–antibody interaction, T-cell interaction with dead cells and foreign-bodies, signaling pathway control, transport of nutrients across a cell-membrane, metabolic processes involved in the degradation and synthesis of biomolecules, muscle contraction and cellular process regulation. There are a huge number of protein–protein interactions involved among the proteins in a human-body.

A signaling pathway involves many protein–protein interactions. A protein–protein interaction may change the 3D-configuration of the interacting proteins, unfold the protein-structures exposing protected domains and split a protein changing its functionality. The cascade of protein–protein interactions in signaling pathways are responsible for cell-growth, post-translational changes in the three-dimensional confirmation, proliferation of the proteins, metabolism and other processes.

There are many types of proteins involved in signaling pathways such as *G-protein*, nuclear factor *NF-kB, Ras, Raf, Mek, Mapk* and so on. *NF-kB* is a protein-complex that controls the transcription of DNA and plays a key role in regulating the immune response to infection. *Ras* is a family of proteins involved in growth, differentiation and survival of the cells. *Raf, Mek and Mapk* play a key role in cellular metabolism, and their absence can cause abnormal development of liver, vascular and neuronal tissues.

Protein–protein interactions have been studied in the wet-labs using many techniques. The networks of protein–interactions have been identified using: 1) wet-lab experiment and 2) text-mining tools to perform a literature search on reported experiments. Under the assumption, that protein–protein interactions can be studied by aligning the amino-acid sequences of the proteins, amino-acid subsequences involved in protein–protein interaction can be derived.

The problems in deriving the networks of protein–protein interactions are: 1) temporal nature of protein–protein interactions depend on existing concentrations and/or the presence of other biomolecules; 2) the amount of interaction depends upon the affinity of the amino-acids involved in the interactions; 3) multiple domains in proteins resulting into more than one functionality depending on the 3D-configuration; 4) more than one interaction of proteins and 5) lack of available wet-lab data.

Protein-interaction networks are modeled as a graph such that proteins are vertices, and interactions are edges. A pair of interacting proteins are connected using an *interaction-edge*. Most proteins have few interactions. Hence, the number of edges connected to a vertex are limited. However, some proteins have multiple interactions and act like a hub in a protein-interaction network.

Computational prediction of protein–protein interaction network is based upon many techniques that involve: 1) the conservation of the proteins' proximity across multiple genomes; 2) protein-fusion in another genome and 3) a common subset of proteins involved in the network of multiple genomes. In all these cases, the assumption is that there is a natural pressure to conserve the common protein–protein interaction networks. Multiple machine learning techniques have been used to study protein–protein interactions such as clustering, support vector machine, Markov model, Bayesian classifiers and decision trees. Protein–protein interactions exhibit dependencies. The knowledge of these dependencies improves the modeling of interaction-networks.

Protein-interaction networks have been used to study many genetic and metabolic diseases related to autoimmune abnormality, Alzheimer's disease, Schizophrenia and cancer.

10.1.5 Protein–DNA Interactions

Many types of proteins bind to DNA to regulate the transcription or the repair of the DNA. For example, transcription-factors bind to DNA and transcription-proteins to regulate the DNA-transcription to mRNA; *DNA-polymerase* synthesizes the DNA molecules; *helicases* are used to unbind the DNA double-strand by breaking down the hydrogen bonds between nucleotides in opposite strands. Some proteins bind to double-stranded DNA, while others bind to single-stranded DNA. The protein-domains that recognize DNA-patterns to bind are called DNA-binding domains.

10.1.6 Structure and Function

Protein is a multidomain and possibly multifunctional entity. Protein-functions change based upon the dynamic three-dimensional configuration that depends upon the protein–protein or protein–DNA interactions. A protein may interact with more than one protein in vicinity, with DNA and protein at the same time and with transcription-factors that enhance/repress the transcription-rate. An authentic three-dimensional structure of a protein is established using X-ray crystallography. The database of exact structures of proteins is limited. Instead, the three-dimensional structures of proteins are computationally approximated using protein modeling software based upon overall energy-minimization and/or computationally joining the structure of protein-domains available in X-ray crystallography databases. The process of computationally joining the domain structures is called *threading*.

In an unfolded form, protein is a sequence of amino-acids. However, function of a protein differs from the linear sequence of amino-acids. Amino-acid sequences are first folded into a combination of secondary structures that further folded into a dynamic three-dimensional configuration.

There are two types of secondary structures: α-*helix* and β-*sheet*. *Alpha-helix* is a right-handed helical structure that is maintained using a hydrogen-bond between N–H group and C=O group of four amino-acid residue occurring earlier. *Beta-sheet* is pleated and consists of three to ten polypeptides-strands. A strand in a β-*sheet* is connected to the adjacent strand in a lateral direction using hydrogen-bonds.

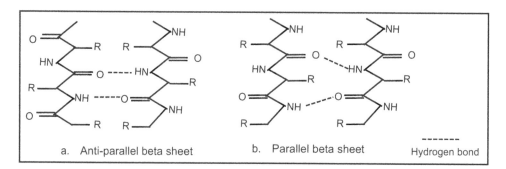

FIGURE 10.7 Beta-sheet binding and orientations: (a) Antiparallel beta; (b) Parallel beta-sheet

A β-*sheet* has two types of orientations: *parallel* and *antiparallel*. In the *parallel orientation*, N-termini of successive strands are oriented in the same direction as illustrated in Figure 10.7a. In the *antiparallel orientation*, the N-terminus of a strand is aligned to C-terminus of the adjacent strand as illustrated in Figure 10.7b. An individual strand may show a combined version of parallel and antiparallel orientation, which is less stable.

Figure 10.8 Illustrates a three-dimensional structure of a hemoglobin molecule that carries iron responsible for red-blood cells in the blood. It has four alpha-helical structures, each holding one heme-binding unit.

10.1.6.1 Domain conservation and functionality

In the absence of the availability of actual three-dimensional structures, the best technique to approximate the functionality of a newly sequenced protein is to match against a database of protein sequences with known domains. If the function involves a specific domain, then exact matching of that domain will identify the similarity of functionality.

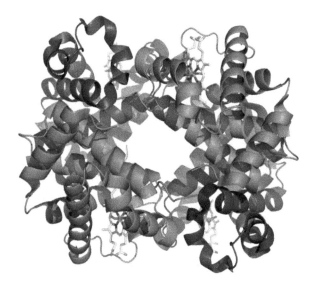

FIGURE 10.8 Three-dimensional model of a hemoglobin molecule involving four alpha-helixes (Image courtesy Richard Wheeler (Zephyris) 2007 created from PDB Enzyme, distributed under GNU Free Document License and Creative Common Attribution – share 3.0, Available at http://www.infotopia.info/articles/images/111/11170.png.htm).

There is a natural pressure to conserve the subsequence of a protein corresponding to the domains because disturbing the amino-acid pattern of the functional-domains will alter the function. Hence, a way to recognize the functional-regions of a protein sequence is to identify the most conserved regions. A conserved region ensures that the three-dimensional folding structure of the protein does not alter the overall protein-function. The amino-acids in the conserved regions need not be the same. Instead, their biochemical and biophysical properties should be conserved. For example, the amino-acids leucine and isoleucine both are hydrophobic as described in Table 10.1. If the domain requires to be hydrophobic, then substituting *leucine* by *isoleucine* will retain its property.

Despite the inherent limitations in annotating the functionality of a newly sequenced protein from protein-sequences, sequence-based matching has been used in predicting the functionality. The basic premise is that sequence matching may not be a necessary or sufficient condition. However, it enhances the probability of identifying similar function in newly sequenced proteins. There are many approaches to annotate a newly sequenced protein such as 1) motif-based annotation; 2) homology-based annotation and 3) approximate dynamic string matching technique described in Sections 2.3.3.

Dynamic matching technique computes sequence-similarity based upon popular matrices such as BLOSUM (BLOcks SUbstitution Matrix) and PAM (Protein Alignment Matrix). These matrices are based upon matching of the biochemical and biophysical properties of amino-acids. The match-score is positive for amino-acids having similar properties and negative for dissimilar properties. A specific popular substitution matrix is BLOSUM62 (see Table 10.3). As shown in Tables 10.2 and 10.3, similar amino-acid such as *leucine* and *isoleucine* have a similarity-core of +2; dissimilar amino-acids such as *leucine* and *histidine* have a similarity score of −4. There are many variations of BLOSUM and PAM matrices such as BLOSUM-50, BLOSUM-62, BLOSUM-80, PAM-30 and PAM-70.

TABLE 10.3 BLOSUM-62 matrix

	A	R	N	D	C	E	Q	G	H	I	L	K	M	F	P	S	T	W	Y	V
A	4																			
R	−1	5																		
N	−2	0	6																	
D	−2	−2	1	6																
C	0	−3	−3	−3	9															
E	−1	1	0	0	−3	5														
Q	−1	0	0	2	−4	2	5													
G	0	−2	0	−1	−3	−2	−2	6												
H	−2	0	1	−1	−3	0	0	−2	8											
I	−1	−3	−3	−3	−1	−3	−3	−4	−3	4										
L	−1	−2	−3	−4	−1	−2	−3	−4	−3	2	4									
K	−1	2	0	−1	−3	1	1	−2	−1	−3	−2	5								
M	−1	−1	−2	−3	−1	0	−2	−3	−2	1	2	−1	5							
F	−2	−3	−3	−3	−2	−3	−3	−3	−1	0	0	−3	0	6						
P	−1	−2	−2	−1	−3	−1	−1	−2	−2	−3	−3	−1	−2	−4	7					
S	1	−1	1	0	−1	0	0	0	−1	−2	−2	0	−1	−2	−1	4				
T	0	−1	0	−1	−1	−1	−1	−2	−2	−1	−1	−1	−1	1	−1	1	5			
W	−3	−3	−4	−4	−2	−2	−3	−2	−2	−3	−2	−3	−1	1	−4	−3	2	11		
Y	−2	−2	−2	−3	−2	−1	−2	−3	2	−1	−1	−2	−1	3	−3	−2	−2	2	7	
V	0	−3	−3	−3	−1	−2	−2	−3	−3	3	1	−2	1	−1	−2	−2	0	−3	−1	4

10.1.6.2 *Sequence matching*

Despite its limitations, sequence-matching has been used by the researchers to predict the structural and functional similarity of proteins and DNA sequences using the database of well-studied sequences. There are many approaches to match a pair of nucleotide or amino-acid sequences. Some popular approaches are: 1) by matching the conserved subsequences using dynamic warping technique similar to dynamic time-warping technique used in Section 2.7.2.1; 2) starting with small matching-subsets of potential matching seeds and extending the seed so the probability of matching remains high and 3) dynamic string matching as explained in Section 2.3.3.1. The first and third techniques are similar with one exception. The first technique uses no gap-penalty and matches the segments with amino-acids having similar biochemical and biophysical properties.

In the first technique, the exact matching amino-acid characters will give a high value in the diagonal cells. However, the presence of additional characters or the absence of certain characters distorts the exact match. The matching value from a BLOSUM matrix can fill in the difference between two nonmatching characters, and the smallest values along the diagonals are picked to align the characters in two sequences using Equation 2.38. The matching follows Equation 10.1.

$$match\_score(i,0) = -\infty \ where \ 0 < i < m$$
$$match\_score(0,j) = -\infty \ where \ 0 < j < n$$

$$match\_score(i,j) = max \begin{pmatrix} match\_score(i-1,j-1) + similarity - score(seq_i^1, seq_j^2) \\ match\_score(i-1,j) + similarity - score(seq_i^1, seq_j^2) \\ match\_score(i,j-1) + similarity - score(seq_i^1, seq_j^2) \end{pmatrix} \quad (10.1)$$

After filling in the matrix, the traversal starts with the *cell(m, n)* and moves toward the *cell (0, 0)* by moving to the cell having a maximum value between *cell(i − 1, j)*, *cell(i − 1, j − 1)* and *cell(i, j − 1)*. The patterns along the diagonal descent show the matching patterns; movement in the vertical or horizontal direction is equivalent to insertion/deletion of amino-acid (or nucleotide) characters. An example of this alignment technique is illustrated in Example 10.4.

Example 10.4: Aligning using first scheme

Consider two amino-acid sequences "AFCRITAACQ" and "AFFCRVTALIRFRACQ." The first technique match will look like Figure 10.9. Note that additional inserted characters "LIRFR" have been ignored in the matching process. The matching segments are {"AF," "CR (I|V)TA," "ACQ"}. Note that the amino-acid "I" (isoleucine) and "V" (valine) have a positive matching-score. These subsequences are conserved (common) in both the sequences.

The second scheme adds gap-penalties for the inserted/deleted amino acid characters in addition to similarity-score matrices. The gap-penalty is added when the previous cell is vertical or horizontal in the alignment matrix. This scheme is known as *Smith–Waterman algorithm*, and is given by Equation 10.2. The scheme supports two types of alignment: *global* and *local*. In global alignment, the gap-penalty for the missing characters is small and the complete alignment is used. In local alignment, the gap-penalty is kept high, and matching of the segment is suspended after the cumulative similarity score goes below a threshold-value. Local alignment

FIGURE 10.9   Matching two sequences using dynamic warping technique with no gap penalty

	A	F	C	R	I	T	A	A	C	Q
Q	0	2	14	22	24	34	39	43	49	58
C	1	5	18	22	27	35	40	44	53	49
A	1	7	9	25	28	36	40	44	44	43
R	-3	9	9	26	31	36	36	36	36	36
F	-2	12	10	21	34	37	37	37	39	41
R	0	6	12	24	34	36	39	38	41	44
I	1	9	15	19	37	35	40	44	44	41
L	2	9	16	22	33	37	41	45	45	43
A	3	9	18	23	31	38	42	46	46	45
T	-1	11	20	26	32	38	38	38	37	36
V	-1	10	21	27	33	33	33	33	32	30
R	-1	11	22	30	27	26	25	24	26	29
C	0	14	25	22	21	20	20	20	29	25
F	0	16	14	11	10	11	9	7	5	3
F	2	10	8	5	5	6	4	6	6	5
A	4	2	2	1	0	0	4	8	8	7

FIGURE 10.10   Dynamic matching without penalty gaps to identify a set of matching patterns

derives a sequence of multiple closely matching subsequences with limited gaps. While, global alignment includes long stretches of nonmatching characters and longer gaps. Example 10.5 illustrates it.

$$match\_score(i, j) = max \begin{cases} match\_score(i-1, \ j-1) + similarity\_score(Seq_i^1, Seq_j^2) \\ match\_score(i-1, \ j) + gap \ penalty \\ match\_score(i, \ j-1) + gap \ penalty \\ otherwise \ 0 \end{cases} \quad (10.2)$$

Example 10.5: Aligning sequences with penalty-gaps and scoring matrix

The matrix in Figure 10.10 describes the pattern-matching of the two amino-acid sequences "AFCRITAACQ" and "AFFCRVTALIRFRACQ" using similarity scores of BLOSUM62 matrix. The corresponding alignment is shaded by gray color. The cumulative matching score without gap-penalty is 58. There are seven indels (insertions or deletions). Assuming a gap-value of "g" for each indel, the cumulative score would be $58 - 7g$ ($g > 0$). For example, for $g = -2$, the cumulative score would be $58 - 7 \times 2 = 44$. The corresponding aligned subsequences are given in Figure 10.11.

The third technique is BLAST (**B**asic **L**ocal **A**lignment **S**earch **T**ool). The technique starts with matching smaller trimers (three characters long subsequences), and then extends the match using similarity-score matching matrix such as *BLOSUM-62* so the matching-score remains above a threshold-value. For DNA matching, nucleotide-matrix is used (matching nucleotide score = +5, and a mismatch score of −4 or −3). Further extension of a segment is suspended if the matching score drops below a threshold-value. This

FIGURE 10.11   Aligning the sequences using BLOSUM-62 similarity-score matrix

process is repeated for every trimer unless it merges with another extended subsequence. The scheme is fast and is used in finding *homologs* – similar sequences. After identifying the higher matching segments, the probability of the two sequences being aligned is computed using probability theory. If the probability meets the user-specified probability, then two sequences are considered similar. Example 10.6 illustrates it. Besides these popular techniques, there are other techniques such as FASTA- and HMM-based matching.

Example 10.6: BLAST scheme

Consider the same two amino-acid sequences "AFCRITAACQ" and "AFFCRVTALIRFRACQ." Consider matching these two sequences using BLAST based seed technique. First, three amino acid long matching subsequences are identified. Here, the matching trimer subsequences are: "CRI," (or "CRV"), "RIT" (or "RVT"), "ITA" (or "VTA") and "ACQ." These trimer subsequences are extended in both the directions and merged to give two local sequences {"CR(I|V)TA," "ACQ"}.

10.1.6.3 Multifunctional proteins

Most proteins are multifunctional and interact with more than one type of proteins. Some examples are *kinases* and *arrestin*. *Kinase* is a large class of enzymes responsible for phosphorylation – transfer of a phosphate group from a high-energy ATP molecule or another substrate. *Arrestin* is a small class of regulatory proteins that bind to *G-protein-coupled receptors* (*GPCR*). *GPCR* is a large family of trans-membrane receptors outside a cell to sense biomolecules.

Multifunctional proteins play a major role in disease identification. There are many ways to study the role of multifunctional genes: 1) knocking down the protein; 2) downregulating the gene-expression corresponding to a multifunctional protein and 3) domain-level regulation of a protein. Knocking down a protein shuts down the pathway related to the functions associated with the multifunction protein. However, often, an alternate pathway is activated that compensates for the activity of the original pathways. In such a case, the high-level phenotype functional equivalent remains unaltered making it difficult to assess the functionality of the protein.

siRNA or miRNA knock out proteins with shorter half-life and degrades more proteins in addition to the intended proteins causing side-effects. Proteins can also be silenced by changing the key conserved amino-acids in the functional-domain. For example, replacing a conserved *alanine* by *phenylalanine* removes the kinase-activity from the kinases.

10.2 PROTEOEMICS AND PATHWAYS

There are two types of pathways: *metabolic* and *cell signaling*. *Metabolic pathways* have two major classes of reactions: 1) *catabolic* – breaking down complex molecules into simpler molecules; 2) *anabolic* – synthesizing complex molecules from simpler molecules. In addition, reactions store and release energy in a biochemical form. *Cellular respiration* is an example of *catabolic reaction*. In *cellular respiration*, sugar is imported in the cells, and energy is released after the reaction.

Signaling pathways are a cascade of protein–protein interactions induced by external conditions and autoimmune response to foreign bodies or sometimes to own body cells. Signaling pathways have many functions in the body such as 1) communications between the cells; 2) identifying and tagging foreign-bodies; 3) attacking and killing foreign-bodies; 4) cleaning up the dead cells; 5) hormone regulation and 6) activating latent / alternate pathways in case of emergencies or altered external conditions.

10.2.1 Metabolic Pathways

Majority of energy is acquired from the food we eat. We ingest biomolecules comprising protein, carbohydrate and fat. *Fat* is stored in the muscles and is released in the form of complex carbohydrates to acquire needed energy. Body stores energy in *ATP (Adenosine Triphosphate)*. There are three systems to produce energy in a human-body: 1) *phosphagen* that produces *ATP* from *creatinine phosphate* in muscles during intense activities; 2) *anaerobic glycolysis* utilizing glucose in the cytoplasm and 3) *aerobic glycolysis* that transforms carbohydrates and fats in the presence of oxygen.

There are many components of a metabolic pathway: 1) chemical molecules, called *substrates*, at the input-end of the reactions; 2) *products* at the output-end of a reaction; 3) *enzymes* that regulate the speed of metabolic reactions in a metabolic pathway; 4) *coenzymes* whose presence enhances the enzymes' activities. *Enzymes* are a class of proteins that regulate a specific biochemical reaction. The specificity enables a cell to produce desired metabolites without altering other reactions. Enzymes' active sites bind to substrates and alter three-dimensional configuration of substrates to promote reaction-rate. The enzyme itself remains unaltered during the reaction.

In humans, there are many classes of metabolic pathways: 1) alcohol metabolism; 2) carbohydrate and sugar metabolism; 3) cell cycle and mitosis; 4) drug metabolism; 5) lipid and fatty acid metabolism; 6) neurotransmitter metabolism; 7) nucleotide and nucleoside metabolism; 8) peptide hormone metabolism; 9) protein and amino acid metabolism; 10) steroid metabolism and 11) vitamin and coenzyme metabolism besides many other metabolisms such as iron metabolism, nitric oxide metabolism, nitrogen metabolism, selenium metabolism, sulfur metabolism, reversible hydration of carbon dioxide, benzopyrene metabolism and porphyrin metabolism. A complete list of metabolic pathways in different classes is given in Tables 10.4a and 10.4b.

While all metabolic pathways are needed, there are seven major metabolic pathways: 1) glycolysis–glucose oxidation to derive ATP; 2) citric acid cycle (Kerb's cycle) – acetyl-CoA oxidation to get GTP and other intermediates; 3) Oxidative phosphorylation – disposal of the electrons released by glycolysis and citric acid cycle; 4) pentose phosphate pathway – synthesis of pentoses needed for anabolic reactions; 5) urea cycle – disposal of NH_4^+ in less toxic forms; 6) fatty acid β-oxidation – breaking down fatty acid into acetyl-CoA to be used by Krebs' cycle and 7) gluconeogenesis–glucose synthesis from precursors used by the brain. Excess diet amino-acids are converted into precursors of glucose, fatty acids and ketones. Amino acid conversion removes an amine-group through the formation of urea for excretion through urine.

Neurons use glucose as the energy source. Liver maintains a steady concentration of glucose in the bloodstream through a regulated operation of glucose generation and degradation. Liver is also the site of urea synthesis. Muscles use glucose, fatty acids, ketone-bodies and amino acids as the source of energy. Muscles also store creatine-phosphate needed by ADP (Adenosine Diphosphate) → ATP (Adenosine Triphosphate) conversion for energy release without the use of glucose. Kidney can also perform glucogenesis and release glucose in the bloodstream. It is also responsible for the excretion of urea and other undesired electrolytes. Excess nitrogen is removed using joint actions of liver and kidney.

The sugar level in a body is regulated using two hormones: *insulin* and *glucagon*. Both these hormones are synthesized in the pancreas. *Insulin* is released in blood when the sugar-levels are high. *Insulin* stimulates glucose uptake by muscles, glycogen synthesis, triglyceride synthesis, inhibiting glycogen synthesis and promoting glucose degradation. *Glucagon* is released when the sugar-level is very low, and its effect is the opposite of insulin.

10.2.1.1 Modeling metabolic pathways

Metabolic pathways are modeled as a network of reactions. Important information associated with the pathways is: 1) the structure and connectivity of substrates and reactions, 2) rate of the reactions in terms of substrate consumption, metabolite production and energy release-rate and 3) metabolic-level regulation to study cell-reaction and switching to alternate pathways in the absence or low concentration of the preferred enzyme.

TABLE 10.4 List of metabolic pathways

CLASS	COMPONENTS
Alcohol metabolism	1) Ethanol oxidation; 2) inositol phosphate metabolism
Carbohydrate and Sugar metabolism	1) Ascorbate and aldarate metabolism; 2) Chondroitin sulfate and dermatan sulfate metabolism; 3) citric acid cycle; 4) eicosanoid metabolism; 5) fructose metabolism; 6) galactose metabolism; 7) glucose and energy metabolism; 8) glycogen breakdown; 9) glycolysis and gluconeogenesis; 10) glycosaminoglycan degradation; 11) heparan sulfate biosynthesis and metabolism; 12) keratan sulfate and keratin metabolism; 13) mannose metabolism; 14) nepathalene metabolism; 15) pentose phosphate pathway; 16) pyruvate metabolism; 17) respiratory electron transport; 18) sorbitol degradation; 19) starch and sucrose metabolism
Cell cycle and mitosis	1) Cell cycle; 2) mitosis: G1-G1/S phases; 3) mitosis G2-G2/M phases; 4) mitosis: M-M/G1 phases; 5) mitotic metaphase and anaphase
Drug metabolism	1) 4-hydroxy benzoate biosynthesis; 2) abacavir metabolism; 3) acetaminophen metabolism; 4) butirosin and neomycin biosynthesis; 5) caffeine metabolism; 6) codeine and morphine metabolism; 7) nicotine metabolism; 8) tamoxifen metabolism
Lipid and fatty acid metabolism	1) Arachidonic acid metabolism; 2) bile acid and bile salt metabolism; 3) biosynthesis of unsaturated fatty acids; 4) cholesterol biosynthesis; 5) ether lipid metabolism; 6) fatty acid beta oxidation; 7) fatty acid biosynthesis; 8) fatty acid elongation; 9) fatty acid omega oxidation; 10) fatty acid, triacylglycerol and ketone body metabolism; 11) glycerolipid metabolism; 12) glycerophospholipid metabolism; 13) glycosphingolipid biosynthesis; 14) glyoxylate and dicarboxylate metabolism; 15) HETE and HPETE biosynthesis and metabolism; 16) leukotriene metabolism; 17) linoleic metabolism; 18) lipoic acid metabolism; 19) lipids and lipoprotein metabolism; 20) oleate metabolism; 21) palmitate biosynthesis; 22) peroxisomal lipid metabolism; 23) phospholipid metabolism; 24) propanoate metabolism; 25) prostaglandin biosynthesis; 26) sphingolipid metabolism; 27) stearate biosynthesis
Neurotransmitter metabolism	1) acetylcholine synthesis; 2) anandamide degradation; 3) dopamine degradation; 4) serotonin metabolism
Nucleotide and nucleoside metabolism	1) Amino sugar and nucleotide sugar metabolism; 2) purine metabolism; 3) pyrimidine metabolism; 4) wybutosine metabolism
Peptide hormone metabolism	1) Catecholamine biosynthesis; 2) Melatonin degradation; 3) Noradrenaline and adrenaline degradation; 4) Peptide hormone metabolism; 5) Regulation of insulin secretion; 6) Renin-angiotensin system
Protein metabolism	1) Alanine, aspartate and glutamate metabolism; 2) arginine and proline metabolism; 3) aryl amine metabolism; 4) asparagine degradation; 5) beta-alanine metabolism; 6) butanoate metabolism; 7) citrulline metabolism; 8) collagen biosynthesis; 9) creatine metabolism; 10) D-arginine and D-ornithine metabolism; 11) D-Glutamine and D-glutamate metabolism; 12) glutamine metabolism; 13) glutathione metabolism; 14) glycine, serine and threonine metabolism; 15) histamine metabolism; 16) histidine metabolism; 17) lysine metabolism; 18) metabolism of proteins; 19) methionine and cysteine metabolism; 20) one-carbon metabolism; 21) phenylalanine metabolism; 22) putrescine degradation; 23) spermine and spermidine degradation; 24) taurine and hypo taurine metabolism; 25) tRNA aminoacylation; 26) tryptophan metabolism; 27) tyrosine metabolism; 28) urea cycle; 29) valine, leucine and Isoleucine degradation
Steroid metabolism	1) Androgen biosynthesis; 2) estradiol biosynthesis; 3) estrogen metabolism; 4) estrone metabolism; 5) glucocorticoid and mineralocorticoid metabolism; 6) metabolism of steroid hormones and Vitamin D; 7) steroid hormone biosynthesis; 8) thyroid hormone synthesis

(Continued)

TABLE 10.4 *(Continued)*   List of metabolic pathways

CLASS	COMPONENTS
Vitamin and coenzyme metabolism	1) Nicotinate and nicotinamide metabolism; 2) biotin metabolism; 3) folate metabolism; 4) NAD metabolism; 5) pantothenate and CoA biosynthesis; 6) propionyl-CoA catabolism; 7) retinoate biosynthesis; 8) retinol metabolism; 9) riboflavin metabolism; 10) thiamine metabolism; 11) ubiquinol biosynthesis; 12) Vitamin A and carotenoid metabolism; 13) Vitamin B6 metabolism; 14) Vitamin digestion and absorption
Other metabolism	1) Iron metabolism; 2) nitric oxide metabolism; 3) nitrogen metabolism; 4) reversible hydration of CO_2; 5) selenium metabolism; 6) sulfur metabolism; 7) benzopyrene metabolism; 8) porphyrin metabolism

Source: © BioMagResBank – Biological Magnetic Resonance Databank, Available at http://www.bmrb.wisc.edu/data_library/Genes/Metabolic_Pathway_table.html, published with written permission.

The structure and connectivity of the substrates and reactions are modeled as a directed acyclic graph (DAG): vertices are the substrates/products, and the reactions are the edges. Each edge is associated with an enzyme that acts as a catalyst in the reaction. A vertex can have more than one in-degrees and out-degrees. Multiple in-degree means that a biochemical compound is being generated by multiple reactions, and multiple out-degree means that the same biomolecule is involved as the source in many chemical reactions. Figure 10.12 illustrates the graph-based modeling of a metabolic pathway for *glycolysis/gluconeogenesis*.

Example 10.7: Graph-based modeling of metabolic pathways

Glycolysis is the process of converting *glucose* into *pyruvate* and generating small amounts of ATP (energy) and NADH (reducing power). It is a central pathway that produces many precursor metabolites: *glucose-6P, fructose-6P, glycerone-P, glyceraldehyde-3P, glycerate-3P, phosphoenolpyruvate* and *pyruvate. Acetyl-CoA*, another important precursor metabolite, is produced by oxidative decarboxylation of *pyruvate*. Gluconeogenesis is a synthesis-pathway of glucose from noncarbohydrate precursors. It is a reversal of *glycolysis* with minor variations.

The nodes show various substrates and products. The edges illustrate the reactions. The small rectangles show the enzymes involved in the reactions. Enzymes have a nested nomenclature based on their classification. The big rectangles show the connection of a pathway to other pathways.

10.2.1.2 Modeling metabolic reactions

A metabolic reaction depends upon many factors: 1) concentration of the substrate; 2) concentration of the enzyme; 3) presence of other related enzymes that work on the same substrate; 4) concentration of the end-product and 5) environmental factor such as PH-value, salinity and so on. The assumption is that the reaction occurs in a steady state. In a steady state, the concentration of each substrate remains invariant. Each substrate is modeled as a variable, and the rate of change of the concentration is modeled using partial differential equations.

The study of reaction-rate in a large metabolic network is called *metabolomics*. Multiple techniques and their combinations have been used to model the network of reaction-rates: 1) metabolic flux analysis; 2) system of partial differential equations; 3) time-course measurements and clustering; 4) correlation between reactions creating a correlation-network and 5) Gaussian Graph modeling (GGM) based upon multivariate Gaussian distribution.

Metabolic flux analysis measures the rate of reaction, and the enzyme concentration required for the optimum rate of reaction. The correlation-network measures both direct and indirect changes due to a chain of reactions.

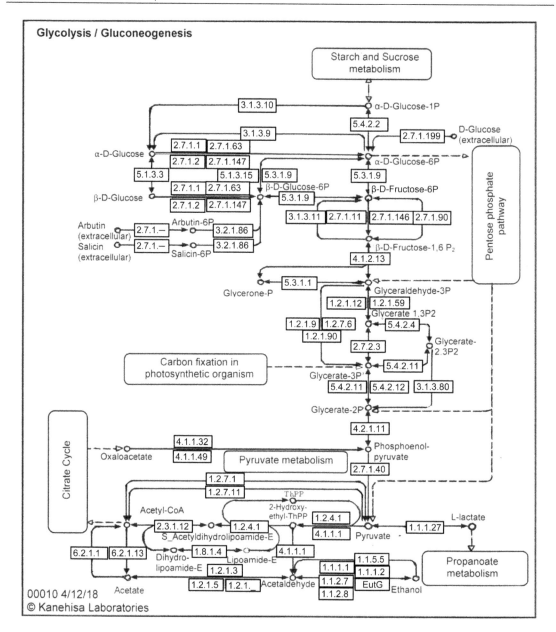

FIGURE 10.12   An illustration of a metabolic pathway for glycolysis/glucogenesis (Pathway Image courtesy Japanese KEGG (Kyoto Encyclopedia of Genes and Genomes) database https://www.genome.jp, published with written permission).

A GGM is a directed acyclic graph: each edge represents pair-wise correlation between each variable (substrate concentration). It uses regression analysis to predict the concentration of the products (molecule being produced) affected by the substrates (input molecule) considering that all the other variables do not affect the outcome. The major difference between correlation-network and GGM is that GGM-based modeling approximates reversible reactions better. If reversible reactions occur, GGM measurements show high pair-wise correlation between substrates involved in reversible reactions.

10.2.1.3 Modeling flux analysis

Metabolic flux is defined as the difference of the forward reaction and the reverse reaction. In an equilibrium state, when the forward reaction is the same as the reverse reaction, magnetic flux is zero. The metabolic flux in a reaction is described using many parameters: 1) the activity of the enzyme catalyzing the reaction, 2) the properties of the enzyme, 3) the metabolites' concentration affecting enzyme activities and 4) the catalytic constants and reaction-coefficients. The study of the metabolic flux is considered important because abnormality in the metabolic flux causes many episodes of cancer.

The reaction rate depends upon the enzyme binding to the substrate. Using partial differential equations, it can be expressed as being proportional to the product of the reactants. The reaction-rate and various coefficients are shown in Equation 10.3. In the reaction, E is the enzyme; S is the substrate; P is the product; κ^{for} is the forward reaction-coefficient; κ^{rev} is the reverse reaction-coefficient and κ^{cat} is the turnover number – the maximum number of substrate molecules converted to product molecule per second.

$$E + S \underset{k^{for}}{\overset{k^{rev}}{\rightleftharpoons}} ES \xrightarrow{k^{cat}} E + P \tag{10.3}$$

The rate of the product formation $\frac{d[P]}{dt}$ is given by Equation 10.4, where *[P]* denotes the concentration of the product P. Under the equilibrium condition, the forward reaction-rate is equal to reverse reaction-rate. Using simplification, the rate of product-formation is modeled as Equation 10.4, where K^d (dissociation constant) is the ratio $\kappa^{rev}/\kappa^{for}$ and V^{max} (*maximum reaction-rate*) is given by $\kappa^{cat} \times [E]_0$. The term $[E]_0$ is the original concentration of the enzyme. The term $[S]$ is the concentration of the substrate.

$$V = \frac{d[P]}{dt} = \frac{V_{max} \times [S]}{K_d + [S]} \tag{10.4}$$

Michaelis–Menten equations (see Equations 10.5 and 10.6) are used to study the kinetics of enzyme's reaction. The rate of reaction is measured at different substrate concentration values. The *Michaelis constant K_m* is the substrate concentration at which the reaction-rate is 50% of the maximum reaction-rate at the saturation point. The flux J through the linear pathway of n unimolecular reversible reaction is given by *Michaelis–Menten equations*. The substrates in the chain-reactions are denoted as S_i ($1 \le i \le n + 1$) with S_1 as the initial substrate and S_{n+1} as the final substrate. The equilibrium constants $K_{1 \to j}$ is the product of reaction constants from reaction one to reaction $S_{j-1} \to S_j$. The coefficient K_j^m is the *Michaelis constant* for the jth reaction. The maximum velocity of the jth reaction V_j^{max} is given by $[E]_j \times \kappa_j^{cat}$, and the coefficient A_j (see Equation 10.7) is equal to $\frac{\kappa_j^{cat} \times K_{1 \to j}}{K_j^m}$.

$$J = \left(S_1 - \frac{S_{n+1}}{K_{1 \to n+1}} \right) \Big/ \left(\sum_{j=1}^{j=n} \frac{K_j^m}{V_j \times K_{1 \to j}} \right) \tag{10.5}$$

$$J = \left(S_1 - \frac{S_{(n+1)}}{K_{1 \to (n+1)}} \right) \Big/ \left(\sum_{j=1}^{j=n} \frac{K_j^m}{K^{cat} \times K_{1 \to j} \times E_j} \right) \tag{10.6}$$

The set of enzymes' concentration that optimizes the metabolic flux is given by $\{E_1^{opt}, E_2^{opt} \ldots, E_n^{opt}\}$, and each E_i^{opt} is described by Equation 10.7, where $E^{total} = \Sigma_{i=1}^{i=n} E_i$ is the total concentration of enzymes.

$$\frac{E_i^{opt}}{E^{total}} = \left(\frac{1}{\sqrt{A_i}} \right) \Big/ \left(\sum_{j=1}^{j=n} \frac{1}{\sqrt{A_j}} \right) \tag{10.7}$$

It is difficult to model a network of pathways that includes a multitude of single chains of metabolic reactions interconnected with cell signaling pathways. In this network of reactions, alternate pathways contribute toward the production and consumption of intermediate substrates. This increases the number of parameters to be modeled.

10.2.2 Cell Signaling Pathways

Signaling pathways are present in all living organisms, including bacteria and eukaryotes. Bacterial signaling pathways are used to: 1) activate latent pathways in response to extreme environmental conditions such as a change in PH-value, excessive temperatures, pressure, saline concentration, antibiotics, humidity level; 2) activate peers for virulence and 3) generate a coordinated response.

The human-body has many types of cell signaling pathways such as: 1) cell-growth; 2) programmed cell-death; 3) cytoskeleton and extracellular matrix; 4) disease drivers, EMT (Epithelial-Mesenchymal Transition for wound healing) and angiogenesis (generation of blood-vessels); 5) epigenetics (changes in the organism due to gene-expression variations); 6) immune cell signaling and 7) neurodegenerative signaling.

Programmed cell-death is the automated death of an inactive cell. There are two types of cell-death: *apoptosis* and *autophagy*. *Apoptosis* occurs due to the lack of survival-factors. *Autophagy* is a natural degradation that disassembles dysfunctional components. *Cytoskeleton* gives a cell its shape and facilitates its movement using protein fibers. *Extracellular matrix* pathways are involved in cell-binding, cell-migration, proliferation and differentiation. The extracellular matrix is made of proteoglycans, water, minerals and fibrous proteins. *Proteoglycans* have a protein core surrounded by long chains of *glycosaminoglycans* – a starch like molecule. *Angiogenesis* is the formation of new blood-vessels. *Epigenetic* is the heritable changes caused by gene-expression variations with no change in genome sequence.

Table 10.5 describes a subset of signaling pathways.

TABLE 10.5 A subset of cell signaling pathways

PATHWAY CLASS	CELL SIGNALING PATHWAYS
Cell growth and viability	1) mTOR signaling; 2) P53 signaling (DNA damage signaling pathway); 3) MapK/Erk in growth and differentiation pathway; 4) autophagy signaling; 5) P13 kinase/Akt signaling.
Programmed cell-death	1) Death receptor signaling; 2) regulation of apoptosis; 3) necrotic cell death.
Cytoskeleton/extraellular matrix	1) Adherens junction signaling; 2) regulation of actin dynamics; 3) microtubule dynamics.
Disease drivers and growth signals	1) Nuclear receptor signals; 2) P53 signaling; 3) receptor tyrosine kinase signaling; 4) hippo signaling; 5) TGF-β/Smad signaling; 6) Wnt/β-catenin signaling; 7) notch signaling.
Angiogenesis and EMT	1) Angiogenesis signaling; 2) contribution of ECM and cytoskeletal factors to EMT; 3) contribution of soluble factors to EMT.
Epigenetics	1) ATP-dependent chromatin modeling complexes; 2) histone, lysine and methylation pathway; 3) nuclear receptor signaling; 4) epigenetics writers and erasers of histones; 5) DNA methylation.
Immune cell signaling	1) B-cell receptor signaling; 2) immune checkpoint signaling in the tumor microenvironment; 3) T-cell receptor signaling; 4) sting signaling; 5) necrosis factor-κβ signaling; 6) IL6 receptor signaling; 7) inflammasome signaling.
Metabolic reprogramming	1) Hypoxia signaling; 2) phosphoinositide signaling; 3) glutamine metabolism signaling; 4) autophagy signaling; 5) Warburg effect; 6) energy metabolism; 7) P13 kinase/Akt signaling; 8) insulin and glucose signaling; 9) mTOR signaling.
Neurodegenerative signaling	1) Dopamine signaling pathway in Parkinson's disease; 2) Alzheimer's disease signaling pathway.

Source: http://www.cellsignal.com. Information courtesy of Cell Signaling Technology, Available at https://www.cellsignal.com/contents/science/cst-pathways/science-pathways, published with written permission.

Example 10.8

Figure 10.13 illustrates a cell signaling pathway for *insulin receptors*. *Insulin* is the major hormone controlling critical energy-functions such as glucose and lipid metabolism. The pathway involves protein–protein interactions between multiple types of proteins such as *kinases, phosphatases, receptors, transcription-factors, enzymes, G-protein, deacetylase, acetylase, GTPase, pro-apoptotic, prosurvival* and *caspase*. In addition, *insulin* signaling inhibits *gluconeogenesis* in the liver, through disruption of CREB/CBP/mTORC2 binding.

A signaling pathway is connected to other signaling and metabolic pathways such as *apoptosis* (cell-death), *lipolysis, gluconeogenesis, glycogen synthesis, fatty acid and cholesterol synthesis, glucose and lipid metabolism and protein synthesis and growth.* Disruption in this complex chain of protein–protein interactions negatively affects the associated pathways that may show up as a disease.

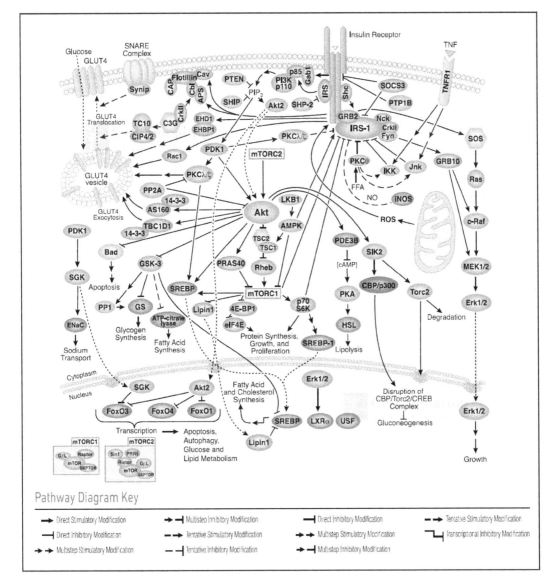

FIGURE 10.13 An illustration of cell signaling pathway for insulin receptor signaling (Image courtesy of Cell Signaling Technology, website: http://www.cellsignal.com, published with written permission).

10.3 GENOME ANALYSIS TOOLS

Genome analysis tools play a major role in understanding the structure and function of genes, gene-variations, pathway variations and facilitate identifying possible causes of genetic diseases. The tools include: 1) identifying the sequence of nucleotides making a genome; 2) identifying gene-sequences in a genome sequence; 3) identifying the coregulated genes in gene-groups called *operons*; 4) predict three-dimensional folding-structure of proteins; 5) identifying protein-domains and their properties; 6) identifying the protein-DNA binding patterns; 7) identifying the protein-protein binding patterns; 8) deriving metabolic pathways in newly sequenced genomes; 9) deriving cell signaling pathways using a combination of artificial intelligence technique such as clustering and microarray analysis; 10) simulation of metabolic flux using partial differential equations; 11) derivation of microRNAs and 12) analysis of variations of genes using multiple techniques such as SNP (Single-Nucleotide Polymorphism), GWAS (Genome-Wide Association Studies), SAGE (Serial Analysis of Gene Expressions), etc.

10.3.1 BLAST Search for Similar Genes/Proteins

BLAST is a sequence-similarity based technique that identifies similar sequences, both nucleotide sequences as in genes and amino-acid sequences, from a large database of DNA-sequences or protein-sequences. It is based upon probabilistically matching largest matching subsequences. BLAST has many applications such as deriving the gene-functionality and annotating the proteins in newly sequenced genomes.

BLAST uses similarity-matrices to match different types of sequences. For example, BLASTN is used to match the nucleotide sequences using nucleotide-matrix. BLASTP matches the amino-acid sequences using BLOSUM or PAM matrices. In addition, there are versions that build their own similarity-matrix by database comparisons. It also supports other similarities based tasks such as 1) identifying possibly conserved domains; 2) identifying receptors and 3) aligning nucleotide and protein sequences.

10.3.2 SNP (Single-Nucleotide Polymorphism)

SNP (Single-Nucleotide Polymorphism pronounced as "Snip") studies the variations of single nucleotides in genes. The rationale is that even a single nucleotide variation in the nucleotide sequence of a functionally significant domain can affect the gene-functionality resulting into a disease. There are many examples of such diseases such as *sickle-cell anemia* - a disease caused by deformity and less oxygen carrying capability of red blood cells. A combination of many SNPs may also correspond to a disease-state.

SNPs are the single base-pair positions where multiple sequence-alternatives exist in the gene-pool so the percentage of such alleles (alternatives) is more than *one percent*. Normally, the regular base-pairing nucleotides is A $\Longleftrightarrow$ T (or T $\Longleftrightarrow$ A) and G $\Longleftrightarrow$ C (or C $\Longleftrightarrow$ G) in a double-stranded DNA helix. In SNPs, T may be substituted by G or C, A may be substituted by G or C, G may be substituted by A or T, or C may be substituted by A or T.

Since the third nucleotide of every codon has many-to-one mapping to the same amino-acid (see Table 10.1), mostly a change in the third nucleotide character of a codon does not alter the corresponding amino-acid. However, change in first two nucleotides of a codon causes the corresponding amino-acid to change. This alteration may cause the change in three-dimensional folding-structure of protein and the functionality resulting into a possible disease.

The change in nucleotide may occur in the protein-coding region or the noncoding region. A change in the protein-coding region may cause a change in the amino-acid as described in the previous paragraph. However, changes in the noncoding region may cause its capability to bind to transcriptase and/or polymerase effecting the gene-expression capability. This affects the concentration of protein being produced that can affect both metabolic pathways for enzymes and cell signaling pathways. Such SNPs that result into amino-acid sequence variation are called *nonsynchronous SNP* (*nsSNP*) and can cause a drug to become dysfunctional. *nsSNPs* are significant targets for developing new drugs.

It is impossible for wet-lab researchers to identify function altering nsSNP. Thus, bioinformatics tools are used to identify potentially functionality changing nsSNP before applying wet-lab techniques. There are three popular approaches for bioinformatics-based SNP analysis: 1) sequence-based approach; 2) 3D-structure-based approach and 3) combination of sequence-based and structure-based approach. Sequence-based approaches are *similarity-based* and *conservation-based*. However, they cannot predict the alteration in functionality since they cannot infer changes in 3D-configuration. 3D-structure-based approach approximates 3D-configuration and lacks accuracy. The combined approach performs much better.

Many bioinformatics tools have been used to identify the functionally significant nsSNP such as: 1) SIFT (Sorting Intolerant from Tolerant); 2) MAPP (Multivariate Analysis of Protein Polymorphism); 3) PANTHER; 4) Parepro; 5) PhD-SNP (Predictor of Human Deleterious SNP) ; 6) SNPS&GO; 7) PolyPhen; 8) SNPS3D; 9) LS-SNP; 10) SNPeffect; 11) SNAP; 12) PMUT; 13) SAPRED; 14) MutPred and 15) MuD. The approach combines the machine learning techniques such as HMM, neural net and support vector machines along with multiple sequence-analysis, 3D-structure matching and matching of the biochemical properties such as hydrophobicity to identify a base-pair in a conserved domain that will be deleterious to the overall 3D-configuration and the functionality of the protein. Conserved domains are considered potentially vital for the functionality of the protein. Hence, any variation is a candidate for nsSNP.

For example, *SIFT* is a multiple-sequence alignment tool to identify the variations in a conserved section. *MAPP* is a multiple-sequence alignment tool that decides the aberrations based upon variation in the biochemical properties of conserved domains. *PANTHER* uses multiple-sequence alignment and HMM-based statistical analysis to derive the variation in the conservation score. *Parepro* uses support vector machine to study the detrimental effect of base-pair change on the overall functionality of a protein. *PhD-SNP* also exploits SVM-based classifier to predict whether change in a local base-pair will cause a deleterious effect on the protein-function. *Polyphen* combines the sequence-based analysis with the 3D-structure matching of a protein sequence.

10.3.3 Linkage Analysis and GWAS

There are three approaches to study nucleotide patterns associated with diseases: 1) studying candidate genes associated with diseases based upon localization using *linkage analysis*; 2) studying the set of genes associated with variation of gene-expression in diseased states using microarray analysis and 3) genome-wide association studies.

Linkage analysis has been traditionally used for hunting the genes of a disease – mutated nucleotides in a gene responsible for a disease. Since descendants from a common ancestor have a high probability of inheriting genes in the same chromosome and in the same loci, it is easy to hunt for the mutations in genes responsible for a disease. However, linkage analysis does not predict the overall heritability of the disease. It only provides a potential. In many common diseases, more than one aberrations in multiple genes cause a disease, and this may be missed in linkage analysis.

One approach to understand common disease genetically is to understand the correlation of more than one SNP associated with the disease. Such an analysis is called *Genome-Wide Association of Sequences*.

10.3.4 SAGE (Serial Analysis of Gene Expressions)

SAGE is a technique to study the gene-expressions by tagging the mRNA using a nine base-pair unique nucleotide sequence. It produces a snapshot of different mRNA population in a sample. For each type of mRNA (protein-template), a unique tag is used. This is done by clipping the mRNA using an *anchoring*

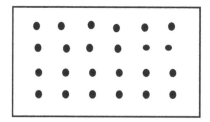

FIGURE 10.14 A matrix of gene-array to study gene-expression

enzyme and then attaching the nine base pair sequence using a *tagging enzyme*. This way, the gene-expression for every protein can be measured. It also allows for the identification of abnormal gene changes and gene-expression profiling. The major application of SAGE is in identifying the gene-expressions in cancer. By comparing the gene-expressions in cancerous cells and normal cell, cancer-specific genes are identified.

10.3.5 CAGE (CAP Analysis of Gene Expressions)

As the amount of sequenced genomes have increased, it has become important to tag them to identify in the newly sequenced genomes and determine the transcription start-sites on a genome-wide scale. CAGE is a mRNA-based tagging scheme that studies and catalogs a short subsequence of mRNA at 5′ end. It also helps in establishing the transcription for different genes in the regulatory network. The typical length of CAP tags is around *20–30* nucleotides.

10.3.6 Microarray Analysis of Gene-Expressions

Microarray analysis was the first scheme to study the signaling pathway using a combination of wet-lab techniques and computational techniques. It is based upon studying the gene-expression of the cells in a normal environment and comparing it with the gene-expressions in the disease condition. In a disease state, certain genes in the signaling pathways are expressed differently suggesting their involvement.

The process of microarray analysis creates a cluster of a maximum amount of gene-mRNAs into a matrix format on an etched glass-plate as illustrated in Figure 10.14 and allow them to interact with the abnormal cells. *Hybridization events* are detected using fluorescent dyes. In the hybridization process, CDNA (cloned DNA) binds to the DNA-probes on microarray slides. Image analysis and clustering techniques are used to study variations in gene-expressions. The experiments are repeated multiple times on different samples of tissues to get statistically significant results.

The data is statistically studied using many artificial intelligence techniques such as decision trees, support vector machines, K-means clustering, hierarchical clustering and Bayesian network. In a micro-array analysis, all the genes with varying expressions cannot be considered important as biomarkers for a disease such as cancer because many genes depend strongly on each other. This requires dimension reduction techniques such as LDA (linear discriminant analysis) to reduce the number of genes in the set.

10.4 GENOME AND PROTEOME ANALYSIS

To find disease, deviations of the functionality have to be identified. The deviations are caused by: 1) gene mutations; 2) insertion and deletion of genes resulting into pathway variations; 3) horizontal transfer of genes (gene jumping) between bacterial species; 4) changes in the biome inside the body causing imbalance and promoting opportunistic bacteria to become pathogenic and 5) aberrations in signaling pathways due to interaction of the antigens with proteins involved in signaling pathway. In addition, bacterial cells become resistant to antibiotics and grow unabated.

To analyze the cause of disease, computational analysis of genes, gene-groups, insertion/deletion of genes, gene-fusion, gene-duplication, proteins and pathways are important. Genes, gene-groups and pathways are derived and analyzed using computational techniques to study their variations in a disease-state. This characterization helps in diagnosing the diseases and facilitates drug discovery.

10.4.1 Identifying Genes

One of the major problems after assembling the genome sequence is to identify various genes using computational techniques. One approach is to identify the start and stop codons and perform similarity-based matching against the known genes in the databases and the potential coding-region between the start and stop codons. In bacteria, identifying the start and stop region is relatively easy and more accurate because regulatory-region and TATA-box occur just before the coding-region. However, it is not so straightforward in eukaryotes as regulatory-region need not be next to the coding-region.

Many artificial intelligence techniques are used to predict the genes such as HMM (Hidden Markov model), ANN (artificial neural network) and generalized probabilistic models to derive 75%–80% accuracy. *GENSCAN* and *GlimmerHMM* are two popular software tools. The GENSCAN software analysis is based upon analyzing various features such as gene-density, regional composition of "C + G," known tags, known coding-regions, known splicing-sites and homology-based search and a generalized Hidden Markov model. A generalized HMM emits complete feature-set at every state that improves the probability of prediction.

10.4.2 Finding Gene-Groups

Genes can have duplicates with somewhat varying functionality, fuse with each other to form a bigger gene with many domains controlled by the same regulatory region or be next to each other regulated by a common promoter. Such gene-groups are identified using a combination of homology analysis (gene similarity using BLAST) and bipartite-graph based matching. Bipartite-graph is a pair of sets of nodes having a special property: an edge (V_1, V_2) has V_1 as an element of the first set, and V_2 is an element of the second set. Besides this property, each genome is modeled as an ordered set of genes, and similarity-score between two genes derives the edges between genes in two genomes. A cluster of edges in a user-defined neighborhood derives the matching-group of genes in two genomes as shown in Figure 10.15. This clustering is derived using a moving window-based analysis.

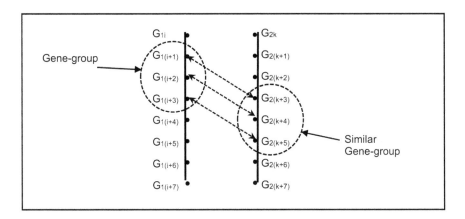

FIGURE 10.15 Identifying gene-groups using Blast-based similarity search and graph matching

10.4.3 Deriving Metabolic Pathways

A metabolic pathway is modeled as a network of biochemicals being consumed and produced with enzymes (proteins) acting as catalysts. The biochemicals being consumed are the source-vertices; the biochemicals being produced are destination-vertices, and the reactions are the edges.

By knowing the proteins and biochemical compounds being produced and consumed, such a network can be derived using model-based reasoning. The subsets of proteins involved in known metabolic pathways are used to build the corresponding pathways in newly sequenced genomes. A missing enzyme will cause either another similar enzyme to substitute for it, or the needed biochemical compound will be supplied externally. However, substitution of an enzyme by another or the presence (or absence of) coenzyme changes the reaction-rate that may not be healthy for a human-body.

This graph-theoretic approach of deriving metabolic pathways can give ambiguous pathways as many enzymes may have the same input and/or output. To reduce this ambiguity, many approaches have been taken: 1) use of the gene-groups to identify the enzymes involved in the pathway; 2) studying the biochemical reaction-rate and 3) using pathways of the evolutionary close genome as a model to study the pathway of the newly sequenced genome.

10.4.4 Deriving Disease-Related Pathways

Microarray analysis helps in the gene-expression analysis. However, most diseases are caused by aberrations in metabolic or signaling pathways. The variations are caused by insertion (or deletion) of genes, a change in reaction-rates due to variations in gene-expressions and mutations within the genes. The cluster analysis of gene-expression data derives the set of genes that are co-expressed during the disease-state. The information about how these genes are involved in an affected signaling pathway cannot be determined in a straightforward manner by simple gene-expression analysis. The relationship between the expressed genes in a signaling pathway involves the study of a sequence of protein–protein interactions and timing behaviors.

10.4.4.1 Correlation-based analysis

The coexpressed set of genes is derived using: 1) similarity: identifying the coexpressed gene-pairs based upon expression correlation and 2) clustering: building hierarchical clusters of the coexpressed gene-pairs based upon similarity analysis illustrated in Figure 10.16a. *Gene 1* and *Gene 2* are strongly correlated across the datasets; genes 3 and 4 are strongly correlated across the datasets.

High correlation shows close functional relationships between a gene-pair. Weaker similarity below a threshold shows functional links that are activated or deactivated in a similar manner under the similar conditions. Based upon the similarity score, a hierarchical cluster (see Figure 10.16b) is made of the gene-expressions that derives the pathway module of the expressed genes as illustrated in Figure 10.16c.

A positive correlation is defined as the variation in the same direction; a negative correlation is defined as the variation in the opposite direction. While analyzing the correlations, both positive and negative correlations are counted as shown in Figures 10.16b and 10.16c. The dashed boxes in Figure 10.16b show a negative correlation between the gene-pairs, and boxes with solid lines show positive correlations between the gene-pairs. Similarly, the dashed edges between the proteins a and c show a negative correlation, and the solid line between the proteins b and d shows the positive correlation.

10.4.4.2 Bayesian network-based analysis

An alternative approach to build pathways is based on maximizing the joint-probability distribution of Bayesian networks that uses mutual information and entropy. As described in Section 3.7.1, Bayesian network is a directed acyclic graph involving N variables where nodes are the variables, and edges show the

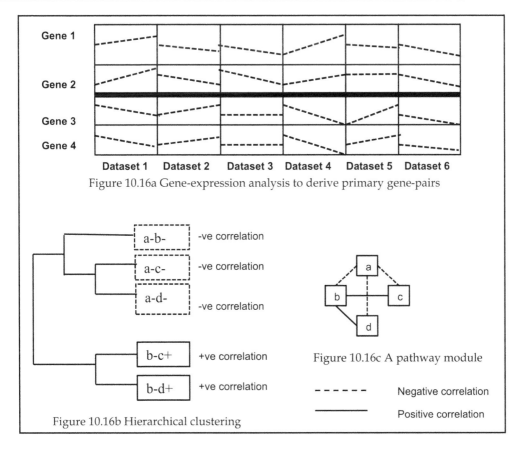

Figure 10.16a Gene-expression analysis to derive primary gene-pairs

Figure 10.16c A pathway module

Figure 10.16b Hierarchical clustering

FIGURE 10.16   A gene-expression analysis scheme for deriving signaling pathway modules: (a) Gene-expression analysis to derive primary gene-pairs; (b) Hierarchical clustering; (c) A pathway module

probability of occurrences of an event given the set of parent conditions. The joint probability distribution in a Bayesian network is given by Equation 10.8.

$$Pr(X_1, X_2, ..., X_n) = \prod_{i=1}^{i=n} Pr(X_i \mid parent(X_i)) \tag{10.8}$$

The mutual information $MI(X, Y)$ between two variables (nodes) is calculated by Equation 10.11, where $En(X)$ is the entropy of the variable X, and $En(X|Y)$ is the conditional entropy of the variable X given the variable Y. The entropy $En(X)$ is given by Equation 10.9, and the conditional entropy $En(X|Y)$ is given by Equation 10.10.

$$En(X) = \sum_{i=1}^{i=n} -Pr(X_i) log(Pr(X_i)) \tag{10.9}$$

$$En(X \mid Y) = \sum_{i=1}^{i=n} \sum_{j=1}^{j=n} -Pr(X = x_i, Y = y_i) \times log(Pr(X = x_i \mid Y = y_i)) \tag{10.10}$$

$$MI(X, Y) = En(X) - En(X \mid Y) \tag{10.11}$$

The approach is to derive the Bayesian network that describes the maximum posterior joint-probability. The pathway module is first split into submodules. To split the graph-module in submodules, the number of nodes connected to other nodes is counted and sorted in the decreasing order. Then the submodules are selected by finding all the nodes connected to the nodes with highest connectivity, until all the nodes are consumed. After splitting the modules into submodules, the combination of parent–child relationships is tried that would maximize the overall probability.

To derive the signaling pathway, mutual information is derived by the maximum similarity of the gene-expression change in two nodes as described in Figure 10.16a. The nodes (genes) are connected if there is a strong gene-expression similarity, and this process is done in a greedy way between a pair of nodes until all the nodes are connected by at least one edge. This forms the undirected pathway module.

Despite the identification of possible pathway modules, the directionality of the paths and transient timing diagrams of gene-expressions in signaling pathways are difficult to establish. Parent–child relationship between the nodes is derived using experimental data from wet-lab data, literature search, and prior knowledge of the pathway in other similar organisms.

10.4.5 Host–Pathogen Interactions

Pathogenesis (causing disease) is a complex phenomenon. A bacterium under different circumstances and environment can become pathogenic based upon host–pathogen interactions. A human-body is a carrier of pathogens that become dominant under a specific environmental condition, which varies from individual to individual depending upon their *microbiome* and immune system response. *Microbiome* is the set of environmental conditions within a human-body where a cell lives. Important portals for human–pathogen interactions are: 1) digestive tract, including mouth and gut; 2) nasal cavity and 3) wounds. In addition, once the pathogens are in the fluid system such as blood they can travel in any part of the body.

The proteins characterized as common virulence-factors include: 1) *adhesins* – bacterial cell-surface proteins that allow the pathogen to get attached to host cells; 2) *toxins* – an antigenic substance produced by the pathogens toxic to the human-cells; 3) *invasins* – bacterial-cell proteins that penetrate and damage host human-cell and are responsible for the spread of pathogens; 4) *secretion systems* – protein complexes that secrete virulence-factors to invade the host cells and 5) *iron uptake system* – proteins involved in taking iron from the environment.

Host–pathogen interactions' study involves: 1) horizontal gene-transfer using plasmid based gene-jumping; 2) antigen–receptor interaction (protein–protein interaction); 3) antigen–antibody interaction (protein–protein interaction); 4) release of toxins by the pathogens; 5) consumption of required substrate by the bacterial-cells causing a biochemical imbalance in a human-body and 6) evading immune system response using external capsule, pumps inside microbial cells that reduce the concentration of administered antibiotics from the microbial cell, genetic mutation to reduce adverse interactions and so on.

10.4.6 Biomarker Discovery

Another aspect to the gene-expression analysis is the discovery of biomarkers associated to a disease-state. An ideal biomarker is a biochemical compound or protein with significant over-expression or reduced expression due to a disease-state, and this variation is specific to the disease-state. Identification of a proper biomarker will assist diagnosing a disease and facilitating drug discovery.

In biomarker discovery, K-means clustering is used to identify clusters of genes with significant gene-expression variations. The clusters of genes are ranked using support vector machines, and clusters with lower ranking are removed. The remaining clusters are merged to give a resulting group of genes that vary significantly in the disease-state. This group forms a potential biomarker for the disease.

Every gene-expression that shows significant changes is not suitable for a biomarker. Some genes in the same pathway show strong correlation. Based upon a correlation-matrix analysis, a representative gene is picked from a set of strongly correlated genes; rest of strongly correlated genes are removed. This gives a minimal set of biomarkers. A biomarker should be easily accessible. For instance, bodily fluids such as urine and blood are a good source to search for biomarkers.

10.4.7 Case Studies in Disease Discovery

Diseases can be of many types such as *genetic diseases* or *infectious diseases*. Genetic diseases can be identified using GWAS study, microarray analysis and linkage analysis. However, infectious diseases may not affect genes. Rather, they affect the pathways and reaction-rates altering the production of the biomolecules in the body and the gene-expression. Complex diseases have multiple causes and rarely involve the abnormalities of a single gene.

GWAS and linkage analysis based studies are good for predicting the susceptibility of the genetic diseases. GWAS is a genetic level information. Its effect on the variations in pathways, gene-expression and protein–protein interaction remains unclear partly due to the lack of complete knowledge of miRNA and gene-silencing and lack of complete knowledge of various pathways and their timing-diagrams.

GWAS is not always useful in diseases caused by a disruption in pathways (both metabolic and signaling pathways) and protein–protein interactions that include foreign-body invasion unless the foreign-body gets embedded into human-genome or invokes a specific immune response. Many viral diseases affect genome of affected humans because viral-genome uses the human-cell replication system to survive and reproduce. Besides current technology does not support very accurate nucleotide profiling of genomes of individual humans. Besides, there are many false-positives in GWAS-based identification that cannot be curated due to the lack of wet-lab confirmations. Another problem is of prioritizing the genes that are likely to be relevant for specific human diseases.

Diseases that cause immune response from the body can be predicted by studying the genome. However, it can only tell the susceptibility of human-body to specific diseases or past-occurrence of specific diseases and does not predict the recurrence. Besides, current human genome sequencing is not sufficiently accurate nor individual-specific to study nucleotide level variations.

An important aspect is to identify the diseases early before it is perceptible to human-sensors. Early detection reduces mortality, pain and the cost of long-term care. Following sections describe certain case studies in identifying cancer – a class of nonreversible terminal disease in the advanced stages. Early detection has been facilitated by microarray analysis and GWAS.

The major issue with microarray analysis is the small number of samples and more than one associated features of a gene. Unless unwanted features are filtered, clustering gives a high percentage of false-positives. Another issue is the lack of an accurate gene–gene correlation matrix that shows dependency between the coexpressed genes.

10.4.7.1 Prostate cancer detection using GWAS

Prostate cancer is the second most common cancer in men worldwide. It varies based on geographic region and ethnicity. The most contributing factors are: age, family history and ethnicity. Linkage analysis identified 8q24 as a significant prostate cancer risk region. Based upon GWAS analysis, 63 novel precursors and chromosomal loci are identified. These precursors come from various long and short arms of 20 chromosomes: one-to-twelve, fourteen-twenty and twenty-three. A membrane protein called CD96 interacts with cancer associated miRNA. Other suspected genes are *MSMB* (*micro-semino-protein-beta protein*) gene, *LMTK2* (*serine/threonine-protein-kinase enzyme*) gene and *KLK3* (*prostate-specific antigen precursor protein*) gene.

10.4.7.2 Breast cancer metastasis profiling

The recurrence of breast cancer and metastasis is a major problem. Microarray analysis of gene-expressions has been used to derive the cluster of genes that act as a signature to predict the recurrence of breast cancer and metastasis. The comparison between the expressed genes of the ten-years surviving patients and the deceased patients identified 27 genes whose expression differed significantly. These genes are located in chromosomes 1, 8, 16 and chromosome X. The patients with the gain in gene-expressions in chromosome one and loss in the gene-expressions in chromosome 16 had a better prognosis. The patients with the loss in gene-expressions in chromosome eight had a poor prognosis.

10.4.7.3 Liver cancer detection

The liver develops cancer through two types of infection HBV (Hepatitis B infection) and HCV (Hepatitis C infection). HBV is caused by Hepatitis B virus and is transmitted through a bodily fluid such as semen or blood contaminated through interaction with the contaminated fluid of another person. HCV is a viral infection caused by the Hepatitis C virus that causes liver inflammation and spreads through contaminated blood. Upregulation and downregulation of many genes vary for HBV and HCV. Identifying these genes separates the infection and act as a biomarker for cancer.

A microarray analysis placed expressed-genes in four categories: 1) significantly upregulated (gene-expression increases); 2) significantly downregulated; 3) insignificant gene-expression variations and 4) genes that are not expressed. A total of 165 genes (or ESTs) were upregulated, and 170 genes (or ESTs) were downregulated. EST (Expressed Sequence Tag) is a short subsequence of genes used to derive the gene-expressions of cloned DNA (cDNA). A total of 19 known genes and 21 ESTS had a separate gene-expression between HBV and HCV making them biomarkers for HBV and HCV. Additional genes were identified that were upregulated or downregulated in cancer improving the knowledge of genes involved in carcinogenesis. The techniques used to study and identify these genes were: 1) hierarchical clustering based upon edit-distance and 2) filtering the gene-expressions using a threshold value that separated significant upregulation or downregulation.

10.5 DRUG DEVELOPMENT

The drugs can be of many types: 1) drug killing or disrupting the foreign invading body; 2) drug boosting the immune system; 3) drug blocking the invading foreign-body to interact with host-protein and 4) drug countering the upregulated or downregulated gene-expressions in a disease-state. Most of the antibacterial drugs belong to the first category. They remove or reduce the transcription and reproduction capability of bacteria, or reduce the capability of the bacteria to form the cell-membranes.

Antibodies, protein with Y-shaped structure, prevent protein–protein interactions between the foreign-proteins (antigens) and the surface cell-receptor of the host by binding. Vaccines improve the immune system by activating the immune-system pathways by injecting dead or diluted version of bacterial (or viral) virulent genes. The body itself produces the antibodies in response to a perceived attack by the foreign-body and strengthens the immune system.

An important aspect of a drug is its ability to be partially specific to harmful strains of bacteria or virus. A nonspecific drug will interact with other useful bacteria, may attack human-cells or produce harmful biomolecule that may cause side-effects. While side-effects of many drugs cannot be avoided, by adjusting the dosage, the harmful effect called *toxicity* can be reduced.

Developing drugs is a long process and is divided into multiple stages. The first stage is to identify the genes and/or pathways that cause the disease. Once identified, the specific reactions or protein–protein interactions that disrupt the regular pathway and/or reaction-rates are identified. After the identification of the receptor-proteins that interact with antigens, the corresponding antibodies are developed using a combination of these techniques: 1) identifying from a library of antibodies (or small biomolecules) that interact with the receptor-proteins; 2) 3D modeling of proteins to modify antibodies (or small molecules) for enhanced protein-protein interactions with the receptors; 3) synthesis of the improved antibodies; 4) multistage clinical trials starting with smaller animals and 5) trial of the drug on volunteering human-subjects and get it approved by FDA (Federal Drug Association) for the use if the toxicities (undesired harmful effects) are minimal, and the dosages are small enough for humans.

There are many issues for drug development: 1) the animal model does not translate effectively to the human clinical trials due to differences in genome and physiology; 2) the efficacy of the drug varies based upon ethnicity, gender, age and region; 3) drugs may have significant toxicity making them useless for real use and 4) the cost of testing is in billions of dollars. The use of gene-expression analysis and GWAS has reduced the cost of drug-discovery and promises the cure of many diseases that were considered incurable. Analysis of gene-expression data has helped in studying the correlation between the diseases and a *phenotype disease network* (PDN) which relates two or more diseases that influence each-other or exhibit dependencies. Gene-expression data has also been used to develop and study disease–disease, disease–drug and drug–drug interactions. These networks have been used to repurpose the drug for diseases sharing similar gene-expressions.

10.5.1 Structure and Function of Antibodies

Antibodies are Y-shaped multidomain proteins generated by the immune system to protect the cell-surface proteins within the host-cells from binding to the invading pathogens' antigens. Multiple disulfide-bonds (S–S bonds) provide stability to its structure. The forked part of the antibody has two parallel strands, and the outer strand binds to antigens; the inner strand is required for stability of the structure. Antibodies circulate through the blood and bind to the antigens thus neutralizing their effect. Antibodies have two types of regions: 1) constant region and 2) variable region. Antigen-binding region is at the tip of the light chain arm (outer arm) of the antibody as illustrated in Figure 10.17. These areas are called *complementary determining region* (CDR) because the antigen-binding site is complementary to the antigens. The remainder of the region is called the *framework region* (FR). The hinge region is flexible to support an angular change to the structure when seeking and binding the antigens.

The function of an antibody is to identify antigens and invoke effector functions that would neutralize the antigen causing the death (or neutralization) of pathogens. There are many types of antibodies such as *IgA* (immunoglobulin A), *IgG1, IgG2, IgG3, IgG4* and *IgM*. The antibodies crosslink large antigens using many epitopes thus inhibiting viral and bacterial infection.

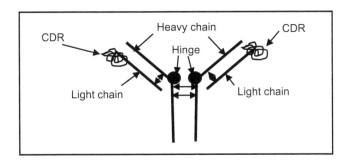

FIGURE 10.17   A schematic of an antibody structure

10.5.2 Virtual Screening in Drug Design

Virtual screening is a bioinformatics technique to prune the library of compounds using computational structure-based similarity analysis under the assumption that similar chemical structure of a molecule represents similar function. *Docking methods* is a popular technique to prune the molecules that do not match. Docking uses matching based on the shape, orientation, protein flexibility (or rigidity), translational and rotational motion and affinity of the amino-acid sequences of CDRs and antigens.

Shape-based technique matches the surface of the binding-site of the protein with the surface of the binding-site of the antigen. Shape matching algorithm uses six degree of freedom: three translational and three rotational for trying various orientations.

A search algorithm is used to identify the CDR based upon the above parameters, and a scoring function is used to compute the match. Higher similarity score above a threshold selects the candidate molecules. The different types of search algorithms used are: 1) systematic; 2) stochastic and 3) simulation based.

Systematic algorithm generates possible binding confirmations by exploring all degrees of freedom. There are three major types of systematic search methods: 1) exhaustive search; 2) fragment-based methods and 3) conformational ensemble. In *exhaustive search methods*, possible translations and rotations of rotatable bonds are performed, and the corresponding binding score is computed. The problem with the exhaustive search is that it is computationally very expensive as the size of the binding area increases. In the *fragment-based method*, the binding area is divided into multiple rigid fragments, then the binding conformation is incrementally extended one fragment at a time. In the *conformational ensemble method*, pregenerated conformations are matched and ranked using binding-energy scores.

Stochastic algorithm makes random changes in the conformational, translational and rotational space of the ligand. Random changes are accepted or rejected based upon a probabilistic criterion such as Boltzmann probability function used in simulated annealing method. Other schemes used for finding the best conformational spaces are evolutionary algorithms and swarm optimization methods.

10.6 IMMUNOINFORMATICS – VACCINE DEVELOPMENT

There are many bacterial and viral diseases which are cured by boosting immune system using vaccines. Some of the most common ones are illustrated in Table 10.6. In addition, many vaccines are under development against leukemia (blood cancer), breast cancer, liver cancer and cancer of other organs.

The basis of the immune system is the ability to identify the genes within the human-body and foreign proteins. The invading pathogens (bacteria and virus) are recognized and attacked, and their growth is inhibited. Immune system uses blood circulatory system and lymphatic circulatory system to protect the body against foreign-bodies. The immune system also possesses long-term immunological memory that once triggered can protect a body against the same or similar pathogen for a long time. Vaccines use the memory of the immune system to quickly activate the immune defense in the future against the actual attack by the foreign body that could be bacteria or virus. The immune system uses a complex signaling pathway to tag foreign-bodies that are killed by neutrophils (white blood cells) circulating in the blood. Neutrophils also release more signaling molecules such as *chemokines* and *cytokines* that bind to foreign-bodies and attract macrophages and natural killer cells.

Macrophages and NK cells (natural killer cells) detect infected cells and destroy them using apoptosis. In addition, bone-marrow produces *B-cells* that produce antibodies after being activated. Another type of important cell involved in immunity is called a *T-cell*, which is produced by

TABLE 10.6   Common diseases for which vaccine has proved useful

	DISEASE	MORTALITY AND MORBIDITY	OTHER SIDE-EFFECTS ON SURVIVORS
1	Diphtheria	Bacteria attack heart, kidney or nervous system, 20% death rate among children	Respiratory failure
2	Hemophilus Influenza	Life-threatening air passage blockage 5% death rate without antibiotics	30% survivors have hearing loss
3	Hepatitis A		Fever, extreme tiredness, a loss of appetite, nausea, vomiting, stomach pain, yellow skin, contagious
4	Hepatitis B	Chronically infected for life 300 times more risk of liver cancer	Infected women have a 90% chance to pass on to fetus
5	Human Papilloma virus (HPV)	Causes 70% of cervical cancer; 90% of anal cancer; 40% of genital; 10% of oral and throat cancer	
6	Influenza ("flu")	Most deaths in infants and elderly above 65	Chronic medical conditions such as heart disease
7	Measles	Common cause of death in infants	Permanent damage to nervous system
8	Meningitis	Brain and spinal cord infection. Mortality rate: 12% for meningitis and 40% if blood stream is infected	20% have hearing loss due to permanent damage to nervous system.
9	Mumps		5% of infected have hearing loss; 15% have heart muscle irritation; A small percentage of testes or ovary inflammation
10	Pertussis (Whooping cough)	Death among infants	Can crack ribs, collapsed lungs, hernia or bleeding in the brain. 5% of children have a nervous system damage; a small percentage have seizures
11	Pneumococcal disease	Infection in blood stream; 20% death among infants	Brain infection; irritation of heart muscles; lung abscess; lung collapse
12	Polio	Paralysis in 1% of the children; 2%–5% of death	
13	Rotavirus		Severe diarrhea among children
14	Rubella	Premature delivery, fetal death	80% of birth have defects such as deafness, eye, heart, nervous system defects and/or mental retardation
15	Tetanus	Severe muscle spasm and 10% death. Combined with pneumonia 50%–70% death rate	Difficulty in eating and swallowing
16	Varicella (chickenpox)	33% fetus death if a pregnant woman is infected. Adult death rate is 25/100,000	Skin rashes and fluid-filled blisters

Source: Centers for Disease Control and Prevention, Available at https://www.cdc.gov/vaccines/index.html, public domain

bone-marrow but matures in the *thymus* – a small gland between the sternum and the lungs at the base of the neck. It would attract or activate macrophages or kill the infected cells directly. The immune system also can evolve and produce a quick response to kill the same foreign-body invasion providing natural immunity.

A bacterium has three layers of protection: 1) cell-capsule, an outer layer, that protects bacteria against macrophages; 2) cell-wall that protects bacteria against death due to interaction with water or bodily fluid and 3) secretion of a toxic substance called *exotoxins* that damage local tissues and also remote organs using the blood circulatory system. The task of the immune system macrophages is to break these protective layers and kill the foreign cells.

Viruses are of many types and mutate much faster than bacteria. For example, influenza virus ("common flu") mutates quickly. In addition, virus-genomes get embedded in human-genome for replication. If the genome of the invading virus evolves quickly, it is difficult for the antibodies to bind to the different strains of antigens present in the viruses. There are two types of variations: 1) *antigenic drift* caused by natural mutation and evolution and 2) *antigenic shift* caused by the recombination of two strains. Antigenic shifts are dangerous and can cause epidemics.

Many types of vaccines are used. The traditional approach was to use *attenuated virulent genes* (disease-causing genes in pathogens) that are too weak to cause the disease but strong enough to invoke the immune response to counter the disease. Traditional vaccines can be: 1) live attenuated vaccine; 2) inactivated vaccine; 3) component vaccines; 4) VLP (virus-like particle) vaccine; 5) liposomal vaccines; 6) adjuvanted vaccines; 7) conjugate vaccine and 8) toxoid vaccine.

Live attenuated vaccines work on the principle that if the pathogen is placed in another species, then only a small fraction of it mutates and survives. However, its effect is attenuated for humans. This process of attenuation is called *passaging*. After many rounds of *passaging*, the pathogen is attenuated enough so it cannot infect humans. However, it can generate a sufficient immune system response to prepare the human-body for the actual pathogen attack.

Inactivated vaccines are inactivated microorganisms that are safe against active infection. However, antigens in the microorganisms cause immune system to respond and make antibodies. Inactivation is achieved using chemicals such as formaldehyde, formalin, heat or radiation. The immunogenic response can cause a mild reaction such as redness, swelling, fever and so on.

Component vaccines contain only the antigen-related part of the microorganisms and are sufficient for an immunogenic response. However, the absence of complete microorganisms may cause a weaker immunogenic response. Variations of component vaccines are *virus-like particles* (*VLP*), which have a similar structure like viruses that are easily recognized by the immune system.

Adjuvant vaccines use a chemical such as emulsified oil or aluminum hydroxide along with the dead or attenuated microorganism to invoke the immunological response. Using adjuvants augments the immune system against the bacteria that coat there outer surface to make them unrecognizable by the immune system. In the presence of adjuvants, antigens bind to other proteins making them recognizable by the immune system. However, it can cause serious immunogenic reaction, an ulcer or muscle-lesions.

Toxoid vaccines are produced by removing the toxin capability of bacteria and injecting in the human-body to generate an immune system response. Two examples of toxoid vaccines are vaccines against diphtheria and tetanus.

Another approach is the development of a *synthetic peptide vaccine* based upon identifying the dominant epitopes on the surface of antigens and embed these antigens into an inactivated viral or plasmid carrier so immune system can develop an antibody against these epitopes. It is important to identify dominant epitopes that can be easily recognized by the immune system. The vaccines produced in this way are much safer and have fewer side-effects or toxic response.

The identification of virulent genes (antigens) and their persistent inactivation by the immune system is the key idea in the design of the vaccines. The attenuated virulent gene can be inserted into a viral genome or a plasmid and injected into the human-body. *Plasmids* are an independent unit in a genome of a microorganism that can self-replicate without the presence of a genome.

The role of the bioinformatics in the vaccine development is in 1) the identification of the antigens – the set of potential virulent genes and pathogenic islands; 2) identifying the receptors; 3) identifying the epitopes – the regions in the surface-receptors that bind and 4) enhancing the affinity of the antibodies using 3D modeling of the antibodies and antigens.

10.6.1 Identifying Target Genes

Virulent genes have been identified using comparative genomics, proteomics and transcriptomics (gene-expression analysis) and the "G + C" (guanine and cytosine) composition analysis. G + C composition of gene changes in an adverse environment such as extreme temperature, acidity and so on and indicates the change through the survival mechanism and pathways in the pathogen. Despite great progress in the bioinformatics methods, the algorithms are based upon approximate modeling and probabilistic sequence-based matching and suffer from limited accuracy suggesting only potential candidates.

Comparative genomics is based upon comparing the genomes of a pathogenic strain with a non-pathogenic strain within the similar class of bacteria and deriving the set of genes and gene-groups that vary in the pathogenic and the nonpathogenic strains. The assumption is that: 1) newly inserted genes are caused by the *horizontal transfer* of genes (using plasmids) from some other genome for the survival of the strain in a different environment and 2) genes mutate significantly changing their functionality. The set of altered (or inserted) genes or gene-groups help pathogens to survive but cause them to a malfunction resulting into a disease.

The genes are matched using strong homology-based matching techniques such as BLAST, Smith–Waterman alignment and bipartite graph-matching technique described in Section 10.4.3. Those genes and/or gene-groups in pathogenic strains that do not have corresponding homologous genes (or gene-groups) in the nonpathogenic strain are considered potential *virulent genes* and the region of the genome that contains multiple such genes is a potential *pathogenicity island*. Homology-based search of virulent genes from other pathogenic strains in a pathogenic strain also identifies the potential virulent genes. The genome-based search for virulent genes is also called *reverse vaccinology*.

Despite identification of potentially virulent genes, the pathogenicity of a bacterium is due to many reasons: 1) deletion of a conserved gene; 2) disruption in the pathway due to deletion of an enzyme in the metabolic pathway; 3) change in the environmental condition invoking a latent pathway in the bacteria and 4) mutation within a gene changing the behavior of the corresponding protein (or enzyme). The deletion of conserved genes is derived using a combination of pair-wise comparison of many genomes, and homology-based search as described in Section 10.4.3. Disruption of pathways is derived partially by identifying the missing genes in the pathway or the corresponding gene-groups.

10.6.2 Epitope Analysis

There are two types of epitopes: 1) linear epitopes and 2) discontinuous epitopes as illustrated in Figure 10.18. *Linear epitopes* have one continuous binding segment of amino-acids and are easy to predict. *Discontinuous epitopes* are a set of several small segments of the antigen brought together due to 3D folding structure of the antigens. There are multiple databases for epitope analysis such as AntiJen, IEDB, BciPep, Los Alamos HIV Database and Protein Databank.

The computational tools for epitope analysis include: 1) hydrophobicity analysis of different protein domains; 2) flexibility within the protein structure; 3) protein secondary structure prediction and 4) loop and side structure analysis in proteins. The analysis of linear epitopes is easier and can be identified experimentally. To predict linear sequence, antigen sequences are fragmented into small subsequences, and their bindings to antibodies are analyzed. However, only 10% of the epitopes are linear, and linear epitopes give a weak immunogenic response.

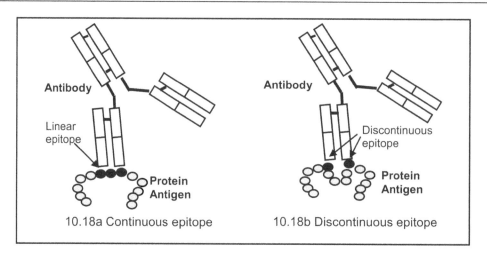

FIGURE 10.18 Types of epitopes: (a) Continuous epitope; (b) Discontinuous epitope

The *Major-Histocompatibility-Complex (MHC)* is a large class of the gene-family involved in the immune system. The proteins encoded by MHC-complex are known as MHC-proteins and play a major role in antigen-binding. The antigens first bind to MHC-proteins, and then are recognized by T-cell receptors that activate the T-cells cloning, and leads to an immune system response. MHC molecules are two types: *MHC I* and *MHC II*. MHC I binds peptides in a narrow range length; MHC II allows a binding for varied length usually between *9* and *25* residues. There is a segment of nine amino-acid long essential for MHC binding to antigens. The prediction of these nine amino-acids is based upon deriving the amino-acid segments with strong binding affinity and is approximated using many artificial intelligence techniques such as neural networks, ant-colony optimization (metaheuristic search), iterative learning, clustering using microarray data and particle-swarm optimization (metaheuristic search). Binding-affinity should be strong for the proper immune-system activation.

Quantitative analysis of affinity is derived using *Position Specific Scoring Matrix (PSSM)*. *Binding-affinity* is approximated as the sum of the affinity-score of various matchings of the amino-acids under the assumption that each binding is independent of the other. *PSSM optimization* is the key to identifying the optimum affinity prediction.

PSSM is calculated by matching and aligning k sequences using a two-step process as given in Equations 10.12–10.14. Equation 10.12 shows the *fractional frequency* f_{ij} of amino-acids at each position by first aligning them then dividing the occurrence of amino-acid i at position j by the number of peptide sequences k. Equation 10.13 calculates the propensity of the amino-acid i at location j by dividing the fractional-frequency at that location by the *background frequency* $f(i)$ of the amino-acid i.

The *expected background frequency* of an amino-acid is calculated by multiplying the probability of nucleotides in the codon and adding the probabilities of each codon in an amino-acid. The resulting probability is multiplied by a correcting factor of 1.057. The frequencies of the nucleotides A, U, G and C occurring are 0.0303, 0.022, 0.0261 and 0.0217, respectively. The correcting factor is used to compensate for the three stop-codons. The expected background frequency of amino-acids is close to actual observed frequency except *Arginine*, which shows a deviation from the probabilistic calculations.

$$F_{ij} = \frac{\textit{cumulative-frequency of amino-acid i at location j}}{k} \tag{10.12}$$

$$P_{ij} = \frac{F_{ij}}{f(i)} \textit{ where } f(i) \textit{ is the background frequency of the amino-acid i.} \tag{10.13}$$

$$\textit{PSSM score } w_{ij} = log_2 \ (F_{ij}) \tag{10.14}$$

Example 10.9: Background frequency of an amino-acid

Calculate the expected background frequency of the amino acid *lysine?*

The amino acid *lysine* has two codons: AAA and AAG. The corresponding mRNA equivalents are AAA and AAU. The random expectation of its frequency is: $1.057 \times (0.303^3 + 0.303^2 \times 0.022) = 1.057 \times (0.028 + 0.020) = 1.057 \times 0.048 = 0.051$.

PSSM is represented by a $m \times 20$ matrix where m is the window-size, and 20 is types of amino acids. The value w_{ij} corresponds to the binding-affinity score of an amino acid j at the location i. For an MHC-peptide-sequence $Psq = \{A_1, ..., A_n\}$, let the experimentally derived affinity be aff^{exp}. There are $n - m + 1$ possible matching subsequence. The overall cumulative binding score is given by Equation 10.15. The value b_{ij} is a binary value equal to 1 if the amino-acid j matches the amino-acid in the peptide sequence at the position i; otherwise, it is 0. The value w_{ij} is the expected match-score calculated using Equation 10.14. There are $n - m + 1$ such scores: one for each matching subsequence of m amino acids. The predicted affinity is given by the matching subsequence that gives the maximum score as given in Equation 10.16.

$$subsequence\ score = \sum_{i=k}^{i=k-m+1} \sum_{j=1}^{j=20} b_{ij} w_{ij} + w_0 \tag{10.15}$$

$$aff^{pred} = max(subsequence\ score_i)|_{i \in \{1, ..., n-m+1\}} \tag{10.16}$$

A total of $(n - m + 1)$ subsequence-scores are calculated. Each score is treated as a particle, and an optimization function f is derived by Equation 10.17 where the first term is the average error between experimentally derived values and the predicted values, and the second-term f_2 regulates the complexity of the weights in PSSM. The parameter σ is used to balance the two terms. The equations for the average-error and the second-term f_2 are given in Equations 10.18 and 10.19.

$$f = average\ error + \sigma \times f_2 \tag{10.17}$$

$$average\ error = \frac{\sqrt{\sum_{i=1}^{i=m} (aff^{exp} - aff^{pred})^2}}{m} \tag{10.18}$$

$$f_2 = \frac{\sum_{i=k}^{i=k-m+1} \sum_{j=1}^{j=20} |w_{ij}|}{20 \times m} \tag{10.19}$$

Metaheuristic search algorithms such as particle swarm optimizations are used to minimize the value of the optimization function f. There are other factors that affect the binding affinity such as length of the peptide, peptide flanking residues, matching biochemical properties that can be incorporated in the analysis to enhance the accuracy.

Most epitopes have 3D discontinuous structure and require 3D-folding methods for an accurate analysis and prediction. This also requires the analysis and prediction of the surface exposed region, hydrophilic (or hydrophobic) analysis, polarity analysis of peptide segments, 3D conformational change during protein-protein interaction of antigen and host-protein interaction, glycosylation information and so on.

Protein structure prediction is possible using a combination of three-dimensional structures of protein domains from an X-ray crystallography database, homology-based prediction of similar structures and 3D modeling software based upon energy minimization. However, except for X-ray crystallography structure, other techniques are based upon approximate modeling.

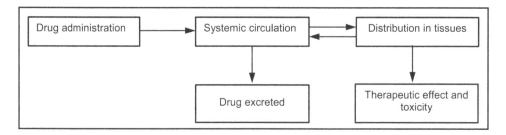

FIGURE 10.19   Overall steps in pharmacokinetics

10.7 PHARMACOKINETICS AND PHARMACODYNAMICS

Pharmacokinetics is the study of effect of the body-system to the drug administered to a human-body. *Pharmacodynamics* is the study of the effect of the drug on a human-body. It involves both the therapeutic effect and adverse effect. After a drug intake, it is *absorbed in the blood-stream, distributed to the tissues, metabolized* and *excreted* from the body. An overall schematic is illustrated in Figure 10.19.

The effect of the drug-metabolism is both positive and negative. The negative effect of the drug is called *toxicity*. For a drug to be effective, its concentration level should be above a threshold. The concentration reaches to a peak level when it is fully distributed, and after that it decreases slowly and exponentially with a half lifetime. If the drug remains too long in the body system, it can cause excessive toxicity. Hence, concentration and duration of excretion of the drug from the body system have to be carefully managed for the drug to be effective with reduced toxicity.

Drugs are administered in five ways: 1) injection; 2) externally on the skin surface; 3) respiration; 4) rectal and 5) oral. The therapeutic effect of drug is stronger if the administration results into quick absorption (such as intravenous injections) and distribution to the target-site. Most drugs are taken orally and are absorbed gastro-intestinally. Absorption relies on the passage through the lipid bilayer and diffusion. A drug with high molecular weight, donating/accepting a high number of hydrogen bonds is poorly absorbed. Lipid-soluble drugs diffuse across the membranes easily. A fraction of the drug reaches the systemic circulation system. The ratio of the drug reaching the systemic circulation and administered drug is called *bioavailability*. The bioavailability of intravenous drug is near 1.0. The bioavailability of the oral drugs is much less than 1, and peaks with a delay as shown in Figure 10.20.

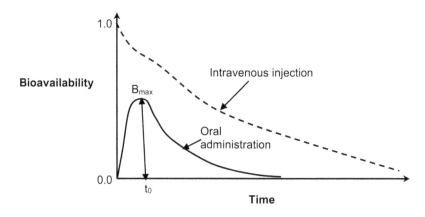

FIGURE 10.20   An illustration of bioavailability of intravenous and oral medication

10.7.1 Drug Distribution and Excretion

Orally administered drugs, after being absorbed in the gastrointestinal system, move to the liver where they are mixed with the systemic circulation system because the blood passes through the liver. The increase in concentration of the drug increases the positive effect along with the associated toxicity. The effectiveness of a medication saturates beyond a specific concentration. The dosage for a medication depends on: 1) the pharmacodynamics studies of the drug that decides the drug-efficacy; 2) bioactivity measurement of the protein–protein interaction and 3) reported toxicity.

The drug concentration after reaching the peak decays exponentially while the distribution saturates above a certain concentration and does not contribute the enhancement of the drug's therapeutic effect. Increasing concentration beyond that value only increases the toxicity without added benefit.

The bioavailability decays due to the reduced availability of the drug in the blood-plasma caused by drug-excretion from the system. Bioavailability depends on: 1) the ratio of the administered drug and the total blood-plasma volume, 2) time taken to distribute the drug to tissues and 3) drug removal from the body using excretion of the body-fluids. The reduction in the concentration is modeled as an exponential decay given by Equation 10.20 where B is the bioavailability at time t, B^{max} is the peak bioavailability, B^T is the threshold of the bioavailability for the drug to be effective, k is the decay-rate constant, t is the duration of the drug in the system and t_0 is the time taken to reach the peak bioavailability.

$$B = B^{max} e^{-k(t - t_0)}$$ (10.20)

Assuming that t_0, the time taken to distribute the drug in the system and reaching the peak availability is relatively smaller than the duration of the drug in the system, the time the drug remains effective is approximated by Equation 10.21 where B^{Th} is the Bioavailability threshold.

$$t^{effective} = \frac{1}{k} ln \left(\frac{B_{max}}{B^{Th}} \right)$$ (10.21)

10.7.2 Drug Efficacy and Toxicity

The efficacy (effectiveness) of a drug is enhanced by increasing the dosage until it saturates at the cost of an increase in the toxicity. It is important to perform drug efficacy and toxicity analysis. The statistically significant data is collected, and a regression analysis is done by studying the effect of the drug-dosage and is therapeutic effect along with the toxicity as a function of time. An effective drug has a significant difference between the drug-effectiveness and the drug-toxicity as illustrated in Figure 10.21.

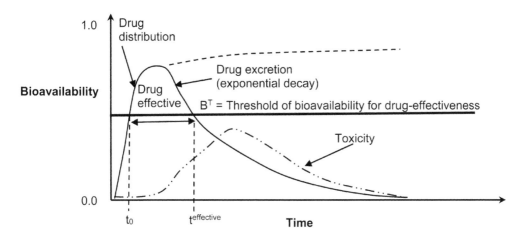

FIGURE 10.21   An illustration of drug effectiveness duration and toxicity

The toxicity should decease as the drug is excreted from the system in a reasonable time. The dosage is adjusted to maintaining the bioavailability above the threshold for effective therapy during the period of the treatment. Different patients respond differently to the same dosage because of the variability of the receptors needed for the effective drug-response and the body-weight. The dosage based upon a statistical study only provides an approximation for the dosage.

10.7.3 Drug–Drug Interactions

Administration of two or more medications simultaneously can cause drug–drug interaction (DDI) that adversely affects and reduces the effectiveness of the administered drugs. It may also change the toxicity of the drug by increasing the duration a drug is in the body. The major cause of drug–drug interaction is the change in metabolism. Drug's action may be inhibited if the metabolic activity of a specific drug is slowed down. Distribution of the drugs may be affected without altering the concentration of a drug in the blood system to affect membrane transportation system.

Drug–drug interactions are hard to predict because of the lack of complete knowledge of drug-metabolism and their effect on various pathways. Many known common medications that affect elderly populations are anticoagulants, antihypertensives and anti-inflammatories. It is very difficult to document the drug–drug interactions because of different combinations of medications and dosages taken by patients. Clinical trials usually recruit a small number of volunteers, and the adverse effect symptoms may not be statistically significant to be reported. Genomic variation of patients also plays a major role in drug–drug interactions.

10.7.3.1 Techniques to study drug–drug interactions

Informatics approaches collect a large amount of data about patients and correlate the data using clustering techniques and regression analysis techniques to predict adverse effects. The sources are: 1) scientific studies; 2) scientific articles and abstracts and 3) adverse events reporting systems such as healthcare professionals and drug companies. The current research is in building large DDI knowledge-bases, extraction of the adverse effects from the reported literatures, data-mining of the adverse events and combining pharmacogenomic and chemoinformatic data. DDI relation extraction is treated as a classification task.

10.8 SUMMARY

The human system and microorganism system have many similarities in the blueprint with subtle variations. The basic building block of components in a cell is an arrangement of four nucleotides: *adenine* (*A*), *thymine* (*T*), *guanine* (*G*) and *cytosine* (*C*). The nucleotides "*A*" and "*T*" are complementary and are bounded by two hydrogen bonds; "*G*" and "*C*" are complementary and are bonded with three hydrogen bonds. DNA is made of nucleotides. Genome stays in the compressed form of the double-helix structure. Genome of higher order organisms (other than bacteria) is further packed in multiple chromosomes.

The higher-level structure comprises protein-mediated pathways involved in a biochemical reaction to consume and produce required biochemicals and signal the information or actuate another reaction. There are two types of pathways: *metabolic* and *signaling*. Metabolic pathways break up molecules to simpler molecules and synthesize complex molecules needed by a body. Enzymes are proteins that act as catalysts for biochemical reactions.

The machinery to convert DNA to protein involves: 1) transcription – conversion of DNA to mRNA, 2) translation – conversion of mRNA to a sequence of amino-acids using ribosomes and 3) maturation to three-dimensional structure of proteins. A cell has multiple coordinating functions such as replication, defense against foreign bodies, energy generation, signaling using protein–protein interaction, breaking up or the synthesis of biochemicals, protection against extreme environmental conditions, intake of nutrients, cell-repair and so on. Proteins have multiple domains and have 3D dynamic configuration that can change with interaction with other proteins. Change in the 3D configuration of a protein may alter its functionality.

Genes become defective with aberrant function due to random mutations caused by the variations in the microbiome conditions such as a change in temperature, pressure, acidity/alkaline nature, nutrient, water, etc. In addition, alternative protein-domains may get activated and protein–protein interactions may be altered. Despite a strong built-in gene-repair mechanism, all mutations are not repaired. Gene-alterations are caused in the post-transcription process due to variations in miRNA. Metabolic pathways get altered due to gene insertions and deletions changing the biochemical reactions and reaction-rate.

Pathways are modeled as directed graphs. Graphs modeling metabolic pathways have substrates and biochemical products as the vertices and the chemical reactions as edges. Protein-interaction networks are modeled as graphs where proteins are vertices and protein-interactions are edges. A protein may interact with more than one protein in a protein-interaction network.

Metabolic reactions depend upon the concentration of the substrates, concentration of the product concentration of the enzymes, environment-factors and the presence of other enzymes. The study of metabolic reactions is called metabolomics. The study simulates the reactions using a combination of techniques such as partial differential equations, time-based clustering, correlation between multiple reactions, regression analysis and Gaussian graph modeling using a multivariate Gaussian distribution. Metabolic flux, the difference between the forward and backward reaction, is modeled using multiple parameters related to enzyme properties, catalytic constants and reaction-coefficients.

A disease can be caused due to abnormalities that disrupt the normal pathways in a human-body. The cause may be genetic variation, an attack by a foreign-body such as a bacterium or a virus, auto-immune disease and cell-death or mutations caused by persistent exposure to extreme conditions and/or radiation. The host–pathogen interaction occurs through *protein–protein interaction* of the foreign-body and the cell-surface receptors in the host-cells. This changes the three-dimensional configuration of the protein and may alter its functionality altering the pathway that results in an abnormal cell-behavior.

A disease is removed by killing the foreign cells using immune system, stopping the foreign cells from interacting with surface receptors of host cells by using antibodies that bind to antigens, making them inactive, arresting the reproduction of the foreign-bodies, weakening their membranes or extra-cellular matrix, or removing their capability to defend themselves against an act by the host immune system.

Bioinformatics has contributed significantly in sequencing genomes, identifying genes and predicting their functionality using homology-based search. Homology-based search uses probabilistic string matching to derive the similar string-segments between genes with known structure/function. Under the assumption that high sequence similarity is equivalent to a similar function (and/or structure), pair-wise genome comparisons have been used to identify the gene-functions, gene-groups and similarity and difference between two genomes. Pair-wise comparisons of genomes have also been used to derive coexpressed gene-groups, conserved genes, horizontal transfer of genes, pathogenicity islands and virulent genes. BLAST and its variations are popular tools for homology-based search. Other sequence comparison tools are based upon dynamic programming such as Smith–Waterman algorithm.

One of the major roles of bioinformatics is to derive the gene-sequence and its function, gene-groups and their roles in metabolic pathways. Gene-sequence is identified using multiple machine learning techniques such as HMM, ANN and generalized probabilistic models. The gene-groups are derived using pair-wise genome comparison that is modeled as a bipartite graph matching. Gene-function is derived by using BLAST-based search with a high level of match under the assumption that sequence-similarity maps to functional similarity.

Besides sequence comparison tools, bioinformatics has also contributed to study of genome-wide presence of SNPs and the prediction of genetic diseases due to SNPs. Most genome sequences have "A" to "T" or "G" to "C" bindings. However, in SNP, this combination is altered at a single nucleotide. This change affects the concentration of the protein being produced that can affect metabolic pathways and cell signaling pathways. Genome-wide studies of SNPs and their various combinations can be analyzed and associated with diseases using GWAS. SAGE (Serial Analysis of Gene Expressions) is used to study the change in gene-expressions by tagging mRNA using a nine base-pair sequence. CAGE is a mRNA-based tagging scheme that catalogs a short subsequence of mRNA at $5'$ end.

Another approach to derive the diseases caused by a change in transcription of the genes to proteins is by an intelligent analysis using microarray data. Microarray is a matrix form of gene-mRNA that interacts with the abnormal cells to derive changes in the gene-expressions. This analysis has been used to derive disease-related pathways by identifying strong correlation between gene-expressions and building a hierarchical cluster of coexpressed genes. The directionality of the pathway is derived using wet-lab data.

Host–pathogen interaction is a complex phenomenon involving factors such as gene transfers between two bacteria, antigen–receptor interaction, antigen–antibody interaction, release of toxins by the pathogens and evasion of the pathogens from the immune system. Gene-expression analysis has been used for biomarker discovery by identifying a minimal cluster of highly correlated uniquely coexpressed gene-clusters. GWAS and linkage analysis have been used to predict susceptibility to the genetic diseases. However, infectious diseases and the disruptions in the signaling pathways caused by the foreign-body invasions cannot be predicted by GWAS and require gene-expression analysis.

Drugs can be of many types. Antibacterial drugs kill or arrest the growth of the pathogen by limiting the reproduction capability or breaking down their protection mechanisms. Vaccines boost human immune system by activating immune system pathways. Antibodies act as a drug by disrupting the potential bindings between antigen and host-cells' surface-protein. Antibodies are Y-shaped proteins where the tip of the side-arms interact and bind to the antigens. Antibody design requires the analysis for effective epitopes (the binding tip of the antibody arms). The process involves an exhaustive search of the known epitope library, 3D affinity analysis, shape matching of the 3D structure of the epitopes with the complementary shape of the antigens and optimizing the binding-energy score.

There are two types of epitopes: linear and discontinuous. Linear epitopes are easy to analyze as the binding sequence is contiguous. The discontinuous epitopes bind the adjacent space due to 3-dimensional structure of the epitopes and are difficult to analyze. Matrix-based PSSM technique is used to derive the amino-acid subsequence with the best affinity prediction for antigen–antibody binding.

Pharmacokinetics and pharmacodynamics are the studies associated with the behavior of the body on the drug and therapeutic effect of the drug on the body, respectively. After a drug is taken, it gets into the bloodstream, and gets distributed in the tissue-cells. After the drug is metabolized, the therapeutic effect takes place for a limited duration when the drug concentration is above the therapeutic-threshold. Increase in the drug concentration increases the effect to a certain extent. However, it saturates beyond a concentration. The adverse effect of the drug increases with the increase in the dosage and duration of the drug in the body. To reduce the toxicity, the drug is excreted from the body in a reasonable amount of time. The concentration of the drug depends on three factors: initial dosage of the drug, volume of the blood-plasma and the constant of the exponential decay of the concentration of the drug in the blood-plasma due to drug-excretion.

10.9 ASSESSMENT

10.9.1 Concepts and Definitions

3D structure; adhesin; adjuvant vaccine; alpha-helix; amino-acid; anabolic pathway; anaphase; antibiotics; antigen; antibody; antigen-antibody interaction; apoptosis; arrestin; ATP; attenuated vaccine; autophagy; B-cell; bacteria; balanced flux analysis; beta-sheet; bioavailability; bipartite-graph matching; BLAST; BLOSUM; CAGE; CAP; catabolic pathway; cell signaling pathways; chromosome;

codon; conjugate vaccine; conserved domain; continuous epitopes; cytokinesis; discontinuous epitope; disease-disease network; disease-related pathways; DNA; domain; double strand; drug distribution; drug-efficacy; drug-drug interaction; drug enhancers; drug-toxicity; dynamic programming; enzymes; epitope; evolution; exon; flux analysis; gap penalty; Gaussian graph modeling; gene-expression; gene-expression analysis; gene-group; gene structure; genomics; glucose pathway; epitope; eukaryotes; FASTA; flux analysis; gene; gene-expression; gene-groups; gene-repair; genome; genome comparison; GGM; GTP; GWAS; helicase; homologs; host-pathogen interaction; hybridization; hydrogen bond; hydrophilic; hydrophobic; immune system; insulin; interaction; intron; invasion; iron uptake system; kinase; linear epitope; linkage analysis; macrophage; maturation; metabolic flux; metabolic pathway; metabolism; metabolomics; MHC; microarray analysis; meiosis; Michaelis–Menten equation; microbiome; miRNA; mitosis; mitotic spindle; mRNA; mutation; nsSNP; nucleotide; open-reading frame; operons; PAM; passaging; pathogen; pathogenicity island; plasmid; polar; polymerase; polypeptide; Pribnow Box; prokaryotic; promoter; protein; protein-DNA interaction; protein-interaction network; protein–protein interaction; proteomics; PSSM; receptor; regulation; repressors; ribosome; rRNA; RNA; SAGE; secondary structure; sex chromosome; sequence alignment; sequence matching; signaling pathway; SiRNA; similarity score; Smith–Waterman alignment; SNP; splicing; substrate; surface receptor; T-cell; TATA box; telophase; tertiary structure; toxicity; toxin; toxoid vaccine; transcription; transcription factor; transcriptomics; translation; transmembrane protein; transposon; tRNA; vaccine; vesicle; Virtual screening; virus; virulent genes.

10.9.2 Problem Solving

10.1 Given a nucleotide subsequences "aug aua agc cac aca ucc ccu uca gau acc cca agg aau," give the corresponding amino-acid sequence using Figure 10.2.

10.2 After inserting a nucleotide "c" in the fourth position in the sequence in Problem 10.1, again derive the amino-acid sequence for the new nucleotide subsequence using Figure 10.2. Note that the insertion will cause the frameshift in the nucleotide sequence.

10.3 Derive the complement of the nucleotide sequence in Problem 10.1.

10.4 Assume that the 13th element in the Problem 10.1 can mutate to one of the three other nucleotides. For each change in the sequence, give the corresponding change in the amino-acid sequence. Repeat the problem if the 15th element in the nucleotide subsequence is altered.

10.5 Analyze the amino acid subsequence derived from the Problem 10.2 using Table 10.1 to identify the different regions of biochemical properties.

10.6 Write a program to automatically derive the biochemical domains in a nucleotide subsequence having similar properties.

10.7 Write a program to derive all the trimers in an amino acid sequence starting from the TATA box.

10.8 Change the program to discover the codons in the nucleotide sequence and convert them to amino acid sequence using the Table 10.1.

10.9 Using the nucleotide-matching matrix as given in Table 10.7, align the two nucleotide sequences using the dynamic programming. The sequences are:

atgccgccaccc and *aggaccggccg*

TABLE 10.7 Nucleotide-matching matrix

	A	G	C	T
A	5	−4	−4	−4
G	−4	5	−4	−4
C	−4	−4	5	−4
T	−4	−4	−4	5

> MKTAYIAKQR QISFVKSHFS RQLEERLGLI
> MWDVIDLSRW QFALTALYHF LFVPLTLGLI
> MRKSVIAIII IVLVVLYMSV FVVKEGERGI
> MNTIFSARIM KRLALTTALC TAFISAAHAD

FIGURE 10.22 Four protein subsequences for Problem 10.11

10.10 Write a program to match two pairs of amino-acid sequences using a dynamic programming technique as described in Figure 10.10 using BLOSUM-62 similarity-matrix as given in Table 10.3. Store BLOSUM-62 matrix as a file that is loaded into a two-dimensional array using user-based interaction.

10.11 A conserved domain is the amino-acid subsequence that is the same after the multiple sequence-alignments. Multiple sequence-alignments are derived using combinations of pair-wise sequence alignments, and the hierarchical tree based upon similarity of the two amino-acid subsequences. Amino-acid subsequences are matched using BLOSUM-62 matrix (see Table 10.3) and using the dynamic programming technique given by Equation 10.1. The technique is to perform pair-wise alignment first to derive the edit-distance between the sequences. This distance is used to build the hierarchical tree. This hierarchical tree is used to align the sequence pairs iteratively so that the resulting sequences are aligned with the next nearest sequence. The process is repeated until all the sequences are aligned. Perform the process for the four amino acid subsequences given in Figure 10.22 to derive multiple sequence alignment and conserved subsequences (if any) that have more than nine amino-acids.

10.12 Amino-acid subsequences are also matched by the biochemical properties. Amino-acids are mapped to the corresponding biochemical properties, and then biochemical properties are matched. For the amino-acid subsequences, given in the Problem 10.11, derive the biochemical properties of the sequences as given in Table 10.1 and match the amino-acid sequences based on hydrophobicity/hydrophilicity giving $+1$ if they match and -1 if they do not match. Based upon the matching score, draw a hierarchical graph.

10.13 Given the gene subsequences, $<g_{11}, g_{12}, g_{13}, g_{14}, g_{15}>$ and $<g_{21}, g_{22}, g_{23}, g_{24}, g_{25}>$ and the matrix of the matching score in Table 10.8, derive gene-groups using a neighborhood-window of size 3. Note that the gene-group keeps expanding, and the adjacent genes are coalesced. Find the direction of the matching: increasing order of genes or inverse order. Assume that the threshold for high-level similarity is 92. Identify the genes which may have similar functions using a threshold of 86.

10.14 For the following nucleotide sequence, give at least ten possible SNPs. Most popular SNPs have $C \Longleftrightarrow T$ or $A \Longleftrightarrow G$ variations.

atgccgccaccccggaccggccg

TABLE 10.8 Matching of the two gene subsequences in a genome

	g_{21}	g_{22}	g_{23}	g_{24}	g_{25}
g_{11}	71	78	89	45	67
g_{12}	65	89	46	58	73
g_{13}	54	70	99	67	56
g_{14}	73	54	68	94	88
g_{15}	78	68	80	67	69

TABLE 10.9   Gene-expression percent change in micro-array for Problem 10.15

171	78	89	45	67
65	89	46	58	73
54	70	174	67	62
172	54	68	88	123
125	68	118	116	115

10.15 A gene-expression variation percentage ratio is illustrated in Table 10.9. Assuming that a threshold of ±20% change from the original is significant, which set of genes are affected by the disease. Classify genes into clusters, which are almost equally affected by the disease. Assume that two genes which are expressed within a variation of ∓4% belong to the same cluster. First derive the most strongly correlated cluster. A gene can belong to more than two clusters probabilistically. Show the Venn diagram for the clusters. For the genes belonging to two soft-clusters, derive the probabilities of belonging to the clusters.

10.16 Assume that a person's cognitive function reduces with age after the age of 60 using the equation *cognitive- capability* $= -2 \times (age - 60) + 100$ on a scale of *100...0*. Assume that the person needs at least *90%* of the cognitive capability to be socially active. At what age should he take medication for cognitive enhancement? Show your computation. (*Use your knowledge of straight lines.*)

10.17 Calculate the expected frequencies of the amino acids *isoleucine, proline* and *histidine* using the knowledge of codons in the amino acids and the knowledge of the frequency of the comprising nucleotides. As described in Example 10.9, the expected frequency is calculated as a cumulative sum of the probabilities of natural occurrence of nucleotides. Use Table 10.1 to derive the comprising codons for each amino acid.

10.18 Make a PSSM for the four amino acid subsequences given in problem 10.11. First align the multiple subsequences using a combination pair-wise alignment and make a hierarchical cluster based upon matching score. Use multiple sequence-alignments as in Problem 10.11. Use Equations 10.12–10.14 to compute the PSSM.

10.19 The bioavailability threshold of a medication is *2.0* micrograms/ml. A human-body has around *5700* ml of blood. Assuming quick diffusion, initial peak bioavailability is given by (dosage/blood-volume). If the half-life time of the medication in a human-body is four hours, and the decay is exponential, what should be the dosage of the medication, and how frequently should it be taken? Keep the dosage (not more than *40.0* mg) and frequency reasonable (not more than six). Show the equation and the computation you used. Plot the decay of the drug for a 24-hour period on a graph paper.

10.9.3 Extended Response

10.20 Explain the gene-transcription process and the role of enhancers in the transcription.

10.21 Explain the gene-translation process to protein.

10.22 Explain the role of miRNA and siRNA on the mRNA and protein formation. How is siRNA different from miRNA?

10.23 Explain mitosis and meiosis and their differences.

10.24 Explain protein–protein interaction and its role in biochemical pathways.

10.25 Explain the function of metabolic and signaling pathways.

10.26 Read about at least one metabolic pathway from the Internet related to human disease and explain.

10.27 Read about at least one signaling pathway from the Internet related to cancer and explain.

10.28 Explain dynamic pair-wise sequence alignment and the role of a gap-penalty.

10.29 Explain the concept of conserved domains and their relationship to gene functionality.

10.30 Explain the approaches of deriving metabolic and signaling pathways.

10.31 Explain the issues involved and limitations in deriving metabolic pathways using bioinformatics approaches.

10.32 Explain the issues involved and limitations in deriving signaling pathways using bioinformatics techniques.

10.33 Explain the issues involved in GWAS and SNP-base analysis in disease discovery.

10.34 Explain microarray analysis techniques and its limitations to study gene-expressions.

10.35 Explain flux analysis modeling and kinetics of enzyme reaction.

10.36 Read about the metabolic pathways in Table 10.4 and explain at least five metabolic pathways including at least one of the carbohydrate and sugar pathways.

10.37 Read literature about five signaling pathways in Table 10.5 and discuss them.

10.38 Explain Michaelis–Menten equation for studying kinetics of enzyme reactions using equations.

10.39 Explain different types of vaccines and give at least one example of each type of vaccine.

10.40 Discuss various gene-expression analysis techniques to derive signaling pathways.

10.41 Explain the operons.

10.42 Explain the principal of immune response, and how does a vaccine boost the immune system.

10.43 Explain the role of bioinformatics in the vaccine development.

10.44 Explain the role and mechanism of antibodies to protect against the foreign-bodies.

10.45 Explain the role and an algorithm to derive pathogenicity islands and virulent genes using homology-based comparative genomics.

10.46 Perform a literature search and explain the algorithms for epitope analysis.

FURTHER READING

10.1 Bansal, A. K., "Bioinformatics in Microbial Biotechnology - A Mini Review," *Microbial Cell Factories*, 4(9), June 2005, DOI: 10.1186/1475-2859-4-19.

10.2 Rocha, Miguel, and Ferreira, Pedro G, *Bioinformatics Algorithms: Design and Implementation in Python*, Academic Press, Cambridge, MA, USA, 2018.

Genome and Cell Structure

10.3 Alberts, Bruce, Johnson, Alexander, Lewis, Julian, Morgan, David, Raff, Martin, Roberts, Keith, and Walter, Peter, *Molecular Biology of the Cells*, 6th edition, Garland Science, Taylor and Francis group, New York, NY, USA, 2015.

10.4 Jo, Bong-Seok, and Choi, Sun Shim, "Introns: The Functional Benefits of Introns in Genomes," *Genomics Informatics*, 13(4), 2015, 112–118, DOI: 10.5808/GI.2015.13.4.112.

Gene Structure

10.5 Archer, Irene M., Harper, Peter S., Wusteman, Fredrick S., "Multiple Forms of Iduronate 2-Sulphate Sulphatase in Human Tissues and Body Fluids," *Biochimica et Biophysica Acta (BBA) – Protein Structure and Molecular Enzymology*, 708(2), November 1982, 134–140, DOI: 10.1016/0167-4838(82)90213-8.

10.6 Gebert, Luca F. R., and MacRae, Ian J., "Regulation of microRNA Function in Animals," *Nature Reviews/ Molecular Cell Biology*, 20, January 2019, 21–37.

10.7 Wong, Nathan, and Wang, Xiaowei, "miRDB: An Online Resource for microRNA Target Prediction and Functional Annotation," *Nucleic Acids Research*, 43(D1), January 2015, D146–D152, DOI: 10.1093/nar/gku1104.

Protein–Protein Interactions

10.8 Blas, Jose R., Segura, Joan, and Fernandez-Fuentes, Narcis, "Computational Tools and Databases for the Study and Characterization of Protein Interactions," In *Protein-Protein Interactions- Computational and Experimental Tools* (editors: Weibo Cai, and Hao Huang), InTech, Rijeka, Croatia, 2012, 379–404.

10.9 Chang Tien-Hao, "Computational Approaches to Protein-Protein Interactions," In *Protein-Protein Interactions – Computational and Experimental Tools*, (editors: Weibo Cai and Hao Huang), InTech, Rijeka, Croatia, March 2012, 231–246.

10.10 Guerich, Vsevolod V., and Gurevich, Eugenia V., "Analyzing the Roles of Multi-functional Proteins in Cells: the Case of Arrestins and GRKs," *Critical Reviews in Biochemistry and Molecular Biology*, 50(5), October 2015, 440–452.

10.11 Hase, Takeshi, and Niimura, Yoshihito, "Protein-Protein Interaction Networks, Structures, Evolution and Application to Drug Design," *Protein-Protein Interactions*, Computational and Experimental Tools, (Editors: Cai, Weibo and Hong, Hao), InTechOpen, March 2012, 405–426, available at https://www.intechopen.com/books/protein-protein-interactions-computational-and-experimental-tools/protein-protein-interaction-networks-structures-evolution-and-application-to-drug-design, DOI: 10.5772/36665.

10.12 HPRD Human Protein Reference Database, Available at http://www.hprd.org, Accessed February 10, 2019.

10.13 Ji, Junghong, Zhang, Aidong, Liu, Chunnian, Quan, Xiaomei, and Liu, Zhijun, "Survey: Functional Module Detection from Protein-Protein Interaction Networks," *IEEE Transactions on Knowledge and Data Engineering*, 26(2), February 2014, 261–277.

10.14 Mitchell, Alex, Chang, Hsin-Yu, Daugherty, Louise, Fraser, Matthew, Hunter Sarah, Lopez, Rodrigo, et al., "The InterPro Protein Families Database: The Classification Resource after 15 Years," *Nucleic Acid Research*, 43(Database Issue), D213–D221, January 2015, DOI: 10.1093/narlgkul1243.

10.15 Rivas, Javier D. L., Fontanillo, Celia, "Protein-Protein Interactions Essentials: Key Concepts to Building and Analyzing Interactome Networks," *PLoS Computational Biology*, 6(6), June 2010, e1000807. DOI: 10.1371/journal.pcbi.1000807.

10.16 Salwinski, Lukasz, Miller, Christopher S., Smith, Adam J., Petitt, Frank K., Bowie, James U., and Eisenburg, David, "The Database of Interacting Proteins,: 2004 Update," *Nucleic Acids Research*, 32(Database Issue), January 2004, D449–D451, DOI: 10.1093/nar/gkh086.

Structure and Function

10.17 Bahamish, Hesham A.A., Abdullah, Rosni, and Salam, Rosalina A., "Protein Tertiary Structure Prediction Using Artificial Bee Colony Algorithm," In Proceedings of the IEEE Third Asia International Conference on Modeling and Simulation, Bali, Indonesia, May 2009, 258–263, DOI: 10.1109/AMS.2009.47.

10.18 Becker, Emmanuelle, Robisson, Benoît, Chapple, Charles E., Guénoche, Alain, and Brun, Christine, "Multifunctional Proteins Revealed by Overlapping Clustering in Protein Interaction Network," *Bioinformatics*, 28(1), January 2012, 84–90.

10.19 Dorn, Márcio, Silva, Mariel B. E., Buriol, Lucioana S., and Lamb, Luis C., "Three-Dimensional Protein Structure Prediction: Methods and Computational Strategies," *Computational Biology and Chemistry*, 53, 2014, 251–276, DOI: 0.1016/j.compbiolchem.2014.10.001.

10.20 Marcotte, Edward M., Pellegrinl, Matteo, Thompson, Michael J., Yeates, Todd O., and Eisenberg, David, "A Combined Algorithm for Genome Wide Prediction of Protein Function," *Nature*, 402, November 1999, 83–85.

Proteomics and Pathways

Metabolic Pathways

10.21 Berman, H. M., Henrick, K., Nakamura, H., and Markley, J. L., "The Worldwide Protein Data Bank (wwPDB): Ensuring a Single, Uniform Archive of PDB Data," *Nucleic Acids Research*, 35 (Database Issue), January 2007, D301–D303.

10.22 Kanehisa, Minoru, Sato, Yoko, Furumichi, Miho, Morishima, Kanae, and Tanabe, Mao, "New Approach for Understanding Genome Variations in KEGG," *Nucleic Acids Research*, 47(D1), January 2019, D590–D595, DOI: 10.1093/nar/gky962.

10.23 Kanehisa, Minoru, Furumichi Miho, Tanabe, Mao, Sato, Yoko, and Morishima, Kanae, "KEGG: New Perspectives on Genomes, Pathways, Diseases and Drugs," *Nucleic Acids Research*, 45(D1), January 2017, D353–D361, DOI: 10.1093/nar/gkw1092.

10.24   Kanehisa, Minoru, and Goto, Susumu, "KEGG: Kyoto Encyclopedia of Genes and Genomes," *Nucleic Acids Research*, 28(1), January 2000, 27–30.

Modeling Metabolic Reactions

10.25   Krumsiek, Jan, Suhre, Karsten, Illig, Thomas, Adamski, Jerzy, and Theis, Fabian J., "Guassian Graphical Modeling Reconstructs Pathway Reactions for High-throughput Metabolomics Data," *BMC Systems Biology*, 5(21), 2011, DOI: 10.1186/1752-0509-5-21.

10.26   Pang, Guangchang, Xie, Junbo, Chen, Qingsen, and Hu, Zhihe, "Energy Intake, Metabolic Homeostasis, and Human Health," *Food Science and Human Wellness*, 3(3–4), September–December 2014, 89–103, DOI: 10.1016/j.fshw.2015.01.001.

Modeling Flux Analysis

10.27   Fiévet, Julie B., Dillmann, Christine, Curien, Gilles, and De Vienne, Dominique, "Simplified Modeling of Metabolic Pathways for Flux Prediction and Optimization: Lesson from an In Vitro Reconstruction of the Upper Part of Glycolysis," *The Biochemical Journal*, 396(Pt 2), June 2006, 317–326, DOI: 10.1042/BJ20051520.

Cell Signaling Pathways

10.28   Leicht, D. T., Balan, Vitaly, Kaplun, Alexander, Singh-Gupta, Vinita, Kaplun, Ludmila, Dobson, Melissa, and Tzivion, Guri, "Raf Kinases: Function, Regulation and Role in Human Cancer," *Biochimica et Biophysica Acta*, 1773(8), August 2007, 1196–1212.

10.29   Palsson, Bernhard O., *Systems Biology: Properties of Reconstructed Networks.* Cambridge University Press, New York, NY, USA, 2006.

Genome Analysis Tools

Blast Search for Similar Genes

10.30   Altschul, Stephen F., Gish, Warren, Miller, Webb, Myers, Eugene W., and Lipman, David J., "Basic Local Alignment Search Tools," *Journal of Molecular Biology*, 215(3), October 1990, 403–410.

10.31   Edgar, Robert C., "Local Homology Recognition and Distance Measures in Linear Time using Compressed Amino-acid Alphabets," *Nucleic Acids Research*, 32(1), January 2004, 380–385.

10.32   Li, W., and Godzik, A., "Cd-hit: A Fast Program for Clustering and Comparing Large Sets of Proteins or Nucleotide Sequences, *Bioinformatics*, 22(13), 2006, 1658–1659.

10.33   Langmead B, and Salzberg S., "Fast Gapped-Read Alignment with Bowtie 2," *Nature Methods*, 9, 2012, 357–359.

SNP (Single Nucleotide Polymorphism)

10.34   Brookes, Anthony J., "The Essence of SNPs," *Gene*, 234(2), August 1999, 177–186.

10.35   Chen, Carla C-M., Schwender, Holger, Keith, Jonathan, Nunkesser, Robin, Mengersen, Kerrie, and Macrossan, Paula, "Method for Identifying SNP Interactions: A Review on Variations of Logic Regression, Random Forest and Bayesian Logistic Regression," *IEEE/ACM Transactions on Computational Biology and Bioinformatics*, 8(6), November/December 2011, 1580–1591.

10.36   Kim, Sobin, and Misra, Ashish, "SNP Genotyping: Technologies and Biomedical Applications," *Annual Review of Biomedical Engineering*, 9, August 2007, 289–320, DOI: 10.1146/annurev.bioeng. 9.060906.152037.

10.37   Mah, James T. L., Low, Esther S. H., and Lee, Edmund, "In Silico SNP Analysis and Bioinformatics Tools: A Review of the State-of-the Art to Aid Drug Discovery," *Drug Discovery Today*, 16(17/18), September 2011, 800–809, DOI: 10.1016/j.drudis.2011.07.005.

Linkage Analysis and GWAS

10.38   Hirschhorn, Joel N., and Daly, Mark J., "Genome-wide Association Studies for Common Diseases and Complex Traits," *Nature Reviews Genetics*, 6(2), February 2005, 95–108, DOI: 10.1038/nrg1521.

10.39   Visscher, Peter M., Wray, Naomi R., Zhang, Qian, Sklar, Pamela, and McCarthy, Mark I., "Ten Years of GWAS Discovery: Biology, Function and Translation," *The American Journal of Human Genetics*, 101(1), July 2017, 5–22, DOI: 10.1016/j.ajhg.2017.06.005.

SAGE (Serial Analysis of Gene Expressions)

10.40 Yamamoto, Mikio, Wakatsuki, Toru, Hada, Akiyuki, and Kyo, Akihide, "Use of Serial Analysis of Gene Expression (SAGE) Technology," *Journal of Immunological Methods*," 250(1–2), May 2001, 45–66, DOI: 0.1016/S0022-1759(01)00305-2.

CAGE (CAP Analysis of Gene Expressions)

10.41 De Hoon, Michael, and Hayashizaki, Yoshihide, "Deep CAP Analysis Gene Expression (CAGE): Genome-wide Identification of Promoters, Quantification of their Expression, and Network Interference," *Biotechniques*, 44(5), April 2008, 627–628, 630, 632, DOI: 10.2144/000112802.

Microarray Analysis of Gene Expressions

10.42 Bolón-Canedo, Verónica, Sánchez-Maraño, Noelia, Alonso-Betanzos, Amparo, Benítez, José M., and Herrera, Francisco, "A Review of Microarray Datasets and Applied Feature Selection Methods," *Information Sciences*, 282(5), October 2014, 111–135.

10.43 Ding, Chris, and Peng, Hanchuan, "Minimum Redundancy Feature Selection from Microarray Gene Expression Data," *Journal of Bioinformatics and Computational Biology*, 2(3), April 2005, 185–205.

10.44 McLachlan, Geoffrey J., Bean, Richard W., and Peel David, "A Mixture Model-Based Approach to the Clustering of Microarray Expression Data," *Bioinformatics*, 18(3), March 2002, 413–422.

Genome and Proteome Analysis

Identifying Genes and Gene-Groups

10.45 Burge, Chris, and Karlin, Samuel, "Prediction of Complete Gene Structures in Human Genomic DNA," *Journal of Molecular Biology*, 268(1), May 1997, 78–94, DOI: 10.1006/jmbi.1997.0951.

10.46 Majoros, W. H., Pertea, M., and Salzberg, S. L., "TigrScan and GlimmerHMM: Two Open-Source ab initio Eukaryotic Gene-Finders," *Bioinformatics*, 16(20), November 2004, 2878–2879, DOI: 10.1006/jmbi.1997.0951.

10.47 Bansal, Arvind K., "An Automated Comparative Analysis of Seventeen Complete Microbial Genomes," *Bioinformatics*, 15(11), November 1999, 900–908.

Deriving Metabolic Pathways

10.48 Bansal, Arvind K., and Woolverton, Christopher, "Applying Automatically Derived Gene-Groups to Automatically Predict and Refine Microbial Pathways," *IEEE Transactions of Knowledge and Data Engineering*, 15(4), July/August 2003, 883–894.

Deriving Disease-Related Pathways

10.49 Chen, Xue-wen, Anantha, Gopalkrishna, and Wang, Xinkun, "An Effective Structure Learning Method for Constructing Gene Networks," *Bioinformatics*, 22(11), November 2006, 1367–1374, DOI: 10.1093/bioinformatics/btl090.

10.50 Cooper, Gregor F., and Herskovits, Edward, "A Bayesian Method for the Induction of Probabilistic Networks from Data," *Machine Learning*, 9(4), October 1992, 309–347.

10.51 Xu, Min, Kao, Ming-Chih J., Nunez-Iglesias, Juan, Nevins, Joseph R., West, Mike, and Zhou, Xianghong J., "An Integrative Approach to Characterize Disease-Specific Pathways and their Coordination: A Case Study in Cancer," *BMC Genomics*, 9(Supplemental I: S12), March 2008, DOI: 10.1186/1471-2164-9-S1-S12.

Host–Pathogen Interactions

10.52 Casadevall, Arturo, and Pirofski, Liise-Anne, "Host-Pathogen Interactions: Basic Concepts of Microbial Commensalism, Colonization, Infection, and Disease," *Infection and Immunity*, 68(12), December 2000, 6511–6518.

Biomarker Discovery

10.53 Yousef, Malik, Ketany, Mohamed, Manevitz, Larry, Showe, Louise C., and Showe, Michael K, "Classification and Biomarker Identification using Gene Network Modules and Support Vector Machines," *BMC Bioinformatics*, 10, October 2009, 337, DOI: 10.1186/1471-2105-10-337.

Case Studies in Disease Discovery

10.54   Arnold, Roland, Boonen, Kurt, Sun, Mark G. F., and Kim, Philip M., "Computational Analysis of Interactomes: Current and Future Perspectives for Bioinformatics Approaches to Model Host-Pathogen Interaction Space," *Methods*, 57, June 2012, 508–518, DOI: 10.1016/j.ymeth.2012.06.01.

10.55   Frasca, Marco, "Gene2DisCo: Gene to Disease Using Disease Commonalities," *Artificial Intelligence in Medicine*, 82, September 2017, 34–46, DOI: 10.1016/j.artmed.2017.08.001.

Prostate Cancer Detection Using GWAS

10.56   Benafif, Sarah, Kote-Jarai, Zsofia, and Elles, Ropsalind A., "A Review of Prostate Cancer Genome-Wide Association Studies (GWAS)," *Cancer, Epidemiology, Biomarkers and Prevention*, 27(8), 845–857, August 2018, DOI: 10.1158/1055-9965.EPI-16-1046.

Breast Cancer Metastasis Profiling

10.57   Mittempergher, Lorenza, Saghatchian, Mahasti, Wolf, Denise M., Michiels, Stefan, Canisius, Sander, and Dessen, Philippe, et. al., "A Gene Signature for Late Distant Metastasis in Breast Cancer Identifies a Potential Mechanism of Late Recurrences," *Molecular Oncology*, 7, 2013, 987–999.

10.58   Möllerström, Elin, Delle, Ulla, Danielsson, Anna, Parris, Toshima, Olsson Björn, Karlsson, Per, and Helou, Khalil, "High-Resolution Genomic Profiling to Predict 10-year Overall Survival in Node-Negative Breast Cancer," *Cancer Genetics and Cytogenetics*, 198(2), April 2010, 79–89.

Liver Cancer Detection

10.59   Okabe, Hiroshi, Satoh, Seiji, Kato, Tatsushi, Kithara, Osamu, Yanagawa, Renpei, Yamoka, Yoshio, Tsunoda, Tatsuhiko, Furukawa, Yoichi, and Nakamura, Yusuke, "Genome-wide Analysis of Gene Expression in Human Hepatocellular Carcinomas using cDNA Microarray: Identification of Genes Involved in Viral Carcinogenesis and Tumor Progression," *Cancer Research*, 61(5), March 2001, 2129–2137.

Drug Development

10.60   Bansal, Arvind K., "Role of Bioinformatics in the Development of Anti-Infective Therapy," *Expert Review of Anti-Infective Therapy*, 6(1), January 2008, 51–63.

10.61   Brown, David K., and Bishop, Özlem B., "Role of Structural Bioinformatics in Drug Discovery by Computational SNP Analysis: Analyzing Variations at Protein Level," *Global Heart*, 12(2), June 2017, 151–161.

10.62   Buchan, Natalie S., Rajpal, Deepak K., Webster, Yue, Alatorre, Carlos, Gudivada, Ranga C., Zheng, Chengyi, Sanseau, Philippe, Koehler, Jacob, "The Role of Translational Bioinformatics in Drug Discovery," *Drug Discovery Today*, 16(9/10), May 2011, 426–434.

10.63   Kortagere, Sandhya (editor), *In Silico Models for Drug Discovery, Methods in Molecular Biology*, Springer Science + Business Media, New York, NY, USA, 2013, 993.

10.64   Strømgaard, Kristian, KrogsgaardLarsen, Povl, and Madsen, Ulf, *Textbook of Drug Design and Discovery*, CRC Press (Taylor & Francis Group), Boca Raton, FL, USA, 2017.

10.65   Young, David C., *Computational Drug Design: A Guide for Computational and Medicinal Chemists*, John Wiley and Sons, Hoboken, NJ, USA, 2009.

Structure and Function of Antibodies

10.66   Carter, Paul J., "Potent Antibody Therapeutics by Design," *Nature Reviews Immunology*, 6, May 2006, 343–356, DOI: 10.1038/nri1837.

10.67   Edelman, Gerald M., "Antibody Structure and Molecular Immunology," *Science*, 180(4088), May 1973, 830–839.

10.68   Huang, Sheng-You, and Zou, Xiaokin, "Advances and Challenges in Protein-Ligand Docking," *International Journal of Molecular Science*, 11(8), August 2010, 3016–3034, DOI: 10.3390/ijms11083016.

Virtual Screening in Drug Design

10.69   Lill, Markus, "Virtual Screening in Drug Design," In *Silico Methods for Drug Discovery* (editor: Kortagere, Sandhya), Methods in Molecular Biology, 993, Springer Science + Business Media, New York, NY, USA, 2013, 1–12, DOI: 10.1007/978-1-62703-342-8_1.

Immunoinformatics – Vaccines Development

10.70 Abbas, Abdul K., Litchtman, Andrew H., and Pillai, Shiv, *Cellular and Molecular Immunology*, 8th edition, Elsevier Saunders, Philadelphia, PA, USA, 2014.

10.71 Atkinson W, Wolfe S, and Hamborsky J, (editors), *Epidemiology and Prevention of Vaccine-Preventable Diseases*, 12th edition, Centers for Disease Control and Prevention, Public Health Foundation, Washington, DC, USA, 2011.

10.72 Bowman, Brett N., McAdam, Paul R., Vivona, SDandro, Zhang, Jin X., Luong, Tiffany, Belew, Richard K., Sahota, Harpal, Guiney, Donald, Valafar, Faramarz, Fierer, Joshua, and Woelk, Chreistopher H., "Improving Reverse Vaccinology with a Machine Learning Approach," *Vaccine*, 29(45), October 2011, 8156–8164, DOI: 10.1016/j.vaccine.2011.07.142.

10.73 He, Yongqun, Rappuoli, Rino, De Groot, Anne S., and Chen, Robert T., "Emerging Vaccine Informatics," *Journal of Biomedicine and Biotechnology*, 2010, Article Id 218590, June 2011, DOI: 10.1155/2010/218590.

10.74 Janeway, Charles A., Travers, Paul, Walport, Mark, and Shlomchik, Mark J., *Immunobviology: The Immune System in Health and Disease*, 9th edition, Garland Services, New York, 2017.

10.75 Lew-Tabor, A. E., and Valle, Rodriguez M., "A Review of Reverse Vaccinology Approaches for the Development of Vaccines Against Ticks and Tick-Borne Diseases," *Ticks and Ticks Borne Diseases*, 7(4), June 2016, 573–585, DOI: 10.1016/j.ttbdis.2015.12.012.

10.76 Meinke, Andreas, Henics, Tamas, and Nagy, Eszter, "Bacterial Genomes Pave the Way to Novel Vaccines," *Current Opinion in Microbiology*, 7(3), 2004, 314–320, DOI: 10.1016/j.mib.2004.04.008.

10.77 Nauta, Jozef, "Basic Concepts of Vaccine Immunology," Chapter 1 in *Statistics in Clinical Vaccine Trials* (editor: Nauta, Jos), Springer-Verlag, Berlin/Heidelberg, Germany, 2011, 1–12, DOI: 10.1007/978-3-642-14691-6.

10.78 Sirskyz, Danylo, Diaz-Mitoma, Francisco, Golshani, Ashkan, Kumar Ashok, and Azizi, Ali, "Innovative Bioinformatics Approaches for Developing Peptide-Based Vaccines Against Hypervariable Viruses," *Immunology and Cell Biology*, 89(1), January 2011, 81–89, DOI: 10.1038/icb.2010.65.

Identifying Target Genes

10.79 Casadevall, Arturo, and Pirofski, Liise-Anne, "Host-Pathogen Interactions: Redefining the Basic Concepts of Virulence and Pathogenicity.," *Infection Immunity*, 67(8), August 1999, 3703–3713.

10.80 Hacker, Jörg, and Kaper, James B., "Pathogenicity Islands and the Evolution of Microbes," *Annual Review of Microbiology*, 54, October 2000, 641–679.

Epitope analysis

10.81 He, Linglin, and Zhu, Jiang, "Computational Tools for Epitope Vaccine Design and Evaluation," *Current Opinion in Virology*, 11, April 2015, 103–112, DOI: 10.1016/j.coviro.2015.03.013.

10.82 He, Yongqun, Rappuoli, Rino, De Groot, Anne, and Chen, Robert T., "Emerging Vaccine Informatics," *Journal of Biomedicine and Biotechnology*, 2010, Article Id: 218590, December 2010, DOI: 10.1155/2010/218590.

10.83 Ponomorenko, Julia, Bui, Huynh-Hoa, Li, Wei, Fusseder, Nicholas, Bourne, Philip E., Sette, Alessandro, and Peter Bjoern, "ElliPro: A New Structure-Based Tool for the Prediction of Antibody Epitopes," *BMC Bioinformatics*, 9: 514, December 2008, DOI: 0.1186/1471-2105-9-514.

10.84 Zhang, Wen, Liu, Juan, and Niu, Yanqing, "Quantitative Prediction of MHC-II Binding Affinity Using Particle Swarm Optimization," *Artificial Intelligence in Medicine*, 50(2), October 2010, 127–312, DOI: 10.1016/j.artmed.2010.05.003.

Pharmacokinetics and Pharmacodynamics

10.85 Rowland, Malcolm, and Tozer, Thomas N., *Clinical Pharmacokinetics and Pharmacodynamics*, Lippincott, Williams and Wilkins, Philadelphia, PA, USA, 2011.

Drug–Drug Interactions

10.86 Dmitriev, Alexander V., Karasev, Dmitry A., Lagunin, Alexey, and Rudik, Anastasia V., "Prediction of Drug-Drug Interactions Related to Inhibition or Induction of Drug-Metabolizing Enzymes," *Current Topics in Medicinal Chemistry*, 19(5), January 2019, DOI: 10.2174/1568026619666190123160406.

10.87 Ferdousi, Reza, Safdari, Reza, and Omidi, Yadollah, "Computational Prediction of Drug-Drug Interactions based on Drugs Functional Similarities," *Journal of Biomedical Informatics*, 70, June 2017, 54–64, DOI: 10.1016/j.jbi.2017.04.021.

10.88 Percha, Bethany, and Altman, Russ B., "Informatics Confronts Drug-Drug Interactions," *Trends in Pharmacological Sciences*, 34(3), March 2013, 178–184.

End-User's Emotion and Satisfaction

11

Contributed by Leon Sterling

BACKGROUND CONCEPTS

Section 1.4.4 Privacy and Security; Section 1.6.1.4 Maintaining Patients' Privacy; Section 1.6.10 Biosignal Processing; Section 1.6.12 Pervasive Healthcare; Section 2.2 Digitization of Sensor Data; Section 2.5 Modeling Multimedia Feature Space; Section 2.9 Middleware for Information Exchange; Section 2.10 Human Physiology; Chapter 4 Healthcare Data Organization; Section 5.7 Medical Image Archival and Transmission; Section 6.2 Modeling Medical Imaging Information using DICOM; Chapter 9 Pervasive Healthcare

The potential for health-informatics software to help practitioners improve health outcomes for patients is enormous. The purpose of health-informatics software is to improve the patients' quality of life and surviv-ability, optimize the utilization of resources, improve patients' recovery time, reduce the cost of patient care, improve people's mobility and improve future care with the discovery of new patterns of diseases, diagnosis, drugs and treatments. Many classes of experts are involved in developing health-informatics software, including experts from various domains whose knowledge is captured to develop health-informatics software and end-users of the software. End-users include physicians, specialist doctors, patients, nurses, pharmacists and occupational therapists. Each user-class has its own requirements and needs the software to behave in an optimized way to utilize the users' time and other resources in a user-friendly and meaningful way.

The effective utilization of health-informatics software depends on the advocacy, adoption and appropriation of the software by a wide range of stakeholders, with diverse abilities and motivations. The emotional aspects of user-interaction with the software and convenience are important—arguably essential—for utilization but are routinely ignored by software engineers. Any software that is not intu-itively simple to understand, asks more from the users and lacks awareness of medical conditions of patients that could prohibit them from interacting in a programmed way will lose engagement from the users. The literature indicates that the emotional expectations of the end-user are critical if users are to appropriate technology meaningfully into their lives.

Interactive software needs the involvement of the end-users so that it meets their emotional needs and provides the feeling of satisfaction. Software designers and programmers are often unaware of the stress they can cause to the end-users by being very technical and by expecting too much irrelevant information from the end-users, and/ or being confused in their requests. The focus of software developers and pro-grammers is on the system development, information safety and security. Often, they ignore the experien-tial aspects of interacting with the software, including user-friendliness in the software-interface, and the stress that interacting with inappropriate and/or confusing software can cause. Further, they do not try to communicate a feeling of human warmth and reassurance. Often, the medical software replaces humans in the process of restructuring and efficient utilization of resources. While the software can handle the information and computational processing capabilities of the human agents, they are still incapable of showing emotional intelligence, warmth and concerns for the patients' condition that is communicated by the human agents to the patients.

This chapter discusses the need for incorporating the emotions, need for satisfaction and feeling of well-being of end-users, especially patients, during the design and implementation phase, and describes a model for the same. Several research projects in the health domain with a component of building software are discussed. For each, emotional factors have been considered during requirements elicitation and design. The research projects include a screening test for patients suffering from depression, a system to promote recovery and self-management for people with psychosis, helping people with sleep disorders and emergency well-being checks for elderly people living on their own.

11.1 NEED FOR END-USERS' EMOTIONAL WELL-BEING AND SATISFACTION

To develop software, the requirements arising from emotional needs differ from regular requirements. For example, consider the requirement of security in an e-commerce system. The usual understanding of security is technical, involving encryption for secure communication. However, the technical aspects are invisible to most users. Users must perceive that: 1) the e-commerce system is secure when they use it; 2) the system is user-friendly and does not prompt them for irrelevant personal security-related questions, or else they are unlikely to keep using it. Technical efficiency is an important criterion. However, the perception of security and user-friendliness is paramount.

Feeling secure has an emotional component and needs to be explicitly catered for when designing the workflows involving interactions between a user and a software-system. Incorporating emotion in a design refers to three levels of cognitive and visceral processing associated with the product: 1) "look and feel"; 2) behavioral processing associated with product characteristics such as performance and usability and 3) reflective processing associated with its impact on self-image and satisfaction.

In a case study of an emergency alarm system for elderly people, despite being well-engineered and highly reliable, the system was perceived as a failure by the users because it did not meet the required emotional goals. Many elderly people choose not to wear their emergency alarm pendant due to the emotional stigma attached to it. As another example, a game must be fun for a player to adopt it. Ensuring a player is engaged and having fun while playing the game is part of the requirements and design process not covered by standard concepts such as ease of use and efficient response-time. Technology is only valuable and adaptable if it addresses and fulfills people's emotional needs and satisfaction such as experiencing fun, feeling engaged and feeling valued. An automated system such as health-informatics software should provide the same personal touch and warmth as good nursing-care would. Focusing on technical issues while ignoring emotional factors fails to get technology adapted despite its long-term advantages.

Another example is to provide reminders for patients with Alzheimer's disease to take their medication. A technically sound software system expects patient to press a keyboard in response to the question "Have you taken the medication today?" However, the patient may forget to take the medication, press the button after taking the medication or press the button even without taking the medication. The software needs to be intuitive to test different possibilities without asking too many questions and call for human help if doubt occurs without overburdening the patient.

The instincts of most health software developers are to focus on the technical aspects: be it the collection of clinical data from doctor and nurse; clinical data analysis and/or communication of health information between the lab to doctor, lab to a patient or between two healthcare providers. Besides technical aspects, satisfaction of emotional needs as perceived by the users' needs to be treated is a crucial end goal. Our earlier research and research by others have shown that even if the emotional satisfaction was incorporated as one of the goals, the uniform process did not allow sufficient focus on emotional aspects. Emotional goals have to be separated from other quality requirements, with greatly increased stakeholder participation and engagement in the software development and use.

11.2 MODELING EMOTION DURING SOFTWARE DESIGN

The software system has to essentially model the patients, their medical capabilities, past trainings, behavior patterns, expectations, limitations and constraints in medical and social terms and needs to address this model in user-interaction in an intuitive human-like manner. The goals are multifaceted, encompassing functional goals, quality goals and emotional goals. Models are described at three levels of abstraction: 1) a motivation level where models give the overall picture of what is trying to be achieved by the system; 2) a design level where models specify the types of software agents needed to perform required tasks and 3) an implementation level suitable for developers to specify communication and user-interface features, for example.

Consider elderly physically-challenged persons living independently. A software system has to be developed for a "wellbeing check" and "emergency alarm system" to interact with them and help them communicate with their relatives and neighbors in the case of emergency situations. At the same time, it should not take away their independence. Figure 11.1 describes a motivational scenario for the emergency care domain. The scenario has four components: a name, a description, a quality description, and an emotional description. The scenario, quality and emotional descriptions are self-explanatory.

The system has three components: 1) a daily well-being check; 2) an emergency alarm system and 3) a communication-unit to inform relatives and neighbors in the case of emergency. The communication system is automatically triggered in case an emergency system is activated either manually or when a fall is detected by the system. The daily "well-being check" is initiated from a monitor using a button. The older person at home is required to push the button at a set-time each day between a specific time-period. If the button is not pushed, the caregiver and responsible relatives receive a phone call from the

Scenario	Provide technology-supported care
Scenario Description	Neighbor and relatives keep an eye on the elderly person to ensure that everything is fine. Technology supported care is added to monitor the person's wellbeing. The technology supported care consists of a keep-in-touch activity that the elderly persons perform periodically, and an emergency alarm activated by the person.
Quality description	The system must be accessible to an elderly person, and be invisible to others interacting with her. The system must be responsive and flexible to adapt to the specific needs of the person. It must fit her lifestyle.
Emotional Description	The relatives want to feel reassured that the elderly person is well. The elderly person wants to feel that system s integrated in her life, is robust, and under their control. Using the system should give her added feeling of independence. By using the system, the elderly person should feel both physically and emotionally safe, and cared for by the caregivers and/ or relatives without being overburdened.

FIGURE 11.1 Motivational scenario for technology-enhanced care supporting well-being

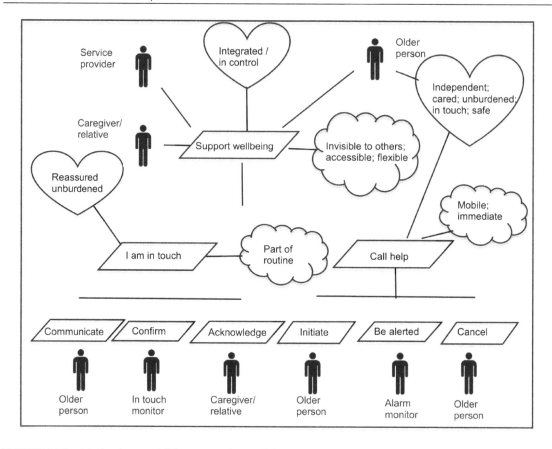

FIGURE 11.2   Motivation model for supporting well-being

communication system to inquire, visually monitor and physically check the patient. To enhance mobility of the patients, emergency alarm systems enforce wearing a pendant around patients' necks. The pendant has a button that is pressed in an emergency such as having fallen in the house and being unable to get up.

Figure 11.2 contains a motivational model for the overall system. There are four types of entities in the model: 1) functional requirements of the system, 2) agents who interact with the system, 3) quality goals and 4) emotional goals. In Figure 11.2, parallelograms denote the functional requirements; human figures denote the agents; cloud-shapes denote the quality goals and heart-shapes are emotional goals.

The functional goal "support well-being" has two subgoals: 1) "I am in touch" and 2) "call help." The subgoal "I am in touch" abstracts the action pressing-down the button on a monitor provided by the emergency services' provider to indicate that the elderly person living independently is ok. It includes the subgoal "communicate" where the elderly person communicates with the monitoring service. The monitor confirms that the elderly person has communicated, and the caregiver/relative needs to acknowledge the communication sent by the monitor. The second subgoal "call help" involves an initiation by the elderly person or automated fall-detector, which triggers behaviors on the alarm-monitor. The model depicts a requirement that the person can cancel the call. The exact details of the acknowledgement are specified after an elicitation and the design process.

The three heart-shapes represent emotional requirements. Elderly persons want to feel in control, have their own health in their hands and have the well-being integrated into their lives. They want to feel independent, yet cared for. They want to feel safe with early respectful identification of issues, in touch with their caregivers, and not be a burden on family and friends. The caregiver/relative wants to be reassured that the older person's well-being is being supported without feeling a large burden on themselves.

These are not usually expressed requirements, but they are easily elicited by asking people how they want to feel in their interactions with the system.

Five roles are depicted in the motivational model: *elderly person, care provider/ relative, service provider, in-touch monitor and alarm monitor.* It will be determined during the design process what agents are needed for each role. There are three clouds representing quality requirements: 1) routine monitoring with "I am in touch"; 2) invisible to others and flexible support and 3) mobile-based immediate communication to caregiver/relatives with "call help."

Motivational models have been developed for the case studies to be presented in the next section. The modeling process engenders useful conversations and allows people to agree on a high-level of understanding. The models are deliberately not fully prescribed.

11.3 COLLECTING INFORMATION AND PROVIDING SUPPORT

If extensive information is conveyed in an unfocused way, the end-user accessing the site becomes confused and frustrated. Common mistakes by website developers are: 1) embedding important information at the bottom of the page or in an inaccessible place; 2) failing to highlight the most critical items and 3) using overly small-fonts, sparking poor emotional reactions. People are less inclined to use such websites and exhibit frustration when forced to do so, which affects the usefulness of the site. Interacting with a website or health-informatics software is not an emotionally neutral experience. Making the interaction fun is emotionally positive and can encourage people to engage more with the website or software. For sites dispensing advice or seeking information, creating a positive emotional experience is likely to increase the effectiveness of the information.

Positive and satisfying interaction between end-users and medical software can improve the quality of care, enhance patients' participation in their own health care and increase the efficiency of healthcare encounters. The potential is achieved if a positive emotional experience is engendered for multiple stakeholders. Positive engagement improves the probability of stakeholders heeding the information and following the instructions. An example of positive interaction is when a specialist doctor (or a patient) can intuitively mark the related abnormality on a visual rendering of a human part instead of giving a whole lot of textual descriptions.

User-friendly software makes people comfortable, while the technology that requires too many steps to learn frustrates and confuses end-users. If patients do not understand the technology or its use, they can feel incompetent, distressed and disconnected. Patients feel supported if the software and the associated technology are perceptually understandable, easy to use and gives them a feeling of independence yet being taken care of when needed.

11.3.1 Study I: Software for Insomnia

The first case study is an app called "Sleep-e" to help people with insomnia. The app development was part of a larger project about technology-delivered interventions for improving sleep. The project had three components: 1) development of an online treatment program for insomnia; 2) evaluation of the online treatment program for insomnia and 3) the design of a supportive app to help monitor sleep disorders. A key purpose of the Sleep-e app was to advise people having sleep difficulties during the night, with a list of strategies and relaxation exercises that people can try to help them get back to sleep.

The software-development task was given to a group comprising design students and computer science students. The software was much more engaging due to the design-students in the group.

The design-students were more imaginative and perceptive about the end-user experience. The design-students demonstrated a greater concern than the software students about the emotional experience of the users by applying product-design principles. The design-students thought about: 1) colors (or lack thereof) to relax the people and 2) clear fonts to make it easy for people to follow instructions during the night and a reassuring layout.

The software has been tested and correlated with data about sleep generated from a Fitbit, reviewed by experts and provided via the Mental Health Online clinic (https://www.mentalhealthonline.org.au/). In usability testing with 12 individuals, Sleep-e was easy to understand, helpful and easy to navigate.

11.3.2 Study II: Software for Prescreening for Severity of Depression

The next project involves improving health outcomes for people with depression. In a research project funded by the National Health and Medical Research Council of Australia, the research-team is performing randomized, clinical trials as to whether patients with depression achieve better outcomes by interacting with electronic resources than through standard consultations with general practitioners. Currently, there is a mismatch between the patients' needs and the treatment received. Patients with subthreshold or mild depression, who are likely to recover spontaneously, are often treated with anti-depressant medication. On the other hand, patients with severe symptoms rarely receive minimally adequate treatment.

A prescreening tool was designed and implemented on an iPad to assess potential participants in the study as to whether their mental health symptoms are at low risk, moderate risk or high risk for developing into serious depression. A motivational model was built that incorporated relevant emotional goals for depression. The tool consists of a set of questions that the medical team considered is a good indication of depression, based on the analysis of data collected over a decade.

The designs included icons relating to each question on the app and an assortment of ways of presenting results. Various design-concepts engendered different reactions from the focus group participants. One of the design-concepts was preferred. The iPad app was subsequently designed and developed so that interactions with the tool were positive. It needed to be sufficiently appealing to engage people who are unaware that they are at risk of depression and/or have negative attitudes toward the diagnosis of depression. It also needed to deliver the results in a way that does not lead to panic but to discussion with a general practitioner.

The software application was piloted through two focus groups with eight potential end-users. None of the participants had to be instructed in how to use the app. The participants gave positive feedback. Participants indicated that the app could aid their self-reflection, provide hope and motivate them to seek treatment.

Patients reported that they wanted the app to make them feel supported, comfortable and that the information felt relevant and important, and they wanted to see their results. General practitioners wanted to have confidence that the app provided reliable professional information, useful for improving depression-care. The researchers wanted the app to be scientifically reliable, trustworthy and useful for improving depression-care. Participants expressed a desire for a positive treatment recommendation through a message tailored to their individual results. Their emotional needs were catered for their interaction with the app. One finding was that the system needed to provide feedback to the participants about their well-being.

An interesting issue uncovered in the focus groups was the conflicting views on conveying information about risk. The doctors wished to convey the risk that depression could be overcome by seeking treatment instead of doing nothing. Several design images were tried, but ultimately were confusing to participants. Because of the interactions with the focus groups, depiction of risk within the iPad app was abandoned.

11.3.3 Study III: Website for Empowering Psychosis Patients

The third project was developed to support people suffering from psychosis. The goals of the project were: 1) to enable the delivery of an intervention to support mental health recovery and 2) to make patients feel empowered, connected and hopeful about having a meaningful life. These are emotional outcomes.

Personal recovery is an important target for persons with persisting psychotic disorders. The recovery refers to a satisfying and contributing life despite ongoing symptoms and disability. Contacts with peers, especially about sharing lived experience, is cited as facilitative of recovery. The project facilitated recovery-based digitally supported intervention for people suffering from psychosis.

A website was developed to be used on a tablet computer by health-workers to structure therapeutic discussions about personal recovery. Site content was developed from video interviews of people with lived experience of psychosis discussing how they had navigated issues within their own recovery based on the *Connectedness—Hope—Identity—Meaning—Empowerment* (CHIME) model of recovery.

The role of the development-team was to help the psychologists in the elicitation of emotion-informed requirements, modeling and validation of the platform. Many focus groups consisting of consumers and mental-health workers were conducted. The psychologists were initially skeptical but were pleasantly surprised at the responses when they asked the focus group participants how they felt when using the platform. By explicitly addressing emotions, better results addressing the requirements were produced compared to a related project in another area of mental-health.

The SMART system does not replace counseling. Ideally, a counselor works with the recovering psychotic to use the app. The app is helpful as an enhanced form of support, especially in the absence of the counselor, instead of as a substitute for the counselor.

11.3.4 Study IV: Emergency Alarm Services for Elderly Persons

Emergency-alarm systems are designed to support people, particularly elderly adults who live independently despite having some physical or mental-health issues such as dementia, diabetes, heart problems or osteoarthritis. These systems are designed as a bundle of two components: 1) an *emergency-alarm* to alert the caregivers and relatives about the need for physical assistance by the elderly persons and 2) a *well-being check* to ensure that the elderly person is doing well.

The emergency-alarm mechanism is typically implemented as a pendant or wristband that is worn all times by an elderly person due to the sudden and random nature of hazardous episodes. The device has a push-button that immediately connects to the service-provider for assistance. The *well-being check* consists of a base-station connected to a landline. The aged person is required to press a button on the base-station daily during a specified period to inform the service-provider of their well-being. In the absence of the well-being signal in a timely manner, the service-provider calls the elderly person to: 1) ensure their well-being and 2) find out the occurrence of any hazard that requires assistance.

Emergency-alarm systems are designed with reliability and robustness as the primary concerns, ignoring the emotional needs of elderly persons or any consideration of aesthetics. Many end-users have serious emotional concerns about the use of the pendants. For instance, some elderly users refuse to wear the pendant in public because they would be perceived as old, frail and dependent. These devices are quite complex for elderly users and are based on the expectation that aged persons behave rationally with full awareness of time. However, elderly persons may suffer from dementia, Alzheimer's disease and episodes of regular forgetfulness. A daily call to check if they are well is irritating and burdensome. Due to forgetfulness or dementia, they may press the button at the wrong time. Pressing the button at a specific period during the day could become agonizing to some elderly persons and might not serve to meet their emotional needs.

Based upon ethnographic studies interviewing elderly persons and their caregivers, the emergency alarm system has been redesigned so that the emergency-button is separated from the well-being check.

The well-being check is performed by the elderly persons interacting with a digital photo-frame that takes care of their emotional needs without stressing them.

A motivational model was built by extending the model in Figure 11.2. The emotional goals such as feeling independent and in control have been incorporated during this project. An app called "Touchframe" was built and is available in the Apple app-store.

11.3.5 Study V: A Tele-Health Hearing Aid System

Hearing-aids are important for elderly persons living independently at home. Being unable to hear people at the door and not being able to respond to phone calls has negative consequences for people's well-being. However, there is a stigma attached to hearing aids, in stark contrast to spectacles. Elderly persons with hearing aids are perceived as frail, dependent and old by the public.

This section sketches a hearing aid service offered by "Blamey Saunders Hears," an innovative and award-winning small company in Melbourne, Australia. A potential user can test her/his hearing at home using the "Blamey Saunders hearing test" for free online. The test, launched in 2013, uses real words — not tones, which hearing-tests for the last 70 years have used — and tells a patient about deficiencies in hearing based upon misspelled words.

The hearing-aids are adjusted based on the results of an at-home test, and they require minimal further tuning. The hearing aids connect via cords to a widget which then communicates via Bluetooth with an app on a smartphone. The users adjust their hearing to suit noise levels for their current environment and save the settings for similar surroundings. Previously, users would have had to carry a programming-box and laptop computer, or visit an audiologist to have their hearing-aids adjusted. Now, the hearing-aid is instead tuned with the assistance of a mobile phone. It is an example of telehealth.

11.4 SUMMARY

Consideration of the emotions of end-users is an important goal in designing and implementing healthcare software. Several research projects in the health domain with a component of building software have been discussed. For each, emotional factors have been considered during the requirements' elicitation and design. It is an advantage that the agent metaphor covers both people and software. Deferring the decision about the roles played by people and software until requirements are better understood and several design concepts have been contemplated can lead to better outcomes. It is also important that the successful systems should be developed in collaboration with the end-users: patients, nurses, caregivers and doctors.

Software developers and programmers have to perceive the emotional needs of the end-users. Incorporating additional error-checks and unrealistic security and safety features in their software only confuses, stresses, and frustrates the end-users reducing the chance of software adoption. The end-users, including caregivers, relatives and patients, should abstractly understand and contribute to the system models and architectures. End-users' participation has received insufficient attention as analysts and programmers have solely developed the model, architecture and technical details without sufficient input from the end-users.

Another point to note is that applications like SMART discussed in Section 11.3.3 are intended to enhance the relationship between case-workers and patients rather than replace the case-workers. Seeing technology enhances people relationships rather than replace people will enhance the interaction with the technology, will be emotionally satisfying and increase adoption.

ACKNOWLEDGMENTS

These case studies are based upon research by the author in conjunction with his colleagues from Swinburne University of Technology and the University of Melbourne. The research was supported from multiple grants. The emergency services research was supported through the Smart Services CRC and the Australian Research Council with the Discovery Grant DP130102660. The sleep-e app was supported by a grant from the Barbara Dicker Foundation. The depression study was supported by NHMRC, and the SMART program was supported from the Victorian Government Mental Illness Research Fund.

11.5 ASSESSMENT

11.5.1 Concepts and Definitions

Adoption; background emotion; beliefs; consumer willingness; emotional goal; functional requirements; motivation model; quality goal; requirement engineering; social emotion; user satisfaction.

11.5.2 Extended Response

1. Explain various goals and components of health-informatics software.
2. Explain the need of emotional goals in health-informatics software.
3. Read the articles on motivational needs and requirements engineering and write a two-page report on its impact on healthcare software development.
4. Discuss various aspects of expressed emotions of patients and elderly adults in response to newly introduced healthcare technology and software.
5. Design a system that takes care of emotional responses of an elderly patient suffering from dementia to make sure that she has taken medications taking care of her emotional needs.
6. Design a system that takes care of an elderly person suffering from Parkinson's disease to make sure that she has no serious falls while taking care of her emotional needs.

FURTHER READING

11.1 Australian Standards, http://www.e-health.standards.org.au/ABOUTIT014/WhatisHealthInformatics.aspx, Accessed December 21, 2016.

Need for End-User's Emotion and Satisfaction

11.2 Anderson, Catherine L., and Agarwal, R., "The Digitization of Healthcare: Boundary Risks, Emotion, and Consumer Willingness to Disclose Personal Health Information," *Information Systems Research*, 22(3), April 2011, 469–490.
11.3 Dam'asio, Antonio, *The Feeling of What Happens: Body and Emotion in the Making of Consciousness*, Harcourt Brace, New York, NY, USA, 1999.

11.4   Insel, Thomas R., and Wang, Philip S., "The STAR*D Trial: Revealing the Need for Better Treatments," *Psychiatric Service*, 60(11), November 2009, 1466–1467, DOI: 10.1176/appi.ps.60.11.1466.

11.5   Norman, Donald A., *Emotional Design: Why We Love (or Hate) Everyday Things*, Kindle edition, Basic Books, New York, NY, USA, 2005.

11.6   Ramos, Oao I., Berry, Daniel M., and Carvalho, Joao A., "The Role of Emotion, Values, and Beliefs in the Construction of Innovative Work Realities", In *Proceedings of First International Conference on Soft-Ware: Computing in an Imperfect World* (editors: Bustard, D. Liu, W., and Sterritt, R., Springer, LNCS 2311, Belfast, Northern Ireland, April 2002, 300–314.

11.7   Wood, Stacy L., and Moreau, Page C., "From Fear to Loathing? How Emotion Influences the Evaluation and Early Use of Innovations," *Journal of Marketing*, 70(3), July 2006, 44–57.

11.8   Zeelenberg, Marcel, Nelissen, Rob M.A., Breugelmans, Seger M., and Pieters, Rik, "On Emotion Specificity in Decision-Making: Why Feeling is for Doing," *Judgment and Decision Making*, 3(1), January 2008, 18–27.

Modeling Emotions During Software Design

11.9   Lopez-Lorca, Antonio A., Miller, Tim, Pedell, Sonja, Mendoza, Antonette, Keirnan, Alen, and Sterling, Leon, "One Size Doesn't Fit All: Diversifying "The User" Using Personas and Emotional Scenarios," In Proceedings of the 6th International Workshop on Social Software Engineering, Hong Kong, China, November 2014, 25–32.

11.10   Miller, Tim, Pedell, Sonja, Sterling, Leon, Vetere, Frank, and Howard, Steve, "Understanding Socially Oriented Roles and Goals through Motivational Modelling," *Journal of Systems and Software*, 85(9), September 2012, 2160–2170.

11.11   Ramos, Isabela, Berry, Daniel M., and Carvalho, João Á., "Requirements Engineering for Organizational Transformation", *Journal of Information and Software Technology*, 47(5), May 2005, 479–495.

11.12   Ramos, Isabela, and Berry, Daniel M., "Is Emotion Relevant to Requirements Engineering?" *Requirements Engineering*, 10(3), 2005, 238–242.

11.13   Mollon, Brent, Chong, Jaron J.R, Holbrook, Anne M, Sung, Melani, Thabane, Lehana, and Foster, Gary, "Features Predicting the Success of Computerized Decision Support for Prescribing: A Systematic Review of Randomized Controlled Trials," *BioMed Central Medical Informatics and Decision Making*, 9(1), February 2009, DOI: 10.1186/1472-6947-9-11, Available at https://www.ncbi.nlm.nih.gov/pmc/articles/PMC2667396/pdf/1472-6947-9-11.pdf, Accessed on February 13, 2017.

11.14   Proynova, Rumanya, Paech, Barbara S., Koch, Hilary, Wicht, Andreas, and Wetter, Thomas, "Use of Personal Values in Requirements Engineering – A Research Preview," In *Proceedings of the 16th International Working Conference on Requirements Engineering: Foundations for Software Quality* (REFSQ 2010), Essen, Germany, July 2010, LNCS 6182, (editors: Roel Wieringa and Anne Persson), Springer Verlag, Berlin, Germany, 17–22.

11.15   Proynova, Rumanya, Paech, Barbara S., Koch, Hilary, Wicht, Andreas, and Wetter, Thomas, "Investigating the Influence of Personal Values on Requirements for Health Care Information Systems," In Proceedings of the 3rd ACM Workshop on Software Engineering in Health Care (ICSE 2011), Honolulu, Hawaii, USA, May 2011, 48–55.

Collecting Information and Providing Support

11.16   Sterling, Leon and Taveter, Kuldar, *Art of Agent-Oriented Modelling*, MIT Press, Cambridge, MA, USA, 2009.

Software for Pre-Screening for Depression Severity

11.17   Davidson, Sandra K., Harris, Meredith G., Dowrick, Christopher F., Wachtler, Caroline A., Pirkis, Jane, and Gunn, Jane M., "Mental Health Interventions and Future Major Depression among Primary Care Patients with Subthreshold Depression." *Journal of Affective Disorder*, 177, May 2015, 65–73, DOI: 10.1016/j.jad.2015.02.014.

11.18   Herrman, Helen E., Patrick, Donald L., Diehr, Paula K.H., Martin, M.L., Fleck, M., Simon, Gregory E., and Buesching, Don P., "Longitudinal Investigation of Depression Outcomes in Primary Care in Six Countries: The LIDO Study. Functional Status, Health Service Use and Treatment of People with Depressive Symptoms," *Psychological Medicine*, 32(5), May 2002, 889–902.

Website for Empowering Psychosis Patients

11.19 Thomas, Neil, Farhall, John, Foley, Fiona, Leitan, Nuwan D., Villagonzalo, Kristi-Ann, et al., "Promoting Personal Recovery in People with Persisting Psychotic Disorders: Development and Pilot Study of a Novel Digital Intervention," *Frontiers in Psychiatry*, 7, Article 196, December 2016, DOI: 10.3389/fpsyt.2016.00196, Available at https://www.ncbi.nlm.nih.gov/pmc/articles/PMC5179552/pdf/fpsyt-07-00196.pdf.

11.20 Thomas, Neil, Farhall, John, Foley, Fiona, Rossell, Susan, Castle, David, Ladd, Emma, et al., "Randomized Controlled Trial of a Digitally Assisted Low Intensity Intervention to Promote Personal Recovery in Persisting Psychosis: SMART-Therapy Study Protocol," *BMC Psychiatry*, 16(1), Article 312, September 2016, DOI: 10.1186/s12888-016-1024-1, Available at https://bmcpsychiatry.biomedcentral.com/articles/10.1186/s12888-016-1024-1.

Emergency Alarm Services for Elderly Persons

11.21 Sterling, Leon, Curumsing, Maheshwaree K., Lopez-Lorca, Antonio, Miller, Timothy, and Vasa, Rajesh, "Viewpoint Modelling with Emotions: A Case Study," *International Journal of People Oriented Programming*, 4 (2), July 2015, 25–53.

11.22 Miller, Timothy, Pedell, Sonja, Lopez-Lorca, Antonio A., Mendoza, Antonette, Sterling, Leon, and Keirnan, Alen, "A. Emotionally-Driven Models for People-Oriented Requirements Engineering: The Case Study of Emergency Systems," *Journal of Systems & Software*, 105, July 2015, 54–71.

11.23 Mendoza, Antonette, Miller, Tim, Pedell, Sonja, and Sterling, Leon, "The Role of Users' Emotions and Associated Quality Goals on Appropriation of Systems: Two Case Studies," *24th Australasian Conference on Information Systems (ACIS)*, RMIT University, Melbourne, Australia, 2013, Available at http://citeseerx.ist.psu.edu/viewdoc/download?doi=10.1.1.717.7674&rep=rep1&type=pdf, Accessed August 19, 2018.

A Tele-Health Hearing Aid-System

11.24 Hears, Blamey S., web sites (https://www.blameysaunders.com.au/) and (http://www.scienceinpublic.com.au/media-releases/backgrounder#blamey), Accessed on August 19, 2018.

Conclusion 12

Computational Health Informatics is an outcome of the desire to improve the life-expectancy and general well-being of humans in the last few decades. Before the 1950s, healthcare was more about managing the abnormalities. However, since 1950, human society has seen many revolutionary changes that have impacted our healthcare system. Some of these are: 1) improvements in hygienic living conditions; 2) availability of antibiotics and vaccines; 3) availability of purified drinkable water; 4) availability of improved agricultural output reducing malnutrition; 5) better health education; 6) proliferation of health technology and 7) exponential growth in research and technology resulting in multiple beneficial technologies such as X-rays, magnetic resonance imaging, optics (including laser techniques), computers and computational techniques, molecular technologies such as genome sequencing and gene-splicing, display devices, miniaturization in manufacturing, surgical techniques and equipment that can substitute and sustain human-organs and long-distance data-communications and their integration.

Each of these sciences and related technologies has improved with a record-breaking pace, and their integration has become intertwined and sophisticated. It is this integration of science and technology in the last few decades that has given rise to the modern fields of health-informatics systems. Computers and optics have infiltrated every technology, providing unprecedented automation and miniaturization making many devices portable and affordable to an extent that was not imaginable 20 years ago. Some examples are miniaturized wearable devices, implants, miniaturized cameras in capsules for photographing digestive systems, portable low-energy X-ray and genome-sequencing devices, mobile phones, portable ECG machines and so on.

This integration has become the key source for the 20-year increase in life-expectancy over the past three decades. The technology is being adopted by the developing world, giving them a leapfrog capability in health-management. The increase in life-expectancy and the availability of information technology has exposed society to many health issues that went undetected in earlier periods when there was a lower life-expectancy or were not a matter of concern since these diseases occurred primarily among those above the age of sixty. These diseases include dementia, neuromuscular diseases, heart diseases, strokes and Alzheimer's disease.

The shortfall of resources has become a key motivation for the improvement in health-informatics. Another motivating factor for modern health-informatics is the increased mobility of people in society because of a revolution in transportation during the last 50 years. The availability of rapid transportation along with an improved life-expectancy has made people more mobile than ever, even at an age of 60 or more. This has started a new field of pervasive health care where a person can be monitored at home or away while in transit.

The revolution in computational science such as computer-miniaturization, fast and compact terabyte level memory-storage capability, formats to store multimedia-objects and multidimensional data have enabled storage, retrieval, comparison, transmission and statistical analysis of a huge amount of data. This has lowered the level of duplication of expensive tests and led to revolution in data analytics and data-mining for automated discovery of knowledge. Combined with the automated genome sequencing, there is an upcoming revolution in the discovery of diseases, biomarkers and drugs, along with vaccine development. These developments have started an emerging field of personalized medicine and evidence-based treatment.

12.1 EVOLUTION OF HEALTH INFORMATICS

Health-informatics was paper-based in the early 1970s and then converted into local relational databases in the mid-1980s when inexpensive secondary storage became available. However, the database lacked nested structures, multimedia and object-based programming. Medical images and patient information were stored in paper and emulsion-gelatin. Networking and email capability were limited. There was no Internet, limiting information-exchange capability. However, bulk data was stored in secondary storage for future retrieval.

In the early 1990s (1990–1995), many major changes occurred. The Internet was adapted for commercial use. Large-scale gigabyte-level secondary hard disks became available. The size of computers shrunk to the desktop-level. Much automated software for diagnostics started appearing. The size of imaging machines such as X-ray machines, ultrasound machines, MRI machines, spectrum analyzers, electronic microscopes, ECG machines and EEG machines kept shrinking. The signal-recording became faster and more accurate. Computers started getting networked. Thus, image and data were transferred from one hospital-unit to another. However, there was a dearth of image and video storage formats. There was no standardization of the various medical data exchange formats. There was a lack of web-based delivery of data, and a lack of miniaturized medical devices.

The late 1990s (1995–1999) saw major changes in the development and adoption of health-informatics with the support of Congress, international government agencies, and computer science agencies such as IEEE and NEMA. Many standards were developed for information exchange including HL7 and DICOM. Image compression was incorporated into image formats. Video formatting standards such as JPEG and MPEG were developed. Lossy and lossless variations of JPEG such as JPEG2000 were developed. Many vendors started developing tools for automated data-analysis. Object-oriented databases were developed by the vendors. The trend toward miniaturization and wearable devices started. The hospital-units started collaborating. Security of data to protect patients' confidentiality and privacy became important. Collaborative surgery and diagnostics among experts across hospitals became possible. Effort was also made for robotics surgery. However, the tools being developed had no common interface, and interoperability was still a problem.

The early 2000s (1999–2005) saw a big change in the development of indexing structures and techniques for multimedia and temporal databases. Datamarts and data warehouses were created to store a huge amount of data. Datamarts of different hospitals and other organizations began to become integrated using the techniques based upon the master-patient index, virtual medical records and role-based ports. Cloud-based storage also became possible. The memory cost dropped to support terabytes of data. Laptops and tablets became a reality. Computational speed increased further for faster data-mining of a huge amount of data. Secure transmission of high-resolution images using a combination of lossy and lossless compression became possible. This started a new field of pervasive health care. Telemedicine, remote health care and teleconsultation became possible. However, this rapid change in medical informatics started interfering with physicians' activities as many tasks were being automated, and physicians were being forced to use the computer-archived information to improve overall efficiency. Some examples are the use of tablets and desktops for patient information entry and retrieval of patient information. Paper archives were removed. Skilled medical staff found the change unnecessary.

End-user and machine interfaces lacked human-like intelligence, which created serious drawbacks when compared to earlier human-based data processing. There were many reasons such as: 1) the lack of knowledge of artificial intelligence techniques by the programmers; 2) additional questions and checks by the programmers to ascertain verification using traditional keyboard-based questions; 3) lack of high-quality natural language and text-based interfaces and 4) the lack of visual-interfaces due to the lack of high-resolution touchscreen technologies.

Since 2006, a good deal of intelligent software tools have been implemented and installed by healthcare organizations despite their slow adoption by the medical experts. This period also saw the use of artificial intelligence and clinical domain-specific systems increasing in the healthcare system.

Cell phones, wearable and ECG devices became portable; miniaturized wearable devices became available for remote health care. Telemedicine and teleconsultation became common.

Complex simulation that used intelligent techniques became available for training medical practitioners. Intelligent inferring of images and data facilitated automated detection of cancers and other abnormalities. Using intelligent techniques in data and process analysis is making life convenient for medical practitioners. Database integration is well established, and the information retrieval and interaction have become web-based. Physicians have started experimenting with virtual and augmented reality to visualize and interact with 3D regions of interest before performing surgery.

There is also a significant improvement in genomics and proteomics analysis that is already revolutionizing vaccine development, the development of visualizing the three-dimensional interactions of molecules, a large-scale simulation of microbiome in the gut, and understanding the biochemical nature of the brain.

With increased informatics technology and seamless integration, a huge amount of data has become available that needs to be archived and managed. A typical hospital processes up to billions of documents every year. This has started a quest for an automated text-analysis systems to identify the themes and topics occurring in the text documents.

The future will belong to intelligent automated data acquisition, seamless integration between remote-care and automated data-acquisition, the development of an intelligent patient-sensitive software that will care for human emotion and human-convenient multimedia-interfaces, simulation of vital organs and large-scale data-mining and process-mining. The focus will also change in predictive medication and cures through the analysis of biomarkers by wearable and implanted molecular chips that will show the condition long before it becomes untreatable.

12.2 EVOLUTION OF STANDARDS

With the growth in computational health-informatics, a clear need for standards arose to share the information unambiguously among healthcare providers and various organizations. The standards had to be developed for health procedures, lab procedures, disease-codes, treatments, personal information representation, communication protocol, medical data and image encoding, wireless communication formats for remote-care, data-acquisition from the sensor devices and information-exchange.

The major problems in the development of standards are many: 1) the lack of previous knowledge of standards; 2) individual vendors promoting their own innovations; 3) the need for multiple parties such as vendors, healthcare providers, insurance companies and technical committees to agree; 4) the slowness of healthcare providers to adopt new standards because of embedded adoption and training costs and the lack of resources to implement; 5) cultural barriers in language and data representations; 6) the lack of compatibility between the old standards and the proposed new standards and 7) the lack of realization of the limitations of the old standards when implemented to capture real scenarios.

Despite these limitations, multiple standards have been evolving in medical data-representation, information-exchange, ontology of medical terms and technological standards such as wireless communications. Different collaborating committees representing multiple stakeholders have been created who regularly update the standards. Some prominent standards that are evolving are SNOMED CT, ICD, HL7, PACS, DICOM, IHE, telemedicine standards, clinical classification and terminology. Besides these standards, underlying technologies such as XML, XML-based languages, database implementations and query languages are also evolving. This two-layer evolution makes the overall standardization daunting and causes many incompatibilities.

An interesting issue concerning the switch from paper-based records to electronic medical records is to ascertain the authenticity of older paper-based records when converted to electronic medical records. Healthcare providers have preferred to store the scanned version of the documents besides their conversion to an electronic format.

The evolution in standards has accelerated as the adoption of the computerized healthcare system speeds up. Earlier, the standards were developed on small datasets to promote the success of the implementations. Once proven, the scope of the coding and classification has increased. For example, ICD has evolved to ICD 11 from ICD 9 in recent years, and HL 7 3.x is replacing HL7 2.x. This evolution is not without pains because the underlying computational technology in HL7 2.x differs from HL7 3.x, which is object-oriented. This difference creates significant backward incompatibility.

Another problem with multiple standards is the incompatibility between the standards of information exchange. Information related to many domains overlaps. For example, information in SNOMED overlaps with information in ICD. In such cases, there has to be an interface to facilitate the interpretation of information in one standard by the other. A similar overlap problem is between HL7 and DICOM. HL7 is meant for clinical data exchange, and DICOM is for image-exchange. Thus, HL7 and DICOM should be interfaced. Currently, HL7 document refers to the DICOM documents and does not carry image information.

12.3 ELDERLY CARE AND ADOPTION

One motivation for electronic healthcare is to have cost-effective quality healthcare management of the rapidly aging population around the world. The percentage of the elderly population will rise from the current 16% to 24% (in USA) by the year 2030 burdening the existing healthcare resources.

The major issues in elder care are: 1) the growth of the elderly population is faster than the resources and infrastructure being developed, even with remote care; 2) the current technology development is not suitable for satisfying the emotional needs of the end-user; 3) the technology does not address the needs of gender variations; 4) technology does not address the loneliness, isolation and depression faced by the elderly patients; 5) elderly persons are impaired, and lack the desire to learn complex new technology. The limitations on physical ability vary and include hearing loss, vision loss, muscle loss, decreased mobility, physical pain, deformity affecting mobility and chronic diseases. The limitation on mental ability is caused by dementia, Alzheimer's and the loss of memory.

Current improvement in technology is focused on the availability and analysis of data, connectivity of data sources and end-users, technical models of data communication and intelligent data analytics. It is assumed that the developed technology will be adopted. While it is true for the literate younger generation, the same cannot be said about elderly persons. Besides, elderly persons face a challenge with remembering, learning, free movement, the embarrassment of feeling dependent on caregivers for the use of equipment and the violation of privacy by home-monitoring systems hindering their adoption.

Elderly persons face additional challenges such as dementia, problems with motor control, decreased listening and vision capabilities, decreased cognitive capabilities and the random occurrence of Alzheimer's disease episodes. This requires sociotechnical and socioeconomic analysis of elderly patients and their challenges for intelligent technology development, which will give them the feeling of independence without overburdening them with learning the new technology. The technology should be at the back end, and the front end should be simple to use.

Virtual connectivity and support provided by the Internet are plausible sources of reducing isolation and loneliness. It results in increased social connectivity, and can mitigate the challenge of loss of mobility. However, web-based and Internet-based technologies presume that the end-user will be able to log into the computer and surf the websites to extract the information and connect to the healthcare systems. In reality, a large percentage of the elderly population has little knowledge of computer usage and finds it a challenge to use computer-based technology. Such people can benefit from an extended version of voice-activated front-end like "Alexa," "Google Home" or "Amazon Echo" that can understand the patients as humans and are in addition linked to the Internet for data-transfer and retrieval to caregivers and healthcare providers.

12.4 HEALTH INFORMATICS IN DEVELOPING COUNTRIES

Population growth and economic resources are uneven between developed countries and developing countries. The population growth in developing countries is much faster than in the developed world. Despite the percentage of the elderly population being lower in developing countries compared with the percentage in developed countries, the availability of health-care and information infrastructure for secure information transmission is still developing, limited and uncoordinated. Because of economic inequity, there is more than one type of infrastructure. In urban areas, the part of the system geared to the affluent fraction of society is well-developed. However, rural areas and government-supported healthcare systems lack the facilities and infrastructure. The percentage of rural areas and the lower middle-class population availing government healthcare system is much larger.

The healthcare standards and data-standards for information-exchange and archiving are not followed in developing countries due to the lack of infrastructure and resources. Health care consists of a combination of independent healthcare providers not following shared common standards, leading to duplication of information and lab-tests. The heterogeneity of data is widespread at multiple levels: local unit level and community clinic level, district level, state or province level and national level. The interface between national and international level is nonexistent.

Developing countries also lack 1) reliable communication technology; 2) continuous power sources to run the equipment and computers; 3) infection-free environment; 4) availability of recent hardware and software; 5) resources in local clinics and 6) medical data for health-related decision making. Most of the houses are connected through private Internet Service Providers (ISP), which charge based upon time or data-transferred. In such cases, continuous connectivity and data-logging becomes expensive.

Despite these limitations, the usefulness of the mobile phone-based messaging to monitor the patients, electronic health-records to retrieve information and the use of complex decision support systems to augment and enhance the availability of medical health care has been growing. The protocols and decision support systems have also helped follow the standard procedures to treat patients in remote clinics. A major source of health-related information transmission to local communities is the development of local tele-centers that broadcast medical information to the community. Tele-centers have also been used to transmit information to the hospitals on a demand basis. These tele-centers are operated, where possible, using solar-powered equipment and satellite links to make them independent of nonavailability or non-continuous availability of the land-based power and communication infrastructure.

12.5 CURRENT STATUS OF TECHNOLOGY ADOPTION

The current status of technology adoption is divided into many aspects such as implementing electronic health records, medical information storage, information-exchange networks, automatic prescription filling, implementation of decision support systems for reducing errors in diagnosis by healthcare providers, entering clinical notes, automated lab result dissemination to healthcare providers and patients, pervasive technologies and remote care. Adoption of technology has been increasing in the developed world 10 to 15% every year because of government incentives and enforcement. It varies from 60% to 80% for different services, with automated prescription transmission to pharmacies being readily adopted.

The implementation of health information technologies have both advantages and disadvantages. The advantages are long-term storage of data, removal of duplication of tests, use of past data for better diagnosis and prognosis, increased collaboration between healthcare providers, improved monitoring, improved information sharing with patients, the use of decision support systems and other computer-assisted software in reducing errors in diagnosis and health management. The disadvantages are: increased financial cost of implementation and training the staff, incompatibility between the implemented standards, incompatibility between existing paper-based systems and electronic systems and a lack of human-friendly software that imposes new behavior-patterns on physicians as well and patients whose emotional needs are often ignored in the software-development process.

The review of the implementation literature shows that even the developed nations have mostly implemented electronic health records. However, they lag in the implementation of the health information-exchange systems. Implementation of health information systems is expensive and financially draining. The major expense is to keep up with the ever-changing technology and training the staff lacking the required skill-set. There is also a problem with the lack of the information scientists and trained software programmers to implement and maintain the required software infrastructure. Because of this, the quality of implementation is heterogeneous.

There is also a problem in training the end-users to use the technology due to existing low levels of education and the lack of familiarity with the technology and implemented systems. It is also difficult to improve the perception and motivation of a technology being used. A technology that does not match the people's existing needs, and the available infrastructure is less likely to be supported by clinicians. This mismatch generates sufficient resistance to adoption by the clinicians. Clinicians and end-users are also apprehensive of the errors in using the new technology that may negatively affect patients' health recovery.

12.6 FUTURE DEVELOPMENT

The future development in healthcare informatics will occur on multiple fronts: 1) technological advancement in wearable mobile technology with improved broadband capabilities; 2) incorporation of advanced intelligent interfaces that relieve patients and caregivers from learning technical details of smart home equipment and wearable technologies; 3) intelligent genome analysis tools for better biomarkers and drug development; 4) improved government policies to enhance adoption of healthcare protocols and technology; 5) propagation of healthcare technology in the developing world with more flexibility in the standards and 6) incorporation of HIPAA in electronic health records.

The technological advancement will continue in miniaturized intelligent noninvasive and minimally invasive sensors that can: 1) transmit useful physiological information about the human body to a central hub that can transmit the information to the caregivers and healthcare providers; 2) automatically release drugs in a patient's body based upon limited intelligent analysis of biosensor data to relieve the emergency conditions and 3) provide guidance to elderly persons for navigating and performing daily activity for improved independence without hurting their self-confidence. This will include the development of ubiquitous placement of health equipment and the integration with intelligent interfaces that connect to the Internet. The improvement in healthcare technology will also continue in two emerging areas: *social robotics, virtual and augmented reality*. *Virtual reality* is a computational visualization technique that gives a realistic interaction capability with an object or scene. *Augmented reality* provides a seamless integration with the actual reality and the virtual object such that the interacting person can put and interact with virtual objects in the real environment and interact with them like real objects.

Social robotics will emerge as a major area where intelligent machines integrated with health equipment will provide companionship to the elderly person in the absence of caregivers and inform caregivers in the case of emergencies. Virtual and augmented reality, when integrated with 3D radiology techniques, will help visualize the internal organs and their functioning for better diagnosis, assistive surgery and design of new drug molecules by understanding and analyzing epitope interactions.

Intelligent machines will assist many functions such as mobility, emergency symptom analysis, activity-pattern analysis, including movement analysis, providing help with daily activities such as bathroom activities, medication-related activities, reading a story or playing music (TV), handling basic commands given by elderly persons or recovering patients, and automatically analyzing the situation based upon daily-activity pattern data. Some of these assistive functions are already incorporated in emerging intelligent interfaces such as Alexa, Echo and Google Home. However, these devices are still limited to a human voice command and lack the interface with smart homes that enables them to collect and analyze activity-patterns automatically.

Wearable and embedded smart devices would be common as biosensors and drug-actuators for monitoring and regulating release of the drugs for chronic conditions. Some conditions are sugar control of diabetic patients, anxiety attacks and arrhythmic heart-conditions. New standards and biosignal analysis techniques will be developed for automated regulation. Implementing automated data-collection and drug-release will require secure transmission of data and commands to reduce hacking of the devices. The technology will also require a human-friendly natural language interface to explain the emergency conditions to the caregivers and the patients. Thus, it would have onboard intelligent analysis capability to be automatically personalized to the patient's condition and commands and the caregivers' commands with appropriate HIPAA privileges.

One of the major problems faced by elderly persons is navigation inside the dwelling and also outside the dwelling, such as daily walks or visiting nearby friends and shops. The pattern of movements collected will have to be transferred to the central hub that has intelligent learning and analysis technique to analyze the deviations for collision avoidance, avoid falling, and provide reminders for any activity to be performed to avoid instantaneous forgetfulness.

Current progress toward the development of intelligent genome analysis tools will continue as the memory and computational power of the computer increases. This growth will assist the development of more personalized treatment, effective drugs and vaccines. The virtual reality development will be integrated with drug and vaccine development activities to facilitate the interaction between the chemists involved and the 3D-structure of molecules to study the bindings between antibody and epitopes. Intelligent techniques will also be used to develop biomarkers for genetic diseases such as cancer.

The developed world will continue to forge forward with governmental agencies taking the initiative to develop policies, standards and incentives to improve healthcare decision making, promote patient-centric care, improve the quality and safety of medical management and promote the use of new clinical decision support systems for improved accuracy of diagnosis and prognosis.

One of the major problems in implementing healthcare technologies in the developing world is to restructure the existing clinical systems to bring in innovations. These changes require significant financial resources. Hence, progress will be slower than expected, and the developing world will go through many hybrid systems that need to be interfaced with the state-of-the art healthcare systems of the developed world. The development of such software interfaces will require significant governmental commitment and investments. It will also require progressive policies that work in phases instead of sudden leapfrog movements. Thus, there will be an infrastructure and technological divide between the healthcare technology in the developed world and developing world despite regular progress in technology and its adoption.

FURTHER READING

Evolution of Health Informatics

12.1 Haux, R., "Health Information Systems – Past, Present, Future," *International Journal of Medical Informatics*, 75(3–4), March–April 2006, 268–281, DOI: 10.1016/j.ijmedinf.2005.08.002.

12.2 Luo, Jake, Gopukumar, Deepika, and Zhao, Yiqing, "Big Data Application in Biomedical Research and Health Care: A Literature Review," *Biomedical Informatics Insights*, 8, 2016, 1–10.

12.3 Mihalas, George I., "Evolution of Trends in European Medical Informatics," *ACTA Informatica Medica*, 22(1), February 2014, 37–43, DOI: 10.5455/aim.2014.22.37-43.

12.4 Peek, Niels, Combi, Carlo, Marin, Roque, and Bellazzi, Ricardo, "Thirty Years of Artificial Intelligence in Medicine (AIME) Conferences: A Review of Research Themes," *Artificial Intelligence in Medicine*, 65(1), September 2015, 61–73, DOI: 10.1016/j.artmed.2015.07.003.

12.5   Peterson, Hans E., "From Punched Cards to Computerized Patient Records," *IMIA Yearbook of Medical Informatics*, 15(1), January 2006, 180–186, DOI: 10.1055/s-0038-1638483.

12.6   Shortliffe, Edward H., and Cimino, James J. (editors), *Biomedical Informatics*, 4th edition, Springer, London, 2014.

Evolution of Standards

12.7   Carroll, Randy, Cnossen, Rick, Schnell, Mark, and Simons, David, "Continua: An Interoperable Personal Healthcare Ecosystem," *IEEE Pervasive Computing*, 6(4), October–December 2007, 90–94, DOI: 10.1109/MPRV.2007.72.

12.8   Kart, First, Moser, Louise E., and Melliar-Smith, P. Michael, "E-Healthcare System Using SOA," *IT PRO*, 10(2), March/April 2008, 24–30, DOI: 10.1109/MITP.2008.22.

Elderly Care and Adoption

12.9   Cotton, Sheila R., Ford, George, Ford, Sherry, and Hale, Timothy M., "Internet Use and Depression among Retired Older Adults in the United States: A Longitudinal Analysis," *Journal of Gerontology, Series B: Psychological Science and Social Sciences*, 69(5), September 2014, 763–771, DOI: 10.1093/geronb/gbu018.

12.10 Golant, Stephen M., "A Theoretical Model to Explain the Smart Technology Adoption Behaviors of Elder Consumers (ElderAdopt)," *Journal of Aging Studies*, 42, August 2017, 56–73, DOI: 10.1016/j.jaging.2017.07.003.

12.11 Gottesman, Omri, Kuivaniemi, Helena, Tromp, Gerard, Faucett, W. Andrew, Li, Rongling, et al., "The Electronic Medical Records and Genomics (Emerge)Network: past, present and Future, *Genetics in Medicine*, 15(10), October 2013, 761–771, DOI: 10.1038/gim.2013.72.

12.12 Hirsch, Tad, Forlizzi, Jodi, Hyder, Elaine, Goetz, Jennifer, Kurtz, Chris, and Stroback, Jacey, "The ELDer Project: Social and Emotional Factors in the Design of Eldercare Technologies," In *Proceedings of the conference on universal usability*, Arlington, VA, USA, November 2000, 72–80, DOI: 10.1145/355460.355476.

12.13 Hoof, Joost V., Kort, Helianthe S. M., Rutten, Paul G. S., and Duijnstee, Mia S. H, "Aging-in-Place with the use of Ambient Intelligence Technology: Perspectives of Older Users, *International Journal of Medical Informatics*, 80(5), March 2011, 310–331, DOI: 10.1016/j.ijmedinf.2011.02.010.

12.14 Koch, Sabine, and Hägglund, Maria, "Health Informatics and the Delivery of Care to Older People," *Maturitas*, 63(3), July 2009, 195–199, DOI: 10.1016/j.maturitas.2009.03.023.

12.15 Liu, Lili, Stroulia, Eleni, Nikolaidis, Ioanis, and Miguel-Cruz, Antonio, "Smart Homes and Home Health Monitoring Technologies for Older Adults: A Systematic Review," *International Journal of Medical Informatics*, 91, July 2016, 44–59, DOI: 10.2016/ijmedinf.2016.04.007.

12.16 Weiner, Michael, Callahan, Christopher M., Tierney, William M., Overhage, Joseph M., Mamlin, Burke, Dexter, Paul R., and McDonald, Clement J., "Using Information Technology to Improve Health Care of Older Adults," *Annals of Internal Medicine*, 139(5_part_2), 2003, 430–436, DOI: 10.7326/0003-4819-139-5_Part_2-200309021-00010.

12.17 Yuseif, Salifu, Soar, Jeffrey, and Hafeez-Baig, Abdul, "Older People, Assistive Technologies, and the Barriers to Adoption: A Systematic Review," *International Journal of Medical Informatics*, 94, October 2016, 112–116, DOI: 10.1016/j.ijmedinf.2016.07.004.

Health Informatics in Developing Countries

12.18 Braa, Jørn, Hanseth, Ole, Heywood, Arthur, Mohammed, Woinshet, and Shaw, Vincent, "Developing Health Information Systems in Developing Countries: The Flexible Standards Strategy," *MIS Quarterly*, 31(2), June 2007, 381–402, DOI: 10.2307/25148796.

12.19 Lucas, Henry, "Information and Communication Technology for Future Health Systems in Developing Countries," *Social Science and Medicine*, 66(10), May 2008, 2122–2132, DOI: 10.1016/j.socsscimed.2008.01.033.

Current Status of Technology Adoption

12.20 Ben-Assuli, Ofir, "Electronic Health Records, Adoption, Quality of Care, Legal and Privacy Issues and their Implementation in Emergency Departments," *Health Policy*, 119(3), March 2015, 287–297, DOI: 10.1016/j.healthpol.2014.11.014.

12.21 Ryan, Jamie, Doty, Michelle M., Abrams, Melinda K., and Riley, Pamela, "The Adoption and Use of Health Information Technology by Community Health Centers, 2009-2013," *The Commonwealth Fund Publication 1746*, 10, May 2014,

12.22 Sligo, Judith, Gauld, Robin, Robert, Vaughan, and Villa, Luis, "A Literature Review for Large-scale Health Information System Project Planning, Implementation and Evaluation," *International Journal of Medical Informatics*, 97, 2017, 86–97, DOI: 10.1016/j.ijmedinf.2016.09.007.

12.23 *Update on the Adoption of Health Information Technology and Related Efforts to Facilitate the Electronic Use and Exchange of Health Information*, Report to Congress, February 2016, Available at https://www.healthit.gov/sites/default/files/Attachment_1_-_2-26-16_RTC_Health_IT_Progress.pdf, Accessed June 30, 2019.

Future Development

12.24 *AHRQ Health Information Technology Division's 2017 Annual Report*, AHRQ Publication No. 18-0028-EF, April 2018, Available at https://healthit.ahrq.gov/sites/default/files/docs/page/2017-ahrq-hit-annual-report.pdf, Accessed June 29, 2019.

12.25 Bates, David W., and Bitton, Asaf, "The Future of Health Information Technology in the Patient-centered Medical Home," *Health Affairs*, 29(4), April 2010, 614–621, DOI: 10.1377/hlthaff.2010.0007.

12.26 Broadbent, Elizabeth, Stafford Rebecca, and MacDonald, Bruce, "Acceptance of Healthcare Robots for the Older Population: Review and Future Directions," *International Journal of Social Robotics*, 1(4), November 2009, 319–330, DOI: 10.1007/s12369-009-0030-6.

12.27 Collins, Francis S., Green, Eric D., Guttmacher, Alan E., and Guyer, Mark S., "A Vision for the Future of Genomic Research: A Blueprint of Genomic Era," *Nature*, 422(6934), April 2003, 835–847, DOI: 10.1038/nature01626.

12.28 Gorini, Alessandra, and Riva, Giuseppe, "Virtual Reality in Anxiety Disorders: The Past and the Future," *Expert Review of Neurotherapeutics*, 8(2), 2008, 215–233, DOI: 10.1586/14737175.8.2.215.

12.29 Lányi, Cecilia S., "Virtual Reality in Healthcare," In *Intelligent Paradigms for Assistive and Preventive Healthcare*, Studies in Computational Intelligence Series (Editors: Ichalkaranje, Nikhil, Ichalkaranje, Ajita, and Jain, Lakhmi C), Springer, Berlin, Germany, Vol. 19, April 2006, 87–116.

12.30 McGowan, Julie J., Cusack, Caitlin M., and Bloomrosen, Meryl, "The Future of Health IT Innovation and Informatics: A Report from AMIA's 2010 Policy Meeting," *Journal of American Informatics Association*, 19(3), May-June 2012, 460–467, DOI: 10.1136/amiajnl-2011-000522.

12.31 McLoy, Roiry, and Stone, Robert, "Virtual Reality in Surgery," *BMJ*, 323(7318), October 2001, 912–915, DOI: 10.1136/bmj.323.7318.912.

12.32 Scopelliti, Massimiliano, Giuliani, Maria V., and Fornara, Ferdinando, "Robots in a Domestic Setting: A Psychological Approach," *Universal Access in the Information Society*, 4(2), December 2005, 146–155, DOI: 10.1007/s10209-005-0118-1.

12.33 Sittig, Dean F., Wright, Adam, Osheroff, Jerome A., Middleton, Blackford, Teich, Jonathan M., Ash, Joan S., et al., "Grand Challenges in Clinical Decision Support," *Journal of Biomedical Informatics*, 41(2), April 2008, 387–392, DOI: 10.1016/j.jbi.2007.09.003.

12.34 Stark, Zornitza, Dolman, Lena, Manolio, Teri A., Ozenberger, Brad, Hill, Sue L., Caulfied, Mark J., et al., "Integrating Genomics into Healthcare: A Global Responsibility," *The American Journal of Human Genetics*, 104(1), January 2019, 13–20, DOI: 10.1016/j.ajhg.2018.11.014.

12.35 Tinker Anthea, and Lansley Peter, "Introducing Assistive Technology into the Existing Homes of Older People: Feasibility, Acceptability, Costs and Outcomes," *Journal of Telemedicine and Telecare*, 11 (Supplement 1), July 2005, 1–3, DOI: 10.1258/1357633054461787.

12.36 Young, James E., Hawkins, Richard, Sharlin, Ehud, and Igarashi, Takeo, "Toward Acceptable Domestic Robots: Applying Insights from Social Psychology," *International Journal of Social Robot*, 1(1), January 2009, 95–108, DOI: 10.1007/s12369-008-0006-y.

Appendix I: Websites for Healthcare Standards

STANDARD	WEBSITE
DICOM	https://www.dicomstandard.org/
HIPAA	https://www.hhs.gov/hipaa/index.html
HL7 (Health Level 7)	https://www.hl7.org/
ICD	https://www.who.int/classifications/icd/en/
ICD 11	https://icd.who.int/en/
Indoor GML	https://www.opengeospatial.org/standards/indoorgml
KML	https://www.opengeospatial.org/standards/kml
LOINC	https://loinc.org/
medDRA	https://www.meddra.org/
MICS (Monitoring, Information-exchange and Control…)	https://standards.ieee.org/standard/1547_3-2007.html
SNOMED	https://www.snomed.org/
WBAN (Wireless Body Area Network)	https://standards.ieee.org/standard/802_15_6-2012.html
WPAN (Wireless Personal Area Network)	https://standards.ieee.org/standard/802_15_4x-2019.html
UWB (Ultra-Wide Band)	https://www.etsi.org/technologies/radio/ultra-wide-band
VMD (Virtual Medical Devices) standards	https://standards.ieee.org/standard/11073-10207-2017.html
WMTS (Wireless Medical Telemetry Service)	http://www.ashe.org/wmts/
WSDL (Web Services Description Language)	https://www.w3.org/TR/wsdl.html
XML (eXtended Markup Language)	https://www.w3schools.com/xml/
XML SOAP	https://www.w3schools.com/xml/xml_soap.asp
Zigbee	https://www.zigbee.org/

Appendix II: Healthcare-Related Conferences and Journals

The list is a rich subset of international conferences and journals that publish research articles in computational health informatics. The titles are based upon the conferences and journals listed at the back of each chapter. Conferences and journals are grouped based on broad categories. There may be more than one category corresponding to a chapter in the book. The conferences are separated from journals. Individual conferences (or journals) titles are separated by ";".

ARTIFICIAL INTELLIGENCE AND KNOWLEDGE MODELING IN HEALTH CARE

Journals

Annals of Mathematics and Artificial Intelligence; Artificial Intelligence; Artificial Intelligence in Medicine; Decision Support Systems; Engineering Applications of Artificial Intelligence; IEEE Transactions of Knowledge and Data Engineering; IEEE Transactions on Neural Networks; IEEE Transactions on Systems, Man, and Cybernatics; Machine Learning; International Journal of Intelligent Systems; Neural Networks; Proceedings of the IEEE Inst. Electrical and Electronics Engineering; Procedia Computer Science.

BIOINFORMATICS/BIOMARKERS/BIOSCIENCE/ BIOTECHNOLOGY/GENOMICS IN HEALTHCARE

International Conferences

ACM Conference on Bioinformatics, Computational Biology, and Health Informatics; IEEE International Conference of the EMBS.

Journals

Annual Review of Biomedical Eng.; Biochimica et Biophysica Acta; Bioinformatics; Biology, Biomarkers and Prevention; BMC Bioinformatics; Function and Translation; Biotechniques; BMC Genomics; BMC Systems Biology; Cancer, Cancer Genetics and Cytogenetics; Cancer Research;

Epidemiology, Gene; Genomics, Genomics Informatics; IEEE/ACM Transactions on Computational Biology and Bioinformatics; IEEE Transactions on Knowledge and Data Engineering; Information Sciences; International Journal of Bioscience and Biotechnology; Journal of Bioinformatics and Computational Biology; Journal of Biomedical Informatics; Journal of Immunological Methods; Journal of Molecular Biomarkers and Diagnosis; Methods; Microbial Cell Factories; Molecular Oncology; Nature Reviews Genetics; Nature Reviews/Molecular Cell Biology; Nucleic Acids Research; Pathologie Biologie; PLoS Computational Biology; Protein-protein Interactions; Proteomics and Bioinformatics; The Biochemical Journal.

BIOSIGNALS ANALYSIS/IMAGING/ MEDICAL IMAGE ANALYSIS

International Conferences

Data Compression Conference; International Conference on Digital Signal Processing; IEEE International Conference on Advanced Information Networking and Applications (AINA); International Conference on Engineering in Medicine and Biology (EMBS); International Conference on Health Informatics and Medical Systems (HIMS); Signal Processing.

Journals

American Association of Medical Physics; Biological Cybernatics; Biomedical Engineering Online; Biomedical Signal Processing and Control; Computerized Medical Image Processing and Graphics; Current Medical Imaging Reviews; EURASIP Journal on Advances in Signal Processing; IEEE Multimedia; IEEE Signal Processing Magazine; IEEE Transactions on Biomedical Engineering; IEEE Transactions on Consumer Electronics; IEEE Transactions on Imaging Processing; IEEE Transactions on Medical Imaging; IEEE Transactions on Signal processing; IEEE Transaction on Systems, Man and Cybernatics; IEEE Transactions on Ultrasonics, Ferroelectrics, and Frequency Control; Journal of Digital Imaging; Journal of Medical Physics; Journal of Medical Systems; Medical Engineering and Physics; Medical Image Analysis; Medical Physics; Nature; Pattern Recognition; Proceedings of the IEEE; Physiology Measurements; Signal Processing; Statistical Methods in Medical Research; Ultrasonic Imaging; Ultrasound in Medicine and Biology.

BRAIN ABNORMALITIES/NEUROTHERAPY

Journals

Alzheimer's and Dementia; Journal of Clinical Neurophysiology; Journal of the Neurophysiology; NeuroImage; Neuropsychiatric Disease and Treatment; Stroke and Vascular Neurology; The Scientific World Journal.

CLINICAL DATA ANALYTICS/BIOSTATISTICS/ CLINICAL DECISION SUPPORT SYSTEMS

Journals

ACM Conference on Bioinformatics; BMJ; Computational Biology, and Health Informatics; American Journal of Health-Systems in Pharmacy; Annual Reviews in Public Health; Biometrics; Biometrika; British Journal of Cancer; Contemporary Clinical Trials; Controlled Clinical Trials; Decision Support Systems; Journal of American Medical Association (JAMA); Journal of Biomedical Informatics; Journal of Clinical Epidemiology; Journal of Methodology; Movement Disorders; Neuropsychiatric Disease and Treatment; Pediatrics; Risk Management and Health Care Policy; Physiology Measurements; Statistics in Medicine.

CARDIOLOGY/COMPUTERS IN CARDIOLOGY

Journals

American Journal of Cardiology; Applied Soft Computing; Circulation; Clinical Cardiology; Europace; IEEE Engineering in Medicine and Biology Magazine; IEEE Transactions on Biomedical Engineering; Journal of American College of Cardiology; Journal of Electrocardiography; Journal of Electrocardiology; Physica Medica; The Annals of Thoracic Surgery; The Journal of Emergency Medicine.

CLINICAL DATA AND PROCESS MINING

International Conferences

International Conference on Computer Systems and Applications (AICCSA).

Journals

Expert Syst Applications; IMIA Yearbook Medical Informatics; International Journal of Medical Informatics; Journal of Biomedical Informatics; Nature Reviews/Genetics; Prog Pediatrics Cardiology.

DATA MODELING AND DATABASE SEARCH

International Conferences

ACM SIGMOD International Conference on Management of Data; ACM SIGSPATIAL International Conference in Advances in Geographic Information System; annual ACM-SIAM Symposium on Discrete algorithms (SODA); Computers in Biology and Medicine; Data Cleaning and Object Consolidation; Federated Conference on Computer Science and Information Systems; IEEE International Conference on Data Mining; IEEE International Conference on Intelligent Information Hiding and Multimedia Signal Processing; IEEE International Conference on Pattern Recognition, International Conference on Informatics and Mobile Engineering; International Computer Software and Applications Conference; International Conference on Very Large Data Bases; Conference on Retrieval for Image and Video Databases; International Symposium Advances in Spatial and Temporal Databases; SPIE: Storage and Retrieval for Image and Video Databases; VLDB Conference.

Journals

ACM Computing Surveys; ACM International Conference in Multimedia; ACM Transactions on Database Systems; Communications of the ACM; Computer Methods and Programs in Biomedicine; Computational Medical Image and Graphics; Computational Methods and Programs in Biomedicine; Computerized Medical Imaging and Graphics; International Journal of Computer Applications; International Journal of Computational Models and Algorithms in Medicine; International Journal of Computer Assisted Radiology and Surgery; International Journal of Computer Science and Information Technologies; IEEE Transactions on Knowledge and Data Engineering; IEEE Transactions on Multimedia; IEEE Transactions on Software Engineering; Journal of the American Medical Informatics Association (JAMIA); Journal of Digital Imaging; Journal of Medical Image Analysis; Pattern Recognition; The VLDB Journal.

DOMAIN-SPECIFIC HEALTHCARE LANGUAGE

Journals

Clinical Chemistry; Computers and Biomedical Research; International Journal of Clinical Monitoring and Computing; International Journal of Medical Informatics; Journal of American Medical Informatics Association (JAMIA); Journal of Biomedical Informatics.

DRUG DEVELOPMENT, IMMUNOLOGY, EPITOPE ANALYSIS AND VACCINES

Journals

Anesthesia and Intensive Care Medicine; Computational and Structural Biotechnology Journal; Computes Rendus Chimie; Current Topics in Medicinal Chemistry; Current Opinion in Microbiology; Current Opinion in Virology; Drug Discovery Today; Expert Review of Anti-Infective Therapy; Global Heart; Immunology and Cell Biology; Infection and Immunity; Infection and Immunity; INtechopen; International Journal of Molecular Science; Journal of Biomedicine and Biotechnology; Journal of Immunological Methods; Nature Reviews; Nature Reviews Genetics; Nature Reviews Immunology; Trends in Pharmacological Sciences; Vaccine.

ELECTRONIC MEDICAL RECORDS

Journals

Communications of the ACM; Health Informatics – An International Journal; Communications of the ACM; Journal of the American Medical Informatics Association (JAMIA).

ELDERLY CARE AND END-USER SATISFACTION

International Conferences

ACM Workshop on Software Engineering in Health Care (ICSE); International Conference on Soft-Ware: Computing in an Imperfect World; International Working Conference on Requirements Engineering: Foundations for Software Quality (REFSQ).

Journals

American Journal of Critical care; Frontiers in Psychiatry; BioMed Central Medical Informatics and Decision Making; Information Systems Research; International Journal of People Oriented Programming; International Journal of Social Robot; Journal of Aging Studies; Journal of Affective Disorder; Journal of Information and Software Technology; Journal of Marketing; Journal of Systems and Software; Maturitas; Psychiatric Service; Psychological Medicine; Requirements Engineering.

EMG

Journals

Biological Procedures Online; Computers in Biology and Medicine; IEEE Transactions on Rehabilitation Engineering; Journal of Computing; Journal of Electromyography and Kinesiology; Neuroscience Letters; Medical Engineering and Physics.

DISEASE PATHWAYS, IMMUNE SYSTEM AND GENETIC DISEASES

Journals

Artificial Intelligence in Medicine; BMC Systems Biology; Current Opinion in Genetics and Development; Genetic Epidemiology; Infection and Immunity; International Journal of Biological Sciences; Journal of Biomedical Informatics; Medical Physics; Nature Reviews Genetics; Nucleic Acids Research.

GENOME SEQUENCING AND ANALYSIS

Journals

Bioinformatics; BMC Genomics; Journal of Bioinformatics and Computational Biology; Journal of Molecular Biology; Nucleic Acids Research; Nature Methods.

HEALTH CARE

International Conferences

Annual Symposium on Computer Application in Medical Care.

Journals

Advanced Healthcare Material; American Journal of Critical care; Annual Intern Med; Annual Reviews Public Health; Clinics in Dermatology; Diabetes Care; Food Science and Human Wellness; Genetics in Medicine; Health Affairs; Health Policy; Journal of Evaluation and Clinical Practices; The American Journal of Human Genetics; Medical care Research and Review; Nature; Nature Reviews/Genetics; Patient Education and Counseling.

HEALTH INFORMATICS/MEDICAL INFORMATICS

International Conferences

ACM Symposium on Applied Computing; Annual Symposium of American Medical Informatics Association (AMIA).

Journals

ACTA Inform Med.; Applied Clinical Informatics; Applied Medical Informatics; Biomedical Informatics Insights; BMJ; Clinics in Laboratory Medicine; Communications of the ACM; Computers Informatics Nursing; Health Informatics Journal; IMIA Yearbook of Medical Informatics; International Journal of Functional Informatics and Personalised Medicine; International Journal of Medical Informatics; IT PRO; Journal of American Informatics Association; Journal of American Medical Association (JAMA); Journal of Biomedical Information; Journal of Biomedical Informatics; Journal of Information Science; Journal of Pathology Informatics; Methods of Information in Medicine; MIS Quarterly; The Journal of Systems and Software; Information Systems.

HEALTH-INFORMATION EXCHANGE/ MEDICAL IMAGE TRANSMISSION

International Conferences

Conference on Mobile Computing in Medicine; Free and Open Source Conference; Annual Symposium of American Medical Informatics Association (AMIA).

Journals

IBM Systems Journal; IEEE Journal on Selected Areas in Communications; IEEE Micro; Journal of American Medical Informatics Association (JAMIA); International Journal of Medical Informatics; RadioGraphics.

HEALTHCARE STANDARDS, INTERFACES, INTEROPERABILITY AND VISUALIZATION

International Conferences

ACM International Conference of Software Engineering; Australasian Workshop on Advances in Ontologies; Australian Symposium on the ACSW Frontiers; Foundations and Trends in Human–Computer Interaction; IEEE Symposium on Computer Based Medical Systems; International Workshop on Managing Interoperability and Complexity in Health Systems; SPIE International Society for Optical Engineering.

Journals

Acta Informatica in Medicine; Clinical Chemistry; Clinical Radiology; Computer; Computerized Medical Imaging and Graphics; Computer Standards and Interfaces; European Journal of Radiology; Image and Vision Computing; IEEE Signal Processing Letters; IEEE Transactions on Knowledge and Data Engineering; International Journal of Functional Informatics and Personalized Medicine; Journal of the American Health Information Management Association; Journal of American Medical Informatics Association (JAMIA); Journal of Biomedical Semantics; Journal of Digital Imaging; Journal of Medical Image Analysis; RadioGraphics; SPIE Medical Imaging 2002: PACS and Integrated Medical Information Systems.

MEDICINE

Journals

New England Journal of Medicine; Ultrasound in Medicine and Biology.

PERVASIVE CARE AND REMOTE MONITORING/TELEMEDICINE

International Conferences

Annual Conference of IEEE Communications Society on Sensor and Ad Hoc Communications and Networks (SECON); IEEE Symposium of the Computer-Based Medical Systems.

Journals

Annals of Biomedical Engineering; Communications of the Association for Information Systems; IEEE Antennas and Propagation Magazine; IEEE Network; IEEE Pervasive Computing; IEEE Transactions on Systems, Man, and Cybernatics; International Journal of Medical Informatics; International Journal of Telemedicine Applications; IEEE Transactions on Information Technology in Biomedicine; Journal of Medical Association (JAMA); Journal of Telemedicine and Telecare; Pervasive and Mobile Computing; Surgical Innovation; Journal of Medical Association (JAMA); Telemedicine and E-Health.

RADIOLOGY AND IMAGING

Journals

Abdominal Imaging; Canadian Association of Radiologist Journal; Clinical Radiology; European Journal of Radiology; Expert Opinion in Medical Diagnostics; IEEE Transactions on Ultrasonics, Ferroelectrics, and Frequency Control; International Journal of Dental Clinics; Journal of Medical Ultrasonics; Journal of Medicinal Physics; Magnetic Resonance Imaging; Optometry and Vision Science; Radiology; Theranostics; Ultrasonic Imaging.

RFID, LOCALIZATION, MONITORING AND TRACKING

International Conferences

International Conference of IEEE EMBS.

Journals

Age and Aging; IEEE Transactions of Information Technology in Biomedicine; Pervasive and Mobile Computing; PLoS One.

SECURITY AND PRIVACY/HIPAA/ WATERMARKING OF IMAGES

International Conferences

IEEE Symposium on Security and Privacy; International Conference of Engineering in Medicine and Biology Society (EMBC).

Journals

Computerized Medical Imaging and Graphics; IEEE Transactions on Information Technology in Biomedicine; International Journal of Computational Models and Algorithms in Medicine; International Journal of Medical Informatics; Journal of American Medical Informatics Association (JAMIA); Journal of the American Statistical Association; Journal of Digital Imaging; Journal of Medical Systems.

SOCIOECONOMIC STUDY AND ADOPTION

Journals

International Journal of Information Management; International Journal of Medical Informatics; Procedia – Social and Behavioral Sciences; Social Science and Medicine; Univers Access Inf Soc.

SENSOR NETWORKS, WIRELESS NETWORKS AND STANDARDS IN HEALTH CARE

International Conferences

High Confidence Medical Device Software and Systems Workshop; IEEE SoutheastCon; International Conference on Convergence Information Technology (ICCIT); International Conference of the IEEE Engineering in Medicine and Biology Society; IEEE International Symposium on Applied Sciences in Biomedical and Communication Technologies (ISABEL); International Workshop on Wearable and Implantable Body Sensor Networks.

Journals

Ad Hoc Networks; Advanced Materials; Computer Networks; IEEE Antennas and Propagation Magazine; IEEE Network; IEEE Transactions on Biomedical Engineering; IEEE Transaction of Cloud Computing; IEEE Transactions on Information Technology in Biomedicine; IEEE Transactions on Systems, Man, and Cybernatics; Information Fusion; International Conference of Body Sensor Networks; Journal of Medical Systems; Journal of Neuroengineering and Rehabilitation; Journal of Wireless Networks; Multimedia Tools and Applications; Physiological Measurement; Proceedings of the IEEE; Sensors and Actuators A Physical; Sensors and Actuators; Transaction of the Institute of the Measurement and Control.

Appendix III: Health Informatics Related Organizations

AGENCY	WEBSITE
American Association for the Study of Liver Diseases	https://www.aasld.org/
American Brain Foundation	https://www.americanbrainfoundation.org/
American Health Care Association (AHCA)	https://www.ahcancal.org
American Health Information Association (AHIMA)	https://www.ahima.org/
American Heart Association	https://www.heart.org/
American Liver Foundation	https://liverfoundation.org/
American Lung Association	https://www.lung.org/
American Medical Informatics Association (AMIA)	https://www.amia.org/
American Neurological Association	https://myana.org/
American Nursing Informatics Association (ANIA)	https://www.ania.org/
Association of Computing Machinery (ACM)	https://www.acm.org/
Association for Electrical Equipment and Medical Imaging (NEMA)	https://www.nema.org
European Biobank for Medical Research Innovation (BBMRI-ERIC)	http://www.bbmri-eric.eu/
European Clinical Research Infrastructure Network (ECRIN)	https://www.ecrin.org/
European Innovative Medicines Initiative (IMI)	https://www.imi.europa.eu/
European Institute for Translational Medicine (EATRIS)	https://eatris.eu/
European Life Science Infrastructure for Biological Information (ELIXIR)	https://elixir-europe.org/
European Bioinformatics Institute (EMBL-EBI), UK	https://www.ebi.ac.uk/
European Molecular Biology Laboratory (EMBL), Germany	https://www.embl.org/
Institute of Electrical and Electronics Engineers (IEEE)	https://www.ieee.org/
Japanese GenomeNet Database Resources	https://www.genome.jp/
Japanese GenomeNet Bioinformatics Tools Resource	https://www.genome.jp/en/gn_tools.html
National Institute of Health (NIH), USA	http://www.nih.gov
National Center of Biotechnology Information (NCBI), USA	http://www.ncbi.nlm.nih.gov
National Human Genome Research Institute (NHGRI), USA	http://research.nhgri.nih.gov
National Cancer Institute (NCI), USA	https://www.cancer.gov/
National Cancer Research Institute (NCRI), UK	https://www.ncri.org.uk/
National Heart, Lung and Blood Institute, USA	https://www.nhlbi.nih.gov/
National Institute of Aging, USA	https://www.nia.nih.gov/
National Institute of General Medical Sciences, USA (NIGMS)	https://www.nigms.nih.gov/
National Institute of Neurological Disorder and Strokes	https://www.ninds.nih.gov/
National Kidney Foundation	https://www.kidney.org/
National Library of Medicine	https://www.nlm.nih.gov/
National Network of Libraries of Medicine	https://nnlm.gov/
National Pharmaceutical Association	http://nationalpharmaceuticalassociation.org

Appendix IV: Health Informatics Database Resources

NAME	DESCRIPTION / WEBSITE / CITATION
Collection of Multiple Databases	
Physionet	Collection of Clinician databases, ECG and EMG waveforms with different disease conditions. http://www.physionet.org Citation: Goldberger, Ary L., Amaral, Luis A. N., Glass, Leon, Hausdorff, Jeffrey M., Ivanov, Plamen Ch, Mark, Roger G., Mietus, Joseph E., Moody, George B., Peng, Chung-Kang, and Stanley, H. Eugene, "PhysioBank, PhysioToolkit, and PhysioNet: Components of a New Research Resource for Complex Physiologic Signals," *Circulation*, 101(23):e215–e220 http://circ.ahajournals.org/content/101/23/e215.full, June 2000.
Clinical Databases	
MIMIC	Clinical database. https://mimic.physionet.org/ Citation: Johnson, Alistair E. W., Pollard, Tom J., Shen, Lu, Lehman, Le-wei H., Feng, Mengling, Ghassemi, Mohammad, Moody, Benjamin et al., "MIMIC III – A Freely Accessible Critical Care Database," *Scientific Data*, 3, Article Number: 160035, 2016.
ECG Databases	
CU	Creighton University ECG database. https://physionet.org/physiobank/database/cudb/ Citation: Nolle, Floyd M., Badura, F. K., Catlett, J. M., Browser, R. W., and Sketch, M. H., "Crei-Gard, A New Concept in Computerized Arrythmia Monitoring Systems," *Computers in Cardiology*, 13, 1986, 515–518.
ECG View II	Website: http://www.ecgview.org Citation: Kim, Young-Gun, Shin, Dahye, Park, Man Young, Jeon, Min Seok, Yoon, Dukyong, and Park, Rae Woong, "ECG-ViEW II, A Freely Accessible Electrocardiogram Database," *PLoS One*, 12(4), 2017, DOI: 10.1371/journal.pone.0176222.
MIT BIH	Website: https://physionet.org/physiobank/database/mitdb/ Citation: Moody, George B., and Mark, Roger G., "The impact of the MIT-BIH Arrhythmia Database," *IEEE Engineering in Medicine and Biology*, 20(3), May–June 2001, 45–50.
PTB	Website: https://physionet.org/physiobank/database/ptbdb/ Citation: Bousseljot, Ralf-Dieter, Kreiseler, Dieter, and Schnabel, Allard, "Nutzung der EKG-Signaldatenbank CARDIODAT der PTB über das Internet," *Biomedizinische Technik*, 40(1), January 1995, S317.

(Continued)

NAME	DESCRIPTION / WEBSITE / CITATION
Genomes and Protein (and Related Disorders) Databases	
BLAST	A powerful search engine based upon similarity based search to annotate newly sequenced genes and proteins. Website: https://blast.ncbi.nlm.nih.gov/Blast.cgi
CDD	A database of conserved domains in protein families. Website: https://www.ncbi.nlm.nih.gov/Structure/cdd/cdd.shtml
Pfam	Protein family database. Website: https://pfam.xfam.org/
Ensemble	A genome browser for vertebrate genomes for comparative genomics. Website: https://useast.ensembl.org
EPD	A eukaryote promoter database. Website: https://epd.epf.ch
Human Protein Atlas	A database of proteins in human genome, their sequences, functions and genetic disorders. Website: http://www.proteinatlas.org Citation: Uhlén Mathias, Fagerberg, Linn, Hallström, Björn, Lindskog, Cecilia, Oksvold, et al., "Tissue-based map of the human proteome," *Science*, 347(6220), January 2015, DOI: 10.1126/science.1260419.
InterPro	A database showing classification of protein families. Website: https://www.ebi.ac.uk/interpro/beta/
OMIM	Online Mendelian Inheritance in Man – An online catalog of human genes and genetic disorders. Website: https://www.omim.org
Prosite	A database of protein families. Website: https://prosite.expasy.org
UniProt	A database of proteins and their functions. Website: https://www.uniprot.org/
Cell Signaling and Metabolic Pathways Databases	
Kegg	A database of wiring diagrams of molecular reactions including metabolism, cell signaling, immune system and cancer-related pathways. Website: https://www.genome.jp/kegg/pathway.html
PharmGKB	An evidence based database of pathways involving pharmacokinetics and pharmacodynamics of drugs. Website: https://www.pharmgkb.org/pathways
Reactome	A database of protein pathways. Website: https://www.reactome.org
WikliPathways	A database of biological pathways. Website: https://www.pharmgkb.org/pathways
Drug-Related Databases	
Drugbank	A detailed database of drug and drug targets. Website: https://www.drugbank.ca
Epitome	A database of epitopes. Website: http://www.rostlab.org/services/epitome/

Appendix V: Selected Companies in Healthcare Industry

NAME	DESCRIPTION / WEBSITE / CITATION
Electronic Health Record	
Athenahealth	For hospitals and multi-physician practices. Website: http://www.athenahealth.com
Cerner	Integrated clinical and financial solutions. Website: http://www.cerner.com
CureMD	All in one cloud-based platform. Website: http://www.cureMD.com
eClinicWorks	For small size practices. Website: http://www.eClinicalWorks.com
EPIC	For mid and large size practices and hospitals. Website: https://www.epic.com/
GEHealthcare	For large size integrated management. Website: https://www.gehealthcare.com/en-us
Practice Fusion	Integrated software for 1–10 doctors; free web-based HER. Website: https://www.practicefusion.com/electronic-health-record-ehr/
Artificial Intelligence/Knowledge Discovery/Data Archival	
Cannon Medical Systems	Integrated Artificial Intelligence for medical imaging. Website: https://us.medical.canon/
DeepMind Health (Google)	Artificial intelligence and mobile tools. Website: https://deepmind.com/applied/deepmind-health/
IBM	Artificial intelligent techniques, knowledge bases, data mining. https://www.ibm.com/watson/health/
Janssen (Johnson and Johnson)	Artificial Intelligence in drug discovery. Website: https://www.jnj.com
Microsoft	Artificial intelligent techniques, cybersecurity, virtual reality, cloud-based archival. Website: https://www.microsoft.com/en-us/enterprise/health
Siemens	Artificial intelligence in medical diagnosis. Website: https://www.siemens-healthineers.com/en-us/
Zebra Medical Vision	Artificial Intelligence in radiology-based diagnosis. Website: https://www.zebra-med.com
Medical Imaging and Imaging-Based Diagnostics	
Fujifilm	Digital mammography; X-ray imaging, films and diagnostic systems; optical devices. Website: https://www.fujifilm.com/
Cannon Medical Systems	Angiography; Computed tomography; MRI system; Ultrasound systems; X-ray machines. Website: https://us.medical.canon/
Carestream Healthcare	Mammography picture archiving and communication; mammography printing and film; cardiology-based systems; dental imaging system. Website: https://www.carestream.com/en/us

(Continued)

NAME	DESCRIPTION / WEBSITE / CITATION
General Electric	MRI system; mammography; molecular imaging; nuclear imaging; PET system; radiopharmacy; preclinical imaging; fluoroscopy systems. Website: https://www.gehealthcare.com/en-us
Hitachi Medical	MRI system; Computed tomography; Ultrasound; Digital radiography; Information technology. Website: https://www.hitachimed.com/
Philips	Advanced molecular imaging; Computed tomography; EEG neuroimaging; X-ray systems; MRI systems; Fluoroscopy; Radiography; Ultrasound. Website: https://www.usa.philips.com/healthcare
Shimadzu	Angiography; Mobile X-ray; Fluoroscopy; Fluorescence imaging; Radiography. Website: https://www.shimadzu.com/med/
Siemens Healthcare	Angiography; Computed tomography; Fluoroscopy; Imaging for radiation therapy; MRI; mammography system; Imaging software; Molecular imaging system (PET scanner). Website: https://www.siemens-healthineers.com/en-us/medical-imaging

Wearable and Monitoring Devices and Others

Alphabet/Verily (A Google company)	ECG measuring watches; fall detecting shoes; smart lens; miniaturized continuous glucose meters; retinal imaging for retinopathy. Website: https://verily.com
Fitbit	Smart watch-like device for heart rate, activity measurement. http://www.fitbit.com
Meditronic	Multiple monitoring devices for heart, brain, nerve and diabetes; pulse oximetry; cerebral oximetry; surgical imaging systems, bone grafting; multiple surgical devices, dialysis related equipment, etc. Website: https://www.medtronic.com
Nokia	Distributed cloud-based archival and image transmission and robot assisted surgery; real-time AI-based data analytics; smart wireless/wired connectivity between devices. Website: https://www.nokia.com
Philips	Diagnostic ECG; Telehealth; patient monitoring; Respiratory care; Emergency care; advanced visualization. Website: https://www.usa.philips.com
PolarH10	Chest-strap with Bluetooth connectivity to measure heart rate. Website: http://www.polar.com
Quardiocore	Wearable ECG devices wirelessly connected to the smart phone. Website: https://www.getqardio.com/
Somnomedics	Home and ambulatory wearable devices for ECG, sleep and blood pressure. Website: https://somnomedics.eu/
Vitaljacket	Makes T-shirt for monitoring ECG and other vital signs. Website: http://www.vitaljacket.com

Drug Discovery and Immunotherapy (Diseases covered and Websites)

Abbot	Parkinson's disease management; heart-valves; cardiology; monitoring diseases such as HIV, cancer, heart failure; generic drugs; diagnostics. Website: http://www.abbott.com
Abbvie	Drugs for rheumatology; dermatology; gastroenterology; oncology; neurodegenerative disorders; Alzheimer's, Parkinson, HIV, Hepatitis C and general medicine. Website: https://www.abbvie.com

(Continued)

NAME	DESCRIPTION / WEBSITE / CITATION
Alexion Pharmaceuticals	Hemolysis; hypophosphatasia; enzyme deficiency; central nervous system disorder; bacterial infection. Website: http://www.alexion.com
Amgen	Oncology (metastatic breast cancer)Osteoporosis'; migraine prevention; anemia; leukemia; chronic heart failure; rheumatoid arthritis; osteoporosis; melanoma; myeloma; low platelet count; etc. Website: https://www.amgen.com
AstraZeneca	Metabolism; oncology; respiratory disorders; Inflammation and autoimmunity; neuroscience; infection; vaccines. Website: https://www.astrazeneca.com
Bayer	Cardiology and women's healthcare; oncology; hematology; ophthalmology; radiology contrast enhanced diagnostics; dermatology; digestive health; allergy; etc. Website: http://www.bayer.com
Biogen	Neurodegenerative diseases including Alzheimer's disease; epilepsy; stroke; gene therapy; spinal muscular atrophy. Website: http://www.biogen.com
Boehringer Ingelheim	Lung cancer; diabetes; pulmonary fibrosis; atrial fibrillation; COPD; asthma; myocardial infarction; stroke prevention; HIV/AIDS; Hypertension; sleep disorder; Parkinson's disease; etc. Website: http://www.boehringer-ingelheim.com
Bristol-Myers	HIV; drug resistant bacteria; atrial fibrillation; thrombosis; pulmonary embolism; stroke; myeloma; leukemia; inflammation; cancer; blood thinner, statin for heart related issues; etc. Website: http://www.bms.com
CelGene	Arthritis; cancer; myeloma; psoriasis; myelofibrosis; etc. Website: https://www.celgene.com
Eli Lily	Arthritis; bone muscle; cancer; cardiovascular; diabetes; migraine. Website: https://www.lilly.com
Glaxo SmithKline	Antibiotics; Asthma; COPD; diabetes; epilepsy; hyperplasia; migraine; inflammation; migraine; prostate hyperplasia; pulmonary hypertension; vaccines; etc. Website: https://gsksource.com/pharma/content/gsk/source/us/en/brands.html
Gilead	HIV/AIDS; Liver diseases; hematology/oncology; cardiovascular; inflammation/respiratory. Website: https://www.gilead.com/science-and-medicine/medicines
Janssen	Alzheimer's, arthritis; cancer; dental diseases; diabetes; HIV; mental diseases (depression and schizophrenia); lupus; pulmonary hypertension; tuberculosis; vaccines; etc. Website: https://www.janssen.com
Merck	Bacterial and fungal infection; seasonal allergy; asthma; cancer therapy; post cardiac surgery problems; diabetes, hypertension; inflammation; neuromuscular disorder; HIV; skin infections; sleeping disorders; vaccines; venoms; women's osteoporosis; etc. Website: https://www.merck.com
Novartis	Alzheimer's disease; arthritis; asthma; atherosclerosis; cancer; cell and gene therapy; liver diseases; migraine; multiple sclerosis; psoriasis; retinal diseases; sickle cell anemia; etc. Website: https://www.novartis.com
Pfizer	Bacterial infection; Asthma; diabetes; kidney diseases; congestive heart failure; hypertension; inflammation; metabolic diseases; pulmonary congestion; ventricular fibrillation; etc. Website: www.pfizer.com
Roche	Anxiety disorders; arthritis; cancer and after-effects; chronic heart failure; HIV; hypertension; infections; macular degeneration; malaria; multiple sclerosis; obesity; organ rejection; osteoporosis; Parkinson's disease; pulmonary fibrosis; viral infections. Website: http://www.roche.com

Index

Milton Keynes UK
Ingram Content Group UK Ltd.
UKHW051928141024
449569UK00027B/1402